Bin Ding • Jianyong Yu
Editors

Electrospun Nanofibers for Energy and Environmental Applications

Springer

Editors
Bin Ding
Jianyong Yu
State Key Laboratory for Modification
of Chemical Fibers and Polymer Materials
College of Materials Science
and Engineering
Donghua University
Shanghai, China

Nanomaterials Research Center
Modern Textile Institute
Donghua University
Shanghai, China

ISSN 1571-5744 ISSN 2197-7976 (electronic)
ISBN 978-3-642-54159-9 ISBN 978-3-642-54160-5 (eBook)
DOI 10.1007/978-3-642-54160-5
Springer Heidelberg New York Dordrecht London

Library of Congress Control Number: 2014936170

Printed on acid-free paper

Springer is part of Springer Science+Business Media (www.springer.com)

Preface

Energy and environment head the list of top global problems facing society in the twenty-first century. Nanotechnology is responding to these challenges by designing and fabricating functional nanofibers optimized for energy and environmental applications. The route towards these nano-objects is based primarily on electrospinning: a highly versatile method that allows the fabrication of continuous fibers with diameters down to a few nanometers. In recent years, breakthroughs in electrospun nanofibers, especially in their applications in the energy and environmental sectors, have opened up the possibility of moving beyond our current options to ensure sustainable energy and preserve our environment for the future.

This book, *Electrospun Nanofibers for Energy and Environmental Applications*, is a collection of 20 chapters contributed by the world's leading experts in the electrospinning field. Each chapter covers a special subject that falls within three general areas: Introduction, energy applications, and environmental applications, corresponding to the three parts of the book.

Part I Introduction first provides a general overview of the electrospinning technique, including the history, basic principles, theoretical background, and controlling parameters. This is followed by a detailed introduction to electrospun nanofibers, involving the materials classes as well as the morphologies and structures of nanofibers. Finally, a brief analysis of nanofibers used for advanced energy and environmental applications indicates their substantial and promising impact.

Part II Electrospun nanofibers for energy applications covers the main developments in the use of nanofibers in energy conservation and storage. Each chapter reports on recent contributions of nanofibers in selected areas of energy, such as fuel cells, electrodes of lithium-ion batteries, separators for lithium-ion batteries, dye-sensitized solar cells, organic and hybrid solar cells, super capacitors, hydrogen storage, thermal storage, and carbon dioxide capture.

Part III Electrospun nanofibers for environmental applications focuses on the potential applications of nanofibers in environmental monitoring and management, which are mainly concerned with the ever-growing issues of water and air pollution. These chapters present how functional nanofibers help to improve the environment in contexts such as sensors, air and liquid filtration, structurally-colored fibers, fibers with photocatalytic, self-cleaning and adsorbent properties, oil spill clean-up, and electromagnetic shielding, examining them in depth.

This book is intended to provide a balanced treatment of the various topics of electrospun nanofibers in energy and environmental-related areas with both a general overview and detailed discussions to suit a broader audience. We sincerely hope it will pique your interest in electrospun nanofibers and their fascinating applications in energy and environmental related areas.

China
November, 2013

Bin Ding
Jianyong Yu

Contents

Part I
Introduction

Chapter 1
Electrospun Nanofibers: Solving Global Issues

Yang Si, Xiaomin Tang, Jianyong Yu, and Bin Ding

Abstract Energy and environment will head the list of top global issues facing society for the next 50 years. Nanotechnology is responding to these challenges by designing and fabricating functional nanofibers optimized for energy and environmental applications. The route toward these nano-objects is based primarily on electrospinning: a highly versatile method that allows the fabrication of continuous fibers with diameters down to a few nanometers. The mechanism responsible for the fiber formation mainly includes the Taylor Cone theory and flight-instability theory, which can be predicted theoretically and controlled experimentally. Moreover, the electrospinning has been applied to natural polymers, synthetic polymers, ceramics, and carbon. Fibers with complex architectures, such as ribbon fiber, porous fiber, core-shell fiber, or hollow fiber, can be produced by special electrospinning methods. It is also possible to produce nanofibrous membranes with designed aggregate structure including alignment, patterning, and two-dimensional nanonets. Finally, the brief analysis of nanofibers used for advanced energy and environmental applications in the past decade indicates that their impact has been realized well and is encouraging, and will continually represent a key technology to ensure sustainable energy and preserve our environment for the future.

Y. Si • X. Tang • J. Yu • B. Ding (✉)
State Key Laboratory for Modification of Chemical Fibers and Polymer Materials, College of Materials Science and Engineering, Donghua University, Shanghai 201620, China

Nanomaterials Research Center, Modern Textile Institute, Donghua University, Shanghai 200051, China
e-mail: binding@dhu.edu.cn

B. Ding and J. Yu (eds.), *Electrospun Nanofibers for Energy and Environmental Applications*, Nanostructure Science and Technology,
DOI 10.1007/978-3-642-54160-5_1,

1.1 Energy and Environment Global Issues

Energy and environment head the list of top global problems facing society in the twenty-first century. According to the World Energy Outlook 2011 published by International Energy Agency (IGA), the global energy demand will have a dramatic increase of 40 % between 2009 and 2035, and the fossil fuels—coal, natural gas, and petroleum—will continue to be the primary sources used to meet the energy demands of humanity (Fig. 1.1a) [1–3]. However, these resources are finite and the regeneration would take millions of years; the concerns over fossil fuel supplies are often the direct or indirect causes of regional and global conflicts [4, 5]. Furthermore, the production, transmission, and use of fossil fuels also lead to environmental degradation. Combustion of carbon-based fossil fuels generates not only air pollutants, for example, sulfur oxides and heavy metals, but also CO_2, the infamous greenhouse gas widely believed to be the culprit of global climate change (Fig. 1.1b), which make the world on a trajectory that results in a level of emissions consistent with a long-term average temperature increase of more than 3.5 °C by 2035 [1, 6–8]. One of the solutions to this energy challenge relies upon increasing the efficiency in production, transmission, and utilization of the remaining fossil fuels while reducing their negative impacts to the environment. On the other hand, technologies and infrastructures have to be developed or improved in preparation for the smooth transition to the alternative and renewable energy sources, that is, nuclear power, solar energy, wind power, geothermal energy, biomass and biofuels, hydropower, etc. [4, 9–11].

Environmental pollution has become another major global concern due to rapid growth of industrialization, urbanization, and modern agricultural development. Technological innovations and advancements in products and processes in industries have given rise to new products and new pollutants in abundant level which are above the self-cleaning capacity of the environment [12–14]. Pollution is the introduction of contaminants into an environment that causes instability, disorder, harm, and discomfort to the ecosystem, that is, physical systems or living organisms. Water and air represent two major environmental systems where the most pressing

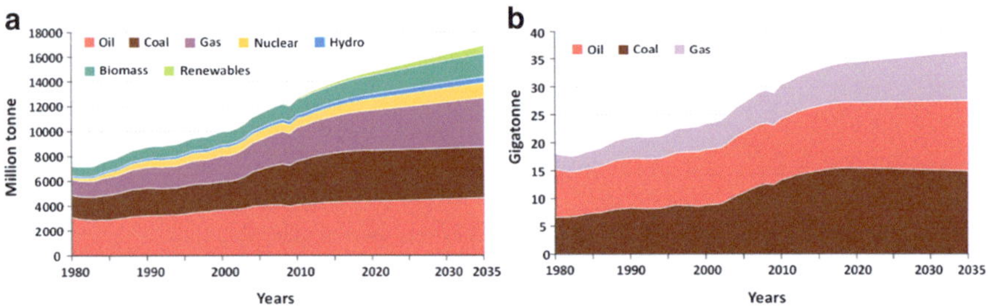

Fig. 1.1 (**a**) World primary energy demand by fuel, 1980–2035. (**b**) Energy-related CO_2 emissions by fuel, 1980–2035 (Reprinted with permission from The International Energy Agency [1]. © 2011 IGA)

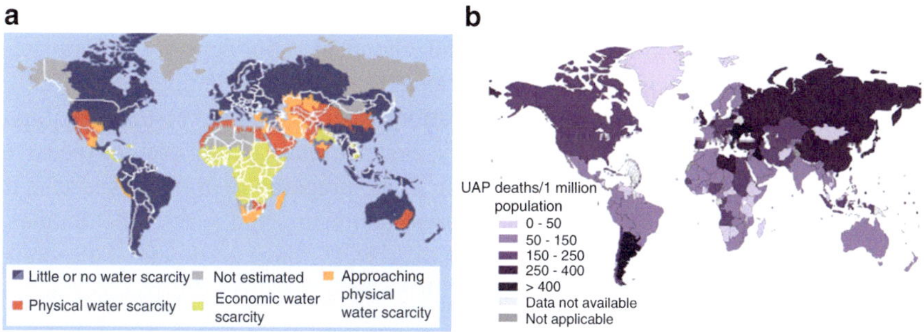

Fig. 1.2 (**a**) The map showing global physical and economic water scarcity. (**b**) The global map showing the death attribute to the urban air pollution (**a**: Reprinted with permission from The Water Project. © 2011 The Water Project. **b**: Reprinted with permission from WHO. © 2011 WHO)

environmental pollution issues persist [15–17]. As People Action International states, water is essential for human life, development, and environment, but it is a finite and vulnerable resource which has quantitative limitations and qualitative vulnerability (Fig. 1.2a) [18, 19]. It is estimated that more than 50 % of nations in the world will face freshwater stresses or shortages by 2025 and will increase to 75 % by 2075. Thus, water pollution and dwindling freshwater supplies are becoming a critical global issue due to increasing population, economic growth, and climate change [2, 12]. Similarly, air is another environmental system that is of concern. Air pollution is the contamination of the indoor or outdoor environment by any chemical, physical, or biological agent that modifies the natural characteristics of the atmosphere. The major air pollutants are particulate matter, nitrogen dioxide, sulfur dioxide, and carbon monoxide [20–22]. Motor vehicles, household combustion devices, and industrial activity are the main man-made sources of outdoor air pollution, causing respiratory and other diseases. The World Health Organization states that 2.4 million people die each year from causes directly attributable to air pollution, with 1.5 million of these deaths attributable to indoor air pollution, as shown in Fig. 1.2b [23–25].

Given the recognized threats to the world's collective energy security and environment, the focus must be redirected, as quickly as possible, toward addressing these critical challenges and driving global research to develop technology and devices for clean energy conversion, storage and conservation, and a clean environment—water and air pollution abatement [2, 26]. These demands for such high-performance materials have led to increasing attention in advanced functional nanosized materials. Among those existing nanostructures, one-dimensional (1D) nanostructures are of particular interest currently because of their unique applications. 1D semiconducting nanowires, nanofibers, and nanotubes and also patterning of materials can be used to enhance charge transport and as building blocks in energy and electronic devices because of their confinement effects. Water and air purification or removal of contaminants can be effectively achieved using nanofiber membranes [2, 9, 27]. Up to now, substantial strategies have been established to generate 1D nanostructures including electron-beam or focused-ion-beam writing, lithography, hydrothermal, chemical vapor deposition, template-directed, self-assembly, interface synthesis technique, electrospinning, etc. [27–29]. However, most of these methods have limitations, such as material restrictions, high cost, and high process complexity. Recently, electrospinning, a simple, inexpensive technique, has attracted significant attention in the preparation of nanomaterials. It has been used to make nanofibers, nanotubes, nanobelts, and porous membranes, which have unique properties applicable to a wide range of fields. This book highlights the potential and application of electrospun fibrous materials for energy and environmental applications.

1.2 Overview of the Electrospinning

Electrospinning is currently the most effective technique that allows the fabrication of continuous fibers with diameters down to a few nanometers, which shares the characteristics of both electrospraying and conventional solution dry spinning. The method can be applied to synthetic and natural polymers, polymer alloys, as well as to metals and ceramics [30, 31]. Fibers with complex architectures, such as core-shell fibers, helix fibers, porous fibers, and hollow fibers, can be produced by special electrospinning methods. It is also possible to produce structures ranging from single fibers to ordered arrangements of fibers. Since the beginning of this century, researchers all over the world have been reexamining the electrospinning process, which is not only employed in university laboratories but also increasingly being applied in industry [28–30].

1.2.1 History of Electrospinning

The history of electrospinning is a rather complicated story. The first record of the electrostatic attraction of a liquid was reported by William Gilbert in the late sixteenth century. He observed that when a suitably electrically charged piece of amber was brought near a droplet of water, it would form a cone shape and small droplets which would be ejected from the tip of the cone [32, 33]. Following, the development of the electricity had reached a sufficient level to provide inductive generation of high voltages. In 1745 Georg Mathias Bose described the formation of aerosols formed by high electric potentials on the surfaces of droplets. In 1885 Lord Rayleigh described the amount of charge required for the deformation of droplets, providing the first theoretical explanation of electrospinning [31, 32, 34].

The process of electrospinning was firstly patented by Gooley in 1900. He proposed four types of indirectly charged spinning heads, a conventional head, a coaxial head, an air-assisted model, and a spinneret featuring a rotating distributor. In 1914 Zeleny et al. published work on the behavior of fluid droplets at the end of

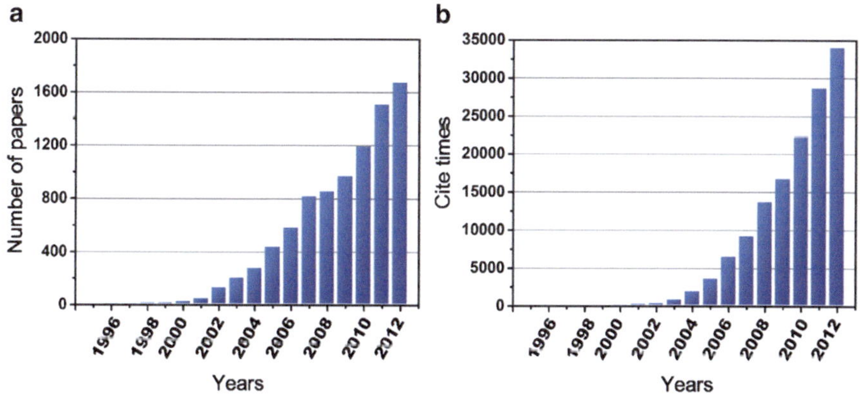

Fig. 1.3 (**a**) Numbers of publications and (**b**) cited times from 1996 to 2012 with the keyword of "electrospinning" (Source from ISI web of science)

metal capillaries. His effort began the attempt to mathematically model the behavior of fluids under electrostatic forces [32, 35–37]. Further developments toward commercialization were made by Anton Formhals and described in a sequence of patents at 1934 for the fabrication of textile yarns. His spinning process consists of a movable thread collecting device to collect threads in a stretched condition, like that of a spinning drum in conventional spinning. Meanwhile, electrospinning from a melt rather than a solution was patented by C. L. Norton in 1936 using an air blast to assist fiber formation [32–34].

It took a long time until others picked up on these early reports. In 1969 Sir Geoffrey Ingram Taylor produced the theoretical underpinning of electrospinning. Taylor's work contributed to electrospinning by mathematically modeling the shape of the cone formed by the fluid droplet under the effect of an electric field. This characteristic droplet shape is now known as the Taylor Cone [27, 30, 38]. In 1971 Baumgarten reported on electrospinning of acrylic microfibers, and Larrondo and Manley reported in a series of papers on electrospinning of polymer melts. In the 1970s also some attempts at commercialization were undertaken [32, 38]. Simm, from the Bayer company, submitted a series of patents on electrospinning of plastics. A variety of electrospinning setups were suggested in early electrospinning setups that have some similarities to recent efforts. The first technical application for electrospinning was suggested for the nonwoven industry [26, 34, 39].

Academia picked up electrospinning in the 1990s due to the notable work of Reneker and co-workers (at the University of Akron). Since then, more and more groups started to enter the field of electrospinning driven by the growing interest in nanomaterials, which resulted in a dramatic increase of publications from very few per year to 1675 publications in 2012 (Fig. 1.3). With such a huge impact, electrospinning is certainly among the very hot topics in materials science. The popularity of the electrospinning process can be realized by the fact that over 500 universities and research institutes worldwide are studying various aspects of the

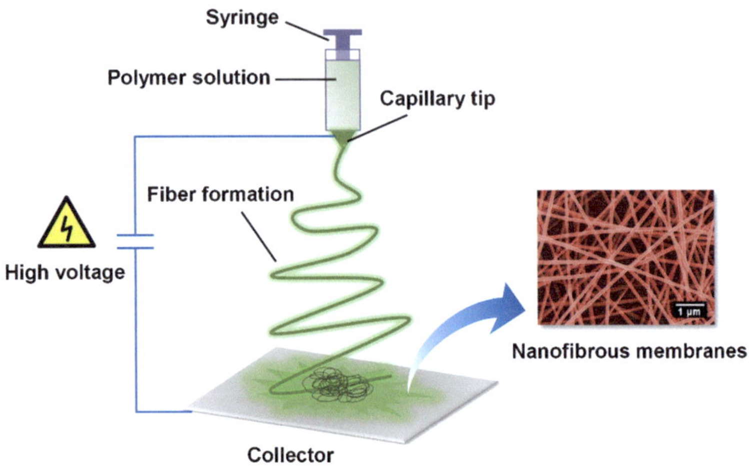

Fig. 1.4 Schematic illustration of the basic setup for electrospinning. The *inset* shows a typical SEM image of the PA6 nanofibrous membranes deposited on the collector

electrospinning process and the electrospun nanofibers, and also the number of patents for applications based on electrospinning has grown in recent years. Some companies such as eSpin Technologies, NanoTechnics, Elmarco Ltd., and KATO Tech are actively engaged in reaping the benefits of the unique advantages offered by electrospinning, while companies such as Donaldson Company and Freudenberg have been using this process for the last two decades in their air filtration products [32, 35, 40, 41].

1.2.2 Basic Principles

A typical electrospinning apparatus consists of three major components: a high-voltage power supply, a spinneret (a metallic capillary tip), and a collector (a grounded conductor). A schematic illustration of the basic setup for electrospinning is shown in Fig. 1.4. The syringe is filled with a melt or blend polymer solution, and a high voltage (typically 10–50 kV) is applied between the syringe tip and a collector, and the solution can be fed through the spinneret at a constant and controllable rate with the use of a syringe pump [30, 35, 42]. Electrospinning is fundamentally different from air or other mechanically driven spinning techniques in that the extrusion force is generated by the interaction between the charged polymer fluid and an external applied electric field. When high voltage is applied, the solution becomes highly charged, and as a result, the solution droplet at the tip will experience two major types of forces, the electrostatic repulsion force and surface tension. Under the action of these electrostatic interactions, a conical fluid structure called the Taylor Cone is formed at the tip of the syringe [30, 43, 44].

At a critical voltage, the repulsive force of the charged polymer overcomes the surface tension of the solution, and a charged jet erupts from the tip of the Taylor Cone, which moves toward the counter electrode and becomes narrower in the process. On the way to the counter electrode, the solvent evaporates (or the melt solidifies), and solid fibers with diameters ranging from micrometers to nanometers are often deposited as a randomly oriented, nonwoven mat; see the inset of Fig. 1.4 for the SEM image of a typical sample [35, 43, 45]. With the use of this relatively simple and straightforward method, more than 100 different types of materials have already been processed as fibers with diameters ranging from several micrometers to a few nanometers.

1.2.3 Theoretical Background

Electrospinning is a straightforward and inexpensive process that produces continuous nanofibers through an electrically charged jet of polymer solution. Although the setup for electrospinning is extremely simple, the physics behind it is very complex and very new to the scientific community, which involves the understanding of electro-statistics, fluid rheology, and polymer solution properties [30, 46]. In the past, the formation of ultrathin fibers by electrospinning was often ascribed to the splitting or splaying of the electrified jet as a result of repulsion between surface charges. Recent experimental observations indicated that the formation process is mainly controlled by the interaction of several physically unstable processes in electric field [31, 32, 37]. Currently, theories on electrospinning mainly include two moieties by observing the phenomena during the fiber formation: (1) liquid droplet under high voltage, Taylor Cone theory; and (2) jet in flight-instability theory.

The first step toward fiber initiation consists in the formation of a fluid droplet at the tip of the syringe, considering a droplet is positioned inside a capacitor. The interaction of the charged fluid with the external electric field gives rise to deformations of its shape relative to the field-free case, as shown in Fig. 1.5a, that is, the droplet assumes a prolate shape [31, 32, 47]. With increasing electric field the shape of the droplet becomes increasingly longer. When the voltage is increased to the critical potential and any further increase will destroy the equilibrium, the liquid body acquires a conical shape, with a half angle of 49.3°, referred to as the Taylor Cone [48, 49]. Recently, Yarin and Reneker [32, 50, 58] modified the Taylor Cone theory based on the experimental data, and they indicated that the Taylor Cone corresponds essentially to a specific self-similar solution, whereas nonself solutions exist which do not tend toward the Taylor Cone. Thus, the Taylor Cone does not represent a unique critical shape: another shape exists which is not self-similar. Moreover, they also proved that as a liquid surface develops a critical shape, its configuration approaches the shape of a cone with a half angle of 33.5°, rather than the previous angle of 49.3° as shown in Fig. 1.5b [47, 51].

Current research on electrospinning indicated that after the charged jets eject from the Taylor Cone in a nearly straight line during the process of electrospinning,

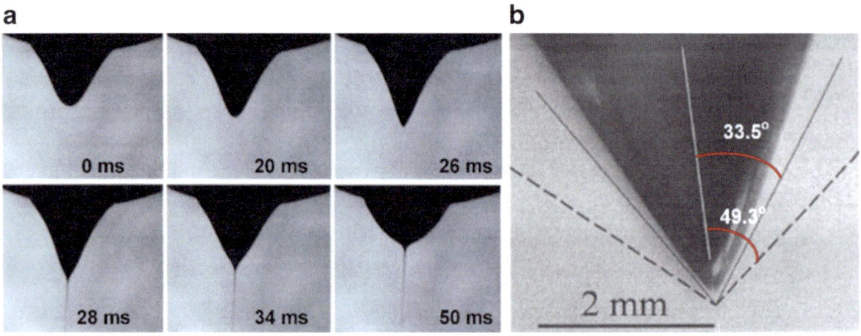

Fig. 1.5 (**a**) Deformation of a pending droplet by electric fields for various times following the application of a high electric potential close to a critical one. (**b**) Critical droplet shape observed for a pendant drop (**a**: Reprinted with permission from Reneker and Yarin [58]. © 2008, Elsevier. **b**: Reprinted with permission from Reneker et al. [47]. © 2001 AIP Publishing LLC)

the jets usually bend into a complex path, during which electric forces stretch and thin them by large ratios, as shown in Fig. 1.6a [32, 52, 53]. In some cases, splaying of the electrified jet might also be observed, though it was never a dominant process during spinning. To observe the different processes controlling the transition from the polymer solution to the final fiber, the experimental techniques, such as high-speed video analysis, tracer-particle tracking techniques, the analysis of interference colors indicating jet diameters, and other techniques, are often used [30–32]. Reneker et al. [54–56] studied the straight part of the jet by optical observation, that is, looking at the interference colors immediately allows the variation of the diameter of the jet to be followed as the distance from the tip increases, as shown in Fig. 1.6b. They revealed that the jet diameter decreases typically, by a factor of about 4 over a distance of 10 cm, which corresponds to the reduction of the cross section by a factor of about 20. It is obvious that a significant part of this decrease comes from the evaporation of the solvents and that a further contribution originates from longitudinal deformations of the jet induced by electric forces [32, 57, 58].

As shown in Fig. 1.6a, the distance away from the die of the jet is no longer able to follow a straight path in the direction of the counter electrode. Following, the jets will bend and turn sideways, and then it begins to perform spiraling, looping motions. Finally, in each loop the jet becomes thinner and elongated as the loop diameters increase. The envelope of these loops that is apparently in electrospinning observed by the naked eye resembles a cone with its opening oriented toward the counter electrode [55, 58]. Reneker et al. [54, 57, 58] treated the charged liquid jet as a system of connected, viscoelastic dumbbells and provided a good interpretation for the formation of bending instability. They demonstrated that the prototypical instantaneous position of the path of an electrospinning jet contained three successive electrical bending instabilities, as shown in Fig. 1.6c. The elongation of each segment by the electric forces caused by the charge carried with the jet continued. As the diameter of the jet decreased, the path of the jet

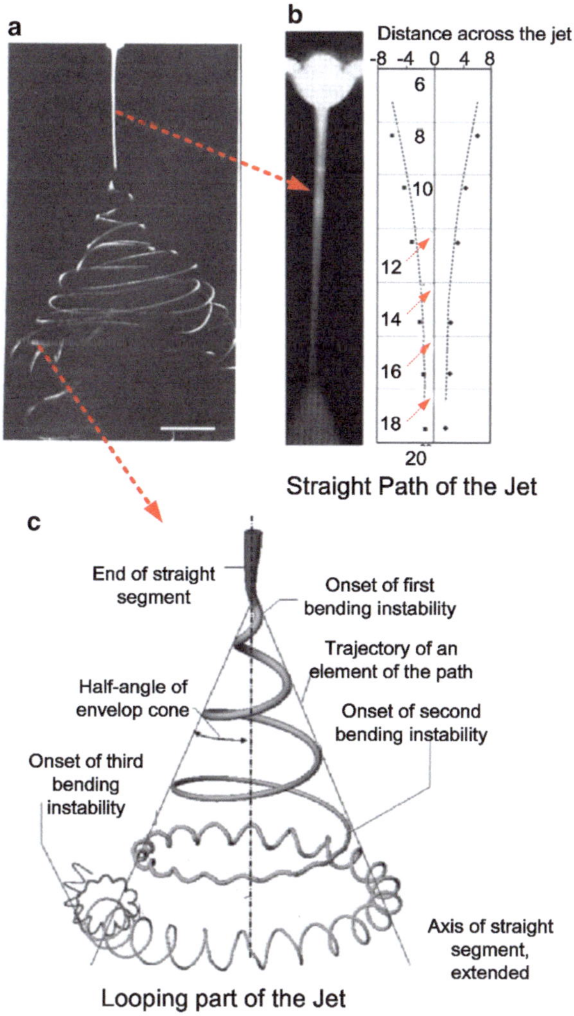

Fig. 1.6 (**a**) Experimental observation of the electrospinning process. (**b**) Linear part of the jet and the relevant results on the diameter variation along the jet as analyzed via interference colors. (**c**) Schematic drawing of the looping part of the jet showing a sequence of bending instabilities (**a** and **b**: Reprinted with permission from Wendorff et al. [32]. © 2012 Wiley-VCH. **c**: Reprinted with permission from Reneker and Yarin [58]. © 2008, Elsevier)

again became unstable, and a new, smaller diameter electrical bending instability developed [31, 32, 55]. A succession of three or more smaller diameter bending instabilities was often observed before the jet solidified. As a result, such jets acquired fractal-like configurations, and their length increased enormously as their cross-sectional diameter decreased to a fraction of a micron. After several turns were formed, a new electrical bending instability formed a smaller coil on a turn of the larger coil. The turns of the smaller coil transformed into an even smaller coil, and so forth until the elongation stopped, usually by solidification of the thin jet [32, 35, 45, 58]. Thanks to these instabilities, nanofibers with diameters down to a few nanometers can be generated stably and without decomposition of the jet into droplets. The typical stretching ratios of fiber-forming processes during electrospinning are in the range of 10^5, and stretching rates are up to 10^5 s^{-1}. Such values are not accessible with other methods, such as fiber extrusion followed by mechanical stretching [32, 35, 55, 58].

All these studies provide a better understanding of the mechanism responsible for electrostatic spinning process. More importantly, these characteristic processes can be predicted theoretically and controlled experimentally in such a way that a richness of fiber structures becomes accessible in a highly controlled way based on guidelines put forward by theoretical treatments.

1.2.4 Controlling Parameters on Electrospinning

The electrospinning process is controlled by many parameters, which mainly comprises of three parts: solution parameters, process parameters, and ambient parameters. Solution parameters include viscosity, conductivity, molecular weight, and surface tension, and process parameters include applied electric field, tip to collector distance, and feeding or flow rate. Each of these parameters significantly affects the morphology and structure of obtained electrospun nanofibers, and we can get nanofibers of desired morphology and diameters by proper manipulation of these parameters [27, 35]. In addition, ambient parameters encompassing the humidity and temperature of the surroundings also play a significant role in determining the structure of electrospun nanofibers. The summarized various parameters and their effects on fiber morphology are listed in Table 1.1 [27].

1.2.4.1 Solution Parameters

Polymer Concentration and Viscosity

The polymer concentration is the key factor to affect the chain entanglement; thus, the solution viscosity increased regularly with the increasing of polymer concentration. Although other factors such as molecular weight, solvent property, and temperature could also affect the viscosity, the polymer concentration is critical to control the viscosity during electrospinning, and the fiber formation of electrospinning process requires a minimum solution concentration [27, 35]. Generally, as at low concentrations beads are formed instead of fibers; meanwhile, at high concentrations the formation of continuous fibers is prohibited because of the inability to maintain the flow of the solution at the tip of the needle resulting in the formation of larger fibers [9, 35]. Figure 1.7 presents the detailed effect of polymer concentration on the morphology of polystyrene (PS) fibers [59]. Figure 1.7a shows that numerous beads were presented among fibers upon low concentration. With the increasing of concentration, the beaded fiber significantly decreased, and the shape of the beads changed from spherical to spindle-like,

Table 1.1 Electrospinning parameters and their effects on fiber morphology

Parameters	Effect on fiber morphology
Solution parameters	
Polymer concentration and viscosity	Low beads generation, high increase in fiber diameter, disappearance of beads. Increase in fiber diameter with increase of concentration
Conductivity	Decrease in fiber diameter with increase in conductivity
Surface tension	No conclusive link with fiber morphology, high surface tension results in instability of jets
Processing parameters	
Applied voltage	Decrease in fiber diameter with increase in voltage
Tip to collector distance	Generation of beads with too small and too large distance, minimum distance required for uniform fibers
Feed rate	Decrease in fiber diameter with decrease in flow rate, generation of beads with too high flow rate
Ambient parameters	
Humidity	High humidity results in circular pores on the fibers
Temperature	Increase in temperature results in decrease in fiber diameter

and finally uniform fibers with increased diameters were formed because of the higher viscosity resistance (Fig. 1.7b, c). The high concentration solution generated uniform fibers with no beads could be observed, as shown in Fig. 1.7d, which was due to the high entanglement of polymer chain. Moreover, the diameter of fibers from high concentration (30 wt%) was one order of magnitude larger than that of low concentration (5 wt%), indicating the remarkable increase of fiber diameter toward the concentration increase. Researchers have attempted to find a relationship between solution concentration and fiber diameter, and they found a power-law relationship, that the fiber diameter increases with the increasing of the solution concentration [27, 60, 61]. Similar results have also been reported about other polymer fibers such as polyurethane (PU), polylactide (PLA), polyvinyl chloride (PVC), polyamide6 (PA6), and chitosan (CS) [60, 62–65], which indicated the significant role of polymer concentration and viscosity in controlling the structure of electrospun fibers.

Conductivity

For electrospinning process to be initiated, the solution must gain sufficient charges such that the repulsive forces within the solution are able to overcome the surface tension of the solution. Subsequent stretching or drawing of the electrospinning jet is also dependent on the ability of the solution to carry charges [35]. Typically, with

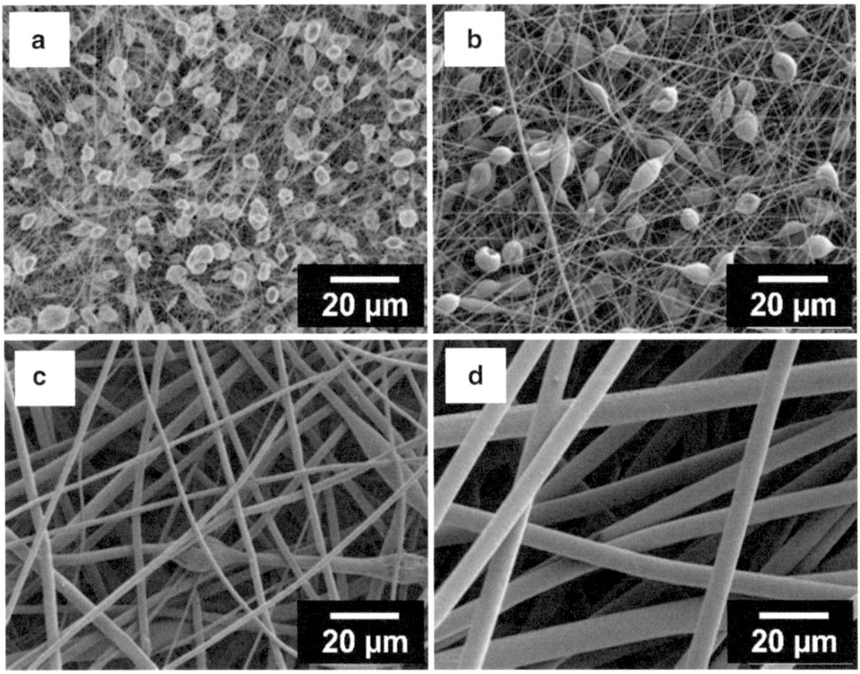

Fig. 1.7 FE-SEM images of electrospun PS fibers from various concentrations of (**a**) 5 wt%, (**b**) 10 wt%, (**c**) 20 wt%, and (**d**) 30 wt% (Reprinted with permission from Lin et al. [59]. © 2010, American Chemical Society)

the increase of electric conductivity of the solution, there is a significant decrease in the diameter of the electrospun nanofibers. Whereas with low conductivity of the solution, which results in insufficient elongation of a jet by electric force to produce uniform fiber, and beads may also be observed [27]. Previous theoretical study has also reported that the diameter of charged jets is in proportion to the third power of solution conductivity [66]. Solution conductivity is mainly determined by the polymer type, the solvent sort, and the salt. Fong et al. [67] have showed that introducing NaCl salts could greatly increase the conductivity of polyethylene oxide (PEO)/water solution and obtain uniform fibers with low diameter, and the beaded fibers were decreased as well. Zhang et al. [68] have demonstrated that the diameter of electrospun polyacrylonitrile (PAN) nanofibers decreased from 214 to 159 nm, which was due to the obvious increase of conductivity from 1.53 to 10.5 mS/cm. Natural polymers are generally polyelectrolytic in nature, for example, gelatin. The ions increase the charge-carrying capacity of the jet, thereby subjecting it to higher tension with the applied electric field. Thus, the fiber-forming ability of the gelatin is less as compared to the synthetic ones. Wang et al. [69] have reported the effect of ions by adding ionic salt on the morphology and diameter of electrospun gelatin fibers and found that with the addition of NaCl salts (0.1 wt%), the diameter of fibers (90 nm) was nearly one order of magnitude less than the pristine fibers (730 nm), and the fiber diameter uniformity enhanced as well. The strategy of increasing the solution conductivity by the use of salt addition has also been used for other polymers such as PLA, PA6, polyacrylic acid (PAA), and others [62, 70, 71]. With the use of salts, the uniformity of fibers increases and there is a decrease in beads generation.

Surface Tension

In electrospinning, the charges on the polymer solution must be high enough to overcome the surface tension of the solution. As the solution jet accelerates from the tip of the source to the collector, the solution is stretched, while surface tension of the solution may cause the solution to break up into droplets [27, 32, 35]. Generally, the high surface tension of a solution inhibits the electrospinning process because of instability of the jets and the generation of sprayed droplets. Zeng et al. [72] have studied the effect of surface tension on the structure of PLA fibers, and the experiment result showed that the stretch effect of charged jets increased obviously toward the decrease of surface tension by adding surfactant, finally generating nanofibers with high uniformity. The formation of droplets, beads, and fibers depends on the surface tension of solution, and a lower surface tension of the spinning solution helps electrospinning to occur at a lower electric field. These phenomena have been also found in the electrospinning process of PS, PVC, PEO, and CS [63, 65, 67, 73]. In summary, surface tension determines the upper and lower boundaries of the electrospinning window if all other variables are held constant.

1.2.4.2 Processing Parameters

Applied Voltage

A crucial element in electrospinning is the application of a high voltage to the solution. The high voltage will induce the necessary charges on the solution and, together with the external electric field, will initiate the electrospinning process when the electrostatic force in the solution overcomes the surface tension of the solution. Generally, both high negative and positive voltages of more than 6 kV are able to cause the solution drop at the tip of the needle to distort into the shape of a Taylor Cone during jet initiation [9, 27, 35]. However, the affection of the applied voltages on the diameter of electrospun fibers is a little controversial, which is based on the properties of polymer solution. Fong et al. [67] reported that the beaded PEO fiber structure was enhanced obviously with the voltage increase from 5 to 18 kV, and the fiber diameter decreased as well. A possible explanation for this result could be that the higher voltage causes greater stretching of the solution due to the greater columbic forces in the jet as well as a stronger electric field, which lead to the generation of beaded structure and also reduction in the fiber diameter. Similar behavior of applied voltage on fiber structure is also observed for polyvinyl

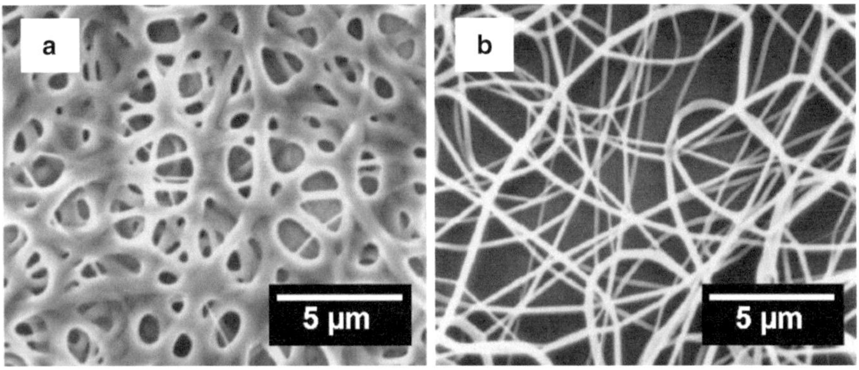

Fig. 1.8 FE-SEM images of electrospun PA6 nanofibers with the tip to collector distance of (**a**) 0.5 cm and (**b**) 2 cm (Reprinted with permission from Buchko et al. [78]. © 1999, Elsevier)

alcohol (PVA), PLA, poly ε-caprolactone (PCL), and PS [62, 74–76]. Moreover, it is worth to note that the voltage influences fiber diameter, but the level of significance varies with the polymer solution concentration and on the distance between the tip and the collector.

Tip to Collector Distance

The flight time as well as the electric field strength will affect the electrospinning process and the resultant fibers. Varying the distance between the tip and the collector will have a direct influence on both the flight time and the electric field strength. During fiber formation, the charged jet must be allowed enough time for most of the solvents to be evaporated. With a low tip to collector distance, the jet will have a shorter distance to travel before it reaches the collector plate, and the electric field strength will also increase at the same time; thus, maybe there is not enough time for the solvents to evaporate [27, 28, 35]. Lee et al. [77] have found that a minimum distance is required to give the fibers sufficient time to dry before reaching the collector; otherwise, with distances that are either too close or too far, beads have been observed. Buchkoa et al. [78] have reported that the extremely low distance (0.5 cm) may cause the fibers to merge where they contact to form junctions resulting in inter- and intralayer bonding (Fig. 1.8), which was due to the excess solvent in nanofibers. On the other hand, the longer distance means that there is a longer flight time for the solution to be stretched before it is deposited on the collector, and obvious increase in fiber diameter could be observed which was due to the decrease in the electrostatic field strength resulting in less stretching of the fibers [27, 43]. Summarily, there is an optimum distance between the tip and collector which favors the evaporation of solvent for each electrospinning system.

Feed Rate

The feed rate will determine the amount of solution available for electrospinning. For a given voltage, there is a corresponding feed rate if a stable Taylor Cone is to be maintained. When the feed rate is increased, there is a corresponding increase in the fiber diameter or beads size [35]. Hartman et al. [79] have demonstrated that the diameter of charged jet (D) was related to the feed rate (Q) with the relationship of $D \sim Q^{0.48}$, which was due to the bending instability of polymer jets. Moreover, there is always a minimum flow rate of the spinning solution. It has been observed that the fiber diameter and the pore diameter increase with an increase in the polymer flow rate in the case of PS fibers and by changing the flow rate, indicating that the morphological structure can be slightly changed [27, 80].

1.2.4.3 Ambient Parameters

Humidity

The humidity of the electrospinning environment may have an influence on the polymer solution during electrospinning. At high humidity, it is likely that water condenses on the surface of the fiber when electrospinning is carried out under normal atmosphere. As a result, this may have an influence on the fiber morphology especially polymer dissolved in volatile solvents [35, 81]. Casper et al. [82] have showed detailed affection of humidity on the porous structure of PS nanofibers, as presented in Fig. 1.9. It is clearly shown that the porous structure is significantly enhanced with the increase of humidity, and the size and depth of the relevant pores increased as well. Further increasing the humidity, the depth, diameter, and number of the pores start to saturate. Moreover, the humidity will also determine the rate of evaporation of the solvent in the solution. At a very low humidity, a volatile solvent may dry very rapidly; thus, the solvent evaporation rate may be faster than the removal of the solvent from the tip of the needle. As a result, the electrospinning process may only be carried out for a few minutes before the needle tip is clogged.

Temperature

The temperature of the solution has both the effect of increasing its evaporation rate and reducing the viscosity of the polymer solution, thus enabling the polymer solution which was spinnable under room temperature to be electrospun easily. For example, the hyaluronic acid is a natural polyelectrolyte that is hard to be electrospun under room temperature [27]. Li et al. [83] used the high electrospinning temperature that was higher than the gel temperature of hyaluronic acid (40 °C) and thus successfully fabricated hyaluronic acid composite nanofibers with uniform morphology. Similar method was also suitable for the fabrication of other natural

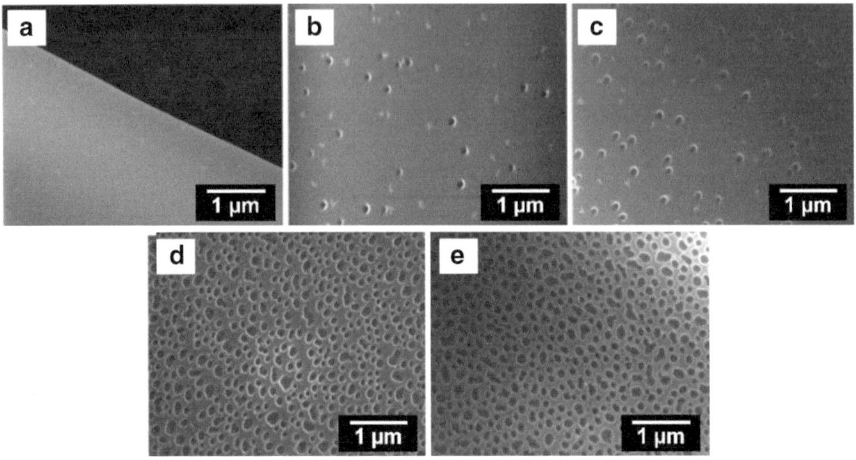

Fig. 1.9 FE-SEM images of PS fibers electrospun under varying humidity: (**a**) <25 %, (**b**) 31–38 %, (**c**) 40–45 %, (**d**) 50–59 %, and (**e**) 60–72 % (Reprinted with permission from Casper et al. [82]. © 2004 American Chemical Society)

polymers such as alginate and gelatin [84]. Demir et al. [60] have demonstrated that higher temperature could enable the electrospun polyurethane fibers to have a more uniform diameter, which could be attributed to the lower viscosity of the solution and greater solubility of the polymer in the solvent which allows even greater stretching of the solution. Consequently, increased polymer molecules mobility due to increased temperature also allows the columbic torte to stretch the solution further. However, when the biological substances such as enzymes and proteins are added to the solution, the use of high temperature may cause the nanofibers to lose their functionality.

1.3 Electrospun Nanofibers

1.3.1 Materials Classes

There are numerous types of materials that could be used for electrospinning, and individual material properties must be considered depending on its applications. The electrospinning process may be modified so as to yield electrospun fiber with the desired morphology and properties. In the production of ceramic fibers, post processes are required after the fibers are electrospun. Up to now, more than 200 types of electrospun nanofibers have been fabricated, including natural polymer, synthetic polymer, ceramic, and carbon. Thus, it is important to have a basic understanding of the different groups of materials before selecting the most appropriate electrospun fibers for specific applications.

Table 1.2 Different natural polymers used in electrospinning

Materials	Molecular weight	Solvent	Diameter (nm)	References
Cellulose acetate	100,000	Acetone/DMAc	100–1,200	[85]
Chitosan	70,000–700,000	TFA	58–260	[86]
Chitin	91,000	HFIP/PBS	~130	[87]
Hyaluronic acid	1,500,000	DMF/water	50–700	[88]
Silk fibroin	26,000–35,000	Formic acid	1,600–1,700	[89]
Fibrinogen	–	PBS/ethanol	185–190	[90]
Wheat protein	–	HFIP	100–1,000	[91]
Gelatin	–	TFE	100–340	[92]
Elastin	81,000	Water	200–3,000	[93]
Zein	35,000	Water/ethanol	~500	[94]
Dextran	64,000–76,000	Water	290–1,950	[95]
Deoxyribonucleic acid	124,000–186,000	Water/ethanol	–	[96]

DMAc dimethylacetamide, *TFA* trifluoroacetic acid, *HFIP* 1,1,1,3,3,3-hexafluoro-2-propanol, *DMF* dimethylformamide, *PBS* phosphate buffer saline, *TFE* trifluoroethanol

1.3.1.1 Natural Polymer

One of the greatest potentials in electrospun fiber is in the area of bioengineering. For many biomedical applications, the materials used have to be biocompatible; thus, natural polymers have a distinct advantage over synthetic materials. Since most natural polymer can be degraded by naturally occurring enzymes, it can be used in applications where temporary implants are desired or in drug release. It is also possible to control the degradation rate of the implanted polymer by chemical cross-linking or other chemical modifications, thus allowing greater versatility in the design of the implant [27, 30]. To date, the reported natural polymers used in electrospinning are summarized in Table 1.2, which mainly include polysaccharides and proteins.

Cellulose is one of nature's most abundant polymers that has received much attention in research to modify its physical and/or chemical structure to improve its properties and broaden its applications. Frey et al. [97] reported the fabrication of electrospun cellulose nanofibers from waste cellulose materials, suggesting a promising way to recycle these biomass resources. In 2006, Viswanathan et al. [98] first used nonvolatile ionic liquid as solvent for the electrospun of cellulose, and after removal of the residual ionic liquid in fibers by ethanol, the cellulose nanofibers were obtained. As shown in Fig. 1.10, the relevant fibers have a smooth surface with high adhesion structure, which could be attributed to the high viscosity and nonvolatile property of ionic liquid solution. Although several studies have demonstrated the preparation of cellulose nanofibers by directly using cellulose as materials, the strong hydrogen bond in cellulose made it hard to be dissolved by common solvent; thus, the use of nonvolatile ionic liquid solvent limited the controlling of fibrous structure. Alternatively, cellulose derivatives such as cellulose acetate, ethyl cellulose, and hydroxypropyl methyl cellulose which have good

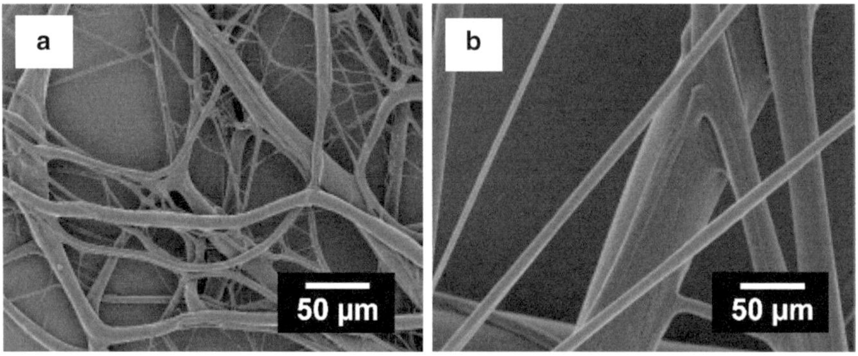

Fig. 1.10 (**a**) FE-SEM image and (**b**) high magnification FE-SEM image of electrospun cellulose nanofibers (Reprinted with permission from Viswanathan et al. [98]. © 2006 American Chemical Society)

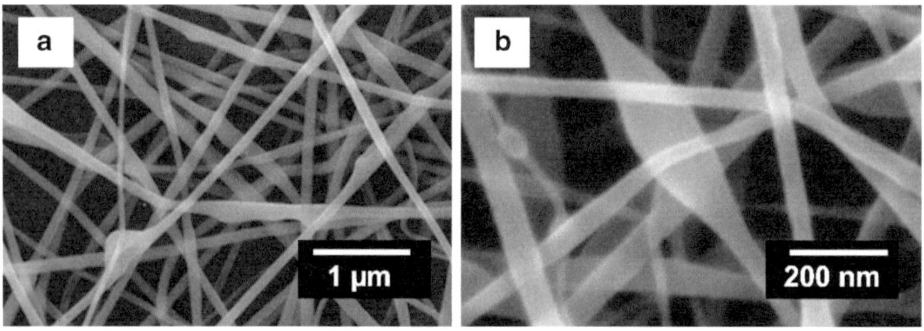

Fig. 1.11 FE-SEM images of nanofibers electrospun from (**a**) N. clavipes silk and (**b**) B. mori silk solution (Reprinted with permission from Zarkoob et al. [100]. © 2004, Elsevier)

solubility are better raw materials for the preparation of cellulose nanofibers. Therefore, most cellulose nanofibers were obtained by the hydrolysis of electrospun cellulose derivative nanofibers [99].

Silk is a well-described natural fiber obtained from silkworms (Bombyx mori and Antheraea mylitta) and spiders and has been used in textile industries for thousands of years. Generally, silk proteins are of two kinds, namely, hydrophobic fibroin and hydrophilic sericin. Silk fibroin protein has been extensively studied as one of the promising materials for biomedical applications due to the distinctive biological properties including good biocompatibility, biodegradability, and minimal inflammatory reaction and excellent mechanical properties [27, 35]. In 2004, Zarkoob et al. [100] first reported the preparation of silk electrospun nanofibers by dissolving silk protein in HFIP solution. As shown in Fig. 1.11, the obtained silk nanofibers had the fiber diameter of 100–200 nm, which was significant lower than that of natural silk fiber. Moreover, the silk nanofibers with annealing showed similar crystal structure with the original silk fibroin and had better mechanical strength. The high surface area to volume ratio of the silk nanofiber could encourage cell attachment, growth, and proliferation, thus implying as a prominent material for tissue engineering application.

Hyaluronic acid (HA) is another naturally occurring polysaccharide commonly found in the specialized tissues such as synovial fluid, dermis, and cartilage. This molecule plays an important role among the interstitial proteins of the extracellular matrix, providing important biological and tissue engineering. Generally, it was difficult to form uniform size fibers from HA nanofibers using electrospinning due to the high viscosity, surface tension, and strong water retention ability of the HA solution [27, 29, 38]. The water retention ability might lead to the fuse of nanofibers electrospun on the collector due to the insufficient evaporation of solvents in electrospinning. In 2004, Um et al. [101] showed several approaches to prepare a HA solution that had sufficient molecular entanglement in a rapidly evaporated solvent while still maintained low viscosity and surface tension by using the electroblowing technique. Thus the HA nanofibers can be successfully fabricated by controlling the rate and the temperature of air blow. Consequently, the high surface area to volume ratio of the HA nanofibrous membrane would make it attractive in applications such as tissue scaffolds, wound dressings, and artificial blood vessels.

1.3.1.2 Synthetic Polymer

To date, there are many polymers that have been electrospun including custom-made polymers. Listed in Table 1.3 are commonly used synthetic polymers and the corresponding molecular weight and solvent used that is able to yield fibers without beads.

There are water-soluble polymers such as PVA, PAA, polyvinylpyrrolidone (PVP), polyacrylamide (PAM), polyethylene glycol (PEG), and polyethyleneimine (PEI). These polymers can be directly electrospun using water as solvent, which avoid the treatment of toxic organic solvent vapor. On the other hand, these water-soluble polymer nanofibers are easy to swelling or collapse upon moisture; thus, the obtained nanofibers should be kept in special dry environment to prevent the fiber structure from destroying.

Besides water-soluble polymers, more synthetic polymers are insoluble in water and should be dissolved in organic solvent for electrospinning, such as polylactic-co-glycolic acid (PLLA), PCL, polybutylene succinate-co-butylene terephthalate (PBST), polyhydroxyalkanoates (PHA), polybutylene succinate (PBS), PAN, polysulfone (PSF), polyimide (PI), polyethylene-co-vinyl alcohol (PEVA), PU, polypyrrole (PPy), polyoxymethylene (POM), PS, polymethyl methacrylate (PMMA), PVC, polyvinylidene fluoride (PVDF), polytrimethylene terephthalate (PTT), polybutylene terephthalate (PBT), poly *N*-vinylcarbazole (PVK), polymeta-phenylene isophthalamide (PMIA), polyethylene terephthalate (PET), polycarbonate (PC), polybenzimidazole (PBI), polymer vinyl acetate (PVAc), polyvinyl butyral (PVB), and polyethylene-co-vinyl acetate (PEV).

Table 1.3 Different synthetic polymers used in electrospinning

Materials	Molecular weight	Solvent	Diameter (nm)	References
PVA	~65,000	Water	50–250	[102]
PAA	~450,000	Water	~100	[103]
PVP	85,000–95,000	Ethanol	300–2,400	[104]
PAM	~6,000,000	Water	200–2,000	[105]
PEG	~10,000	DMAc/acetone	~1,500	[106]
PEI	~215,000	Methanol	460–1,200	[107]
PLLA	100,000–300,000	DCM	600–1,000	[108]
PCL	~80,000	Acetone	1,000–1,500	[109]
PBST	~20,000	Isopropyl alcohol/chloroform	100–700	[110]
PHA	–	Chloroform	100–1,000	[111]
PBS	200,000–300,000	Chloroform	~1,000	[112]
PAN	~75,000	DMF	50–1,300	[113]
PSF	–	DMAc/acetone	300–400	[80]
PI	–	DMF	1,000–4,000	[114]
EVAH	–	Alcohol/water	200–8,000	[115]
PEO	~400,000	Water	~300	[116]
PU	~30,000	DMAc/DMF	500–2,000	[117]
PPy	~62,300	Water	70–300	[118]
POM	–	HFP	100–1,000	[119]
PS	~350,000	Butanone	2,000–3,000	[120]
PMMA	~350,000	DCM/DMF	~2,000	[121]
PVC	~81,000	DMF/THF	~400	[122]
PVDF	~275,000	DMAc/acetone	100–400	[123]
PTT	–	TFA/MC	200–600	[124]
PBT	–	TFA/DCM	200–800	[125]
PVK	~1,100,000	DMF	200–600	[126]
PMIA	–	DMAc/LiCl ionic liquid	100–200	[127]
PET	18,000–25,000	TFA/dichloromethane	1,000–2,000	[128]
PC	~27,000	DMF/THF	100–500	[129]
PBI	–	DMAc	200–300	[130]
PVAc	~140,000	Ethanol	~700	[131]
PVB	70,000–100,000	Water/ethanol	900–1,500	[132]
PEVA	–	Methylene dichloride	~1,000	[133]

1.3.1.3 Ceramics

Ceramic nanofibers can be made from electrospinning of the ceramic precursors and followed by sintering of the electrospun fibers to derive ceramic fibers. Ceramics are materials that are composed of both metallic and nonmetallic elements and commonly exist as compounds of oxides, nitrides, and carbides. While most ceramics are crystalline, there are common noncrystalline ceramics like window glass which is made of silicon dioxide. As the atomic bonding is either ionic or covalent, there are no "free" electrons in ceramics, making them excellent insulators [35, 43, 134]. The strong ionic and covalent bonding gives the complex

Table 1.4 Different types of ceramic nanofibers fabricated by electrospinning

Materials	Precursor	Solvent	Diameter (nm)	References
TiO_2	Titanium tetraisopropoxide/PVP	Acetic acid	~200	[135]
SiO_2	Tetracthyl orthosilicate/PVA	Water	200–300	[136]
ZnO	Zinc acetate/PVA	Water	~100	[137]
In_2O_3	Indium nitrate/PVP	DMF/ethanol	60–160	[138]
Al_2O_3	Aluminum chloride/PVP	Water	~1,000	[139]
CuO	Copper acetate/PVA	Water	~60	[140]
NiO	Nickel acetate/PVA	Water	50–150	[141]
GeO_2	Germanium isopropoxide/PVAc	Isopropanol	~325	[142]
V_2O_5	Vanadium oxide isopropoxide/PVAc	Water/ethanol	~500	[143]
Mn_2O_3	Manganese acetate/PVP	Ethanol	38–80	[144]
ZrO_2	Zirconium oxide chloride/PVP	Water/ethanol	~40	[145]
SnO_2	Stannous chloride/PVAc	DMF	100–150	[136]
SiC	Tetraethyl orthosilicate/PVP	Ethanol	~2,000	[146]
$CoFe_2O_4$	$Co(NO_3)_2/Fe(NO_3)_3$/PVP	DMF	~320	[147]

structure of ceramics many advantages such as high-temperature stability, resistance to chemical attacks, and adsorption of foreign atoms, which have wide applications in electronic and optical nanodevices, chemical and biological sensors, catalysis and electrocatalysis, superhydrophobic surfaces, environment, energy, and biomedical fields [26, 28, 30]. To date, dozens of ceramic nanofibers have been successfully fabricated by using electrospinning, as summarized in Table 1.4.

Just like the pristine ceramic materials, the strong ionic and covalent bonding also gives these ceramic nanofibers high brittleness. Typically, although the ceramic nanofibers exhibited continuous fiber structure under microscale SEM measurement, most of them showed the powdered macro morphology due to the weak mechanical properties, which significantly limited the practical applications. In 2010, Ding et al. [148] for the first time reported the preparation of flexible SiO_2 nanofibers by electrospinning. As shown in Fig. 1.12a, b, the as-prepared SiO_2 nanofibers had a fiber diameter of 300–500 nm with robust flexibility. After the FAS modification, the SiO_2 nanofibers showed promising amphiphobicity with the water contact angle of 154° and oil contact angle of 144° (Fig. 1.12c). Additionally, the fluorinated inorganic fibrous mats exhibited a high heat resistance; they kept their amphiphobicity even after the annealing treatment at 450 °C for 30 min (Fig. 1.12d). Potential applications of the fluorinated fibrous mats include high-temperature filtration, selective filtration, and self-cleaning coatings.

1.3.1.4 Carbon

Carbon fibers were first prepared from rayon fibers by the US Union Carbide Corporation and the US Air Force Materials Laboratory in 1959. With the development of nanotechnology in fiber fields, carbon nanofibers (CNFs) gradually attracted

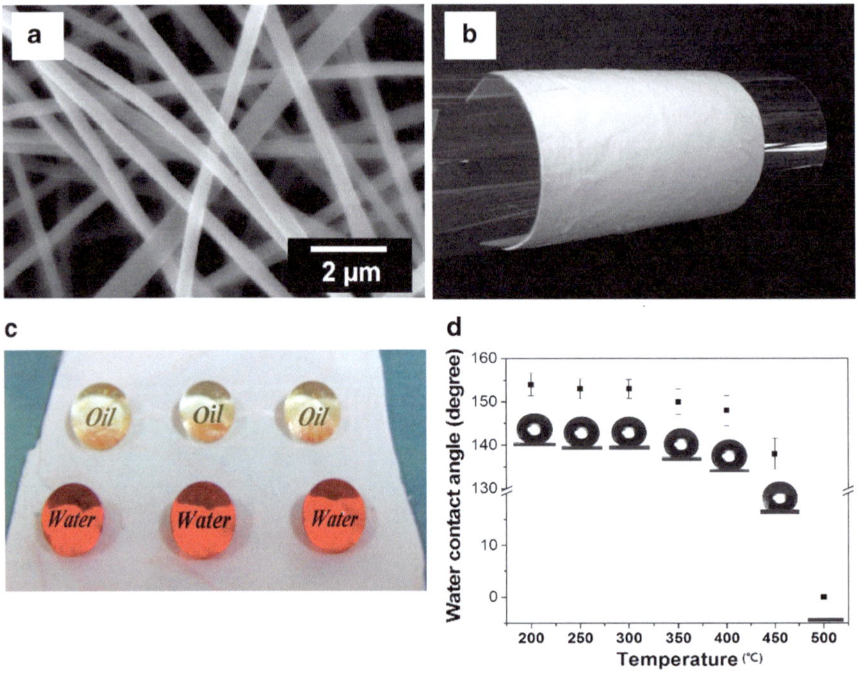

Fig. 1.12 (**a**) FE-SEM images of SiO_2 nanofibers. (**b**) Optical image of flexible SiO_2 nanofibrous membranes when bent with a PET film. (**c**) Several water and dodecane droplets placed on the FAS-modified SiO_2 membranes showing superhydrophobicity and oleophobicity. (**d**) Water contact angle and the corresponding shapes of water droplets for the FAS-modified SiO_2 membrane annealed at different temperatures (Reprinted with permission from Guo et al. [148]. © 2010, American Chemical Society)

much attention after the discovery of carbon nanotubes by Iijima in 1991 [42, 149]. The CNFs are of great interest, owing to their high aspect ratio, large specific surface area, high-temperature resistance, and good electrical/thermal conductivities, and have various applications such as catalyst carriers, electrochemical probes, energy storage electrodes, supercapacitors, high-temperature filters, and adsorbents [35, 150]. Electrospinning is a versatile process to obtain continuous carbon nanofibers at low cost. Generally, the most reported method is the preparation of electrospun-derived CNFs from PAN precursor as PAN is a well-known precursor for the traditional microsized carbon fibers. Other precursors, such as pitch, PVDF, PAA, PI, cellulose, PBI, PVA, and polybenzoxazine (PBZ), have also been reported. In particular, highly rigid polymers (e.g., PAA, PI, PBI, and PBZ), without the need of the time-consuming and costly stabilization process, are of great importance as precursors giving potentially large energy saving [29, 150, 151].

The performance of the resultant electrospun CNFs is highly dependent on the chemical and physical natures of the precursor nanofibers, which is because the internal and surface defects of precursors may be easily transferred into the resultant CNFs leading to performance deterioration. Typically, stabilization and carbonization merely refine and perfect the as-spun structure without changing the initial morphology significantly. Therefore, the careful control of the electrospinning parameters, such as solution properties (e.g., concentration, solvent type, conductivity, temperature, and surface tension), process parameters (e.g., applied voltage, spinneret diameter, polymer flow rate, collecting distance, and collector type), as well as ambient conditions (i.e., temperature and humidity), is crucial to obtain high-performance precursor nanofibers [30, 149, 151].

1.3.2 Morphologies and Structures

Electrospun nanofibers are featured with very small diameters, extremely long length, large surface area per unit mass, and small pore size. Various materials such as polymers, ceramics, and carbon can be used to electrospin uniform fibers with well-controlled sizes, compositions, and morphologies. Different nanofiber structures and morphologies can be obtained via control of the processing conditions, and these structure regulations could be mainly divided into two categories: single fiber structure and aggregate structure.

1.3.2.1 Single Fiber Structure

In the majority of cases, the goal of electrospinning will be the production of nanofibers uniform in diameter along their complete length and with a circular cross-sectional shape. In fact, electrospinning tends to produce just such fibers in the range of polymer solution concentrations in which droplet formation is suppressed, which is because the geometries of the dies, of the applied field, and thus of the deformation of the droplets are in the majority of cases axially symmetric [28, 32, 35]. The single fiber structure is of significant importance as far as the packing of the fibers is concerned in nonwovens, that is, the total porosity. Furthermore, the structure is of significance with respect to the internal specific surface as controlled by geometric effects and also with respect to the shape of the pores controlled by intersecting fibers as well as with respect to the flow of fluids or gases around the fibers, that is, the permeation properties [29, 32, 43].

Generally, it is important to know how to fabricate fibers with circular cross sections at will, and the aim of electrospinning for particular applications is in general to obtain just these particular types of fibers. On the other hand, other types of structure of fibers may be beneficial for specific functionality and applications. To date, by carefully controlling the preparing conditions, a variety of specific fiber structures originating from the complex self-assembly processes intrinsic in electrospinning have been fabricated, such as ribbon fiber, helix fiber, porous fiber, necklace-like fiber, core-shell fiber, and hollow fiber, as shown in Fig. 1.13.

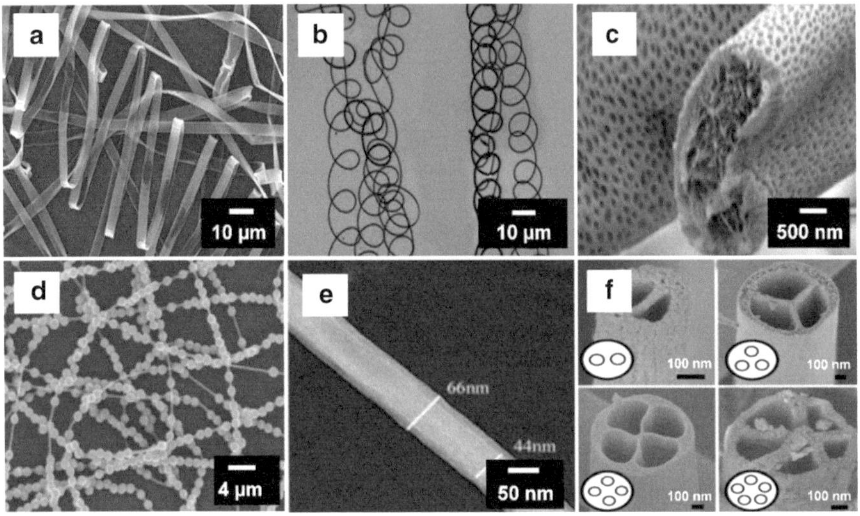

Fig. 1.13 Various types of single fiber structure: (**a**) PEI ribbon fiber, (**b**) PA6 helix fiber, (**c**) PS porous fiber, (**d**) PVA/SiO_2 necklace-like fiber, (**e**) PSU/PEO core-shell fiber, and (**f**) TiO_2 hollow fiber (**a**: Reprinted with permission from Koombhongse et al. [152]. © 2001 John Wiley & Sons, Inc. **b**: Reprinted with permission from Han et al. [57]. © 2007, Elsevier. **c**: Reprinted with permission from Lin et al. [59]. © 2010, American Chemical Society. **d**: Reprinted with permission from Jin et al. [154]. © 2010, American Chemical Society. **e**: Reprinted with permission from Sun et al. [155]. © 2003 WILEY-VCH Verlag GmbH& Co. KGaA, Weinheim. **f**: Reprinted with permission from Han et al. [57]. © 2007, Elsevier)

The ribbon fiber is a frequent observation in electrospinning using high concentration polymer solution, which is characterized by a flat rectangular cross-sectional area (Fig. 1.13a) [152]. The diameters of the band-type fibers increase with increasing polymer concentration in a similar way as observed for fibers with circular cross-sectional areas discussed previously [32]. The helix fiber is usually fabricated via two main strategies: one is using bicomponent electrospinning that the two polymer components have different elasticity and another is the side-by-side electrospinning method. The helix structure endows the fibrous membranes with high porosity and flexibility, which could further enhance the application performance (Fig. 1.13b) [57, 153]. Moreover, introducing porous structure into nanofibers leads to even higher surface area and larger pore volumes, thereby resulting in further improvements in relevant applications. Ding et al. [59] reported the preparation of highly porous PS fibers by the control of solvent evaporation character, as shown in Fig. 1.13c, which was due to the solvent-induced phase separation during electrospinning. The necklace-like fiber is usually fabricated by adding inorganic nanoparticle in polymer solution (Fig. 1.13d) [154], and the fiber structure could be controlled by changing the nanoparticle size and concentration. The core-shell fibers are commonly prepared by the specific core-shell electrospinning (Fig. 1.13e) [155], and similarly, the hollow fiber is obtained by the removal of the relevant core component by using physical or chemical methods (Fig. 1.13f) [156].

1.3.2.2 Aggregate Structure

The deposition of nanofibers on a counter electrode or on a specific planar substrate such as copper screen, papers, and textiles is characterized by the aggregate structure, including random orientation structure, alignment structure, patterning structure, and two-dimensional (2D) nanonets [28]. The aggregate structure is mainly affected by two factors: one is the processes controlling the transition from the fluid jet emerging from the die to the solid fiber at rest on the substrate and another is the shape of the counter electrode [29, 32]. Generally, the aggregate structure will influence the packing density of the fibrous membranes and the pore dimensions and shapes, but they might also give rise to local anisotropies of fiber orientations and pore shapes [30, 35].

In the majority of cases, there will be a random fiber orientation within the deposition plane, and the fibers that are deposited tend to be straight, at least for a length that is much larger than the diameter of the fibers. As shown in Fig. 1.14a, the representative FE-SEM image of the PAN nanofibrous membranes indicates randomly oriented three-dimensional (3D) nonwoven membranes with a uniform diameter of 200–300 nm, revealing a typical random deposition aggregate structure of electrospinning nanofibers [28, 151]. Moreover, alignment nanofibers that are completely straight within the deposition plane become accessible by using specific counter electrode configuration. One method is the use of two parallel electrodes in the deposition plane giving rise to a parallel deposition of the fibers, which is due to the anisotropic field imposed on the jet. Another method for the preparation of alignment nanofibers is the use of high-speed rotating roller as collecting electrodes; the strong stretching of the roller makes the fiber to be aligned among the rotating orientation, as shown in Fig. 1.14b, the highly aligned PMIA nanofibers [157]. Similarly, various types of patterning aggregate structure of nanofibers could be fabricated by controlling the collecting electrodes, including materials, sharps, mode of motions, etc. For example, the uniform mesh-like PLLA patterning nanofibers were fabricated by using electroconductive collector with woven structure, as shown in Fig. 1.14c [158].

Electrospun nanofibrous membranes typically have fiber diameters in the range of 100–500 nm, but properties such as surface area and porosity become more significant when the fiber diameter falls below 50 nm. Great efforts have been made to study the effects of basic processing parameters on the structure and morphology of electrospun fibers, but the objective of reducing fiber diameter to 50 nm was rarely achieved [28]. In 2006, a major breakthrough was reported by Ding et al. for the direct formation of a large number of netlike structured nanowires with small diameters (~20 nm) which distribute among conventional

Fig. 1.14 Different aggregate structures of electrospun nanofiber: (**a**) PAN nonwoven nanofibers, (**b**) PMIA alignment nanofiber, (**c**) PLLA patterning nanofiber, and (**d**) PA6 two-dimensional nanonets (**a**: Reprinted with permission from Ren et al. [151]. © 2012, Royal Society of Chemistry. **b**: Reprinted with permission from Wang et al. [157]. © 2013, Royal Society of Chemistry. **c**: Reprinted with permission from Zhang and Chang [158]. © 2007 Wiley-VCH Verlag GmbH & Co. KGaA, Weinheim. **d**: Reprinted with permission from Wang et al. [160]. © 2011, Royal Society of Chemistry)

electrospun fibers [159]. This new type of aggregate structure was termed as "nanonets," which was obtained by employing the novel one-step "electronetting" process. Figure 1.14d shows the FE-SEM images of PA6 nanonets, and the primary electrospun fibers acted as a support for the nanonets comprising interlinked ultrathin nanowires with a diameter of 18 nm [160]. This structure matched well with that of a typical reticular soap bubble with a clear ideal and weighted Steiner-tree structure, in which three neighboring nanowires form a three-way junction with angular symmetry. The possible formation mechanism of nanonets concerns a combined effect of instabilities and rapid phase separation on charged droplets, which has been discussed in our previous reports. The instability of the Taylor Cone induced by the high electric field results in the formation of the electrospray droplets. During the flight of charged droplets from the capillary tip to the collector, the microsized droplet was distorted and expanded into a thin film due to the comprehensive effects of the forces acting on it. The occurrence of rapid phase separation on the splitting film and obeying the minimal energy principle may lead to the formation of such an unexpected structure [45]. To date, a variety of types of nanonets including PA6, PVA, PAA, PMIA, PAA, Gelatin, PU, and CS have been fabricated [28, 45, 69, 160–162], which would be particularly useful for applications such as ultrafiltration, ultrasensitive sensors, catalyst supports, etc.

1.4 Electrospun Nanofibers in Energy and Environmental Applications

As introduced above, much effort has been applied to electrospun polymer nanofibers to comprehend the fundamental phenomena of fabrication process as well as the physical and chemical properties from the material science viewpoint. Consequently, researchers are able to flexibly control the fiber component, morphology, and aggregate structure, which have provided much impetus to the dream of realizing the applications of electrospun nanofibers [2, 30, 35]. Particularly, electrospinning nanofibers have provided important evidence of the potential roles in solving the energy and environmental issues, which are the most important factors that influence the society in the twenty-first century.

In view of the cleaning energy, researchers have been investigating opportunities such as using solar energy more efficiently and cost effectively for generation of electricity, efficient energy conservation and storage, and novel electrochemical devices [2, 4]. Electrospinning technology responds to these challenges by designing, fabricating, and evaluating functional nanofibers optimized to apply in fuel cells, lithium ion batteries, dye-sensitized solar cells, organic and hybrid solar cells, supercapacitors, hydrogen storage, thermal storage, and carbon dioxide capture. As shown in Fig. 1.15a, the major applications in energy are lithium ion batteries,

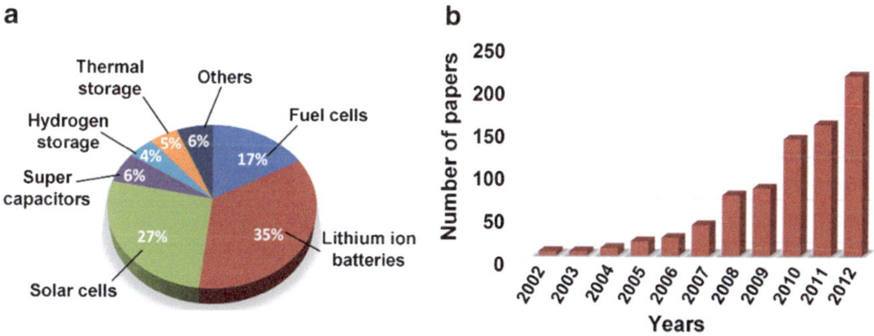

Fig. 1.15 Statistics on the literature published on the applications of nanofibers in energy field. (**a**) Distribution of the relevant energy applications. (**b**) Numbers of publications from 2002 to 2012 (Source from ISI web of science)

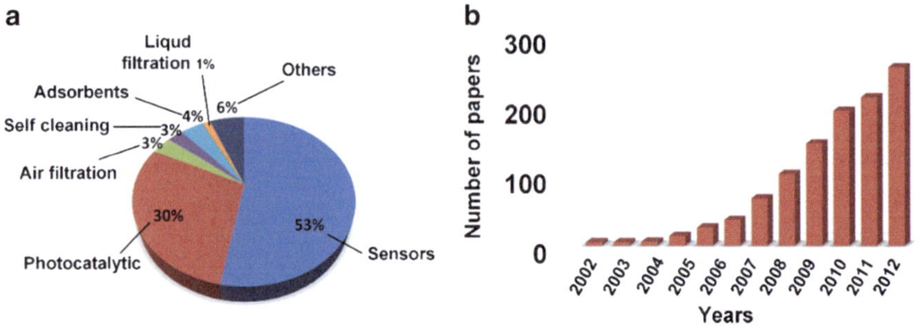

Fig. 1.16 Statistics on the literature published on the applications of nanofibers in environmental field. (**a**) Distribution of the relevant environmental applications. (**b**) Numbers of publications from 2002 to 2012 (Source from ISI web of science)

solar cells, and fuel cells, by providing nanofiber electrodes and separators. The common challenges are how to improve the performance, conversion efficiency, energy/power density, and discharge rate and at the same time extend the lifetime with reversibility and reduce the production and operation costs. In particular, rapid Li^+, e^-, and H^+ transport are essential for achieving desired performance in high-efficiency solar cells, high energy density batteries, and high-efficiency fuel cells, respectively [2, 9]. In addition, the applications of nanofibers in novel energy storage device including supercapacitors, hydrogen storage, and thermal storage are becoming of interest as well, particularly when considering the use of them as power source for electric device or engineering equipment. Consequently, Fig. 1.15b illustrates the recent remarkable growth in the number of papers published on the application of electrospun materials in energy conversion devices, indicating that the relevant applications have been realized well and the perspective is encouraging.

The need of clean water and air is an ever-growing problem for underdeveloped countries and will be a major stressor on modern economies in the future. To meet this environmental challenge, the nanofibrous membranes were widely studied for the applications in sensors, photocatalytic, air filtration, liquid filtration, oil spill cleanup, self-cleaning, adsorbents, and electromagnetic shielding, as shown in Fig. 1.16a. The environmental monitoring is the precondition for the handling of the environmental pollution; thus, the sensors are the major application of nanofibers in relevant fields, including the various types of nanofiber sensors toward numerous analytes in air or water [17, 28]. Moreover, the environmental pollution control also involves the treatment of polluted air and water; thus, the functional nanofibers could be used in air filtration, liquid filtration, oil spill cleanup, self-cleaning, adsorbents, etc. A brief analysis of nanofibers used for advanced environmental applications over the past 10 years (Fig. 1.16b) has indicated that their impact is substantial, which also revealed that the promising nanofibrous materials will continually represent a key technology that provides a solution to pollution in water and air environments.

Acknowledgments This work is supported by the National Basic Research Program of China (973 Program, 2011CB606103 and 2012CB525005), the National Natural Science Foundation of China (No. 51173022, U1232116, 51273038, and 51322304), and the Program for New Century Talents of the University in China.

References

1. The International Energy Agency (2011) World energy outlook 2011. http://www.worldenergyoutlook.org. Accessed 13 May 2011
2. Thavasi V, Singh G, Ramakrishna S (2008) Electrospun nanofibers in energy and environmental applications. Energy Environ Sci 1(2):205. doi:10.1039/b809074m
3. De Oliveira MCK, Teixeira A, Vieira LC, de Carvalho RM, de Carvalho ABM, do Couto BC (2012) Flow assurance study for waxy crude oils. Energy Fuel 26(5):2688–2695. doi:10.1021/ef201407j
4. Moniz EJ (2010) Nanotechnology for the energy challenge. Wiley-VCH, Weinheim
5. Serrano E, Rus G, Garcia-Martinez J (2009) Nanotechnology for sustainable energy. Renew Sust Energy Rev 13(9):2373–2384. doi:10.1016/j.rser.2009.06.003
6. Painter P, Williams P, Mannebach E (2010) Recovery of bitumen from oil or tar sands using ionic liquids. Energy Fuel 24(2):1094–1098. doi:10.1021/ef9009586
7. Lee M, Bae J, Lee J, Lee S, Hong S, Wang ZL (2011) Self-powered environmental sensor system driven by nanogenerators. Energy Environ Sci 4(9):3359. doi:10.1039/c1ee01558c
8. Koopmans R, Duyvendak JW (1995) Political construction of the nuclear energy issue and its impact on the mobilization of anti-nuclear movements in Western Europe. Soc Probl 42(2):235–251
9. Dong Z, Kennedy SJ, Wu Y (2011) Electrospinning materials for energy-related applications and devices. J Power Sources 196(11):4886–4904. doi:10.1016/j.jpowsour.2011.01.090
10. Pietrosemoli L, Rodríguez Monroy C (2013) The impact of sustainable construction and knowledge management on sustainability goals. A review of the Venezuelan renewable energy sector. Renew Sust Energy Rev 27:683–691. doi:10.1016/j.rser.2013.07.056
11. Mascal M, Nikitin EB (2008) Direct, high-yield conversion of cellulose into biofuel. Angew Chem Int Ed 120(41):8042–8044. doi:10.1002/anie.200801594
12. Wiesner MR, Bottero JY (2007) Environmental nanotechnology: applications and impacts of nanomaterials. McGraw Hill Professional, New York
13. Wheeler D, Elkington J (2001) The end of the corporate environmental report? Or the advent of cybernetic sustainability reporting and communication. Bus Strat Environ 10(1):1–14. doi:10.1002/1099-0836(200101/02)10:1
14. Webster PJ, Holland GJ, Curry JA, Chang HR (2005) Changes in tropical cyclone number, duration, and intensity in a warming environment. Science 309(5742):1844–1846. doi:10.1126/science.1116448
15. Herzog F (2005) Agri-environment schemes as landscape experiments. Agric Ecosyst Environ 108(3):175–177. doi:10.1016/j.agee.2005.02.001
16. Gleick P, Cooley H, Cohen MJ, Morikawa M, Morrison J, Palaniappan M (2006) The world's water 2008–2009: the biennial report on freshwater resources. Island Press, Washington, DC
17. Gv L, Duffy SJ (2005) Environmental chemistry: a global perspective, vol 2. Oxford University Press, London
18. McIntyre BD (2009) IAASTD International Assessment of Agricultural Knowledge, Science and Technology for Development: Global report. Island Press, Washington, DC
19. Vörösmarty CJ, McIntyre P, Gessner MO, Dudgeon D, Prusevich A, Green P, Glidden S, Bunn SE, Sullivan CA, Liermann CR (2010) Global threats to human water security and river biodiversity. Nature 467(7315):555–561. doi:10.1038/nature09440

20. Pope CA III, Dockery DW (2006) Health effects of fine particulate air pollution: lines that connect. J Air Waste Manage Assoc 56(6):709–742. doi:10.1016/S0140-6736(00)02653-2
21. Nel A (2005) Air pollution-related illness: effects of particles. Science 308(5723):804–806. doi:10.1126/science.1108752
22. Volkamer R, Jimenez JL, San Martini F, Dzepina K, Zhang Q, Salcedo D, Molina LT, Worsnop DR, Molina MJ (2006) Secondary organic aerosol formation from anthropogenic air pollution: rapid and higher than expected. Geophys Res Lett 33(17):L17811. doi:10.1029/2006GL026899
23. Krewski D, Jerrett M, Burnett RT, Ma R, Hughes E, Shi Y, Turner MC, Pope CA III, Thurston G, Calle EE (2009) Extended follow-up and spatial analysis of the American Cancer Society study linking particulate air pollution and mortality, vol 140. Health Effects Institute, Boston
24. Marshall JD, Brauer M, Frank LD (2009) Healthy neighborhoods: walkability and air pollution. Environ Health Perspect 117(11):1752. doi:10.1289/ehp.0900595
25. Anderson HR (2009) Air pollution and mortality: a history. Atmos Environ 43(1):142–152. doi:10.1016/j.atmosenv.2008.09.026
26. Ramakrishna S, Fujihara K, Teo WE, Yong T, Ma Z, Ramaseshan R (2006) Electrospun nanofibers: solving global issues. Mater Today 9(3):40–50. doi:10.1016/s1369-7021(06)71389-x
27. Bhardwaj N, Kundu SC (2010) Electrospinning: a fascinating fiber fabrication technique. Biotechnol Adv 28(3):325–347. doi:10.1016/j.biotechadv.2010.01.004
28. Ding B, Wang M, Wang X, Yu J, Sun G (2010) Electrospun nanomaterials for ultrasensitive sensors. Mater Today 13(11):16–27. doi:10.1016/s1369-7021(10)70200-5
29. Ding B, Yu J (2011) Electrospinning and nanofibers. China Textile & Apparel Press, Beijing
30. Greiner A, Wendorff JH (2007) Electrospinning: a fascinating method for the preparation of ultrathin fibers. Angew Chem Int Ed 46(30):5670–5703. doi:10.1002/anie.200604646
31. Wang ZM (2008) One-dimensional nanostructures, vol 3. Springer, Heidelberg
32. Wendorff JH, Agarwal S, Greiner A (2012) Electrospinning: materials, processing, and applications. Wiley-VCH, Weinheim
33. Tucker N, Stanger JJ, Staiger MP, Razzaq H, Hofman K (2012) The history of the science and technology of electrospinning from 1600 to 1995. J Eng Fibers Fabr Spec Issue-Fibers 7:63–73
34. De Vrieze S, De Clerck K (2009) 80 years of electrospinning. In: International conference on latest advances in high-tech textiles and textile-based materials, Ghent University. Department of Textiles, pp 60–63
35. Ramakrishna S (2005) An introduction to electrospinning and nanofibers. World Scientific, Singapore
36. Subbiah T, Bhat G, Tock R, Parameswaran S, Ramkumar S (2005) Electrospinning of nanofibers. J Appl Polym Sci 96(2):557–569. doi:10.1002/app.21481
37. Teo W, Ramakrishna S (2006) A review on electrospinning design and nanofibre assemblies. Nanotechnology 17(14):R89. doi:10.1088/0957-4484/17/14/R01
38. Sill TJ, von Recum HA (2008) Electrospinning: applications in drug delivery and tissue engineering. Biomaterials 29(13):1989–2006. doi:10.1016/j.biomaterials.2008.01.011
39. Iimura K, Oi T, Suzuki M, Hirota M (2010) Preparation of silica fibers and non-woven cloth by electrospinning. Adv Powder Technol 21(1):64–68. doi:10.1016/j.apt.2009.11.006
40. Luo CJ, Stoyanov SD, Stride E, Pelan E, Edirisinghe M (2012) Electrospinning versus fibre production methods: from specifics to technological convergence. Chem Soc Rev 41(13):4708. doi:10.1039/c2cs35083a
41. Jirsak O, Petrik S (2010) Needleless electrospinning-history, present and future. In: Proceedings of the 7th international conference-TEXSCI 2010, Czech Republic, Liberec, 6–8 September
42. Si Y, Ren T, Li Y, Ding B, Yu J (2012) Fabrication of magnetic polybenzoxazine-based carbon nanofibers with Fe_3O_4 inclusions with a hierarchical porous structure for water treatment. Carbon 50(14):5176–5185. doi:10.1016/j. carbon .2012.06.059

43. Li D, Xia Y (2004) Electrospinning of nanofibers: reinventing the wheel? Adv Mater 16(14):1151–1170. doi:10.1002/adma.200400719
44. Yang L, Raza A, Si Y, Mao X, Shang Y, Ding B, Yu J, AlDeyab SS (2012) Synthesis of superhydrophobic silica nanofibrous membranes with robust thermal stability and flexibility via in situ polymerization. Nanoscale 4(20):6581–6587. doi:10.1039/c2nr32095a
45. Wang X, Ding B, Sun G, Wang M, Yu J (2013) Electro-spinning/netting: a fascinating strategy for the fabrication of three-dimensional polymer nano-fiber/nets. Prog Mater Sci 58(8):1173–1243. doi:10.1016/j.pmatsci.2013.05.001
46. Li Z, Wang C (2013) Introduction of electrospinning. Springer, Heidelberg
47. Reneker D, Kataphinan W, Theron A, Zussman E, Yarin A (2002) Nanofiber garlands of polycaprolactone by electrospinning. Polymer 43(25):6785–6794. doi:10.1016/s0032-3861(02)00595-5
48. Higuera F (2003) Flow rate and electric current emitted by a Taylor Cone. J Fluid Mech 484:303–327. doi:10.1017/S0022112003004385
49. Xie J, Wang CH (2007) Encapsulation of proteins in biodegradable polymeric microparticles using electrospray in the Taylor Cone-jet mode. Biotechnol Bioeng 97(5):1278–1290. doi:10.1002/bit.21334
50. Reznik SN, Yarin AL, Theron A, Zussman E (2004) Transient and steady shapes of droplets attached to a surface in a strong electric field. J Fluid Mech 516:349–377. doi:10.1017/s0022112004000679
51. Yarin AL, Koombhongse S, Reneker DH (2001) Taylor Cone and jetting from liquid droplets in electrospinning of nanofibers. J Appl Phys 90(9):4836. doi:10.1063/1.1408260
52. Reneker DH, Yarin AL, Fong H, Koombhongse S (2000) Bending instability of electrically charged liquid jets of polymer solutions in electrospinning. J Appl Phys 87(9):4531. doi:10.1063/1.373532
53. Yarin A, Kataphinan W, Reneker D (2005) Branching in electrospinning of nanofibers. J Appl Phys 98(6):064501–064512. doi:10.1063/1.2060928
54. Thompson C, Chase G, Yarin A, Reneker D (2007) Effects of parameters on nanofiber diameter determined from electrospinning model. Polymer 48(23):6913–6922. doi:10.1016/j.polymer .2007.09.017
55. Han T, Yarin AL, Reneker DH (2008) Viscoelastic electrospun jets: initial stresses and elongational rheometry. Polymer 49(6):1651–1658. doi:10.1016/j.polymer.2008.01.035
56. Reneker D, Yarin A, Zussman E, Xu H (2007) Electrospinning of nanofibers from polymer solutions and melts. Adv Appl Mech 41:43–346. doi:10.1016/S0065-2156(07)41002-X
57. Han T, Reneker DH, Yarin AL (2007) Buckling of jets in electrospinning. Polymer 48(20):6064–6076. doi:10.1016/j.polymer.2007.08.002
58. Reneker DH, Yarin AL (2008) Electrospinning jets and polymer nanofibers. Polymer 49(10):2387–2425. doi:10.1016/j.polymer.2008.02.002
59. Lin JY, Ding B, Yu JY, Hsieh Y (2010) Direct fabrication of highly nanoporous polystyrene fibers via electrospinning. ACS Appl Mater Interfaces 2(2):521–528. doi:10.1021/am900736h
60. Demir MM, Yilgor I, Eea Y, Erman B (2002) Electrospinning of polyurethane fibers. Polymer 43(11):3303–3309. doi:10.1016/S0032-3861(02)00136-2
61. Deitzel J, Kleinmeyer J, Dea H, Beck Tan N (2001) The effect of processing variables on the morphology of electrospun nanofibers and textiles. Polymer 42(1):261–272. doi:10.1016/S0032-3861(00)00250-0
62. Zong X, Kim K, Fang D, Ran S, Hsiao BS, Chu B (2002) Structure and process relationship of electrospun bioabsorbable nanofiber membranes. Polymer 43(16):4403–4412. doi:10.1016/S0032-3861(02)00275-6
63. Lee KH, Kim HY, La YM, Lee DR, Sung NH (2002) Influence of a mixing solvent with tetrahydrofuran and N, N-dimethylformamide on electrospun poly (vinyl chloride) nonwoven mats. J Polym Sci Part B: Polym Phys 40(19):2259–2268. doi:10.1002/polb.10293
64. Mit-uppatham C, Nithitanakul M, Supaphol P (2004) Effects of solution concentration, emitting electrode polarity, solvent type, and salt addition on electrospun polyamide-6 fibers: a preliminary report. Macromol Symp 216:293–300, Wiley Online Library

65. Geng X, Kwon OH, Jang J (2005) Electrospinning of chitosan dissolved in concentrated acetic acid solution. Biomaterials 26(27):5427–5432. doi:10.1016/j.biomaterials.2005.01.066
66. Baumgarten PK (1971) Electrostatic spinning of acrylic microfibers. J Colloid Interface Sci 36(1):71–79. doi:10.1016/0021-9797(71)90241-4
67. Fong H, Chun I, Reneker D (1999) Beaded nanofibers formed during electrospinning. Polymer 40(16):4585–4592. doi:10.1016/S0032-3861(99)00068-3
68. Zhang C, Yuan X, Wu L, Han Y, Sheng J (2005) Study on morphology of electrospun poly (vinyl alcohol) mats. Eur Polym J 41(3):423–432. doi:10.1016/j.eurpolymj.2004.10.027
69. Wang X, Ding B, Yu J, Yang J (2011) Large-scale fabrication of two-dimensional spider-web-like gelatin nano-nets via electro-netting. Colloids Surf B Biointerfaces 86(2):345–352. doi:10.1016/j.colsurfb.2011.04.018
70. Kim B, Park H, Lee SH, Sigmund WM (2005) Poly (acrylic acid) nanofibers by electrospinning. Mater Lett 59(7):829–832. doi:10.1081/E-ENN-120013552
71. Barakat NA, Kanjwal MA, Sheikh FA, Kim HY (2009) Spider-net within the N6, PVA and PU electrospun nanofiber mats using salt addition: novel strategy in the electrospinning process. Polymer 50(18):4389–4396. doi:10.1016/j.polymer.2009.07.005
72. Zeng J, Xu X, Chen X, Liang Q, Bian X, Yang L, Jing X (2003) Biodegradable electrospun fibers for drug delivery. J Control Release 92(3):227–231. doi:10.1016/S0168-3659(03)00372-9
73. Lee K, Kim H, Bang H, Jung Y, Lee S (2003) The change of bead morphology formed on electrospun polystyrene fibers. Polymer 44(14):4029–4034. doi:10.1016/S0032-3861(03)00345-8
74. Ding B, Kim HY, Lee SC, Lee DR, Choi KJ (2002) Preparation and characterization of nanoscaled poly(vinyl alcohol) fibers via electrospinning. Fiber Polym 3(2):73–79. doi:10.1002/polb.10191
75. Jarusuwannapoom T, Hongrojjanawiwat W, Jitjaicham S, Wannatong L, Nithitanakul M, Pattamaprom C, Koombhongse P, Rangkupan R, Supaphol P (2005) Effect of solvents on electro-spinnability of polystyrene solutions and morphological appearance of resulting electrospun polystyrene fibers. Eur Polym J 41(3):409–421. doi:10.1016/j.eurpolymj.2004.10.010
76. Jeun JP, Lim YM, Nho YC (2005) Study on morphology of electrospun poly (caprolactone) nanofiber. J Ind Eng Chem 11(4):573–578
77. Min BM, Lee G, Kim SH, Nam YS, Lee TS, Park WH (2004) Electrospinning of silk fibroin nanofibers and its effect on the adhesion and spreading of normal human keratinocytes and fibroblasts in vitro. Biomaterials 25(7):1289–1297. doi:10.1016/j.biomaterials.2003.08.045
78. Buchko CJ, Chen LC, Shen Y, Martin DC (1999) Processing and microstructural characterization of porous biocompatible protein polymer thin films. Polymer 40(26):7397–7407. doi:10.1016/S0032-3861(98)00866-0
79. Hartman R, Brunner D, Camelot D, Marijnissen J, Scarlett B (2000) Jet break-up in electrohydrodynamic atomization in the cone-jet mode. J Aerosol Sci 31(1):65–95. doi:10.1016/S0021-8502(99)00034-8
80. Yuan X, Zhang Y, Dong C, Sheng J (2004) Morphology of ultrafine polysulfone fibers prepared by electrospinning. Polym Int 53(11):1704–1710. doi:10.1002/pi.1538
81. Lin J, Tian F, Shang Y, Wang F, Ding B, Yu J (2012) Facile control of intra-fiber porosity and inter-fiber voids in electrospun fibers for selective adsorption. Nanoscale 4(17):5316–5320. doi:10.1039/c2nr31515g
82. Casper CL, Stephens JS, Tassi NG, Chase DB, Rabolt JF (2004) Controlling surface morphology of electrospun polystyrene fibers: effect of humidity and molecular weight in the electrospinning process. Macromolecules 37(2):573–578. doi:10.1021/ma0351975
83. Li J, He A, Zheng J, Han CC (2006) Gelatin and gelatin-hyaluronic acid nanofibrous membranes produced by electrospinning of their aqueous solutions. Biomacromolecules 7(7):2243–2247. doi:10.1021/bm0603342
84. Nie H, He A, Wu W, Zheng J, Xu S, Li J, Han CC (2009) Effect of poly (ethylene oxide) with different molecular weights on the electrospinnability of sodium alginate. Polymer 50(20):4926–4934. doi:10.1016/j.polymer.2009.07.043

85. Deng L, Young RJ, Kinloch IA, Zhu Y, Eichhorn SJ (2013) Carbon nanofibres produced from electrospun cellulose nanofibres. Carbon 58:66–75. doi:10.1016/j.carbon.2013.02.032
86. Schiffman JD, Schauer CL (2007) Cross-linking chitosan nanofibers. Biomacromolecules 8(2):594–601. doi:10.1021/bm060804s
87. Park KE, Kang HK, Lee SJ, Min B-M, Park WH (2006) Biomimetic nanofibrous scaffolds: preparation and characterization of PGA/chitin blend nanofibers. Biomacromolecules 7(2):635–643. doi:10.1021/bm0509265
88. Ji Y, Ghosh K, Shu X, Li B, Sokolov J, Prestwich G, Clark R, Rafailovich M (2006) Electrospun three-dimensional hyaluronic acid nanofibrous scaffolds. Biomaterials 27(20):3782–3792. doi:10.1016/j.biomaterials.2006.02.037
89. Hang Y, Zhang Y, Jin Y, Shao H, Hu X (2012) Preparation of regenerated silk fibroin/silk sericin fibers by coaxial electrospinning. Int J Biol Macromol 51(5):980–986. doi:10.1016/j.ijbiomac.2012.08.010
90. Wnek GE, Carr ME, Simpson DG, Bowlin GL (2003) Electrospinning of nanofiber fibrinogen structures. Nano Lett 3(2):213–216. doi:10.1021/nl025866c
91. Woerdeman DL, Ye P, Shenoy S, Parnas RS, Wnek GE, Trofimova O (2005) Electrospun fibers from wheat protein: investigation of the interplay between molecular structure and the fluid dynamics of the electrospinning process. Biomacromolecules 6(2):707–712. doi:10.1021/bm0494545
92. Huang ZM, Zhang YZ, Ramakrishna S, Lim CT (2004) Electrospinning and mechanical characterization of gelatin nanofibers. Polymer 45(15):5361–5368. doi:10.1016/j.polymer.2004.04.005
93. Huang L, McMillan RA, Apkarian RP, Pourdeyhimi B, Conticello VP, Chaikof EL (2000) Generation of synthetic elastin-mimetic small diameter fibers and fiber networks. Macromolecules 33(8):2989–2997. doi:10.1021/ma991858f
94. Yao C, Li X, Song T (2007) Electrospinning and crosslinking of zein nanofiber mats. J Appl Polym Sci 103(1):380–385. doi:10.1002/app.24619
95. Ritcharoen W, Thaiying Y, Saejeng Y, Jangchud I, Rangkupan R, Meechaisue C, Supaphol P (2008) Electrospun dextran fibrous membranes. Cellulose 15(3):435–444. doi:10.1007/s10570-008-9199-3
96. Shin MK, Kim SH, Si J, Kim SI, Kim SJ, Kim BJ, So I (2008) The effect of DNA on mechanical properties of nanofiber hydrogels. Appl Phys Lett 93(17):171903. doi:10.1063/1.3009204
97. Kim CW, Frey MW, Marquez M, Joo YL (2005) Preparation of submicron-scale, electrospun cellulose fibers via direct dissolution. J Polym Sci Part B: Polym Phys 43(13):1673–1683. doi:10.1002/polb.20475
98. Viswanathan G, Murugesan S, Pushparaj V, Nalamasu O, Ajayan PM, Linhardt RJ (2006) Preparation of biopolymer fibers by electrospinning from room temperature ionic liquids. Biomacromolecules 7(2):415–418. doi:10.1002/polb.20475
99. Liu H, Hsieh YL (2002) Ultrafine fibrous cellulose membranes from electrospinning of cellulose acetate. J Polym Sci Part B: Polym Phys 40(18):2119–2129. doi:10.1002/polb.10261
100. Zarkoob S, Eby RK, Reneker DH, Hudson SD, Ertley D, Adams WW (2004) Structure and morphology of electrospun silk nanofibers. Polymer 45(11):3973–3977. doi:10.1016/j.polymer.2003.10.102
101. Um IC, Fang D, Hsiao BS, Okamoto A, Chu B (2004) Electro-spinning and electro-blowing of hyaluronic acid. Biomacromolecules 5(4):1428–1436. doi:10.1021/bm034539b
102. Ding B, Kim HY, Lee SC, Lee DR, Choi KJ (2002) Preparation and characterization of nanoscaled poly (vinyl alcohol) fibers via electrospinning. Fibers Polym 3(2):73–79. doi:10.1002/polb.10191
103. Kim B, Park H, Lee SH, Sigmund WM (2005) Poly (acrylic acid) nanofibers by electrospinning. Mater Lett 59(7):829–832. doi:10.1016/j.matlet.2004.11.032
104. Li X, Lin L, Zhu Y, Liu W, Yu T, Ge M (2013) Preparation of ultrafine fast-dissolving cholecalciferol-loaded poly (vinyl pyrrolidone) fiber mats via electrospinning. Polym Compos 34(2):282–287. doi:10.1002/pc.22402

105. Lu P, Hsieh YL (2009) Organic compatible polyacrylamide hydrogel fibers. Polymer 50(15):3670–3679. doi:10.1016/j.polymer.2009.05.040
106. Chen C, Wang L, Huang Y (2007) Electrospinning of thermo-regulating ultrafine fibers based on polyethylene glycol/cellulose acetate composite. Polymer 48(18):5202–5207. doi:10.1016/j.polymer.2007.06.069
107. Khanam N, Mikoryak C, Draper RK, Balkus KJ (2007) Electrospun linear polyethyleneimine scaffolds for cell growth. Acta Biomater 3(6):1050–1059. doi:10.1016/j.actbio.2007.06.005
108. Inai R, Kotaki M, Ramakrishna S (2005) Structure and properties of electrospun PLLA single nanofibres. Nanotechnology 16(2):208–213. doi:10.1088/0957-4484/16/2/005
109. Hong JK, Xu G, Piao D, Madihally SV (2012) Analysis of void shape and size in the collector plate and polycaprolactone molecular weight on electrospun scaffold pore size. J Appl Polym Sci 128(3):1583–1591. doi:10.1002/app.38326
110. Wang L, Liu YS, Mo LF, Liu FJ, Xu L (2011) Electrospun nanoporous poly(butylenes succinate-co-butylene terephthalate) nonwoven mats. Adv Mater Sci Eng 2011:1–3. doi:10.1155/2011/654360
111. Yu BY, Chen CR, Sun YM, Young TH (2009) The response of rat cerebellar granule neurons (rCGNs) to various polyhydroxyalkanoate (PHA) films. Desalination 245(1–3):639–646. doi:10.1016/j.desal.2009.02.031
112. He JH, Liu Y, Xu L, Yu JY (2007) Micro sphere with nanoporosity by electrospinning. Chaos Soliton Fract 32(3):1096–1100. doi:10.1016/j.chaos.2006.07.045
113. Basu S, Gogoi N, Sharma S, Jassal M, Agrawal AK (2013) Role of elasticity in control of diameter of electrospun PAN nanofibers. Fiber Polym 14(6):950–956. doi:10.1007/s12221-013-0950-5
114. Zhu J, Wei S, Chen X, Karki AB, Rutman D, Young DP, Guo Z (2010) Electrospun polyimide nanocomposite fibers reinforced with core-shell Fe-FeO nanoparticles. J Phys Chem C 114(19):8844–8850. doi:10.1021/jp1020033
115. Kenawy ER, Layman JM, Watkins JR, Bowlin GL, Matthews JA, Simpson DG, Wnek GE (2003) Electrospinning of poly (ethylene-co-vinyl alcohol) fibers. Biomaterials 24(6):907–913. doi:10.1016/S0142-9612(02)00422-2
116. Deitzel J, Kleinmeyer J, Hirvonen J, Beck Tan N (2001) Controlled deposition of electrospun poly (ethylene oxide) fibers. Polymer 42(19):8163–8170. doi:10.1016/s0032-3861(01)00336-6
117. Kang YK, Park CH, Kim J, Kang TJ (2007) Application of electrospun polyurethane web to breathable water-proof fabrics. Fiber Polym 8(5):564–570. doi:10.1007/BF02875881
118. Chronakis IS, Grapenson S, Jakob A (2006) Conductive polypyrrole nanofibers via electrospinning: electrical and morphological properties. Polymer 47(5):1597–1603. doi:10.1016/j.polymer.2006.01.032
119. Luo H, Huang Y, Wang D (2013) Confined crystallization of POM in the CA-nanotubes fabricated by coaxial electrospinning. Eur Polym J 49(6):1424–1436. doi:10.1016/j.eurpolymj.2013.02.037
120. Tao R, Zhu X, Zuo YT, Fan LJ, Wang B, Xin JT (2013) Grafting modification of electrospun polystyrene fibrous membranes via an entrapped initiator in an acrylic acid aqueous solution. J Appl Polym Sci 127(5):4102–4109. doi:10.1002/app.36797
121. Bae HS, Haider A, Selim KMK, Kang DY, Kim EJ, Kang IK (2013) Fabrication of highly porous PMMA electrospun fibers and their application in the removal of phenol and iodine. J Polym Res 20(7):158. doi:10.1007/s10965-013-0158-9
122. Zhu X, Jiang X, Cheng S, Wang K, Mao S, Fan LJ (2009) Preparation of high strength ultrafine polyvinyl chloride fibrous membrane and its adsorption of cationic dye. J Polym Res 17(6):769–777. doi:10.1007/s10965-009-9368-6
123. Damaraju SM, Wu S, Jaffe M, Arinzeh TL (2013) Structural changes in PVDF fibers due to electrospinning and its effect on biological function. Biomed Mater 8(4):045007. doi:10.1088/1748-6041/8/4/045007

124. Zhang B, Li C, Chang M (2009) Curled poly(ethylene glycol terephthalate)/poly(ethylene propanediol terephthalate) nanofibers produced by side-by-side electrospinning. Polym J 41(4):252–253. doi:10.1295/polymj.PJ2008270
125. Guo S, Ke Q, Wang H, Jin X, Li Y (2012) Poly (butylene terephthalate) electrospun/melt-blown composite mats for white blood cell filtration. J Appl Polym Sci 128(6):3653–3655. doi:10.1002/app.38423
126. Gao L, Li C (2010) Preparation and photoluminescence properties of electrospun nanofibers of C60/PVK. J Lumin 130(2):236–239. doi:10.1016/j.jlumin.2009.08.015
127. Tang X, Si Y, Ge J, Ding B, Liu L, Luo W, Zheng G, Yu J (2013) In-situ polymerized superhydrophobic and superoleophilic nanofibrous membranes for gravity driven oil–water separation. Nanoscale. doi:10.1039/C3NR03937D
128. Ke H, Cai Y, Wei Q, Xiao Y, Dong J, Hu Y, Song L, He G, Zhao Y, Fong H (2013) Electrospun ultrafine composite fibers of binary fatty acid eutectics and polyethylene terephthalate as innovative form-stable phase change materials for storage and retrieval of thermal energy. Int J Energ Res 37(6):657–664. doi:10.1002/er.2888
129. Shawon J, Sung C (2004) Electrospinning of polycarbonate nanofibers with solvent mixtures THF and DMF. J Mater Sci 39(14):4605–4613. doi:10.1023/B:JMSC.0000034155.93428.ea
130. Walmsley RS, Hlangothi P, Litwinski C, Nyokong T, Torto N, Tshentu ZR (2013) Catalytic oxidation of thioanisole using oxovanadium (IV)-functionalized electrospun polybenzimidazole nanofibers. J Appl Polym Sci 127(6):4719–4725. doi:10.1002/app.38067
131. Park JY, Lee IH, Bea GN (2008) Optimization of the electrospinning conditions for preparation of nanofibers from polyvinylacetate (PVAc) in ethanol solvent. J Ind Eng Chem 14(6):707–713. doi:10.1016/j.jiec.2008.03.006
132. Zakaria SM, Sharif Zein SH, Othman MR, Jansen JA (2013) Hydroxyapatite nanoparticles: electrospinning and calcination of hydroxyapatite/polyvinyl butyral nanofibers and growth kinetics. J Biomed Mater Res A 101A(7):1977–1985. doi:10.1002/jbm.a.34506
133. Sanders EH, Kloefkorn R, Bowlin GL, Simpson DG, Wnek GE (2003) Two-phase electrospinning from a single electrified jet: microencapsulation of aqueous reservoirs in poly (ethylene-co-vinyl acetate) fibers. Macromolecules 36(11):3803–3805. doi:10.1021/ma021771l
134. Lu XF, Wang C, Wei Y (2009) One-dimensional composite nanomaterials: synthesis by electrospinning and their applications. Small 5(21):2349–2370. doi:10.1002/smll.200900445
135. Li D, Xia Y (2003) Fabrication of titania nanofibers by electrospinning. Nano Lett 3(4):555–560. doi:10.1002/jbm.a.34293
136. Dharmaraj N, Kim CH, Kim KW, Kim HY, Suh EK (2006) Spectral studies of SnO_2 nanofibres prepared by electrospinning method. Spectrochim Acta A-Mol Biomol Spectrosc 64(1):136–140. doi:10.1016/j.saa.2005.07.007
137. Melo EF, Alves KGB, Junior SA, Melo CP (2013) Synthesis of fluorescent PVA/polypyrrole-ZnO nanofibers. J Mater Sci 48(10):3652–3658. doi:10.1007/s10853-013-7159-2
138. Zheng W, Lu X, Wang W, Li Z, Zhang H, Wang Y, Wang Z, Wang C (2009) A highly sensitive and fast-responding sensor based on electrospun In_2O_3 nanofibers. Sens Actuat B-Chem 142(1):61–65. doi:10.1016/j.snb.2009.07.031
139. Zhang P, Jiao X, Chen D (2013) Fabrication of electrospun Al_2O_3 fibers with CaO–SiO_2 additive. Mater Lett 91:23–26. doi:10.1016/j.matlet.2012.09.077
140. Lee SS, Bai H, Liu Z, Sun DD (2013) Novel-structured electrospun TiO_2/CuO composite nanofibers for high efficient photocatalytic cogeneration of clean water and energy from dye wastewater. Water Res 47(12):4059–4073. doi:10.1016/j.watres.2012.12.044
141. Guan H, Shao C, Wen S, Chen B, Gong J, Yang X (2003) Preparation and characterization of NiO nanofibres via an electrospinning technique. Inorg Chem Commun 6(10):1302–1303. doi:10.1016/j.inoche.2003.08.003
142. Viswanathamurthi P, Bhattarai N, Kim HY, Khil MS, Lee DR, Suh EK (2004) GeO_2 fibers: preparation, morphology and photoluminescence property. J Chem Phys 121(1):441. doi:10.1063/1.1755666

143. Viswanathamurthi P (2003) Vanadium pentoxide nanofibers by electrospinning. Scr Mater 49(6):577–581. doi:10.1016/s1359-6462(03)00333-6
144. Li Y, Gong J, He G, Deng Y (2011) Synthesis of polyaniline nanotubes using Mn_2O_3 nanofibers as oxidant and their ammonia sensing properties. Synth Met 161(1–2):56–61. doi:10.1016/j.synthmet.2010.10.034
145. Jing N, Wang M, Kameoka J (2005) Fabrication of ultrathin ZrO nanofibers by electrospinning. J Photopolym Sci Technol 18(4):503–506. doi:10.1021/nl034039o
146. Liu HA, Balkus KJ (2009) Electrospinning of beta silicon carbide nanofibers. Mater Lett 63(27):2361–2364. doi:10.1016/j.matlet.2009.08.009
147. Sangmanee M, Maensiri S (2009) Nanostructures and magnetic properties of cobalt ferrite ($CoFe_2O_4$) fabricated by electrospinning. Appl Phys A 97(1):167–177. doi:10.1007/s00339-009-5256-5
148. Guo M, Ding B, Li X, Wang X, Yu J, Wang M (2009) Amphiphobic nanofibrous silica mats with flexible and high-heat-resistant properties. J Phys Chem C 114(2):916–921. doi:10.1021/jp909672r
149. Liu C-K, Lai K, Liu W, Yao M, Sun R-J (2009) Preparation of carbon nanofibres through electrospinning and thermal treatment. Polym Int 58(12):1341–1349. doi:10.1002/pi.2669
150. Si Y, Ren T, Ding B, Yu J, Sun G (2012) Synthesis of mesoporous magnetic Fe_3O_4@ carbon nanofibers utilizing in situ polymerized polybenzoxazine for water purification. J Mater Chem 22(11):4619–4622. doi:10.1039/C2JM00036A
151. Ren T, Si Y, Yang J, Ding B, Yang X, Hong F, Yu J (2012) Polyacrylonitrile/polybenzoxazine-based Fe_3O_4@carbon nanofibers: hierarchical porous structure and magnetic adsorption property. J Mater Chem 22(31):15919–15927. doi:10.1039/C2JM33214K
152. Koombhongse S, Liu W, Reneker DH (2001) Flat polymer ribbons and other shapes by electrospinning. J Polym Sci Part B: Polym Phys 39(21):2598–2606. doi:10.1002/polb.10015
153. Lin T, Wang H, Wang X (2005) Self-crimping bicomponent nanofibers electrospun from polyacrylonitrile and elastomeric polyurethane. Adv Mater 17(22):2699–2703. doi:10.1002/adma.200500901
154. Jin Y, Yang D, Kang D, Jiang X (2009) Fabrication of necklace-like structures via electrospinning. Langmuir 26(2):1186–1190. doi:10.1021/la902313t
155. Sun Z, Zussman E, Yarin AL, Wendorff JH, Greiner A (2003) Compound core-shell polymer nanofibers by co-electrospinning. Adv Mater 15(22):1929–1932. doi:10.1002/adma.200305136
156. Zhao Y, Cao XY, Jiang L (2007) Bio-mimic multichannel microtubes by a facile method. J Am Chem Soc 129(4):764–765. doi:10.1021/Ja068165g
157. Wang X, Si Y, Wang X, Yang J, Ding B, Chen L, Hu Z, Yu J (2013) Tuning hierarchically aligned structures for high-strength PMIA–MWCNT hybrid nanofibers. Nanoscale 5(3):886–889. doi:10.1039/C2NR33696K
158. Zhang D, Chang J (2007) Patterning of electrospun fibers using electroconductive templates. Adv Mater 19(21):3664–3667. doi:10.1002/adma.200700896
159. Ding B, Li CR, Miyauchi Y, Kuwaki O, Shiratori S (2006) Formation of novel 2D polymer nanowebs via electrospinning. Nanotechnology 17(15):3685–3691. doi:10.1088/0957-4484/17/15/011
160. Wang N, Wang X, Ding B, Yu J, Sun G (2012) Tunable fabrication of three-dimensional polyamide-66 nano-fiber/nets for high efficiency fine particulate filtration. J Mater Chem 22(4):1445–1452. doi:10.1039/C1JM14299B
161. Hu J, Wang X, Ding B, Lin J, Yu J, Sun G (2011) One-step electro-spinning/netting technique for controllably preparing polyurethane nano-fiber/net. Macromol Rapid Commun 32(21):1729–1734. doi:10.1002/marc.201100343
162. Yang SB, Wang XF, Ding B, Yu JY, Qian JF, Sun G (2011) Controllable fabrication of soap-bubble-like structured polyacrylic acid nano-nets via electro-netting. Nanoscale 3(2): 564–568. doi:10.1039/C0NR00730G

Part II
Electrospun Nanofibers for Energy Applications

Chapter 2
Electrospun Nanofibers for Design and Fabrication of Electrocatalysts and Electrolyte Membranes for Fuel cells

Zhan Lin, Yingfang Yao, and Xiangwu Zhang

Abstract In the past decades, in response to the energy needs of modern society and emerging ecological concerns, the pursuit of novel, low-cost, and environmentally friendly energy conversion and storage systems has raised significant interest. Among these systems, fuel cells have gained much attention for their high efficiency and high power density, with low greenhouse gas emission. As one of the most promising and versatile fabrication methods for one-dimensional mesostructured nanomaterials composed of organic, inorganic, metallic, or hybrid components prepared as randomly or orientedly arranged continuous nanofibrous mats with possibilities of ordered internal morphologies such as core-sheath, hollow, or porous fibers, or even multichanneled microtubes, electrospinning has been widely investigated to fabricate electrocatalysts and electrolyte materials applied in fuel cells because of their dimensional, directional, and compositional flexibility. In this chapter, the application of electrospun nanofibers for the specific design and fabrication of different components is reviewed in detail. Particular progresses with the use of electrospun nanofibers on improved fuel cell performance, such as power density, ionic conductivity, interfacial resistance, and chemical stability, as well as mechanical strength are emphasized, which, as we hope, can trigger further development and evolution of fuel cells as one potential energy conversion device and system.

Zhan Lin and Yingfang Yao contributed equally to this chapter.

Z. Lin • Y. Yao • X. Zhang (✉)
Fiber and Polymer Science Program, Department of Textile Engineering, Chemistry and Science, North Carolina State University, Raleigh, NC 27695-8301, USA
e-mail: xiangwu_zhang@ncsu.edu

B. Ding and J. Yu (eds.), *Electrospun Nanofibers for Energy and Environmental Applications*, Nanostructure Science and Technology,
DOI 10.1007/978-3-642-54160-5_2,

2.1 Introduction

Fuel cells are electrochemical cells where the chemical energy of the fuel was converted into electricity for power generation with high efficiency [1, 2]. Industrial purified hydrogen and air are often used in fuel cells to eliminate any pollution or emission, which is known as proton exchange membrane fuel cells (PEMFCs). In a typical PEMFC, a steam of hydrogen is delivered to the anode side of the membrane electrode assembly (MEA) [3, 4]. At the anode, it is catalyzed by platinum (Pt) and split into protons and electrons. This oxidation half-cell reaction is represented as follows:

$$H_2 \rightarrow 2H^+ + 2e^- \qquad E^0 = 0V_{SHE} \tag{2.1}$$

The protons formed in this reaction permeate through the proton exchange membrane (PEM). However, the electrons travel along an external load circuit to the cathode side of the MEA and provide electric power along the way. At the cathode, a stream of oxygen is meanwhile delivered, and the oxygen molecules react with the protons permeating through the PEM and the electrons arriving through the external circuit to form water molecules. The reduction half-cell reaction is represented as follows:

$$1/2O_2 + 2H^+ + 2e^- \rightarrow H_2O \quad E^0 = 1.229V_{SHE} \tag{2.2}$$

The full-cell reaction is

$$H_2 + 1/2O_2 \rightarrow H_2O \quad E^0 = 1.229V_{SHE} \tag{2.3}$$

The scheme of a PEMFC can be seen in Fig. 2.1.:

Fuel cells can also use hydrocarbon as fuels although they are associated with local carbon dioxide emissions. There is great interest in direct methanol fuel cells (DMFCs) for portable electronic applications; however, the major development of fuel cells is focused on stationary and transportation usage. Over the past two decades, intense research and development endeavors were triggered by the need to integrate alternative energy carriers and thus energy conversion systems to improve energy sustainability, increase lifetime, and reduce cost [5, 6]. These efforts are currently dedicated to advanced performance and robustness, focusing on all components of membrane electrode assembly: electrocatalysts (lower catalyst loading, increase specific activity, and reduce cost), catalyst support materials (stabilize different support materials to redox reactions), and electrolyte membranes (increase conductivity at high temperature and low relative humidity, improve mechanical strength, and chemical stability) [7, 8].

As one of the versatile and widely spread technologies to produce 1-D nanofibers, electrospinning has been extensively investigated as the potential materials applied in both the catalyst materials and electrolyte membranes [9, 10]. With the properties

Fig. 2.1 The schematic diagram of a PEMFC

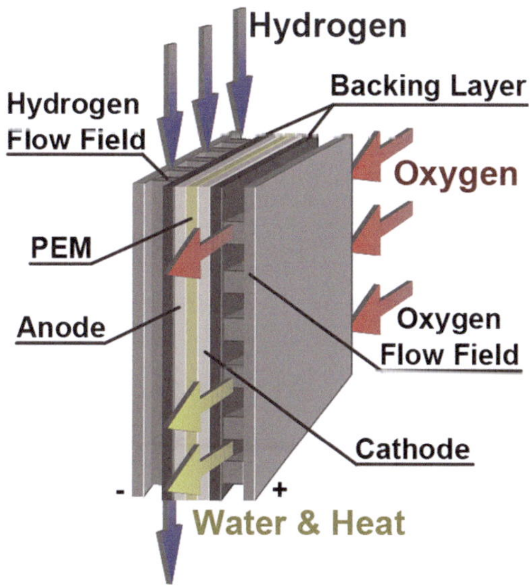

of high aspect ratio and wide material diversity, electrospun nanofibers generally exhibit high conductivity, large specific surface area, and excellent mechanical and chemical stability.

2.2 Electrospun Catalysts

Catalyst material is the key component in a fuel cell; and it facilitates electrode reactions [11]. Commercial fuel cells typically use Pt nanoparticles as the electrocatalyst; however, the Pt nanoparticle electrocatalyst suffers from low activity and poor durability. To address the problems, electrospun catalysts were developed with high activity, good poisoning resistance, and excellent stability. Table 2.1 shows the applications of various electrospun catalysts in fuel cells. The synthesis of catalytic electrospun nanofibers of PtRh and PtRu was first reported by Kim et al. [12]. As shown in Fig. 2.2, the diameter of the nanofibers was within 50 nm. The mass activity of PtRu nanofibers was better than that of highly dispersed PtRu nanoparticles on carbon. The enhanced electrical properties and pertinent interface formation with nanofiber catalysts are believed to contribute to the improved performance. The centimeter-long nanofibers of Pt by electrospinning were fabricated by Shui et al. [16]. These Pt nanofibers showed the diameter of 5–20 nm, with the fiber formation rather than beads dependent on the type of carrier polymer (e.g., PVP), which was removed during heat treatment. There was no need for a catalyst support in the fuel cell electrode by using these Pt nanofibers. Shui et al. continued to report Pt-Fe alloy nanofibers prepared by electrospinning and dealloying for oxygen reduction reaction [15]. The overall nanofiber diameter was

Table 2.1 Applications of electrospun catalysts in fuel cells

Fuel cells	Materials	Microstructure	Advantages	References
DMFC	PtRh and PtRu	D 50 nm, MA 355 and 449 mA mg^{-1}	Electrocatalytic properties enhanced by 30 % relative to PtRu/C	[12]
	Pt	D 50–70 nm, MA 47 mA mg^{-1}	Electrocatalytic properties on MOR	[13]
PEMFC	Pt	D 20–40 nm	Higher ORR activity by 50 %	[14]
	Pt-Fe	D 10–20 nm	2.3 times higher than that of conventional Pt/C catalysts	[15]
Others	Pt	D 5–20 nm	Freestanding Pt electrocatalysts	[16]
	Pd	MA 322 mA mg^{-1}	Simple and cost-effective synthesis of nanofibers	[17]
	Pt and PtRh	D 25–35 nm	High power density	[18]

Microstructure descriptions include diameter (*D*) and/or mass activity (*MA*)

10–20 nm and the ligament diameter only 2–3 nm. These porous long nanowires interweaved to form a self-supporting network with a high specific activity, which was 2.3 times higher than that of conventional Pt/C catalysts. There are a few other reports on the electrospun catalysts; however, comparing to other methods such as chemical vapor deposition, an extremely fine fiber diameter (<10 nm) by electrospinning is difficult to achieve.

2.3 Electrospun Nanofiber-Supported Catalysts

The requirements of the support material for electrodes, such as high electronic conductivity, large specific surface area, and excellent electrochemical and chemical stability, are crucial for both the performance and durability of fuel cells. In addition, the metallic catalyst loading should be reduced due to the high cost of catalyst materials. The dispersion and size distribution of catalyst nanoparticles are optimized for maximum exploitation and rapid kinetics, which provide a way to decrease the metal loading. Table 2.2 summarizes electrospun nanofibers from carbon and other materials such as TiO_2 [24, 33] and conducting polymers [38, 39] as the supports for electrode catalysts in different fuel cells.

2.3.1 Carbon Nanofiber-Supported Catalysts

Carbon nanomaterials have been considered as suitable catalyst supports in fuel cells [40–45] because of their unique graphite properties combined with flexible

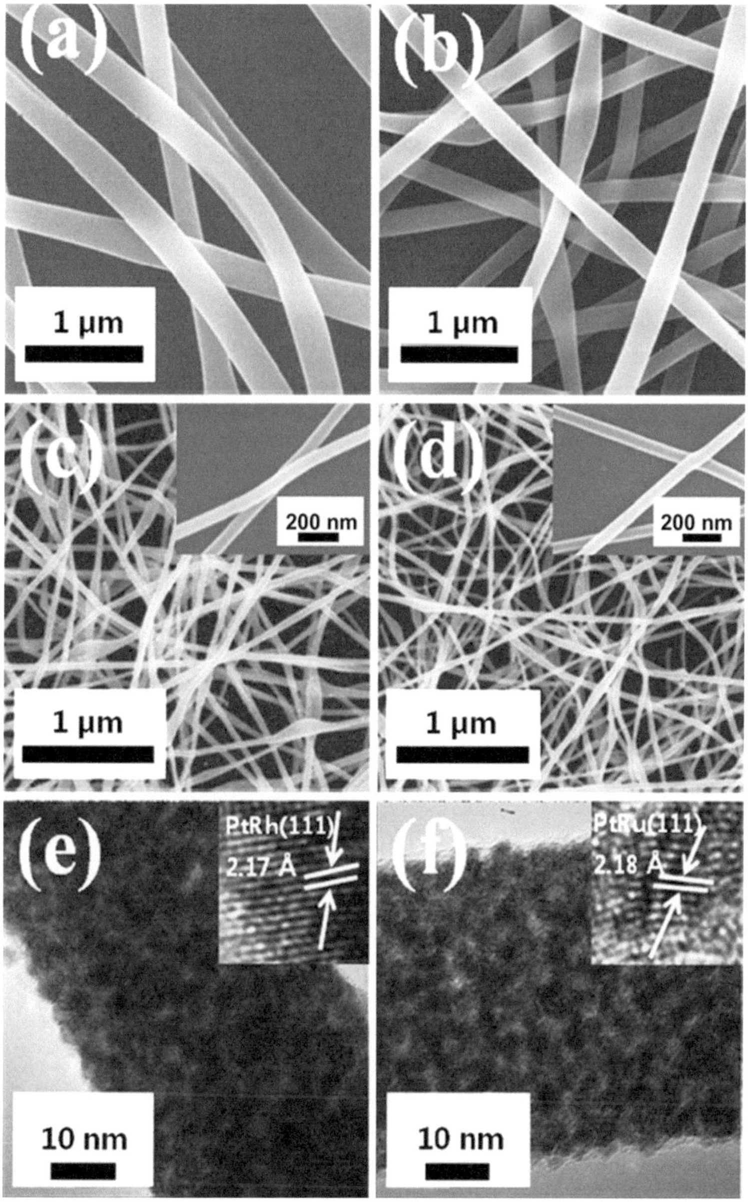

Fig. 2.2 FESEM images of as-spun (**a**) PtRh and (**b**) PtRu precursor/PVP and bimetallic nanofibers of (**c**) PtRh and (**d**) PtRu; *insets* of (**c**) and (**d**) show the diameter of bimetallic nanofibers. HRTEM images of (**e**) PtRh and (**f**) PtRu nanofibers; *insets* of (**e**) and (**f**) are inverse fast Fourier transform image of PtRh and PtRu nanofibers (Reprinted from Kim et al. [12], Copyright 2008, with permission from Elsevier)

Table 2.2 Electrospun nanofiber-supported catalysts for fuel cells

Fuel cells	Materials	Microstructure	Advantages	References
DMFC	Pt/CFM	FD ~150 nm, PD 50–200 nm, MA 185 mA mg^{-1} Pt	Carbon fibrous mats improve the performance of catalyst	[19]
	Pt, Au/CFM	FD 740 nm, PD 7.5 and 4.2 nm, MA 450 mA mg^{-1} Pt	Better stability and smaller charge transfer resistance	[20]
	PtRu/1-aminopyrene-CNFs	FD 100–300 nm, PD 3.1–3.6 nm, MA 32.7 m^2 g^{-1}	High active surface area and improved performance	[21]
	Electrodeposited Pt/CNFs	FD 200–300 nm, PD 20 nm	Pt/CNFs with a Pt loading of 0.170 mg cm^{-2} has the highest activity	[22]
	Pt/CNFs	FD 150 nm, MA 98 mA cm^{-2}	Higher electrocatalytic activity, more stability, larger exchange current, and smaller charge transfer resistance	[23]
	Pt (NP and NW)/TiO_2	FD 150–300 nm, PD ~7 or 150 nm, MA 56.2 mA mg^{-1} Pt	The additional catalytic facets on the nanofibers can enhance both catalytic ability and robustness	[24]
	Pt/TiO_2	FD 30–60 nm, PD ~5 nm	Highly dispersed Pt nanoparticles without agglomeration	[25]
	PdO-Co_3O_4	FD 300–500 nm	Nafion/PdO-Co_3O_4/GCE exhibits the greatest electrocatalytic activity toward methanol electrooxidation	[26]
	Pt-PAN fibers			
	Pt-anchored electrospun WO_3	FD 200–400 nm	The peak current density of Pt nanotubes anchored to WO_3 nanofibers is 4.5 times higher than that of their spherical counterparts	[27]

PEMFC	Pt/e-CNFs	FD 250 nm, PD ~2.6–2.9 nm, MA 302 mA mg^{-1} Pt	The performance of the MEA prepared by e-CNF was excellent, owing to the morphology and the enhanced electrical conductivity	[28]
	PtPd/CNFs	FD ~250 nm, PD 29–64 nm	Pt-Pd/CNFs possess good performance in the electrocatalytic reduction of oxygen	[29]
	Pt/Nb-doped TiO_2	FD 50–100 nm, PD 5 nm, MA 17.1 $m^2\ g^{-1}$	Nb-doped TiO_2 retained 60 % of the electrochemically active surface area after 1,000 cycles	[30]
	$CrN\text{-}Cr_3C_2$	FD 200–500 nm	Conductivity studies revealed that the fiber systems are suitable for electrochemical applications	[31]
Others	Pt/Nb-TiO_2	FD 10 nm, PD ~3 nm	Excellent catalytic activity for oxygen reduction	[32]
	Pd layer/TiO_2	FD 115 nm, PD ~60 nm	Promote glycerol electrooxidation in alkaline medium	[33]
	PtxAu100-x/CNF	FD 700 nm, PD ~120 nm	Highly efficient electrocatalytic oxidation of formic acid	[34]
	Ni/YSZ	FD 300 nm	The peak power density is twice of that with the conventional anode	[35]
	FeCo-CNF	FD 100–200 nm	FeCo-CNF catalysts possess the comparable electrocatalytic activity and better tolerance to ethanol crossover than Pt/C in the oxygen reduction	[36]
	CNF-based biofilms	FD ~1,000 nm	The bioelectrocatalytic anode current density is shown to reach values of up to 30 A m^{-2}, which represents the highest reported values for electroactive microbial biofilms	[37]

Microstructure descriptions include fiber diameter (*FD*), particle diameter (*PD*), and/or mass activity (*MA*)

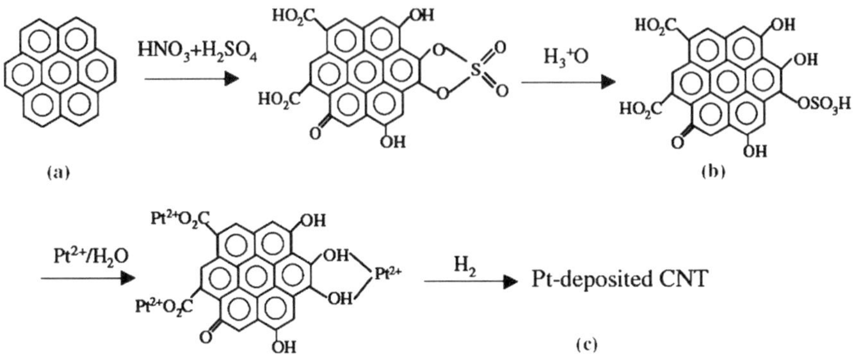

Fig. 2.3 Scheme of the mechanism of Pt deposition via the oxidation of $HNO_3 + H_2SO_4$ on CNFs (Reprinted with the permission from Yu et al. [51]. Copyright 1998 American Chemical Society)

structures. Among various carbon nanomaterials, electrospun carbon nanofibers (CNFs) are promising support materials for catalysts in fuel cells, due to their unique one-dimensional structure, high electronic and thermal conductivities, and good electrochemical stability [46–48]. However, CNFs are chemically inert and it is necessary to activate their graphitic surface because the deposition, distribution, and size of Pt or Pt alloy nanoparticles strongly depend on the surface properties of CNFs [49, 50]. Until now, many methods, such as electroless deposition and electrodeposition, have been carried out to prepare electrospun CNF-supported catalysts for the alcohol oxidation and oxygen reduction in fuel cells.

2.3.1.1 Electroless Deposition

Electroless deposition, also known as chemical deposition, is the most widely used method in preparing highly dispersed Pt nanoparticles on carbon supports for fuel cells. Figure 2.3 shows the typical preparation procedure of the chemical deposition method for Pt/CNF catalysts [51]. First, the aromatic conjugate ring structure of the CNFs was modified by an oxidation process of extremely aggressive reagents (HNO_3 or H_2SO_4 or their mixture). During that process, various surface functional groups such as quinoid (═O), hydroxy (–OH), and carboxyl (–COOH) were produced. These functional groups can supply defect sites for the Pt nanoparticle deposition. After adding Pt salt and a reductive agent such as H_2, HCHO, or ethylene glycol, Pt^{2+} ions can be reduced and deposited on the defect sites created during the acid treatment.

This procedure was adopted by many scientists working on the preparation of electrospun nanofiber-supported catalysts. Steigerwalt et al. reported a multistep deposition and reactive decomposition of a precursor molecule containing 1 Pt and 1 Ru atom on herringbone graphitic carbon nanofibers (GCNFs) [52]. The resultant nanocomposites had a total metal content of 42 wt% with a bulk Pt/Ru atomic ratio of 1:1, and metal alloy nanoclusters had average particle sizes of 6–7 nm.

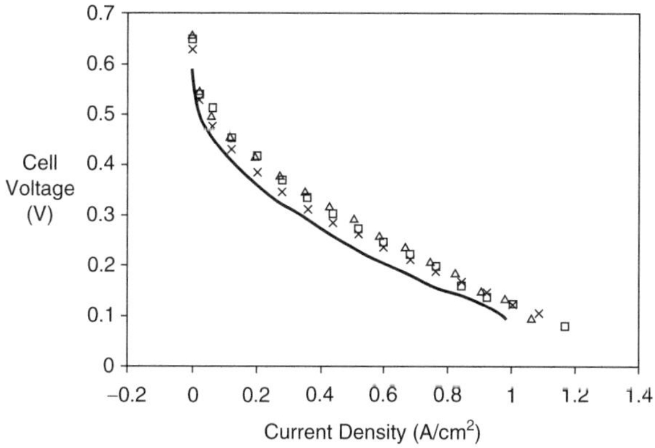

Fig. 2.4 *I-V* curves comparing the performance of Pt-Ru nanocomposites at a loading of 1.5 mg cm^{-2} with (*triplicate*) and without (*solid line*) CNF supports as anode catalysts in DMFCs (Reprinted from Guo et al. [53], Copyright 2006, with permission from Elsevier)

Comparative testing of those nanocomposites and unsupported Pt-Ru colloid with similar surface area and catalyst particle size revealed a 50 % increase in electrochemical performance for the Pt-Ru/GCNF nanocomposites. Guo et al. reported CNFs as supports to prepare highly dispersed Pt-Ru catalysts for DMFCs [53]. The results showed that the diameter of Pt-Ru nanoparticles was around 5 nm in Pt-Ru/CNF composites, and the CNF-supported catalysts gave better performance than commercial catalysts when the current density was higher than 50 mA cm^{-2} in spite of low methanol oxidation peak current density (Fig. 2.4). This improvement is ascribed to the large specific surface and unique structural properties of CNFs. More recently, a microwave-heated polyol process was employed to synthesize carbon-supported metal catalysts for DMFCs [54–56]. Unlike the conventional heating strategy, the microwave heating process can accelerate the reduction of the metal ions on the surface of carbon supports and reduce the temperature and concentration gradients in the solution, resulting in a more uniform environment for the nucleation and growth of metal particles [57]. Tsuji et al. prepared Pt-Ru alloy nanoparticles supported on CNFs within a few minutes by using a microwave-polyol method [58]. Three types of CNFs with very different surface structures, such as platelet, herringbone, and tubular ones, were used as carbon supports. The methanol fuel cell activities of Pt-Ru/CNF catalysts were in the order of platelet > tubular > herringbone, and the methanol fuel cell activities of PtRu/CNFs measured at 60 °C were 1.7–3.0 times higher than those of a standard Pt-Ru catalyst loaded on carbon black Vulcan XC72R, which is shown in Fig. 2.5. Since CNF surfaces are always smooth and chemically inert, harsh oxidative methods are used to produce active sites to allow catalyst nanoparticles to be deposited on and anchored to the surface of CNFs. However, such treatments can potentially destroy the integrity of the CNFs, which result in their low electronic conductivity and poor cyclability. A method based on mild activation for electrospun CNFs and subsequent deposition can be utilized, that is, electrodeposition.

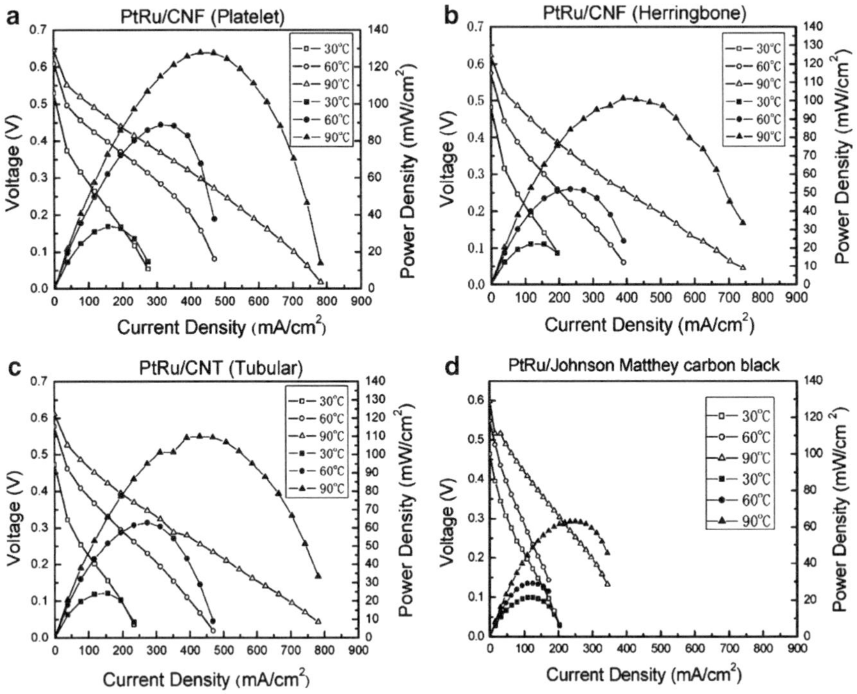

Fig. 2.5 Electrocatalytic activity of Pt-Ru catalysts loaded on platelet CNFs (**a**), herringbone CNFs (**b**), tubular CNFs (**c**), and carbon black Vulcan XC72R (**d**). *Open* and *solid symbols* represent voltage and power density, respectively (Reprinted with the permission from Tsuji et al. [58]. Copyright 2007 American Chemical Society)

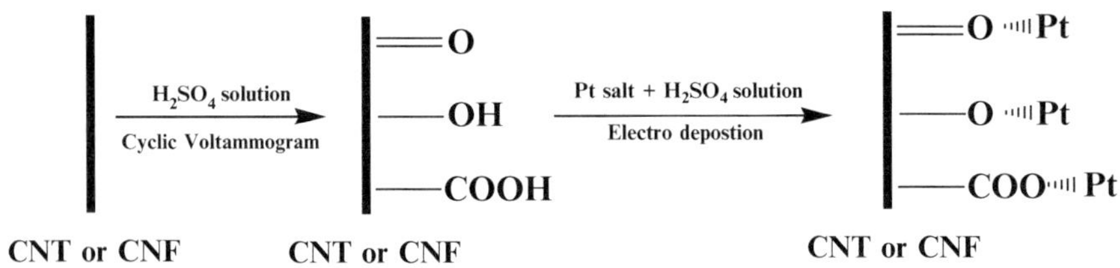

Fig. 2.6 Schematic diagram illustrating the procedure of electrodepositing Pt nanoparticles onto a CNT or CNF

2.3.1.2 Electrodeposition

Electrodeposition, which is used for attaining Pt particles onto the carbon supports, was developed by several research groups to improve the Pt utilization and reduce Pt loading [59–62]. The synthesis procedure for Pt/CNF catalysts using electrodeposition is shown in Fig. 2.6, which is similar to that used in the electroless deposition. In the electrodeposition, CNFs are first cycled in a sulfuric acid solution

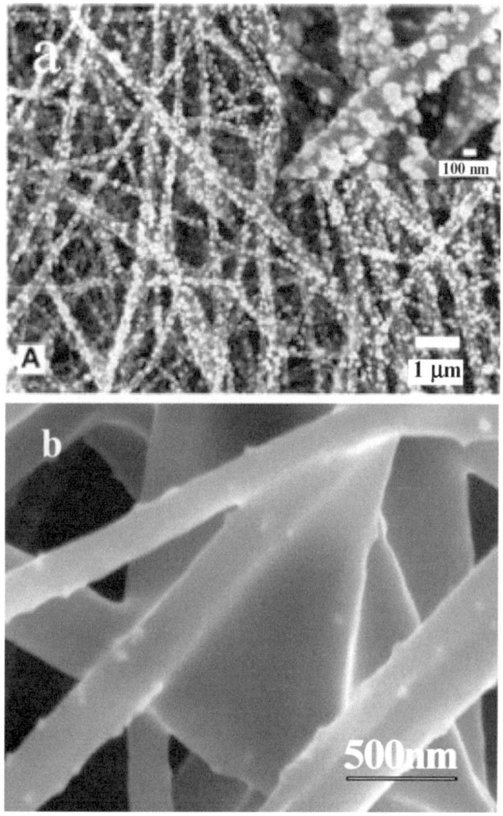

Fig. 2.7 SEM images of as-prepared Pt/CNF electrode using multi-cycle CVs in 2.0 M $H_2PtCl_6 + 0.5$ M H_2SO_4 solutions (**a**) (Reprinted from Li et al. [19], Copyright 2008, with permission from Elsevier) and Pt/CNFs prepared with the electrodeposition potentials −0.2 V (vs. Ag/AgCl/4.0 M KCl) in 5.0 mM $H_2PtCl_6 \cdot xH_2O + 1.0$ M H_2SO_4 solutions (**b**) (Reprinted from Lin et al. [65], Copyright 2009, with permission from Elsevier)

between −0.7 and +1.2 V to oxidize the surface. During that process, various surface functional groups such as quinoid (=O), hydroxy (–OH), and carboxyl (–COOH) were produced, which can supply defect sites for the Pt nanoparticle deposition [63]. For instance, Wang et al. have successfully prepared Pt/CNT catalysts by electrodepositing Pt nanoparticles on CNTs in a $H_2SO_4 + H_2PtCl_6$ solution [64]. The results showed that although the Pt loading was 0.2 mg cm^{-2} on the CNT surface, the average diameter of the Pt particle was much larger (i.e., 25 nm) compared to the commercially available Pt/C catalysts, and the fuel cell performance was also lower than that of the conventional catalyst at a comparable Pt loading. Li et al. used the cyclic voltammogram method to deposit Pt particles onto CNFs, but the diameters of deposited Pt nanoclusters were too large (50–200 nm) for practical fuel cell applications (Fig. 2.7a) [19]. In comparison, Lin et al. reported the preparation and characterization of Pt/CNFs by the electrodeposition of smaller Pt nanoparticles ($\leq$55 nm) onto electrospun CNFs under different potentials (Fig. 2.7b) [65]. The Pt/carbon composite nanofibers possessed the properties of high active surface area and fast electron transfer rate, which led to a good performance toward the electrocatalytic oxidation of methanol.

2.3.2 *Other Nanofiber-Supported Catalysts*

Except for carbon, polymers and other materials were also explored as alternatives for supporting catalysts in fuel cells. Electrospun polyaniline (PANI) nanofibers were fabricated by Liu et al. as the support for Pt. The diameter of the PANI nanofibers was 60–80 nm, which supplied a high surface area to disperse the nanoparticles [66]. These electrospun PANIs also exhibited a porous network for gas flow and good electrical conductivity, which made them an excellent support for electrode catalysts in fuel cells. The electrocatalytic activity of PANI/Pt was found to be much higher than that of bulk Pt electrodes; for example, the current density value was 24.7 versus 7.5 mA cm^{-2} mg^{-1} at 0.68 V. Xia et al. electrospun anatase TiO_2 nanofibers followed by the deposition of Pt nanoparticles with 2–5 nm in diameter [24]. The electrochemical activity and durability for anatase TiO_2 nanofiber-supported Pt nanostructures for direct methanol oxidation were much higher compared with those of commercial Pt/C catalysts. The improvement in catalytic activity and durability results from the synergistic effect of the underlying anatase surface and the Pt nanostructures with well-defined facets. Novel conductive or semiconductive oxides such as TiO_x [67, 68], SiO_2 [69], and MnO_x [70] were also electrospun as catalyst supports for fuel cell electrodes with enhancement in the catalytic performance. Metal oxides are superior to conventional catalyst supports due to their exceptional chemical stability and corrosion resistance at the potentials and pH in fuel cell operations. Such metal oxide-supported catalyst systems showed promising results in terms of catalyst performance and durability and also the tolerance to methanol, compared with conventional Pt/C electrodes [71, 72].

2.4 Electrospun Electrolyte Membranes

As a versatile technique for producing nanofibrous mats, electrospinning has been utilized for the development of composite fuel cell membranes for the purpose of modifying the proton exchange membranes and improving their performance. For example, non- or less-proton-conductive component, acting as mechanical reinforcement, can be incorporated into a porous matrix, generally Nafion®, for higher mechanical properties or lower methanol crossover. An overview on this type of studies was shown in Table 2.3.

Alternatively, electrospun functionalized nanofibers with high proton conductivity could also be incorporated into polymer matrices, in which perfluorosulfonated polymers, such as Nafion®, were selected as the matrix polymers. With the filled proton-conducting component fibers, the proton conductivity of composite membranes could be greatly enhanced. The relative reports are shown in Table 2.4.

There was a general thought that the proton conductivity is highly dependent on the concentration of the protonated groups or the ion-exchange capacity (IEC), that is, the volume fraction of the proton conductive component. For example, with the

Table 2.3 Electrospun non- or less-conductive component as mechanical reinforcements

Matrix polymer	Filler nanofibers	Remark	References
Nafion®	Polyvinyl alcohol	PVA was cross-linked with glutaraldehyde and had lower proton conductivity and lower methanol crossover	[73–76]
Nafion®	Polyvinylidene fluoride	17 wt% PVDF was electrospun with the solvent of 1:1 w/w acetone/dimethylacetamide. The composite membranes had lower conductivity but better properties of MEA performance and methanol crossover	[77]
Nafion®	Sulfonated poly(ether sulfone)	The composite membranes had a triple-layer structure of Nafion-SPES/Nafion-Nafion membrane. The composite membranes had a relative lower conductivity but better fuel cell performance and lower methanol crossover	[78]
Sulfonated poly(2,6-dimethyl-1,4-phenylene oxide) (SPPO)	Cross-linked bromomethylated sulfonated polyphenylene oxide (BPPO)	The composite membranes had lower conductivity of 0.03–0.07 S/cm compared with SPPO of 0.09–0.15 S/cm but much lower water swelling ratio and improved fuel cell performance	[79, 80]
Perfluorosulfonic acid prepared in laboratory	Polyethersulfone loaded with $CaCO_3$	The composite membranes had highly increased break strength and Young's modulus	[81]
660 EW perfluorosulfonic acid (PFSA) from 3 M Company	Polyphenylsulfone	The composite membrane with 70 vol% PFSA and 30 vol% polyphenylsulfone had lower conductivity but lower water swelling ratio	[82, 83]
Sulfonated poly(arylene ether sulfone) (SPAES)	Sulfonated silicate-nanoencapsulated polyimide (sPI)	The composite membrane had relatively comparable conductivity with that of SPAES but better mechanical strength with SPAES	[84, 85]
PBI	Polybenzoxazine (PBz)-modified polybenzimidazole (PBI)	The composite membranes showed enhanced mechanical properties, twofold higher proton conductivity at 160 °C, and 34 % increased power density	[86]
Nafion®	SiO_2-coated polyetherimide	The composite membranes showed high mechanical properties, much lower water swelling, and lower proton conductivity	[39]
Nafion®	Poly(phenyl sulfone)	The composite membranes exhibited higher Young's modulus	[82, 83]

Table 2.4 Electrospun proton-conducting fiber fillers for PEM applications

Matrix polymer	Filler nanofibers	Proton conductivity (S/cm)	References
Poly(diallyldimethylammonium chloride) (PDDA), poly(styrene sulfonic acid, sodium salt) (PSS) LBL self-assembly	Sulfonated poly(fluorenyl ether ketone) (SPFEK) and polyether sulfone (PES) co-electrospun fibers sandwiched with PES beaded nanofibers	0.06 (80 °C, 100 % RH), $E_a = 1.30$ KJ/mol	[87]
Norland Optical Adhesive (NOA63)	Perfluorosulfonic acid polymer (PFSA, 3 M Corporation)/poly(acrylic acid) (PAA)/sulfonated octaphenyl polyhedral silsesquioxane (sPOSS) with weight ratio of 65: 5: 35	0.498 (120 °C, 90 % RH)	[88]
UV-cured NOA63	Sulfonated poly(arylene ether sulfone)	~0.10 (25 °C, in water) for 80 vol% fibers	[88, 89]
Sulfonated poly(arylene ether sulfone)	Sulfonated polyhedral oligomeric silsesquioxane	0.094 (30 °C, 80 % RH)	[88, 89]
UV-cross-linked polyurethane (PU)	Sulfonated poly(arylene ether sulfone) (sPAES)/sulfonated polyhedral oligomeric silsesquioxane (sPOSS) (w/w = 60/40)	0.37 (80 °C, 80 % RH)	[90]
UV-cross-linked NOA63	Perfluorosulfonic acid polymers from 3 M Corporation (825 EW and 733 EW polymers)	~0.1 (80 °C, 80 % RH, 733 EW PFSA with 70 % fiber volume fraction)	[91]
Nafion®	SiO_2/sulfonated polyether ether ketone	0.077 (90 °C, 100 % RH)	[92]
N/A	Nafion® nanofibers	~0.12 (30 °C, 90 % RH)	[93–95]
N/A	Sulfonated poly(ether ether ketone ketone) (SPEEKK)	~0.08 (25 °C, 100 % RH)	[96]
Cross-linked poly(2-acrylamido-2-methylpropanesulfonic acid)	Sulfated zirconia (S-ZrO_2)	0.34 (100 °C, 80 % RH)	[97]
Nafion®	Sulfonated polystyrene (S-PS)	0.18 (80 °C, 100 % RH)	[98]
Nafion®	S-ZrO_2	0.31 (80 °C, 100 % RH)	[99]
Lab-made sulfonated polyimide: NTDA-BDSA-r-APPF	NTDA-BDSA-r-APPF	1.3 (80 °C, 98 % RH, fiber axial direction) 0.036 (fiber perpendicular direction)	[100]

addition of cross-linked PVA nanofibers, the proton conductivity of Nafion/PVA composite membranes was reduced to 0.025 S/cm compared with 0.096 S/cm for Nafion 117. This value could be further reduced to around 0.01 S/cm with IEC value reduced from 0.58 to 0.33 meq/g, compared with the IEC value of 0.91 meq/g for Nafion 117 [73–76]. Similar phenomenon was evident with the use of Nafion/PVDF [77], Nafion/SiO_2-coated polyetherimide composite membranes [39], etc., as the proton exchange membranes. In addition, with the filling of long-range one-dimensional nanofibrous materials, a series of other properties of the membranes were significantly improved, such as the mechanical properties [73–76], hydrolytic stability [79], reduced methanol crossover [73–76], and so forth.

With the in-depth comprehension of the working principle of ion transport in proton exchange membranes, for example, Nafion®, quantitative and qualitative modeling studies have been conducted for supporting the optimization of not only the composition of the membranes but also the operating conditions, thereby yielding higher efficiencies and power densities. In terms of microscopic models, there have been many models based on statistical mechanics, molecular dynamics, and macroscopic phenomena applied to the microscopic structure of the membranes. These models provide a fundamental understanding of processes like diffusion and conduction in the membranes on a microscopic scale including the effect of small perturbations, such as nonhomogeneity of pores, electric fields on transport, and the introduction of small-scale structural effects. The basis for all these models is the description of the microscopic structure of the polymer that was proposed in the early 1980s by Gierke and Hsu [101]. In their work, the authors correlated the experimental data through geometric and phenomenological relationships for the swelling of the polymer caused by the uptake of water and its effect on the diffusion coefficient of water in the membrane pores. The correlation evolved by analyzing the data taken under different operating conditions led to the formulation of a widely accepted description of the polymeric membranes in terms of an inverted micellar structure, in which the ion-exchange sites are separated from the fluorocarbon backbone, and thus forming spherical clusters (pores). These clusters are connected by short narrow channels. The model was hence termed as "cluster network" model. When the membrane is dry, an average cluster has a radius of about 1.8 nm and it contains about 26 $-SO_3^-$groups distributed on the inner pore surface, as illustrated in Fig. 2.8a. In the swollen state, the diameter increases to about 4 nm and the number of fixed SO_3^- groups goes up to about 70. Under these conditions, each pore is filled with about 1,000 water molecules and the connecting channels have a diameter and a length of about 1 nm. According to this theory, there is a critical amount of water available in the membrane, below which proton transport is extremely difficult due to the absence of extended pathways. Above or near the threshold, the conductivity (σ) is as follows:

$$\sigma = \sigma_0(c - c_0)^n \tag{2.4}$$

where c is the volume fraction of the aqueous phase, c_0 the percolation threshold for the water content in the membrane, n a universal constant which depends on the

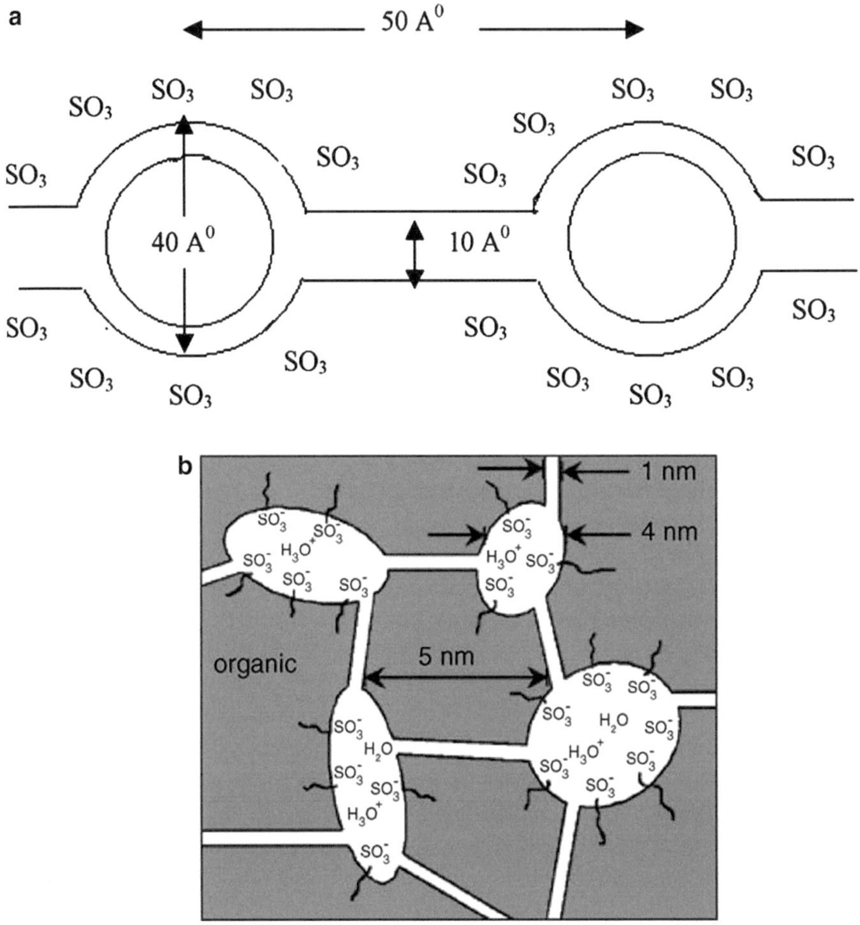

Fig. 2.8 Schematic diagrams of (**a**) cluster network model (Reprinted with the permission from Yeager and Eisenberg [101], Copyright 1982 American Chemical Society) and (**b**) random cluster network model (Reprinted with permission from Eikerling et al. [102]. Copyright 1997 American Chemical Society)

dimensions of the system, and σ_0 a factor related to the molecular interactions that can only be computed from specific microscopic models. A comparison between experimental results and percolation theory showed a good agreement.

Based on the cluster network model, further studies propose an interpretation of the percolation properties of proton conductivity as a function of water content by using a "random network model" [102], which is a modification of the "cluster network model." This model includes an intermediate region wherein the side chains ending with pendant sulfonic acid groups, which are bonded to the perfluorinated backbones, tend to form cluster within the overall structure of the material resulting in the formation of hydrated regions. Unlike the "cluster network model," the

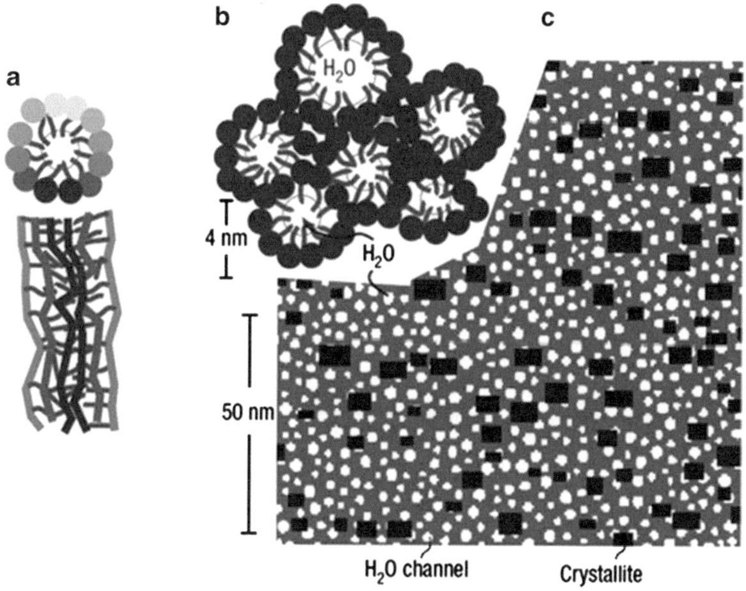

Fig. 2.9 (**a**) Two views of an inverted-micelle cylinder, with the polymer backbones on the outside and the ionic side groups lining the water channel. Shading is used to distinguish chains in front and in the back. (**b**) Schematic diagram of the approximately hexagonal packing of several inverted-micelle cylinders. (**c**) Cross sections through the cylindrical water channels (*white*) and the Nafion crystallites (*black*) in the noncrystalline Nafion matrix (*dark gray*) (From Schmidt-Rohr and Chen [105]. Reprinted by permission of Nature Publishing Group)

hydrated regions in this model are distributed randomly in the polymer matrix, which facilitates quicker transport of protons upon the rotation of these side chains. In this case, although the hydrated regions drift apart, the traverse motion of protons through the membranes is possible, as shown in Fig. 2.8b. Haubold et al. verified through small-angle X-ray scattering (SAXS) that the "random network model" could be applied to Nafion [103]. James et al. deployed the atomic force microscopy (AFM) technique and found the "random network model" to be generally acceptable on Nafion 117, which is a commonly available Nafion membrane, typically used in the region of 9–34 % RH [104]. Based on the recent model (inverted-micelle cylinder model) of Nafion (Fig. 2.9), the diameter of the ionic channel is 1.8–3.5 nm, with an average diameter of 2.4 nm, while the length of the channel is at the level of tens of nanometers, with the hydration degree of 20 vol% (11 wt%) water.

According to the modeling of proton exchange membranes, the transport of hydrated protons depends not only on the amount of protogenic functional groups but also on the effective phase separation and appropriate arrangement of ionic clusters in the membrane. Inspired by the ion channel structure, 1D nanomaterials should be one of the most promising candidates, including sulfonate single-wall carbon nanotubes [106, 107], sulfonate nano- or micro-channels [108–110], self-assembled nanomaterials [111], and electrospun nanofibers [82, 100, 112–114].

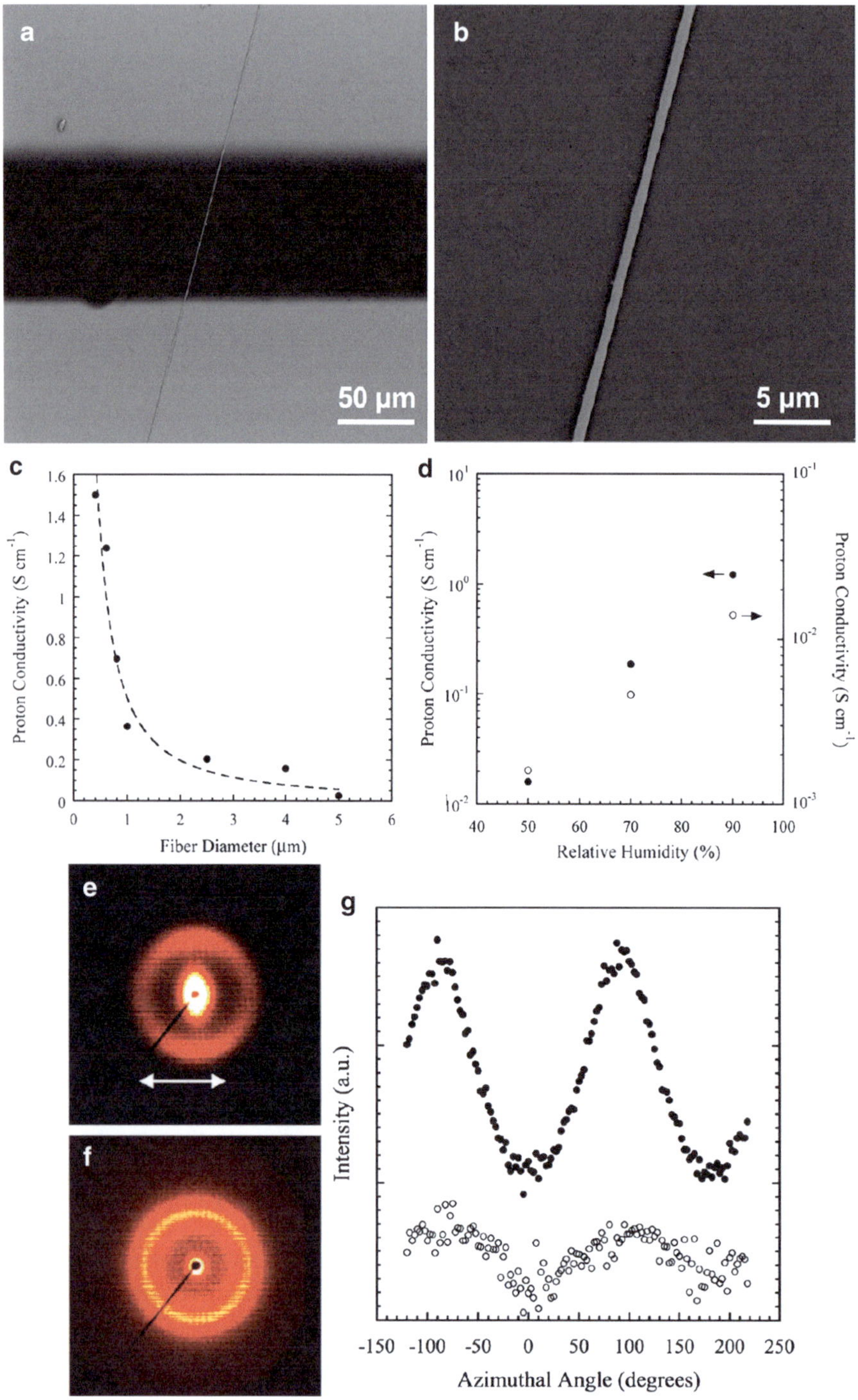
a
50 μm
b
5 μm
c
Proton Conductivity (S cm⁻¹)
Fiber Diameter (μm)
d
Proton Conductivity (S cm⁻¹)
Relative Humidity (%)
Proton Conductivity (S cm⁻¹)
e
f
g
Intensity (a.u.)
Azimuthal Angle (degrees)

It was shown that even with inert polymer matrix, such as UV-cross-linked NOA63 or PU for mechanical reinforcement, the composite membranes with electrospun functionalized nanofibers could still offer a significantly higher in-plane proton conductivity than that of commercial Nafion membranes [77, 83, 114–116], which was possibly due to the continuous network of proton-conducting nanofibrous paths which span the membrane thickness. It was also reported that because of the high shear force and fast drying process during electrospinning, the electrospun Nafion nanofibers show obvious higher scattering maxima of $q \sim 1.65\ nm^{-1}$ on 2D X-ray patterns than that of $1.58\ nm^{-1}$ for Nafion cast films, indicating the anisotropic alignment of ion aggregates in nanofibers and isotropic ion aggregates in cast films. The inner alignment of ionic structure can significantly improve proton conductivity. As shown in Fig. 2.10, with the fiber diameter reducing to 400 nm, the proton conductivity along the fiber axis can be as high as 1.5 S/cm [93–95]. Interestingly, when composite membranes were filled with well-aligned sulfonated nanofibers, the in-plane conductivity was different between the axial and perpendicular directions, verifying the uniaxial enhancement of proton transport in functionalized polymers [100].

Besides the electrospun functional polymers, ceramic nanofibers, such as solid superacid like S-ZrO_2, can also be used as the nanofibrous fillers. During the formation of hybrid membranes, hydrophilic attractions between the sulfate groups on S-ZrO_2 fibers and sulfonic acid groups in Nafion lead to the aggregation of ionic clusters onto the organic-inorganic interfaces. The increased amount of interfacial protogenic groups forms long-distance ionic pathways along the S-ZrO_2 fibers that ensure effective proton transport through the entire membranes, as shown schematically in Fig. 2.11. As a result, the in-plane proton conductivity can also be highly improved.

2.5 Summary and Outlook

Nanofibers for fuel cells can potentially play a key role in meeting global energy demands in the coming century due to their unique structures and high surface area to volume ratios [117, 118]. Among many nanofiber fabrication methods,

Fig. 2.10 Properties of electrospun Nafion nanofibers: (**a**) SEM image of a single high-purity Nafion nanofiber (99.9 wt% Nafion) bridging two electrodes; (**b**) enlarged image of (**a**); (**c**) proton conductivity (at 30 °C, 90 % RH) versus fiber diameter for high-purity Nafion nanofibers measured on individual nanofibers; (**d**) humidity-dependent proton conductivity (30 °C) of a single high-purity Nafion nanofiber (*solid symbols*) and a cast Nafion film (*open symbols*); 2D X-ray scattering patterns (at 86 % RH, 25 °C) of (**e**) macroscopically aligned high-purity Nafion nanofibers (*arrows* indicating longitudinal direction of fibers) and (**f**) Nafion cast film; (**g**) X-ray scattering intensity as a function of azimuthal angle for macroscopically aligned high-purity Nafion nanofibers (*solid symbols*) and a cast Nafion film (*open symbols*) (Reprinted with permission from Tamura and Kawakami [100]. Copyright 2010 American Chemical Society)

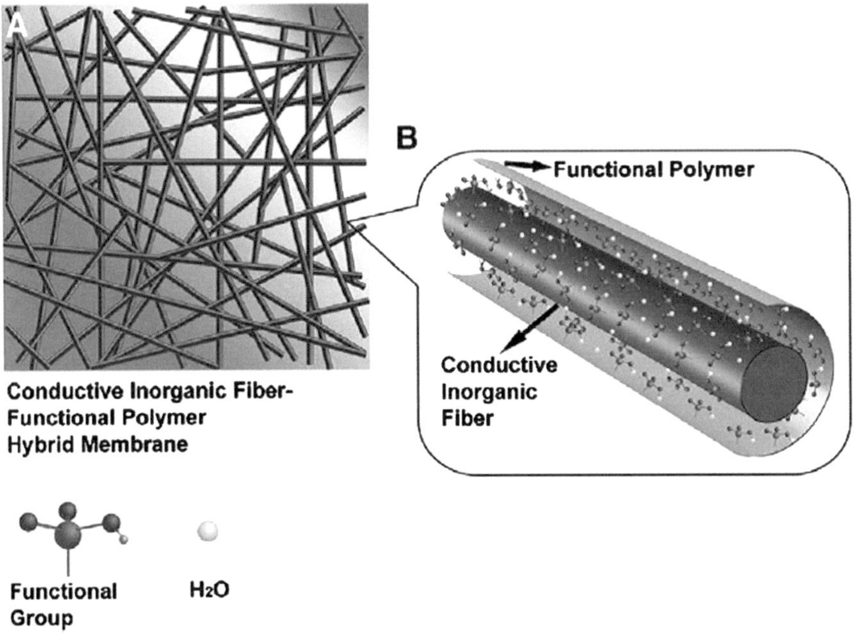

Fig. 2.11 Schematic diagrams of (**a**) conductive inorganic fiber/polymer hybrid membrane and (**b**) the long-range ionic channel between the conductive fiber and functional polymer matrix (Reprinted from Yao et al. [97], Copyright 2011, with permission from Elsevier)

electrospinning is arousing the most intense interest due to its comparative low cost and high production rate for practical energy-related applications. It is a relatively feasible method which provides great control over the fiber morphology, such as the homogeneity, porosity, and a variety of structures including core-shell, hollow, and aligned fibers [119].

There are still several challenges left to be overcome before the large-scale usage of electrospun materials in fuel cell applications. For nanofibrous applications in catalysts and supports, the developments are further needed to reduce fiber diameters and improve the diameter uniformity for higher electrocatalytic ability and lower cost. The process parameters affecting the electrospinning technique are still not fully understood; thus, there are still limitations on the types of catalysts and support materials synthesized by electrospinning. In the applications of composite membranes, nanofibers have been widely investigated as either mechanical or stability reinforcing component or proton conductor. With the further investigation of electrospinning technology, solutions with sufficient viscosity can be electrospun to nanofibrous networks [120]. Other fibrous structures composed of various materials, such as solutions of self-assembled small molecules, can also be electrospun to nanofibers and used as the filler of proton exchange membranes [121]. Furthermore, as the fiber mats generally lie in the composite membranes, while the in-plane proton conductivity increased, the through-plane conductivity was still on a relatively low level, which however was the actual proton transporting direction. Further work on electrospun nanofibers used in proton exchange membranes needs to focus on the progress of through-plane proton conductivity and the improvement of fuel cell performance. Addressing these concerns requires dedicated, renewed, intense, and extensive research on nanofibers from electrospinning, which will make their scale-up become practical for fuel cells from an industrial perspective in the near future.

References

1. Mehta V, Cooper JS (2003) Review and analysis of PEM fuel cell design and manufacturing. J Power Sources 114(1):32–53
2. Liu HS, Song CJ, Zhang L, Zhang JJ, Wang HJ, Wilkinson DP (2006) A review of anode catalysis in the direct methanol fuel cell. J Power Sources 155(2):95–110
3. Yu Y, Li H, Wang HJ, Yuan XZ, Wang GJ, Pan M (2012) A review on performance degradation of proton exchange membrane fuel cells during startup and shutdown processes: causes, consequences, and mitigation strategies. J Power Sources 205:10–23
4. Zhang SS, Yuan XZ, Hin JNC, Wang HJ, Friedrich KA, Schulze M (2009) A review of platinum-based catalyst layer degradation in proton exchange membrane fuel cells. J Power Sources 194(2):588–600
5. Chandan A, Hattenberger M, El-Kharouf A, Du SF, Dhir A, Self V, Pollet BG, Ingram A, Bujalski W (2013) High temperature (HT) polymer electrolyte membrane fuel cells (PEMFC) – a review. J Power Sources 231:264–278
6. Li XL, Faghri A (2013) Review and advances of direct methanol fuel cells (DMFCs) part I: Design, fabrication, and testing with high concentration methanol solutions. J Power Sources 226:223–240
7. Cavaliere S, Subianto S, Savych I, Jones DJ, Roziere J (2011) Electrospinning: designed architectures for energy conversion and storage devices. Energy Environ Sci 4(12):4761–4785
8. Dong ZX, Kennedy SJ, Wu YQ (2011) Electrospinning materials for energy-related applications and devices. J Power Sources 196(11):4886–4904
9. Szentivanyi AL, Zernetsch H, Menzel H, Glasmacher B (2011) A review of developments in electrospinning technology: new opportunities for the design of artificial tissue structures. Int J Artif Organs 34(10):986–997
10. Panda PK (2008) Ceramic nanofibers by electrospinning technique – a review. Trans Indian Ceram Soc 66(2):65–76
11. Debe MK (2012) Electrocatalyst approaches and challenges for automotive fuel cells. Nature 486(7401):43–51
12. Kim YS, Nam SH, Shim HS, Ahn HJ, Anand M, Kim WB (2008) Electrospun bimetallic nanowires of PtRh and PtRu with compositional variation for methanol electrooxidation. Electrochem Commun 10(7):1016–1019
13. Kim JM, Joh HI, Jo SM, Ahn DJ, Ha HY, Hong SA, Kim SK (2010) Preparation and characterization of Pt nanowire by electrospinning method for methanol oxidation. Electrochim Acta 55(16):4827–4835
14. Kim HJ, Kim YS, Seo MH, Choi SM, Cho J, Huber GW, Kim WB (2010) Highly improved oxygen reduction performance over Pt/C-dispersed nanowire network catalysts. Electrochem Commun 12(1):32–35
15. Shui JI, Chen C, Li JCM (2011) Evolution of nanoporous Pt-Fe alloy nanowires by dealloying and their catalytic property for oxygen reduction reaction. Adv Funct Mater 21(17): 3357–3362

16. Shui JL, Li JCM (2009) Platinum nanowires produced by electrospinning. Nano Lett 9(4):1307–1314
17. Su L, Jia WZ, Schempf A, Ding Y, Lei Y (2009) Free-standing palladium/polyamide 6 nanofibers for electrooxidation of alcohols in alkaline medium. J Phys Chem C 113(36):16174–16180
18. Kim HJ, Kim YS, Seo MH, Choi SM, Kim WB (2009) Pt and PtRh nanowire electrocatalysts for cyclohexane-fueled polymer electrolyte membrane fuel cell. Electrochem Commun 11(2):446–449
19. Li MY, Han GY, Yang BS (2008) Fabrication of the catalytic electrodes for methanol oxidation on electrospinning-derived carbon fibrous mats. Electrochem Commun 10(6): 880–883
20. Liu XM, Li MY, Han GY, Dong JH (2010) The catalysts supported on metallized electrospun polyacrylonitrile fibrous mats for methanol oxidation. Electrochim Acta 55(8):2983–2990
21. Lin Z, Ji LW, Krause WE, Zhang XW (2010) Synthesis and electrocatalysis of 1-aminopyrene-functionalized carbon nanofiber-supported platinum-ruthenium nanoparticles. J Power Sources 195(17):5520–5526
22. Lin Z, Ji LW, Zhang XW (2009) Electrocatalytic properties of Pt/carbon composite nanofibers. Electrochim Acta 54(27):7042–7047
23. Li MY, Zhao SZ, Han GY, Yang BS (2009) Electrospinning-derived carbon fibrous mats improving the performance of commercial Pt/C for methanol oxidation. J Power Sources 191(2):351–356
24. Formo E, Peng ZM, Lee E, Lu XM, Yang H, Xia YN (2008) Direct oxidation of methanol on pt nanostructures supported on electrospun nanofibers of anatase. J Phys Chem C 112(27):9970–9975
25. Long Q, Cai M, Li JR, Rong HL, Jiang L (2011) Improving the electrical catalytic activity of Pt/TiO_2 nanocomposites by a combination of electrospinning and microwave irradiation. J Nanopart Res 13(4):1655–1662
26. Zhang YQ, Wang YZ, Jia JB, Wang JG (2012) Electro-oxidation of methanol based on electrospun PdO-Co_3O_4 nanofiber modified electrode. Int J Hydrog Energy 37(23): 17947–17953
27. Zhao ZG, Yao ZJ, Zhang J, Zhu R, Jin Y, Li QW (2012) Rational design of galvanically replaced Pt-anchored electrospun WO_3 nanofibers as efficient electrode materials for methanol oxidation. J Mater Chem 22(32):16514–16519
28. Park JH, Ju YW, Park SH, Jung HR, Yang KS, Lee WJ (2009) Effects of electrospun polyacrylonitrile-based carbon nanofibers as catalyst support in PEMFC. J Appl Electrochem 39(8):1229–1236
29. Lin Z, Ji LW, Toprakci O, Krause W, Zhang XW (2010) Electrospun carbon nanofiber-supported Pt-Pd alloy composites for oxygen reduction. J Mater Res 25(7):1329–1335
30. Bauer A, Lee K, Song CJ, Xie YS, Zhang JJ, Hui R (2010) Pt nanoparticles deposited on TiO_2 based nanofibers: electrochemical stability and oxygen reduction activity. J Power Sources 195(10):3105–3110
31. Garcia-Marquez A, Portehault D, Giordano C (2011) Chromium nitride and carbide nanofibers: from composites to mesostructures. J Mater Chem 21(7):2136–2143
32. Park KW, Seol KS (2007) Nb-TiO_2 supported Pt cathode catalyst for polymer electrolyte membrane fuel cells. Electrochem Commun 9(9):2256–2260
33. Su L, Jia WZ, Schempf A, Lei Y (2009) Palladium/titanium dioxide nanofibers for glycerol electrooxidation in alkaline medium. Electrochem Commun 11(11):2199–2202
34. Huang JS, Hou HQ, You TY (2009) Highly efficient electrocatalytic oxidation of formic acid by electrospun carbon nanofiber-supported PtxAu100-x bimetallic electrocatalyst. Electrochem Commun 11(6):1281–1284
35. Li LP, Zhang PG, Liu RR, Guo SM (2011) Preparation of fibrous Ni-coated-YSZ anodes for solid oxide fuel cells. J Power Sources 196(3):1242–1247
36. Uhm S, Jeong B, Lee J (2011) A facile route for preparation of non-noble CNF cathode catalysts in alkaline ethanol fuel cells. Electrochim Acta 56(25):9186–9190

37. Chen SL, Hou HQ, Harnisch F, Patil SA, Carmona-Martinez AA, Agarwal S, Zhang YY, Sinha-Ray S, Yarin AL, Greiner A, Schroder U (2011) Electrospun and solution blown three-dimensional carbon fiber nonwovens for application as electrodes in microbial fuel cells. Energy Environ Sci 4(4):1417–1421
38. Hong SH, Lee SA, Nam JD, Lee YK, Kim TS, Won S (2008) Platinum-catalyzed and ion-selective polystyrene fibrous membrane by electrospinning and in-situ metallization techniques. Macromol Res 16(3):204–211
39. Lee JR, Kim NY, Lee MS, Lee SY (2011) SiO_2-coated polyimide nonwoven/Nafion composite membranes for proton exchange membrane fuel cells. J Membr Sci 367(1–2): 265–272
40. Liu L, Pu G, Viswanathan R, Fan QB, Liu RX, Smotkin ES (1998) Carbon supported and unsupported Pt-Ru anodes for liquid feed direct methanol fuel cells. Electrochim Acta 43(24):3657–3663
41. Arico AS, Shukla AK, El-Khatib KM, Creti P, Antonucci V (1999) Effect of carbon-supported and unsupported Pt-Ru anodes on the performance of solid-polymer-electrolyte direct methanol fuel cells. J Appl Electrochem 29(6):671–676
42. Nam JH, Jang YY, Kwon YU, Nam JD (2004) Direct methanol fuel cell Pt-carbon catalysts by using SBA-15 nanoporous templates. Electrochem Commun 6(7):737–741
43. Shukla AK, Raman RK, Choudhury NA, Priolkar KR, Sarode PR, Emura S, Kumashiro R (2004) Carbon-supported Pt-Fe alloy as a methanol-resistant oxygen-reduction catalyst for direct methanol fuel cells. J Electroanal Chem 563(2):181–190
44. Salgado JRC, Antolini E, Gonzalez ER (2005) Carbon supported Pt-Co alloys electrocatalysts for as methanol-resistant oxygen-reduction direct methanol fuel cells. Appl Catal B-Environ 57(4):283–290
45. Zeng JH, Lee JY, Zhou WJ (2006) A more active Pt/carbon DMFC catalyst by simple reversal of the mixing sequence in preparation. J Power Sources 159(1):509–513
46. De Jong KP, Geus JW (2000) Carbon nanofibers: catalytic synthesis and applications. Catal Rev-Sci Eng 42(4):481–510
47. Tian ZQ, Jiang SP, Liang YM, Shen PK (2006) Synthesis and characterization of platinum catalysts on muldwalled carbon nanotubes by intermittent microwave irradiation for fuel cell applications. J Phys Chem B 110(11):5343–5350
48. Chan KY, Ding J, Ren JW, Cheng SA, Tsang KY (2004) Supported mixed metal nanoparticles as electrocatalysts in low temperature fuel cells. J Mater Chem 14(4):505–516
49. Yang CW, Hu XG, Wang DL, Dai CS, Zhang L, Jin HB, Agathopoulos S (2006) Ultrasonically treated multi-walled carbon nanotubes (MWCNTs) as PtRu catalyst supports for methanol electrooxidation. J Power Sources 160(1):187–193
50. Jha N, Reddy ALM, Shaijumon MM, Rajalakshmi N, Ramaprabhu S (2008) Pt-Ru/multi-walled carbon nanotubes as electrocatalysts for direct methanol fuel cell. Int J Hydrog Energy 33(1):427–433
51. Yu RQ, Chen LW, Liu QP, Lin JY, Tan KL, Ng SC, Chan HSO, Xu GQ, Hor TSA (1998) Platinum deposition on carbon nanotubes via chemical modification. Chem Mater 10(3): 718–722
52. Steigerwalt ES, Deluga GA, Cliffel DE, Lukehart CM (2001) A Pt-Ru/graphitic carbon nanofiber nanocomposite exhibiting high relative performance as a direct-methanol fuel cell anode catalyst. J Phys Chem B 105(34):8097–8101
53. Guo J, Sun G, Wang Q, Wang G, Zhou Z, Tang S, Jiang L, Zhou B, Xin Q (2006) Carbon nanofibers supported Pt–Ru electrocatalysts for direct methanol fuel cells. Carbon 44(1): 152–157
54. Chen WX, Lee JY, Liu ZL (2002) Microwave-assisted synthesis of carbon supported Pt nanoparticles for fuel cell applications. Chem Commun 21:2588–2589
55. Liu ZL, Lee JY, Chen WX, Han M, Gan LM (2004) Physical and electrochemical characterizations of microwave-assisted polyol preparation of carbon-supported PtRu nanoparticles. Langmuir 20(1):181–187

56. Li X, Chen WX, Zhao J, Xing W, Xu ZD (2005) Microwave polyol synthesis of Pt/CNTs catalysts: effects of pH on particle size and electrocatalytic activity for methanol electrooxidization. Carbon 43(10):2168–2174
57. Liu ZL, Gan LM, Hong L, Chen WX, Lee JY (2005) Carbon-supported Pt nanoparticles as catalysts for proton exchange membrane fuel cells. J Power Sources 139(1–2):73–78
58. Tsuji M, Kubokawa M, Yano R, Miyamae N, Tsuji T, Jun MS, Hong S, Lim S, Yoon SH, Mochida I (2007) Fast preparation of PtRu catalysts supported on carbon nanofibers by the microwave-polyol method and their application to fuel cells. Langmuir 23(2):387–390
59. He ZB, Chen JH, Liu DY, Zhou HH, Kuang YF (2004) Electrodeposition of Pt-Ru nanoparticles on carbon nanotubes and their electrocatalytic properties for methanol electrooxidation. Diamond Relat Mater 13(10):1764–1770
60. Tang H, Chen JH, Yao SZ, Nie LH, Kuang YF, Huang ZP, Wang DZ, Ren ZF (2005) Deposition and electrocatalytic properties of platinum on well-aligned carbon nanotube (CNT) arrays for methanol oxidation. Mater Chem Phys 92(2–3):548–553
61. Chien CC, Jeng KT (2006) Effective preparation of carbon nanotube-supported Pt-Ru electrocatalysts. Mater Chem Phys 99(1):80–87
62. Wee JH, Lee KY, Kim SH (2007) Fabrication methods for low-Pt-loading electrocatalysts in proton exchange membrane fuel cell systems. J Power Sources 165(2):667–677
63. Rasheed A, Howe JY, Dadmun MD, Britt PF (2007) The efficiency of the oxidation of carbon nanofibers with various oxidizing agents. Carbon 45(5):1072–1080
64. Wang C, Waje M, Wang X, Tang JM, Haddon RC, Yan YS (2004) Proton exchange membrane fuel cells with carbon nanotube based electrodes. Nano Lett 4(2):345–348
65. Lin Z, Ji L, Zhang X (2009) Electrodeposition of platinum nanoparticles onto carbon nanofibers for electrocatalytic oxidation of methanol. Mater Lett 63:2115–2118
66. Liu FJ, Huang LM, Wen TC, Gopalan A (2007) Large-area network of polyaniline nanowires supported platinum nanocatalysts for methanol oxidation. Synth Met 157(16–17):651–658
67. Shanmugam S, Gedanken A (2009) Synthesis and electrochemical oxygen reduction of platinum nanoparticles supported on mesoporous TiO_2. J Phys Chem C 113(43): 18707–18712
68. Gojkovic SL, Babic BM, Radmilovic VR, Krstajic NV (2010) Nb-doped TiO_2 as a support of Pt and Pt-Ru anode catalyst for PEMFCs. J Electroanal Chem 639(1–2):161–166
69. Seger B, Kongkanand A, Vinodgopal K, Kamat PV (2008) Platinum dispersed on silica nanoparticle as electrocatalyst for PEM fuel cell. J Electroanal Chem 621(2):198–204
70. Elezovic NR, Babic BM, Radmilovic VR, Vracar LM, Krstajic NV (2009) Synthesis and characterization of MoO_x-Pt/C and TiOx-Pt/C nano-catalysts for oxygen reduction. Electrochim Acta 54(9):2404–2409
71. Xiong L, Manthiram A (2004) Synthesis and characterization of methanol tolerant Pt/TiOx/C nanocomposites for oxygen reduction in direct methanol fuel cells. Electrochim Acta 49(24):4163–4170
72. Saha MS, Banis MN, Zhang Y, Li RY, Sun XL, Cai M, Wagner FT (2009) Tungsten oxide nanowires grown on carbon paper as Pt electrocatalyst support for high performance proton exchange membrane fuel cells. J Power Sources 192(2):330–335
73. Molla S, Compan V, Lafuente SL, Prats J (2011) On the methanol permeability through pristine Nafion (R) and Nafion/PVA membranes measured by different techniques. A comparison of methodologies. Fuel Cells 11(6):897–906
74. Molla S, Compan V, Gimenez E, Blazquez A, Urdanpilleta I (2011) Novel ultrathin composite membranes of Nafion/PVA for PEMFCs. Int J Hydrog Energy 36(16):9886–9895
75. Molla S, Compan V (2011) Polyvinyl alcohol nanofiber reinforced Nafion membranes for fuel cell applications. J Membr Sci 372(1–2):191–200
76. Molla S, Compan V (2011) Performance of composite Nafion/PVA membranes for direct methanol fuel cells. J Power Sources 196(5):2699–2708
77. Choi SW, Fu YZ, Ahn YR, Jo SM, Manthiram A (2008) Nafion-impregnated electrospun polyvinylidene fluoride composite membranes for direct methanol fuel cells. J Power Sources 180(1):167–171

78. Hasani-Sadrabadi MM, Shabani I, Soleimani M, Moaddel H (2011) Novel nanofiber-based triple-layer proton exchange membranes for fuel cell applications. J Power Sources 196(10):4599–4603
79. Yun SH, Woo JJ, Seo SJ, Wu LA, Wu D, Xu TW, Moon SH (2011) Sulfonated poly(2,6-dimethyl-1,4-phenylene oxide) (SPPO) electrolyte membranes reinforced by electrospun nanofiber porous substrates for fuel cells. J Membr Sci 367(1–2):296–305
80. Wu D, Wu L, Woo JJ, Yun SH, Seo SJ, Xu TW, Moon SH (2010) A simple heat treatment to prepare covalently crosslinked membranes from sulfonated poly(2,6-dimethyl-1,4-phenylene oxide) for application in fuel cells. J Membr Sci 348(1–2):167–173
81. Lu PP, Xu ZL, Yang H, Wei YM (2012) Processing-structure-property correlations of polyethersulfone/perfluorosulfonic acid nanofibers fabricated via electrospinning from polymer-nanoparticle suspensions. ACS Appl Mater Interfaces 4(3):1716–1723
82. Ballengee JB, Pintauro PN (2011) Composite fuel cell membranes from dual-nanofiber electrospun mats. Macromolecules 44(18):7307–7314
83. Ballengee JB, Pintauro PN (2011) Morphological control of electrospun Nafion nanofiber mats. J Electrochem Soc 158(5):B568–B572
84. Seol JH, Won JH, Yoon KS, Hong YT, Lee SY (2012) SiO_2 ceramic nanoporous substrate-reinforced sulfonated poly(arylene ether sulfone) composite membranes for proton exchange membrane fuel cells. Int J Hydrog Energy 37(7):6189–6198
85. Seol JH, Won JH, Lee MS, Yoon KS, Hong YT, Lee SY (2012) A proton conductive silicate-nanoencapsulated polyimide nonwoven as a novel porous substrate for a reinforced sulfonated poly(arylene ether sulfone) composite membrane. J Mater Chem 22(4):1634–1642
86. Li HY, Liu YL (2013) Polyelectrolyte composite membranes of polybenzimidazole and crosslinked polybenzimidazole-polybenzoxazine electrospun nanofibers for proton exchange membrane fuel cells. J Mater Chem A 1(4):1171–1178
87. Liu W, Wang SJ, Xiao M, Han DM, Meng YZ (2012) A proton exchange membrane fabricated from a chemically heterogeneous nonwoven with sandwich structure by the program-controlled co-electrospinning process. Chem Commun 48(28):3415–3417
88. Choi J, Wycisk R, Zhang WJ, Pintauro PN, Lee KM, Mather PT (2010) High conductivity perfluorosulfonic acid nanofiber composite fuel-cell membranes. ChemSusChem 3(11):1245–1248
89. Choi J, Lee KM, Wycisk R, Pintauro PN, Mather PT (2008) Nanofiber network ion-exchange membranes. Macromolecules 41(13):4569–4572
90. Choi J, Lee KM, Wycisk R, Pintauro PN, Mather PT (2010) Sulfonated polysulfone/POSS nanofiber composite membranes for PEM fuel cells. J Electrochem Soc 157(6):B914–B919
91. Choi J, Lee KM, Wycisk R, Pintauro PN, Mather PT (2010) Nanofiber composite membranes with low equivalent weight perfluorosulfonic acid polymers. J Mater Chem 20(30): 6282–6290
92. Lee C, Jo SM, Choi J, Baek KY, Truong YB, Kyratzis IL, Shul YG (2013) SiO_2/sulfonated poly ether ether ketone (SPEEK) composite nanofiber mat supported proton exchange membranes for fuel cells. J Mater Sci 48(10):3665–3671
93. Dong B, Gwee L, Salas-de la Cruz D, Winey KI, Elabd YA (2010) Super proton conductive high-purity nafion nanofibers. Nano Lett 10(9):3785–3790
94. Chen H, Snyder JD, Elabd YA (2008) Electrospinning and solution properties of Nafion and poly(acrylic acid). Macromolecules 41(1):128–135
95. Subianto S, Cavaliere S, Jones DJ, Roziere J (2013) Effect of side-chain length on the electrospinning of perfluorosulfonic acid ionomers. J Polym Sci Part A-Polym Chem 51(1):118–128
96. Li XF, Hao XF, Xu D, Zhang G, Zhong SL, Na H, Wang DY (2006) Fabrication of sulfonated poly(ether ether ketone ketone) membranes with high proton conductivity. J Membr Sci 281(1–2):1–6
97. Yao YF, Guo BK, Ji LW, Jung KH, Lin Z, Alcoutlabi M, Hamouda H, Zhang XW (2011) Highly proton conductive electrolyte membranes: fiber-induced long-range ionic channels. Electrochem Commun 13(9):1005–1008

98. Yao YF, Ji LW, Lin Z, Li Y, Alcoutlabi M, Hamouda H, Zhang XW (2011) Sulfonated polystyrene fiber network-induced hybrid proton exchange membranes. ACS Appl Mater Interfaces 3(9):3732–3737
99. Yao YF, Lin Z, Li Y, Alcoutlabi M, Hamouda H, Zhang XW (2011) Superacidic electrospun fiber-Nafion hybrid proton exchange membranes. Adv Energy Mater 1(6):1133–1140
100. Tamura T, Kawakami H (2010) Aligned electrospun nanofiber composite membranes for fuel cell electrolytes. Nano Lett 10(4):1324–1328
101. Yeager HL, Eisenberg A (1982) Perfluorinated ionomer membranes – introduction. ACS Symp Ser 180:1–6
102. Eikerling M, Kornyshev AA, Stimming U (1997) Electrophysical properties of polymer electrolyte membranes: a random network model. J Phys Chem B 101(50):10807–10820
103. Haubold HG, Vad T, Jungbluth H, Hiller P (2001) Nano structure of NAFION: a SAXS study. Electrochim Acta 46(10–11):1559–1563
104. James PJ, McMaster TJ, Newton JM, Miles MJ (2000) In situ rehydration of perfluorosulphonate ion-exchange membrane studied by AFM. Polymer 41(11):4223–4231
105. Schmidt-Rohr K, Chen Q (2008) Parallel cylindrical water nanochannels in Nafion fuel-cell membranes. Nat Mater 7(1):75–83
106. Markovic Z, Jovanovic S, Kleut D, Romcevic N, Jokanovic V, Trajkovic V, Todorovic-Markovic B (2009) Comparative study on modification of single wall carbon nanotubes by sodium dodecylbenzene sulfonate and melamine sulfonate superplasticiser. Appl Surf Sci 255(12):6359–6366
107. Priya BR, Byrne HJ (2008) Investigation of sodium dodecyl benzene sulfonate assisted dispersion and debundling of single-wall carbon nanotubes. J Phys Chem C 112(2):332–337
108. Yameen B, Kaltbeitzel A, Glasser G, Langner A, Muller F, Gosele U, Knoll W, Azzaroni O (2010) Hybrid polymer-silicon proton conducting membranes via a pore-filling surface-initiated polymerization approach. ACS Appl Mater Interfaces 2(1):279–287
109. Yameen B, Kaltbeitzel A, Langer A, Muller F, Gosele U, Knoll W, Azzaroni O (2009) Highly proton-conducting self-humidifying microchannels generated by copolymer brushes on a scaffold. Angew Chem-Int Ed 48(17):3124–3128
110. Yameen B, Kaltbeitzel A, Langner A, Duran H, Muller F, Gosele U, Azzaroni O, Knoll W (2008) Facile large-scale fabrication of proton conducting channels. J Am Chem Soc 130(39):13140–13144
111. Wang H, Xu X, Johnson NM, Dandala NKR, Ji HF (2011) High proton conductivity of water channels in a highly ordered nanowire. Angew Chem-Int Ed 50(52):12538–12541
112. Park AM, Pintauro PN (2012) Alkaline fuel cell membranes from electrospun fiber mats. Electrochem Solid State Lett 15(3):B27–B30
113. Takemori R, Kawakami H (2010) Electrospun nanofibrous blend membranes for fuel cell electrolytes. J Power Sources 195(18):5957–5961
114. Bajon R, Balaji S, Guo SM (2009) Electrospun Nafion nanofiber for proton exchange membrane fuel cell application. J Fuel Cell Sci Technol 6(3)
115. Lei S, Chen DJ, Chen YQ (2011) A surface acoustic wave humidity sensor with high sensitivity based on electrospun MWCNT/Nafion nanofiber films. Nanotechnology 22(26)
116. Zhou CS, Liu Z, Dai JY, Xiao D (2010) Electrospun Ru(bpy)(3)(2+)-doped nafion nanofibers for electrochemiluminescence sensing. Analyst 135(5):1004–1009
117. Sebastian D, Calderon JC, Gonzalez-Exposito JA, Pastor E, Martinez-Huerta MV, Suelves I, Moliner R, Lazaro MJ (2010) Influence of carbon nanofiber properties as electrocatalyst support on the electrochemical performance for PEM fuel cells. Int J Hydrog Energy 35(18):9934–9942
118. Choi J, Lee KM, Wycisk R, Pintauro PN, Mather PT (2009) Nanofiber network ion-exchange membranes for PEM fuel cells. Abstr Pap Am Chem Soc 237
119. Collins G, Federici J, Imura Y, Catalani LH (2012) Charge generation, charge transport, and residual charge in the electrospinning of polymers: a review of issues and complications. J Appl Phys 111(4)

120. McKee MG, Layman JM, Cashion MP, Long TE (2006) Phospholipid nonwoven electrospun membranes. Science 311(5759):353–355
121. Pillay V, Dott C, Choonara YE, Tyagi C, Tomar L, Kumar P, du Toit LC, Ndesendo VMK, du Toit LC, Ndesendo VMK (2013) A review of the effect of processing variables on the fabrication of electrospun nanofibers for drug delivery applications. J Nanomater. doi:10.1155/2013/789289

Chapter 3
Applications of Electrospinning in Design and Fabrication of Electrodes for Lithium-Ion Batteries

Aikifa Raza, Jianyong Yu, Yunyun Zhai, Gang Sun, and Bin Ding

Abstract Lithium-ion battery technology has become very imperative in recent years as these batteries show great promise as power sources in portable electric appliances and vehicles. The development of new materials for lithium-ion batteries is the focus of research in prominent groups in the field of materials science throughout the world. On the other hand, electrospinning is appealing attracting close attention as a versatile fabrication method for one-dimensional organic, inorganic, and hybrid nanomaterials of controlled dimensions as random or oriented continuous nanofibers. The controllable dimensions and compositional flexibility of electrospun nanofibers are increasingly being investigated for the targeted development of electrodes and electrolyte materials in energy conversion and storage devices. We present here a review on the application of electrospinning for the design and fabrication of architectured, nanofibrous materials for electrodes of lithium-ion batteries with particular emphasis on improved energy and power density imparted by performance improvement via better ionic conductivity, cyclability, reversibility, and electrochemical stability of electrospun electrode for lithium-ion batteries.

A. Raza • Y. Zhai
State Key Laboratory for Modification of Chemical Fibers and Polymer Materials, College of Materials Science and Engineering, Donghua University, Shanghai 201620, China

J. Yu • B. Ding (✉)
State Key Laboratory for Modification of Chemical Fibers and Polymer Materials, College of Materials Science and Engineering, Donghua University, Shanghai 201620, China

Nanomaterials Research Center, Modern Textile Institute, Donghua University, Shanghai 200051, China
e-mail: binding@dhu.edu.cn

G. Sun
Nanomaterials Research Center, Modern Textile Institute, Donghua University, Shanghai 200051, China

B. Ding and J. Yu (eds.), *Electrospun Nanofibers for Energy and Environmental Applications*, Nanostructure Science and Technology,
DOI 10.1007/978-3-642-54160-5_3,

3.1 Introduction

Lithium-ion battery is a member of a family of rechargeable battery types in which lithium ions move from anode to the cathode during discharge reaction and move back when charging. Lithium-ion batteries use an intercalated lithium compound as the electrode material, compared to the metallic lithium used in non-rechargeable lithium battery. The first published interest in lithium batteries began in 1958 which eventually led to the development and commercialization of a variety of primary lithium cells. The 1980s brought many attempts to develop rechargeable lithium-ion batteries using lithium insertion compounds as positive electrodes, and the first cells of this type appeared when Exxon and Moli Energy tried to commercialize the Li/TiS_2 and Li/MoS_2 systems, respectively [1]. The last system, developed by Tadiran Inc. in the mid-1990s, was a commercial AA battery that was addressed to power cellular phones. Li/MoS_2 and Li/TiS_2 systems did not take off due to safety problems [2].

From the early stages of research and development of lithium-ion batteries, it was clear that transition metal oxides and sulfides can serve as excellent reversible cathode materials for rechargeable lithium batteries. After the era of the transition metal chalcogenides came the higher voltage metal oxides (where M = Ni, Co, or Mn) [3]. These materials are the basis for the most commonly used cathodes in commercial lithium-ion cells. At about that time the concept of a lithium-ion cell was tested in the laboratory with two insertion electrodes cycling lithium ions between them, thus eliminating the use of a metallic lithium anode [4]. The next decade saw substantial research and development on advanced battery systems based upon the insertion and removal of lithium ions into host compounds serving as both electrodes. The recent lithium-ion battery technology development, which enabled the commercialization of the rechargeable, high energy density batteries that are conquering the market, emerged due to the introduction of graphite as the anode material instead of lithium metal and the use of lithiated transition metal oxide cathode materials.

The three contributors in the electrochemical reactions in a lithium-ion battery are the anode, cathode, and electrolyte (Fig. 3.1). During insertion (or intercalation), ions move into the electrode, while during the reverse process, extraction (or deintercalation), ions move back out. To control the rate of electron transfer, the cathode must be physically and electrically isolated from the anode using an ionically conductive but electrically insulating medium, typically a liquid or polymeric electrolyte [5]. Following are the electrochemical reactions occurring in a lithium-ion battery ($LiCoO_2$ system). The positive electrode half reaction is

$$LiCoO_2 \rightleftharpoons Li_{1-n}CoO_2 + nLi^+ + ne^-$$

The negative electrode half reaction is

$$nLi^+ + ne^- + C \rightleftharpoons Li_nC$$

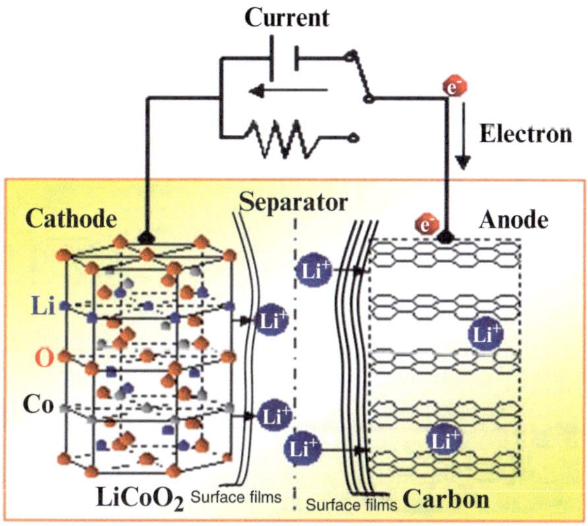

Fig. 3.1 A schematic presentation of the most commonly used lithium-ion battery based on graphite anodes and $LiCoO_2$ cathodes (Reprinted with permission from Ref. [5]. Copyright © 2011, Royal Society of Chemistry)

The overall chemical reaction employed in the current generation of Li-ion batteries is [6, 7]

$$LiCoO_2 + C \rightleftharpoons Li_{1-n}CoO_2 + Li_nC$$

In the fully charged state, lithium is stored in the anode material (e.g., graphite), and during discharge, it migrates from the anode through the electrolyte to the cathode in the form of lithium ions and inserts into the cathode. Simultaneously, electrons are removed from the anode and are transferred through an external circuit, which creates current to power a load, and enter the cathode (reduction). The discharge process continues until the potential difference between the two electrodes becomes too low, at which point the cell is fully discharged. This process is reversed during charging.

The significant features of lithium-ion batteries like high energy density, long cycle life, low gravimetric density, and the flexible design have made them very attractive for academic research and industrial applications [8]. So far, lithium-ion batteries have empowered the development of small and portable electronic devices, but future challenges regarding larger and more demanding applications such as electric or hybrid electric vehicles require significant advances in battery technology as they require fast charging and discharging at high power rates. Thus, more improvements are needed in various aspects such as cycling life, energy density, power, charge/discharge rate, safety, and cost in order to achieve satisfactory performance in these demanding applications [9]. One of the approaches used to improve the efficiency and the durability of lithium-ion batteries is to maintain a low weight, volume, and cost by designing and fabricating nanomaterials and nanostructured materials for use as electrode and electrolyte materials [10–14].

In this regard, the versatility of electrospinning makes them extremely suitable to prepare both these electrodes with the benefits of smaller, nanostructured geometries and architectures and reduced manufacturing costs [15–20]. Electrospinning leads to the formation of ultrafine fibers with diameters in the micrometer to nanometer range, starting from molten polymers, polymers, or hybrids in solution. Electrospun nanofibers present specific properties, such as high surface/volume and aspect ratios, leading to low density and high pore volume, and exceptional mechanical strength. Some excellent reviews describing the technique are available [17, 21–28]. Electrospinning has the advantages of easiness, efficiency, low cost, high yield, and high degree of reproducibility of the obtained materials. Due to its inherent adaptability and applicability, the electrospinning technique can be applied to the preparation of different components of energy materials, from polymer electrolytes to ceramic and metallic electrodes [29, 30]. The aim of this review is then to give a detailed overview of recent advances in the application of materials issued from electrospinning in electrodes of lithium-ion batteries.

3.2 Electrospun Nanofibers as Electrodes of Lithium-Ion Batteries

For highly efficient rechargeable batteries, the electrode materials must possess a high specific charge and charge density, generate a high cell voltage, and have a high reversibility of electrochemical reactions at both electrodes to maintain the specific charge for thousands of charge/discharge cycles. Due to multifaceted advantages, the nanostructured materials have been extensively explored to prepare high-performance lithium-ion battery electrodes [31–34]. Among these the nano-sized fibrous structures not only enhanced the rate capability and power density of the electrodes by significantly increasing the lithium-ion insertion/extraction but also noticeably escalated the electron transport reaction at cathodes. Secondly, the high surface area of electrospun nanofibrous electrodes allows a high electrode/electrolyte contact area, which thus increases the rate capability of the battery. Finally, the nanofibrous electrodes provide better solution to the strain associated with insertion/extraction, thus avoiding the damage of the material structure and improving the cycle life of the cell [28]. By adjusting the electrospinning configuration and the posttreatments, a wide variety of morphologies and architectures such as nanofibers, nanorods, nanowires, nanobelts, hollow, and core-sheath biaxial and triaxial fibers were investigated, as shown in Fig. 3.2 [35–43]. In this regard, electrospinning presents a simple and versatile alternative approach for preparing nanostructured anode and cathode hosts for lithium-ion batteries.

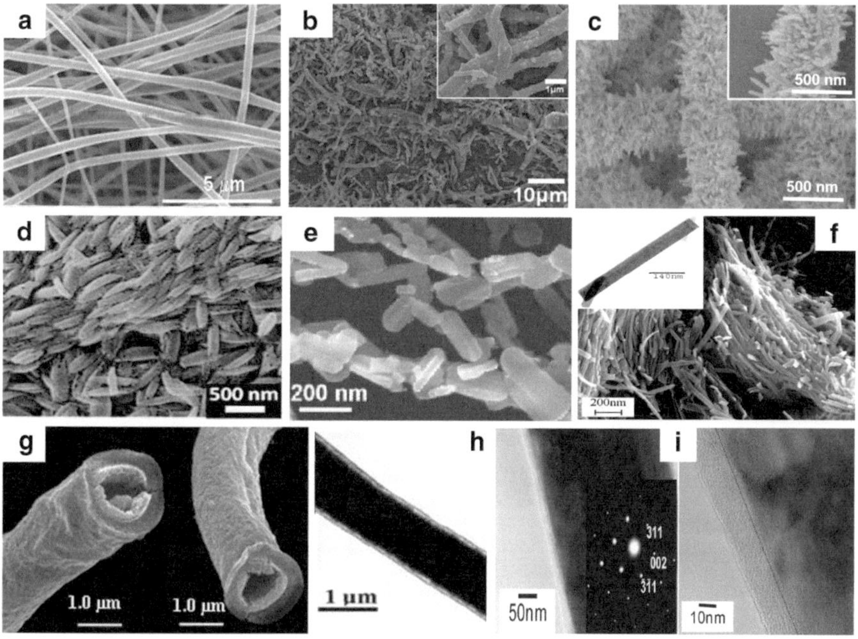

Fig. 3.2 Scanning electron microscopic (*SEM*) images of (**a**) typical smooth CNFs obtained after calcination of PAN nanofibers at 1,000 °C, (**b**) Si/CNF composite nanofibers, (**c**) Fe_2O_3 nanorods grown on CNFs, (**d**) rice grain-shaped TiO_2-CNT (4 wt%) nanostructures, (**e**) ultralong hierarchical vanadium oxide nanowires, (**f**) $H_{0.48}V_4O_{10} \cdot 2H_2O$ nanobelts (inset is the transmission electron microscopic (*TEM*) image of single $H_{0.48}V_4O_{10} \cdot 2H_2O$ nanobelt), and (**g**) hollow coaxial $LiNi_{0.8}Co_{0.1}Mn_{0.1}O_2$-MgO fibers. TEM images of (**h**) core/shell biaxial $LiCoO_2$-MgO fiber and (**i**) triaxial $LiFePO_4$/CNTs/C nanowire ((**a**) Reprinted with permission from Ref. [35]. Copyright © 2006 WILEY-VCH Verlag GmbH & Co. KGaA, Weinheim (**b**) Reprinted with permission from Ref. [36]. Copyright © 2010, Elsevier (**c**) Reprinted with permission from Ref. [37]. Copyright © 2012, Elsevier (**d**) Reprinted with permission from Ref. [38]. Copyright © 2011, Royal Society of Chemistry (**e**) Reprinted with permission from Ref. [39]. Copyright © 2010, American Chemical Society (**f**) Reprinted with permission from Ref. [40]. Copyright © 2009, Elsevier (**g**) Reprinted with permission from Ref. [41]. Copyright © 2008, American Chemical Society (**h**) Reprinted with permission from Ref. [42]. Copyright © 2007, Royal Society of Chemistry (**i**) Reprinted with permission from Ref. [43]. Copyright © 2010, American Chemical Society)

3.2.1 Anode Materials

Cycle life and rate capability of lithium-ion batteries strongly depend on the properties of the anode material, while their electrical storage capacities depend on how much lithium can be held in the said anode. In this section, recent advances in the area of electrospun anodes for lithium-ion batteries are described and summarized (Table 3.1).

Table 3.1 Electrospun anode materials for lithium-ion batteries and the corresponding electrochemical performance

Material	Fiber/NPs diameter nm	Discharge capacity (1st cycle) mA h g^{-1} [rate/C]	Discharge capacity (*n*th cycle) mA h g^{-1} [rate/C]	References
CNFs	200–300	1,000 [0.08]	$n=2$, 500 [0.08]	[35]
	150–300	621 [0.13]	$n=50$, 454 [0.13]	[44]
CNFs + Si NPs	–	1,710 [0.1]	$n=50$, 1,100 [0.13]	[46]
	300–500/50–80	1,060–780 [0.01]	$n=50$, 410–380 [0.01]	[36]
	150–250/30–70	1,070 [–]	$n=50$, 600 [–]	[47]
	150–250/30–70	730 [–]	$n=50$, 445 [–]	[48]
CNFs + Sn NPs	100–1,000/1	816 [–]	$n=20$, 400 [–]	[45]
	150–350/30–40	1211.7 [0.025]	$n=2$, 800 [0.025]	[49]
	–	1,417 [0.2]	$n=100$, 460 [1.4]	[54]
CNFs + Sn/SnO_x NPs	50–500/20–40	1,192 [0.03]	$n=40$, 520 [0.03]	[53]
CNFs/SnO_2/C	–	961 [0.1]	$n=200$, 810 [0.1]	[56]
CNFs + MnO_x NPs	150–200/21–42	650 [–]	500 [–]	[51]
CNFs + αFe_2O_3	175–485	680 [–]	$n=75$, 484 [–]	[57]
CNFs + Fe_2O_3 nanorods	200/20–40	1,250 [–]	$n=50$, 780 [–]	[37]
CNFs + TiO_2	200	220 [–]	$n=100$, 200 [–]	[58]
	–	207 [–]	$n=500$, 146 [–]	[38]
CNFs + Ni NPs	150–200/20	795 [0.13]	$n=50$, 500 [0.13]	[59]
SnO_2	50–80	1,650 [0.1]	$n=50$, 480 [0.13]	[61]
	220	1,650 [0.23]	$n=50$, 807 [0.23]	[62]
$CaSnO_3$	560	1,170 [–]	$n=50$, 648 [–]	[63]
ZnO-SnO_2	–	1,780 [–]	$n=588$, 200 [–]	[64]
$Li_4Ti_5O_{12}$	–	192 [0.5]	$n=30$, 140 [0.5]	[65]
Ag/Au + TiO_2	50, 5–10	180–300 [1]	$n=2$, 160–170 [1]	[66]
Nb/TiO_2	–	140 [–]	23 [–]	[67]
α-Fe_2O_3	250–350	1,700 [0.06]	$n=40$, 1,293 [0.06]	[68]
$ZnFe_2O_4$	80–130	900 [0.06]	$n=30$, 733 [0.06]	[69]
Co_3O_4	200	1,336 [0.5]	$n=40$, 604 [0.5]	[70]

3.2.1.1 Carbon Nanofiber Anodes

Most recent secondary lithium batteries use carbon as its anode, replacing metallic lithium found in primary batteries. Carbon anode provides greater cell life with safety and low cost; however, it also has lower cell voltage, rate capability, and specific charge (theoretical capacity of 372 mA h g^{-1}). For the improvement of these specific properties, scientists have studied various nanofibrous anode materials with improved performance with respect to the above parameters. A comparison of specific capacities of various modified and unmodified carbon nanofiber (CNF) anodes is presented in Fig. 3.3. Kim et al. prepared CNFs by combining electrospinning and thermal treatments. Due to the particular nanotexture, these polyacrylonitrile

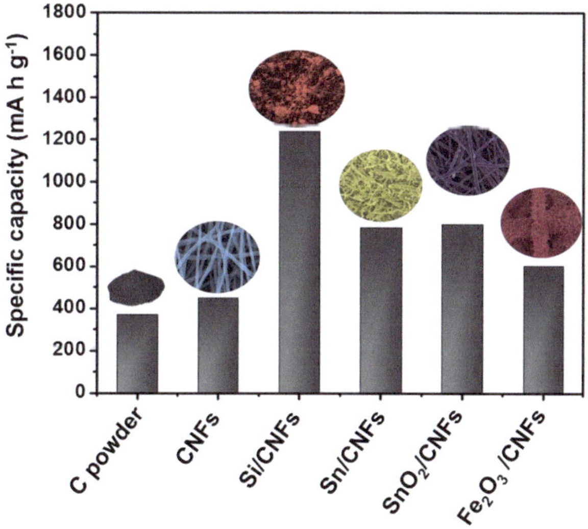

Fig. 3.3 Comparison of specific capacities of carbon powder to the modified nanofibrous carbon anode materials for lithium-ion batteries

(PAN)-derived CNFs exhibited the highest reversible capacity (ca. 450 mA h g^{-1}) and a good rate capability (350 mA h g^{-1} at a charge current of 100 mA h g^{-1}) [35]. CNFs prepared from PAN/polypyrrole also showed a relatively large reversible capacity (ca. 360 mA h g^{-1}) with high rate capability, good cycle performance, and structural integrity after 50 charge/discharge cycles, attributable to their large surface area and extended interface with electrolyte, leading to short pathway for charge and electron transport [44].

Studies on anode materials for lithium-ion batteries also concern electrospun CNFs loaded with metal or metal oxide nanoparticles (NPs). For instance, CNFs have been doped with silicon and tin NPs [45–47]. The fine dispersion of nano-sized particles within CNFs allows them to act as structural buffers, particle stabilizers, and electroactive materials and also eliminates the need for binding or conducting additives in the electrode materials. Wang et al. first used electrospun C/Si nanofiber as an anode material for lithium-ion batteries, and these composite nanofibers exhibited reversible capacity (1,240 mA h g^{-1}) with stable capacity retention even in 40 cycles [46]. Choi et al. studied the effect of the hard confinements of Si NPs on the anodic performance in lithium-ion batteries. The improved cycle stability and reversible capacity of the Si-based anode was found to be attributable to the formation of nanofibrillar networks which favor electronic and ionic transport through C/Si composite nanofibers [36]. Carbonization temperature has also proven to be a key factor for electrospun CNF anode materials since it affects their morphology and electrochemical performance [48]. Yu et al. prepared tin nanoparticle-dispersed carbon (Sn/C) nanofibers by stabilization of electrospun $SnCl_2$/PAN fibers and subsequent carbonization at different temperatures [45, 49]. Sn/C nanofibers at 700 and 850 °C presented the highest charge (785.8 and 811 mA h g^{-1}) and discharge (1211.7 and 993 mA h g^{-1}) capacities due to the unique feature of reticular nanofiber geometries.

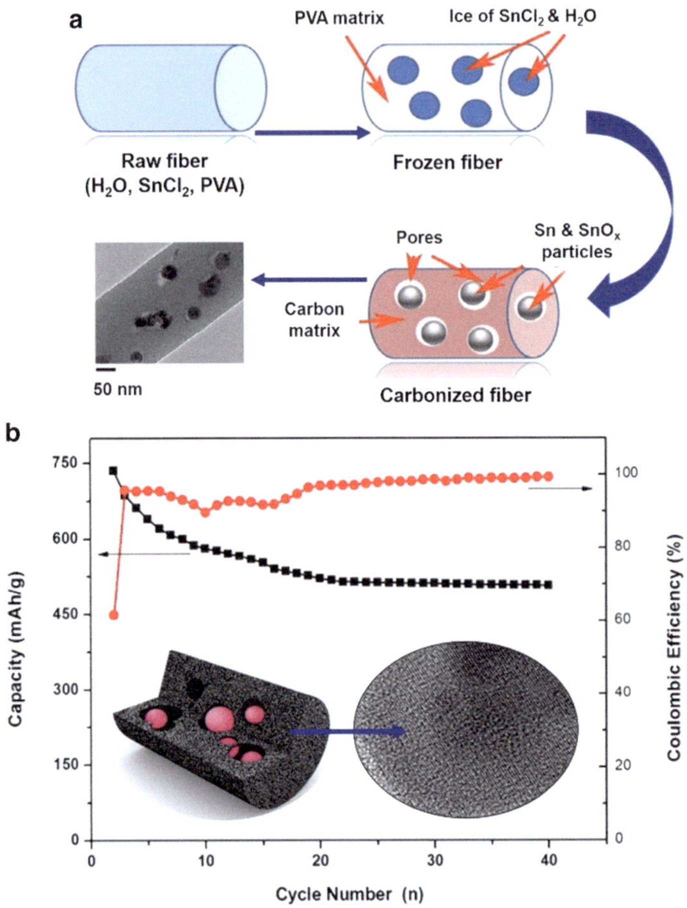

Fig. 3.4 (**a**) In situ synthesis of tin-/tin oxide-encapsulated porous carbon nanofibers. (**b**) Cycle performance and coulombic efficiency during cycling of the tin-/tin oxide-encapsulated porous carbon nanofibrous anode at a current density of 30 mA g^{-1}. Inset is the model of the microstructure of the tin-/tin oxide-encapsulated porous carbon nanofibers and TEM image of the nanoparticle in the pores (Reprinted with permission from Ref. [53]. Copyright © 2011, Elsevier)

Electrospun nanofibrous carbon anodes loaded with transition metal oxide NPs have been widely investigated as anode materials in lithium-ion batteries, because of their high theoretical capacities, safety, nontoxicity, and low cost [50, 51]. One example is the fabrication of C/SnO_2 composite nanofibers with adjustable sheath thicknesses, involving the electrodeposition of SnO_2 on CNFs [52–55]. These composite nanofibrous anodes have delivered nearly 800 mA h g^{-1} discharge capacity at the first cycle at 50 mA g^{-1} current rate, with 69 % capacity retention even after 100 charge/discharge cycles, reflecting major improvement over pure SnO_2 powder-based anodes, as shown in Fig. 3.4 [53]. Recently Kong et al. prepared C/SnO_2/C core/shell/shell hybrid nanofibrous mats via single-spinneret electrospinning followed by carbonization and hydrothermal treatment. The embedded and deaggregated SnO_2 NPs in the carbon phase, which are less than 10 nm in size, provided huge number of reaction sites for lithium ions and ensured complete alloying with them [56]. Similarly, α-Fe_2O_3 nanorods and NPs with an average size of about 20 nm were loaded in CNFs through electrospinning and were used as anode with high reversible capacity of about 604 mA h g^{-1} at 50 mA g^{-1}, improved capacity retention for at least 100 cycles, and enhanced rate performance even at the high current density of 500 mA g^{-1} [37, 57]. Another example is provided by the preparation of TiO_2/C composite nanofibrous anodes with reversible capacity of 206 mA h g^{-1} up to 100 cycles at the current density of 30 mA g^{-1} [38, 58]. The unique improvement of anode performance upon the incorporation of a lithium-inactive metal such as Ni in CNFs is particularly noteworthy [59].

3.2.1.2 Transition Metal Oxide Nanofiber Anodes

The next typology of electrospun anode materials for lithium-ion batteries is represented by pure metal oxide nanofibers, and among them TiO_2 and SnO_2 have been thoroughly investigated [60]. Due to the peculiar porous structure, electrospun SnO_2 fibers consisting of NPs have shown to facilitate electrolyte diffusion, charge transfer, and the mechanical strain due to the volume changes during charge/discharge reactions [61]. The anode has exhibited the specific capacity of 446 mA h g^{-1} after 50 cycles at a 100 mA g^{-1} rate with high rate capability of 477.7 mA h g^{-1}. The performance was additionally improved by exploiting the adaptability of electrospinning and preparing SnO_2 porous nanotubes that have shown a high discharge capacity of 807 mA h g^{-1} after 50 cycles [62]. Li et al. electrospun $CaSnO_3$ nanotubes with eggroll-like morphology with improved lithium storage properties and better cycling performance compared to the nanorod structured $CaSnO_3$. This is ascribed to the large specific surface areas for loading more lithium ions and the exclusive internal hollow space for buffering the volume changes [63]. Recently, Feng et al. reported hierarchical ZnO-SnO_2 composite nanofibers as an anode material for lithium-ion batteries. The initial discharge and charge capacities of the typical ZnO-SnO_2 composite nanofibers were 1,795 and 1,364 mA h g^{-1}, respectively, with high reversible capacity of 588 mA h g^{-1} obtained after 100 cycles at a current density of 50 mA g^{-1}. The higher reversible capacity and good stability can be related to the mesoporous nanostructures of the composite nanofibers, the formation of heterostructures, and the synergistic effect between ZnO and SnO_2 in the nanofibers [64]. Lu et al. prepared three-dimensional architectures stacked from a crossbar array of aligned electrospun fibers of anatase TiO_2 and compared cathodic performance with $Li_4Ti_5O_{12}$ spinel [65]. Lithium-ion battery anodes based on TiO_2 nanofibers loaded with 10 % Au or Ag NPs showed improved performance compared to the bare material. The inserted metallic NPs not only decreased the charge transfer but also enhanced the electronic conductivity of the composite TiO_2. The capacity was improved by 20 % or more compared to bare one-dimensional TiO_2 especially for the material doped with gold [66].

Recently, niobium-doped TiO_2 nanofibrous anode elaborated by electrospinning also presented a higher rate capability in electrochemical cycling experiments than nondoped materials [67].

Chaudhari et al. reported another example of transition metal oxide of hollow-structured α-Fe_2O_3 with a high reversible capacity of 1,293 mA h g^{-1} at a current density of 60 mA g^{-1} [68]. Teh et al. demonstrated the electrochemical properties of electrospun nanowebs consisting of $ZnFe_2O_4$ nanofibers as lithium-ion battery anodes. In addition, they also studied the transition of the one-dimensional $ZnFe_2O_4$ nanofibers to nanorods by altering the viscosity of the precursor solution. As compared to nanorods, a higher reversible capacity of 925 mA h g^{-1} with enhanced cyclability was achieved by electrospun $ZnFe_2O_4$, owing to the interconnected $ZnFe_2O_4$ nanowebs which effectively escalated the ion flux and accommodated the lattice stress generated upon prolonged cycling [69]. Finally, another example of metal oxide anode material attained by electrospinning is tricobalt tetraoxide, known for its high reversible capacity and electrochemical stability. Co_3O_4 electrospun nanofibers presented an initial discharge capacity of 1,336 mA h g^{-1}, and after 40 cycles it was maintained at 604 mA h g^{-1} [70]. The cyclic performance of various transition metal oxide anodes for lithium-ion batteries has been shown in Fig. 3.5.

3.2.2 *Cathode Materials*

In recent years, we have witnessed a growing development of the electrospinning of transition metal oxide with and without lithium processing different phases as cathodes for lithium-ion batteries. In the following section, recent advances in the area of electrospun cathodes for lithium-ion batteries are described and summarized (Table 3.2).

3.2.2.1 Lithium Transition Metal Oxide Nanofibers

Most commercial lithium-ion batteries use $LiCoO_2$ as a cathode material due to its better performance in terms of specific energy density and good cycle life [71]. Nanofibrous $LiCoO_2$ cathodes have been fabricated by electrospinning to achieve a fast solid state diffusion rate due to the short diffusion distance of lithium cations [72]. The electrospun $LiCoO_2$ fibrous cathodes with large surface areas and small pores have exhibited the initial discharge capacity of 182 mA h g^{-1} compared with ca. 140 mA h g^{-1} for conventional powder and film electrodes [73]. However, such cathodes have also suffered from loss of capacity during the charge/discharge process due to the dissolution of cobalt and lithium cations forming Li_2CO_3 and CoF_2 impurities. To increase the stability and cycling performance of nanofibrous cathodes, by using coaxial electrospinning, $LiCoO_2$ fibers and $LiNi_{0.8}Co_{0.1}Mn_{0.1}O_2$ tubes coated with MgO shells were fabricated [41, 42]. The MgO coating reduced the impedance growth by protecting the surface from passive film formation during cycling, and thus, cathode showed improved reversibility and cyclability. After 40 cycles, 90 % of the initial discharge capacity of $LiCoO_2/MgO$ was maintained when compared to 52 % for the uncoated fibrous cathode. Likewise, electrospun core/shell $Li(Ni_{1/3}Co_{1/3}Mn_{1/3})O_2/Li(Ni_{1/2}Mn_{1/2})O_2$ nanofibers exhibited an enhanced cycle stability compared to core $Li(Ni_{1/3}Co_{1/3}Mn_{1/3})O_2$ nanofibers with the loss of 24.5 % of the initial capacity after 40 cycles [74]. $LiCoO_2$ nanofibrous cathode performance was also stabilized by depositing layer of lithium phosphorus oxynitride (LiPON) by radio-frequency sputtering that showed a loss of 0.11 % per cycle after 100 cycles at the discharge rate of 0.05 mA cm^{-2} [75].

$LiFePO_4$ is another promising candidate for cathode material because of its large specific capacity (170 mA h g^{-1}), good thermal stability, high discharge potential, low cost, and safety. However, its low ionic and electronic conductivity lead toward the low rate capability and high impedance in lithium-ion batteries. To avoid these limitations and in order to introduce an electron path, carbon-coated $LiFePO_4$ cathode was used, and these coatings also protected the electrode surface from undesired side reactions [76, 77]. Zhu et al. reported single-crystalline thin carbon-coated $LiFePO_4$ three-dimensional interconnected nanowires with fast lithium-ion transport. The capacity of carbon-coated cathode was close to the theoretical value (169 mA h g^{-1}), and after 100 cycles it was 86 % of the theoretical value (169 mA h g^{-1}) [78]. Recently, Hagen et al. fabricated monoclinic α-$Li_3V_2(PO_4)/C$ composites by combining sol-gel and electrospinning techniques. The nanofibrous self-supported electrodes showed superior cycling stability giving an initial discharge capacity of 124 mA h g^{-1} at 0.2 °C with retention of 95 % after 50 cycles with rate variation in a voltage range of 3–4.3 V, as shown in Fig. 3.6 [79]. This study evaluates an interesting approach toward a simple one-step electrode preparation by electrospinning technique and gives an insight into the structure–electrochemical relationship of electrode materials.

3.2.2.2 Transition Metal Oxide Nanofibers

Vanadium pentoxide with high theoretical capacity (ca.400 mA h g^{-1}) is another example of cathode materials for lithium-ion batteries, but the performance of conventional bulk V_2O_5 powder cathodes is limited by low electronic conductivity, slow lithium diffusion rate, and structural instability upon charge/discharge reactions. To overcome these problems, the use of electrospun nanofibrous V_2O_5 positive electrodes is increasing. Single-phase electrospun V_2O_5 fibers were fabricated with an initial discharge capacity of 320 mA h g^{-1} and excellent coulombic efficiency ($\sim$100 %) throughout 50 charge/discharge cycles. This study highlights once again the importance of the network morphology of electrospun fibers to increase the electroactive surface area, thus improving the accessibility of electrolytes to the cathode material and the overall efficiency of the Li-ion batteries [80].

Recently, Al-inserted electrospun V_2O_5 nanofibers were synthesized via in situ addition of Al^{3+} ions and compared with native V_2O_5 nanofibers. Al-inserted V_2O_5

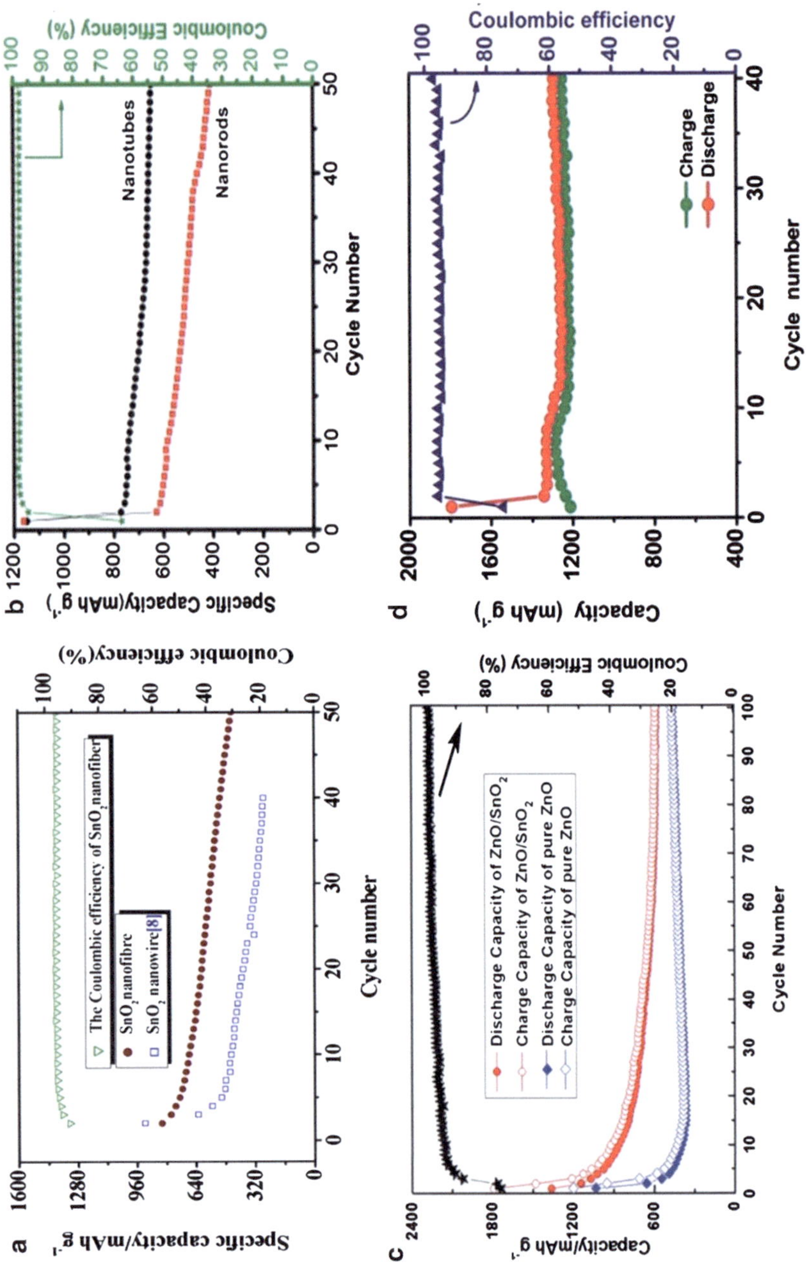
a
Specific capacity/mAh g-1
Coulombic efficiency(%)
Cycle number
The Coulombic efficiency of SnO2 nanofiber
SnO2 nanofibre
SnO2 nanowire[8]
b
Specific Capacity(mAh g-1)
Coulombic Efficiency (%)
Cycle Number
Nanotubes
Nanorods
c
Capacity/mAh g-1
Coulombic Efficiency (%)
Cycle Number
Discharge Capacity of ZnO/SnO2
Charge Capacity of ZnO/SnO2
Discharge Capacity of pure ZnO
Charge Capacity of pure ZnO
d
Capacity (mAh g-1)
Coulombic efficiency
Cycle number
Charge
Discharge

Fig. 3.5 (**a**) The cycling performance from the second cycle to the 50th cycle of the SnO_2 nanofibers and SnO_2 nanowires at 100 mA g^{-1}. (**b**) Cycling performance of $CaSnO_3$ nanotubes and $CaSnO_3$ nanorods at a current density of 60 mA g^{-1}. (**c**) Capacity retention and coulombic efficiency of the galvanostatic test for the ZnO-SnO_2 nanofiber electrode and the cyclic performance of the pure ZnO nanofiber electrode. (**d**) Cycling performance of the electrode made from α-Fe_2O_3 hollow fibers at a current density of 60 mA g^{-1}((**a**) Reprinted with permission from Ref. [61]. Copyright © 2010, Elsevier (**b**) Reprinted with permission from Ref. [63]. Copyright © 2013, Royal Society of Chemistry (**c**) Reprinted with permission from Ref. [64]. Copyright © 2013, Royal Society of Chemistry (**d**) Reprinted with permission from Ref. [68]. Copyright © 2012, Royal Society of Chemistry)

Table 3.2 Electrospun cathode materials for lithium-ion batteries and the corresponding electrochemical performance

	Fiber/NPs diameter	Discharge capacity (1st cycle)	Discharge capacity (*n*th cycle)	
Material	nm	mA h g^{-1} [rate/C]	mA h g^{-1} [rate/C]	References
$LiCoO_2$	500–2,000	182 [0.14]	$n = 20$, 123 [0.14]	[73]
$LiCoO_2/MgO$	1,000–2,000, shell 10–100	181 [0.14]	n ¼ 40, 163 [0.14]	[42]
LiPON-coated $LiCoO_2$	60–80	129 [–]	$n = 100$, 120 [–]	[75]
$LiNi_{0.8}Co_{0.1}Mn_{0.1}O_2/MgO$	1,000–2,000, 30–60	195 [0.14]	n ¼ 50, 174 [0.14]	[41]
$LiNi_{1/3}Co_{1/3}Mn_{1/3}O_2/LiNi_{1/2}Co_{1/2}Mn_{1/2}O_2$	<5,000, shell 300	192 [0.1]	$n = 5$, 166 [0.1]	[74]
$LiFePO_4/C$	400–1,000	160 [0.06]	$n = 20$, 160 [0.06]	
$C/LiFePO_4/C$	100/shell 2–5	150 [1]	$n = 100$, 146 [1]	[43]
V_2O_5	100–200	275–390 [–]	$n = 50$, 187–201 [–]	[39]
	300–800	300 [0.1]	$n = 20$, 200 [0.1]	[80]
	350	207 [–]	$n = 500$, 146 [–]	[81]

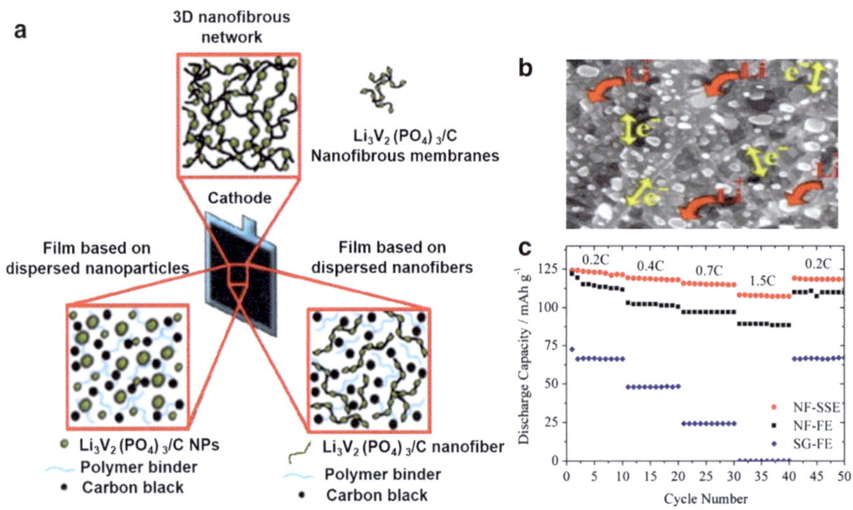

Fig. 3.6 (**a**) Schematic illustration of the three different $Li_3V_2(PO_4)_3/C$ electrodes and (**b**) distribution of $Li_3V_2(PO_4)_3/C$ crystallites on the carbon nanofiber matrix. (**c**) Cycling performance of the respective $Li_3V_2(PO_4)_3/C$ electrodes at different rates in the potential range of 3–4.3 V (Reprinted with permission from Ref. [79]. Copyright © 2013, Elsevier)

nanofibrous cathodes have shown an initial discharge capacity of ~ 250 mA h g^{-1} and improved capacity retention of >60 % after 50 cycles at 0.1 C rate, whereas native V_2O_5 nanofibers showed only ~40 % capacity retention at room temperature (Fig. 3.7) [81]. Viet et al. compared the electrochemical properties and lithium

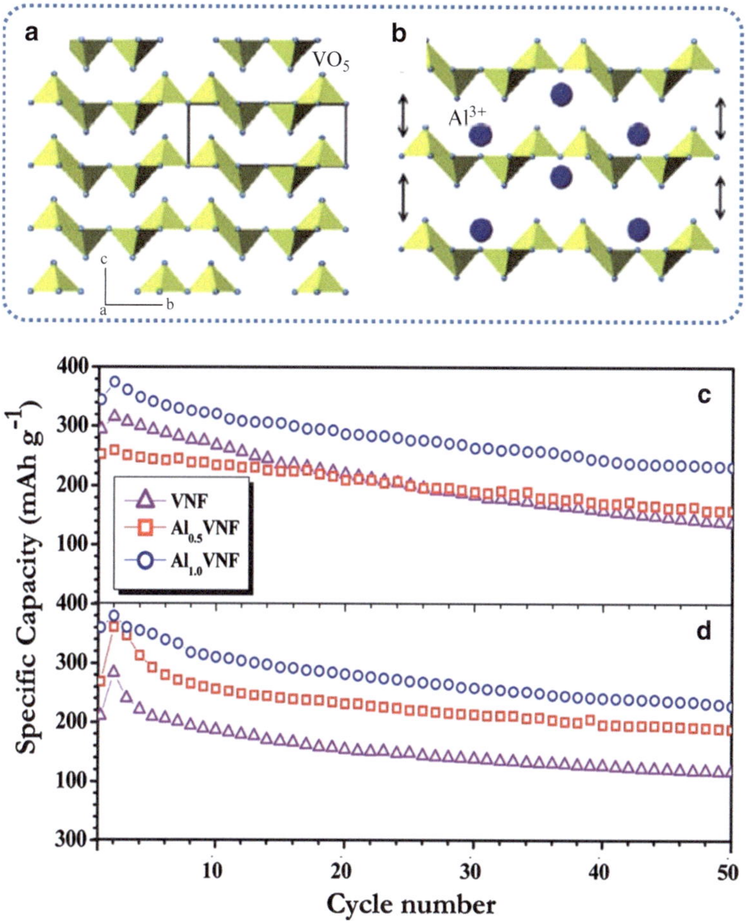

Fig. 3.7 (**a**) V_2O_5 (Shcherbinaite) crystal structure. (**b**) V_2O_5 lattice with possible positions of Al atoms, as indicated by blue spheres, and possible movement of V_2O_5 layers upon intercalation of lithium ions, as indicated by arrows. Plots of specific discharge capacity versus cycle number of V_2O_5NF, $Al_{0.5}V_2O_5$NF, and $Al_{0.5}V_2O_5$NF cycled at (**c**) 0.1 C rate at room temperature and (**d**) 0.1 C rate at 55 °C (Reprinted with permission from Ref. [81]. Copyright © 2012, American Chemical Society)

cycling of various one-dimensional Nb_2O_5 polymorphs obtained through electrospinning followed by annealing treatment. The studies indicated that electrospun monoclinic Nb_2O_5 nuggets compared to pseudohexagonal Nb_2O_5 and orthorhombic Nb_2O_5 could be better candidate for rechargeable 2 V lithium batteries in view of the commercial viability of the electrospinning process [82].

3.3 Concluding Remarks and Future Perspectives

Development of new functional nanomaterials is indispensable to meet the performance and durability targets of energy conversion and storage applications. Apart from other avenues, one-dimensional nanostructured materials have significantly contributed to meeting some of these challenges as the structural control of nanomaterials can be as important as the composition of the materials themselves. As a simple fabrication method, electrospinning is undergoing renewed interest, and recent advances indicate a high potential of electrospun nanofibrous material in energy-related applications. Several challenges remain to be overcome before large-scale usage of electrospun materials in energy-related applications, and this is partly due to the act that the process parameters affecting the electrospinning technique are still not fully understood. Therefore, there are still restrictions on the types of materials that can be synthesized by electrospinning, and it is sometimes difficult to fully eliminate defects such as beads and to obtain completely homogeneous nanofibers. Furthermore, although advances in aligning and bridging adjacent fibers have been made, still further solutions must be found to temper the fragility of inorganic electrospun nanofiber networks particularly for electrodes of lithium-ion batteries, for which removal of the carrier polymer while maintaining fiber integrity and composition is still challenging. Despite these shortcomings, electrospinning has been shown to be capable of producing nanofibrous electrodes of excellent uniformity with more complex architectures such as core/shell and hollow nanofibers by coaxial electrospinning, as well as novel composite materials by the use of co-electrospinning. In conclusion, successful research and development of nanofibrous electrodes for lithium-ion batteries that will combine excellent performance with the necessary features is the opportunity to meet the challenge of moving the world that can utilize sustainable "green" energy sources.

Acknowledgments This work is supported by the National Basic Research Program of China (973 Program, 2011CB606103 and 2012CB525005), the National Natural Science Foundation of China (No. 51173022, U1232116, and 51273038), the Shanghai Nano Special Projects (11 nm0502900), the Shanghai Committee of Science and Technology (No. 12JC1400101), the Huo Yingdong Foundation (131070), and the Program for New Century Talents of the University in China.

References

1. Ohzuku T, Brodd RJ (2007) An overview of positive-electrode materials for advanced lithium-ion batteries. J Power Sources 174(2):449–456. doi:10.1016/j.jpowsour.2007.06.154
2. Mengeritsky E, Dan P, Weissman I, Zaban A, Aurbach D (1996) Safety and performance of Tadiran TLR-7103 rechargeable batteries. J Electrochem Soc 143(7):2110–2116. doi:10.1149/1.1836967

3. Mizushima K, Jones PC, Wiseman PJ, Goodenough JB (1980) $LixCoO_2$ $(0 < x < -1)$: a new cathode material for batteries of high energy density. Mater Res Bull 15(6):783–789, doi: http://dx.doi.org/10.1016/0025-5408(80)90012-4
4. Lazzari M, Scrosati B (1980) A cyclable lithium organic electrolyte cell based on two intercalation electrodes. J Electrochem Soc 127(3):773–774. doi:10.1149/1.2129753
5. Etacheri V, Marom R, Elazari R, Salitra G, Aurbach D (2011) Challenges in the development of advanced Li-ion batteries: a review. Energ Environ Sci 4(9):3243–3262. doi:10.1039/c1ee01598b
6. Choi HC, Jung YM, Noda I, Kim SB (2003) A study of the mechanism of the electrochemical reaction of lithium with CoO by two-dimensional soft X-ray absorption spectroscopy (2D XAS), 2D Raman, and 2D heterospectral XAS-Raman correlation analysis. J Phys Chem B 107(24):5806–5811. doi:10.1021/jp030438w
7. Amatucci GG, Tarascon JM, Klein LC (1996) CoO_2, the end member of the $LixCoO_2$ solid solution. J Electrochem Soc 143(3):1114–1123. doi:10.1149/1.1836594
8. Winter M, Brodd RJ (2004) What are batteries, fuel cells, and supercapacitors? Chem Rev 104(10):4245–4269. doi:10.1021/cr020730k
9. Cabana J, Monconduit L, Larcher D, Rosa Palacin M (2010) Beyond intercalation-based Li-ion batteries: the state of the art and challenges of electrode materials reacting through conversion reactions. Adv Mater 22(35):E170–E192. doi:10.1002/adma.201000717
10. Bruce PG, Scrosati B, Tarascon JM (2008) Nanomaterials for rechargeable lithium batteries. Angew Chem Int Ed 47(16):2930–2946. doi:10.1002/anie.200702505
11. Shukla AK, Kumar TP (2008) Materials for next-generation lithium batteries. Curr Sci 94(3):314–331
12. Hua Kun L, Guo Xiu W, Zaiping G, Jiazhao W, Konstantinov K (2006) Nanomaterials for lithium-ion rechargeable batteries. J Nanosci Nanotechnol 6(1):1–15. doi:10.1166/jnn.2006.103
13. Chen X, Li C, Graetzel M, Kostecki R, Mao SS (2012) Nanomaterials for renewable energy production and storage. Chem Soc Rev 41(23):7909–7937. doi:10.1039/c2cs35230c
14. Cheng F, Chen J (2013) Nanoporous catalysts for rechargeable Li-air batteries. Acta Chim Sinica 71(4):473–477. doi:10.6023/a13010098
15. Ding B, Lin J, Wang X, Yu J, Yang J, Cai Y (2011) Investigation of silica nanoparticle distribution in nanoporous polystyrene fibers. Soft Matter 7(18):8376–8383. doi:10.1039/c1sm05791j
16. Ding B, Ogawa T, Kim J, Fujimoto K, Shiratori S (2008) Fabrication of a super-hydrophobic nanofibrous zinc oxide film surface by electrospinning. Thin Solid Films 516(9):2495–2501. doi:10.1016/j.tsf.2007.04.086
17. Ding B, Wang M, Wang X, Yu J, Sun G (2010) Electrospun nanomaterials for ultrasensitive sensors. Mater Today 13(11):16–27. doi:10.1016/S1369-7021(10)70200-5
18. Hu J, Wang X, Ding B, Lin J, Yu J, Sun G (2011) One-step electro-spinning/netting technique for controllably preparing polyurethane nano-fiber/net. Macromol Rapid Commun 32(21):1729–1734. doi:10.1002/marc.201100343
19. Lin J, Ding B, Yang J, Yu J, Sun G (2012) Subtle regulation of the micro- and nanostructures of electrospun polystyrene fibers and their application in oil absorption. Nanoscale 4(1):176–182. doi:10.1039/c1nr10895f
20. Yang S, Wang X, Ding B, Yu J, Qian J, Sun G (2011) Controllable fabrication of soap-bubble-like structured polyacrylic acid nano-nets via electro-netting. Nanoscale 3(2):564–568. doi:10.1039/c0nr00730g
21. Wang X, Ding B, Sun G, Wang M, Yu J (2013) Electro-spinning/netting: a strategy for the fabrication of three-dimensional polymer nano-fiber/nets. Prog Mater Sci 58(8):1173–1243, doi: http://dx.doi.org/10.1016/j.pmatsci.2013.05.001
22. Lin J, Wang X, Ding B, Yu J, Sun G, Wang M (2012) Biomimicry via electrospinning. Crit Rev Solid State 37(2):94–114. doi:10.1080/10408436.2011.627096

23. Wang X, Ding B, Yu J, Wang M (2011) Engineering biomimetic superhydrophobic surfaces of electrospun nanomaterials. Nano Today 6(5):510–530. doi:10.1016/j.nantod.2011.08.004
24. Li D, Xia YN (2004) Electrospinning of nanofibers: reinventing the wheel? Adv Mater 16(14):1151–1170. doi:10.1002/adma.200400719
25. Kumar PR, Khan N, Vivekanandhan S, Satyanarayana N, Mohanty AK, Misra M (2012) Nanofibers: effective generation by electrospinning and their applications. J Nanosci Nanotechnol 12(1):1–25. doi:10.1166/jnn.2012.5111
26. Ramakrishna S, Fujihara K, Teo WE, Yong T, Ma Z, Ramaseshan R (2006) Electrospun nanofibers: solving global issues. Mater Today 9(3):40–50. doi:10.1016/s1369-7021(06)71389-x
27. Thavasi V, Singh G, Ramakrishna S (2008) Electrospun nanofibers in energy and environmental applications. Energ Environ Sci 1(2):205–221. doi:10.1039/b809074m
28. Dai Y, Liu W, Formo E, Sun Y, Xia Y (2011) Ceramic nanofibers fabricated by electrospinning and their applications in catalysis, environmental science, and energy technology. Polym Adv Technol 22(3):326–338. doi:10.1002/pat.1839
29. Miao J, Miyauchi M, Simmons TJ, Dordick JS, Linhardt RJ (2010) Electrospinning of nanomaterials and applications in electronic components and devices. J Nanosci Nanotechnol 10(9):5507–5519. doi:10.1166/jnn.2010.3073
30. Dong Z, Kennedy SJ, Wu Y (2011) Electrospinning materials for energy-related applications and devices. J Power Sources 196(11):4886–4904. doi:10.1016/j.jpowsour.2011.01.090
31. Dong S, Chen X, Zhang X, Cui G (2013) Nanostructured transition metal nitrides for energy storage and fuel cells. Coord Chem Rev 257(13–14):1946–1956. doi:10.1016/j.ccr.2012.12.012
32. Zhang Q, Uchaker E, Candelaria SL, Cao G (2013) Nanomaterials for energy conversion and storage. Chem Soc Rev 42(7):3127–3171. doi:10.1039/c3cs00009e
33. Guo YG, Hu JS, Wan LJ (2008) Nanostructured materials for electrochemical energy conversion and storage devices. Adv Mater 20(15):2878–2887. doi:10.1002/adma.200800627
34. Manthiram A, Murugan AV, Sarkar A, Muraliganth T (2008) Nanostructured electrode materials for electrochemical energy storage and conversion. Energ Environ Sci 1(6):621–638. doi:10.1039/b811802g
35. Chan K, Kap Seung Y, Kojitna M, Yoshida K, Yong Jung K, Yoong Ahm K, Endo M (2006) Fabrication of electrospinning-derived carbon nanofiber webs for the anode material of lithium-ion secondary batteries. Adv Funct Mater 16(18):2393–2397. doi:10.1002/adfm.200500911
36. Choi HS, Lee JG, Lee HY, Kim SW, Park CR (2010) Effects of surrounding confinements of Si NPs on Si-based anode performance for lithium ion batteries. Electrochim Acta 56(2):790–796. doi:10.1016/j.electacta.2010.09.101
37. Liu Z, Tay SW (2012) Direct growth Fe_2O_3 nanorods on carbon fibers as anode materials for lithium ion batteries. Mater Lett 72:74–77. doi:10.1016/j.matlet.2011.12.083
38. Zhu P, Wu Y, Reddy MV, Nair AS, Chowdari BVR, Ramakrishna S (2012) Long term cycling studies of electrospun TiO_2 nanostructures and their composites with MWCNTs for rechargeable Li-ion batteries. RSC Adv 2(2):531–537. doi:10.1039/c1ra00514f
39. Mai L, Xu L, Han C, Xu X, Luo Y, Zhao S, Zhao Y (2010) Electrospun ultralong hierarchical vanadium oxide nanowires with high performance for lithium ion batteries. Nano Lett 10(11):4750–4755. doi:10.1021/nl103343w
40. Ban C, Chernova NA, Whittingham MS (2009) Electrospun nano-vanadium pentoxide cathode. Electrochem Commun 11(3):522–525. doi:10.1016/j.elecom.2008.11.051
41. Gu Y, Jian F (2008) Hollow $LiNi_{0.8}Co_{0.1}Mn_{0.1}O_2$-MgO coaxial fibers: sol-gel method combined with co-electrospun preparation and electrochemical properties. J Phys Chem C 112(51):20176–20180. doi:10.1021/jp808468x
42. Gu Y, Chen D, Jiao X, Liu F (2007) $LiCoO_2$-MgO coaxial fibers: co-electrospun fabrication, characterization and electrochemical properties. J Mater Chem 17(18):1769–1776. doi:10.1039/b614205b

43. Hosono E, Wang Y, Kida N, Enomoto M, Kojima N, Okubo M, Matsuda H, Saito Y, Kudo T, Honma I, Zhou H (2010) Synthesis of Triaxial $LiFePO_4$ nanowire with a VGCF core column and a carbon shell through the electrospinning method. ACS Appl Mater Interfaces 2(1):212–218. doi:10.1021/am900656y
44. Ji L, Yao Y, Toprakci O, Lin Z, Liang Y, Shi Q, Medford AJ, Millns CR, Zhang X (2010) Fabrication of carbon nanofiber-driven electrodes from electrospun polyacrylonitrile/polypyrrole bicomponents for high-performance rechargeable lithium-ion batteries. J Power Sources 195(7):2050–2056. doi:10.1016/j.jpowsour.2009.10.021
45. Zou L, Gan L, Kang F, Wang M, Shen W, Huang Z (2010) Sn/C non-woven film prepared by electrospinning as anode materials for lithium ion batteries. J Power Sources 195(4):1216–1220. doi:10.1016/j.jpowsour.2009.08.052
46. Wang L, Ding CX, Zhang LC, Xu HW, Zhang DW, Cheng T, Chen CH (2010) A novel carbon-silicon composite nanofiber prepared via electrospinning as anode material for high energy-density lithium ion batteries. J Power Sources 195(15):5052–5056. doi:10.1016/j.jpowsour.2010.01.088
47. Li Y, Guo B, Ji L, Lin Z, Xu G, Liang Y, Zhang S, Toprakci O, Hu Y, Alcoutlabi M, Zhang X (2013) Structure control and performance improvement of carbon nanofibers containing a dispersion of silicon NPs for energy storage. Carbon 51:185–194. doi:10.1016/j.carbon.2012.08.027
48. Nan D, Wang JG, Huang ZH, Wang L, Shen W, Kang F (2013) Highly porous carbon nanofibers from electrospun polyimide/SiO_2 hybrids as an improved anode for lithium-ion batteries. Electrochem Commun 34:52–55, doi: http://dx.doi.org/10.1016/j.elecom.2013.05.010
49. Yu Y, Yang Q, Teng D, Yang X, Ryu S (2010) Reticular Sn nanoparticle-dispersed PAN-based carbon nanofibers for anode material in rechargeable lithium-ion batteries. Electrochem Commun 12(9):1187–1190. doi:10.1016/j.elecom.2010.06.015
50. Zhang P, Guo ZP, Huang Y, Jia D, Liu HK (2011) Synthesis of Co_3O_4/Carbon composite nanowires and their electrochemical properties. J Power Sources 196(16):6987–6991. doi:10.1016/j.jpowsour.2010.10.090
51. Lin Z, Ji L, Woodroof MD, Zhang X (2010) Electrodeposited MnO_x/carbon nanofiber composites for use as anode materials in rechargeable lithium-ion batteries. J Power Sources 195(15):5025–5031. doi:10.1016/j.jpowsour.2010.02.004
52. Ji L, Lin Z, Guo B, Medford AJ, Zhang X (2010) Assembly of carbon-SnO_2 core-sheath composite nanofibers for superior lithium storage. Chem Eur J 16(38):11543–11548. doi:10.1002/chem.201001564
53. Zou L, Gan L, Lv R, Wang M, Huang Z, Kang F, Shen W (2011) A film of porous carbon nanofibers that contain Sn/SnO_x nanoparticles in the pores and its electrochemical performance as an anode material for lithium ion batteries. Carbon 49(1):89–95. doi:10.1016/j.carbon.2010.08.046
54. Meschini I, Nobili F, Mancini M, Marassi R, Tossici R, Savoini A, Focarete ML, Croce F (2013) High-performance Sn@carbon nanocomposite anode for lithium batteries. J Power Sources 226:241–248. doi:10.1016/j.jpowsour.2012.11.004
55. Bonino CA, Ji L, Lin Z, Toprakci O, Zhang X, Khan SA (2011) Electrospun carbon-tin oxide composite nanofibers for use as lithium ion battery anodes. ACS Appl Mater Interfaces 3(7):2534–2542. doi:10.1021/am2004015
56. Kong J, Liu Z, Yang Z, Tan HR, Xiong S, Wong SY, Li X, Lu X (2012) Carbon/SnO_2/carbon core/shell/shell hybrid nanofibers: tailored nanostructure for the anode of lithium ion batteries with high reversibility and rate capacity. Nanoscale 4(2):525–530. doi:10.1039/c1nr10962f
57. Ji L, Toprakci O, Alcoutlabi M, Yao Y, Li Y, Zhang S, Guo B, Lin Z, Zhang X (2012) alpha-Fe_2O_3 Nanoparticle-loaded carbon nanofibers as stable and high-capacity anodes for rechargeable lithium-ion batteries. ACS Appl Mater Interfaces 4(5):2672–2679. doi:10.1021/am300333s

58. Yang Z, Du G, Meng Q, Guo Z, Yu X, Chen Z, Guo T, Zeng R (2012) Synthesis of uniform TiO_2@carbon composite nanofibers as anode for lithium ion batteries with enhanced electrochemical performance. J Mater Chem 22(12):5848–5854. doi:10.1039/c2jm14852h
59. Ji L, Lin Z, Medford AJ, Zhang X (2009) In-situ encapsulation of nickel particles in electrospun carbon nanofibers and the resultant electrochemical performance. Chem Eur J 15(41):10718–10722. doi:10.1002/chem.200902012
60. Deng D, Kim MG, Lee JY, Cho J (2009) Green energy storage materials: nanostructured TiO_2 and Sn-based anodes for lithium-ion batteries. Energ Environ Sci 2(8):818–837. doi:10.1039/b823474d
61. Yang Z, Du G, Feng C, Li S, Chen Z, Zhang P, Guo Z, Yu X, Chen G, Huang S, Liu H (2010) Synthesis of uniform polycrystalline tin dioxide nanofibers and electrochemical application in lithium-ion batteries. Electrochim Acta 55(19):5485–5491. doi:10.1016/j.electacta.2010.04.045
62. Li L, Yin X, Liu S, Wang Y, Chen L, Wang T (2010) Electrospun porous SnO_2 nanotubes as high capacity anode materials for lithium ion batteries. Electrochem Commun 12(10):1383–1386. doi:10.1016/j.elecom.2010.07.026
63. Li L, Peng S, Cheah YL, Wang J, Teh P, Ko Y, Wong C, Srinivasan M (2013) Electrospun eggroll-like $CaSnO_3$ nanotubes with high lithium storage performance. Nanoscale 5(1):134–138. doi:10.1039/c2nr32766j
64. Feng N, Qiao L, Hu D, Sun X, Wang P, He D (2013) Synthesis, characterization, and lithium-storage of ZnO-SnO_2 hierarchical architectures. RSC Adv 3(21):7758–7764. doi:10.1039/c3ra40229k
65. Lu HW, Zeng W, Li YS, Fu ZW (2007) Fabrication and electrochemical properties of three-dimensional net architectures of anatase TiO_2 and spinel $Li_4Ti_5O_{12}$ nanofibers. J Power Sources 164(2):874–879. doi:10.1016/j.jpowsour.2006.11.009
66. Nam SH, Shim HS, Kim YS, Dar MA, Kim JG, Kim WB (2010) Ag or Au nanoparticle-embedded one-dimensional composite TiO_2 nanofibers prepared via electrospinning for use in lithium-ion batteries. ACS Appl Mater Interfaces 2(7):2046–2052. doi:10.1021/am100319u
67. Fehse M, Cavaliere S, Lippens PE, Savych I, Iadecola A, Monconduit L, Jones DJ, Rozière J, Fischer F, Tessier C, Stievano L (2013) Nb-Doped TiO_2 nanofibers for lithium ion batteries. J Phys Chem C 117(27):13827–13835. doi:10.1021/jp402498p
68. Chaudhari S, Srinivasan M (2012) 1D hollow alpha-Fe_2O_3 electrospun nanofibers as high performance anode material for lithium ion batteries. J Mater Chem 22(43):23049–23056. doi:10.1039/c2jm32989a
69. Teh PF, Sharma Y, Pramana SS, Srinivasan M (2011) Nanoweb anodes composed of one-dimensional, high aspect ratio, size tunable electrospun $ZnFe_2O_4$ nanofibers for lithium ion batteries. J Mater Chem 21(38):14999–15008. doi:10.1039/c1jm12088c
70. Ding Y, Zhang P, Long Z, Jiang Y, Huang J, Yan W, Liu G (2008) Synthesis and electrochemical properties of Co_3O_4 nanofibers as anode materials for lithium-ion batteries. Mater Lett 62(19):3410–3412. doi:10.1016/j.matlet.2008.03.033
71. Sun YK, Oh IH, Hong SA (1996) Synthesis of ultrafine $LiCoO_2$ powders by the sol-gel method. J Mater Sci 31(14):3617–3621. doi:10.1007/bf00352769
72. Chen LJ, Liao JD, Chuang YJ, Hsu KC, Chiang YF, Fu YS (2011) Synthesis and characterization of PVP/$LiCoO_2$ nanofibers by electrospinning route. J Appl Polym Sci 121(1):154–160. doi:10.1002/app.33499
73. Gu YX, Chen DR, Jiao ML (2005) Synthesis and electrochemical properties of nanostructured $LiCoO_2$ fibers as cathode materials for lithium-ion batteries. J Phys Chem B 109(38):17901–17906. doi:10.1021/jp0521813
74. Yang Z, Cao C, Liu F, Chen D, Jiao X (2010) Core-shell $Li(Ni_{1/3}Co_{1/3}Mn_{1/3})O_{-2}$/$Li(Ni_{1/2}Mn_{1/2})O_2$ fibers: synthesis, characterization and electrochemical properties. Solid State Ion 181(15–16):678–683. doi:10.1016/j.ssi.2010.03.032
75. Lu HW, Yu L, Zeng W, Li YS, Fu ZW (2008) Fabrication and electrochemical properties of three-dimensional structure of $LiCoO_2$ fibers. Electrochem Solid State 11(8):A140–A144. doi:10.1149/1.2932054

76. Gabrisch H, Wilcox JD, Doeff MM (2006) Carbon surface layers on a high-rate $LiFePO_4$. Electrochem Solid State 9(7):A360–A363. doi:10.1149/1.2203309
77. Sisbandini C, Brandell D, Gustafsson T, Nyholm L (2009) The mechanism of capacity enhancement in $LiFePO_4$ cathodes through polyetheramine coating. J Electrochem Soc 156(9):A720–A725. doi:10.1149/1.3155451
78. Zhu C, Yu Y, Gu L, Weichert K, Maier J (2011) electrospinning of highly electroactive carbon-coated single-crystalline $LiFePO_4$ nanowires. Angew Chem Int Ed 50(28):6278–6282. doi:10.1002/anie.201005428
79. Von Hagen R, Lepcha A, Song X, Tyrra W, Mathur S (2013) Influence of electrode design on the electrochemical performance of $Li_3V_2(PO_4)_3/C$ nanocomposite cathode in lithium ion batteries. Nano Energ 2(2):304–313. doi:10.1016/j.nanoen.2012.10.002
80. Cheah YL, Gupta N, Pramana SS, Aravindan V, Wee G, Srinivasan M (2011) Morphology, structure and electrochemical properties of single phase electrospun vanadium pentoxide nanofibers for lithium ion batteries. J Power Sources 196(15):6465–6472. doi:10.1016/j.jpowsour.2011.03.039
81. Cheah YL, Aravindan V, Madhavi S (2012) Improved elevated temperature performance of Al-Intercalated V_2O_5 electrospun nanofibers for lithium-ion batteries. ACS Appl Mater Interfaces 4(6):3270–3277. doi:10.1021/am300616k
82. Le Viet A, Reddy MV, Jose R, Chowdari BVR, Ramakrishna S (2010) Nanostructured Nb_2O_5 polymorphs by electrospinning for rechargeable lithium batteries. J Phys Chem C 114(1): 664–671. doi:10.1021/jp9088589

Chapter 4
Electrospun Fibrous Membranes as Separators of Lithium-Ion Batteries

Shuiliang Chen, Wan Ye, and Haoqing Hou

Abstract Electrospinning is known to be an effective technology for producing ultrathin fibers with diameters ranging from several nanometers to several micrometers. Electrospun fibrous membranes show advantages of high porosity, small and interconnected open pore size, high surface area, and high permeability of gas, and thus are employed as separators in lithium-ion batteries (LIBs). In this chapter, the electrospun pure polymer, polymer blend, and polymer/inorganic fibrous membranes for separators in LIBs with liquid or gel electrolytes are summarized and highlighted.

4.1 Introduction of LIB

Lithium-ion battery (LIB) is an important member of the rechargeable battery family, in which lithium ions move from the negative electrode to the positive electrode during discharge and reverse when charging. LIBs show excellent characteristics of the best energy density and extremely slow loss of charge when not in use. Thus, LIB is one of the most popularly used rechargeable batteries for portable electronics in recent years [1]. Besides being used as power source in consumer electronics such as computers, mobile phones, instruments, and so on, LIBs also have growing applications for military, electric vehicle, and aerospace applications.

The LIB mainly consists of three parts, positive electrode (or anode), negative electrode (or cathode), and separator which is placed between the two electrodes, as shown in Fig. 4.1. The LIBs usually use an intercalated lithium compound as the electrode material, which is different from the metallic lithium used in the non-rechargeable lithium battery. During discharge (Fig. 4.1b), lithium ions (Li^+) carry

S. Chen • W. Ye • H. Hou (✉)
Department of Chemistry and Chemical Engineering, Jiangxi Normal University, Nanchang, China
e-mail: haoqing@jxnu.edu.cn

B. Ding and J. Yu (eds.), *Electrospun Nanofibers for Energy and Environmental Applications*, Nanostructure Science and Technology,
DOI 10.1007/978-3-642-54160-5_4,

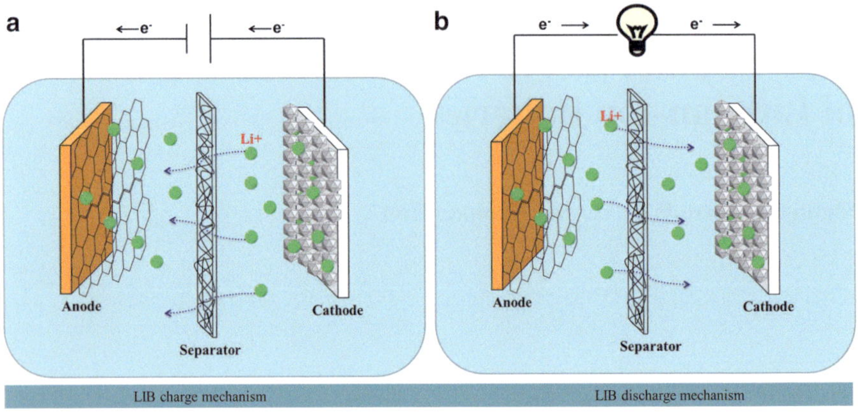

Fig. 4.1 Charge (**a**) and discharge (**b**) mechanism of lithium-ion battery

the current from the negative to the positive electrode, through the nonaqueous electrolyte and separator diaphragm. While during charge (Fig. 4.1a), an overvoltage (a higher voltage but of the same polarity than that produced by the battery) is applied by an external electrical power source (the charging circuit), which forces the current to pass in the reverse direction. The lithium ions then migrate from the positive to the negative electrode, where they become embedded to the porous electrode material through a process known as intercalation.

4.2 Introduction of Separators in LIB

The separator is one of the key components in the LIBs; its main function is to keep the positive and negative electrodes apart to prevent electrical short circuits and at the same time allow rapid transport of ionic charge carriers that were needed to complete the circuit during the passage of current in an electrochemical cell. According to the electrolytes used, the LIBs here are mainly classified into the following three types: (a) a liquid-type battery using liquid electrolytes (LEs), (b) a gel-type battery using gel polymer electrolytes (GPEs), and (c) a solid-type battery using inorganic ceramics as electrolyte which ions migrate through the lattice [2].

Generally, the commonly used electrolytes in LIBs were $LiPF_6$ solution in nonaqueous solvents which were stable at high voltage, such as ethylene carbonate (EC), ethyl methyl carbonate (EMC), and diethyl carbonate (DEC). For the LIBs with LEs, porous membranes including microporous plastic films and nonwoven fabrics were used as separators. These separators allowed the liquid electrolytes to flow through and ensured a good ionic conductivity. The microporous plastic films were generally based on polyethylene (PE), polypropylene (PP), and their composite PP/PE/PP. In general, they were prepared by stretch polymer film through a dry or wet process. The nonwoven fabrics were textile products processed directly from random assembled fibers. The nonwoven fabrics

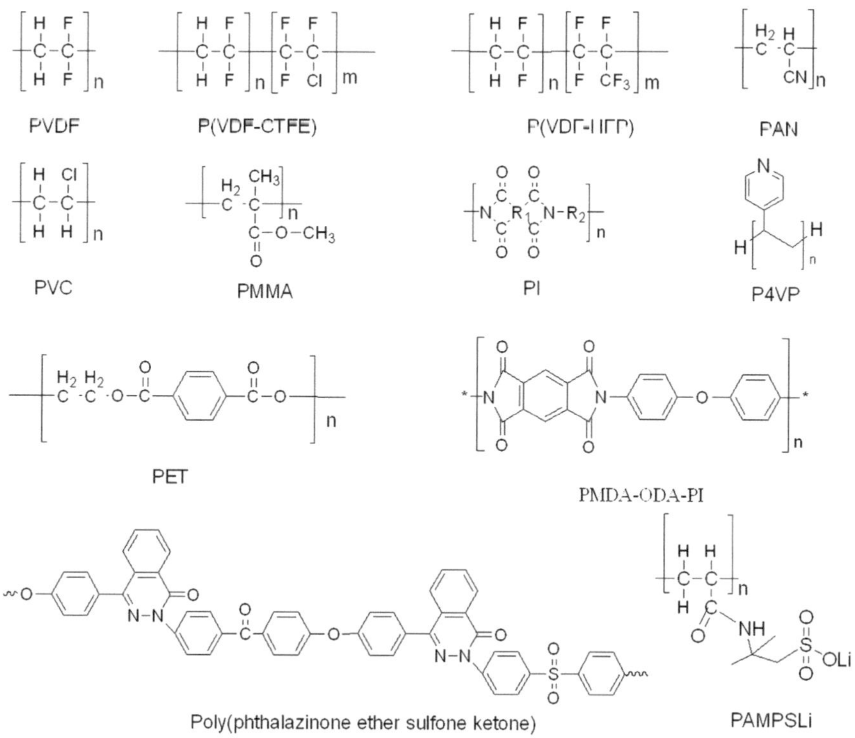

Fig. 4.2 Macromolecular formula of some polymers

generally possessed much higher porosity and lower weight comparing to the microporous membranes. To improve the safety and durability of the LIBs, the liquid electrolytes were incorporated into a polymer membrane to form a GPE membrane. The electrolytes restrained by the polymer chains showed similar mechanism of ionic conduction to that in liquid electrolytes, but exhibited better shape flexibility over liquids. In that case, the GPE membrane served as both electrolyte and separator in LIBs. Polymers of poly(vinylidene fluoride) (PVDF) [3], poly(vinylidene fluoride-co-hexafluoropropylene) (P(VDF-HFP)) [4], polyvinylidene fluoride-co-chlorotrifluoroethylene (P(VDF-CTFE)) [5], polyacrylonitrile (PAN) [6], and poly(methyl methacrylate) (PMMA) [7] had been commonly developed for GPEs. Macromolecular formulas of some polymers are shown in Fig. 4.2. In addition, some inorganic ceramics or solid polymer membranes were also employed as separators in LIBs due to their good intrinsic ionic conductivity and thermal stability.

The most widely used separators in LIBs were the porous polymer membranes. Such membranes were very good electronic insulators and had the capability of conducting ions either by intrinsic ionic transmission or by ion migration through the soaked electrolyte. They were able to minimize any processes that adversely affect the electrochemical energy efficiency of the batteries. Different methods such as solvent casting [8], plasticizer extraction [6, 9], phase inversion [10], and electrospinning [4, 11] had been reported for the preparation of porous polymer membranes. Electrospinning is known to be an effective method for preparing nonwovens with fiber diameters from several micrometers down to several nanometers [12]. Compared to the other methods, it had been reported that porous polymer membranes prepared by electrospinning (denoted as electrospun fibrous membranes, EFMs) showed higher porosity, electrolyte uptake, and ionic conductivity due to the presence of fully interconnected pore structure [11]. In this chapter, the EFMs used for the separators in LIBs were highlighted.

4.3 EFMs for Separators in Liquid Electrolyte LIBs

The traditional polyolefin microporous membranes, such as PP, PE, and PP/PE/PP composite separators, showed disadvantages of low porosity (about 40–55 %), poor thermal stability, and poor wettability due to nonpolarity, which lead to high cell resistance and low rate capability in the LIBs, which severely restricted the electrochemical performance of LIBs. To solve these problems, EFMs with higher porosity, thermal stability, and wettability were developed to be used as the separators in LIBs, including polyethylene terephthalate (PET) [13], polyimide (PI) [14], and poly(phthalazinone ether sulfone ketone) (PPESK) [15]. The high porosity (over 70 %) of the EFMs could greatly increase the ionic conductivity and reduce the cell resistance. Such EFMs were ideal separator candidates for LIBs to achieve high performance, such as large capability, high rate capability, resistance to high temperature, and long cyclic life.

4.3.1 Electrospun PET Fibrous Membrane

PET could be used as the separator material in LIB due to its good thermal stability and resistance to shrinkage, good mechanical properties, and excellent electronic insulation. The PET had a melting point of about 255 °C and degradation temperature of 413 °C (5 % weight loss). Electrospun PET fibrous membrane showed good mechanical properties of tensile strength of 12 MPa and elongation of 40 %. The electrospun PET fibrous membrane had a porosity of 89 % and an electrolyte uptake of 500 %, which were much higher than those of Celgard separators (porosity of 40–55 %, electrolyte uptake of 176 %). It also displayed high ionic conductivity of 2.27×10^{-3} S cm^{-1} at room temperature [13].

4.3.2 Electrospun PI Fibrous Membrane

As one kind of high-performance engineering polymers, aromatic PIs had been electrospun to fibrous nonwoven and developed as separator in LIB. Aromatic PIs

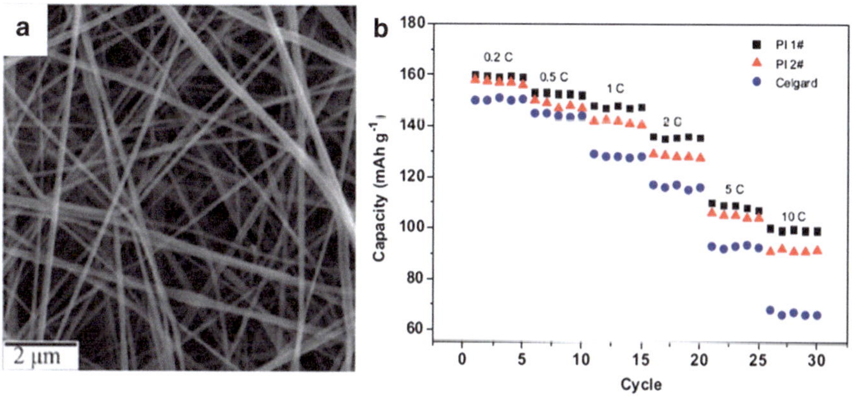

Fig. 4.3 (**a**) SEM image of electrospun PI (1#) fibrous nonwoven. (**b**) Rate capability tests for the cells with Celgard membrane and two types of PI (1#, 40 μm) and PI (2#, 100 μm) fibrous nonwovens with different thickness as separators in LIB, respectively (Reprinted from Miao et al. [14], Copyright 2013, with permission from Elsevier)

showed excellent mechanical properties, low dielectric constant, and inertness to traditional solvents; they were thermally stable at temperatures over 500 °C. Thus, the electrospun aromatic PI fibrous nonwoven could effectively avoid the short circuits caused by the shrinkage of the conventional separators at temperatures higher than 150 °C. Moreover, polar solvents used in LIBs, such as EC and EMC, could be strongly coordinated within the polymer chains due to the presence of the electron donor and acceptor groups of PI, which would largely enhance the retain ability for electrolyte of PI-based battery separators. Aromatic PI is insoluble in common organic and inorganic solvents except for sulfuric acid; thus, it is not able to be directly electrospun. The preparation of electrospun PI fibers was realized by following two steps: (a) electrospinning of its precursor poly(amic acid) (PAA) dissolved in aprotic solvents, such as dimethyl formamide and dimethyl acetamide, to form PAA fibrous nonwovens and (b) conversion of the PAA fibrous nonwovens to PI fibrous nonwoven at a high temperature of over 300 °C [16, 17]. In one of the studies, electrospun PI fibrous nonwovens based on pyromellitic dianhydride and 4, 4′-oxydianiline were prepared by Miao et al. [14] and used as separator in LIB. It did not melt and had a high degradation temperature of over 500 °C. The electrospun PI nonwovens showed no color change and shrinkage under hot oven treatment at 150 °C, demonstrating excellent thermal stability for separators in LIBs. Moreover, the electrospun PI fibrous nonwoven separator exhibited excellent electrolyte wettability due to the high polarity and could guarantee the penetration and high uptake of the electrolyte. Electrochemical tests revealed that the electrospun PI fibrous nonwoven separators were also evaluated to have higher capability of 160 mAh g^{-1} (0.2 C rate), as shown in Fig. 4.3, lower resistance of 90 Ω, and higher rate capability compared to the traditional Celgard separators.

Composite fibrous nonwovens of PI/PET had been prepared by using PET nonwovens as support to collect electrospun PI nanofibers. The PI/PET fibrous

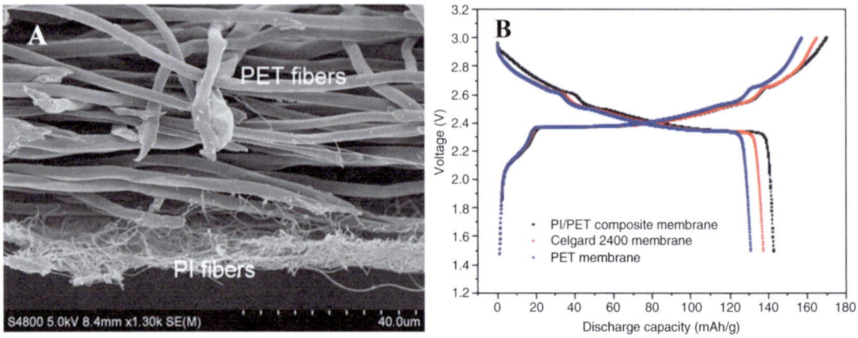

Fig. 4.4 (**a**) Cross-sectional SEM image of composite membrane with electrospun PI nanofibers supported on PET nonwoven. (**b**) Initial charge–discharge curves for the cells with the PET membrane, Celgard 2400 membrane, and PI/PET composite membrane [18] (Reproduced by permission of ECS-The Electrochemical Society)

membrane decreased the porosity, electrolyte uptake, and ionic conductivity of about 10–30 % but showed 400 % higher tensile strength than that of pure PI fibrous nonwovens prepared at low temperature of 250 °C. Moreover, the PI/PET composite fibrous membrane displayed only about 2 % of thermal shrinkage at 180 °C under air atmosphere [18]. The PI/PET fibrous membrane separator showed a capability of 146 mAh g^{-1} with capability retention ratios of 87.5 % after 50 cycles (Fig. 4.4).

4.3.3 Electrospun Poly(phthalazinone ether sulfone ketone) Fibrous Membrane

Poly(phthalazinone ether sulfone ketone) (PPESK) is one class of high performance polymers. Its excellent properties including high strength and thermal stability are due to its aromatic architecture (Fig. 4.2). PPESK could be electrospun to make fibrous membrane and employed as separator in LIBs due to the excellent thermal stability. After thermal treatment at 220 °C for 1 h, the shape of PPESK fabrics was almost unchanged. Moreover, the PPESK nonwovens showed a good wettability with a high electrolyte uptake of 1,210 % and a high electronic ionic conductivity of 3.97 mS cm^{-1}, which were beneficial for improving the rate capability of the cells [15]. Therefore, electrospun PPESK fabrics could be used for developing high-performance lithium-ion battery.

4.4 EFMs for Separators in GPE LIBs

Polymer gel is generally formed by incorporating liquid electrolyte into polymer matrix, which hints that the polymer should show excellent adsorption property toward the liquid electrolytes used in LIB. The ionic conduction mechanism in the polymer gels is very similar to that in the liquid electrolytes, and the polymer gel generally has better shape flexibility over liquid. Typically, liquid electrolyte is restrained by polymer chains to form the polymer gel; thus, the electrolyte can also be restricted by electrospun polymer fibrous membrane to form gel electrospun fibrous membrane and used as GPE separator in LIBs. Gelled electrospun fibrous membrane based on poly(vinylidene fluoride) (PVDF) [11, 19–22], poly(vinylidene fluoride-co-hexafluoropropylene) (P(VDF-HFP) [4], polyvinylidene fluoride-co-chlorotrifluoroethylene (P(VDF-CTFE)) [5, 23], polyacrylonitrile (PAN) [24–28], and blend gel polymers had been developed for GPE separators in LIBs.

4.4.1 Electrospun PVDF Fibrous Membrane

PVDF is one of the commonly used polymer matrix for GPE separator in LIBs because it is known to be chemically and mechanically stable. However, the high crystallinity of the PVDF homopolymer leads to a low ionic conductivity because the crystalline part hinders the migration of Li ions and, thus, results in low charge–discharge capability and poor 1C-rate value. A low rate capability with only about 50 % of the designed capability at high discharge rate (C/1 rate) had been obtained because the residual solvent or non-solvent could not be easily removed in the conventional processing methods, for example, phase inversion. Processing the PVDF by using electrospinning technology could solve the above problems. The electrospun fibrous membrane had high porosity, small pore size of few tens of nanometers to few micrometers, interconnected open pore structure, and large specific surface area. On one hand, the interconnected architecture facilitated the migration of the Li ions through the membrane and thus enhanced the ionic conductivity. On the other hand, the high specific surface area of the PVDF nanofibrous membrane resulted in an excellent wettability and good affinity to the electrolyte. Moreover, the gel phase formed by the swelling of PVDF fibers could cause an enhancement of the electrochemical stability window. The use of electrospun PVDF membrane for separator in GPE LIB was first reported by Chio et al. [11]. It exhibited high electrolyte uptake of 260 %, low cell resistance of 50 Ω, and ionic conductivity of 1.7 mS cm^{-1} (0 °C). The mechanical properties of the electrospun PVDF nanofibrous membranes were enhanced by thermal treatment at 150–160 °C to increase their stability and safety to some extent in LIBs [20]. The enhancement of the mechanical properties of PVDF membrane by thermal treatment was due to the increase of crystallinity. However, the PVDF-based fibrous membranes showed low thermal stability with a melting point of 166 °C similar to that of traditional PE (133 °C) and PP separators (about 163 °C); therefore, they would still face the safety problem in high-power LIBs.

4.4.2 Electrospun P(VDF-HFP) and P(VDF-CTFE) Fibrous Membrane

As two important copolymers of PVDF, the P(VDF-HFP) [4] and P(VDF-CTFE) [23] had been developed for gel polymer electrolyte in LIBs. The introduction of copolymer components was to reduce the crystallinity of the PVDF chain. The reduction of crystallinity could increase the ionic conductivity. Electrospun P(VDF-HFP) and P(VDF-CTFE) fibrous membranes had been proved to show high ionic conductivities in the range of several mS cm^{-1} which was attributed to the easy transportation of the liquid electrolyte through the fully interconnected pore structure of the membrane. For example, the electrospun P(VDF-HFP) fibrous membrane had high ionic conductivities in the range of 4.59 mS cm^{-1}, high electrolyte uptake of 425 % at room temperature, and good electrochemical stability with a potential of over 4.5 V versus Li/Li^{+} [29].

4.4.3 Electrospun PAN Fibrous Membrane

Among these host polymers, PAN-based electrolytes had shown interesting characteristics like high ionic conductivity at room temperature, high thermal stability, resistance to oxidative degradation, and good compatibility with the lithium electrodes, which could minimize the dendrite growth during the charging–discharging process of LIBs [30]. It was reported that the CN groups in PAN could interact with C=O groups of the liquid electrolytes such as propylene carbonate (PC) and ethylene carbonate (EC), as well as with lithium ions [31]. It was found that PAN provided rigidity and other characteristics similar to the PEs with augmented lithium ion conduction. The DSC analysis revealed that the electrospun PAN fibrous membrane did not melt, but exhibited two exothermic peaks at 253 and 284 °C, respectively [24]. It was demonstrated that the electrospun PAN fibrous separator showed much better thermal stability than that of PE and PP separators. Further investigation revealed that the electrospun PAN fibrous membranes were thermally stable at 120 °C, but not sufficiently stable at higher temperature of 150 °C due to the oxidative condition of the charged cathode. Electrochemical tests showed that the PAN-based nanofibrous separators had smaller ionic resistance at higher C-rate than that of PE and PP separators, as shown in Fig. 4.5 [24].

However, electrospun PAN fibrous membranes had undesirable effects, for example, liquid extraction from the gel, which resulted in decreased ionic conductivity of the polymer electrolytes upon long-term storage. Also, the porous PAN membranes were very brittle for the reason that the interaction of adjacent cyanide groups increases the resistance of interior rotation of the main chain and thus decreases the flexibility of the polymer chain [32]. Thus, the separator combination of thermally stable nonwoven frame with the electrospun PAN fibrous membranes is required in high-power battery.

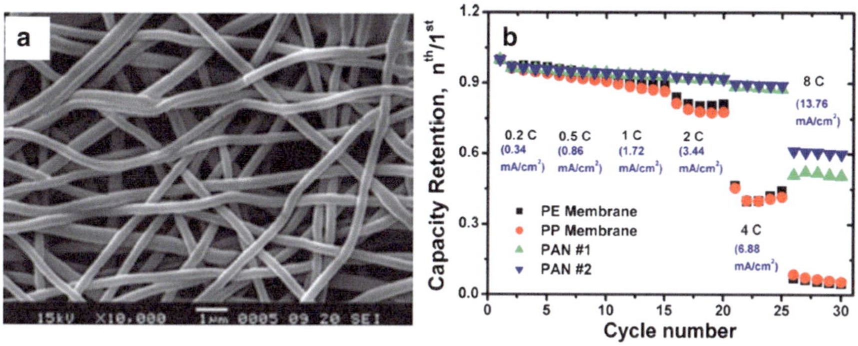

Fig. 4.5 (**a**) SEM image of the electrospun PAN nanofiber-based separator. (**b**) Results of rate capability test for the cells with PE, PP, electrospun PAN #1, and PAN #2 separators, respectively [24] (Reproduced by permission of ECS-The Electrochemical Society)

4.4.4 Electrospun Poly(lithium 2-acrylamido-2-methylpropanesulfonic acid) Fibrous Membrane

Introduction of Li ion to polymer matrix could make an ion conduction polymer and was used as separator materials in LIBs. For example, poly(lithium 2-acrylamido-2-methylpropanesulfonic acid) (PAMPSLi) is a new single-ion conducting polymer, in which anions are covalently attached to the polymer backbones and only cations transport charges [33]. The ion conductivity of single-ion conductors is generally lower than that of the salt-doped polymer electrolytes; the restricted movement of anions suppresses the buildup of concentration gradients [34, 35]. However, the porous architecture of the electrospun fibrous could greatly enhance the migration of Li ion, thus making the electrospun ion conducting polymer fibrous membrane an ideal separator candidate in LIBs.

4.5 EFMs from Polymer Blend for Separators in LIB

The use of EFMs from polymer blends for separators in LIB is to combine the advantages of different polymers, such as mechanical properties, thermal stability, ionic conductivity, and electrochemical stability. Bicomponent and tricomponent fibrous membranes based on PVDF and its copolymers, PAN and PMMA, had been prepared by electrospinning and used as separator in LIBs, including PVDF/PMMA [36], PAN-PVDF [37], PMMA/PVC [38], TPU/PVDF [39, 40], PAN/PMMA/PS [41], PI/PVDF and PI/PAN [42], PVDF/poly(4-vinylpyridine) (P4VP) [43], and nanocrystal cellulose (NCC)/P(VDF-HFP) [44–46]. Their properties and performance in LIBs are summarized in Table 4.1.

Table 4.1 Summary of EFMs as separators in LIBs

Separators	AFD/APS (μm)	Thickness (μm)	Porosity (%)	Ionic cond. at 20 °C (mS cm^{-1})	Stability versus. Li/Li^{+} (V)	Electrolyte uptake/%	Capability /rate (mAh g^{-1})	Retention/ cycles	Strength (MPa)	T_m (°C)	T_d (°C)	References
PET	0.3–0.42	40	89	2.27	5.0	500	147.2/0.1 C	95 %/50	12	255	413	[13]
PMDA-ODA-PI	0.2	40	75	1.242	–	256	160/0.2 C	100 %/100	10.23	>240	500	[14, 18]
PET/PI	0.144	40	65	0.897	–	220	146/0.5 C	87.5 %/50	50.87	>180	–	[18]
PPESK	1.384.23	98	92	3.79	4.47	1,120	100/0.2 C	>95 %/180		>273		[15]
PVDF	0.4/3.28	–	–	1.7*	4.5	260	–	–	1.9	166	–	[11, 19]
PVDF-160	0.51/1.23	–	–	1.6–2.0	–	–	–	–	6.5	–	–	[20]
P(VDF-HFP)	1.1	100–125	–	4.59	>4.5	425	142/0.1 C	96 %/100	6.5	159.5	438	[29]
P(VDF-CTFE)	0.23	–	–	>2.0	>4.0	800	120/0.33 C	–	–	120–180	>455	[5]
PAN	0.35/0.38	35	76	~1.0	>4.7	–	125.9/0.5 C	85 %/200	–	–	253	[24]
PAMPSLi	0.17–1.5	–	–	0.0212	4.4	650–960	–	–	–	–	304	[33]
PVDF/PMMA	–	–	60	3.55	5.1	> 300	–	–	–	–	–	[36]
P(VDF-HFP)/PMMA	0.2–0.35	150–250	–	1.99	>4.1	377	145/0.1 C	92 %150	8	–	–	[47]
PVDF/25 % PAN	0.25	45	85	7.8	5.1	>300	120.4/0.1 C	93 %/150	–	165	–	[37]
10 % PMMA/PVC	0.132	75	79	7.8	4.7	369	140/0.5 C	100 %/100	–	–	–	[38]

50 % TPU/ PVDF	0.57	–	86	3.2	5.0	279	169/0.1 C	86 %/50	–	–	329	[39]
PAN/PMMA/ PS	0.62–0.95	120	84	3.9	4.5	680	120/0.1 C	92.7 %	–	296	288.7	[41]
PVDF/P4VP	0.25–0.6	–	–	3.4	–	100	–	–	21.5	–	–	[43]
Cellulose/ P(VDF-HFP)	0.2/1.66	60–70	78	0.4	5.0	712	100/0.1 C	100/60	8	–	–	[44]
12 % SiO_2/PAN	0.1/1.54	–	92	11	5.0	490	139/0.5 C	89 %/150	–	–	–	[48]
PVDF/ PMMA/ SiO_2	0.333	–	–	2.51–2.95	5.3	360	–	–	–	168.2	–	[49]
TiO_2/PVDF	0.15	–	–	1.4	–	367	148	–	–	170	–	[50]
10 % SiO_2/ P(VDF-HFP)	1.2	150	92	8.06	4.6	592	170/0.1 C	–	–	156.9	450	[29]
10 % LLTO/ PAN	<0.2	–	–	1.95	5.7	–	166/0.2 C	91.7 %/50	–	–	–	[51]
15 % LATP/ PAN	0.18–0.3	40	–	3.4	5.8	520 %	165/0.2 C	93 %/50	–	–	–	[52]

*Represnting at 0 °C
AFD average fiber diameter, *APS* average pore size, T_m melt point, T_d 5 % decomposition temperature

4.5.1 Electrospun Blend PVDF/PMMA Fibrous Membrane

PMMA is one of the host polymers previously used in plasticized polymer electrolytes. It is a common thermoplastic polymer with good chemical properties. Its amorphous structure is beneficial to ionic conduction; thus, the PMMA-based gel electrolytes have shown excellent interfacial stability toward lithium metal. The pendant -$COOCH_3$ is not likely to crystallize around Li ions as PEO does. PMMA-based polymer fibrous membrane exhibited high electrolyte uptake, high ionic conductivity, and good electrochemical stability [53]. PMMA was able to have chemical cross-linking, which could remarkably increase the mechanical strength and the electrolyte solution retention ability of the polymer electrolytes [54, 55]. Also it had been reported that gel electrolytes based on cross-linked PMMA could suppress lithium dendrite formation [54]. The PMMA was used to blend with other polymers, such as PVDF and PAN, to make composite nonwovens by electrospinning technology and used as gel polymer electrolyte separator in LIBs.

The incorporation of PMMA into gel polymer PVDF and its copolymers to make blend fibrous membranes by electrospinning was to increase the mechanical properties and electrolyte uptake and reduced the crystallinity of PVDF-based gel polymer, such as PVDF/PMMA [36] and P(VDF-HFP)-PMMA composite fibrous membrane separator [47]. It had been reported that the crystallinity of P(VDF-HFP) was depressed by the addition of PMMA due to the presence of $COOCH_3$ group and thus exhibited high electrolyte uptake, high ionic conductivity, and good electrochemical stability [53]. The composite membrane P(VDF-HFP)-PMMA showed an electrolyte solution uptake of 377 %. The ionic conductivity of the electrospun P(VDF-HFP)/PMMA fibrous electrolyte reached 2.0 mS cm^{-1} at room temperature. Furthermore, the tensile strength and elongation at break was improved to 30 % by the addition of PMMA component [47].

4.5.2 Electrospun Blend PVDF/PAN Fibrous Membrane

It had been stated above that PAN and PVDF had beneficial characteristics when used individually as host polymers in gel polymer electrolytes. Interestingly, PAN could provide a few important characteristics toward polymer electrolytes which could not be derived from PVDF and vice versa. Compared to PVDF, PAN had better mechanical stability and interfacial characteristics and could avoid dendrite growth [56], while PVDF had good electrochemical stability and affinity for Li ions [57]. Hence, polymer electrolytes prepared from the composites of PVDF and PAN had the synergistic advantages of both PVDF and PAN. Electrospinning conditions like high voltage, fast drying, and fiber formation could induce phase mixing between the partly miscible blend of PVDF and PAN to form a miscible PVDF/PAN fibrous membrane.

It has been reported that electrospun PVDF/PAN composite fibrous membrane could have an electrode uptake of liquid electrolyte (lithium salt in PC) over 300 %

and showed a high ionic conductivity of 7.8 mS cm^{-1} at 25 °C and high anodic stability of 5.1 V (vs. Li). The performance characteristics of PVDF/PAN fibrous membrane as polymer electrolytes in lithium batteries are much superior to the membranes of individual polymer of PAN and PVDF, respectively [37].

4.5.3 Electrospun Tricomponent Blend PAN/PMMA/PS Fibrous Membrane

The third component of PS added to the blend PAN/PMMA is mainly to increase the toughness and thermal stability [41]. Polystyrene (PS) is an amorphous polymer having higher T_g (100 °C) and good thermal stability. Only few reports on electrochemical properties of PS are available, and in most cases the material is used in blend with the addition of other polymer to provide toughness [55]. It is well known that, at higher temperatures, the backbone of PS starts to slide past each other. This makes PS flexible and stretchable. This quality is coupled with the strength of PAN to develop a tough and strong membrane. The aromatic ring in PS also reduces a certain extent of crystallization, thus enhances transportation of Li ions. It meant that the resultant polymer blend membrane had more flexibility and higher ionic conductivity. The blend PAN/PMMA/PS electrolyte separator showed ionic conductivity of about 3.9 mS cm^{-1}. The performance evaluation in coin cells delivered an initial discharge capability of ~120 mAh g^{-1} at 0.1 C with stable cycling properties at room temperature [41].

4.5.4 Electrospun Cellulose/P(VDF-HFP) Composite Fibrous Membrane

Cellulose is a member of the polysaccharide family with advantages of availability in abundance, biodegradability, and low cost. Nanocrystal cellulose (NCC) could be extracted by the acid hydrolysis of cellulose [58]. It had high tensile strength and modulus of elasticity. The modulus of the perfect crystal of native cellulose had been calculated before and was estimated to be 138 GPa [59]. Due to these promising properties, the NCC was an ideal candidate to reinforce polymers used as separators/electrolytes in lithium-ion batteries. In one report, the NCC-reinforced P(VDF-HFP) composite fibrous membrane showed high conductivity of about 3 mS cm^{-1} due to the high uptake of liquid electrolyte of about 712 % inside the porous fiber mat. The cathodic electrochemical stability of the electrospun composite fibrous membrane with 1 M LiTFSI–BMPyrTFSI electrolyte was up to 5 V (vs. Li^+/Li) [44].

Moreover, the addition of cellulose to the P(VDF-HFP) could also increase the thermal stability of the composite porous membrane. It was reported that the bulky

P(VDF-HFP) possessed a melting point at about 145 °C [60]; however, cellulose (3 wt%)/P(VDF-HFP) did not show any obvious endothermic peak below 300 °C. After thermal treatment in oven at 200 °C for 0.5 h, the cellulose/P(VDF-HFP) composite nonwoven exhibited negligible dimensional change [46].

4.6 Electrospun Polymer/Inorganic Composite Fibrous Membrane

The aims of the addition of inert ceramic filler to the electrospun polymer fiber matrix to form polymer/ceramic composite fiber membrane for separator in LIB were on one hand to prevent dimensional changes by thermal deformation at high temperature because of the frame structure of the heat-resistant ceramic powder and on the other hand to increase the ionic conductivity of the membrane due to the intrinsic ionic conductivity of the ceramics. Various ceramic fillers had been incorporated into the electrospun polymer membranes to make composite separators, including aluminum oxide (Al_2O_3) [61], fumed silica (SiO_2) [29, 48, 62], titanium dioxide (TiO_2) [50, 63–65], lithium lanthanum titanate oxide (LLTO) [51], and lithium aluminum titanium phosphate (LATP) [52].

The electrospun polymer/inorganic composite fibrous membranes could be prepared by direct electrospinning of polymer solution blend with inorganic ceramics. It had been reported that electrospun SiO_2/PAN fibrous membrane with 12 wt% SiO_2 exhibited a high conductivity of 1.1×10^{-2} S cm^{-1} due to the large liquid electrolyte uptake (about 490 %). It delivered the discharge capability of 139 mAhg^{-1} as 98 % of theoretical value and still retained a high value of 127 mAh g^{-1} as 89 % at 150th cycle, which was significantly higher than that of pure PAN nanofiber-based electrolyte membranes [48].

In another report, the TiO_2 was grafted onto the PMMA macromolecular chains to form PMMA-g-TiO_2. Then the PMMA-g-TiO_2 was blended with PVDF and electrospun to make composite fibrous membrane. The PMMA-g-TiO_2 could inhibit the crystallization of PVDF during the solidification process and improved the ionic conductivity of the fibrous polymer electrolyte from 2.51 to 2.95 mS cm^{-1} at 20 °C. The electrochemical stability window of the polymer electrolyte was also enhanced due to the presence of PMMA-g-TiO_2 [49].

Ionic-conducting ceramic particles were usually introduced to electrospun PAN fibers to provide the opportunity to obtain fibrous membranes that not only had the basic functions of battery separators but also could conduct Li ions, which significantly improved the battery kinetics. Perovskite-type lithium lanthanum titanate $Li_{3x}La_{2/3x}$ TiO_3(LLTO) [51] and lithium aluminum titanium phosphate (LATP) [52] had been incorporated into fibrous membranes by electrospinning for being used as separator in LIBs. In one study, with the addition of 15 wt% LLTO particles, the liquid electrolyte-soaked composite membrane showed a high ionic conductivity of 1.95 mS cm^{-1} at room temperature. The liquid electrolyte-soaked

LLTO/PAN composite fiber-based membranes exhibited lower interfacial resistance, higher electrolyte uptake, and higher electrochemical stability window than liquid electrolyte-soaked pure PAN fibrous membrane. Liquid electrolyte-soaked LLTO/PAN composite fiber-based membranes also showed good compatibility with lithium electrode. In addition, cells containing liquid electrolyte-soaked LLTO/PAN composite fiber-based electrospun membranes had an initial discharge capability of 166 $mAhg^{-1}$ and a stable cycle performance at 0.2 C rate [51]. In another report, electrospun 15 wt% LATP/PAN composite fiber-based membrane had a high ionic conductivity of 3.4 mS cm^{-1}. Moreover, the first cycle efficiency of cells using this membrane (93 %) was higher than that using commercial Celgard separator (91.9 %) and electrospun PAN fiber-based membrane (92.3 %) [52].

4.7 Electrospun Fibrous/Microporous Membrane Composite

The polyolefin microporous membrane separators commonly suffered from severe thermal shrinkage at higher temperatures, and this problem was fatal for cell safety because it would lead to internal short circuits and finally to fire outbreak or even explosion, especially when cells were exposed to abnormal conditions such as overcharge or overheating. To overcome this drawback, electrospun gel polymer fiber-coated microporous membranes were prepared by using polyolefin microporous membranes as supports to collect electrospun gel polymer fibers, for example, PVDF and its copolymers, for separators in LIBs [23, 66, 67]. Generally, inorganic ceramics were incorporated together with the electrospun gel polymer fibers to the polyolefin microporous membranes. The incorporation of the inorganic ceramics to the porous polyolefin membranes could be realized by the following two methods: coating the membrane by direct electrospinning of polymer solution containing ceramics [61, 68] or sandwiching the inorganic ceramics between electrospun membrane and conventional microporous membrane/nonwovens [69]. Cho et al. had developed a composite nonwoven separator by combining electrospun PAN nonwoven, ceramic powder, and polyolefin nonwoven using a hot roll press method. The ceramic powders were sandwiched between the PAN and polyolefin nonwoven, as shown in Fig. 4.6. The ceramic-containing polyolefin nonwoven had high tensile strength and thermal stability; the fibrous nonwoven could prevent dropping out of the ceramic powder as well as provide small pore size to the separator [69].

4.8 Conclusion

Electrospun fibrous membranes for separator in LIBs have been reviewed. Electrospinning is an effective technology for production of fibrous membrane with fiber diameter of several micrometers to several nanometers. With advantages

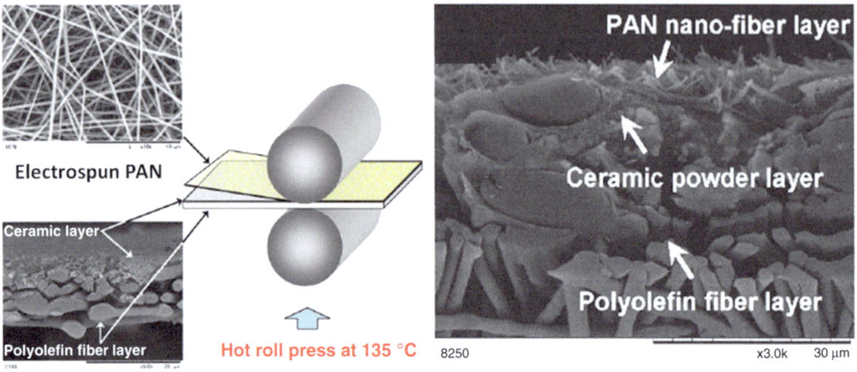

Fig. 4.6 Process for combination of electrospun PAN nanofibers with ceramics and polyolefin nonwoven and cross-sectional SEM image of composite membrane [69] (Reprinted from Miao et al. [14], Copyright 2012, with permission from Elsevier)

of high porosity which contributed to the high ionic conductivity and low cell resistance, the electrospun fibrous membranes showed better performance than the conventional microporous polyolefin membranes. For GPE LIBs, gel polymer was directly electrospun to make gel fibrous membrane for separators. The gel polymer was also blended with other polymers or inorganic ceramics to prepare composite fibrous membrane for separators in LIBs to overcome the disadvantages of the pure gel polymer. For LE LIBs, by selecting of polymer matrix materials with specific characteristics, including thermal resistance and high strength and toughness, the electrospun fibrous membrane-based separator could achieve high battery performance, such as large capability, high rate capability, resistant to high temperature, and long cyclic life.

References

1. Masataka W, Osamu Y (1998) Lithium ion batteries: fundamentals and performance. Wiley, New York (WILEY-VCH). ISBN 3-527-29569-0
2. Arora P, Zhang Z (2004) Battery separators. Chem Rev 104(10):4419–4462. doi:10.1021/cr020738u
3. Kim JR, Choi SW, Jo SM, Lee WS, Kim BC (2004) Electrospun PVdF-based fibrous polymer electrolytes for lithium ion polymer batteries. Electrochim Acta 50(1):69–75. doi:10.1016/j.electacta.2004.07.014
4. Li X, Cheruvally G, Kim J-K, Choi J-W, Ahn J-H, Kim K-W, Ahn H-J (2007) Polymer electrolytes based on an electrospun poly(vinylidene fluoride-co-hexafluoropropylene) membrane for lithium batteries. J Power Sources 167(2):491–498. doi:10.1016/j.jpowsour.2007.02.032
5. Croce F, Focarete ML, Hassoun J, Meschini I, Scrosati B (2011) A safe, high-rate and high-energy polymer lithium-ion battery based on gelled membranes prepared by electrospinning. Energy Environ Sci 4(3):921–927
6. Bohnke O, Frand G, Rezrazi M, Rousselot C, Truche C (1993) Fast ion transport in new lithium electrolytes gelled with PMMA. 1. Influence of polymer concentration. Solid State Ion 66(1–2):97–104, http://dx.doi.org/10.1016/0167-2738(93)90032-X

7. Peramunage D, Pasquariello DM, Abraham KM (1995) Polyacrylonitrile-based electrolytes with ternary solvent mixtures as plasticizers. J Electrochem Soc 142(6):1789–1798. doi:10.1149/1.2044195
8. Wang Z, Tang Z (2003) Characterization of the polymer electrolyte based on the blend of poly(vinylidene fluoride-co-hexafluoropropylene) and poly(vinyl pyrrolidone) for lithium ion battery. Mater Chem Phys 82(1):16–20, http://dx.doi.org/10.1016/S0254-0584(03)00241-4
9. Waugh W, Trissel L, Stella V (1991) Stability, compatibility, and plasticizer extraction of taxol (NSC-125973) injection diluted in infusion solutions and stored in various containers. Am J Health Syst Pharm 48(7):1520–1524
10. Strathmann H, Kock K (1977) The formation mechanism of phase inversion membranes. Desalination 21(3):241–255, http://dx.doi.org/10.1016/S0011-9164(00)88244-2
11. Choi SW, Jo SM, Lee WS, Kim YR (2003) An electrospun poly(vinylidene fluoride) nanofibrous membrane and its battery applications. Adv Mater 15(23):2027–2032. doi:10.1002/adma.200304617
12. Greiner A, Wendorff JH (2007) Electrospinning: a fascinating method for the preparation of ultrathin fibres. Angew Chem Int Ed 46(30):5670–5703. doi:10.1002/anie.200604646
13. Hao J, Lei G, Li Z, Wu L, Xiao Q, Wang L (2013) A novel polyethylene terephthalate nonwoven separator based on electrospinning technique for lithium ion battery. J Membr Sci 428:11–16. doi:10.1016/j.memsci.2012.09.058
14. Miao Y-E, Zhu G-N, Hou H, Xia Y-Y, Liu T (2013) Electrospun polyimide nanofiber-based nonwoven separators for lithium-ion batteries. J Power Sources 226:82–86. doi:10.1016/j.jpowsour.2012.10.027
15. Qi W, Lu C, Chen P, Han L, Yu Q, Xu R (2012) Electrochemical performances and thermal properties of electrospun Poly (phthalazinone ether sulfone ketone) membrane for lithium-ion battery. Mater Lett 66(1):239–241. doi:10.1016/j.matlet.2011.08.042
16. Chen SL, Hu P, Greiner A, Cheng CY, Cheng HF, Chen FF, Hou HQ (2008) Electrospun nanofiber belts made from high performance copolyimide. Nanotechnology 19(1):015604
17. Huang CB, Chen SL, Reneker DH, Lai CL, Hou HQ (2006) High-strength mats from electrospun poly(p-phenylene biphenyltetracarboximide) nanofibers. Adv Mater 18(5):668–671. doi:10.1002/adma.200501806
18. Ding J, Kong Y, Li P, Yang J (2012) Polyimide/poly(ethylene terephthalate) composite membrane by electrospinning for nonwoven separator for lithium-ion battery. J Electrochem Soc 159(9):A1474–A1480. doi:10.1149/2.041209jes
19. Choi SW, Kim JR, Ahn YR, Jo SM, Cairns EJ (2007) Characterization of electrospun PVdF fiber-based polymer electrolytes. Chem Mater 19(1):104–115. doi:10.1021/cm060223+
20. Choi SS, Lee YS, Joo CW, Lee SG, Park JK, Han KS (2004) Electrospun PVDF nanofiber web as polymer electrolyte or separator. Electrochim Acta 50(2–3):339–343. doi:10.1016/j.electacta.2004.03.057
21. Yang C, Jia Z, Guan Z, Wang L (2009) Polyvinylidene fluoride membrane by novel electrospinning system for separator of Li-ion batteries. J Power Sources 189(1):716–720. doi:10.1016/j.jpowsour.2008.08.060
22. Hwang K, Kwon B, Byun H (2011) Preparation of PVdF nanofiber membranes by electrospinning and their use as secondary battery separators. J Membr Sci 378(1–2):111–116. doi:10.1016/j.memsci.2011.06.005
23. Lee H, Alcoutlabi M, Watson JV, Zhang X (2013) Polyvinylidene fluoride-co-chlorotrifluoroethylene and polyvinylidene fluoride-co-hexafluoropropylene nanofiber-coated polypropylene microporous battery separator membranes. J Polym Sci Part B-Polym Phys 51(5):349–357. doi:10.1002/polb.23216
24. Cho TH, Sakai T, Tanase S, Kimura K, Kondo Y, Tarao T, Tanaka M (2007) Electrochemical performances of polyacrylonitrile nanofiber-based nonwoven separator for lithium-ion battery. Electrochem Solid State Lett 10(7):A159–A162. doi:10.1149/1.2730727
25. Cho T-H, Tanaka M, Onishi H, Kondo Y, Nakamura T, Yamazaki H, Tanase S, Sakai T (2008) Battery performances and thermal stability of polyacrylonitrile nanofiber-based nonwoven separators for Li-ion battery. J Power Sources 181(1):155–160. doi:10.1016/j.jpowsour.2008.03.010

26. Huai Y, Gao J, Deng Z, Suo J (2010) Preparation and characterization of a special structural poly(acrylonitrile)-based microporous membrane for lithium-ion batteries. Ionics 16(7): 603–611. doi:10.1007/s11581-010-0431-4
27. Tanaka M, Cho T-H, Nakamura T, Tarao T, Kawabe M, Sakai T (2010) Electrochemical performances of polyacrylonitrile nano-fiber based nonwoven separator for lithium ion battery. Electrochemistry 78(12):982–987
28. Raghavan P, Manuel J, Zhao X, Kim D-S, Ahn J-H, Nah C (2011) Preparation and electrochemical characterization of gel polymer electrolyte based on electrospun polyacrylonitrile nonwoven membranes for lithium batteries. J Power Sources 196(16):6742–6749. doi:10.1016/j.jpowsour.2010.10.089
29. Raghaven P, Choi J-W, Ahn J-H, Cheruvally G, Chauhan GS, Ahn H-J, Nah C (2008) Novel electrospun poly(vinylidene fluoride-co-hexafluoropropylene)-in situ SiO_2 composite membrane-based polymer electrolyte for lithium batteries. J Power Sources 184(2):437–443. doi:10.1016/j.jpowsour.2008.03.027
30. Tsutsumi H, Matsuo A, Takase K, Doi S, Hisanaga A, Onimura K, Oishi T (2000) Conductivity enhancement of polyacrylonitrile-based electrolytes by addition of cascade nitrile compounds. J Power Sources 90(1):33–38, http://dx.doi.org/10.1016/S0378-7753(00)00444-4
31. Abraham KM, Alamgir M (1990) Li^+-conductive solid polymer electrolytes with liquid-Like conductivity. J Electrochem Soc 137(5):1657–1658. doi:10.1149/1.2086749
32. Wu G, Yang H-Y, Chen H-Z, Yuan F, Yang L-G, Wang M, Fu R-J (2007) Novel porous polymer electrolyte based on polyacrylonitrile. Mater Chem Phys 104(2–3):284–287, http://dx.doi.org/10.1016/j.matchemphys.2007.03.013
33. Cui W-W, Tang D-Y (2012) Electrospun poly(lithium 2-acrylamido-2-methylpropanesulfonic acid) fiber-based polymer electrolytes for lithium-ion batteries. J Appl Polym Sci 126(2): 510–518. doi:10.1002/app.36804
34. Kang W-C, Park H-G, Kim K-C, Ryu S-W (2009) Synthesis and electrochemical properties of lithium methacrylate-based self-doped gel polymer electrolytes. Electrochim Acta 54(19):4540–4544, http://dx.doi.org/10.1016/j.electacta.2009.03.050
35. Zygadło-Monikowska E, Florjańczyk Z, Wielgus-Barry E, Paśniewski J (2006) Proton conducting gel polyelectrolytes based on 2-acrylamido-2-methyl-1-propanesulfonic acid (AMPSA) copolymers with polyfunctional monomers: Part I. Anhydrous systems. J Power Sources 159(1):385–391, http://dx.doi.org/10.1016/j.jpowsour.2006.02.036
36. Li X, Cao Q, Wang X, Jiang S, Deng H, Wu N (2011) Preparation of poly(vinylidene fluoride)/poly(methyl methacrylate) membranes by novel electrospinning system for lithium ion batteries. J Appl Polym Sci 122(4):2616–2620. doi:10.1002/app.34401
37. Gopalan AI, Santhosh P, Manesh KM, Nho JH, Kim SH, Hwang C-G, Lee K-P (2008) Development of electrospun PVdF-PAN membrane-based polymer electrolytes for lithium batteries. J Membr Sci 325(2):683–690. doi:10.1016/j.memsci.2008.08.047
38. Jung H-R, Lee W-J (2011) Electrochemical characteristics of electrospun poly(methyl methacrylate)/polyvinyl chloride as gel polymer electrolytes for lithium ion battery. Electrochim Acta 58:674–680. doi:10.1016/j.electacta.2011.10.015
39. Wu N, Cao Q, Wang X, Li X, Deng H (2011) A novel high-performance gel polymer electrolyte membrane basing on electrospinning technique for lithium rechargeable batteries. J Power Sources 196(20):8638–8643. doi:10.1016/j.jpowsour.2011.04.062
40. Wu N, Jing B, Cao Q, Wang X, Kuang H, Wang Q (2012) A novel electrospun TPU/PVdF porous fibrous polymer electrolyte for lithium ion batteries. J Appl Polym Sci 125(4): 2556–2563. doi:10.1002/app.36523
41. Prasanth R, Aravindan V, Srinivasan M (2012) Novel polymer electrolyte based on cob-web electrospun multi component polymer blend of polyacrylonitrile/poly(methyl methacrylate)/polystyrene for lithium ion batteries-Preparation and electrochemical characterization. J Power Sources 202:299–307. doi:10.1016/j.jpowsour.2011.11.057
42. Bansal D, Meyer B, Salomon M (2008) Gelled membranes for Li and Li-ion batteries prepared by electrospinning. J Power Sources 178(2):848–851. doi:10.1016/j.jpowsour.2007.07.070

43. Nah C, Jeong K-U, Lee Y-S, Lee S-H, Kader MMA, Lee HK, Ahn J-H (2013) Polymer electrolyte membranes composed of an electrospun poly(vinylidene fluoride) fibrous mat in a poly(4-vinylpyridine) matrix. Polym Int 62(3):375–381. doi:10.1002/pi.4311
44. Lalia BS, Samad YA, Hashaikeh R (2013) Nanocrystalline cellulose-reinforced composite mats for lithium-ion batteries: electrochemical and thermomechanical performance. J Solid State Electrochem 17(3):575–581. doi:10.1007/s10008-012-1894-1
45. Lalia BS, Samad YA, Hashaikeh R (2012) Nanocrystalline-cellulose-reinforced poly(vinylidenefluoride-co-hexafluoropropylene) nanocomposite films as a separator for lithium ion batteries. J Appl Polym Sci 126:E441–E447. doi:10.1002/app.36783
46. Zhang J, Liu Z, Kong Q, Zhang C, Pang S, Yue L, Wang X, Yao J, Cui G (2012) Renewable and superior thermal-resistant cellulose-based composite nonwoven as lithium-ion battery separator. ACS Appl Mater Interfaces 5(1):128–134. doi:10.1021/am302290n
47. Ding Y, Zhang P, Long Z, Jiang Y, Xu F, Di W (2009) The ionic conductivity and mechanical property of electrospun P(VdF-HFP)/PMMA membranes for lithium ion batteries. J Membr Sci 329(1–2):56–59. doi:10.1016/j.memsci.2008.12.024
48. Jung H-R, Ju D-H, Lee W-J, Zhang X, Kotek R (2009) Electrospun hydrophilic fumed silica/polyacrylonitrile nanofiber-based composite electrolyte membranes. Electrochim Acta 54(13):3630–3637. doi:10.1016/j.electacta.2009.01.039
49. Cui W-W, Tang D-Y, Gong Z-L (2013) Electrospun poly(vinylidene fluoride)/poly(methyl methacrylate) grafted TiO2 composite nanofibrous membrane as polymer electrolyte for lithium-ion batteries. J Power Sources 223:206–213. doi:10.1016/j.jpowsour.2012.09.049
50. Ding Y, Zhang P, Long Z, Jiang Y, Xu F, Di W (2008) Preparation of PVdF-based electrospun membranes and their application as separators. Sci Technol Adv Mater 9(1):015005. doi:10.1088/1468-6996/9/1/015005
51. Liang Y, Ji L, Guo B, Lin Z, Yao Y, Li Y, Alcoutlabi M, Qiu Y, Zhang X (2011) Preparation and electrochemical characterization of ionic-conducting lithium lanthanum titanate oxide/polyacrylonitrile submicron composite fiber-based lithium-ion battery separators. J Power Sources 196(1):436–441. doi:10.1016/j.jpowsour.2010.06.088
52. Liang Y, Lin Z, Qiu Y, Zhang X (2011) Fabrication and characterization of LATP/PAN composite fiber-based lithium-ion battery separators. Electrochim Acta 56(18):6474–6480. doi:10.1016/j.electacta.2011.05.007
53. Rhoo H-J, Kim H-T, Park J-K, Hwang T-S (1997) Ionic conduction in plasticized PVCPMMA blend polymer electrolytes. Electrochim Acta 42(10):1571–1579, http://dx.doi.org/10.1016/S0013-4686(96)00318-0
54. Tatsuma T, Taguchi M, Oyama N (2001) Inhibition effect of covalently cross-linked gel electrolytes on lithium dendrite formation. Electrochim Acta 46(8):1201–1205, http://dx.doi.org/10.1016/S0013-4686(00)00706-4
55. Gray FM, MacCallum JR, Vincent CA (1986) Poly(ethylene oxide) – LiCF3SO3 – polystyrene electrolyte systems. Solid State Ion 18–19(Part 1 (0)):282–286, http://dx.doi.org/10.1016/0167-2738(86)90127-X
56. Tatsuma T, Taguchi M, Iwaku M, Sotomura T, Oyama N (1999) Inhibition effects of polyacrylonitrile gel electrolytes on lithium dendrite formation. J Electroanal Chem 472(2):142–146, http://dx.doi.org/10.1016/S0022-0728(99)00294-6
57. Jiang Z, Carroll B, Abraham KM (1997) Studies of some poly(vinylidene fluoride) electrolytes. Electrochim Acta 42(17):2667–2677, http://dx.doi.org/10.1016/S0013-4686(97)00005-4
58. Lu P, Hsieh Y-L (2010) Preparation and properties of cellulose nanocrystals: rods, spheres, and network. Carbohydr Polym 82(2):329–336, http://dx.doi.org/10.1016/j.carbpol.2010.04.073
59. Zimmermann T, Pöhler E, Geiger T (2004) Cellulose fibrils for polymer reinforcement. Adv Eng Mater 6(9):754–761. doi:10.1002/adem.200400097
60. Ji G-L, Xu Y-Y, Zhu B-K, Zhu L-P (2010) PVDF-HFP membrane prepared via TIPS process as the matrix of gel electrolyte for lithium ion battery. J Macromol Sci Part B 50(2):275–290. doi:10.1080/00222341003651429
61. Lee Y-S, Jeong YB, Kim D-W (2010) Cycling performance of lithium-ion batteries assembled with a hybrid composite membrane prepared by an electrospinning method. J Power Sources 195(18):6197–6201. doi:10.1016/j.jpowsour.2009.11.012

62. Liao YH, Rao MM, Li WS, Yang LT, Zhu BK, Xu R, Fu CH (2010) Fumed silica-doped poly(butyl methacrylate-styrene)-based gel polymer electrolyte for lithium ion battery. J Membr Sci 352(1–2):95–99. doi:10.1016/j.memsci.2010.01.064
63. Xiao W-J, Li Z-H, Huang Z-B, Tan S-T (2007) Preparation of the TiO2/P(VdF-HFP) hybrid fiber microporous film and its electrochemical performance. Acta Chim Sin 65(19):2097–2102
64. Han L, Lu C, Chen P, Qi W, Yu Q (2012) The effects of in situ generated titanium dioxide on the mechanical and electrochemical properties of electrospun polyvinylidene fluoride separator for lithium-ion battery. Acta Polym Sin 11:1319–1325. doi:10.3724/sp.j.1105.2012.12053
65. Yanhuai D, Ping Z, Yong J (2007) Preparation of PVDF/TiO_2 electrospun membranes and its application as separators for lithium-ion batteries. J Fudan Univ Nat Sci 46(5):822, doi:0427-7104(2007)46:5 < 822:popt2e > 2.0.tx;2-#
66. Alcoutlabi M, Lee H, Watson JV, Zhang X (2013) Preparation and properties of nanofiber-coated composite membranes as battery separators via electrospinning. J Mater Sci 48(6):2690–2700. doi:10.1007/s10853-012-7064-0
67. Lee H, Alcoutlabi M, Watson JV, Zhang X (2013) Electrospun nanofiber-coated separator membranes for lithium-ion rechargeable batteries. J Appl Polym Sci. doi:10.1002/app.38894
68. Jeong H-S, Choi E-S, Lee SY (2012) Composition ratio-dependent structural evolution of SiO_2/poly(vinylidene fluoride-hexafluoropropylene)-coated poly(ethylene terephthalate) nonwoven composite separators for lithium-ion batteries. Electrochim Acta 86:317–322. doi:10.1016/j.electacta.2012.03.126
69. Cho T-H, Tanaka M, Ohnishi H, Kondo Y, Yoshikazu M, Nakamura T, Sakai T (2010) Composite nonwoven separator for lithium-ion battery: development and characterization. J Power Sources 195(13):4272–4277, http://dx.doi.org/10.1016/j.jpowsour.2010.01.018

Chapter 5
Electrospun Oxide Nanofibers for Dye-Sensitized Solar Cells

Qinghong Zhang, Yichuan Rui, and Li Zhang

Abstract The Sun continuously irradiates large amount of energy to the Earth, and the power per second irradiation covers thousands of times over the human consumption per year, without concerning about depletion and pollution. It is urgent to develop cheap and efficient devices for turning light into electricity. Unlike silicon-based solar cells which demand highly purified sources and complicated fabrication process, dye-sensitized solar cells owning the merits of low cost, simple manufacturing procedure, and high theoretical energy conversion efficiency have become the promising candidates for the next-generation solar cells (DSSCs). In the past few years, electrospun nanofibers have been introduced into dye-sensitized solar cells. For example, one-dimensional (1D) electrospun TiO_2 and ZnO nanofibers have been widely investigated to be used as photoanode materials in DSSCs, which exhibit more excellent electron transport properties compared with conventional nanoparticle systems with disordered geometrical structures, as well as the potential of large-scale production. Electrospinning method can also be used to prepare polymer gel electrolytes to replace the traditional liquid electrolytes, solving the questions about long-term stability of DSSCs. In this chapter, we present a review on the application of electrospinning for the design and fabrication of architectured, nanofibrous materials for DSSCs, including photoanodes, counter electrodes, and electrolytes.

Q. Zhang (✉) • Y. Rui • L. Zhang
State Key Laboratory for Modification of Chemical Fibers and Polymer Materials, Donghua University, Shanghai 201620, China
e-mail: zhangqh@dhu.edu.cn

B. Ding and J. Yu (eds.), *Electrospun Nanofibers for Energy and Environmental Applications*, Nanostructure Science and Technology,
DOI 10.1007/978-3-642-54160-5_5,

5.1 Introduction

Developing green and renewable energy resources has become an urgent priority for the world with the accelerated depletion of the Earth's fossil fuels. Solar energy which is pollution free and abundant radiating all regions of the globe has been regarded as one of the perfect energy resources. A lot of efforts have been invested in developing highly efficient solar energy conversion technologies, and the most prospective approach is fabricating solar cells. Solar cells are devices converting solar energy into electricity, including silicon-based photovoltaic cells, thin film solar cells, polymer solar cells, and dye-sensitized solar cells (DSSCs). DSSCs, which own the merits of low cost, simple manufacture procedure, and high light-to-electrical energy conversion efficiency, have been attracting global attentions since the pioneering work of O'Regan and Grätzel in 1991 [1–4]. Those cells exhibit impressive high energy efficiency of up to 13 % even if they are based on cheap starting materials and simple techniques [5].

The typical dye-sensitized solar cell contains broadly five components: (1) a mechanical support coated with transparent conductive oxides; (2) the semiconductor porous film composed of oxide, and the most widely researched material is TiO_2; (3) a sensitizer adsorbed onto the surface of the semiconductor; (4) an electrolyte containing a redox mediator; and (5) a counter electrode capable of regenerating the redox mediator like platine [6–8].

The light to electron conversion process in dye sensitization of DSSCs differs fundamentally from the p-n junction-based silicon solar cells. Figure 5.1 shows the

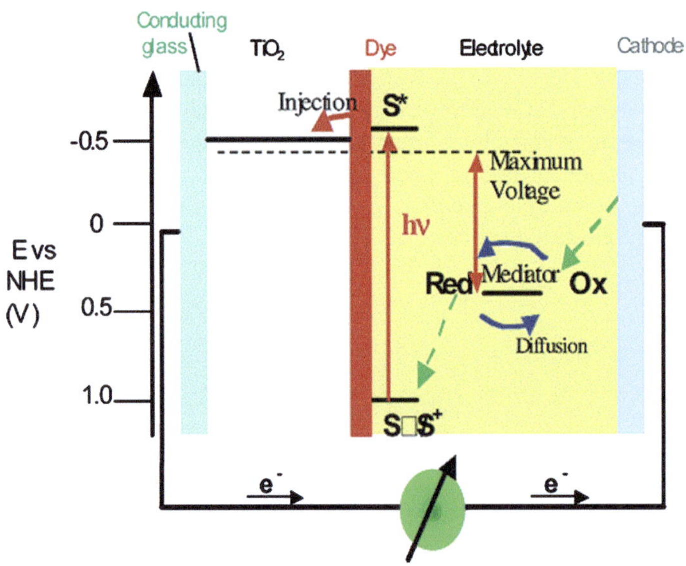

Fig. 5.1 Principle of operation of the dye-sensitized solar cells (Reprinted with the permission from Grätzel [9]. Copyright 2005 American Chemical Society)

operating principles of the dye-sensitized solar cells. The first step is the absorption of a photon by the sensitizer S (Eq. 5.1), leading to the excited sensitizer S* which injects an electron into the conduction band of the semiconductor, leaving the sensitizer in the oxidized state S^+ (Eq. 5.2). The injected electron flows through the semiconductor network to arrive at the back contact and then through the external load to the counter electrode to reduce the redox mediator (Eq. 5.3) which in turn regenerates the sensitizer (Eq. 5.4). This completes the circuit.

Under illumination, the device constitutes a regenerative and stable photovoltaic energy conversion system [6, 7].

$$S_{(\text{adsorbed})} + h\nu \rightarrow S^*_{(\text{adsorbed})} \tag{5.1}$$

$$S^*_{(\text{adsorbed})} \rightarrow S^+_{(\text{adsorbed})} + e^-_{(\text{injected})} \tag{5.2}$$

$$I_3^- + 2e^-_{(\text{cathode})} \rightarrow 3I^-_{(\text{cathode})} \tag{5.3}$$

$$2S^+_{(\text{adsorbed})} + 3I^- \rightarrow 2S_{(\text{adsorbed})} + I_3^- \tag{5.4}$$

Some undesirable reactions resulting in losses in the cell efficiency occur. They are the recombination of the injected electrons either with oxidized sensitizer (Eq. 5.5) or with the oxidized redox couple at the TiO_2 surface (Eq. 5.6).

$$S^+_{(\text{adsorbed})} + e^-_{(\text{titania})} \rightarrow S_{(\text{adsorbed})} \tag{5.5}$$

$$I_3^- + 2e^-_{(\text{titania})} \rightarrow 3I^-_{(\text{anode})} \tag{5.6}$$

Photoexcitation of the sensitizer (S) is followed by electron injection into the conduction band of a semiconductor oxide film. The dye molecule is regenerated by the redox system, which itself is regenerated at the counter electrode by electrons passing through the load. Potentials are referred to the normal hydrogen electrode (NHE) [9].

The photovoltaic performance of the DSSCs can be evaluated in terms of fill factor (FF) and cell efficiency (η) expressed as

$$FF = \frac{V_{\text{max}} J_{\text{max}}}{V_{\text{OC}} J_{\text{SC}}}$$

$$\eta = \frac{V_{\text{OC}} J_{\text{SC}} FF}{P_{\text{in}}} \times 100\,\%$$

where J_{sc} is the short-circuit current density (mA cm^{-2}), V_{oc} is the open-circuit voltage (V), FF is the fill factor, and P_{in} is the incident light power. J_{max} and V_{max} are defined as current and voltage at which maximum power output is obtained.

The *FF* reflects electrical and electrochemical losses occurring during operation of the DSSCs which is mainly dependent on the series resistance of cells, whose value is usually between 0 and 1.

The V_{oc} is determined by the difference between the quasi-Fermi level of TiO_2 and the redox potential of I_3^- ions. The highest theoretical V_{oc} of DSSCs is 0.95 V (the difference between the flat-band potential of TiO_2(−0.7 V) and I^-/I_3^- redox potential (+0.25 V)) [10]. The V_{oc} is also affected largely by the back electron transfer reaction and a lower recombination rate can lead to a higher V_{oc}.

The J_{sc} is a key parameter of DSSCs, determined by the amount of the adsorbed dye molecules, light harvesting of the photoanode, and the diffusion and collection of the generated photoelectrons.

The incident monochromatic photon-to-current conversion efficiency (IPCE), sometimes referred to also as the "external quantum efficiency" (EQE), is another fundamental measurement of the performance of a device. The IPCE value corresponds to the photocurrent density generated in the external circuit under monochromatic illumination of the DSSC divided by the photon flux that strikes the cell [6]. From such an experiment, the IPCE as a function of wavelength can be calculated from the following formula:

$$\mathrm{IPCE}\,(\lambda) = 1240 \frac{J_{sc}\,(\lambda)}{\lambda P_{in}\,(\lambda)}$$

IPCE values provide practical information about the monochromatic quantum efficiencies of a solar cell.

5.2 Photoanodes Contained Electrospun Nanofibers

5.2.1 The Superiorities of DSSC Photoanodes Made by Nanofibers

DSSC photoanodes are generally with the structure of a film of metal oxide on the conductive substrates and the dye sensitizer adsorbed on the surface of metal oxide [11]. The dye in DSSCs generates excitons (bound electron–hole pairs) upon absorption of photons and undergoes dissociation to release them as free electrons and holes. The free electrons are injected into the metal oxide (photoanodes) and transported for collection at the electrode. Electrons form a loop path, so generating the photocurrent. The more the electrons collected, the higher the photocurrent generated and the efficiency of the DSSCs obtained. Researchers have used spin coating [12, 13], screen printing [14], doctor blading [15], and chemical vapor deposition [16] to prepare titania films as a photoelectrode.

The DSSC photoanodes are very important for the optimal light-to-electricity conversion efficiencies. Firstly, charge transport must be enhanced while minimizing charge recombination [17], which can be achieved by controlling the morphology, surface area, porosity, and crystallite size of the photoanode material since these properties could strongly affect the charge transport and recombination processes. For instance, electron transport in photoanode materials such as TiO_2 or ZnO was closely related to the surface states, particle morphology, and interparticle connectivity. Recently, one-dimensional (1D) nanostructured materials such as nanotubes, nanorods, nanobelts, nanowires, and nanofibers have been widely investigated to be used as photoanode materials in DSSC [18–21]. 1D nanofibers exhibit more excellent electron transport properties compared with conventional nanoparticle systems with disordered geometrical structures and interfacial interference in electron transport [22]. It is believed that one-dimensional nanostructures present lower grain boundaries due to their interconnectivity, fine crystallite size, and high surface areas, so electron transport is faster in such structures [23]. The photogenerated electrons could be rapidly transferred to the surface of the photoanode materials and collected by the device, so less recombination of photogenerated electron–holes occurred. The DSSC photoanodes made by nanofibers could maximize the light-harvesting properties of the DSSC device. As it is well known, the dye-sensitized photoanode materials must be excited by the sunlight to generate electrons; for the same photoanode materials, the higher the light utilization, the more electrons generated. 1D inorganic nanofibers with a large surface area (about 50 m^2g^{-1}) and high porosity could absorb the maximum amount of dye sensitizers [24], facilitating the transport of electrons and light utilization. Moreover, the nanofibers behave like a fluffy structure which allows better pore-filling with the electrolyte compared to conventional electrodes made by dense nanoparticles. Such infiltration of a viscous electrolyte into a nanofiber network improves the contact with the inorganic semiconductor and helps the regeneration of oxidized dyes, thus leading to enhanced energy conversion efficiency [25]. However, the photoanode materials should not be too fluffy, or the transportation of photogenerated electrons may be hindered or broken off. In order to further increase the dye loading and the overall efficiency of the solar cell, the electrospun nanofibers have also been shortened into nanorods (by mechanical pressure, grinding, or ultrasonic bath) and DSSCs prepared by using these rods alone or in combination with nanofibers and/or nanoparticles.

5.2.2 TiO_2- and ZnO-Based Nanofiber Photoanodes

Kim et al. prepared TiO_2 nanorods by electrospinning from a solution mixture of titanium *n*-propoxide and poly(vinyl acetate) in dimethyl formamide, as shown in Fig. 5.2 [26]. The as-spun nanorods were collected by the fluorine-doped tin oxide (FTO)-coated glass substrate at a distance of 10 cm. The efficiency of the DSSCs was enhanced by optimizing the nanorod morphology to facilitate charge transport. They demonstrated that the TiO_2 nanorods had an intrinsically higher sensitizer loading capability than conventional TiO_2 nanoparticles and

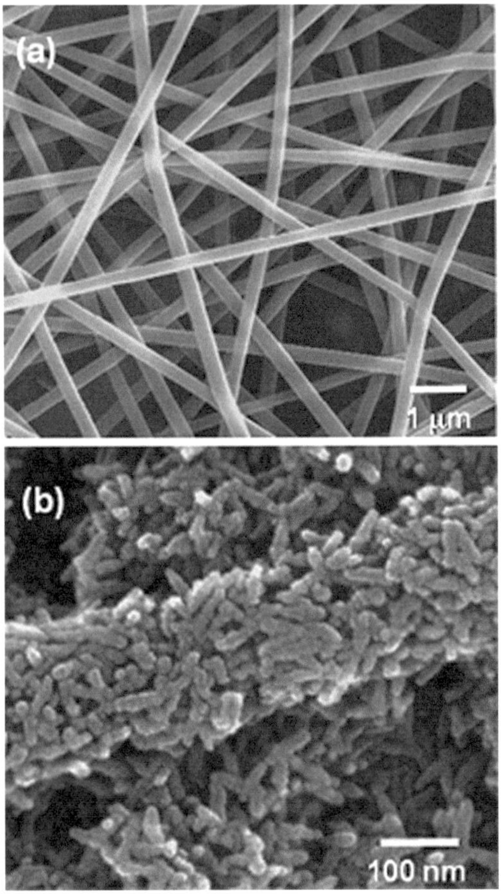

Fig. 5.2 SEM images of electrospun nanofibers: as-spun TiO_2 PVAc nanofibers (**a**) and TiO_2 nanorods after calcination at 400 °C (**b**) (Reprinted with permission from Lee et al. [26]. Copyright 2009 American Chemical Society)

had much slower recombination lifetime compared to conventional nanoparticles. Long electron lifetime in nanorod electrode contributes to the enhanced effective photocarrier collection as well as the conversion efficiency. The electron transport behavior of nanorod photoelectrodes was further improved by $TiCl_4$ post-treatment. The post-treatment reduces the pore volume of nanorod photoelectrodes while improving inter-rod connectivity and enhancing electron diffusion. The electron diffusion coefficient of post-treated nanorod was 51 % higher than that of an untreated one, leading to a charge collection efficiency that was 19 % higher at an incident photon flux of 8.1×10^{16} cm^{-2} s^{-1}. Finally, the efficiency of nanorod-based DSSCs was optimized at a photoelectrode thickness of 14 μm to be achieved of 9.52 %.

Nb-doped anatase TiO_2 nanofibers have been recently explored as photoanode materials for dye-sensitized solar cells [27]. The defects derived from doping play an important role in their photovoltaic properties. The electron diffusion and mobility improved with doping; however, the BET surface area decreased and the charge recombination increased. Balakrishnan et al. studied the electrical and optical properties of electrospun TiO_2–graphene composite nanofibers and its application as DSSC photoanodes [28]. The surface porosity of these samples was found to be in the range of 45–48 %. They used Raman spectroscopy to verify the presence of graphene in TiO_2 nanofibers. Photoluminescence spectroscopy showed excitonic intensity to be lower in graphene–TiO_2 samples indicating that the recombination of photoinduced electrons and holes in TiO_2 can be effectively inhibited in the composite nanofibers. Conductivity measurements showed the mean specific conductance values obtained for TiO_2–graphene composites to be about two times higher values than that of the electrospun TiO_2 fibers. Assembling these TiO_2–graphene fiber composites as photoanodes in dye-sensitized solar cells, an efficiency of 7.6 % was attained.

ZnO has been investigated as an alternative to TiO_2 photoanode material for DSSCs, as it possesses similar electron injection process and bandgap as that of TiO_2 (3.3 eV vs. 3.2 eV for anatase) [29]. In particular one-dimensional ZnO nanomaterials, which show increased conductivity path, have attracted great attention [30]. In spite of the higher charge transport and electron mobility, 1D ZnO has poor chemical stability in acidic dye solutions and low surface area compared to titania nanoparticulate films (and consequently adsorbs a minor amount of dye) [20], resulting in energy conversion efficiencies which are always lower than TiO_2-based ones (6.58 % for nanoparticle-based DSSCs [31] and around 1 % for 1D ZnO-based DSSCs [32]). Many strategies have been applied in order to overcome these drawbacks and again to improve the fiber adhesion on the substrate and consequently to enhance the overall performance of the systems. The use of electrospun ZnO nanofibers in dye-sensitized solar cells was first published in 2007 [22]. To enhance the electrical contact between the nanofibers and the electrode, hot-pressing of the ZnO/PVAc precursor nanofibers was performed before their calcination. This treatment resulted in the formation of dense networks of twisted nanofibers, which further improved the direct conduction paths for efficient electron collection as well as the porosity and surface area. Nevertheless, the efficiency of these systems was still rather low (1.34 %). Ramakrishna et al. treated the as-spun nanofibers using $Zn(OAc)_2$ before calcining them in a high temperature, and a higher energy conversion efficiency of 3.02 % was achieved [33]. Electrochemical impedance spectroscopy results showed the effective suppression of the back reaction between photoelectrons (in the ZnO conduction band) and I_3^- (in the electrolyte) in the $Zn(OAc)_2$-treated systems, leading to the observed improved performance. In order to further enhance their optical and electrical properties, ZnO nanofibers have also been doped with Al [34], which enhances their adhesion to fluorine-doped tin oxide (FTO). This was achieved by introducing an Al-doped ZnO seed layer prepared by electrospinning before the deposition of the Al-doped ZnO electrospun nanofibers. It was shown that nanofibers did not crack nor peel away after the calcination step, as this layer facilitated the release of interfacial tensile stress during calcinations and improved the interfacial adhesion to the conducting substrate. As a result, the efficiency improved when compared to DSSCs without seed layers (0.54–0.55 % vs. 0.03–0.04 %).

In addition to TiO_2 and ZnO nanofibers, 1D nanostructured networks such as carbon nanotubes (CNTs) and silver nanowires also revealed good performances as potential alternatives to be made into photoanode in DSSCs [35, 36]. But they also have inadequacies as CNTs have high junction resistances and limited aspect ratios and the price of silver nanowires is too high. Cu has a high electrical conductivity as Ag. CuO nanofibers were prepared by electrospinning and subsequent calcination of copper acetate/PVA electrospun fibers. Then the CuO nanofibers were subsequently reduced in an H_2 atmosphere [37]. The power efficiency of organic solar cells using copper nanofiber electrode materials was 3.0 %. Cu nanofibers can be easily prepared in a large scale by electrospinning method. Cu is cheap compared with Ag, and CuO has ultrahigh aspect ratios (up to 100,000), very low junction resistances, and great flexibility. The CNTs, Ag, and Cu themselves are not semiconductors and do not work well as the photoanodes of DSSCs; adding some of these highly conductive nanofibers may significantly improve the conductivity and thus also enhance the conversion efficiency of DSSCs.

5.2.3 *Nanofiber–Nanoparticle Composites*

Studies of electron/charge transport in TiO_2 nanofibers, nanowires, and nanorods have been carried out by several research groups, and the results indicate that the electrospun nanofibers and nanorods exhibit high electron diffusion coefficients [38, 39]. Archana et al. found that the electron diffusion coefficient of TiO_2 nanowires prepared by sonication of electrospun nanofibers was orders of magnitude larger than that of TiO_2 nanoparticles [38]. Saji et al. have recently reported that the incorporation of 10 % nanorods into a nanoparticle matrix leads to reduced charge transport resistance [39]. However, it is still controversial to use a composite made of nanoparticles and randomly oriented nanowires, nanofibers, or nanorods to increase the charge transport rate. Wu et al. reported that the effective electron diffusion coefficients in DSSCs based on nanowires only or nanoparticle–nanowire composites, where the nanowires were randomly (non-vertically) oriented, were smaller than those in DSSCs based on TiO_2 nanoparticles alone [40]. Mukherjee et al. studied the electron transport in TiO_2 nanofibers made using a similar method to that used in this study. Their results showed that the electron diffusion length of the TiO_2 nanofibers was similar to that of the TiO_2 nanoparticles but with a lower electron lifetime due to the presence of a large number of surface defects [41]. Tan et al. reported that the use of a photoanode made of a nanowire–nanoparticle composite with small amounts of nanowire ($<$20 wt%) did not contribute to the improvement of electron transport since no nanowire networks were formed [42]. Fujihara et al. reported a solar cell with two TiO_2 layers: a ground electrospun TiO_2 nanoparticle layer on a glass plate and then a TiO_2 nanorod layer [43]. It was found that the devices with a combination of electrospun nanorods and nanoparticles showed improved conversion efficiency over the devices entirely made of nanorods. Chuangchote et al. also suggested that nanofiber-modified nanoparticles are very promising materials for the electrode

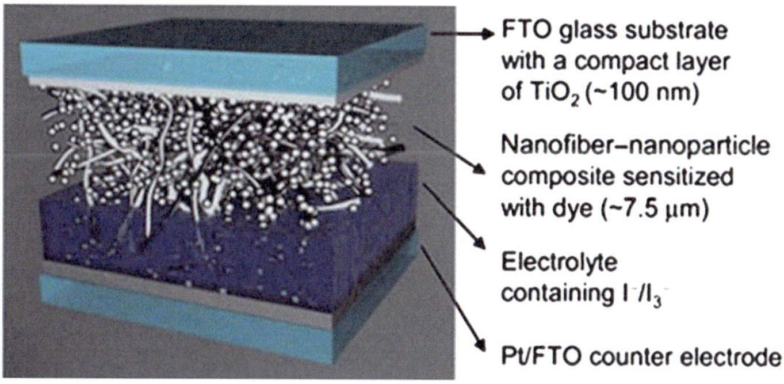

Fig. 5.3 Schematic configuration of a DSSC with TiO_2 nanofiber–nanoparticle composite acting as an innovative photoanode (Reproduced from Joshi et al. [45] by permission of The Royal Society of Chemistry)

structure of DSSCs [44]. In their work, TiO_2 nanofibers were fabricated directly onto thick nanoparticle electrodes by using electrospinning and solgel techniques. The DSSCs comprised of a nanoparticle/nanofiber electrode showed an incident photon-to-current conversion efficiency (IPCE) of 85 % at 540 nm with conversion efficiencies of 8.14 and 10.3 % for areas of 0.25 and 0.052 cm^2, respectively. It is noted that the conversion efficiency of 10.3 % is the highest reported efficiency for electrospun nanofiber DSSCs. Joshi et al. prepared photoanode materials made of electrospun TiO_2 nanofibers and conventional TiO_2 nanoparticles, as shown in Fig. 5.3 [45]. With the same fabrication condition and film thickness, the DSSCs made from the composite with 15 % nanofibers and 85 % nanoparticles by weight demonstrated 44 % higher device efficiency than those made from TiO_2 nanoparticles alone. The photoanode materials made of electrospun TiO_2 nanofibers and conventional TiO_2 nanoparticles are innovative, which noticeably improves the harvesting of light without substantially sacrificing the attachment (uptake) of dye molecules for convenient fabrication of dye-sensitized solar cells with significantly improved efficiency. This study provides new guidance for the development of inexpensive DSSCs using the thinner film of photoanode and the less amount of dye via the incorporation of electrospun TiO_2 nanofibers.

5.2.4 Surface Area and Porosity

It is obvious that the photoanode materials with a high specific surface area and porosity can help to expand the contact area between the dye-adsorbed surface and solar radiation [46–49]. Therefore, in order to further increase the efficiency of the solar cell, it seems to be necessary to develop TiO_2 structure that possesses favorable characteristics such as faster electron transport rate, high porosity, and larger surface area compared to nanoparticles [50–52]. Hwang et al. used a

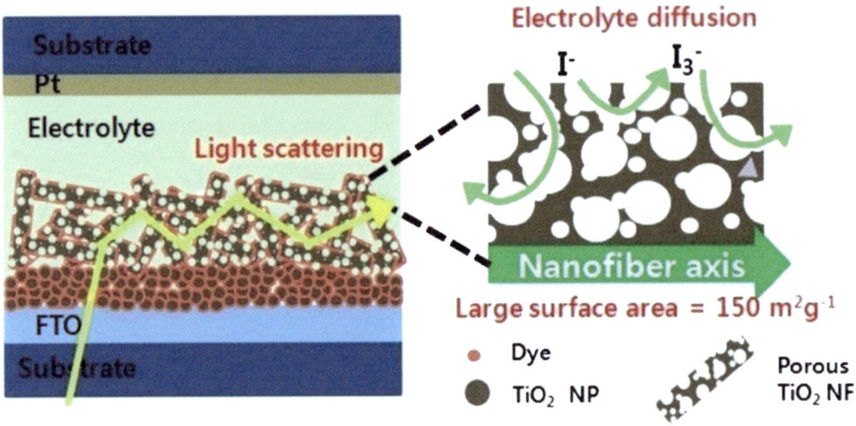

Fig. 5.4 Schematic illustration of the 15–20 nm TiO_2 nanoparticle underlayer, the hierarchical particle TiO_2 nanofibers (HP TiO_2 NFs) scattering overlayer, and the magnified structure of HP TiO_2 NFs (Reprinted with the permission from Hwang et al. [54]. Copyright 2012 American Chemical Society)

modified electrospinning process to create fibers with high specific surface area of 103.3 m^2g^{-1} and maximum porosity of 80.5 %; the nanofibers were fabricated as photoanode and a maximum efficiency of 4.6 % was obtained, which was significantly higher than that of typical TiO_2 nanofibers [53]. Jang et al. designed an architecture of hierarchically porous TiO_2 nanofibers for dye-sensitized solar cell photoanode [54], and their schematic diagram was shown in Fig. 5.4. Hierarchically porous TiO_2 nanofibers were fabricated using a simple electrospinning and etching process with TiO_2/SiO_2 composite nanofibers for high-efficiency dye-sensitized solar cells. TiO_2 nanofibers with different pore sizes (small, large, and hierarchical) were prepared using SiO_2 nanoparticles of various sizes. They claimed that the surface area of the hierarchically porous TiO_2 nanofibers was nine times higher than that of bare TiO_2 nanofibers, which provided sufficient dye adsorption for light harvesting as well as efficient paths for electrolyte contact. They also found that the one-dimensional structure provided efficient light scattering and fast electron transport. As a result, DSSCs exhibited an enhanced current density (J_{sc}) of 16.3 mA cm^{-2} and high photoconversion efficiency (η) of 8.5 %, greater than those of conventional photoelectrodes made of TiO_2 nanoparticles (J_{sc} of 12.0 mA cm^{-2} and η of 6.0 %).

5.2.5 *Aligned Nanofibers to Improve Light Harvesting and Charge Collection*

A further strategy to improve light harvesting and charge collection can be the use of vertical arrays of TiO_2 nanowires. As shown in Fig. 5.5, the electron

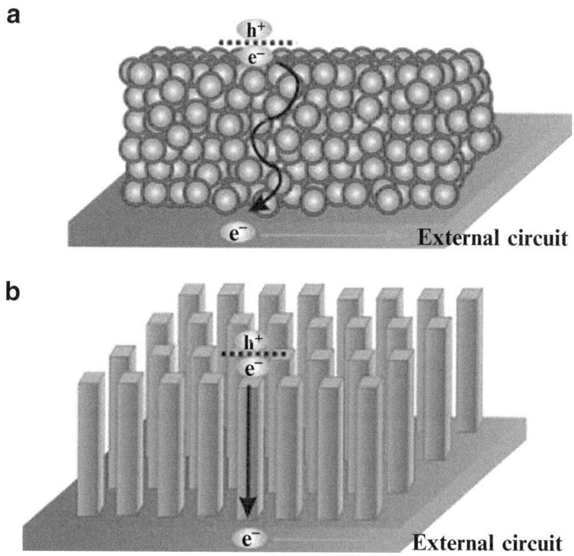

Fig. 5.5 Schematic diagrams of electron pathways in films composed of disordered nanoparticles and oriented nanoarrays (Reprinted from Mu et al. [55]. Copyright 2011 with permission from Elsevier)

transfer rate to the FTO was higher in the aligned nanofibers than that in disordered nanoparticles [55]. The fabrication of such vertically aligned materials by electrospinning followed by a top-down process is very recent [56]. For example, as shown in Fig. 5.6a, Shim et al. demonstrated hybrid solar cells fashioned of planar-aligned TiO_2 nanofiber architectures such as uniaxially aligned nanofibers (Fig. 5.6b) and multiple layers of cross-aligned nanofiber arrays (Fig. 5.6c) with poly[2-methoxy-5-(20-ethyl-hexyloxy)-1,4-phenylenevinylene] (MEH-PPV) [57]. It was found that the active layer composed of the aligned TiO_2 nanofiber array and MEH-PPV polymer showed better photoluminescence quenching (which should be associated with the charge separation efficiency) under similar light absorption as compared to disordered TiO_2 nanofibers. Therefore, as shown in Fig. 5.6e, the photoinduced current (1.28 mA cm^2) and power conversion efficiency of the cells could be significantly improved by at least 50 % under 1-Sun conditions depending on the degree of aligning fibers [57]. In addition, Yan et al. also reported hybrid solar cells based on ordered electrospun TiO_2 and ZnO nanofiber arrays with conjugated poly(3-hexylthiophene) (P3HT), and they found that the organic–inorganic hybrid cells with three cross-aligned layers showed enhanced photovoltaic performance [58, 59]. Moreover, DSSCs with vertical nanofibers of electrospun TiO_2 as a photoelectrode were recently reported [56]. The aligned nanofibrous TiO_2 ribbons were first produced by electrospinning and then erected to vertical nanofibers (Fig. 5.6d) after post-treatment of rolling up, stacking, and cutting. The conversion efficiency, short-circuit current, and open-circuit voltage of the resultant DSSC were measured as 2.87 %, 5.71 mA cm^{-2}, and 0.782 V, respectively (Fig. 5.6f).

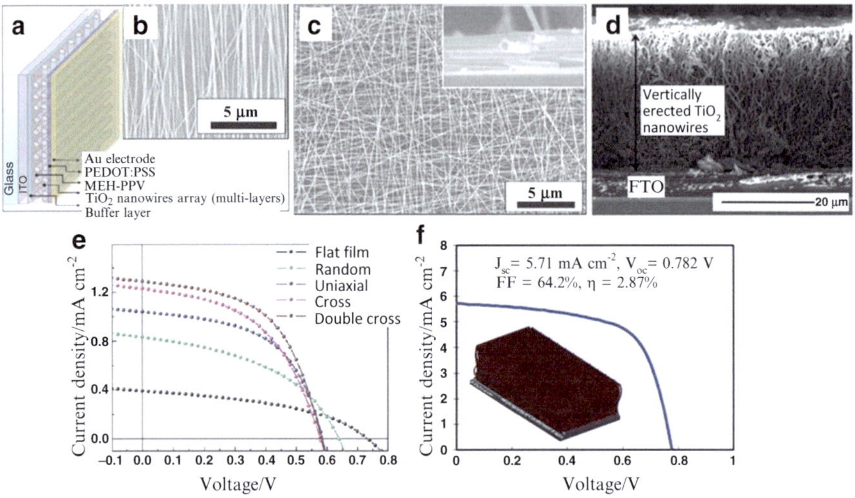

Fig. 5.6 (**a**) Schematic illustration of a hybrid solar cell introduced with aligned TiO_2 or ZnO nanofibers via electrospinning. SEM images of uniaxially aligned TiO_2 nanofibers (**b**), cross-aligned TiO_2 nanofibers (**c**), and vertically erected TiO_2 nanowires (**d**). Current–voltage (J–V) characteristics of (**e**) MEH-PPV: TiO_2 hybrid solar cells depending on different configurations of aligned TiO_2 nanofibers and (**f**) a DSSC based on vertical TiO_2 nanowire array electrode. *Inset* in Fig. 5.6c shows the cross-sectional image of cross-aligned TiO_2 NWs (Reprinted with permission from Krishnamoorthy et al. [56] and Shim et al. [57]. Copyright 2011 The Royal Society of Chemistry and Copyright 2008 American Institute of Physics)

5.2.6 *Morphology of the Photoanode*

Kuang et al. studied the effect of TiO_2 morphology of nanoparticles, nanofibers, hierarchical spheres, and ellipsoid spheres on photovoltaic performance of dye-sensitized solar cells [60]. The DSSC based on the hierarchical anatase TiO_2 sphere photoelectrode shows an overall light-to-electricity conversion efficiency of 9.35 % accompanying a short-circuit current density of 17.94 mA cm^2, an open-circuit voltage of 803 mV, and a fill factor of 0.65, which is much higher than that of nanoparticles (7.37 %), nanofibers (8.15 %), and ellipsoid TiO_2 spheres (7.93 %). Ramakrishna et al. prepared hollow TiO_2 nanofibers with a high surface area of 118 m^{2}g^{-1} [61]. With hollow mesoporous TiO_2 nanofibers as the photoelectrode in dye-sensitized solar cells (DSSC), the solar-to-current conversion efficiency (η) and short-circuit current (J_{sc}) are measured as 5.6 % and 10.38 mA cm^{-2}, respectively, which are higher than those of DSSC made using regular TiO_2 nanofibers under identical conditions (η = 4.2 %, J_{sc} = 8.99 mA cm^{-2}). The improvement in the conversion efficiency is mainly attributed to the higher surface area and mesoporous TiO_2 nanostructure. It facilitates the adsorption of more dye molecules and also promotes the incident photon to electron conversion. Hollow mesoporous TiO_2 nanofibers with close packing of grains and crystals intergrown with each other demonstrate faster electron diffusion and longer electron recombination time than regular TiO_2 nanofibers as well as P25 nanoparticles. Zhang et al. prepared submicrometer-sized rutile TiO_2 nanorod microspheres with diameters of 500–700 nm and with a specific surface area of 63.7 m^2g^{-1} via a salt-assisted hydrothermal method [62]. The DSSC with the scattering overlayer of the special morphological nanorod microspheres exhibited a 39 % increment of cell efficiency (7.32 %) compared with the DSSC without one (5.28 %), and the efficiency was also higher than the DSSC with the same thickness composed of only nanocrystallites (7.14 %). They also prepared ultrafine TiO_2 nanoparticles using the nanofiber as a microsized reactor [63]. The hybrid nanofibers comprised of poly(vinylpyrrolidone) (PVP), and amorphous TiO_2 nanoparticles were prepared via the electrospinning method. When the hybrid nanofibers were treated with water vapor at the temperature of 80 °C or above, the nanofibers acted as microsized reactors in which the amorphous TiO_2 nanoparticles grew and crystallized into ultrafine anatase TiO_2 nanoparticles. The PVP matrix in this microsized reactor could be regarded as a substrate for the reaction and protected the overgrowth of the TiO_2 nanoparticles during the reaction. The ultrafine obtained TiO_2 nanoparticles after dissolving the PVP could be used as the photoanode in DSSCs and high conversion efficiency was expected.

5.2.7 *Scattering Layer*

It has been found that a scattering layer in the photoanode could enhance light harvesting. Such layers could reflect the unabsorbed photons back into the nanowire anode and improve the DSSC efficiency [64]. Wu et al. deposited SnO_2/ZnO nanofibers by electrospinning on the surface of ZnO nanowires and formed a film by methanol vapor treatment followed by high-temperature calcination. J_{sc}, V_{oc}, and FF of the nanofilm/ZnO nanowire DSSCs were all enhanced compared to those of the ZnO nanowire DSSCs. Zhu et al. fabricated the rice grain-shaped TiO_2 by an electrospinning technique as a scattering layer for DSSCs [65]. It is found that the DSSCs based on the rice grain-shaped TiO_2 scattering layer showed a higher efficiency (7.45 %) than the one without a scattering layer (6.44 %). Pan et al. fabricated nest-shaped TiO_2 structures by an electrospinning technique and used them as an effective scattering layer on the top of TiO_2 nanoparticle electrodes in dye-sensitized solar cells [66]. The nest-shaped TiO_2 scattering layer enhanced the photocurrent of DSSCs due to the enhanced light harvesting via the improved light scattering and lower electron transfer resistance. Under one sun illumination (AM 1.5 G, 100 mW cm^2), a high efficiency of 8.02 % was achieved for the cell with a nest-shaped TiO_2 scattering layer, which was an increase of 7.1 % compared to the cell without a scattering layer (7.49 %). We can also cite the use of electrospun materials (e.g., carbon) [67] as templates for the growth of ZnO nanowires. These composites can be used as an anode material for the fabrication of flexible DSSCs.

5.2.8 *Adhesion Problems*

The electrons generated in the TiO_2 or ZnO should transfer to the conductive substrates to form a closed loop. However, one of the issues to be addressed for the electrospun fiber-based DSSCs is the poor adhesion of nanofibrous materials to conductive substrates. It is very crucial to solve this problem to enhance the efficiency of the DSSCs.

Onozuka et al. used a pre-deposited titania layer and a solvent post-treatment to soften the electrospun anatase nanofiber mats to improve their adhesion to the substrate [24]. These treatments improved the photocurrent and the conversion efficiency of the corresponding device, which reached a value of 4.14 %. Mukherjee et al. used TiO_2 pre-deposited film by spin coating and the collected as-spun nanofibers were calcined [41]. They claimed that the electrospun fiber-based DSSCs had a good adhesion to conductive substrates. Chuangchote et al. deposited the TiO_2 electrospun nanofibers on TiO_2 nanoparticles to solve this problem [44]. Such an approach also showed an improvement of the light-harvesting properties and of the overall efficiency (up to 10.3 %) by combining these morphologies. Furthermore, the coexistence of nanofibers or nanorods and nanoparticles enhanced the porosity and thus the penetration of the liquid electrolyte and the amount of the adsorbed dye. Song et al. employed the porous electrospun TiO_2 nanofibers in the quasi-solid-state DSSCs and achieved a photocurrent generation over 90 % of the liquid electrolyte DSSCs [68–70]. To enhance the adhesion and conductivity between the TiO_2 nanofibrous membranes and the substrates, they treated the electrospun TiO_2 membranes with the $TiCl_4$ aqueous solution to implant rutile crystals on the fibers' surface (see Fig. 5.7b). As a result, the photocurrent was found to increase by more than 30 % [69, 71]. Figure 5.7a shows a nanorod-on-nanoparticle configuration developed by Fujihara et al. [43, 72]. Firstly, they mechanically grinded the electrospun TiO_2 nanofibers to yield TiO_2 nanorods. Then they sprayed the TiO_2 nanorods onto a layer of TiO_2 nanoparticles. Figure 5.7b shows a cross section of their resulting electrode. The authors believed this method could potentially solve the poor adhesion problem in DSSCs and therefore achieved an energy conversion efficiency of 5.8 %.

5.3 Counter Electrodes Contained Electrospun Nanofibers

The counter electrode is one of the key components in the DSSCs. The task of the counter electrode (CE) is the reduction of the redox couples used as a mediator in regenerating the sensitizer after electron injection. At present, several different kinds of CEs have been introduced, including platinized transparent CEs, carbon CEs, conductive polymer CEs, and novel compound CEs as carbides and nitrides.

Conventional counter electrodes for DSSCs with iodide/tri-iodide electrolytes can be easily prepared by depositing a thin Pt film onto a transparent conductive oxide (TCO) substrate, usually by thermal deposition of H_2PtCl_6 or vacuum

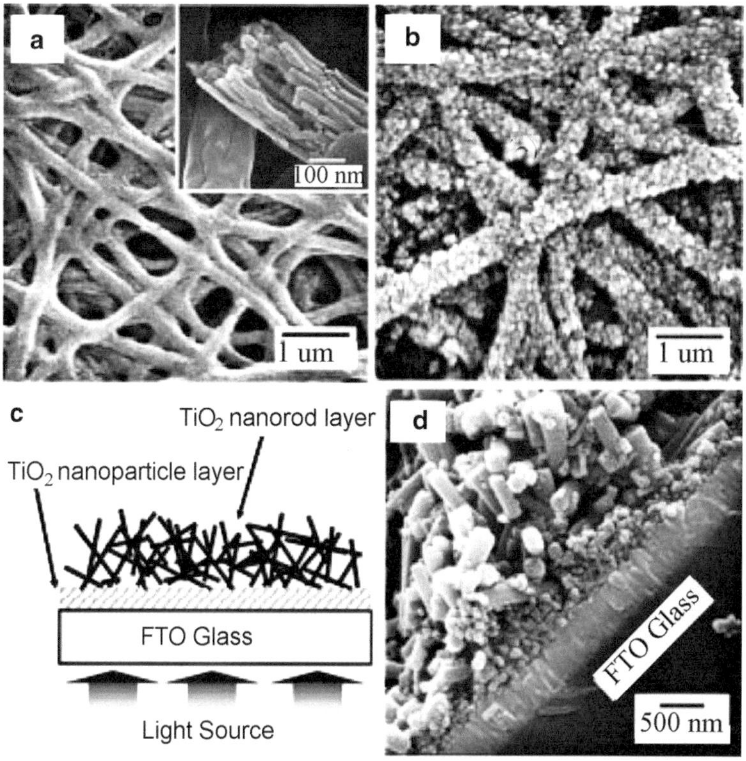

Fig. 5.7 SEM images of anatase TiO_2 electrodes derived from electrospun nanofibers: (**a**) after calcining the as-spun TiO_2/PVAc composite nanofibers (as shown in the inset, each TiO_2 fiber consisted of a bundle of fibrils roughly 20 nm in diameter), (**b**) after treatment with an aqueous $TiCl_4$ solution to deposit an additional layer of rutile to improve the performance, (**C**) schematic of a nanorod-on-nanoparticle electrode prepared by mechanically grinding electrospun TiO_2 nanofibers and then electrospraying the resultant TiO_2 nanorods onto a fluorine-doped tin oxide (FTO) substrate, and (**d**) SEM image of the cross-section of such an electrode

sputtering, due to its high electrocatalytic activity for I_3^- ions reduction. Up to now, Pt is still the most effective catalyst for counter electrodes of DSSCs with conversion efficiencies higher than 11 %. Without Pt, TCO glass is a very poor counter electrode whose charge transfer resistance is as high as $10^6\ \Omega\ cm^2$ in a standard iodide/triiodide electrolyte.

Electrospinning is a simple and versatile nanofabrication method with advantages of low cost, easy process, and the easy formation of films directly on substrates which show the potential for large-scale production; furthermore, it can provide high specific surface area which is beneficial for the reduction of I_3^- ions [73, 74].

To reduce the cost of TCO glass, recently, Lee et al. report a multifunctional platinum nanofiber (PtNF) web produced by electrospinning that can act as a catalyst layer in DSSCs to simultaneously function as a transparent counter electrode, that is, without the presence of an indium-doped tin oxide (ITO) or

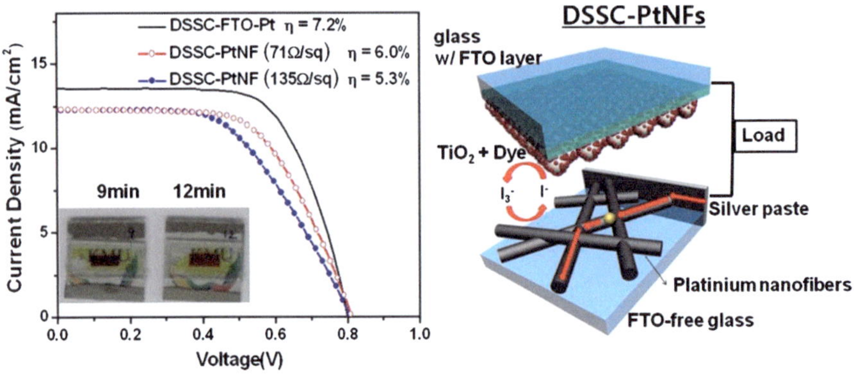

Fig. 5.8 Catalytic, conductive, and transparent platinum nanofiber webs for FTO-free dye-sensitized solar cells (Reprinted with the permission from Kim et al. [75]. Copyright 2013 American Chemical Society)

fluorine-doped tin oxide (FTO) glass. Electrospun PtNFs were straight and had a length of a few micrometers, with a common diameter of 40–70 nm. Each nanofiber was composed of compact, crystalline Pt grains, and they were well fused and highly interconnected, which should be helpful to provide an efficient conductive network for free electron transport and a large surface area for electrocatalytic behavior. When the PtNF web served as a counter electrode in DSSC, a conversation efficiency of 6.0 % was achieved, reaching up to 83 % of the efficiency of the DSSC using a Pt-coated FTO glass as a counter electrode (Fig. 5.8) [75].

Platinum (Pt) has been extensively used as the CE in DSSCs due to its effective catalytic for I_3^- ions. However, as platinum is a noble metal, therefore, low-cost and abundant materials have been developed including carbonaceous materials like graphite, carbon black, carbon nanotubes, and polymers like poly(3,4-ethylenedioxythiophene) (PEDOT) and polystyrene sulfonate (PEDOT-PSS) as potential low-cost replacements for platinum [76–78].

Qiao et al. [79] reported that electrospun carbon nanofibers can be used as a low-cost alternative to Pt counter electrodes in DSSCs. The electrospun carbon nanofibers were relatively uniform with small variations in diameter and an average diameter of approximately 250 nm as shown in Fig. 5.9. The electrospun carbon nanofibers were converted into a paste by adding the POETE binder, followed by grinding and sonication. Although the long carbon nanofibers were broken into micrometer-sized short ones after depositing them as counter electrode onto the FTO, they still show high conductivity of about 164 Sm^{-1}. The charge transfer resistance of the electrospun carbon nanofiber electrode was found to be 0.7 Ω cm^2 which was less than half of that (1.9 Ω cm^2) of the Pt electrode, indicating a high electrocatalytic performance. However, the electrospun carbon nanofiber-based DSSCs showed relatively lower efficiency than that of Pt-based DSSCs because of its higher total series resistance (15.5 vs. 4.8 Ω cm^2) which led to a lower fill factor (FF). In their subsequent research, they soaked the electrospun carbon nanofibers

Fig. 5.9 (**a**) *Top* view SEM image of the electrospun carbon nanofiber (ECN) sheets. (**b**) TEM image of a single typical ECN. SEM image of (**c**) *top* view and (**d**) cross section of ECN counter electrode deposited by doctor blading on an FTO glass substrate (Reprinted with the permission from Joshi et al. [79]. Copyright 2010 American Chemical Society)

into a Pt precursor solution (0.02 M H_2PtCl_6 in anhydrous ethanol) for 6 h; after calcination at 400 °C for 15 min, the electrospun carbon nanofibers were fabricated as a carbon nanofiber/Pt composite counter electrode, and a higher efficiency than that of the conventional Pt cells was achieved [80].

Barakat et al. produced Pd–Co-doped carbon nanofibers by electrospinning of an aqueous colloid composed of Pd NPs, cobalt acetate tetrahydrate, and poly(vinyl alcohol), as Co has the highest activity among the non-noble metal catalysts and Pd is moderately active among the precious metal catalysts. The introduced nanofibers revealed good performance as counter electrode in the dye-sensitized solar cells [81].

Conductive polymers have been widely investigated as promising candidates for the counter electrode materials due to their unique properties, such as relatively high conductivity, good stability, low cost, and easy coating. Conductive polyaniline (PANI) is one of the widely studied conductive polymers. A simple electrospinning method was developed by Peng et al. [82] to directly deposit conductive 10-camphorsulfonic acid (CSA)-doped polyaniline (PANI) blended with polylactic acid (PLA) composite films (PANI CSA–PLA) on conductive substrates.

The average diameter of PANI CSA–PLA nanofibers determined from the SEM images was about 200 nm, with a standard deviation of about 100 nm and length up to several hundred micrometers (Fig. 5.10). The as-electrospun PANI CSA–PLA nanofibers were stacked and bonded with the substrate firmly, resulting in an interconnected nanofibrous network with a thickness of about 2 mm. This network structure was in favor of the infiltration of electrolyte and further promotes the electrocatalytic activity for the reduction of I_3^- ions. The nanofibers were directly deposited on rigid fluorine-doped tin oxide (FTO) and also flexible indium tin oxide-coated polyethylene naphthalate (PEN) substrates to obtain the counter electrodes, which is close to that of sputtered Pt-based DSSCs.

5.4 Electrolytes with Electrospun Nanofibers

Up to now, the most efficient DSSCs are based on nanocrystalline TiO_2 porous film, N719 dye molecules, Pt counter electrode, and liquid electrolytes composed of I^-/I_3^- ion redox couple in volatile organic solvents. Although DSSCs based on liquid electrolytes have already achieved high conversion efficiencies over 11 %, unfortunately, some problems brought by a liquid electrolyte have reduced long-term stability of DSSCs when attempting to put them into practical application. The major drawbacks are the evaporation and leakage of the volatile organic solvents and difficulty in sealing. In order to overcome these problems, many studies have been carried out on alternatives to replace the liquid electrolytes. Solid-state electrolytes are ideal replacements of liquid electrolytes, which show potential of large-scale application for DSSCs by the high-speed roll-to-roll press. Generally, the solid-state electrolyte should primarily conform to the following essential requirements: (1) high ionic conductivity or hole transfer rate to maintain regular regeneration of the dye, (2) good stability, (3) no strong absorption in visible spectrum region, and (4) good physical contact at the TiO_2/dye/electrolyte interface [83–85].

Many groups have focused on the substitution of the liquid electrolyte by solid or gel electrolytes, which could aid in reducing costs and make assembly of dye-sensitized solar cells easier. The development of solid-state electrolytes is generally divided into two main parts: quasi-solid-state electrolyte and solid-state electrolyte.

Solid-state electrolytes include p-type semiconductors (CuI, CuSCN, etc.), polymer electrolytes, or organic hole-transporting materials (HTMs). So far, the most successful organic HTM is 2,2′,7,7′-tetrakis(*N*,*N*- di-4-methoxy-phenylamino)-9,9′-spirobifluorene (spiro-OMeTAD), which leads to an efficiency more than 5 % for solid-state DSSCs with Ru dye-sensitized TiO_2 as photoanode. Although spiro-OMeTAD is the most favorable organic HTM for fabricating solid-state DSSCs, it has some disadvantages such as low hole mobility and high fabrication cost [86]. Polymer electrolytes are solid ionic conductors prepared by the dissolution of salts in a suitable high molar mass polymer containing polyether units. A polymer electrolyte may be defined as a membrane or a similar medium that possesses transport properties similar to liquid ionic media. There are obvious

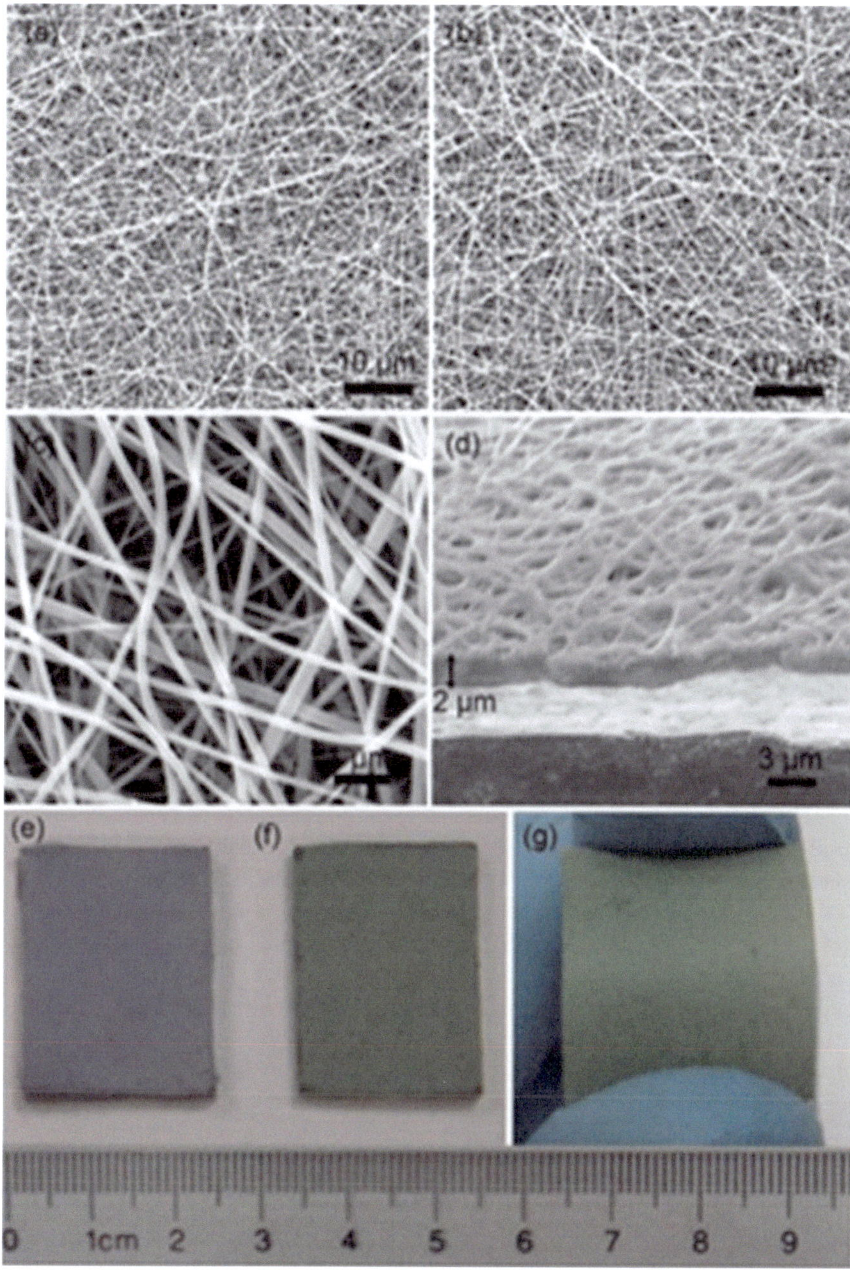

Fig. 5.10 SEM images of the top views (**a**, **b**) of the PANI–PLA and PANI CSA–PLA film on FTO glass, the magnified part (**c**) and cross-sectional view (**d**) of the PANI CSA–PLA nanofiber film, and the photos of the PANI–PLA/FTO (**e**), PANI CSA–PLA/FTO (**f**), and PANI CSA–PLA/PEN (**g**), respectively (Reproduced from Peng et al. [82] by permission of The Royal Society of Chemistry)

advantages with polymer electrolytes such as the absence of internal shorting, leakage of electrolytes, and noncombustible reaction products at the electrolyte surface existing in the liquid electrolytes [87]. Some commonly used polymer electrolytes include poly(ethylene oxide), poly(propylene oxide), poly(acrylonitrile), poly(methyl methacrylate), poly(vinyl chloride), and poly(vinylidene fluoride).

Quasi-solid-state electrolytes include gel polymer electrolytes, ionic liquids, and plastic crystal systems. It is important to distinguish polymer electrolytes and gel polymer electrolytes. In polymer electrolytes, charged cationic or anionic groups are chemically bonded to a polymer chain, while gel polymer electrolytes are solvated by a high dielectric constant solvent and are free to move. In a classical gel electrolyte, polymer and salts are mixed with a solvent, usually having a concentration above 50 wt%, and the role of the polymer is to act as a stiffener for the solvent, creating a three-dimensional network, where cations and anions move freely in the liquid phase [88]. The solid polymer electrolyte includes poly(ethylene oxide) (PEO)-based lithium ion conductors that typically show conductivities of 10^{-8} S cm^{-1}, while the gel polymer electrolytes have semisolid character with much higher ionic conductivities of the order $10^{-5} - 10^{-3}$ S cm^{-1}.

Compared to other kinds of charge transport materials, gel polymer electrolytes are constructed by trapping liquid electrolytes, which usually contain organic solvents and inorganic salts such as acetonitrile (ACN), propylene carbonate (PC), or ethylene carbonate (EC), lithium iodide (LiI), sodium iodide (NaI), and potassium iodide (KI). Gel polymer electrolytes have some advantages, such as low vapor pressure, excellent contacting and filling properties between the nanostructured electrode and counter electrode, higher ionic conductivity compared to the conventional polymer electrolytes, and excellent thermal stability. Thus, the DSSCs based on gel polymer electrolytes have outstanding long-term stability [89].

Up to now, many types of polymer gel electrolytes have been used in quasi-solid-state DSSCs, including polyacrylonitrile (PAN), poly(ethylene oxide) (PEO), polymethylmethacrylate (PMMA), and poly(vinylidene fluoride-co-hexafluoropropylene) (PVDF-HFP). PVDF-HFP exhibits high ionic conductivity and stability at room temperature; however, the complex preparation technology and the poor mechanical strength of these gel polymer-based DSSCs represent a bottleneck to their introduction to the market. To overcome this problem, the electrospinning of such polymers has been performed with the aim of integrating the resulting fibrous, easy-to-obtain, and low-cost materials as electrolytes [23].

An electrospun membrane was prepared from a 16 wt% solution of poly(vinylidene fluoride-co-hexafluoropropylene) (PVDF-HFP) in a mixture of acetone/*N*,*N*-dimethylacetamide (7:3 wt%) at an applied voltage of 12 kV [90]. The electrospun membrane was then activated by immersing it in 0.6 M 1-hexyl-2,3-dimethylimidazolium iodide, 0.1 M LiI, 0.05 M I_2, and 0.5 M 4-tert-butylpyridine in ethylene carbonate/propylene carbonate (1:1 wt%) to obtain the corresponding membrane electrolyte with an ionic conductivity of 10^{-5} S cm^{-1}, at 25 °C. On the basis of this electrospun membrane electrolyte, quasi-solid-state dye-sensitized solar cells were fabricated, which showed an open-circuit voltage of 0.76 V, a fill factor of 0.62, and a short-circuit current density of 15.57 mA cm^{-2} at an incident

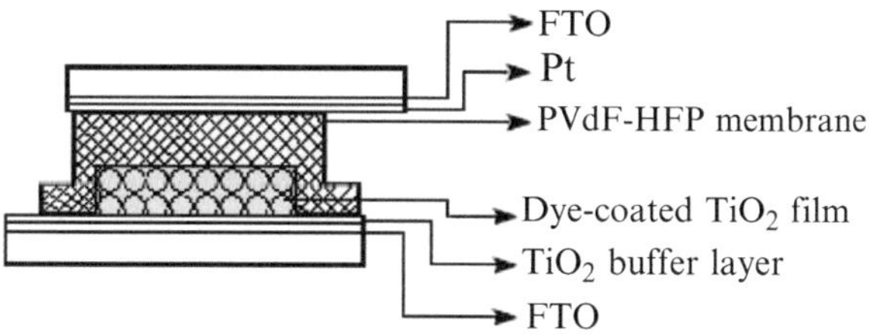

Fig. 5.11 Schematic diagram of a quasi-solid-state DSSC (Reprinted with the permission from Priya et al. [90]. Copyright 2008 American Chemical Society)

light intensity of 100 mW cm^{-2}. This yields a light-to-electricity conversion efficiency of 7.3 %. Moreover, this cell possessed better long-term stability than that fabricated with conventional liquid electrolyte (Fig. 5.11).

Kim's group [91–93] systematically researched DSSCs based on electrospun PVDF-HFP nanofiber films. They found that the photovoltaic performance of DSSC devices, ionic conductivity, uptake, and porosity of the electrospun PVDF-HFP nanofiber films were not influenced by the morphology of nanofibers prepared at various parameter values. The photovoltaic performance of DSSC devices using electrospun PVDF-HFP nanofiber films showed remarkably improved results compared to DSSC devices using spin-coated PVDF-HFP films. In addition, it was found that the photocurrent density and efficiency of DSSC using electrospun PVDF-HFP nanofibers in electrolytes were not necessarily proportional to the ionic conductivity in electrolytes. In their subsequent work, PVDF-HFP and polystyrene (PS) blend nanofibers were prepared. The photovoltaic performance of DSSC devices using electrospun PVDF-HFP/PS (3:1) nanofiber showed remarkably improved results compared to DSSC devices using electrospun PVDF-HFP nanofiber [93]. Because the porosity of electrospun PVDF-HFP/PS blend nanofibers was higher than PVDF-HFP nanofiber, ion transfer occurred well, and regular nanofiber morphology helped the transfer of ions produced by the redox mechanism. They also compared the stability of the electrospun polymer electrolyte-based DSSC and that of the liquid electrolyte-based DSSC through a method of predrilling two holes on the substrate glass. One could see that the solvent was quickly leaked out of the liquid electrolyte-based cell, while no obvious leakage was observed of the electrospun polymer electrolyte-based DSSC as shown in Fig. 5.12.

Jin et al. [94] embedded the liquid crystal (E7) into the electrospun PVDF-HFP nanofiber. Owing to the high ionic conductivity (2.9×10^{-3}) of the E7 embedded on PVDF-HFP polymer gel electrolyte, higher values of current density (14.62 $mAcm^{-2}$) and efficiency (6.82 %) of the quasi-solid-state DSSCs were achieved. The quasi-solid-state DSSC of PVDF-HFP polymer gel electrolyte was found to have a 6.35 % PCE. In addition to that, overall efficiency of the newly prepared DSSCs was 6.82 %, which was nearly equivalent to that of a DSSC made using liquid electrolyte. They believed that DSSCs containing E7 embedded on PVDF-HFP polymer gel electrolyte will find commercial utility.

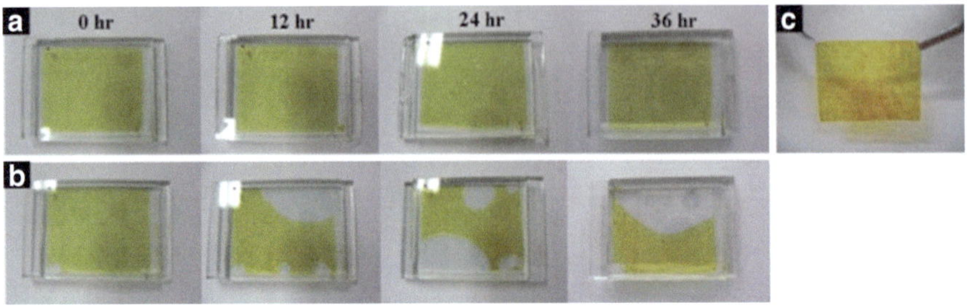

Fig. 5.12 Stability testing of (**a**) the electrospun PVDF-HFP nanofiber electrolyte cells, (**b**) the liquid electrolyte cells, and (**c**) the soaked electrospun PVDF-HFP nanofiber after 36 h (Reprinted from Park et al. [93]. Copyright 2011 with permission from Elsevier)

The above mentioned electrospun nanofibers are all based on fluorinated polymer derivatives. Moon et al. [95] investigated the use of electrospun brominated poly(2,6-dimethyl-1,4-phenylene oxide) (BPPO) which was non-fluorinated polymer backbone structures having polar functional groups for quasi-solid-state electrolytes. It was found that BPPO, which contains ether and bromine functional groups of nonbonding electrons, acts as a Lewis base. This Lewis basicity not only negatively shifted the flat band but also coordinated Li ions in the electrolyte. As a result, by introducing a quasi-solid-state electrolyte, the open-circuit voltage increased by 50 mV and the photon-to-current efficiency was comparable to a conventional liquid electrolyte. Furthermore, the quasi-solid-state DSSCs exhibited enhanced long-term stability, maintaining 97 % of initial efficiency after 7 day under a mild acceleration condition at 50 °C (Fig. 5.13).

Apart from PVDF-HFP nanofiber films, Kim et al. [96] also fabricated poly(vinyl alcohol) (PVA) nanofibers by electrospinning using a PVA solution containing silver nitrate ($AgNO_3$). The Ag nanoparticles were generated after irradiating the as-electrospun PVA/$AgNO_3$ nanofibers with UV light at 310–380 nm. The efficiency of the DSSC devices using the electrospun PVA/Ag nanofibers was slightly higher than that without Ag, because the electrospun PVA/Ag nanofibers facilitated the transfer of the ions as compared with the electrospun PVA nanofibers (Fig. 5.14).

However, the efficiencies of (quasi-)solid-state electrolyte-based DSSCs are still lower (5 %) than the liquid counterpart-based ones. Except for the relatively inferior ionic conductivity, the contact between the electrolyte and the semiconductor oxide is also worse. As conventional TiO_2 photoanode film comprises a three-dimensional highly porous network of spherical interconnected nanoparticles whose inside pores are relatively small and disordered, therefore, it is difficult for those viscous colloidal electrolyte or their precursor to penetrate into the porous titania electrode. New deposition methods are introduced as the use of vacuum techniques has been shown to minimize this effect, leading to a better filling of the porous electrode. Another effective alternative is to control the porous electrode, electrospun TiO_2 nanofiber-based photoanode with ordered and large pores which as we mentioned in the early section could provide well-penetrated channels for those highly viscous polymer electrolytes; thus, a high efficiency is hopeful.

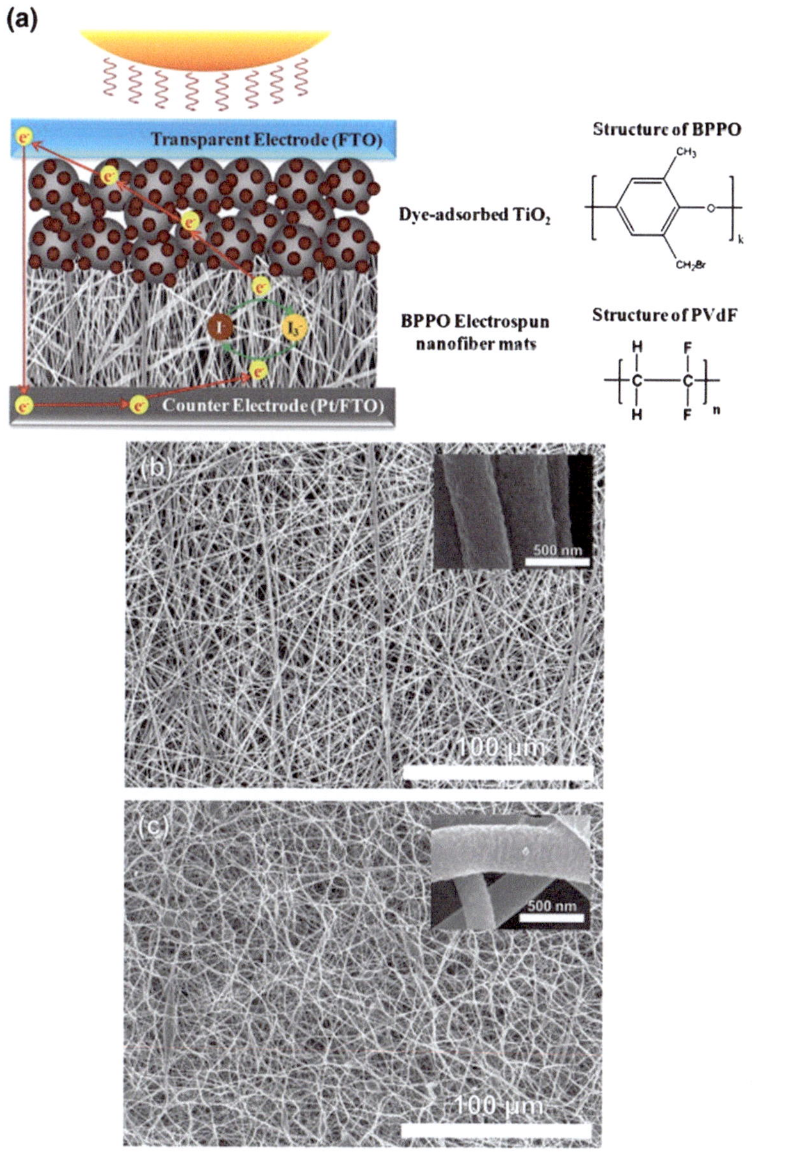

Fig. 5.13 (**a**) Schematic of quasi-solid-state DSSCs in which electrospun nanofiber mats are employed. SEM images of electrospun (**b**) BPPO and (**c**) PVDF nanofiber mats (Reprinted from Seo et al. [95]. Copyright 2011, with permission from Elsevier)

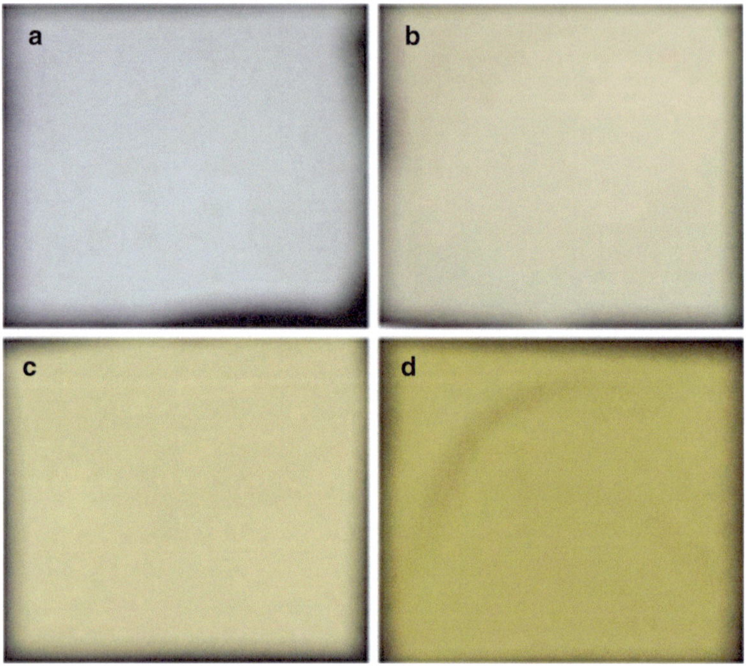

Fig. 5.14 Images of electrospun PVA and PVA/Ag nanofibers. (**a**) electrospun PVA nanofiber, (**b**) electrospun PVA nanofiber containing 1 wt% of Ag, (**c**) 2 wt% of Ag, and (**d**) 5 wt% of Ag (Reprinted from Park et al. [96], with kind permission from Springer Science + Business Media)

5.5 Conclusions

The electrospun nanofibers were intensively investigated in the DSSC components ranging from photoanodes to electrolytes. However, the photoanodes of DSSCs generally have a specific surface area not more than 100 m^2g^{-1} and too many voids between the nanofibers. Both of them make the dye uptake of nanofibers still lower than most of nanoparticle photoanodes, which results in the relative lower photocurrent of electrospun nanofiber DSSCs. Also, the hierarchically porous nanofibers make the poorly adhesive ability between the anode and substrate; the main ways to solve these problems are focused by using the compact layer before the electrospinning and the posttreatment of TiO_2 sol, as well as the compacting technology. Moreover, the nanofibers provide more and larger transportation paths for solid-state electrolytes and make the Pt-free counter electrodes competitive for high-efficiency DSSCs.

Acknowledgments This work was supported by Innovation Program of Shanghai Municipal Education Commission (11ZZ60), the Cultivation Fund of the Key Scientific and Technical Innovation Project (No. 708039), and the Program for Professor of Special Appointment (Eastern Scholar) at Shanghai Institutions of Higher Learning.

References

1. Oregan B, Grätzel M (1991) A low-cost, high-efficiency solar-cell based on dye-sensitized colloidal TiO_2 films. Nature 353(6346):737–740
2. Nazeeruddin MK, Kay A, Rodicio I, Humphrybaker R, Muller E, Liska P, Vlachopoulos N, Grätzel M (1993) Conversion of light to electricity by cis-X_2Bis (2, 2′-bipyridyl-4, 4′dicarboxylate) ruthenium (II) charge transfer sensitizers ($X = Cl^-$, Br^-, I^-, CN^- and SCN^-) on nanocrystalline TiO_2 electrodes. J Am Chem Soc 115(14):6382–6390
3. Grätzel M (2003) Dye-sensitized solar cells. J Photochem Photobiol C 4(2):145–153. doi:10.1016/S1389-5567(03)00026-1
4. Arakawa H, Wang ZS, Kawauchi H, Kashima T (2004) Significant influence of TiO_2 photoelectrode morphology on the energy conversion efficiency of N719 dye-sensitized solar cell. Coord Chem Rev 248(13–14):1381–1389. doi:10.1016/j.ccr.2004.03.006
5. Yella A, Lee HW, Tsao HN, Yi CY, Chandiran AK, Nazeeruddin MK, Diau EWG, Yeh CY, Zakeeruddin SM, Grätzel M (2011) Porphyrin-Sensitized solar cells with cobalt (II/III)-based redox electrolyte exceed 12 percent efficiency. Science 334(6056):629–634. doi:10.1126/science.1209688
6. Nazeeruddin MK, Baranoff E, Grätzel M (2011) Dye-sensitized solar cells: a brief overview. Sol Energy 85(6):1172–1178
7. Hagfeldt A, Boschloo G, Sun LC, Kloo L, Pettersson H (2010) Dye-sensitized solar cells. Chem Rev 110(11):6595–6663. doi:10.1021/Cr900356p
8. Tetreault N, Grätzel M (2012) Novel nanostructures for next generation dye-sensitized solar cells. Energy Environ Sci 5(9):8506–8516
9. Grätzel M (2005) Solar energy conversion by dye-sensitized photovoltaic cells. Inorg Chem 44(20):6841–6851
10. Snaith HJ (2010) Estimating the maximum attainable efficiency in dye-sensitized solar cells. Adv Funct Mater 20(1):13–19. doi:10.1002/adfm.200901476
11. Thavasi V, Singh G, Ramakrishna S (2008) Electrospun nanofibers in energy and environmental applications. Energy Environ Sci 1(2):205–221
12. Willekens J, Adriaenssens GJ, Aernouts T, Geens W, Vanlaeke P, Poortmans J (2004) Probing the subgap absorption of photovoltaic polymer blends by the constant photocurrent method. Thin Solid Films 451:26–28
13. Hagen J, Schaffrath W, Otschik P, Fink R, Bacher A, Schmidt HW, Haarer D (1997) Novel hybrid solar cells consisting of inorganic nanoparticles and an organic hole transport material. Synth Met 89(3):215–220
14. Zhang DS, Ito S, Wada YJ, Kitamura T, Yanagida S (2001) Nanocrystalline TiO_2 electrodes prepared by water-medium screen printing technique. Chem Lett 10:1042–1043
15. Aduda BO, Ravirajan P, Choy KL, Nelson J (2004) Effect of morphology on electron drift mobility in porous TiO_2. Int J Photoenergy 6(3):141–147
16. Nolan MG, Pemble ME, Sheel DW, Yates HM, Yates HM (2006) One step process for chemical vapour deposition of titanium dioxide thin films incorporating controlled structure nanoparticles. Thin Solid Films 515(4):1956–1962
17. Bisquert J, Cahen D, Hodes G, Ruhle S, Zaban A (2004) Physical chemical principles of photovoltaic conversion with nanoparticulate, mesoporous dye-sensitized solar cells. J Phys Chem B 108(24):8106–8118
18. Li ZZ, Rochford C, Baca FJ, Liu JW, Li J, Wu J (2010) Investigation into photoconductivity in single CNF/TiO_2-dye core-shell nanowire devices. Nanoscale Res Lett 5(9):1480–1486
19. Zhu K, Vinzant TB, Neale NR, Frank AJ (2007) Removing structural disorder from oriented TiO_2 nanotube arrays: reducing the dimensionality of transport and recombination in dye-sensitized solar cells. Nano Lett 7(12):3739–3746
20. Lee KM, Hu CW, Chen HW, Ho KC (2008) Incorporating carbon nanotube in a low-temperature fabrication process for dye-sensitized TiO_2 solar cells. Sol Energy Mater Sol Cells 92(12):1628–1633

21. Jiu JT, Isoda S, Wang FM, Adachi M (2006) Dye-sensitized solar cells based on a single-crystalline TiO_2 nanorod film. J Phys Chem B 110(5):2087–2092
22. Kang SH, Choi SH, Kang MS, Kim JY, Kim HS, Hyeon T, Sung YE (2008) Nanorod-based dye-sensitized solar cells with improved charge collection efficiency. Adv Mater 20(1):54–58
23. Cavaliere S, Subianto S, Savych I, Jones DJ, Roziere J (2011) Electrospinning: designed architectures for energy conversion and storage devices. Energy Environ Sci 4(12):4761–4785
24. Onozuka K, Ding B, Tsuge Y, Naka T, Yamazaki M, Sugi S, Ohno S, Yoshikawa M, Shiratori S (2006) Electrospinning processed nanofibrous TiO_2 membranes for photovoltaic applications. Nanotechnology 17(4):1026–1031
25. Song MY, Ahn YR, Jo SM, Kim DY, Ahn JP (2005) TiO_2 single-crystalline nanorod electrode for quasi-solid-state dye-sensitized solar cells. Appl Phys Lett 87(11):113113
26. Lee BH, Song MY, Jang SY, Jo SM, Kwak SY, Kim DY (2009) Charge transport characteristics of high efficiency dye-sensitized solar cells based on electrospun TiO_2nanorod photoelectrodes. J Phys Chem C 113(51):21453–21457
27. Cavaliere S, Subianto S, Chevallier L, Jones DJ, Roziere J (2011) Single step elaboration of size-tuned Pt loaded titania nanofibers. Chem Commun 47(24):6834–6836
28. Madhavan AA, Kalluri S, Chacko DK, Arun TA, Nagarajan S, Subramanian KRV, Nair AS, Nair SV, Balakrishnan A (2012) Electrical and optical properties of electrospun TiO_2-graphene composite nanofibers and its application as DSSC photo-anodes. RSC Adv 2(33):13032–13037
29. Chen HW, Lin CY, Lai YH, Chen JG, Wang CC, Hu CW, Hsu CY, Vittal R, Ho KC (2011) Electrophoretic deposition of ZnO film and its compression for a plastic based flexible dye-sensitized solar cell. J Power Sources 196(10):4859–4864
30. Law M, Greene LE, Johnson JC, Saykally R, Yang PD (2005) Nanowire dye-sensitized solar cells. Nat Mater 4(6):455–459
31. Saito M, Fujihara S (2008) Large photocurrent generation in dye-sensitized ZnO solar cells. Energy Environ Sci 1(2):280–283
32. Kim ID, Hong JM, Lee BH, Kim DY, Jeon EK, Choi DK, Yang DJ (2007) Dye-sensitized solar cells using network structure of electrospun ZnO nanofiber mats. Appl Phys Lett 91(16):163109
33. Zhang W, Zhu R, Liu XZ, Liu B, Ramakrishna S (2009) Facile construction of nanofibrous ZnO photoelectrode for dye-sensitized solar cell applications. Appl Phys Lett 95(4):43304
34. Yun SN, Lim S (2011) Improved conversion efficiency in dye-sensitized solar cells based on electrospun Al-doped ZnO nanofiber electrodes prepared by seed layer treatment. J Solid State Chem 184(2):273–279
35. Kyaw AKK, Tantang H, Wu T, Ke L, Wei J, Demir HV, Zhang QC, Sun XW (2012) Dye-sensitized solar cell with a pair of carbon-based electrodes. J Phys D-Appl Phys 45(16):165103
36. Wang W, Zhao Q, Li H, Wu HW, Zou DC, Yu DP (2012) Transparent, double-sided, ITO-free, Flexible dye-sensitized solar cells based on metal wire/ZnO nanowire arrays. Adv Funct Mater 22(13):2775–2782
37. Wu H, Hu LB, Rowell MW, Kong DS, Cha JJ, McDonough JR, Zhu J, Yang YA, McGehee MD, Cui Y (2010) Electrospun metal nanofiber webs as high-performance transparent electrode. Nano Lett 10(10):4242–4248
38. Archana PS, Jose R, Vijila C, Ramakrishna S (2009) Improved electron diffusion coefficient in electrospun TiO_2 nanowires. J Phys Chem C 113(52):21538–21542
39. Saji VS, Pyo M (2010) Dye sensitized solar cell of TiO_2 nanoparticle/nanorod composites prepared via low-temperature synthesis in oleic acid. Thin Solid Films 518(22):6542–6546
40. Wu JJ, Chen GR, Lu CC, Wu WT, Chen JS (2008) Performance and electron transport properties of TiO_2 nanocomposite dye-sensitized solar cells. Nanotechnology 19(10):105702
41. Mukherjee K, Teng TH, Jose R, Ramakrishna S (2009) Electron transport in electrospun TiO_2 nanofiber dye-sensitized solar cells. Appl Phys Lett 95(1):12101
42. Tan B, Wu YY (2006) Dye-sensitized solar cells based on anatase TiO_2 nanoparticle/nanowire composites. J Phys Chem B 110(32):15932–15938
43. Fujihara K, Kumar A, Jose R, Ramakrishna S, Uchida S (2007) Spray deposition of electrospun TiO_2 nanorods for dye-sensitized solar cell. Nanotechnology 18(36):365709

44. Chuangchote S, Sagawa T, Yoshikawa S (2008) Efficient dye-sensitized solar cells using electrospun TiO_2 nanofibers as a light harvesting layer. Appl Phys Lett 93(3):033310
45. Joshi P, Zhang LF, Davoux D, Zhu ZT, Galipeau D, Fong H, Qiao QQ (2010) Composite of TiO_2 nanofibers and nanoparticles for dye-sensitized solar cells with significantly improved efficiency. Energy Environ Sci 3(10):1507–1510
46. He ZL, Zhu ZF, Li JQ, Zhou JQ, Wei N (2011) Characterization and activity of mesoporous titanium dioxide beads with high surface areas and controllable pore sizes. J Hazard Mater 190(1–3):133–139
47. Zhao X, Liu MH, Zhu YF (2007) Fabrication of porous TiO_2 film via hydrothermal method and its photocatalytic performances. Thin Solid Films 515(18):7127–7134
48. Lee KM, Hsu CY, Chiu WH, Tsui MC, Tung YL, Tsai SY, Ho KC (2009) Dye-sensitized solar cells with a micro-porous TiO_2 electrode and gel polymer electrolytes prepared by in situ cross-link reaction. Sol Energy Mater Sol Cells 93(11):2003–2007
49. Bae KR, Ko CH, Park Y, Kim Y, Bae JS, Yeum JH, Kim IS, Lee WJ, Oh W (2010) Structure control of nanocrystalline TiO_2 for the dye-sensitized solar cell application. Curr Appl Phys 10(3):S406–S409
50. Asagoe K, Suzuki Y, Ngamsinlapasathian S, Yoshikawa S (2007) TiO_2-anatase nanowire dispersed composite electrode for dye-sensitized solar cells. In: Meyer E, Hegner M, Gerber C, Guntherodt HJ (eds) J Phys Confer Ser 61:1112–1116. doi:10.1088/1742-6596/61/1/220
51. Baxter JB, Aydil ES (2005) Nanowire-based dye-sensitized solar cells. Appl Phys Lett 86(5):053114
52. Pavasupree S, Ngamsinlapasathian S, Nakajima M, Suzuki Y, Yoshikawa S (2006) Synthesis, characterization, photocatalytic activity and dye-sensitized solar cell performance of nanorods/nanoparticles TiO_2 with mesoporous structure. J Photochem Photobio A 184(1–2): 163–169. doi:10.1016/j.jphotochem.2006.04.010
53. Jung WH, Kwak N-S, Hwang TS, Yi KB (2012) Preparation of highly porous TiO_2 nanofibers for dye-sensitized solar cells (DSSCs) by electro-spinning. Appl Surf Sci 261:343–352. doi:10.1016/j.apsusc.2012.08.012
54. Hwang SH, Kim C, Song H, Son S, Jang J (2012) Designed architecture of multiscale porous TiO_2 nanofibers for dye-sensitized solar cells photoanode. ACS Appl Mater Interfaces 4(10):5287–5292. doi:10.1021/am301245s
55. Mu QH, Li YG, Zhang QH, Wang HZ (2011) Template-free formation of vertically oriented TiO_2 nanorods with uniform distribution for organics-sensing application. J Hazard Mater 188(1–3):363–368
56. Krishnamoorthy T, Thavasi V, Subodh GM, Ramakrishna S (2011) A first report on the fabrication of vertically aligned anatase TiO_2 nanowires by electrospinning: preferred architecture for nanostructured solar cells. Energy Environ Sci 4(8):2807–2812
57. Shim HS, Na SI, Nam SH, Ahn HJ, Kim HJ, Kim DY, Kim WB (2008) Efficient photovoltaic device fashioned of highly aligned multilayers of electrospun TiO_2 nanowire array with conjugated polymer. Appl Phys Lett 92(18):183107
58. Wu SJ, Tai QD, Yan F (2010) Hybrid photovoltaic devices based on poly (3-hexylthiophene) and ordered electrospun ZnO nanofibers. J Phys Chem C 114(13):6197–6200
59. Tai QD, Zhao XZ, Yan F (2010) Hybrid solar cells based on poly(3-hexylthiophene) and electrospun TiO_2 nanofibers with effective interface modification. J Mater Chem 20(35): 7366–7371
60. Liao JY, He JW, Xu HY, Kuang DB, Su CY (2012) Effect of TiO_2 morphology on photovoltaic performance of dye-sensitized solar cells: nanoparticles, nanofibers, hierarchical spheres and ellipsoid spheres. J Mater Chem 22(16):7910–7918
61. Zhang X, Thavasi V, Mhaisalkar SG, Ramakrishna S (2012) Novel hollow mesoporous 1D TiO_2 nanofibers as photovoltaic and photocatalytic materials. Nanoscale 4(5):1707–1716. doi:10.1039/c2nr11251e
62. Rui YC, Li YG, Zhang QH, Wang HZ (2013) Facile synthesis of rutile TiO_2 nanorod microspheres for enhancing light-harvesting of dye-sensitized solar cells. CrystEngComm 15(8):1651–1656

63. Zhang L, Li YG, Zhang QH, Wang HZ (2013) Formation of the modified ultrafine anatase TiO_2 nanoparticles using the nanofiber as a microsized reactor. CrystEngComm 15(8):1607–1612
64. Wu JJ, Chen YR, Liao WP, Wu CT, Chen CY (2010) Construction of nanocrystalline film on nanowire array via swelling electrospun polyvinylpyrrolidone-hosted nanofibers for use in dye-sensitized solar cells. ACS Nano 4(10):5679–5684
65. Zhu PN, Nair AS, Yang SY, Peng SJ, Ramakrishna S (2011) Which is a superior material for scattering layer in dye-sensitized solar cells-electrospun rice grain- or nanofiber-shaped TiO_2. J Mater Chem 21(33):12210–12212
66. Zhu G, Pan LK, Yang J, Liu XJ, Sun HC, Sun Z (2012) Electrospun nest-shaped TiO_2 structures as a scattering layer for dye sensitized solar cells. J Mater Chem 22(46):24326–24329
67. Unalan HE, Wei D, Suzuki K, Dalal S, Hiralal P, Matsumoto H, Imaizumi S, Minagawa M, Tanioka A, Flewitt AJ, Milne WI, Amaratunga GAJ (2008) Photoelectrochemical cell using dye sensitized zinc oxide nanowires grown on carbon fibers. Appl Phys Lett 93(13):133166
68. Jo SM, Song MY, Ahn YR, Park CR, Kim DY (2005) Nanofibril formation of electrospun TiO_2 fibers and its application to dye-sensitized solar cells. J Macromol Sci Pure A42(11): 1529–1540
69. Song MY, Kim DK, Ihn KJ, Jo SM, Kim DY (2004) Electrospun TiO_2 electrodes for dye-sensitized solar cells. Nanotechnology 15(12):1861–1865
70. Song MY, Kim DK, Ihn KJ, Jo SM, Kim DY (2005) New application of electrospun TiO_2 solid-state dye-sensitized solar electrode to cells. Synth Met 153(1–3):77–80
71. Song MY, Kim DK, Jo SM, Kim DY (2005) Enhancement of the photocurrent generation in dye-sensitized solar cell based on electrospun TiO_2 electrode by surface treatment. Synth Met 155(3):635–638
72. Dai Y, Liu W, Formo E, Sun Y, Xia Y (2011) Ceramic nanofibers fabricated by electrospinning and their applications in catalysis, environmental science, and energy technology. Polym Adv Technol 22(3):326–338. doi:10.1002/pat.1839
73. Li K, Yu Z, Luo Y, Li D, Meng Q (2007) Recent progress of counter electrodes in nanocrystalline dye-sensitized solar cells. J Mater Sci Technol 23(5):577
74. Murakami TN, Grätzel M (2008) Counter electrodes for DSC: application of functional materials as catalysts. Inorg Chim Acta 361(3):572–580. doi:10.1016/j.ica.2007.09.025
75. Kim J, Kang J, Jeong U, Kim H, Lee H (2013) Catalytic, conductive, and transparent platinum nanofiber webs for FTO-Free dye-sensitized solar cells. ACS Appl Mater Interfaces 5(8): 3176–3181. doi:10.1021/am400179j
76. Hong W, Xu Y, Lu G, Li C, Shi G (2008) Transparent graphene/PEDOT-PSS composite films as counter electrodes of dye-sensitized solar cells. Electrochem Commun 10(10):1555–1558. doi:10.1016/j.elecom.2008.08.007
77. Imoto K, Takahashi K, Yamaguchi T, Komura T, Nakamura J, Murata K (2003) High-performance carbon counter electrode for dye-sensitized solar cells. Sol Energy Mater Sol Cells 79(4):459–469. doi:10.1016/s0927-0248(03)00021-7
78. Murakami TN, Ito S, Wang Q, Nazeeruddin MK, Bessho T, Cesar I, Liska P, Humphry-Baker R, Comte P, Pechy P, Grätzel M (2006) Highly efficient dye-sensitized solar cells based on carbon black counter electrodes. J Electrochem Soc 153(12):A2255–A2261. doi:10.1149/1.2358087
79. Joshi P, Zhang L, Chen Q, Galipeau D, Fong H, Qiao Q (2010) Electrospun carbon nanofibers as low-cost counter electrode for dye-sensitized solar cells. ACS Appl Mater Interfaces 2(12):3572–3577. doi:10.1021/am100742s
80. Poudel P, Zhang L, Joshi P, Venkatesan S, Fong H, Qiao Q (2012) Enhanced performance in dye-sensitized solar cells via carbon nanofibers-platinum composite counter electrodes. Nanoscale 4(15):4726–4730. doi:10.1039/c2nr30586k
81. Barakat NAM, Akhtar MS, Yousef A, El-Newehy M, Kim HY (2012) Pd-Co-doped carbon nanofibers with photoactivity as effective counter electrodes for DSSCs. Chem Eng J 211: 9–15. doi:10.1016/j.cej.2012.09.040

82. Peng S, Zhu P, Wu Y, Mhaisalkar SG, Ramakrishna S (2012) Electrospun conductive polyaniline-polylactic acid composite nanofibers as counter electrodes for rigid and flexible dye-sensitized solar cells. RSC Adv 2(2):652–657. doi:10.1039/c1ra00618e
83. Yanagida S, Yu Y, Manseki K (2009) Iodine/iodide-free dye-sensitized solar cells. Acc Chem Res 42(11):1827–1838. doi:10.1021/ar900069p
84. Yum J-H, Chen P, Grätzel M, Nazeeruddin MK (2008) Recent developments in solid-state dye-sensitized solar cells. ChemSusChem 1(8–9):699–707. doi:10.1002/cssc.200800084
85. Li D, Qin D, Deng M, Luo Y, Meng Q (2009) Optimization the solid-state electrolytes for dye-sensitized solar cells. Energy Environ Sci 2(3):283–291. doi:10.1039/b813378f
86. Zhang W, Cheng Y, Yin X, Liu B (2011) Solid-state dye-sensitized solar cells with conjugated polymers as hole-transporting materials. Macromol Chem Phys 212(1):15–23. doi:10.1002/macp.201000489
87. Ileperuma OA (2013) Gel polymer electrolytes for dye sensitised solar cells: a review. Mater Technol 28(1–2):65–70. doi:10.1179/1753555712y.0000000043
88. de Freitas JN, Nogueira AF, De Paoli M-A (2009) New insights into dye-sensitized solar cells with polymer electrolytes. J Mater Chem 19(30):5279–5294. doi:10.1039/b900928k
89. Wang Y (2009) Recent research progress on polymer electrolytes for dye-sensitized solar cells. Sol Energy Mater Sol Cells 93(8):1167–1175. doi:10.1016/j.solmat.2009.01.009
90. Priya ARS, Subramania A, Jung Y-S, Kim K-J (2008) High-performance quasi-solid-state dye-sensitized solar cell based on an electrospun PVdF-HFP membrane electrolyte. Langmuir 24(17):9816–9819. doi:10.1021/la801375
91. Kim J-U, Park S-H, Choi H-J, Lee W-K, Lee J-K, Kim M-R (2009) Effect of electrolyte in electrospun poly(vinylidene fluoride-co-hexafluoropropylene) nanofibers on dye-sensitized solar cells. Sol Energy Mater Sol Cells 93(6–7):803–807. doi:10.1016/j.solmat.2008.09.045
92. Park S-H, Kim J-U, Lee S-Y, Lee W-K, Lee J-K, Kim M-R (2008) Dye-sensitized solar cells using polymer electrolytes based on poly(vinylidene fluoride-hexafluoropropylene) nanofibers by electrospinning method. J Nanosci Nanotechnol 8(9):4889–4894. doi:10.1166/jnn.2008.IC76
93. Park S-H, Won D-H, Choi H-J, Hwang W-P, S-i J, Kim J-H, Jeong S-H, Kim J-U, Lee J-K, Kim M-R (2011) Dye-sensitized solar cells based on electrospun polymer blends as electrolytes. Sol Energy Mater Sol Cells 95(1):296–300. doi:10.1016/j.solmat.2010.04.029
94. Ahn SK, Ban T, Sakthivel P, Lee JW, Gal Y-S, Lee J-K, Kim M-R, Jin S-H (2012) Development of dye-sensitized solar cells composed of liquid crystal embedded, electrospun poly(vinylidene fluoride-co-hexafluoropropylene) nanofibers as polymer gel electrolytes. ACS Appl Mater Interfaces 4(4):2096–2100. doi:10.1021/am3000598
95. Seo S-J, Yun S-H, Woo J-J, Park D-W, Kang M-S, Hinsch A, Moon S-H (2011) Preparation and characterization of quasi-solid-state electrolytes using a brominated poly(2,6-dimethyl-1,4-phenylene oxide) electrospun nanofiber mat for dye-sensitized solar cells. Electrochem Commun 13(12):1391–1394. doi:10.1016/j.elecom.2011.08.018
96. Park S-H, Choi H-J, Lee S-B, Lee S-M, Cho S-E, Kim K-H, Kim Y-K, Kim M-R, Lee J-K (2011) Fabrications and photovoltaic properties of dye-sensitized solar cells with electrospun poly(vinyl alcohol) nanofibers containing Ag nanoparticles. Macromol Res 19(2):142–146. doi:10.1007/s13233-011-0212-6

Chapter 6
Application of Electrospun Nanofibers in Organic Photovoltaics

Surawut Chuangchote and Takashi Sagawa

Abstract This chapter highlights the introduction of organic thin-film solar cells and organic–inorganic hybrid solar cells and the development of electrospun nanofibers for those applications. Organic solar cells need large interfacial surface area for efficient charge separation of excitons, which are normally achieved by thermal annealing. However, this annealing is relatively not compatible with the polymers with low glass transition temperature, not compatible with flexible substrates, and also not compatible with large area devices. An alternative process that yields efficient interface of electron donor and acceptor with continuous phase of each material is necessary. Electrospinning is a good method to fabricate nanofibers of conducting polymers which can be used as electron donor materials of organic solar cells. To improve the efficiencies of organic solar cells, the combination of organic and inorganic semiconductor as organic–inorganic hybrid photovoltaic cells has been extensively investigated to overcome the low mobility of carriers by using nanostructured inorganic materials as electron acceptors and/or electron transporting layers. The use of inorganic semiconductors in the form of electrospun nanofibers as electron transporting layers is a promising alternative way. Nanofiber-based transparent electrodes can also be fabricated by electrospinning with flexible property. These electrodes are very promising to be utilized as flexible substrates for both organic thin-film solar cells and organic–inorganic hybrid solar cells.

S. Chuangchote (✉)
The Joint Graduate School of Energy and Environment, King Mongkut's University of Technology Thonburi, 126 Prachauthit Rd., Bangmod, Tungkru, Bangkok 10140, Thailand
e-mail: surawut_c@jgsee.kmutt.ac.th

T. Sagawa (✉)
Graduate School of Energy Science, Kyoto University, Yoshida Hon-machi, Sakyo-ku, Kyoto 606-8501, Japan
e-mail: sagawa.takashi.6n@kyoto-u.ac.jp

B. Ding and J. Yu (eds.), *Electrospun Nanofibers for Energy and Environmental Applications*, Nanostructure Science and Technology,
DOI 10.1007/978-3-642-54160-5_6,

6.1 Introduction

A photovoltaic cell or solar cell is a semiconductor diode that can convert sunlight into direct current (DC) electricity. Organic solar cell or organic photovoltaic (OPV) cell is a type of solar cell that uses organic materials, especially small organic molecules or conducting organic polymers, for light absorption and charge transport to produce electricity by the photovoltaic effect. A common characteristic of both the small organic molecules and conducting organic polymers used in organic solar cells is that they have large conjugated systems. Carbon atoms covalently bond with alternating single and double bonds form a conjugated system. A delocalized bonding π orbital with a π^* antibonding orbital is possible to form in the conjugated system. The delocalized π orbital is the highest occupied molecular orbital (HOMO), and the π^* orbital is the lowest unoccupied molecular orbital (LUMO). The separation between HOMO and LUMO is considered to be the band gap of organic electronic materials. This band gap is typically in the range of 1–4 eV [1].

When a photon reaches organic semiconductor materials, it will be absorbed. An excited state is then created and confined to an organic molecule or a region of a conducting polymer chain. The excited state can be regarded as an electron-hole pair bound together by electrostatic interactions, the so-called exciton. The exciton is broken up into a free electron-hole pair by an effective field. The effective fields are set up by creating a junction between two dissimilar materials. Effective fields break up excitons by causing the electron to fall from the conduction band of the absorber (the so-called electron donor) to the conduction band of the acceptor molecule (the so-called electron acceptor). It is necessary that the electron acceptor material has a conduction band edge that is lower than that of the electron donor material [2–5]. The most effective junction for an efficient electron-hole pair separation is bulk heterojunction of electron donor and acceptor materials.

Comparing organic solar cells with typical inorganic photovoltaic cells, that is, silicon solar cells or cadmium telluride (CdTe) photovoltaic cells, the power conversion efficiencies of reported organic photovoltaic devices are still relatively low. Such evidence can be attributed to low polymer absorption spectrum and especially low mobility of excited electron or hole in organic semiconductors [6, 7]. To improve the PCE, the combination of organic and inorganic semiconductor as hybrid organic–inorganic photovoltaic cells has been extremely investigated to overcome the low mobility of carriers by using nanostructured inorganic materials as electron transporting layers [8–13].

6.2 Organic Thin-Film Solar Cells

6.2.1 History of Organic Thin-Film Solar Cells

The discovery of organic semiconductors leads to a new field of research in electronic devices, such as organic electroluminescence, thin-film transistors, memory

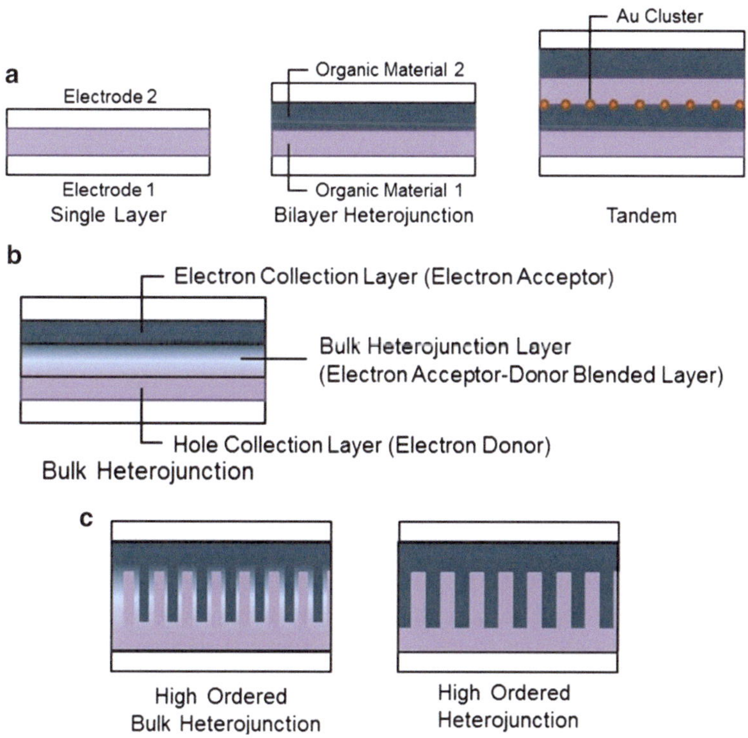

Fig. 6.1 Development of cell structure of organic solar cells; (**a**) primitive device, (**b**) present device, and (**c**) new concept device structures

devices, and solar cells. Organic solar cell is a potential alternative to commercial silicon-based solar cells due to its features like low cost, easy fabrication, environmental benign nature, light weight, and flexibility [14–16].

The history of organic thin-film solar cell started from a simple single-layer cell which is an organic semiconductor layer between two metal electrodes of different work functions (see Fig. 6.1a, left). Power conversion efficiency (PCE) was ranged from 0.001 % to 0.01 %. The single-layer cell based on pure organic material, merocyanine dye, achieved 0.7 % PCE [17]. One breakthrough for the structure of organic solar cells was a report of Tang in 1986, which was a bilayer heterojunction-type cell based on small organic molecules as active layers (perylene as electron donor and phthalocyanine as an electron acceptor) and ITO and Ag as electrodes (see Fig. 6.1a, middle). This cell achieved 1 % PCE [18]. After Tang's report, Hiramoto et al. developed a new cell structure, a tandem structure for organic solar cells, which was a two-stacked bilayer heterostructure (see Fig. 6.1a, right) [19]. A gold ultrathin film acting as a charge recombination part was inserted between stacked bilayer heterostructure. In the report of Hiramoto et al., open-circuit voltage (V_{oc}) was improved to the sum of voltages of upper and bottom layers. Furthermore,

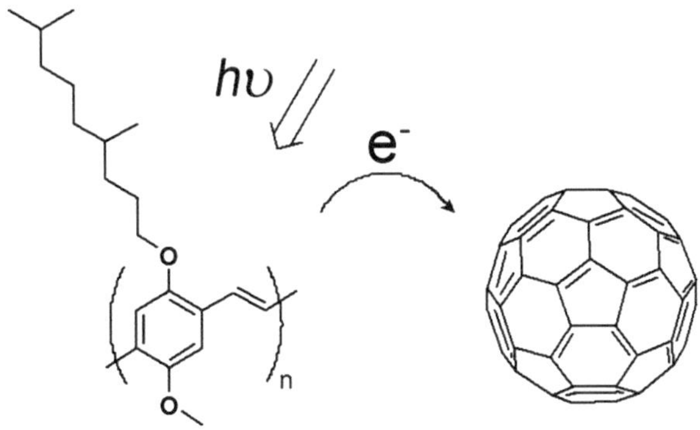

Fig. 6.2 Photoinduced electron transfer from a conducting polymer to fullerene

they also developed the three-layer structure which is composed of electron donor layer, electron acceptor layer, and co-deposited layer of electron donor and acceptor materials (see Fig. 6.1b). This co-deposited layer is the base of the present organic solar cells and similar to the bulk heterojunction (BHJ) concept which is the blend of electron donor and acceptor materials.

Another breakthrough for high-performance organic solar cells is a discovery of ultrafast photoinduced electron transfer from conducting polymer to fullerene and application of fullerene as an *n*-type semiconductor or an electron acceptor. Because of the use of conducting polymer, this type of organic solar cells was also called "polymer solar cells." Studies on the photophysics of poly[2-methoxy-5-(3',7'-dimethyloctyloxy)-1,4-phenylenevinylene] (MDMO-PPV) and fullerene mixture were reported [20, 21]. In these reports, ultrafast, reversible, and metastable-photoinduced electron transfer was observed from a conjugated polymer onto fullerene and it was found that back-electron-transfer process is several orders of magnitude slower than the forward-electron-transfer reaction (see Fig. 6.2). These results suggested that fullerene is a good electron acceptor.

In particular, the introduction of fullerene showed its ability as electron acceptor for performance polymer solar cells. The single-layer device based on conducting polymer was showing the 0.001–0.1 % PCE [22, 23]. Since the discovery of ultrafast photoinduced electron transfer from MDMO-PPV to fullerene, the bilayer cell of conducting polymer/C_{60} was investigated [24, 25]. The heterojunction device showed the rectification ratios on the order of 10^4, while energy conversion efficiency was not high.

Yu et al. reported the BHJ concept using a soluble fullerene derivative, 1-(3-methoxycarbonyl)-propyl-1-phenyl-(6,6)C_{61} (PCBM) (see its structure in Fig. 6.3), as an electron acceptor. A PCE of 2 % was achieved [26]. This BHJ is composed of blended film of mixed *p*- and *n*-type semiconductors, similar to co-deposited layer reported by Hiramoto [19]. In Tang's heterojunction [18], it is hard for all generated excitons to reach *p*-*n* interface, and the generated charge is little. On the other hand, BHJ makes it possible for a generated exciton to reach *p*-*n* junction, and many charge separations occur. A lot of interests on organic thin-film solar cells have been focused since the report of Yu et al. [26]. Until now, high PCEs of more

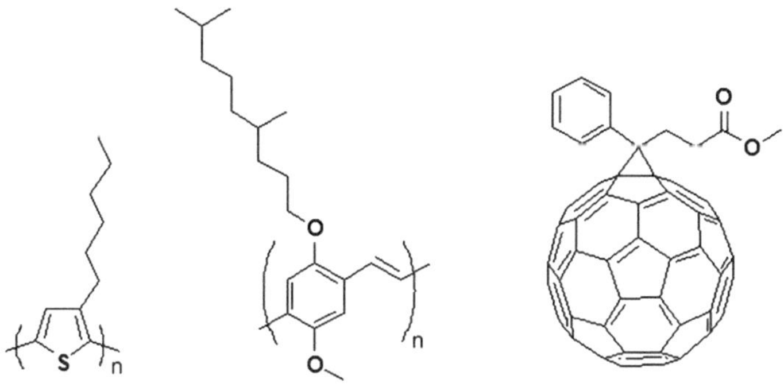

Fig. 6.3 Chemical structures of conducting polymers and a fullerene derivative mostly used in organic polymer cells i.e., P3HT, MDMO-PPV, and PCBM

than 5 % have been achieved in polymer solar cells using poly(3-hexylthiophene-2,5-diyl) (P3HT) (see its structure in Fig. 6.3) as an electron donor and PCBM or other fullerene derivatives as an electron acceptor. Fullerene derivatives are essential materials in order to obtain high energy conversion efficiency in organic polymer solar cells and organic small molecule solar cells. Furthermore, electron donor and acceptor materials and the control method in morphology of layers and charge transporting layers have been developed in order to obtain high energy conversion efficiency of 4–6.5 %.

6.2.2 *Morphology Control of Active Layer*

The performance of organic solar cells depends on the physical interaction of electron donor and electron acceptor phases. It is well known that organic materials have small charge mobility. In heterojunction organic solar cells, exciton within 5–10 nm in *p-n* interface (diffusion length) can be achieved, but exciton in farther distance than exciton diffusion length does not contribute charge generation. In contrast, in BHJ, charge generation can be achieved easily at *p-n* interface (see bilayer heterojunction in Fig. 6.1c; compare with BHJ in Fig. 6.4). Moreover, *p-n* interfacial area increases dramatically in BHJ.

For polymer solar cells, as described above, typically, conjugated low band gap polymer (electron donor) is blended with an electron acceptor to achieve a BHJ layer and fabricated on a transparent conducting oxide (TCO)-coated glass substrate (i.e., tin-doped indium oxide (ITO)). In this system, photons are absorbed by the electron donor, then excitons are generated, diffuse to the interface of the two bulk components, and dissociate. The acceptor receives the electron and transfers to the electrode, while the hole remains within the donor as shown in Fig. 6.4.

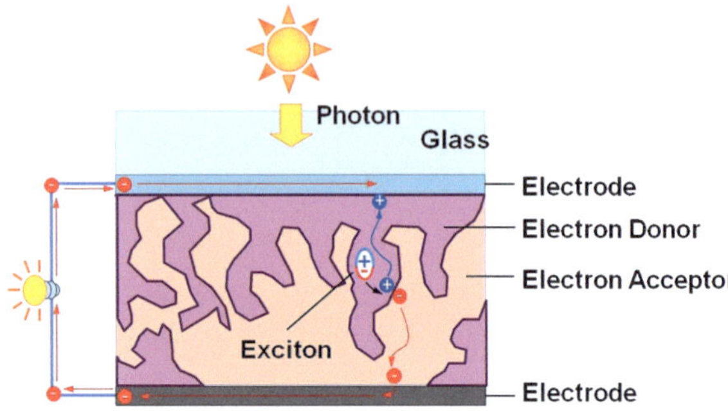

Fig. 6.4 Structure of BHJ organic solar cells

The donor-acceptor components should be the suitable phase size to allow the maximum *p-n* interface for charge generation. This will allow all excitons transfer to *p-n* interface and effective charge transport in the bicontinuous pathway to the electrodes to minimize the recombination of free charges. The morphology of the active layer depends on the interplay between the material property and device fabrication, such as the solvent used, overall concentration of the blend component, deposition technique (spin coating, ink-jet printing, roll-to-roll, etc.), solvent evaporation rate, and temperature.

The solvent in the fabrication of active layer affects the morphology of the active layer and the performance of the device. It was observed that casting the active layer blended from toluene gave the PCE of about 0.9 % with a 1:4 weight ratio of polymer to fullerene. When the solvent was changed to chlorobenzene, the efficiency dramatically improved to 2.5 % [27].

For the film treatment, the most widely studied polymers for BHJ solar cells are MDMO-PPV and P3HT. The film treatment was well investigated in the BHJ based on P3HT/PCBM. Some methods have been prepared for optimization of morphology and donor-acceptor phase in order to obtain high energy conversion efficiency. The most commonly used method is thermal annealing for morphology control. Camaioni et al. reported the improvement of energy conversion efficiency from 0.17 % to 0.6 % on device based on P3HT-fulleropyrrodine at a thermal treatment at 55 °C [28]. Pandinger et al. reported the improvement in energy conversion efficiency from 0.4 % to 3.5 % on device based on P3HT/PCBM [29]. The demonstration by Pandinger et al. stimulated an extensive study on the thermal annealing and high energy conversion efficiency of 4–6 % on the device based on P3HT/PCBM.

Regioregularity (RR), polydispersity index (PDI), and molecular weight also contribute to PCE. Kim et al. investigated P3HT with RR value of 90.7 %, 93 % and 95.2 % [30]. The device made from 1:1 blend of these polymers with PCBM annealed for 2 h at 140 °C is showing PCEs of 0.7 %, 1.8 %, and 2.4 %, respectively. The RR value of 95.4 % polymer shows 4.4 % PCE in 1:1 blend with PCBM after further optimization. The RR value contributes to crystallinity and charge transport property of the polymer phase. They concluded the high RR polymer is needed to achieve high PCE of the cells. Tajima et al. investigated the effect of

RR value in the cell based on MDMO-PPV/PCBM [31]. The prototype MDMO-PPV was regiorandom. They succeeded to synthesize RR-MDMO-PPV by another route. Short circuit photocurrent densities (J_{sc}) and fill factor (FF) of the cells were improved, while V_{oc} decreased. The mobility of RR-MDMO-PPV shows higher value ($3.1 \pm 0.4 \times 10^{-10}$ m^2/V s) than the value of regiorandom-MDMO-PPV mobility ($2.2 \pm 0.4 \times 10^{-11}$ m^2/V s). This study also indicated the need of high RR polymer and development of synthesis method.

In Mw study in BHJ polymer solar cells, Mw of P3HT of about 50,000 is a good value. Higher Mw polymer has high hole-transport property but shows moderate PCE of about 2 %. This result indicated that polymer phase size and shape was suitable for charge generation and charge transport. Furthermore, PDI of polymer study was investigated by Hirons et al. [32]. PCE in the device based on PDI polymer of 2.1 ($M_n = 14{,}800$) was improved from 0.6 % to 3.4 % by thermal annealing at 120 °C. On the other hand, PCE in the device based on the polymer with PDI of 1.4 ($M_n = 12{,}800$) was improved from 0.5 % to 1.9 % by thermal annealing at 120 °C.

The fabrication technique of active layer (electron donor and electron acceptor) also affects the properties and PCE of the fabricated organic solar cells. Thin film of active layer can be prepared by spin coating, solution printing, roll-to-roll, spray coating, etc. In this contribution, fabrications of nanofibers by electrospinning to control the morphology of semiconductor materials and the uses of electrospun nanofibers as active layers or other parts of organic solar cells will be later reviewed.

6.2.3 Recent Progress of PCE of Organic Thin-Film Solar Cells

Researchers at Mitsubishi Chemical Group in Japan used tetrabenzoporphyrin as a solution-processable organic semiconductor that achieves 7.6 % PCE. The Mitsubishi Chemical Group improved upon their earlier work to 10 % PCE with a single-junction device with an area of about 1 cm^2. They also announced their achievement of 11 % PCE (2 mm^2). Construction of tandem solar cell is another promising strategy to achieve high power conversion efficiency. Li et al. reported that incorporation of a new infrared-absorbing polymer material provided by Sumitomo Chemical of Japan into their tandem device made the PCE jump to 10.6 % [14]. The German company Heliatek announced its achievement of 10.7 % PCE in April 2012. Heliatek plans to fabricate solar panels by vacuum deposition in a low-temperature, roll-to-roll process. It does not use a solvent, allowing for better control of crystal growth unlike spin-coating process [15]. It was also reported that 14–20 % PCE is predictable by taking the tandem structure through the numerical calculations [16].

6.3 Organic–Inorganic Hybrid Solar Cells

6.3.1 History of Organic–Inorganic Hybrid Solar Cells

Usually inexpensive organic materials can be easily processed by solution, and their functionality can be tailored by molecular design and chemical synthesis. On the other hand, inorganic semiconductors can be manufactured with various structures. Based on inorganic nanoparticles, their band gap can be tuned, and their absorption/emission spectra can be fitted [33]. Greenham et al. firstly studied hybrid systems containing CdS or CdSe nanoparticles embedded in poly[2-methoxy-5-(2′-ethylhexyloxy)-*p*-phenylenevinylene] (MEH-PPV) [34]. Excitons created upon photoexcitation are separated into free charge carriers very efficiently at interfaces between organic semiconductors and inorganic semiconductor in a hybrid composite thin film. Such structures are often referred later as "hybrid" devices. Crystalline and nanocrystalline inorganic semiconductors have several attributes as electron acceptors, including relatively high electron mobility, high electron affinities, and good physical and chemical stability. Solution-processable nanocrystalline semiconductors that can be prepared in different morphologies offer the potential for a large area interface when combined with a solution-processed organic component.

Hybrid organic polymer-inorganic structures can be prepared in four different ways as shown in Fig. 6.5. The first structure is a planar bilayer structure where an organic layer is deposited on top of an inorganic semiconductor layer. The next structure is a blend of nanocrystals with a polymer where semiconductor nanoparticles and a polymer are deposited from the same solution. The third one is a nanostructured porously inorganic structure where a connected semiconductor

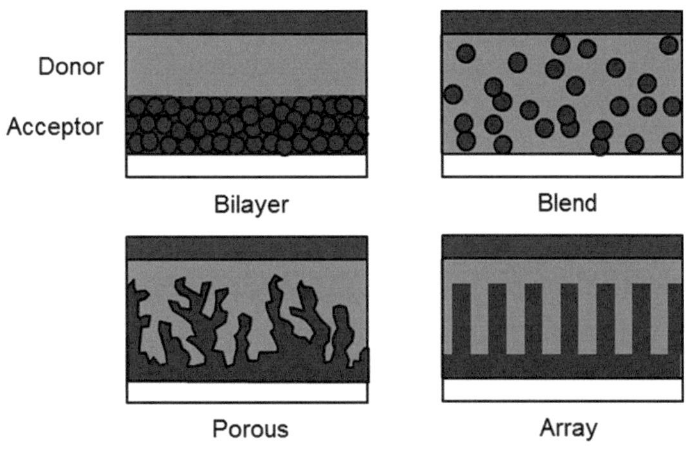

Fig. 6.5 Types of polymer-inorganic (donor-acceptor) heterojunctions; bilayer heterojunction, heterojunction based on nanoparticles dispersed in a donor matrix, heterojunction based on a porous acceptor phase filled with donor, and heterojunction containing vertically arrayed domains

layer is filled with a conjugated polymer. The last one is a vertically aligned inorganic semiconductor array which is filled with a conjugated polymer. In cases of bilayer, porous, and vertically aligned structures, charge separations occurring at the interface are helped to the opposite electrodes by diffusion, while in the blend, charge separations occurring at the interface rely on differences in the work function of the electrodes for direction of the photocurrent. Among various nanostructured inorganic materials, metal oxides, that is, titanium dioxide (TiO_2) and zinc oxide (ZnO), are of interest because of their processability and appropriate band gap [35].

6.3.2 Nanoporous or Nanoparticle Blends for Organic–Inorganic Hybrid Solar Cells

Studies of conjugated polymer-porous metal oxide (i.e., TiO_2) structures have focused on optimization of the layer structure of the device and choice of polymer materials. By optimization of the device thickness and electrode choice, groups achieved photocurrent external quantum efficiency (EQE) values of 10, 23, and 40 % [36–38]. The reduction of the polymer and TiO_2 layer thicknesses to the region of 50–100 nm is a key step for this optimization. In polymer-porous metal oxide structures, the porous metal oxide acceptor layer is normally deposited on top of a TCO substrate (i.e., ITO), the polymer is then fabricated on the metal oxide, and an evaporated metal is deposited on top of the active layers as a hole collecting contact. This polarity is the opposite of most commonly reported organic solar cell structures, so the hybrid cell is also called "inverted organic solar cell."

The work function of ITO (4.5–4.7 eV) is intermediate between typical HOMO and LUMO values for organic solar materials, so ITO can collect either electrons or holes. The problem in hybrid organic–inorganic solar cells is the low value of FF. If the ITO is not insulated from the donor material in the polymer/metal oxide structure, charge leakage occurs, which is the current density-voltage (J-V) characteristic of the device. The opposite electrode should have a sufficiently high work function to collect holes from the donor material. Poor matching of the anode work function and donor HOMO level leads to a sigmoidal J-V characteristic curve of a low conductivity interface and a low fill factor [39]. The power conversion efficiency for a polymer-porous metal oxide structure of 0.58 % was achieved for a multilayer device structure incorporating both these buffer layers and suitable electrode materials [40].

6.3.3 Vertically Aligned Nanostructures for Organic–Inorganic Hybrid Solar Cells

One-dimensional (1D) nanostructured metal oxides are promising electrodes for photovoltaic devices due to their high conductivity as a direct path for charge transport [7]. The 1D nanostructure has been promoted as the traditional *p-n*

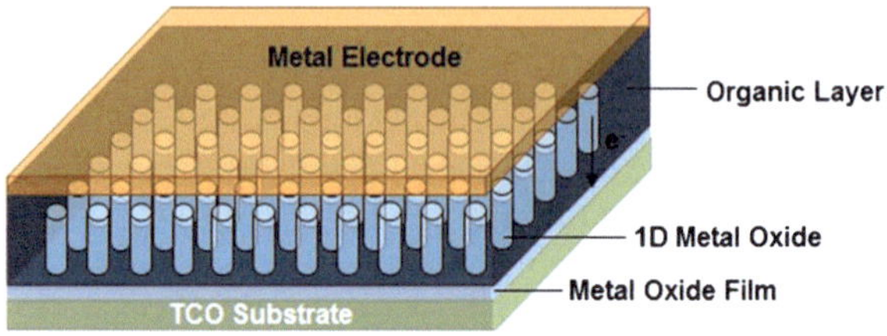

Fig. 6.6 Schematic representation of an organic–inorganic hybrid solar cell based on vertically arrayed 1D metal oxide nanostructured electrodes

junction solar cell performance. Two key constraints for this type of solar cell are (1) materials must be sufficiently thick and pure to absorb solar photon with energy above the material band gap as much as they can and (2) materials must have a high minority carrier diffusion length to effectively collect the photogenerated charge carriers.

An attractive method to improve the light absorption and charge-carrier collection involves high aspect ratio cylindrical absorbers, such as 1D nanowires [40]. To fabricate such solar cells, the following methods are required: (1) preparation of large area arrays that are vertically aligned, (2) fabrication of electrical junctions to such wire arrays, and (3) fabrication of electrical contacts to the backsides of the devices.

In hybrid organic–inorganic solar cells, generally the light-harvesting material is an organic semiconductor. Figure 6.6 shows a schematic representation of a hybrid organic–inorganic solar cell based on vertically arrayed 1D metal oxide nanostructured electrodes.

The applications of vertically arrayed metal oxide nanostructures arise from the need to overcome problems observed at the bulk heterojunction. In the blend type of the hybrid organic–inorganic device, nanoparticles are dispersed within the conjugated polymer matrix. The resulting polymer/metal oxide blend must be annealed to improve nanoparticle connection and to favor electron pathway through the electrode. Nevertheless, these devices are still limited by uncompleted exciton dissociation, due to the isolation of some particles from each other after the thermal annealing treatment. The latter eliminates the possibility of efficient electron transport reducing overall PCE. Figure 6.7 shows the schematic illustrations of the possible path taking place on different nanostructured electrodes made with nanoparticle and nanorod arrays. A nanostructured thin film made of vertical arrays provides a higher interfacial area between the donor and the acceptor materials (polymer/metal oxide, respectively) with highly efficient electron transport pathways.

The ideal nanostructures of the metal oxide could have vertical nanostructures of about 20–40 nm in diameter and the space between them of about 40 nm (the fact that electron diffusion length is in the range of 20 nm). Thus, a dense array of long and thin single crystalline array electrode is of a crucial importance to obtain higher surface and higher electron transport.

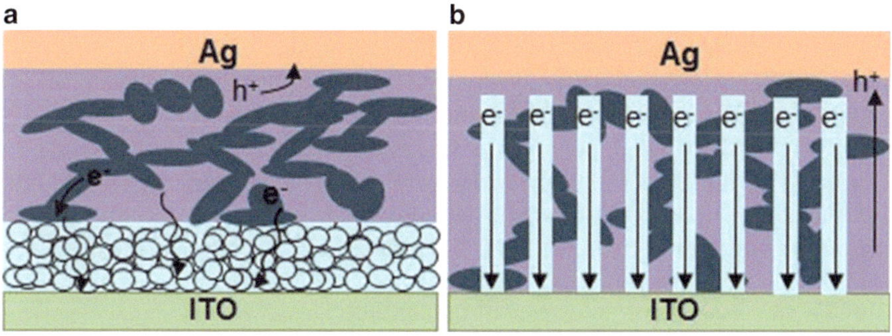

Fig. 6.7 Schematic illustrations of the possible paths taking place on different nanostructured electrodes made from (**a**) nanoparticles and (**b**) nanorod arrays

6.4 Electrospun Nanofibers for Organic Thin-Film Solar Cells and Organic–Inorganic Hybrid Solar Cells

6.4.1 Electrospun Nanofibers for Organic Thin-Film Solar Cells

The BHJ structure in organic solar cells provides a large interfacial surface area for efficient charge separation of excitons to be electrons and holes. Unfortunately, the as-prepared BHJ layers have low charge transport because of the discontinuous and folded topology of the interface of donor and acceptor. Thermal treatment, the so-called thermal annealing, which rearranges the continuous phase of electron donor and acceptor, could improve the charge transport in BHJ [41]. However, thermal annealing is relatively not compatible with the polymers with low glass transition temperature, not compatible with flexible substrates (e.g., transparent plastic substrates), and also not compatible with large area devices. Alternative processes that yield efficient interface of donor and acceptor with continuous phase of each material are necessary. The ideal structures are shown in Fig. 6.1c. Replacing the disordered phase in BHJ with an aligned nanostructured semiconductor to improve charge transport is one of the alternative methods. Electrospinning is a good method to fabricate nanofibers of conducting polymers which can be used as electron donor materials of organic solar cells.

6.4.1.1 Electrospinning of Conducting Polymers

Because of the limitations of solvents and high molecular weight of conducting polymers, only a few polymers such as derivatives of polythiophene (PT), polyaniline (PANI), polypyrrole (PPy), and poly(phenylenevinylene) (PPV) have been electrospun with various associated techniques. Many attempts have been made to overcome the problem of spinnability of conducting polymers by blending the conducting polymers with conventionally spinnable polymers. A derivative of PPV, poly(2-methoxy-5-(2′-ethylhexyloxy)-1,4-phenylenevinylene) (MEH-PPV), has been blended with polystyrene and spun to form ultrafine fibers [42, 43]. For other conducting polymers, it has also been reported that they could be fabricated by blending with polystyrene or poly(ethylene oxide) (PEO) [44–48]. Acid-doped PANI, PANI/PEO, and PANI/polystyrene blends were electrospun into nanofibers with diameters in the 96–275 nm range and electrical conductivities of 0.1^{-33} S/cm [45]. PPy composite fibers have been fabricated by blending with PEO [49]. By increasing the PPy content of the PPy/PEO nanofibers, a higher electrical conductivity has been achieved. Also, the higher concentration of PEO in the initial solution could allow easy stretching of fibers. This provides higher conductive pathways or charge-carrier mobility of PPy molecules along the fibers [50].

6.4.1.2 Modified Electrospinning for Fabrication of Pure Conducting Polymer Fibers

In the above blending systems, nonconducting polymers cannot be easily removed from the blends due to the conducting and nonconducting polymers that can be dissolved in a number of the same solvents. The obtained nanofibers could not be used as electron donor materials in organic solar cells. Therefore, the modified blend has been reported by blending a conducting polymer with a different-solvent extractable polymer in a mixed solvent. The solvent extraction is later carried out to remove nonconducting polymers. For example, MEH-PPV was blended with poly(vinyl pyrrolidone) (PVP), an easily spinnable and easily extractable polymer, in a mixed solvent of 1,2-dichloroethane and chlorobenzene [51] or chlorobenzene and methanol [52–54]. After the removal of PVP, pure MEH-PPV fibers were obtained as a ribbonlike structure aligned with wrinkled surface in fiber direction [53, 54].

Electrospinning of the blends of MEH-PPV and PVP in individually prepared solutions by using a coaxial two-capillary spinneret has been reported [55, 56]. Instead of such polymers, electrospinning of the mixture of polymer precursor and PVP followed by the polymerization had also been reported to prepare uniform nanofibers [57, 58]. Extraction of PVP out from the electrospun fibers in those techniques can be carried out easily, but the design of the spinneret for the first case and the post-thermal conversion of PPV precursor to PPV for the later one are quite complicated.

6.4.1.3 Recent Applications of Electrospun Nanofibers for Organic Thin-Film Solar Cells

We have first reported organic solar cells made of electrospun conducting nanofibers from a derivative of PPV (i.e., MEH-PPV) in 2008 [59]. MEH-PPV

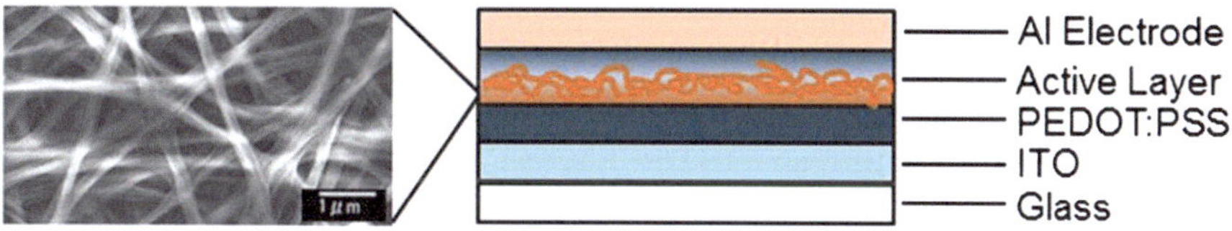

Fig. 6.8 A cell structure and morphology of electrospun fiber-based organic thin-film cells

was blended with PVP for electrospinning. PVP was removed from the composite fibers by Soxhlet extraction [53, 54]. Obtained fibers were applied to organic solar cells [59]. The structure of the cell was glass substrate/ITO/poly(3,4-ethylenedioxythiophene):poly(styrenesulfonate) (PEDOT:PSS)/active layer (MEH-PPV fibers and PCBM)/Al as shown in Fig. 6.8. Fibers were directly spun on the substrate, while PCBM was covered in the active layer by spin coating. Because the conducting polymer was MEH-PPV and MEH-PPV/PCBM was not formed as BHJ in the fibers, PCE of the cells was relatively low (0.2 %).

In 2009, the fabrication of nanofibers from another conducting polymer (a derivative of PPV, that is, poly[2,5-(20-ethylhexyloxy)]-1,4-phenylenevinylene (BEH-PPV)) by electrospinning with PVP blend was reported by Liu et al. [60]. A composite of BEH-PPV/PCBM nanofibers was also successfully fabricated in that report. This study showed the feasibility of fabrication of donor-acceptor composite nanofibers by electrospinning.

Sundarrajan et al. reported the use of electrospinning with coaxial two-capillary spinneret to fabricate solar cells made of P3HT and PCBM composite nanofibers [61]. PVP was used as the shell to form core-shell fibers in which P3HT and PCBM were in the core. PVP shell was extracted by soaking in ethanol. Obtained fibers were applied to organic solar cells with very low PCE of 8.7×10^{-8} %. This is much lower than those of currently reported PCEs. Low conversion efficiency could be attributed to the large diameter of P3HT fibers (1.2 μm) and incompatible core-shell fibers with nonconducting material as shell. The fiber mat morphology seems to interrupt a charge transport process by limited fiber-to-fiber contacts and scattered transport directions along unaligned fibers [62].

Bedford et al. also reported the P3HT/PCBM/polymer composite nanofibers fabricated by co-electrospinning with polycaprolactone (PCL) using a coaxial two-capillary spinneret [63]. PCL was extracted after electrospinning to obtain P3HT/PCBM nanofibers. To overcome the limitation of fiber-to-fiber contact and porous nature of fibers, a P3HT/PCBM solution was coated on top of a P3HT/PCBM fiber by spin coating as a P3HT/PCBM backfill layer. The donor-acceptor interfacial area could be considerably increased. The devices made of nanofibers have relatively high PCE of 4.0 %. The charge extraction was also improved by the addition of a hole transporting layer of PEDOT:PSS.

Again, the design and preparation of the coaxial two-capillary spinneret were complicated. Recently, we have successfully applied electrospinning to fabricate thin nanofibers of P3HT without using the coaxial two-capillary spinneret [64–66].

The P3HT nanofibers exhibited a specific feature of stronger interchain contribution, better crystallinity, and greater optical property than P3HT films. By combining this technique with the backfill layer as reported by Bedford et al. [63], efficient organic solar cells made of electrospun P3HT/PCBM nanofibers should be obtained.

6.4.2 Electrospun Nanofibers for Organic–Inorganic Hybrid Solar Cells

To improve the efficiencies of organic solar cells, the combination of organic and inorganic semiconductor as organic–inorganic hybrid photovoltaic cells has been extensively investigated to overcome the low mobility of carriers by using nanostructured inorganic materials as electron transporting layers [7]. Vertically aligned nanorod or nanotube arrays of metal oxide are interesting electron transporting layers. Hybrid cells fabricated with vertically aligned arrays are expected to provide very high efficiency by reducing the electron-hole recombination and improving the charge transport. However, the organic–inorganic interface in such kind of cells is quite low because the diameter of vertically aligned nanorods or nanotubes in the present researches is still relatively large. Moreover, uncontrolled space in vertically aligned arrays is typically obtained from the solution method. The space is usually in the range of 100 nm [67]. Therefore, the very high-performance cell has not yet done [68, 69]. The use of inorganic semiconductors in the form of electrospun nanofibers as electron acceptors and/or electron transporting layers is a promisingly alternative way.

6.4.2.1 Metal Oxide Nanofibers/Conducting Polymer-Based Devices

Inorganic nanofibers provide high electron transportation in hybrid solar cells. They can also improve the interfacial area of electron donor-acceptor and improve the dissociation of the excitons. An extreme increase in external quantum efficiency (EQE) (~57 % at 515 nm) of the hybrid cells was found when a blend of P3HT/PCBM was added to ZnO nanofibers [70]. It should be noted that these fibers were not fabricated by electrospinning.

Metal oxides, that is, TiO_2 and ZnO, are electrospinnable to form uniform nanofibers, but they cannot be spun from neat precursor solutions. The metal oxide nanofibers are usually electrospun from the solution of metal oxide precursors with a spinnable polymer, that is, PVP, because solutions of metal oxide precursors cannot provide enough entanglement and viscosity for fiber formation. For example, we reported the fabrication of TiO_2 nanofibers by the combination of electrospinning and solgel techniques by using PVP, titanium(IV) butoxide (TiO_2 precursor), and acetylacetone in methanol as a spinning solution [35, 71]. After electrospinning, sintering or the so-called calcination at high temperature (i.e., 400–500 °C) has to be carried out to remove all organic contents and change the phase of metal oxides to crystalline forms [35, 71].

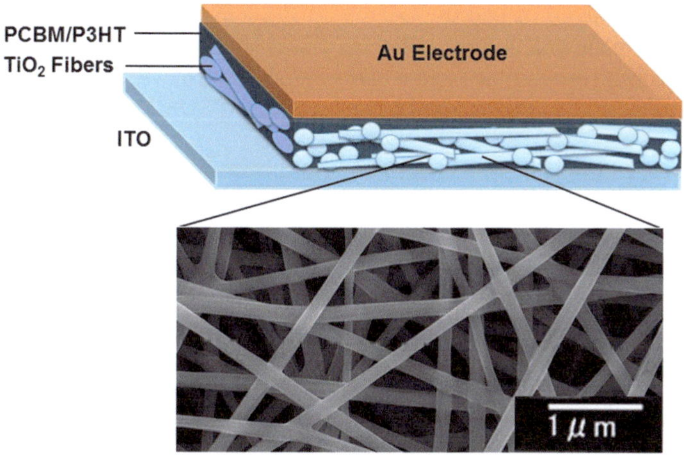

Fig. 6.9 A cell structure and an SEM image of an organic–inorganic hybrid device made of electrospun TiO_2 nanofibers as an electron transporting layer

Shim et al. fabricated the hybrid devices with the active layer composed of multilayer electrospun TiO_2 nanofiber array and MEH-PPV [72]. The nanofibers were randomly, horizontally aligned or cross-aligned on the substrates. Zhu et al. reported the improved adhesion of interconnected random TiO_2 nanofibers on P3HT and the photovoltaic characterizations of the resulted devices [73]. Low PCE of 0.29 % was obtained from the report. Tanveer et al. fabricated hybrid solar cells made of TiO_2 nanofibers/P3HT with a dye (N719) as hybrid composites [74]. Hybrid cells made of ordered electrospun ZnO nanofibers and P3HT were also reported [75]. With an ordered structure (cross-aligned multilayer) of ZnO nanofibers, high interfacial area of ZnO and polymer is relatively high which results in quite high PCE of 0.51 %. In these reports, metal oxide nanofibers were used as electron acceptor in hybrid devices.

In 2011, we demonstrated the use of electrospun TiO_2 nanofibers as an electron transporting layer for hybrid photovoltaic cells [7, 76]. The electron donor and acceptor in our system were typical P3HT and PCBM, respectively [7]. Figure 6.9 shows the cell structure and morphology of hybrid devices made of electrospun TiO_2 nanofibers. With the addition of a hole collecting layer, that is, PEDOT:PSS, between the active layer and Au electrode, PCE of 1.27 % could be achieved [7]. Tanveer et al. reported the improvement of PCE of 2.23 % in the cell made of P3HT/PCBM/ZnO nanofibers by optimization of the thickness of nanofiber network [77].

It can be concluded that the nanofibers provide many advantages in hybrid solar cell applications, including low cost, high charge-carrier mobility, good chemical and mechanical stabilities, and easy control of size and shape [62]. Electrospinning seems to be a simple method to control the porosity and thickness of fiber networks, which can assist complete infiltration of polymers into the networks. Last but not least, it should be noted that the exciton diffusion length in BHJ is about 10 nm or lower in conducting polymers. Hence, closer spacing of adjacent nanofibers is necessary for the increase of exciton dissociation and achievement of larger J_{sc}.

6.4.2.2 Quantum Dots in Hybrid Devices

Quantum dots (QDs) are semiconductor particles of crystals made from chemical compounds from periodic groups II–VI, III–V, or IV–VI, with small diameters. Excitons are confined in all three spatial dimensions of QDs. Multi-exciton generation can be achieved in QDs upon absorption of a single photon. For example, upon absorption of a single photon, crystalline Si and PbSe QDs generate 1.3 and 7 excitons, respectively [78, 79]. A large interfacial area can be obtained from well dispersion of QDs within a conducting polymer [80]. One-dimensional QD nanostructures have been reported for efficient QD-organic–inorganic hybrid system for PV devices. QDs can provide unidirectional electrical transport [10].

In the literature, hybrid solar cells with the active layer composed of CdS-coated cellulose acetate fibers and P3HT were reported by Cortina et al. [81]. Wu et al. reported the enhanced performance of hybrid solar cells made of ZnO nanofibers by modifying the surface of fibers with CdS [82]. To apply more electrospun nanofibers and QDs to hybrid solar cells, the generation of effective interface of electrospun metal oxide nanofibers with QDs before blending with polymer is necessary. Dispersion of QDs in electrospun fibers may be also an alternative way for hybrid applications.

6.4.3 Electrospun Nanofibers for Electron Conducting Substrates

Fabrication of electron conducting materials by electrospinning is possible. Metal nanofibers can be fabricated by co-electrospinning of the metal precursors with electrospinnable polymers similar to the fabrication of metal oxide nanofibers. After the electrospinning heat treatment at a high temperature under a suitable condition is also necessary to converse the untreated fibers to transform to metal fibers. A number of metal nanofibers, for example, Fe, Co, Ni, and Cu, were successfully fabricated by electrospinning [83, 84]. Organic solar cells using copper nanowire networks as transparent electrodes that have a power efficiency of 3.0 %, comparable to devices made with ITO electrodes, were reported [84]. Electrospinning of other electron conducting materials, such as carbon nanotubes, with polymers and the electrical properties of the electrospun products was also reported [85]. These electron conducting nanofibers may apply as photoelectrodes for both organic and hybrid solar cells.

We have demonstrated the fabrication of electrospun ITO nanofibers [86]. The fibers were fabricated by electrospinning of a mixed solution of PVP and ITO precursors. Calcination of the composite fibers at 570 °C resulted in highly crystalline nanofibers and degradation of PVP. The obtained ITO nanofibers showed one-dimensional structure with the diameters about 200 nm. These nanofibers were applied to construct the dye-sensitized solar cells (DSSCs) as an upper layer of the transparent conducting oxide (TCO) substrate (between conventional TCO and the metal oxide photoelectrode). When the ITO nanofibers were applied, enhanced photovoltaic performance was attained as compared with the conventional devices. Properties of ITO nanofibers that are very interesting are the mechanical flexibility and high electrical conductivity. This implies that electrospun ITO nanofibers may be promising photoelectrodes for flexible organic and hybrid solar cells without using rigid TCO glasses.

6.5 Conclusions

Organic solar cell is a type of solar cell that uses organic materials, especially small organic molecules or conducting organic polymers, for light absorption and charge transport to produce electricity by the photovoltaic effect. It needs large interfacial surface area for efficient charge separation of excitons, which is normally achieved by thermal annealing. However, this annealing is relatively not compatible with the polymers with low glass transition temperature, not compatible with flexible substrates, and also not compatible with large area devices. Alternative processes that yield efficient interface of electron donor and acceptor with continuous phase of each material are necessary. Electrospinning is a good method to fabricate nanofibers of conducting polymers which can be used as electron donor materials of organic solar cells. To improve the efficiencies of organic solar cells, the combination of organic and inorganic semiconductor as organic–inorganic hybrid photovoltaic cells has been extensively investigated to overcome the low mobility of carriers by using nanostructured inorganic materials as electron transporting layers. Hybrid cells fabricated with vertically aligned arrays are expected to provide very high efficiency by reducing electron-hole recombination and improving the charge transport. However, the organic–inorganic interface in such kind of cells is quite low because the diameter of vertically aligned nanorods or nanotubes in the present researches is still relatively large. The use of inorganic semiconductors in the form of electrospun nanofibers as electron acceptors and/or electron transporting layers is a promising alternative way. Nanofiber-based transparent electrodes can also be fabricated by electrospinning with flexible property. These electrodes are very promising to be utilized as flexible substrates for both organic and hybrid solar cells. The development of fabrication techniques and new materials in electrospinning to tune up the properties of electrospun products will be the focus for development of nanofiber-based organic and hybrid solar cells.

Acknowledgment The authors thank for all cited reports as the sources of this chapter. The author (S.C.) thanks Building Energy Science and Technology Laboratory (BEST) and Advanced Fuel Processing Laboratory (AFPL) at The Joint Graduate School of Energy and Environment, King Mongkut's University of Technology Thonburi, for the allowed time to review and write this book chapter.

References

1. Rivers PN (ed) (2007) Leading edge research in solar energy. Nova Science Publishers, New York. ISBN 1600213367
2. Halls JJM, Friend RH (2001) Organic photovoltaic devices. In: Archer MD, Hill RD (eds) Clean electricity from photovoltaics. Imperial College Press, London, pp 377–445. ISBN 1860941613
3. Nelson J (2002) Organic photovoltaic films. Curr Opin Solid State Mat Sci 6:87–95. doi:10.1016/S1359-0286(02)00006-2
4. Hoppe H, Sariciftci NS (2004) Organic solar cells: an overview. J Mater Res 19:1924–1945. doi:10.1557/JMR.2004.0252
5. McGehee DG, Topinka MA (2006) Solar cells: pictures from the blended zone. Nat Mater 5:675–676. doi:10.1038/nmat1723
6. Brabec CJ, Sariciftci NS, Hummelen JC (2001) Plastic solar cell. Adv Funct Mater 11:15–26. doi:10.1002/1616-3028(200102)11:1<15::AID-ADFM15>3.0.CO;2-A
7. Chuangchote S, Sagawa T, Yoshikawa S (2011) Electrospun TiO2 nanowires for hybrid photovoltaic cells. J Mater Res 26:2316–2321. doi:10.1557/jmr.2011.167
8. White MS, Olson DC, Shaheen SE, Kopidakis N, Ginley DS (2006) Inverted bulk-heterojunction organic photovoltaic device using a solution-derived ZnO underlayer. Appl Phys Lett 89:143517. doi:10.1063/1.2359579
9. Mor GK, Shankar K, Paulose M, Varghese OK, Grimes CA (2007) High efficiency double heterojunction polymer photovoltaic cells using highly ordered TiO2 nanotube arrays. Appl Phys Lett 91:152111. doi:10.1063/1.2799257
10. Huynh WU, Dittmer JJ, Alivisatos AP (2002) Hybrid nanorod-polymer solar cells. Science 295:2425–2427. doi:10.1126/science.1069156
11. Steim R, Choulis SA, Schilinsky P, Brabec CJ (2008) Interface modification for highly efficient organic photovoltaics. Appl Phys Lett 92:093303. doi:10.1063/1.2885724
12. Cheng Y-J, Hsieh C-H, He Y, Hsu C-S, Li Y (2010) Combination of indene-C60 bis-adduct and cross-linked fullerene interlayer leading to highly efficient inverted polymer solar cells. J Am Chem Soc 132:17381–17383. doi:10.1021/ja108259n
13. Yang T, Cai W, Qin D, Wang E, Lan L, Gong X, Peng J, Cao Y (2010) Solution-processed zinc oxide thin film as a buffer layer for polymer solar cells with an inverted device structure. J Phys Chem C 114:6849–6853. doi:10.1021/jp1003984
14. Li G, Zhu R, Yang Y (2012) Polymer solar cells. Nat Photon 6:153–161. doi:10.1038/nphoton.2012.11
15. Wu C, Murata H (2012) Thinking small for solar. MRS Bull 37:194–195. doi:10.1557/mrs.2012.62
16. Kotlarski JD, Blom PWM (2011) Ultimate performance of polymer: fullerene bulk heterojunction tandem solar cells. Appl Phys Lett 98:053301. doi:10.1063/1.3549693
17. Gosh AK, Feng T (1978) Merocyanine organic solar cells. J Appl Phys 49:5982–5989. doi:10.1063/1.324566
18. Tang CW (1986) Two-layer organic photovoltaic cell. Appl Phys Lett 48:183. doi:10.1063/1.96937
19. Hiramoto M, Suezaki M, Yokoyama M (1990) Effect of thin gold interstitial-layer on the photovoltaic properties of tandem organic solar cell. Chem Lett 19:327–330. doi:10.1246/cl.1990.327

20. Sariciftci NS, Smilowitz L, Heeger AJ, Wudl F (1992) Photoinduced electron transfer from a conducting polymer to buckminsterfullerene. Science 258:1474. doi:10.1126/science.258.5087.1474
21. Morita S, Zakhidov AA, Yoshino K (1992) Doping effect of buckminsterfullerene in conducting polymer: change of absorption spectrum and quenching of luminescence. Solid State Commun 82:249–252. doi:10.1016/0038-1098(92)90636-N
22. Marks RN, Halls JJM, Bradley DDC, Friend RH, Holmes AB (1994) The photovoltaic response in poly(p-phenylene vinylene) thin-film devices. J Phys Condens Matter 6: 1379–1394. doi:10.1088/0953-8984/6/7/009
23. Yu G, Zhang C, Heeger AJ (1994) Dual-function semiconducting polymer devices: light-emitting and photodetecting diodes. Appl Phys Lett 64:1540–1542. doi:10.1063/1.111885
24. Sariciftci NS, Braun D, Zhang C, Srdanov VI, Heeger AJ, Stucky G, Wudl F (1993) Semiconducting polymer-buckminsterfullerene heterojunctions: diodes, photodiodes, and photovoltaic cells. Appl Phys Lett 62:585–587. doi:10.1063/1.108863
25. Roman LS, Mammo W, Petterson LAA, Andersson MR, Inganäs O (1998) High quantum efficiency polythiophene. Adv Mater 10:774–777. doi:10.1002/(SICI)1521-4095(199807)10:10<774::AID-ADMA774>3.0.CO;2-J
26. Yu G, Gao J, Hummelen JC, Wudl F, Heeger AJ (1995) Polymer photovoltaic cells: enhanced efficiencies via a network of internal donor-acceptor heterojunctions. Science 270:1789–1791. doi:10.1126/science.270.5243.1789
27. Shaheen SE, Brabec CJ, Sariciftci NS, Padinger F, Fromherz T, Hummelen CJ (2001) 2.5 % efficient organic plastic solar cells. Appl Phys Lett 78:841. doi:10.1063/1.1345834
28. Camaioni N, Ridolfi G, Casalbore-Miceli G, Possamai G, Maggini M (2002) The effect of a mild thermal treatment on the performance of poly(3-alkylthiophene)/fullerene solar cells. Adv Mater 14:1735–1738. doi:10.1002/1521-4095(20021203)
29. Pandinger F, Rittberger RS, Saricifti NS (2003) Effects of postproduction treatment on plastic solar cells. Adv Funct Mater 13:85–88. doi:10.1002/adfm.200390011
30. Kim Y, Cook S, Tuladhar SM, Choulis SA, Nelson J, Durrant JR, Bradley DDC, Giles M, McCulloch I, Ha C-S, Ree M (2006) A strong regioregularity effect in self-organizing conjugated polymer films and high-efficiency polythiophene: fullerene solar cells. Nat Mater 5:197–203. doi:10.1038/nmat1574
31. Tajima K, Suzuki Y, Hashimoto K (2008) Polymer photovoltaic devices using fully regioregular poly[(2-methoxy-5-(3',7'-dimethyloctyloxy))-1,4-phenylenevinylene]. J Phys Chem C 112:8507–8510. doi:10.1021/jp802688s
32. Hiorns RC, de Bettignies R, Leroy J, Bailly S, Firon M, Sentein C, Khoukh A, Preud'homme H, Dagron-Lartigau C (2006) High molecular weights, polydispersities, and annealing temperatures in the optimization of bulk-heterojunction photovoltaic cells based on poly(3-hexylthiophene) or poly(3-butylthiophene). Adv Funct Mater 16:2263–2273. doi:10.1002/adfm.200600005
33. Weller H (1993) Colloidal semiconductor Q-particles: chemistry in the transition region between solid state and molecules. Angew Chem Int Ed Engl 32:41–53. doi:10.1002/anie.199300411
34. Greenham NC, Peng X, Alivisatos AP (1996) Charge separation and transport in conjugated-polymer/semiconductor-nanocrystal composites studied by photoluminescence quenching and photoconductivity. Phys Rev B 54:17628–17637. doi:10.1103/PhysRevB.54.17628
35. Chuangchote S, Jitputti J, Sagawa T, Yoshikawa S (2009) Photocatalytic activity for hydrogen evolution of electrospun TiO2 Nanofibers. ACS Appl Mater Interfaces 1:1140–1143. doi:10.1021/am9001474
36. Fan Q, McQuillin B, Bradley DDC, Whitelegg S, Seddon AB (2001) A solid state solar cell using sol-gel processed material and a polymer. Chem Phys Lett 347:325–330. doi:10.1016/S0009-2614(01)01003-X
37. Arango AC, Johnson LR, Bliznyuk VN, Schlesinger Z, Carter SA, Hörhold H-H (2000) Efficient titanium oxide/conjugated polymer photovoltaics for solar energy conversion. Adv Mater 12:1689. doi:10.1002/1521-4095(200011)12:22<1689::AID-ADMA1689>3.0.CO;2-9

38. Ravirajan P, Bradley DDC, Nelson J, Haque SA, Durrant JR, Smit HJP, Kroon JM (2005) Efficient charge collection in hybrid polymer/TiO2 solar cells using poly(ethylenedioxythiophene)/polystyrene sulphonate as hole collector. Appl Phys Lett 86:143101. doi:10.1063/1.1890468
39. Nelson J, Kirkpatrick J, Ravirajan P (2004) Factors limiting the efficiency of molecular photovoltaic devices. Phys Rev B 69:035337. doi:10.1103/PhysRevB.69.035337
40. Tsakalakos L, Balch J, Fronheiser J, Korevaar BA, Sulima O, Rand J (2007) Silicon nanowire solar cells. Appl Phys Lett 91:233117. doi:10.1063/1.2821113
41. Chuangchote S, Ruankham P, Sagawa T, Yoshikawa S (2010) Improvement of power conversion efficiency in organic photovoltaics by slow cooling in annealing treatment. Appl Phys Express 3:122302. doi:10.1143/APEX.3.122302
42. Wutticharoenmongkol P, Supaphol P, Srikhirin T, Kerdcharoen T, Osotchan T (2005) Electrospinning of polystyrene/poly(2-methoxy-5-(2′-ethylhexyloxy)-1,4-phenylene vinylene) blends. J Polym Sci B 43:1881–1891. doi:10.1002/polb.20478
43. Chuangchote S, Srikhirin T, Supaphol P (2007) Color change of electrospun polystyrene/MEH-PPV fibers from orange to yellow through partial decomposition of MEH side groups. Macromol Rapid Commun 28:651–659. doi:10.1002/marc.200600769
44. Kahol PK, Pinto NJ (2004) An EPR investigation of electrospun polyaniline-polyethylene oxide blends. Synth Met 140:269–272. doi:10.1016/S0379-6779(03)00370-9
45. MacDiarmid AG, Jones WE Jr, Norris ID, Gao J, Johnson AT Jr, Pinto NJ, Hone J, Han B, Ko FK, Okuzaki H, Llaguno M (2001) Electrostatically-generated nanofibers of electronic polymers. Synth Met 119:27–30. doi:10.1016/S0379-6779(00)00597-X
46. Norris ID, Shaker MM, Ko FK, MacDiarmid AG (2000) Electrostatic fabrication of ultrafine conducting fibers: polyaniline/polyethylene oxide blends. Synth Met 114:109–114. doi:10.1016/S0379-6779(00)00217-4
47. Sun Z, Zussman E, Yarin AL, Wendorff JH, Greiner A (2003) Compound core-shell polymer nanofibers by co-electrospinning. Adv Mater 15:1929–1932. doi:10.1002/adma.200305136
48. Yu JH, Fridrikh SV, Rutledge GC (2004) Production of submicrometer diameter fibers by two-fluid electrospinning. Adv Mater 16:1562–1566. doi:10.1002/adma.200306644
49. Chronakis IS, Grapenson S, Jakob A (2006) Conductive polypyrrole nanofibers via electrospinning: electrical and morphological properties. Polymer 47:1597–1603. doi:10.1016/j.polymer.2006.01.032
50. Lu X, Zhao Q, Liu X, Wang D, Zhang W, Wang C, Wei Y (2006) Preparation and characterization of polypyrrole/TiO2 coaxial nanocables. Macromol Rapid Commun 27: 430–434. doi:10.1002/marc.200500810
51. Zhao Q, Huang Z, Wang C, Zhao Q, Sun H, Wang D (2007) Preparation of PVP/MEH-PPV composite polymer fibers by electrospinning and study of their photoelectronic character. Mater Lett 61:2159–2163. doi:10.1016/j.matlet.2006.08.046
52. Chuangchote S, Sagawa T, Yoshikawa S (2008) Fabrication and optical properties of electrospun organic semiconductor nanofibers from blended polymer solution. Mater Res Soc Symp Proc 1091:1091-AA07-85, 89–94. doi:10.1557/PROC-1091-AA07-85
53. Chuangchote S, Sagawa T, Yoshikawa S (2008) Fabrication and optical properties of electrospun conductive polymer nanofibers from blended polymer solution. Jpn J Appl Phys 47:787–793. doi:10.1143/JJAP.47.787
54. Chuangchote S, Sagawa T, Yoshikawa S (2008) Ultrafine electrospun conducting polymer blend fibers and their photoluminescence properties. Macromol Symp 264:80–89. doi:10.1002/masy.200850413
55. Babel A, Li D, Xia Y, Jenekhe SA (2005) Electrospun nanofibers of blends of conjugated polymers: morphology, optical properties, and field-effect transistors. Macromolecules 38:4705–4711. doi:10.1021/ma047529r
56. Li D, Babel A, Jenekhe SA, Xia Y (2004) Nanofibers of conjugated polymers prepared by electrospinning with a two-capillary spinneret. Adv Mater 16:2062–2066. doi:10.1002/adma.200400606

57. Xin Y, Huang ZH, Yan EY, Zhang W, Zhao Q (2006) Controlling poly(p-phenylene vinylene)/poly(vinyl pyrrolidone) composite nanofibers in different morphologies by electrospinning. Appl Phys Lett 89:053101. doi:10.1063/1.2236382
58. Zhang W, Huang Z, Yan E, Wang C, Xin Y, Zhao Q, Tong Y (2007) Preparation of poly(phenylene vinylene) nanofibers by electrospinning. Mater Sci Eng A 443:292–295. doi:10.1016/j.msea.2006.05.147
59. Chuangchote S, Sagawa T, Yoshikawa S (2008) Fiber-based bulk-heterojunction organic photovoltaic cells. Mater Res Soc Symp Proc 1149:1149-QQ11-04, 25–29
60. Liu HA, Zepeda D, Ferraris JP, Balkus KJ Jr (2009) Electrospinning of poly(alkoxyphenylenevinylene) and methanofullerene nanofiber blends. ACS Appl Mater Interfaces 1:1958–1965. doi:10.1021/am900338w
61. Sundarrajan S, Murugan R, Nair AS, Ramakrishna S (2010) Fabrication of P3HT/PCBM solar cloth by electrospinning technique. Mater Lett 64:2369–2372. doi:10.1016/j.matlet.2010.07.054
62. Cho H, Min S-Y, Lee T-W (2013) Electrospun organic nanofiber electronics and photonics. Macromol Mater Eng 298:475–486. doi:10.1002/mame.201200364
63. Bedford NM, Dickerson MB, Drummy LF, Koerner H, Singh KM, Vasudev MC, Durstock MF, Naik RR, Steckl AJ (2012) Nanofiber-based bulk-heterojunction organic solar cells using coaxial electrospinning. Adv Energy Mater 2:1136–1144. doi:10.1002/aenm.201100674
64. Chuangchote S, Fujita M, Sagawa T, Yoshikawa S (2010) Fabrication and characterizations of poly(3-hexylthiophene) nanofibers. Mater Res Soc Symp Proc 1270:1270-HH14-07, 55–60. doi:10.1557/PROC-1270-HH14-07
65. Chuangchote S, Fujita M, Sagawa T, Yoshikawa S (2010) Electrospun polythiophene nanofibers and their applications for organic solar cells. Mater Res Soc Symp Proc 1303:1303-Y03-28, 63–67. doi:10.1557/opl.2011.402
66. Chuangchote S, Fujita M, Sagawa T, Sakaguchi H, Yoshikawa S (2010) Control of self organization in conjugated polymer fibers. ACS Appl Mater Interfaces 2:2995–2997. doi:10.1021/am1008198
67. Thavasi V, Singh G, Ramakrishna S (2008) Electrospun nanofibers in energy and environmental applications. Energy Environ Sci 1:205–221. doi:10.1039/B809074M
68. Ravirajan P, Peiró AM, Nazeeruddin MK, Graetzel M, Bradley DDC, Durrant JR, Nelson J (2006) Hybrid polymer/zinc oxide photovoltaic devices with vertically oriented ZnO nanorods and an amphiphilic molecular interface layer. J Phys Chem B 110:7635–7639. doi:10.1021/jp0571372
69. Wang GT, Talin AA, Werder DJ, Creighton JR, Lai E, Anderson RJ, Arslan I (2006) Highly aligned, template-free growth and characterization of vertical GaN nanowires on sapphire by metal-organic chemical vapour deposition. Nanotechnology 17:5773–5780. doi:10.1088/0957-4484/17/23/011
70. Olson DC, Piris J, Collins RT, Shaheen SE, Ginley DS (2005) Hybrid photovoltaic devices of polymer and ZnO nanofiber composites. Thin Solid Films 496:26–29. doi:10.1016/j.tsf.2005.08.179
71. Chuangchote S, Sagawa T, Yoshikawa S (2008) Efficient dye-sensitized solar cells using electrospun TiO2 nanofibers as a light harvesting layer. Appl Phys Lett 93:033310. doi:10.1063/1.2958347
72. Shim H-S, Na S-I, Nam SH, Ahn H-J, Kim HJ, Kim D-Y, Kim WB (2008) Efficient photovoltaic device fashioned of highly aligned multilayers of electrospun TiO2 nanowire array with conjugated polymer. Appl Phys Lett 92:183107. doi:10.1063/1.2919800
73. Zhu R, Jiang C-Y, Liu X-Z, Liu B, Kumar A, Ramakrishna S (2008) Improved adhesion of interconnected TiO2 nanofiber network on conductive substrate and its application in polymer photovoltaic devices. Appl Phys Lett 93:013102. doi:10.1063/1.2907317
74. Tanveer M, Habib A, Khan MB (2013) Dependence of organic/inorganic photovoltaic device performance on precursor's concentration used for electrospun TiO2 nanofibers. Nano 8:1350033. doi:10.1142/S1793292013500331

75. Wu S, Tai Q, Yan F (2010) Hybrid photovoltaic devices based on poly (3-hexylthiophene) and ordered electrospun ZnO nanofibers. J Phys Chem C 114:6197–6200. doi:10.1021/jp910921a
76. Chuangchote S, Sagawa T, Yoshikawa S (2011) Electrospun TiO2 nanofibers for organic–inorganic hybrid photovoltaic cells. Mater Res Soc Symp Proc 1359:1359-NN07-09. doi:10.1557/opl.2011.882
77. Tanveer M, Habib A, Khan MB (2012) Improved efficiency of organic/inorganic photovoltaic devices by electrospun ZnO nanofibers. Mater Sci Eng B 177:1144–1148. doi:10.1016/j.mseb.2012.05.025
78. Kolodinski S, Werner JH, Wittchen T, Queisser HJ (1993) Quantum efficiencies exceeding unity due to impact ionization in silicon solar cells. Appl Phys Lett 63:2405–2407. doi:10.1063/1.110489
79. Schaller RD, Sykora M, Pietryga JM, Klimov VI (2006) Seven excitons at a cost of one: redefining the limits for conversion efficiency of photons into charge carriers. Nano Lett 6: 424–429. doi:10.1021/nl052276g
80. Greenham NC, Peng X, Alivisatos AP (1997) Charge separation and transport in conjugated polymer/cadmium selenide nanocrystal composites studied by photoluminescence quenching and photoconductivity. Synth Met 84:545–546. doi:10.1016/S0379-6779(97)80852-1
81. Cortina H, Martínez-Alonso C, Castillo-Ortega M, Hu H (2012) Cellulose acetate fibers covered by CdS nanoparticles for hybrid solar cell applications. Mater Sci Eng B 177: 1491–1496. doi:10.1016/j.mseb.2012.02.014
82. Wu S, Li J, Lo S-C, Tai Q, Yan F (2012) Enhanced performance of hybrid solar cells based on ordered electrospun ZnO nanofibers modified with CdS on the surface. Org Electron 13: 1569–1575. doi:10.1016/j.orgel.2012.04.018
83. Wu H, Zhang R, Liu X, Lin D, Pan W (2007) Electrospinning of Fe, Co, and Ni nanofibers: synthesis, assembly, and magnetic properties. Chem Mat 19:3506–3511. doi:10.1021/cm070280i
84. Wu H, Hu L, Rowell MW, Kong D, Cha JJ, McDonough JR, Zhu J, Yang Y, McGehee MD, Cui Y (2010) Electrospun metal nanofiber webs as high-performance transparent electrode. Nano Lett 10:4242–4248. doi:10.1021/nl102725k
85. Zhang Q, Chang Z, Zhu M, Mo X, Chen D (2007) Electrospun carbon nanotube composite nanofibres with uniaxially aligned arrays. Nanotechnology 18:115611. doi:10.1088/0957-4484/18/11/115611
86. Chuangchote S, Sagawa T, Yoshikawa S (2011) Indium tin oxide nanofibers and their applications for dye-sensitized solar cells. ECS Trans 41:223–229. doi:10.1149/1.3629970

Chapter 7
Application of Nanofibers in Supercapacitors

Kai Wei and Ick Soo Kim

Abstract Recently, rapidly growing demand for high-speed rechargeable energy storage device with high performance has motivated a lot of researchers to develop new materials for supercapacitor. Supercapacitor is a kind of electrochemical device which stores and releases energy at extremely high rate with high power density and long cycling life. However, the energy density of supercapacitor is generally lower than that of batteries. Improving the energy density of supercapacitors while maintaining their high power density and long cycle life has been the key issue in developing future energy storage systems. Nanotechnologies, especially nanofibers, have received a great deal of attentions from various fields, such as medicine, electrical science, and energy resources, where their unique properties contribute to product functionality. As is well known, one of the effective methods of nanofiber fabrication is electrospinning, which has many advantages compared with other conventional methods, such as straightforward, inexpensive, and capability of mass production. In this chapter, the emerging innovative application of electrospun nanofiber products in supercapacitor is discussed, and some recent achievements are also introduced.

7.1 Supercapacitor

A supercapacitor is an electrochemical energy storage device which has high power capabilities, fast charge propagation and charge–discharge processes (in seconds), and long cycle life (more than 100,000 cycles); requires low maintenance; and exhibits low self-discharging [1–6]. As for the advantages of being environmentally friendly, high safety, and can be operated in a wide temperature range with

K. Wei • I.S. Kim (✉)
Faculty of Textile Science and Technology, Shinshu University, Ueda, Nagano 386-8567, Japan
e-mail: weikai@shinshu-u.ac.jp; kim@shinshu-u.ac.jp

B. Ding and J. Yu (eds.), *Electrospun Nanofibers for Energy and Environmental Applications*, Nanostructure Science and Technology,
DOI 10.1007/978-3-642-54160-5_7,

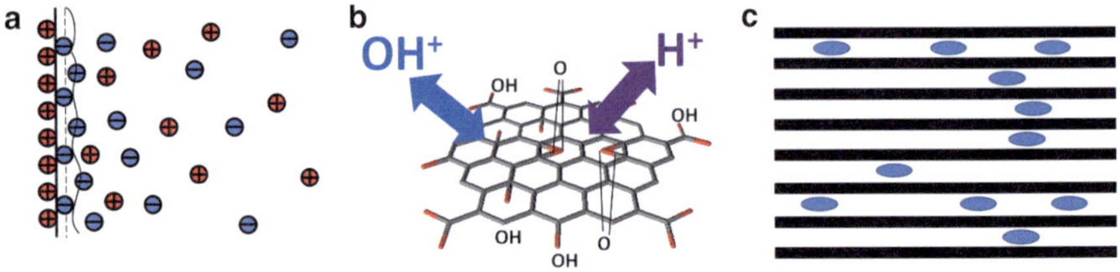

Fig. 7.1 The working principle of supercapacitors. (**a**) Electrical double layer, (**b**) redox reaction on surface, and (**c**) redox reaction in bulk (Reproduced from Liu et al. [14] by permission of John Wiley & Sons Ltd.)

a near-infinite long cycling life, supercapacitors show potential applications in portable electronics (cell phones), memory backup systems, and hybrid cars, where extremely fast charging is a valuable feature [6, 7]. For example, they can be used as uninterruptible power supplies (backup supplies used to protect against power disruption) and load-levelers (backup power for memories, microcomputers, clocks, system boards, etc.).

A supercapacitor unit is usually comprised of two porous carbon electrodes which are immersed in an electrolyte and isolated from electrical contact by a porous separator. The current collectors of metal foil or carbon-impregnated polymers are used to conduct electrical current from each electrode [8]. The separator and the electrodes are impregnated with an electrolyte, which allows ionic current to flow between the electrodes while preventing electronic current from discharging the cell.

Supercapacitor, according to the charge storage mechanism, can be divided into two types. One is the electrochemical double-layer capacitors (EDLCs) based on activated carbons with capacitance proportional to the electrode surface area (the related mechanism is illustrated in Fig. 7.1a). The other, known as pseudocapacitors or redox supercapacitors, uses transition metal oxides or electrical conducting polymers as electrode materials, with the charge storage depending on the fast Faradaic redox reactions (Fig. 7.1b). Electrical double-layer (EDL) supercapacitors are similar to physical dielectric capacitors that are an electrostatic process, but the working principle of pseudocapacitors is a pure electrochemical process [9].

Although EDLs supercapacitors share a similar principle with a dielectric capacitor, where polarization–depolarization takes place at the electrode surface, there are still some big differences. The major difference is that the insulating barrier in a dielectric capacitor is replaced by electrolytes in a supercapacitor [9]. There are mainly two kinds of electrolytes for supercapacitors, aqueous electrolytes and organic electrolytes. In aqueous electrolytes (KOH, H_2SO_4, salt, etc.), supercapacitors have an operating voltage of about 1 V limited by the electrochemical decomposition of water, whereas organic electrolyte-based supercapacitors can be operated in a voltage range of 2.5–3.5 V. An even higher voltage of up to 4.0 V can be achieved for recent ionic liquid-based supercapacitors. Although organic electrolyte-based supercapacitors can store a high overall energy, since the electronic conductivity of organic electrolytes is very low (about 100 times lower than aqueous electrolytes), their power density is usually lower than that of aqueous electrolyte-based supercapacitors.

Another difference is that traditional capacitors rely on a dielectric material to store a charge, while EDLCs rely on the charge of the adsorbed double layer. The capacitance of traditional dielectric capacitors is limited by the thickness of the dielectric material, which separates the charge. The thinnest dielectric materials are 2.5 mm in thickness [10]. However, in the case of EDLCs, charging is achieved by dissociation of ions in the electrolyte salts, resulting in charge separation distances on the order of 1 nm. This arrangement occurs inside the pores of high surface area carbon materials. When the electrode is biased, a double-layer structure is developed with the opposite charge accumulated near the electrode surface. The double-layer thickness is related to the Debye screening length in the modified Gouy–Chapman model. The double-layer capacitance(C) is related to the surface area (A), the effective dielectric constant (e), and the double-layer thickness (d) by an inverse linear relationship ($C = eA/d$). Obviously, the combination of high surface area and small charge separation is necessary for an extremely high capacitance. A typical smooth surface will have a double-layer capacitance of about10–20 μFcm^{-2}, but if a high surface area electrode surface is used, the capacitance can be increased to100 Fg^{-1} for conducting materials that have a specific surface area of 1,000 m^{2}g^{-1} [11]. As a result, a wide range of high surface area carbon materials have been investigated, including activated carbon and multi- and single-walled carbon nanotubes. The capacitance typically ranges from 40 to 140 Fg^{-1} for activated carbon [12] and from 15 to 135 Fg^{-1} for carbon nanotubes [13].

Pseudocapacitive charge storage fundamentally relies on redox reactions between electrode materials and electrolyte ions. The electrical energy is generated by fast Faradaic redox reaction, which produces pseudocapacitance. Depending on the location of these redox reactions, the pseudocapacitive charge storage can be catalogued into surface charge storage and bulk charge storage (Fig. 7.1b, c) [14].

The most commonly used electro-active materials in pseudocapacitive electrodes are metal oxides and conductive polymers. Metal oxides can be divided into two groups: noble metal oxides and cheap metal oxides. One typical candidate for the former is RuO_2, while NiO and MnO_2 are well known for the latter. The most attractive superiority of metal oxides to carbon materials is their much larger capacitance, due to multi-electron transfer during fast Faradaic reactions. RuO_2 shows excellent supercapacitive properties, such as high specific capacitance and very long cycling life, but unfortunately, the high cost of RuO_2 limits its commercial application. The cheaper metal oxides suffer from poor electrical conductivity and poor cycling stability, which precludes the highly desirable performance of these materials. In addition to the above metal compounds, Ir, Co, Mo, Ti, V, Sn, Fe, and other metal oxides can also be used as electrode materials for supercapacitors. Nitrides (tungsten nitride, titanium nitride, etc.) have very good pseudocapacitance as well.

Conducting polymers are another kind of electrode materials capable of delivering Faradaic charge. The applicable conducting polymers for PCs mainly include polyaniline (PANI), poly(3,4-ethylenedioxythiophene) (PEDOT), and polypyrrole

(PPy). Similar to oxide materials, the problem of conducting polymers is their relatively low electrical conductivity and high cost in comparison with carbon materials.

Based on the former description, the performance of supercapacitors is mainly determined by the electrochemical activity and kinetic feature of the electrodes. To improve the power density of supercapacitors, it is crucial to enhance the kinetics of ion and electron transport in electrodes and at the electrode/electrolyte interface. Therefore, electrodes with proper pore structure and good electrical conductivity are highly desirable. On the other hand, by optimizing the structure of electrode materials and designing energy storage devices according to energy storage mechanisms, the energy density of supercapacitors can be increased.

7.2 Electrospun Nanofibers

Electrospun nanofiber has a history of more than 70 years. In 1934, Formhals patented his first invention relating to the process and the apparatus for producing artificial filaments using electrical charge [15]. This technique involves the use of a high voltage to charge the surface of a polymer solution (or melt) droplet and thus to induce the ejection of a liquid jet through a capillary spinneret (Fig. 7.2). Numerous polymer systems, including homopolymers, various kinds of copolymers, blends, and composites, were successfully electrospun [16–19]. From then on, this technique had gradually received more and more attention to prepare for nanosize polymer fibers and numerous polymer systems, including homopolymers, various kinds of copolymers, blends, and composites, were successfully electrospun. Nanofibers produced by electrospinning are of industrial and scientific interest due to their long lengths, small diameters and pores, and high surface area per unit volume which make them an excellent matrix or templates to fabricate various hierarchical nanostructures for sensors, catalysts, remediation, and energy storage applications [20–25].

When electrospun nanofibers are used as the electrodes of a supercapacitor, the nanofibers are immersed in an electrolyte instead of normal carbon material (Fig. 7.3). Electrode materials for supercapacitors mainly include carbon materials, metal oxides, and conducting polymers. Depending on the charge storage mechanism as well as the active materials used, electrospun nanofibers for supercapacitor electrode can be mainly divided into three categories: (1) electrical double-layer capacitors (EDLCs) based on carbon nanofiber with high surface area [26, 27]; (2) pseudocapacitors (PCs) based on conducting polymers such as polyanilines, polypyrroles, and polythiophenes [28–31]; and (3) hybrid capacitors (HCs) based on carbon materials and transition metal oxides such as RuO_2, Fe_3O_4, NiO, and MnO_2 [32–35]. EDLCs are non-Faradaic, PCs are Faradaic, and HCs are a combination of both Faradaic and non-Faradaic reactions. Since the activity of all the nanofiber electrodes depends on both the nature of polymers/oxides and preparation methods, our discussion will focus on the most commonly studied oxides metals and conductive polymers and shed special light on different fabrication methods.

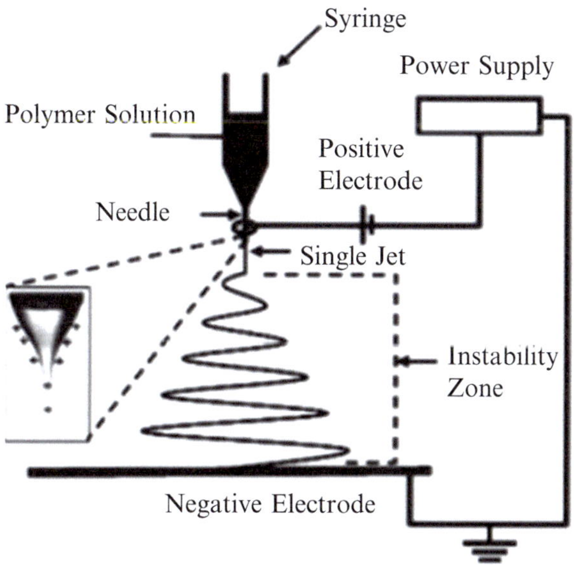

Fig. 7.2 Schematic of electrospinning apparatus

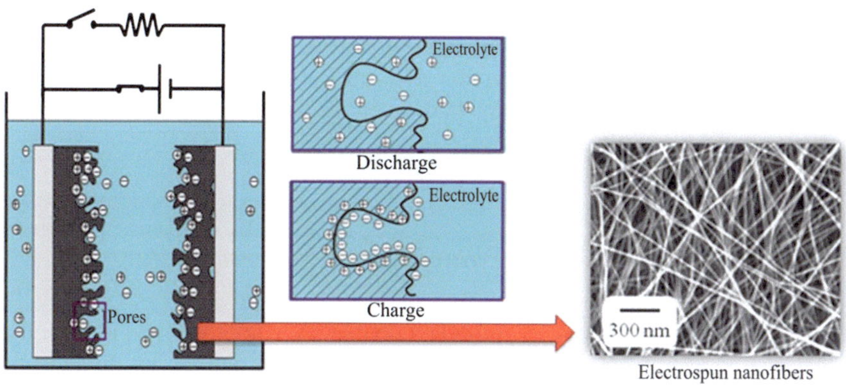

Fig. 7.3 Schematic of a supercapacitor using electrospun nanofibers as electrodes

7.2.1 Electrospun Conducting Polymer Nanofibers for Supercapacitor

Conducting polymers have been extensively studied and widely applied in various organic devices. To improve the performances or extend the functions of the devices, conducting polymers usually have to be nanostructured [36–38]. Electrospun conducting polymer nanofibers have already been fabricated using as supercapacitor electrodes [36–41]. It is well known that polyaniline (PANI) is one of the most useful conducting polymers. Due to its ease of synthesis, good environmental stability, high conductivity, and theoretical specific capacitance, PANI has been widely used in supercapacitor electrode materials [42–47]. Nanostructured PANI can be synthesized through various chemical approaches [48], and they have been applied for fabricating chemical sensors, actuators, memory devices, batteries, and supercapacitors [37, 49, 50].

It is known that poly(amic acid) (PAA), the precursor of polyimide, is soluble in organic solvents and can be easily prepared into uniform fibers with controllable diameters [51]. The COOH groups of PAA endow the electrospun fiber membrane with good hydrophilicity, which makes it an excellent template to capture electropositive molecules through electrostatic attraction in aqueous solutions [52, 53]. Miao and co-workers reported that PAA/PANI core–shell nanofibers with controllable diameters are fabricated by in situ polymerization of aniline using the electrospun PAA fiber membrane as a template, which is followed by removal of the PAA core to form hollow nanofibers of PANI [54]. The hollow structure, thin wall thickness, and orderly pore passages can facilitate the ion diffusion and improve the utilization of the electro-active PANI during the charge–discharge processes, which significantly improve the electrochemical performance of hollow PANI nanofibers. Electrochemical experiments show that a maximum specific capacitance of 601 Fg^{-1} is achieved at 1 Ag^{-1}, suggesting the potential application of hollow PANI nanofibers for supercapacitors. In addition, it is also expected that by using the highly flexible fiber membrane as a template, it will provide possibilities for the facile construction and fabrication of conducting polymers with hollow structures, which may find potential applications in various high-performance electrochemical devices.

7.2.2 Electrospun Carbon Nanofiber for Supercapacitor

Supercapacitors store energy by forming a double layer of electrolyte ions on the surface of conductive electrodes. Supercapacitors are not limited by the electrochemical charge transfer kinetics of batteries and thus can operate at very high charge and discharge rates and can have lifetimes of over a million cycles [55]. However, the energy stored in supercapacitors is currently an order of magnitude lower than that of batteries, which limits their adoption to those applications that require high cycle life and power density. The energy density of existing state-of-the-art supercapacitor devices, which are mainly based on porous activated carbon (AC), is about 4–5 $Whkg^{-1}$, whereas that of lead acid batteries is in the range of 26–34 $Whkg^{-1}$ [2]. A typical AC material, with a Brunauer–Emmett–Teller (BET) specific surface area (SSA) in the range of 1,000–2,000 m^2g^{-1} and a pore size distribution in the range of 2–5 nm, has a gravimetric capacitance of 100–120 Fg^{-1} inorganic electrolytes [56]. Research has thus been focused on increasing energy density without sacrificing cycle life or high power density [57]. An increased capacitance in the organic electrolyte tetraethylammonium tetrafluoroborate (TEA BF4) in acetonitrile (AN) by using carbide-derived carbons (CDCs) has been reported [58]. Metal oxides such as RuO_2, MnO_2 [35], MoO_3 [59], and electronically conducting polymers [28] or their composites have been used to increase specific capacitance via pseudocapacitive redox reactions. Although capacitances of up to 1,300 Fg^{-1} (such as with MnO_2) have been reported in aqueous electrolytes [60], the low electrical conductance, poor compatibility with organic electrolytes, and short cycle life have limited the practical application of these pseudocapacitive materials. Carbon nanomaterials, such as graphene and carbon nanotubes (CNTs), especially single-walled CNTs (SWNTs), have an ideal limit SSA of 1,300 m^2g^{-1} [61], can have high electrical conductance along the tubes, and demonstrate good performance inorganic electrolyte [62]. However, the high cost for mass production of high-quality SWNTs is a challenge for the commercialization of SWNT-based supercapacitors. As a result, electrospun carbon nanofiber has been considered as a potential electrode material for supercapacitor as it can be mass produced. Furthermore, as we know, a supercapacitor unit is usually comprised of two porous carbon electrodes, as a result, carbon nanofibers, which has high surface area and conductivity, is one the most popular and important electrode materials for supercapacitors.

Carbon nanofibers have received growing interests due to their unique chemical and physical properties, depending upon their size, surface area, and shape [63, 64]. Recently, electrospinning has been shown to be a simple but powerful technique for the preparation of carbon nanofibers by carbonization [65, 66] of the electrospun polymer nanofibers at high temperature and could be used to prepare CNFs with sub-micrometer diameters as well as some tens of nanometers to 500 nm [67, 68]. Certainly, PAN is a well-known precursor for the conventional microsized carbon fibers [69, 70]; therefore, many attempts were made to prepare the electrospun-derived carbon nanofibers from pure polyacrylonitrile (PAN) [71–77] as PAN is a well-known precursor for the conventional microsized carbon fibers.

One recent example to prepare electrospun carbon nanofiber with high surface area is that Chan Kim and co-workers [78] have reported synthesis and characterization of high porous carbon nanofibers with hollow cores through the thermal treatment of electrospun copolymeric nanofiber webs, in that the porous carbon nanofibers are clearly seen by using SEM and FE-SEM. In Fig. 7.4, the images are polymeric nanofibers containing two polymer phases of different composition of PAN and PMMA (5:5, 7:3, and 9:1). Figure 7.4d–f shows SEM images of the cross section of the thermally treated carbon fibers; the hollow cores in a single carbon fiber are clearly visible. Interestingly, the number of hollow cores increases with an increase in PMMA concentration.

In addition, Kai Wei and co-workers were also able to tailor the surface area and pore size of the nanofibers simply by changing the blend ratio of the polymers. When the ratio of PAN and PMMA is 7:3, PMMA forms the discontinuous phase and PAN forms the continuous phase. After fiber formation (without change of phase) by electrospinning, the elongated rodlike phase was removed within a single

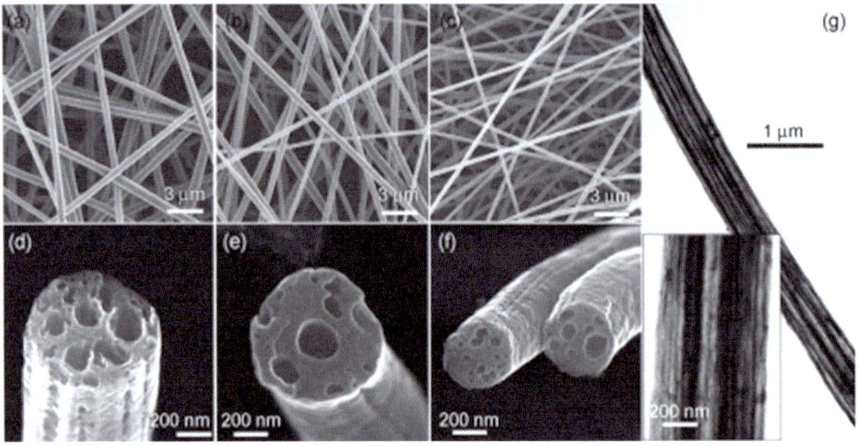

Fig. 7.4 (**a–c**) Macromorphology of electrospun polymeric nanofibers containing two polymer phases. PAN:PMMA = (**a**) 5:5, (**b**) 7:3, and (**c**) 9:1. (**d–f**) Cross-sectional field-emission SEM images of thermally treated nanofibers at 10008C. PAN:PMMA = (**d**) 5:5, (**e**) 7:3, and (**f**) 9:1. (**g**) TEM images of sample (**d**) (Reproduced from Kim et al. [78] by permission of John Wiley & Sons Ltd.)

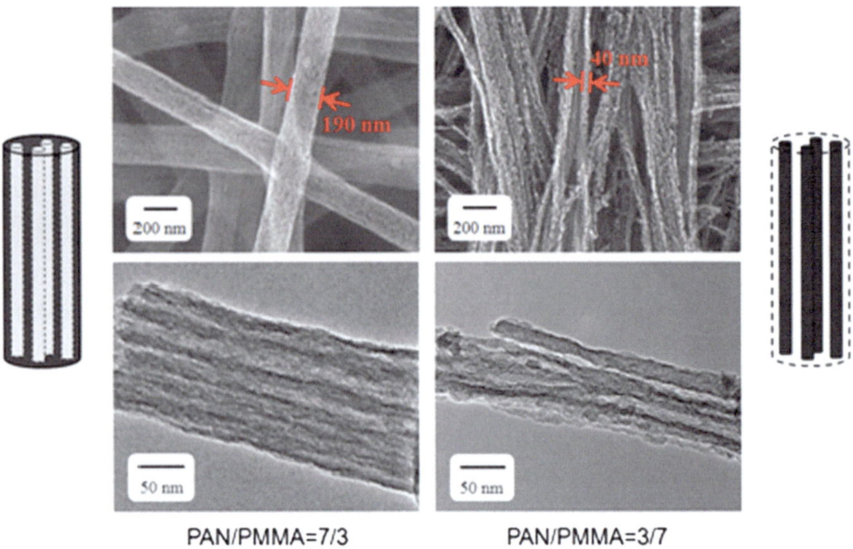

Fig. 7.5 SEM images of electrospun carbon nanofibers with polymer ratio of PAN:PMMA = 7:3 and 3:7

nanofiber through thermal treatment. On the contrary, if the ratio of PAN and PMMA is adjusted to 3:7, PAN forms the discontinuous phase and PMMA forms the continuous phase. Very thin carbon nanofibers (below 50 nm) with high surface area can be obtained after thermal treatment (Fig. 7.5).

These high porous carbon nanofiber webs are promising candidates for utilizing as efficient electrodes due to their low density and expected high electrical conductivities along the plane.

7.2.3 Hybrid Electrospun Nanofiber for Supercapacitor

Recently, the development of high value-added one-dimensional (1D) electrospun nanofibers and structures has tremendously increased over the last decade. Today, a wide variety of nano-objects (such as metal nanoparticles [79–81], clays [82], carbon nanotubes [83, 84], ceramics [85], etc.) can be immobilized onto nanofibers, which bring new materials' properties and potential applications in a broad range of areas such as electronics, medicine, sensor, and controlled release technology. Among which, carbon nanofiber incorporated with metal oxides has been successfully fabricated as a supercapacitor electrode.

Well known that, due to the attractive structural and electrical/mechanical properties, carbon nanotubes (CNTs) are ideal supporting material for metal nanoparticles (NPs). Carbon nanofibers (CNFs), also have the similar mechanical strength and electrical properties to CNTs and the diameter varying from some tens of nanometers to 500 nm [81, 82], are also suitable to be used as metal oxides supporter.

Indeed, extensive research and development in the twenty-first century is directed toward the development of novel hybrid nanomaterials with tunable properties. The purpose of such formulations is to achieve unique and superior properties that cannot be achieved using only a single component of the hybrids. In this way, one can combine the properties of an organic component (polymers, biomolecules) such as mechanical toughness and flexibility with the hardness and thermal stability of the inorganic component (carbon, metals, ceramics, and glass) into a single system [86, 87]. The new nanofibrous products based on the immobilization of functional nanomaterials onto electrospun nanofibers have recently attracted significant attention from both the academic and industrial sectors [88–90]. The main strategies for developing new nanostructured nanofibers were mostly carried out by direct electrostatic spinning of various polymers with molecular building blocks or polymer solutions (or polymer melts) with well-dispersed nano-objects [79–85], by co-electrospinning core–shell polymer nanofiber using a single-nozzle technique [91, 92], and by mimicking natural self-assembled process to produce hierarchically ordered self-assembled nanofibers [15].

Besides lots of efforts in developing these new nanofibers, tailoring the surface chemistry and function of electrospun nanofibers is also of utmost importance for the application of these materials: for example, modification of surface properties such as chemical composition, functionality, and charge as well as biocompatibility is essential for the utilization of the resulting nanofibers in specific applications. As former described, after the preparation of carbon nanofibers using PAN as the precursor, several approaches, including wet-chemical reduction [93, 94],

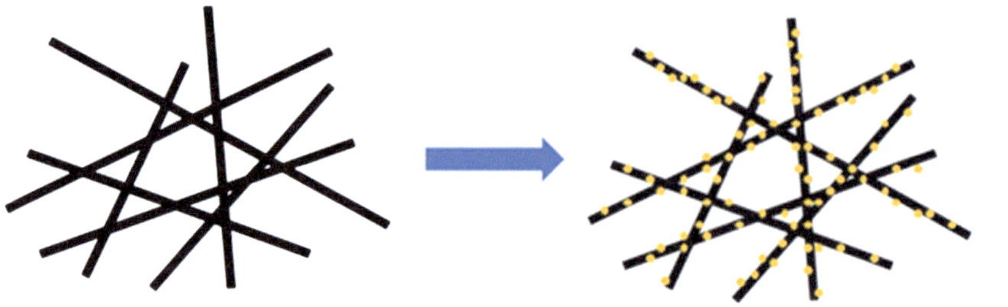

Fig. 7.6 Schematic illustration for fabricating hybrid carbon/metal oxides nanofiber

electrodeposition [95, 96], and metal particles decoration/incorporation [97, 98], have been developed to synthesize metal NP/CNT or NP/CNF nanocomposites. Using different methods (e.g., electrodeposition, wet-chemical reduction, and metal particles decoration/incorporation), different types of metal NPs, such as Au, Pt, Pd, Ag, Co, Rh, Ti, and Cu, have been assembled on these carbon nanomaterials. These metal NP/CNT or NP/CNF nanocomposites have shown great promises in catalysis [99, 100], fuel cells [101, 102], and highly sensitive chemical/biological sensing applications [103, 104].

Especially for supercapacitor electrode, combination of metal oxides with highly conductive carbon nanofiber has been studied intensively. Figure 7.6 shows the illustration for preparing hybrid carbon/metal oxides nanofiber. In these composite electrodes, the carbon substrate acts as a highly porous conducting network that enables a good access of ions and electrons to the metal oxides.

Mu et al. prepared highly dispersed Fe_3O_4 nanosheets on one-dimensional carbon nanofibers through the combination of electrospinning and a solvent–thermal process. Cyclic voltammetry (CV) and galvanostatic charge–discharge measurement in 1 M Na_2SO_3 electrolyte were evaluated to show the electrochemical performances of the CNFs/Fe_3O_4 nanocomposites as the electrode materials for supercapacitors. The samples showed excellent capacitance behavior at different scan rates. Furthermore, the results show that the CNFs/Fe_3O_4 nanocomposites electrode exhibited a higher specific capacitance compared to pure Fe_3O_4 and excellent long cycle life along with 91 % specific capacitance retained after 1,000 cycle tests which shows great potentials for high-performance supercapacitors. They explained the possible mechanism of the high capacitive behavior could be ascribed to the high electrical conductivity and the one-dimensional properties of the CNFs in CNFs/Fe_3O_4 nanocomposites, which could decrease the charge transfer resistance of the Fe_3O_4. Meanwhile, the high specific surface area and high-level exposure of the Fe_3O_4 nanosheets on the surface of the CNFs increased the electrochemical utilization of Fe_3O_4 [105].

Zhi et al. reported the highly conductive carbon nanofiber/MnO_2 coaxial cables in which individual electrospun carbon nanofibers are coated with an ultrathin MnO_2 layer. In this work, PAN was also used as a precursor to prepare CNFs by electrospinning. During the electrospinning process, iron acetylacetonate (AAI) was used to enhance the electronic conductivity; furthermore, the addition of AAI can also enlarge the specific surface area of the CNF. The diameter of synthesized CNFs is around 200 nm, an around 4 nm thick MnO_2 sheath was formed surrounds the carbon CNFs, and nano-whiskers grow radically outward from the sheath in view of the cross section of the coaxial cables, giving a high specific surface area of MnO_2. The as-obtained composites with hierarchical structures exhibit good cycling stability, high energy density (80.2 Wh kg^{-1}), and high power density (57.7 kW kg^{-1}) which indicates that high electronic conductivity of the electrode material is crucial to achieving high power and energy density for pseudosupercapacitors [106].

Nitrogen-doped porous carbon nanofiber, which is another kind of hybrid nanofiber for supercapacitor, exhibits great potential as an efficient electrode material for supercapacitors with favorable electrochemical performance, that is, a relatively high specific capacity and excellent rate performance when compared with the most popular N-containing carbon materials, as well as cycling stability.

Although most of the porous materials exhibit large capacitances, the electrical conductivity suffers from a decrease with increasing porosity due to noncompatibility of conductive pathways or oxygen-containing functional groups [107, 108], which largely limits the power capacity. Thus, it is necessary for the development of carbon materials with high surface area, energy density, specific conductivity properties, and, of course, long cycle life [109].

It has been reported that certain nitrogen doping seems to be the most promising method for enhancing capacity [108, 110], surface wettability of the materials [111], and electronic conductivity [112] while maintaining the super cycle ability [113–116].

Recently, Chen et al. [117] have developed a facile route to prepare the high-content nitrogen-doped porous CNFs containing graphitic frameworks which can be used for the active material of supercapacitor electrodes. Such unique features of composition and structure endow the nitrogen-doped carbon nanofibers with good capacitive performance, that is, with a high electrochemical capacitance (202.0 Fg^{-1} at the current density of 1.0 Ag^{-1}), good retention capability (81.7 % capacitance retention at current densities up to 30 Ag^{-1}), large power capability (89.57 $kWkg^{-1}$), and good cycling stability (3,000 cycles). Furthermore, all of the precursors used in the process can be synthesized on large scale or are commercially available, which will be suitable for scale up production. The results demonstrate that the nitrogen-doped carbon nanofibers might be a promising candidate for the electrode material of high-performance supercapacitors.

Yu and co-workers demonstrate a novel structure based on graphene/MnO_2 nanostructured microfibers [118]. In this study, solution-exfoliated graphene nanosheets with 5 nm thickness were coated on porous microfibers, serving as conductive three-dimensional (3D) frameworks for subsequent controlled electrodeposition of MnO_2 nanomaterials. With further controlled electrodeposition of pseudocapacitive MnO_2 nanomaterials, the hybrid graphene/MnO_2-based

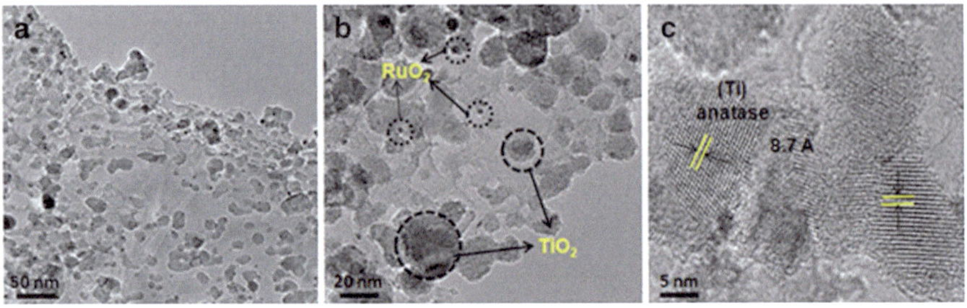

Fig. 7.7 TEM images (**a–c**) of GNS-RuO_2 NPs-TiO_2 NPs

microfibers web yields high capacitance performance with specific capacitance up to 315 F/g achieved. Furthermore, they fabricated a capacitors device using graphene/MnO_2 microfibers web as the positive electrode and single-walled carbon nanotubes (SWNTs) microfibers as the negative electrode in an aqueous Na_2SO_4 electrolyte solution. This capacitor exhibited promising characteristics with a maximum power density of 110 $kWkg^{-1}$, energy density of 12.5 $Whkg^{-1}$, and excellent cycling performance of 95 % capacitance retention over 5,000 cycles.

Although in this work microfibers textile, not nanofiber, was used as the substrate, we consider that instead of microfibers, CNFs and conductive polymer nanofibers can also be used as the electrode substrate which will even increase the conductivity. Wei and co-workers have successfully obtained graphene nanosheets (GNS) incorporated with RuO_2 and TiO_2 nanoparticles (Fig. 7.7). If these composite graphene nanosheets coated on porous nanofibers, excellent supercapacitor electrode can be expected.

Wu et al. prepared composite films from chemically converted graphene and polyaniline nanofibers (PANI-NFs) by vacuum filtration the mixed dispersions of both components [119]. This method provided a new route to graphene-based nanocomposites. The composite film has a layered structure, and PANI-NFs are sandwiched between graphene layers.

The conductivity of this film was measured to be as high as 5.5×10^2 S m^{-1}, which is one order higher than that of pure PANI-NF film. The symmetric supercapacitor device using these films was successful as both electrodes exhibited a high capacitance of 210 Fg^{-1} at 0.3 Ag^{-1}, and this capacitance can be maintained for about 94 % as the discharging current density was increased from 0.3 to 3 Ag^{-1}. Furthermore, this capacitor exhibited excellent cycling performance of a high capacitance of 155 Fg^{-1} was maintained at current density of 3 Ag^{-1}after 800 charging–discharging cycles. Moreover, it is mechanically stable and has a high flexibility; thus, it can be bent into large angles or be shaped into various desired structures. These performances are much higher than those of the supercapacitors based on pure graphene or PANI-NF films, mainly due to the synergic effect of both components.

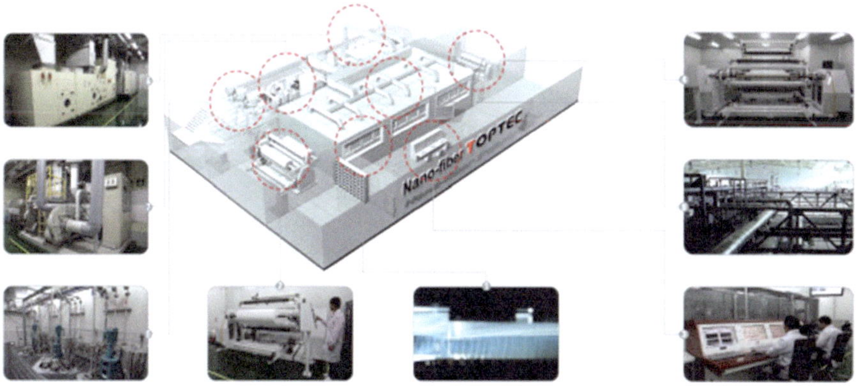

Fig. 7.8 The core technologies of Toptec's Nano-MPI (Nano-Mass Production Innovation)

7.3 Mass Production of Electrospinning Nanofiber

Despite of the potential application in supercapacitors mentioned above, the application of nanofibers has been limited due to the difficulty for mass production because of its peculiarly low process throughput. In 2010, Kim et al. and TOPTEC, a company specialized in factory automation system, completed the development of multifunctional nanofibers production system of nanofiber membranes, named as Nano-Mass Production Innovation (Nano-MPI). Figure 7.8 shows the multi-nozzle spinning part of the machine being commercialized by TOPTEC Company. The device uses upward direction of electrospinning in order to eliminate polymer droplets eventually falling from conventional down-oriented electrospinning elements. This system can provide stable and continuous production from various polymer recipes with 1.8 m width at speed of 80 m/min. It also has automatic QC system and economical heat, ventilation, and air conditioning (HVAC) and regenerative thermal oxidizer (RTO) system. Using this system, large scale of electrospun nanofibers can be obtained, and the practical application of nanofibers in supercapacitors can also be expected.

7.4 Prospect

Supercapacitor is a very attractive alternative to address the rapid increasing global energy consumption. Electrospun nanofibers are being considered for a candidate for supercapacitor electrode where their unique properties contribute to product functionality. These properties include high surface area, small diameter of the fiber, and potentials to incorporate active chemicals. Thus, it is reasonable to believe the electrospun nanofiber-based electrodes may throw a light on the development of supercapacitor.

References

1. Burke A (2000) Ultra capacitors: why, how, and where is the technology. J Power Sources 91:37–50
2. Burke A (2007) R&D considerations for the performance and application of electrochemical capacitors. Electrochim Acta 53(3):1083–1091
3. Reddy N, Xu H, Yang Y (2011) Unique natural-protein hollow-nanofiber membrane produced by weaver ants for medical applications. Biotechnol Bioeng 108:1726–1733
4. Chen X, Wei S, Yadav A, Patil R, Zhu J, Ximenes R, Sun L, Guo Z (2011) Poly(propylene)/carbon nanofiber nanocomposites: ex situ solvent-assisted preparation and analysis of electrical and electronic properties. Macromol Mater Eng 296:434–443
5. Robert RM, Betar MG, Carl VT, Yang SH (2011) All-carbon-nanofiber electrodes for high-energy rechargeable LiO_2 batteries. Energy Environ Sci 4:2952–2958
6. Zhang LL, Zhou R, Zhao XS (2010) Graphene-based materials as supercapacitor electrodes. J Mater Chem 20:5983–5992
7. Yan J, Wei T, Fan Z, Qian W, Zhang M, Shen X, Wei F (2010) Preparation of graphene nanosheet/carbon nanotube/polyaniline composite as electrode material for super capacitors. J Power Sources 195:3041
8. Stoller MD, Park S, Zhu Y, An J, Ruoff RS (2008) Graphene-based ultracapacitors. Nano Lett 8(10):3498–3502
9. Conway BE (1999) Electrochemical super capacitors: scientific fundamentals and technological applications. Plenum Publishers, New York
10. Yang Z, Zhang J, Kintner MCW, Lu X, Choi D, Lemmon JP, Liu J (2011) Electrochemical energy storage for green grid. Chem Rev 111(5):3577–3613
11. Kotz R, Carlen M (2000) Principles and applications of electrochemical capacitors. Electrochim Acta 45:2483–2498
12. Frackowiak E (2007) Carbon materials for supercapacitor application. Phys Chem Chem Phys 9:1774–1785
13. Pico F, Rojo JM, Sanjuan ML, Anson A, Benito AM, Callejas MA, Maser WK, Martinez MT (2004) Single-walled carbon nanotubes as electrodes in supercapacitors. J Electrochem Soc 151(6):A831–A837
14. Liu C, Li F, Ma LP, Cheng HM (2010) Advanced materials for energy storage. Adv Mater 22:E28–E62
15. Jun HW, Yuwono V, Paramonov SE, Hartgerink JD (2005) Enzyme mediated degradation of peptide-amphiphile nanofiber networks. Adv Mater 17:2612–2617
16. Huang ZM, Zhang YZ, Kotaki M, Ramakrishna S (2003) A review on polymer nanofibres by electrospinning and their applications in nanocomposites. Compos Sci Technol 63:2223–2253
17. Kim KO, Seo YA, Kim BS, Yoon KJ, Khil MS, Kim HY, Kim IS (2011) Transition behaviors and hybrid nanofibers of poly(vinyl alcohol) and polyethylene glycol-POSS telechelic blends. Colloid Polym Sci 289:863–870
18. Kimura N, Kim HK, Kim BS, Lee KH, Kim IS (2010) Molecular orientation and crystalline structure of aligned electrospun nylon-6 nanofibers: effect of gap size. Macromol Mater Eng 295:1090–1096
19. Sato H, Kim KO, Kim HK, Kim BS, Enomoto Y, Kim IS (2010) Fabrication of PVA-$BaSO_4$ hybrid nanofibers and dispersion of $BaSO_4$ particles via ultrasonic electrospinning. Fibers Polymers 11:1123–1127
20. Zhang M, Shao C, Guo Z, Zhang Z, Mu J, Cao T, Liu Y (2011) Hierarchical nanostructure of copper (II) phthalocyanine on electrospun TiO_2 nanofibers: controllable solvothermal-fabrication and enhanced visible photocatalytic properties. ACS Appl Mater Interfaces 3:369–377
21. Miao YE, Wang RY, Chen D, Liu ZY, Liu TX (2012) Electrospun self-standing membrane of hierarchical SiO_2@γ-AlOOH (Boehmite) core/sheath fibers for water remediation. ACS Appl Mater Interfaces 4:5353–5359

22. Fang X, Xiao S, Shen M, Guo R, Wang S, Shi X (2011) Fabrication and characterization of water-stable electrospun polyethyleneimine/polyvinyl alcohol nanofibers with super dye sorption capability. New J Chem 35:360–368
23. Wang X, Wang J, Si Y, Ding B, Yu JY, Sun G, Luo W, Zheng G (2012) Nanofiber-net-binary structured membranes for highly sensitive detection of trace HCl gas. Nanoscale 4:7585–7592
24. Ren T, Si Y, Yang J, Ding B, Yang X, Hong F, Yu JY (2012) Polyacrylonitrile/polybenzoxazine-based Fe_3O_4@carbon nanofibers: hierarchical porous structure and magnetic adsorption property. J Mater Chem 22:15919–15927
25. Wang X, Si Y, Wang X, Yang J, Ding B, Chen L, Hu Z, Yu JY (2013) Tuning hierarchically aligned structures for high-strength PMIA–MWCNT hybrid nanofibers. Nanoscale 5: 886–889
26. Conway BE (1999) Electrochemical supercapacitors: scientific fundamentals and technological applications. Kluwer Academic/Plenum, New York
27. Huang J, Sumpter BG, Meunier V (2008) Theoretical model for nanoporous carbon. Supercapacitors. Angew Chem Int Ed 47:520–524
28. Rudge A, Davey J, Raistrick I, Gottesfeld S, Ferraris JP (1994) Conducting polymers as active materials in electrochemical capacitors. J Power Sources 47:89–107
29. Fusalba F, Gouerec P, Villers D, Belanger D (2001) Electrochemical characterization of polyaniline in nonaqueous electrolyte and its evaluation as electrode material for electrochemical supercapacitors. Electrochem Soc 148:A1–A6
30. Frackowiak E, Khomenko V, Jurewicz K, Lota K, Beguin F (2006) Supercapacitors based on conducting polymers/nanotubes composites. J Power Sources 153(2):413–418
31. Mastragostino M, Arbizzani C, Soavi F (2001) Polymer-based supercapacitors. J Power Sources 812:97–98
32. Brousse T, Toupin M, Dugas R, Athouel L, Crosnier O, Belanger D (2006) Crystalline MnO as possible alternatives to amorphous compounds in electrochemical supercapacitors. J Electrochem Soc 153:A2171–A2180
33. Cottineau T, Toupin M, Delahaye T, Brousse T, Belanger D (2006) Nanostructured transition metal oxides for aqueous hybrid electrochemical supercapacitors. Appl Phys A 82:599–606
34. Wang DW, Li F, Cheng HM (2008) Hierarchical porous nickel oxide and carbon as electrode materials for asymmetric supercapacitor. J Power Sources 185:1563–1568
35. Wu NL (2002) Nanocrystalline oxide supercapacitors. Mater Chem Phys 75:6–11
36. Tran HD, Li D, Kaner RB (2009) One-dimensional conducting polymer nanostructures: bulk synthesis and applications. Adv Mater 21:1487–1499
37. Li D, Huang JX, Kaner RB (2009) Polyaniline nanofibers: a unique polymer nanostructure for versatile applications. Acc Chem Res 42:135–145
38. Wan MX (2008) A template-free method towards conducting polymer nanostructures. Adv Mater 20:2926–2932
39. Schultze JW, Karabulut H (2005) Application potential of conducting polymers. Electrochim Acta 50:1739–1745
40. Li C, Bai H, Shi GQ (2009) Conducting polymer nanomaterials: electrosynthesis and applications. Chem Soc Rev 38:2397–2409
41. Lu GW, Hong WJ, Tong L, Bai H, Wei Y, Shi GQ (2008) Drying enhanced adhesion of polythiophene nanotubule arrays on smooth surfaces. ACS Nano 2:2342–2348
42. Bai H, Shi GQ (2007) Gas sensors based on conducting polymers. Sensors 7:267–307
43. Kang ET, Neoh KG, Tan KL (1998) Polyaniline: a polymer with many interesting intrinsic redox states. Prog Polym Sci 23:277–324
44. Ryu KS, Kim KM, Park NG, Park YJ, Chang SH (2002) Symmetric redox supercapacitor with conducting polyaniline electrodes. J Power Sources 103:305–309
45. Ghosh S, Inganas O (1999) Conducting polymer hydrogels as 3D electrodes: applications for supercapacitors. Adv Mater 11:1214–1218
46. Yan X, Tai Z, Chen J, Xue Q (2011) Fabrication of carbon nanofiber-polyaniline composite flexible paper for supercapacitor. Nanoscale 3:212–216

47. Domingues SH, Salvatierra RV, Oliveira MM, Zarbin AJG (2011) Transparent and conductive thin films of graphene/polyaniline nanocomposites prepared through interfacial polymerization. Chem Commun 47:2592–2594
48. Bhadra S, Khastgir D, Singha NK, Lee JH (2009) Progress in preparation, processing and applications of polyaniline. Prog Polym Sci 34:783–810
49. Huang JX, Kaner RB (2004) Nanofiber formation in the chemical polymerization of aniline: a mechanistic study. Angew Chem Int Ed 43:5817–5821
50. Huang JX, Virji S, Weiller BH, Kaner RB (2003) Polyaniline nanofibers: facile synthesis and chemical sensors. J Am Chem Soc 125:314–315
51. Huang C, Wang S, Zhang H, Li T, Chen S, Lai C, Hou H (2006) High strength electrospun polymer nanofibers made from BPDA-PDA polyimide. Eur Polym J 42:1099–1104
52. Carlberg B, Ye LL, Liu J (2011) Surface-confined synthesis of silver nanoparticle composite coating on electrospun polyimide nanofibers. Small 7:3057–3066
53. Yang SQ, Wu DZ, Qi SL, Cui GH, Jin RG, Wu ZP (2009) Fabrication of highly reflective and conductive double-surface-silvered layers embedded on polymeric films through all-wet process at room temperature. J Phys Chem B113:9694–9701
54. Miao YE, Fan W, Chen D, Liu TX (2013) High-performance supercapacitors based on hollow polyaniline nanofibers by electrospinning. ACS Appl Mater Interfaces 5(10):4423–4428
55. John RM, Patrice S (2008) Electrochemical capacitors for energy management. Science 321:651–652
56. Fernández JA, Morishita T, Toyoda M, Inagaki M, Stoeckli F, Centeno TA (2008) Performance of mesoporous carbons derived from poly(vinyl alcohol) in electrochemical capacitors. J Power Sources 175:675–679
57. Simon P, Gogotsi Y (2008) Materials for electrochemical capacitors. Nat Mater 7:845–854
58. Chmiola J, Yushin G, Gogotsi Y, Portet C, Simon P, Taberna PL (2006) Anomalous increase in carbon capacitance at pore sizes less than 1 nanometer. Science 313:1760–1763
59. Brezesinski T, Wang J, Tolbert SH, Dunn B (2010) Ordered mesoporous alpha-MoO_3 with iso-oriented nanocrystalline walls for thin-film pseudocapacitors. Nat Mater 9:146–151
60. Toupin M, Brousse T, Belanger D (2004) Charge storage mechanism for MnO_2 electrode used in aqueous electrochemical capacitor. Chem Mater 16:3184–3190
61. Ali IN, Satoshi Y, Kazufumi K, Takeo Y, Don NF, Hiroaki H, Motoo Y, Sumio I, Kenji H (2010) Extracting the full potential of single-walled carbon nanotubes as durable supercapacitor electrodes operable at 4 V with high power and energy density. Adv Mater 22:E235–E241
62. Don NF, Kenji H, Takeo Y, Tatsuki H, Yuhei H, Yozo K, Osamu T, Hiroaki H, Motoo Y, Sumio I (2006) Shape-engineerable and highly densely packed single-walled carbon nanotubes and their application as super-capacitor electrodes. Nat Mater 5:987–994
63. Burda C, Chen XB, Narayanan R, El-Sayed MA (2005) The chemistry and properties of nanocrystals of different shapes. Chem Rev 105:1025–1102
64. Xia YN, Yang PD, Sun YG, Wu YY, Mayers B, Gates B, Yin YD, Kim F, Yan HQ (2003) One-dimensional nanostructures: synthesis, characterization, and applications. Adv Mater 15: 353–389
65. Ko F, Gogotsi Y, Ali A, Naguib N, Ye H, Yang G, Li C, Willis P (2003) Electrospinning of continuous carbon nanotube-filled nanofiber yarns. Adv Mater 15:1161–1165
66. Kim C, Yang KS, Kojima M, Yoshida K, Kim YJ, Kim YA, Endo M (2006) Fabrication of electrospinning-derived carbon nanofiber webs for the anode material of lithium-ion secondary batteries. Adv Funct Mater 16:2393–2397
67. Reneker DH, Chun I (1996) Nanometre diameter fibres of polymer, produced by electrospinning. Nanotechnology 7:216–223
68. Huang CB, Chen SL, Reneker DH, Lai CL, Hou HQ (2006) High-strength mats from electrospun poly(p-phenylene biphenyltetracarboximide) nanofibers. Adv Mater 18(5): 668–671
69. Donnet JB, Bansal RC (1990) Carbon fibers. Marcel Dekker, New York
70. Peebles LH (1995) Carbon fibers: formation, structure, and properties. CRC, Boca Raton
71. Chun I, Reneker DH, Fong H, Fang XY, Dietzel J, Tan NB, Kearns KJ (1999) Carbon nanofibers from polyacrylonitrile and mesophase pitch. J Adv Mater 31:36–41

72. Wang Y, Serrano S, Santiago-Aviles JJ (2002) Conductivity measurement of electrospun PAN-based carbon nanofiber. J Mater Sci Lett 21:1055–1057
73. Kim C, Yang KS (2003) Electrochemical properties of carbon nanofiber web as an electrode for supercapacitor prepared by electrospinning. Appl Phys Lett 83:1216–1219
74. Wang Y, Santiago-Aviles JJ, Furlan R, Ramos I (2003) Pyrolysis temperature and time dependence of electrical conductivity evolution for electrostatically generated carbon nanofibers. IEEE Trans Nanotechnol 2:39–43
75. Wang Y, Serrano S, Santiago-Aviles JJ (2003) Raman characterization of carbon nanofibers prepared using electrospinning. Synth Met 138:423–427
76. Gu SY, Ren J, Vancso GJ (2005) Process optimization and empirical modeling for electrospun polyacrylonitrile (PAN) nanofiber precursor of carbon nanofibers. Eur Polym J 41:2559–2568
77. Lai D, Xia Y (2004) Electrospinning of nanofibers: reinventing the wheel. Adv Mater 16(14):1151–1170
78. Kim C, Jeong YI, Nhu-Ngoc BT, Yang KS, Kojima M, Kim YA, Endo M, Lee JW (2007) Synthesis and characterization of porous carbon nanofibers with hollow cores through the thermal treatment of electrospun copolymeric nanofiber webs. Small 3:91–95
79. Patel AC, Li S, Wang C, Zhang W, Wei Y (2007) Electrospinning of porous silica nanofibers containing silver nanoparticles for catalytic applications. Chem Mater 19:1231–1238
80. Shin MK, Kim SS, Kim SJ, Kim SK, Lee H (2006) Reinforcement of polymeric nanofibers by ferritin nanoparticles. Appl Phys Lett 88:193901
81. Christopher D, Xin L, David Z, Xianyan W, Ferdinando FB, James W, Lynne AS, Jayant K (2003) Metal oxide-coated polymer nanofibers. Nano Lett 3(1):143–147
82. Ji Y, Li B, Ge S, Sokolov JC, Rafailovich MH (2006) Structure and nanomechanical characterization of electrospun PS/clay nanocomposite fibers. Langmuir 22:1321–1328
83. McCullen SD, Stevens DR, Roberts WA, Ojha SS, Clarke LI, Gorga RE (2007) Morphological, electrical, and mechanical characterization of electrospun nanofiber mats containing multi-walled carbon nanotubes. Macromolecules 40(4):997–1003
84. Wei K, Xia JH, Kim BS, Kim IS (2011) Multiwalled carbon nanotubes incorporated bombyx Mori silk nanofibers by electrospinning. J Polym Res 18(4):579–585
85. Kim CK, Kim BS, Sheikh FA, Lee US, Khil MS, Kim HY (2007) Amphiphilic poly(vinyl alcohol) hybrids and electrospun nanofibers incorporating polyhedral oligosilsesquioxane. Macromolecules 40:4823–4828
86. Davis SR, Brough AR, Atkinson AJ (2003) Formation of silica/epoxy hybrid network polymers. J Non-Cryst Solids 315:197–205
87. Chattopadhyay DK, Raju KVSN (2007) Structural engineering of polyurethane coatings for high performance applications. Prog Polym Sci 32:352–418
88. Caruso RA, Schattka JH, Greiner A (2001) Titanium dioxide tubes from sol–gel coating of electrospun polymer fibers. Adv Mater 13:1577–1579
89. Hou H, Ge JJ, Zeng J, Li Q, Reneker DH, Greiner A, Cheng SZD (2005) Electrospun polyacrylonitrile nanofibers containing high concentration of well-aligned multiwall carbon nanotubes. Chem Mater 17(5):967–973
90. Patel AC, Li S, Yuan JM, Wei Y (2006) In situ encapsulation of horseradish peroxidase in electrospun porous silica fibers for potential biosensor applications. Nano Lett 6:1042–1046
91. Sun Z, Zussman E, Yarin AL, Wendorff JH, Greiner A (2003) Compound core-shell polymer nanofibers by co-electrospinning. Adv Mater 15(22):1929–1932
92. Bazilevsky AV, Yarin AL, Megaridis CM (2007) Co-electrospinning of core-shell fibers using a single nozzle technique. Langmuir 23(5):2311–2314
93. Yoon B, Ma C (2005) Microemulsion-templated synthesis of carbon nanotube-supported Pd and Rh Nanoparticles for catalytic applications. J Am Chem Soc 127:17174–17175
94. Li XL, Liu YQ, Fu L, Cao LC, Wei DC, Wang Y (2006) Efficient synthesis of carbon nanotube–nanoparticle hybrids. Adv Funct Mater 16(18):2431–2437
95. Day TM, Unwin PR, Macpherson JV (2007) Factors controlling the electrodeposition of metal nanoparticles on pristine single walled carbon nanotubes. Nano Lett 7:51–57

96. Quinn BM, Dekker C, Lemay SG (2005) Electrodeposition of noble metals nanoparticles on carbon nanotubes. J Am Chem Soc 127:6146–6147
97. Qu LT, Dai LM, Osawa E (2006) Shape/size controlled synthesis of metal nanoparticles for site selective modification of carbon nanotubes. J Am Chem Soc 128:5523–5532
98. Choi HC, Shim M, Bangsaruntip S, Dai H (2002) Spontaneous reduction of metal ions on the sidewalls of carbon nanotubes. J Am Chem Soc 124:9058–9059
99. Endo M, Kim YA, Ezaka M, Osada K, Yanagisawa T, Hayashi T, Terrones M, Dresselhaus MS (2003) Selective and efficient impregnation of metal nanoparticles on cup-stacked-type nanofibers. Nano Lett 3(6):723–726
100. Van der Lee MK, Van Dillen AJ, Bitter JH, De Jong KP (2005) Deposition precipitation for the preparation of carbon nanofiber supported nickel catalysts. J Am Chem Soc 127: 13573–13582
101. Tian ZQ, Jiang SP, Liang YM, Shen PK (2006) Synthesis and characterization of platinum catalysts on multiwalled carbon nanotubes by intermittent microwave irradiation for fuel cell applications. J Phys Chem B 110:5343–5350
102. Cao L, Scheiba F, Roth C, Schweiger F, Cremers C, Stimming U, Fuess H, Chen L, Zhu W, Qiu X (2006) Novel nano-composite $Pt/RuO_2{\bullet}xH_2O/CNT$ catalysts for DMFC. Angew Chem Int Ed 45(32):5315–5319
103. Hrapovic S, Liu Y, Male KB, Luong JHT (2004) Electrochemical biosensing platforms using platinum nanoparticles and carbon nanotubes. Anal Chem 76:1083–1088
104. Yang MH, Yang YH, Liu YL, Shen GL, Yu RQ (2006) Platinum nanoparticles-doped sol-gel/carbon nanotubes composite electrochemical sensors and biosensors. Biosens Bioelectron 21:1125–1131
105. Mu J, Chen B, Guo Z, Zhang M, Zhang Z, Zhang P, Shao C, Liu Y (2011) Highly dispersed Fe_3O_4 nanosheets on one-dimensional carbon nanofibers: synthesis, formation mechanism, and electrochemical performance as supercapacitor electrode materials. Nanoscale 3: 5034–5040
106. Zhi M, Manivannan A, Meng F, Wu N (2012) Highly conductive electrospun carbon nanofiber/MnO_2 coaxial nano-cables for high energy and power density supercapacitors. J Power Sources 208:345–353
107. Li L, Liu E, Shen H, Yang Y, Huang Z, Xiang X, Tian Y (2011) Charge storage performance of doped carbons prepared from polyaniline for supercapacitors. J Solid State Electrochem 15:175–182
108. Hulicova-Jurcakova D, Seredych M, Lu GQ, Bandosz TJ (2009) Combined effect of nitrogen- and oxygen-containing functional groups of microporous activated carbon on its electrochemical performance in supercapacitors. Adv Funct Mater 19:438–447
109. Zhao L, Fan LZ, Zhou MQ, Guan H, Qiao S, Antonietti M, Titirici MM (2010) Nitrogen-containing hydrothermal carbons with superior performance in supercapacitors. Adv Mater 22:5202–5206
110. Jeong HM, Lee JW, Shin WH, Choi YJ, Shin HJ, Kang JK, Choi JW (2011) Nitrogen-doped graphene for high-performance ultracapacitors and the importance of nitrogen-doped sites at basal planes. Nano Lett 11:2472–2477
111. Guo H, Gao Q (2009) Boron and nitrogen co-doped porous carbon and its enhanced properties as supercapacitor. J Power Sources 186:551–556
112. Lee YH, Lee YF, ChangKH HCC (2011) Synthesis of N-doped carbon nanosheets from collagen for electrochemical energy storage/conversion systems. Electrochem Commun 13:50–53
113. Hulicova-Jurcakova D, Kodama M, Shiraishi S, Hatori H, Zhu ZH, Lu GQ (2009) Nitrogen-enriched nonporous carbon electrodes with extraordinary supercapacitance. Adv Funct Mater 19:1800–1809
114. Qian HS, Yu SH, Luo LB, Gong JY, Fei LF, Liu XM (2006) Synthesis of uniform Te@carbon-rich composite nanocables with photoluminescence properties and carbonaceous nanofibers by the hydrothermal carbonization of glucose. Chem Mater 18:2102–2108

115. Hu B, Wang K, Wu LH, Yu SH, Antonietti M, Titirici MM (2010) Engineering carbon materials from the hydrothermal carbonization process of biomass. Adv Mater 22:813–828
116. Liang HW, Guan QF, Chen LF, Zhu Z, Zhang WJ, Yu SH (2012) Macroscopic-scale template synthesis of robust carbonaceous nanofiber hydrogels and aerogels and their applications. Angew Chem Int Ed 51:5101–5105
117. Chen LF, Zhang XD, Liang HW, Kong MG, Guan QF, Chen P, Wu ZY, Yu SH (2012) Synthesis of nitrogen-doped porous carbon nanofibers as an efficient electrode material for supercapacitors. ACS Nano 6(8):7092–7102
118. Yu GH, Hu LB, Vosgueritchian M, Wang HL, Xie X, James RM, Cui X, Cui Y, Bao Z (2011) Solution-processed graphene/MnO_2 nanostructured textiles for high-performance electrochemical capacitors. Nano Lett 11:2905–2911
119. Wu Q, Xu Y, Yao Z, Liu A, Shi G (2010) Supercapacitors based on flexible graphene/polyaniline nanofiber composite films. ACS Nano 4(4):1963–1970

Chapter 8
Applications of Bulk and Nanostructured Polyaniline in Hydrogen Storage

Sesha S. Srinivasan

Abstract Over the past few years, the need for a clean and renewable fuel has sharply risen. This is due to increasing fossil fuel costs and the desire to limit or eliminate harmful by-products which are created during the burning of these fuels. Hydrogen is the most abundant element in the universe and can be used in either fuel cells or traditional internal combustion engines to produce energy with no harmful emissions. One of the main obstacles facing the implementation of a hydrogen economy is its storage. Classical methods of storage involve either high and unsafe pressures or liquid storage involving a large amount of energy.

Two alternative hydrogen storage methods are investigated – physisorption, which is the weak chemical bonding to a material, as well as chemisorption, which is a strong chemical bond of hydrogen to a host material. Polyaniline, a conducting polymer, is investigated both in its bulk form and in nanostructured forms, more precisely nanofibers and nanospheres, to store hydrogen via physisorption. It is found that the bulk form of polyaniline can store only approximately 0.5 wt% hydrogen, which is far short of the 6 wt% required for practical applications. Nanofibers and nanospheres, however, have been developed, which can store between 4 and 10 wt% of hydrogen at room temperature with varying kinetics.

8.1 Introduction

This chapter investigates hydrogen storage via physisorption in polyaniline (PANI). PANI is a conducting polymer that has been well characterized. Its chemical structure is shown in Fig. 8.1. The emeraldine base form of PANI consists of quinoid and benzenoid rings. This means that the theoretical hydrogen capacity

S.S. Srinivasan (✉)
College of Innovation and Technology, Florida Polytechnic University, 4700 Research Way, Lakeland, FL 33805, USA
e-mail: ssrinivasan@floridapolytechnic.org

B. Ding and J. Yu (eds.), *Electrospun Nanofibers for Energy and Environmental Applications*, Nanostructure Science and Technology,
DOI 10.1007/978-3-642-54160-5_8,

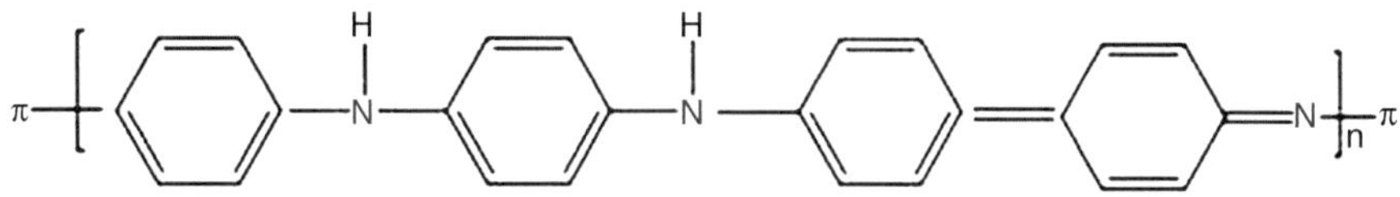

Fig. 8.1 Chemical structure of emeraldine base polyaniline

of polyaniline is approximately 6 wt%. While this is too low to exceed the DOE targets, it is not the chemisorption aspect of PANI that is of interest, but instead the physisorption aspect.

To date, there has been very little research into hydrogen storage using polymers. Since polyaniline is rather simple to synthesize in a laboratory setting, while being relatively inexpensive, it was decided to investigate polyaniline in its bulk form as well as in nanostructured form with various additives for hydrogen storage. This nanocomposite material consists of a polyaniline matrix, which can be functionalized by either catalytic doping or incorporation of nanovariants. It was reported that polyaniline could store as much as 6–8 wt% of hydrogen [1, 2], which, however, another team of scientists could not reproduce [2]. Yet another recent study revealed that a hydrogen uptake of 1.4–1.7 wt% hydrogen has been reported for polymers of intrinsic microscopy [3].

Most hydrogen storage investigations of polymers that have been conducted were performed at 77 K and only involved pure physisorption, mainly by increasing the surface area of the polymer. One form of hydrogen storage in polymers is to use polymeric foam for hydrostatic pressure retainment of hydrogen [4]. In this method of storing hydrogen, the ideal tank contains spherical cells that act as microscopic pressure vessels, ideally in a homogeneous manner. However, no significant amount of hydrogen has been experimentally proved to be stored. As of the time of writing this dissertation, only theoretical work has been performed on hydrogen storage in polymer foams.

Hypercrosslinked polymers, such as styrenic polymer produced from polyvinylbenzylchloride, have also been investigated for their hydrogen sorption properties [5]. It was found that approximately 3 wt% hydrogen could be stored at 15 bar, although a temperature of 77 K was required to ensure that the weak hydrogen bonds created through physisorption did not break, thereby releasing the hydrogen. Polychloromethylstyrene-co-divinylbenzene [6], another hypercrosslinked polymer, was shown to reversibly store 1.6 wt% hydrogen at a temperature of 77 K and a pressure of 12 bar. Again, the surface area of the material played an important part in storing the hydrogen, as is expected for physisorption materials. The hydrogen storage mechanism was purely physisorption and lacked any chemisorption of hydrogen, an attribute that is needed to get away from the low temperatures required for pure physisorption materials.

Another polymer class that has been investigated for its hydrogen storage properties are polymers of intrinsic microporosity, known as PIMs. PIMs are polymers that have very large surface areas, generally in the range of 500–900 m^2/g, and are generally prepared by using a benzodioxane formation reaction between suitable monomers. It has been shown, for example, that HATN-network-PIM can store about 1.6 wt% hydrogen at a pressure of 10 bar and a temperature of 77 K [3, 7]. While this value is higher than that of metal organic frameworks, it is still too low for practical use and of course requires a temperature which is not only impractical but also requires a large amount of energy.

Polyaniline is a conductive polymer, with conductivity on the order of 10°S/cm. This is higher than that of typical nonconducting polymers but much lower than that of metals [8]. In addition to its conductivity, polyaniline in its emeraldine base (EB) form is very simple and inexpensive to polymerize. It is because of this simplicity that it was chosen as a matrix material for the nanocomposite structure discussed in this chapter.

Conducting polymer nanostructures combine the advantages of organic conductors and low-dimensional systems having interesting physicochemical properties [9–11] and useful applications [12–14]. Though many controversial results were reported in terms of hydrogen uptake [15–18] in polymer nanocomposites, there are still a number of parameters, tailor-made properties, surface morphologies, and their correlation with hydrogen sorption behavior to be investigated before these materials can be commercially deployed for onboard hydrogen storage. Similarly, nanotubes [18], nanofibers [19], and nanospheres [20] have attracted more interest because of their novel properties and wide potential application for nanometer-scale engineering applications. It is known that the nanofibrallar morphology significantly improves the performance of polyaniline in many conventional applications involving polymer interactions with its environment [21]. This leads to faster and more responsive chemical sensors [22], new organic/polyaniline nanocomposites, and ultrafast nonvolatile memory devices [23].

A graphical comparison of polymer structures investigated to date is shown in Fig. 8.2. It can be seen that there has not been a lot of work performed on polymeric hydrogen storage and that the work that has been reported to date mostly is performed at low temperatures of 77 K. The polypyrrole and polyaniline with high capacity that are shown in that figure were later refuted by other teams of scientists, as previously mentioned. The polyaniline structures investigated in the following sections are all evaluated for their hydrogen performance between room temperature and 125 °C, which correlates approximately with the temperature range required by the DOE for practical hydrogen storage.

8.2 Bulk Polyaniline

The first form of the emeraldine base polyaniline (PANI) to be investigated was bulk polyaniline. Bulk simply refers to the nanostructure of PANI, indicating that the PANI is used as-synthesized in its non-nanostructured form. This means that the surface morphology was not adjusted into nanofibrous form. After some baseline characterization and hydrogen sorption experiments, filler materials could be added to the PANI. The addition of filler materials to the matrix material could be used to

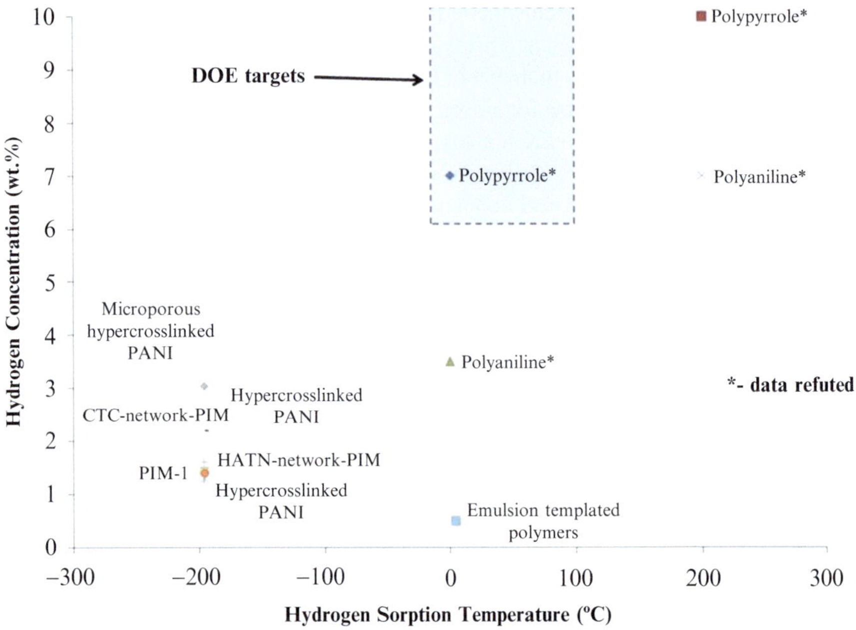

Fig. 8.2 Previous work on hydrogen storage in polymers, mainly in polymers of intrinsic microscopy (*PIM*)

tailor an ideal nanocomposite material for hydrogen storage. Several different filler materials in various concentrations were added to the nanocomposite, and its effect on the characteristics of the material was determined. A schematic representation of the use of filler materials and the mechanism of hydrogen storage for bulk polyaniline can be seen in Fig. 8.3.

Since polyaniline is composed primarily of hydrogen and carbon atoms, it possesses a large number of both primary and secondary hydrogen bonding sites. Figure 8.3 shows a combination of both primary and secondary bonding sites as follows:

1. Hydrogen gas enters the hydrogen storage tank containing the polyaniline matrix material as hydrogen molecules in gaseous form.
2. Any added materials, such as SnO, then react with hydrogen molecule to break it up into hydrogen ions to facilitate the bonding to the host matrix.
3. The hydrogen ions then bond either to filler materials such as Fig. 8.3a carbon nanotubes, Fig. 8.3b fullerenes, or simply to Fig. 8.3c the host material through either primary atomic bonds, where impurities or vacancies are present, or other primarily bonded hydrogen through weak secondary hydrogen bonding.
4. By reducing the hydrogen pressure on the material or heating up the host matrix, the hydrogen is then released from storage.
5. Finally, after recombining to hydrogen molecules, the hydrogen is used either in an internal combustion engine or in a fuel cell.

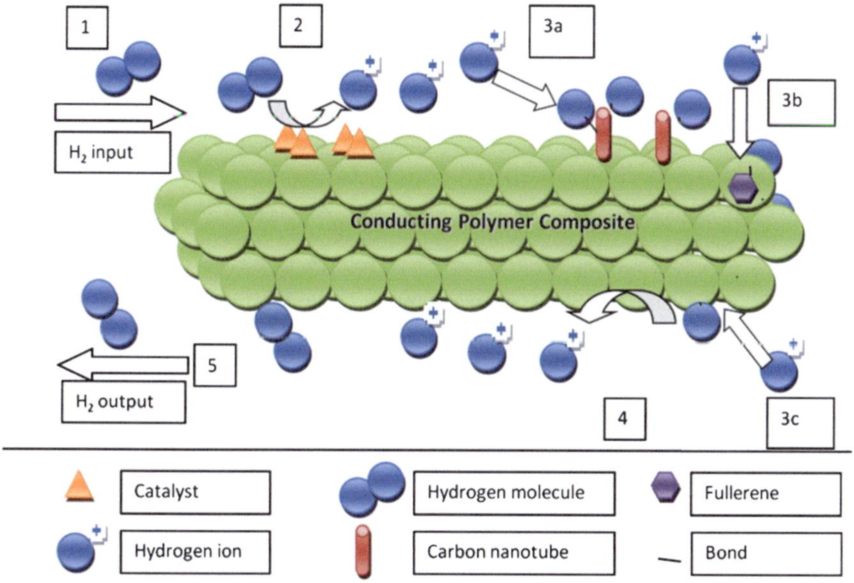

Fig. 8.3 Hydrogen sorption mechanism of standard polyaniline composite material

8.2.1 Synthesis and Characterizations of Bulk Polyaniline

The polyaniline used for this experiment was synthesized by following a well-established method for the synthesis of the emeraldine base (EB) form of polyaniline [8]. 5.71 g of ammonium persulfate was dissolved in a 100 mL beaker with 50 mL of deionized water. Five milliliters of aniline was mixed with approximately 35 mL of deionized water. These two solutions were then slowly mixed with 10 mL of 37 % HCl acid in a large beaker. After the temperature rises to approximately 45 °C, the polymerization is considered finished. The sample, however, is magnetically stirred for approximately 24 h to ensure complete polymerization. After magnetic stirring, the sample was vacuum filtered through an 11 μm filter, after which it was washed twice with HCl to ensure termination of any bonds and methanol. The sample was then vacuum dried at 100 °C for approximately 24 h. While most samples were polymerized at room temperature, which was approximately 23 °C, a few samples, which will be discussed separately, were polymerized at approximately 0 °C by placing the chemicals in a freezer for several hours before polymerization and then mixing the chemicals inside an ice bath-cooled beaker.

The so-called filler materials were added during the polymerization process, so as to ensure uniform distribution within the sample. Some of the filler materials added to the nanocomposite material included SnO_2 (obtained from Sigma-Aldrich in nanopowder form), multiwall carbon nanotubes (with a purity of at least 60 % before purification), as well as aluminum powder (obtained from Riedel-de Haën as fine powder with a purity greater than 93 %). These materials were added in various concentrations, which will be discussed in the individual results section. All materials were added in terms of weight percent based on the expected yield.

The as-synthesized nanocomposite samples were characterized before and after hydrogen sorption employing Fourier transform infrared spectroscopy (FTIR), differential scanning calorimetry (DSC), thermogravimetric analysis (TGA), as well as scanning electron microscopy (SEM). FTIR, which was performed on a PerkinElmer Spectrum One, was utilized to ensure the proper polymerization of the sample as well as determine any chemical effect a filler material might have had. TGA was carried out on a TA SDT Q600, while DSC measurements were made with a TA DSC Q10P. TGA was employed mainly to determine the useful temperature range of the sample material, as well as to determine any hydrogen that might have bonded to the material chemically. The DSC measurement was used to determine any phase transitions that might occur as a result of the heat ramping process. Finally, the SEM was utilized to determine any surface morphological changes of the materials, such as the formation of clusters as a result of hydrogen sorption.

As described, the polyaniline-based nanocomposite materials were synthesized according to a standardized method. Initially, solely polyaniline in its emeraldine base form was synthesized. This sample is referred to as PANI-STD. It was decided to include multiwall carbon nanotubes into the polyaniline matrix material, as these have been shown to increase the conductivity of the conducting polymer [24, 25]. This sample is referred to as PANI-MWCNT. Due to the availability and relatively low cost of multiwall carbon nanotubes, as compared to single wall carbon nanotubes, it was decided to only use multiwall carbon nanotubes, though there has been significant research on hydrogen storage in single wall carbon nanotubes. The hydrogen sorption measurements for carbon nanotubes, however, are usually performed at a temperature of 77 K [26].

10 wt% tin oxide was also added to the matrix material, since it was expected to act as a catalyst material in the capture of hydrogen, breaking the hydrogen molecule into ions. This sample is referred to as PANI-SnO_2 throughout this chapter. Finally, fine aluminum powder was added to the matrix material during polymerization to determine any effect of the aluminum on hydrogen sorption. This sample is referred to as PANI-Al. It was thought that the aluminum would form alane or alanate, thus increasing the hydrogen sorption of the nanocomposite. All samples were prepared with varying concentrations of the filler materials. Therefore, the samples will be denoted not only by their abbreviation, but also labeled with the appropriate concentration of the material added to the nanocomposite, in weight percent.

8.2.1.1 FTIR Characterization Results

FTIR spectrum analysis was performed to identify the bonding environment of the as-synthesized materials and to determine whether the filler materials chemically reacted with the polyaniline matrix material or simply physically mixed with the polyaniline.

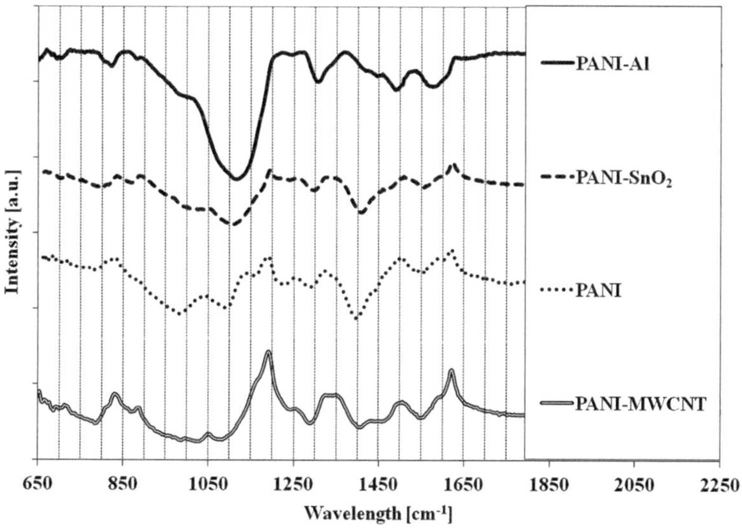

Fig. 8.4 FTIR comparing standard polyaniline (PANI-STD), polyaniline with 10 wt% SnO_2, polyaniline with 10 wt% MWCNT, and polyaniline with 30 wt% aluminum powder

Figure 8.4 shows a comparison of standard polyaniline, polyaniline with multiwall carbon nanotubes, polyaniline with SnO_2, and finally polyaniline with aluminum powder. All four samples exhibit the same benzenoid and quinoid ring vibrations around 1,500 cm^{-1} and 1,600 cm^{-1}, respectively.

Figure 8.4 shows that the polyaniline that was synthesized is in fact polyaniline in its emeraldine base [27]. Additionally, all samples exhibit peaks around 1,306 cm^{-1}, which is attributed to the π-electron delocalization induced in the polymer by protonation. The peak around 1,375 cm^{-1} is attributed to the C–N stretch in the base. The peak around 822 cm^{-1} is due to C–H bond out of plane [28]. It can also be seen that the addition of tin oxide mainly caused a physical rather than a chemical reaction between the polyaniline and the tin oxide, as the FTIR spectrum is essentially unchanged. The addition of multiwall carbon nanotubes to the polyaniline matrix material also shows more of a physical mixing rather than a chemical mixing. This explains the results that are obtained in the hydrogen sorption measurements, to be discussed in the next section. The addition of aluminum to the sample, however, clearly shows a chemical reaction between the polyaniline and the aluminum powder. The large peak around 1,150 cm^{-1} is a sign of increased conductivity and was described as the "electron-like band" [28].

8.2.1.2 Thermogravimetric Results

Once the successful polymerization of the sample was confirmed, a TGA measurement was performed on the sample. The sample was heated from room temperature

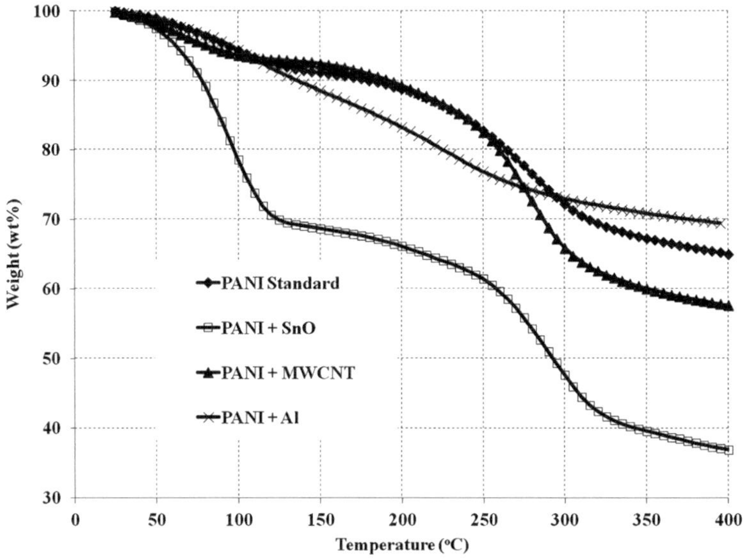

Fig. 8.5 Thermogravimetric analysis of standard polyaniline, polyaniline with 10 wt% multiwall carbon nanotubes, polyaniline with 10 wt% tin oxide, and polyaniline after hydrogen sorption measurements

up to 400 °C. The main purpose for performing the TG analysis was to determine the maximum temperature possible for hydrogen sorption measurements, as well as to confirm the useful thermal range of the samples.

It can be seen in Fig. 8.5 that all samples are stable up to at least 200 °C. The initial weight loss in the samples at around 100 °C is simply due to moisture contained within the nanocomposites. It is interesting to note that the sample with the multiwall carbon nanotubes exhibits a higher weight loss than the other samples. This could be explained by the fact that the multiwall carbon nanotubes were not of 100 % purity, but instead contained soot or other forms of carbon, which were burned off during the temperature increase.

While care was taken to purify the carbon nanotubes as much as possible before synthesis of the nanocomposite material by using a combination of ultrasonication, washing with deionized water, drying in air, and washing with HCl [29], some impurities must have remained in the sample, causing the observed drop. Additionally, the PANI-MWCNT sample contained a higher amount of moisture, as it was observed that this sample readily absorbed moisture, as the sample was handled and stored in atmospheric conditions.

8.2.1.3 Scanning Electron Microscopy Results

Figure 8.6 shows some SEM images of standard polyaniline before and after hydrogen sorption measurements. Figure 8.6a, b clearly shows the rough surface

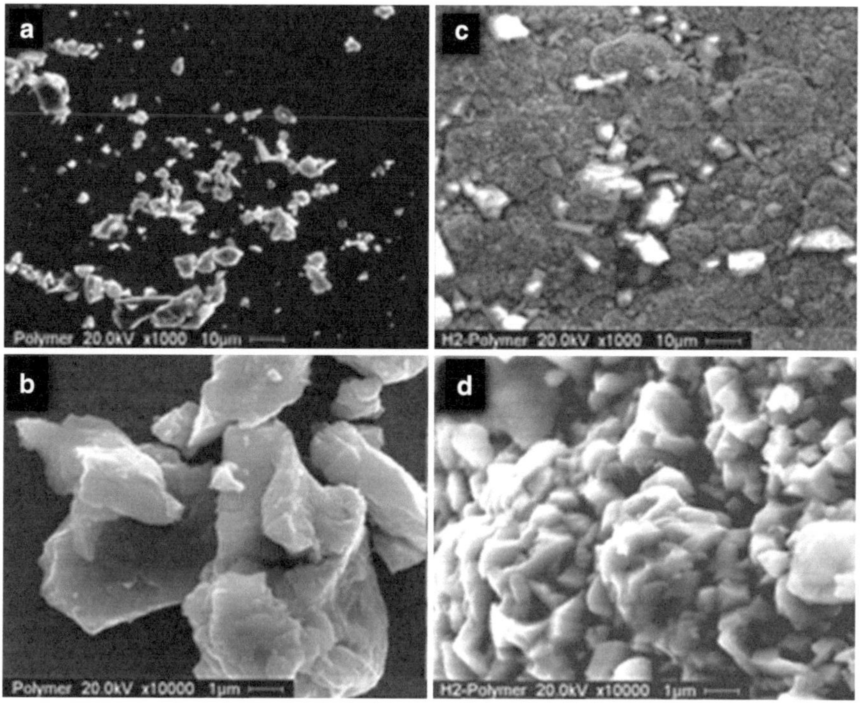

Fig. 8.6 Scanning electron microscopy images of standard polyaniline (**a** and **b**) before and (**c** and **d**) after hydrogen interaction indicating a ballooning caused by hydrogen cycling

of the standard polyaniline. The images show a rocklike surface structure, with sharp edges. Figure 8.6c, d, on the other hand, reveals that hydrogen interacted with the polyaniline during the PCT hydrogen sorption measurements. The interaction of hydrogen with the polyaniline caused a type of ballooning effect, in which the hydrogen seeped into the material and was stored in small pockets. This indicates that the hydrogen is stored physically, rather than chemically.

8.2.1.4 Hydrogen Sorption Measurements

The isothermal volumetric measurements were carried out by HyEnergy's PCT Pro 2000 sorption equipment. This fully automated Sievert's type instrument uses an internal PID-controlled pressure regulator with maximum pressure of 170 bar. It also includes five built-in and calibrated reservoir volumes of 4.66, 11.61, 160.11, 1021.30, and 1169.80 mL. The volume calibration without and with the sample was performed at a constant temperature with an accuracy of ±1 °C using helium. The software subroutines for hydrogen purging cycles, leak test, kinetics, PCT, cycling, etc., were performed by the HyDataV2.1 program. The data collected for each run was analyzed using the Igor Pro 5.03 program with a built-in HyAnalysis Macro.

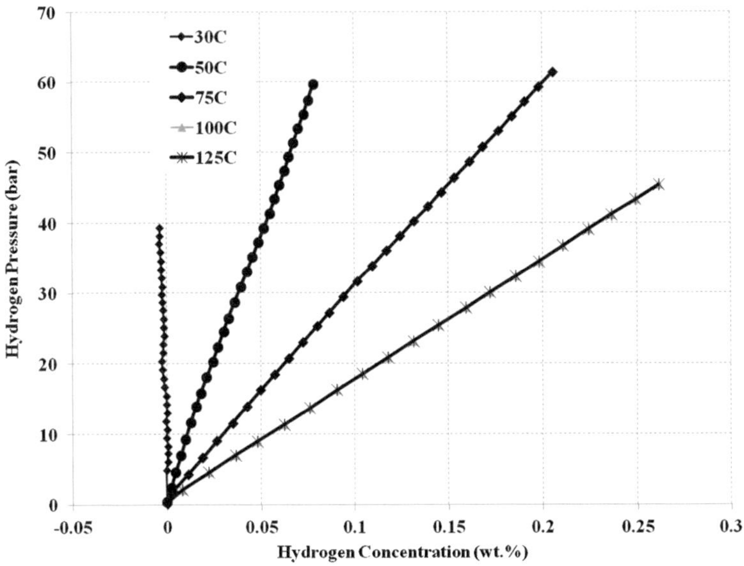

Fig. 8.7 Hydrogen sorption measurements of standard polyaniline at different temperatures

The amount of material loaded for hydrogen sorption measurements corresponds to approximately 2 g. Before the experiment was run, the volume of the sample was determined at the appropriate temperature. The absorption measurement was generally performed at 60 bar of hydrogen pressure at temperatures varying from 25 °C up to 125 °C. Desorption measurements have been carried out under vacuum.

The hydrogen sorption of polyaniline at approximately 77 K [1], was not reproducible elsewhere [2]. This low temperature, however, is not feasible for use in mobile applications. It would simply require too much energy to cool and maintain the hydrogen storage material at such a low temperature. It was therefore decided to conduct the hydrogen sorption measurements between ambient temperature and 125 °C. The absorption of hydrogen was carried out at 60 bar hydrogen pressure, while the sample was initially just below ambient pressure. For desorption measurements, the sample was held at the pressure resulting from the absorption measurement, while the reservoir, into which the hydrogen was to flow, was evacuated.

Figure 8.7 shows the hydrogen sorption measurements of standard polyaniline. The pressure was raised in aliquots from 0 bar hydrogen pressure to 60 bar hydrogen pressure at different temperatures. It can be seen that the standard polyaniline does not absorb any hydrogen at room temperature. At 50 °C, however, the sample can store approximately 0.075 wt% hydrogen at 60 bar. As the temperature is increased, the sample can store more hydrogen. At 125 °C, the standard polyaniline, as synthesized, is capable of storing 0.35 wt% hydrogen.

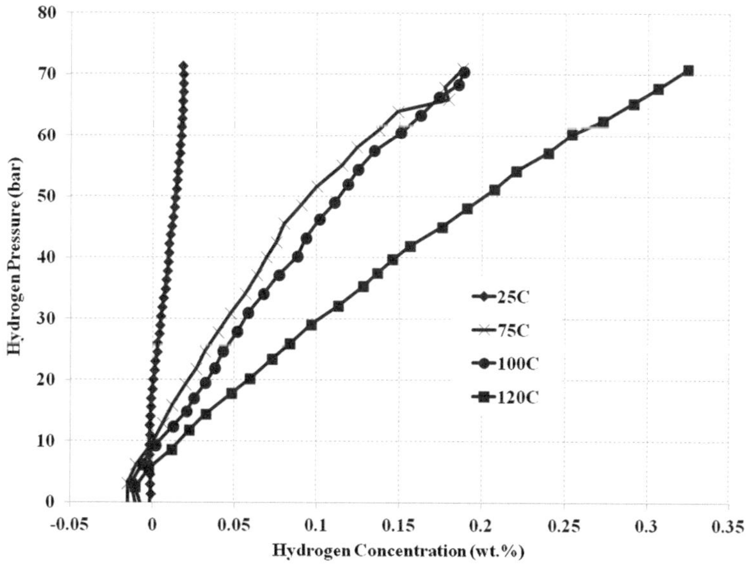

Fig. 8.8 Hydrogen sorption measurements of polyaniline with 10 wt% SnO_2 at different temperatures and increasing pressure

When tin oxide was added to the polyaniline, no change in the hydrogen sorption capabilities was observed, regardless of the amount of tin oxide, as seen in Fig. 8.8. It is thought that the lack of hydrogen sorption increase is due to the measurement temperatures chosen. From hydrogen gas sensing research, it is known that a temperature of around 300 °C is optimum for hydrogen to bind with SnO_2 [26]. The measurements followed the same hydrogen sorption behavior as observed for the standard PANI sample seen in Fig. 8.7. An increase in pressure and temperature meant an increase in hydrogen sorption, provided the temperature of sorption was above 30 °C. Regardless of the amount of SnO_2 added to the polyaniline matrix, there was no change in hydrogen sorption; hence, only the 10 wt% sample has been presented here.

Similarly, the amount of multiwall carbon nanotubes showed no effect on the hydrogen sorption capabilities of the material. Figure 8.9 shows the hydrogen sorption results of the polyaniline sample with 10 wt% multiwall carbon nanotubes. It can be seen that the same trend is evident – a higher temperature and a higher pressure means more hydrogen can be stored. The multiwall carbon nanotubes cause only a very slight increase in the hydrogen sorption capabilities, which can be accredited to an increase in the porosity caused by the inclusion of nanotubes. Again, a minimum temperature of just above 30 °C is needed for the material to absorb hydrogen.

Figure 8.10 shows the hydrogen sorption measurements performed on the polyaniline sample with 30 wt% aluminum powder. It can be seen that this nanocomposite material exhibits a much higher hydrogen sorption than the other

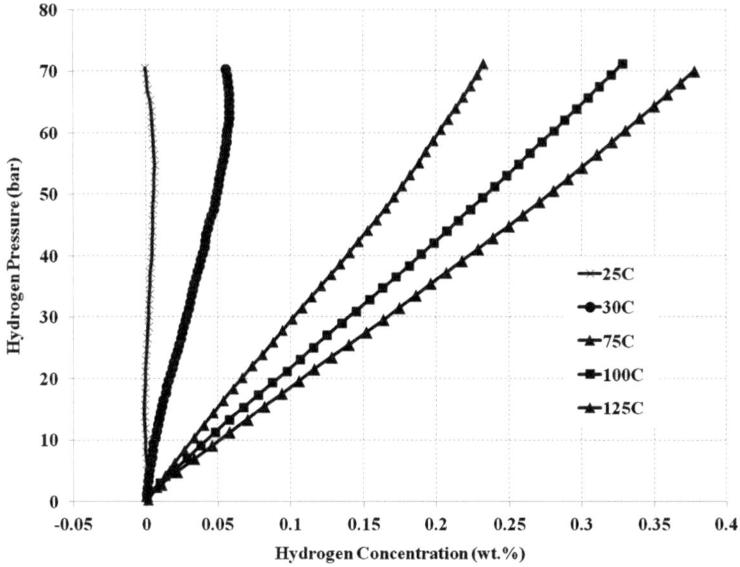

Fig. 8.9 Hydrogen sorption measurements of polyaniline with 10 wt% multiwall carbon nanotubes at different temperatures and increasing pressure

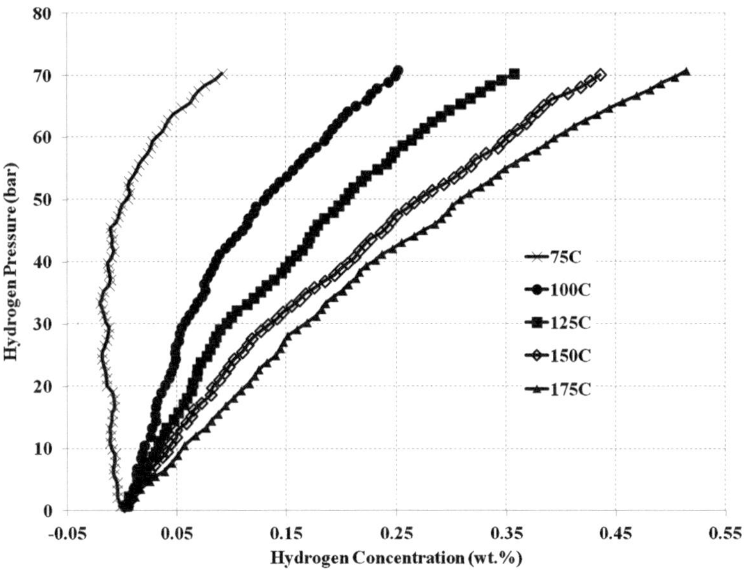

Fig. 8.10 Hydrogen sorption measurements of polyaniline with 30 wt% fine aluminum powder at different temperatures and increasing pressure

samples. It is interesting to note that hydrogen sorption begins at 75 °C, above approximately 50 bar hydrogen pressure, as compared to 30 °C for most other samples. The exact effect of the aluminum on the hydrogen sorption performance of the nanocomposite material is still unclear, as well as the effect of different concentrations of aluminum. It is, however, clear that no hydrogen alanate is formed from this interaction.

8.2.2 Bulk PANI Summary

All experiments were conducted to mimic the US Department of Energy's goals as closely as possible. Hence, it was decided to keep the hydrogen sorption temperature range below 130 °C and above 0 °C, which is reasonably close to DOE targets. Hydrogen sorption measurements were not performed below 0 °C, as it is the author's opinion that it is simply not feasible for use in automotive applications to store hydrogen at such low temperatures, especially considering the freezing effects of fuel cells that would result from such a low temperature.

It has been shown that a polyaniline nanocomposite material can store just under 0.5 wt% hydrogen at 175 °C. The inclusion of multiwall carbon nanotubes has little effect on the hydrogen sorption capabilities of the material, though the increase in porosity of the material causes a slight increase in the hydrogen sorption capabilities of the material. The carbon nanotubes cannot, however, be seen as the primary hydrogen storage material. SEM analysis shows that the hydrogen is mainly physically absorbed, in forms of small pockets of hydrogen. The inclusion of tin oxide has been shown to have no effect on the hydrogen sorption capabilities of the material either. Finally, the addition of aluminum to the nanocomposite matrix has shown the greatest effect on the material's hydrogen capacity. The increase in hydrogen sorption capabilities caused by the inclusion of fine aluminum powder, as well as the exact chemical interaction between the aluminum and the polyaniline, is still unclear. The use of bulk polyaniline, however, is not a feasible or practical option for hydrogen storage.

8.3 PANI Nanospheres

Since the bulk polyaniline showed promise in that it was able to reversibly sorb hydrogen at slightly elevated temperatures, though with a low capacity, it was decided to investigate various nanostructures of polyaniline. As was previously mentioned, a higher surface area allows for more hydrogen bonding sites. Additionally, smaller particle sizes also mean faster kinetics, at least theoretically, since the hydrogen can diffuse through the material with more ease. Hence, the first nanostructured polyaniline that was investigated was PANI nanospheres.

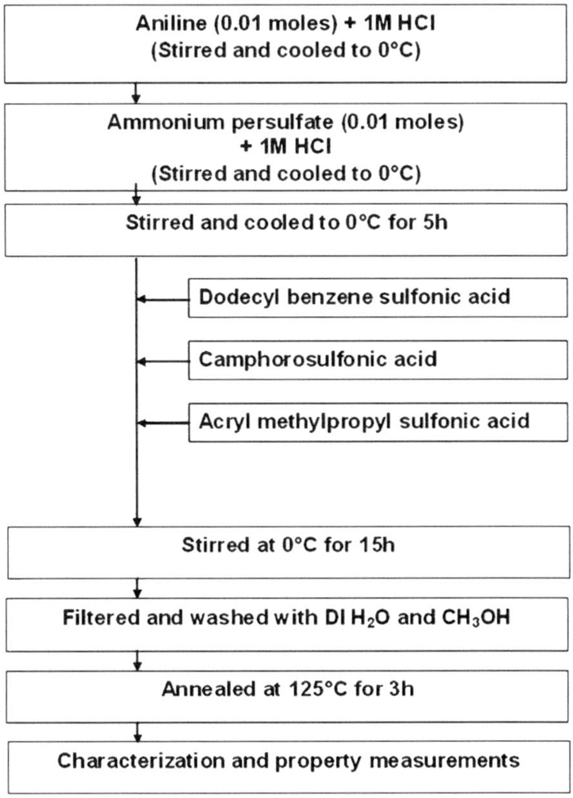

Fig. 8.11 Synthesis of PANI nanospheres

8.3.1 Synthesis and Characterizations of Polyaniline Nanospheres

The polyaniline nanospheres (PANI-NS) were synthesized by oxidative polymerization of aniline monomer at 0 °C in an ice bath using ammonium persulfate as the oxidant in the presence of surfactant. Aniline, ammonium persulfate, Polyvinylpyrrolidone, cetyl ammonium bromide, and camphorsulfonic acid are used as received from Sigma-Aldrich. Camphorsulfonic acid surfactant as the dopant and ammonium persulfate as the oxidant were used in the present synthesis of polyaniline nanospheres (see flow chart in Fig. 8.11). Calculated quantities of aniline monomer (0.005 mol) were mixed with 50 mL of distilled water and stirred using magnetic stirrer for 10 min. Meanwhile, calculated quantities of surfactant (0.75 mol) and oxidant (0.005 mol) were dissolved separately in distilled water and stirred for 10 min in an ice bath. The surfactant solution was first added to the aniline monomer aqueous solution, and then the previously cooled oxidant solution was added dropwise after which the mixture was allowed to react for 10 h in an ice bath. The precipitate was filtered and washed several times with distilled water and

Fig. 8.12 SEM of PANI nanospheres at a magnification of 25,000

methanol to terminate the polymerization reaction and then dried in vacuum at room temperature for 24 h. Later, the vacuum-dried precipitate was annealed at 100 °C for 1 h. The dried polyaniline was characterized and tested for the hydrogen uptake and release measurements.

8.3.1.1 Scanning Electron Microscopy Results

The microstructure of the as-synthesized PANI nanospheres was studied by Hitachi S800 scanning electron microscope (SEM). A fixed working distance of 5 mm and a voltage of 5–25 kV were used. Sample preparation for the SEM measurement was carried out inside the glove box by covering the sample holder with Parafilm® for minimal exposure to oxygen while transferring it to the secondary emission chamber. EDAX Genesis software was used to analyze the SEM images.

It was found that the synthesis, as previously outlined, produced a virtually 100 % yield of polyaniline nanospheres with an average diameter of 100 nm, as can be seen in Fig. 8.12. The nanospheres are not independent of each other, but instead are agglomerated. This is to be expected as the nanospheres are produced without templating and are simply formed through a chemical method. However, the nanospheres are very uniform and possess a better morphology than the bulk PANI previously discussed.

8.3.1.2 FTIR Characterization Results

The quinoid and benzenoid bond stretches of the PANI-NS were compared to the bulk PANI with a PerkinElmer Spectrum One FTIR spectrometer. The PANI-NS samples were pelletized and sealed in a specially designed KBr cell for infrared

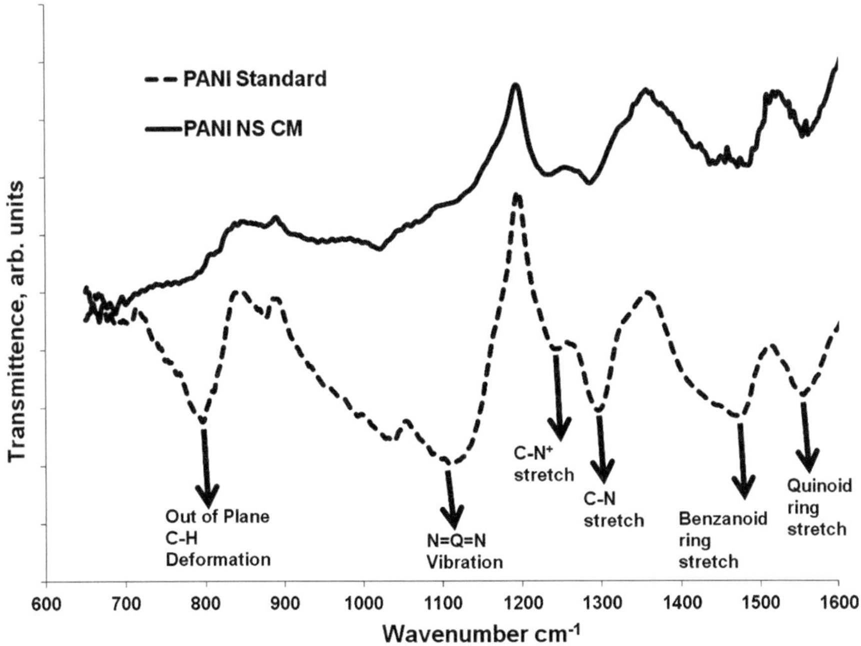

Fig. 8.13 FTIR comparison of bulk PANI with PANI nanospheres

measurements to prevent any further moisture uptake. As Fig. 8.13 indicates, the FTIR spectrum of the PANI nanospheres is virtually identical to the spectrum of the bulk counterpart. The relative intensities of the peaks to each other remain unchanged, with the exception of the out of plane C–H deformation peak observed around 800 cm^{-1}. All other peaks, most importantly the characteristic quinoid and benzenoid peaks, are virtually identical, therefore confirming the formation of the emeraldine base polyaniline.

8.3.1.3 Hydrogen Sorption Results

The volumetric hydrogen sorption measurements are of paramount importance in understanding the hydrogen storage behavior of PANI-NS. Room temperature hydrogen absorption was executed at a high pressure (H_2 ~80 bar) with a pre-calibrated reservoir. These isothermal volumetric measurements were carried out by HyEnergy's PCT Pro 2000 sorption equipment. This fully automated Sievert's type instrument uses an internal PID-controlled pressure regulator with maximum pressure of 170 bar. It also includes five built-in and calibrated reservoir volumes of 4.66, 11.61, 160.11, 1021.30, and 1169.80 mL, of which the 160.11 mL reservoir was used, providing for a roughly 15:1 volumetric ratio of the hydrogen volume to available sample volume. The volume calibration with and without the sample

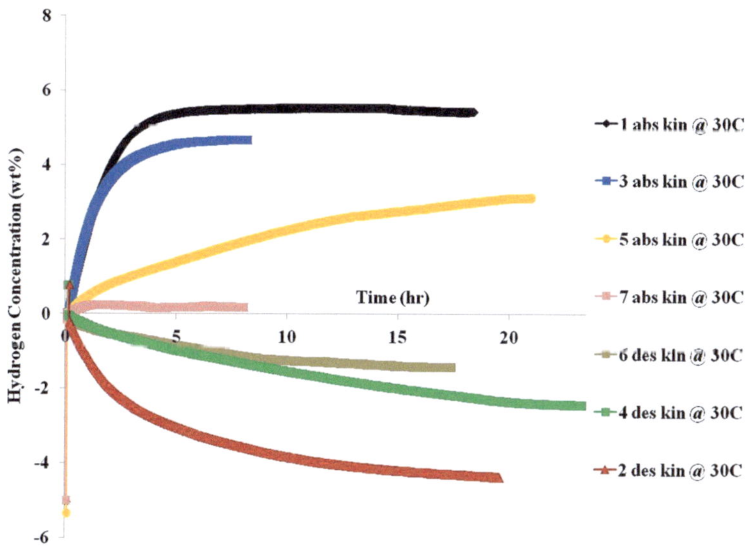

Fig. 8.14 Absorption and desorption kinetics of PANI nanospheres at 30 °C

was performed at a constant temperature with an accuracy of ±1 °C using helium. The software subroutines for hydrogen purging cycles, leak test, kinetics, PCT, and cycling were performed by the HyDataV2.1 Lab-View program. The data collected for each run were analyzed using the Igor Pro 5.03 program with a built-in HyAnalysis Macro.

As with the bulk polyaniline sample, it was decided to perform hydrogen sorption measurements close to room temperature. Since the nanospheres do not contain any hydrogen after synthesis, with the exception of the chemically bonded hydrogen that is part of the polyaniline chemical structure, the first hydrogen measurement performed was absorption. The sample was kept under vacuum and then exposed to 80 bar of hydrogen pressure. Each absorption measurement was then followed by a desorption measurement at 30 °C and given enough time to reach an equilibrium. The results of the kinetic absorption and desorption measurements are shown in Fig. 8.14.

The first hydrogen absorption measurement at 30 °C revealed an uptake of more than 5.5 wt%, a significant improvement over the bulk polyaniline. However, the kinetics of the hydrogen uptake was extremely slow as it took almost 5 h to absorb the full 5.5 wt%. The speed of hydrogen release during the second kinetic measurement was even slower, though, as it took nearly 20 h to release just over 4.5 wt%, meaning that approximately 1 wt% was irreversibly stored in the polymer nanostructure. Consecutive absorption and desorption cycles showed that the amount of hydrogen that was absorbed was always equal to the amount that was released during the previous desorption cycle. However, the amount desorbed continuously decreased as did the kinetics. After only three absorption and

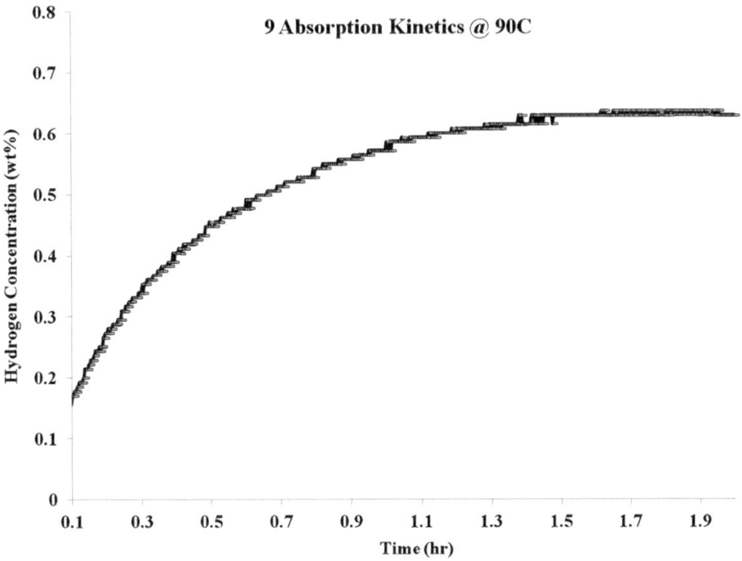

Fig. 8.15 Absorption kinetics of PANI nanospheres at 90 °C

desorption cycles, the amount of hydrogen that was absorbed was well under 1 wt%. Since the previous work with the bulk PANI showed that a higher temperature also meant a higher capacity, an absorption kinetic measurement at 90 °C was performed. The results of this kinetic measurement, shown in Fig. 8.15, indicate that only approximately 0.6 wt% of hydrogen could be reversibly sorbed. This data is in agreement with the best case scenario of the bulk polyaniline previously discussed.

The slow kinetics of the hydrogen absorption indicate that chemisorption is taking place, rather than physisorption, as physisorption is generally characterized by fast, almost instantaneous, hydrogen release or uptake. This is further confirmed by the slow desorption of hydrogen, which is to be expected since the temperature of measurement is relatively low, which means that there is not enough energy to break the hydrogen bonds quickly enough. Furthermore, since the nanospheres are not separated, but instead are agglomerated, there is a larger hydrogen diffusion path which in turn also means slower kinetics.

8.3.1.4 Hydrogen Cycling Effects on PANI Nanospheres

Since the capacity of the nanospheres decreased from 5.5 to 0.6 wt% with consecutive hydrogen cycling, the nanospheres were investigated using SEM. A representative image can be seen in Fig. 8.16. Some remnants of the nanospheres can be seen, though it is evident that the nanostructure was significantly deteriorated. The few remnants of the nanospherical structure that are visible further show that the average diameter has increased to approximately 200 nm. This swelling confirms

Fig. 8.16 SEM image of PANI nanospheres after hydrogen cycling

that the PANI-NS chemically interacted with the hydrogen, thereby causing the increase in particle size. Additionally, the nanostructure was broken down by the repeated cycling, which is not due to any temperature effects that might cause the breakdown of the structure. Some microcracks can also be seen in the SEM image, which is a typical sign of hydrogen cycling in materials as the hydrogen diffuses through the material, thereby creating these cracks. The PANI-NS morphology observed after hydrogen cycling is reminiscent of the bulk PANI morphology. Since the nanospheres are virtually identical to the bulk PANI, as evidenced by the FTIR analysis, the deterioration of the nanostructure explains why the final hydrogen capacity is close to that of the bulk PANI samples.

8.3.2 PANI Nanospheres Summary

Emeraldine base polyaniline nanospheres were successfully synthesized with almost 100 % yield and an average diameter of 100 nm. The FTIR characterization confirms that the chemical structure of the PANI is virtually identical to the bulk PANI previously investigated. It was found that the nanospheres have an initial uptake of approximately 5.5 wt% hydrogen with slow kinetics requiring approximately 5 h for full absorption. The desorption kinetic measurement at 30 °C, however, shows that only 4.5 wt% of hydrogen is released again, leaving approximately 1 wt% contained in the PANI nanospheres. With each further absorption and desorption, the capacity diminished further, leaving more hydrogen in the structure. The rate of hydrogen absorption and desorption also decreased with each cycle, indicating that the material essentially became saturated with hydrogen through chemisorption, as evidenced by the slow kinetics, while making hydrogen diffusion more difficult. SEM analysis of the nanospheres after hydrogen cycling reveals that the nanostructure of the material deteriorated as a result of chemical bonding of hydrogen to the PANI material. Swelling of the few remaining nanospheres was also observed, as were microcracks within the structure, evidence of hydrogen diffusion in the material. Hydrogen cycling at higher temperatures, as was done for the bulk PANI, revealed no further significant hydrogen uptake or release, but instead essentially confirmed the hydrogen capacity of bulk PANI.

8.4 Chemically Grown PANI Nanofibers

Nanofibers with diameters of tens of nanometers appear to be an intrinsic morphological unit that was found to "naturally" form in the early stage of chemical oxidative polymerization of aniline. In conventional polymerization, nanofibers are subject to secondary growth of irregularly shaped particles that form the final granular agglomerates. The key to producing pure nanofibers is to suppress secondary growth. Based on this, two methods (interfacial polymerization and rapidly mixed reactions) have been developed that can readily produce pure nanofibers by slightly modifying the conventional chemical synthesis of polyaniline without the need for any template or structural directing material. With this nanofiber morphology, dispensability and processibility of polyaniline are now greatly improved. On the other hand, the template synthesis method is an effective way to grow the nanofibers of various conducting polymers [30, 31]. The preparation conditions and their effect on morphology, size, and electrical properties of nanofibers have been reported previously in literature [32].

8.4.1 Synthesis and Characterizations of Polyaniline Nanofibers

The polyaniline nanofibers (PANI-NF) were synthesized by oxidative polymerization of aniline monomer at 0 °C in an ice bath using ammonium persulfate as the oxidant in the presence of surfactant. Aniline, ammonium persulfate, dodecyl benzene sulfonic acid, acrylmethylpropyl sulfonic acid, and camphorsulfonic acid are used as received from Sigma-Aldrich. Sulfonic acid-based surfactants as the dopant and ammonium persulfate as the oxidant were used in the present synthesis of polyaniline nanofibers (see flow chart in Fig. 8.17). Calculated quantities of aniline monomer (0.05 mol) were mixed with 50 mL of distilled water and stirred using magnetic stirrer for 10 min. Meanwhile, calculated quantities of surfactant (0.75 mol) and oxidant (0.05 mol) were dissolved separately in distilled water and stirred for 10 min in an ice bath. The surfactant solution was first added into the aniline monomer aqueous solution and then previously cooled oxidant solution dropwise, and the mixture was allowed to react for 15 h in an ice bath.

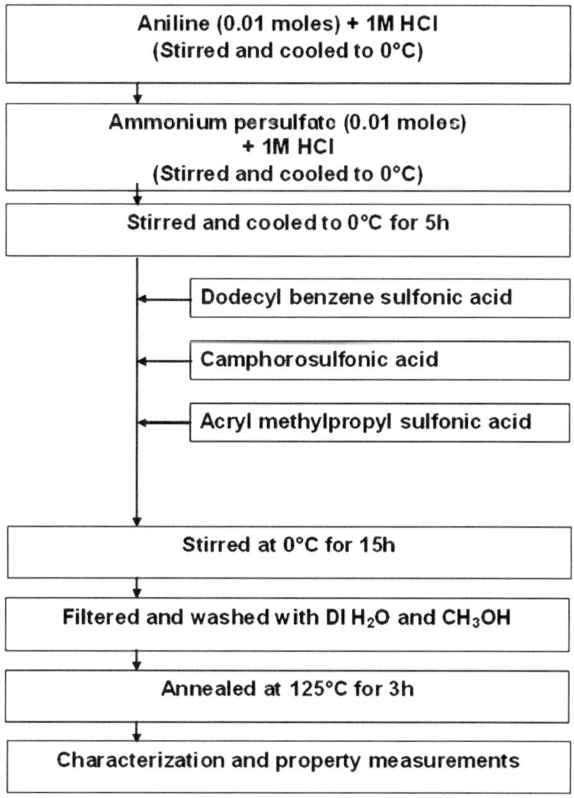

Fig. 8.17 Flow chart for the synthesis of polyaniline nanofibers prepared in an aqueous medium with different surfactants and using ammonium persulfate as oxidizing agent

The precipitate was filtered and washed several times with distilled water and methanol to terminate the polymerization reaction and then dried in vacuum at room temperature for 24 h. Later, the vacuum-dried precipitate was annealed at 125 °C for 3 h. The dried polyaniline was characterized and tested for the hydrogen uptake and release measurements.

8.4.1.1 FTIR Characterization Results

The quinoid and benzenoid bond stretches of the PANI-NF were compared via PerkinElmer Spectrum One FTIR spectrometer. The PANI-NF samples were pelletized and sealed in a specially designed KBr cell for infrared measurements. Fourier transform infrared (FTIR) spectra of standard PANI and PANI-NF prepared from chemical templating method are shown in Fig. 8.18.

The major bonding environment remains unchanged for both standard and nanofibrous polyaniline structures. The presence of two bands in the vicinity of 1,500 cm^{-1} and 1,600 cm^{-1} is assigned to the nonsymmetric C6 ring stretching modes. The higher frequency vibration at 1,600 cm^{-1} is for the quinoid rings, while

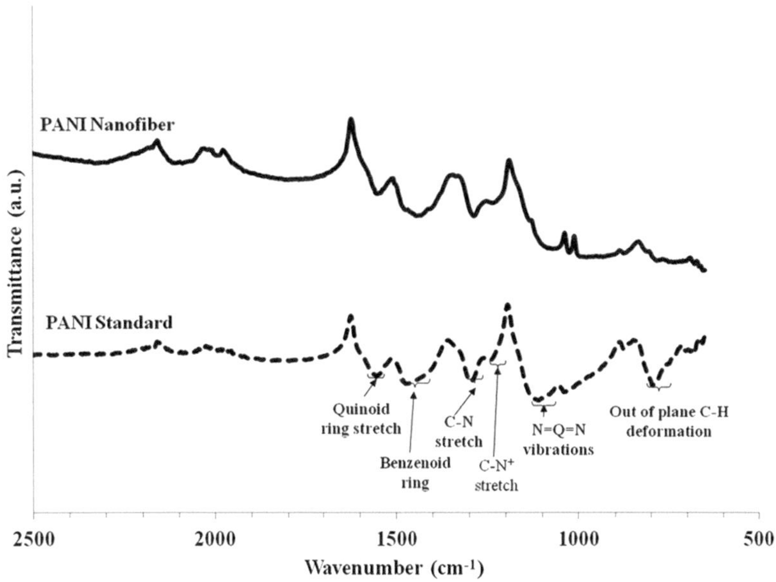

Fig. 8.18 FTIR spectra of PANI nanofiber and standard sample indicating that the major bonding environment remains unchanged for both standard and nanofibrous polyaniline structures

the lower frequency mode at 1,500 cm^{-1} depicts the presence of benzenoid ring units. Furthermore, the peaks at 1,250 and at 800 cm^{-1} are assigned to vibrations associated with the C–N stretching vibration of aromatic amine out of plane deformation of C–H of 1,4 disubstituted rings. The aromatic C–H bending in the plane (1,167 cm^{-1}) and out of the plane (831 cm^{-1}) for a 1,4 disubstituted aromatic ring indicates a linear structure. Independent of surfactant, complete formation of polyaniline has been observed always.

8.4.1.2 Scanning Electron Microscopy Results

The microstructure of the as-synthesized PANI nanofibers was studied by Hitachi S800 scanning electron microscope (SEM). A fixed working distance of 5 mm and a voltage of 5–25 kV were used. Sample preparation for the SEM measurement was carried out inside the glove box by covering the sample holder with Parafilm® for minimal exposure to oxygen while transferring it to the secondary emission chamber. EDAX Genesis software was used to analyze the SEM images.

It is clearly discernible from the scanning electron micrograph, shown in Fig. 8.19, that the density of nanofiber formation remains rather constant, irrespective of the surfactant used during the synthesis of PANI-NF. Furthermore, it can be seen from Fig. 8.19 that the average nanofiber diameter is approximately 250 nm. It is important to note that the surface of the nanofibers is rather rough, which is typical

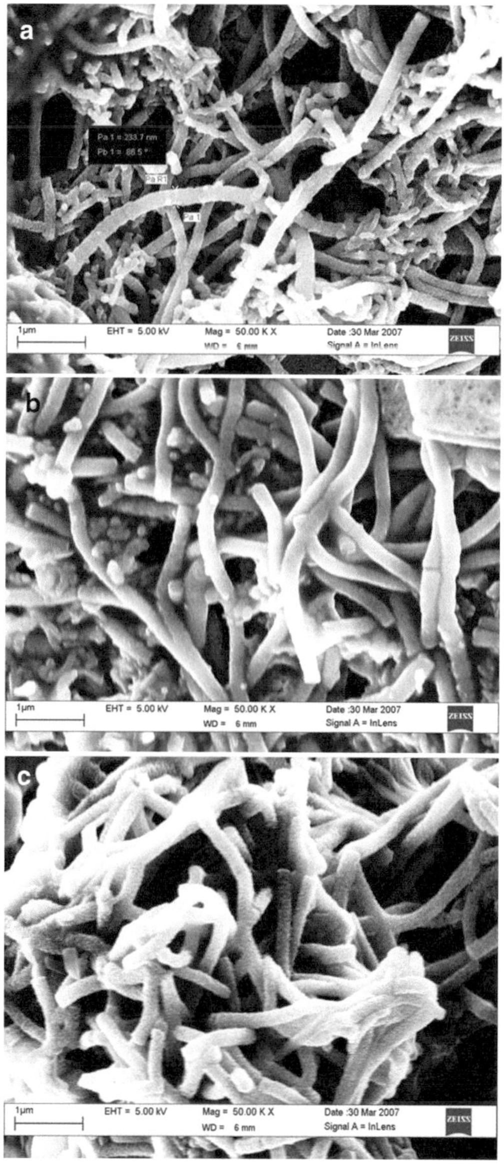

Fig. 8.19 Scanning electron micrographs of polyaniline nanofibers grown at room temperature in aqueous medium with different surfactants (**a**) dodecylbenzenesulfonic acid, (**b**) acrymethylpropyl sulfonic acid, and (**c**) camphorsulfonic acid using ammonium persulfate as oxidizing agent

of chemical nanofiber growth. Additionally, the nanofibers are rather inconsistent in length and overall appear to be fragmented. However, there is no evidence of bulk PANI observed, meaning that there is essentially a 100 % yield of nanofibers. The nanofibers that were synthesized using camphorsulfonic acid, as seen in Fig. 8.19c, were analyzed for their hydrogen sorption behavior because of their slightly rougher surface, as it was thought that they possess more potential hydrogen binding sites.

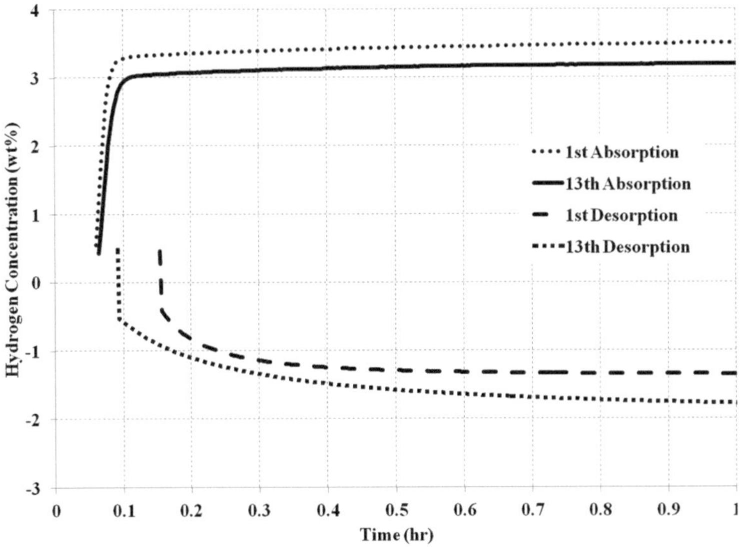

Fig. 8.20 Hydrogen absorption and desorption kinetics of PANI-NF in the 1st and 13th cycle

8.4.1.3 Hydrogen Sorption Results

The volumetric hydrogen sorption measurements are of paramount importance in understanding the hydrogen storage behavior of PANI-NF. Room temperature hydrogen absorption was executed at a high hydrogen pressure of approximately 80 bar with a pre-calibrated reservoir of 160.11 mL, providing for roughly a 15:1 volumetric ratio of hydrogen to sample volume.

Figure 8.20 demonstrates the initial hydrogen uptake of PANI-NF with respect to time. From this figure it is seen that rapid absorption was achieved (i.e., 95 % of total capacity 3–4 wt% in less than 10 min) by the nanofibrous matrix. The absorbed hydrogen was then desorbed against 1 bar and approximately 1–2 wt% hydrogen was released at room temperature (see Fig. 8.20).

In order to confirm and ensure the effective hydrogen sorption, the pressure-composition-temperature (PCT) measurements were carried out at room temperature for the PANI-NF from the 2nd cycle to the 6th cycle as shown in Fig. 8.21a, b. Interestingly, a distinct plateau pressure region was observed, which is usually an identifier for effective hydride formation of chemisorption, around 30 bar (up to 1.5 wt% hydrogen uptake), and a linear region (up to 2 wt%) with total absorption capacity around 3.5 wt%. In the consecutive hydrogenation cycles (3rd–6th), this plateau pressure region diminishes; nevertheless, the hydrogen absorption capacity remains the same as shown in Fig. 8.21a. The initial plateau region observed most likely corresponds to chemisorption of hydrogen into the material. The hydrogen initially bonds via physisorption to the nanofibers, but hydrogen also fills any vacancies and unterminated bonds that arose during the synthesis of the

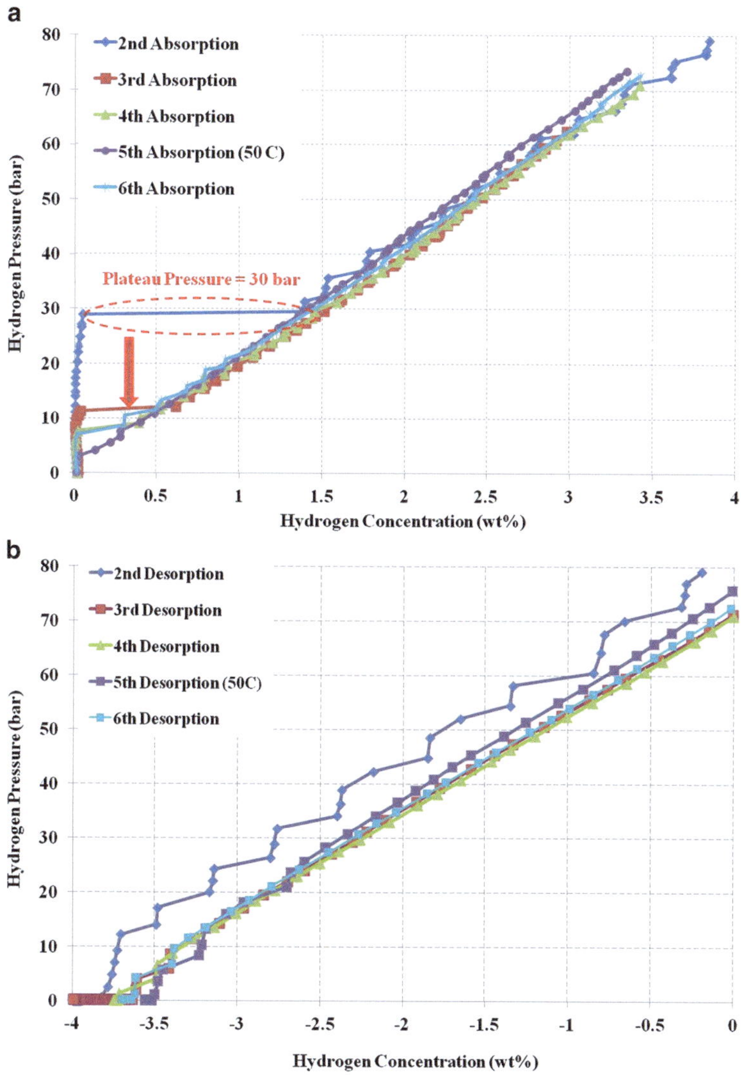

Fig. 8.21 Pressure-composition isotherms (PCT) of PANI-NF at room temperature from 2nd to 6th (**a**) absorption (**b**) desorption cycles

nanofibers, which is very common for chemically synthesized polymers. The fact that no desorption plateau region is observed further confirms hydrogen sorption via chemisorption, as room temperature does not provide the hydrogen with enough energy to break the hydrogen bonds produced through chemisorption. The reduction in the plateau pressure, as well as the reduction in the amount of hydrogen that is actually chemically sorbed during consecutive cycles, indicates that the nanofibers become saturated and terminated with hydrogen. Since hydrogen is still reversibly

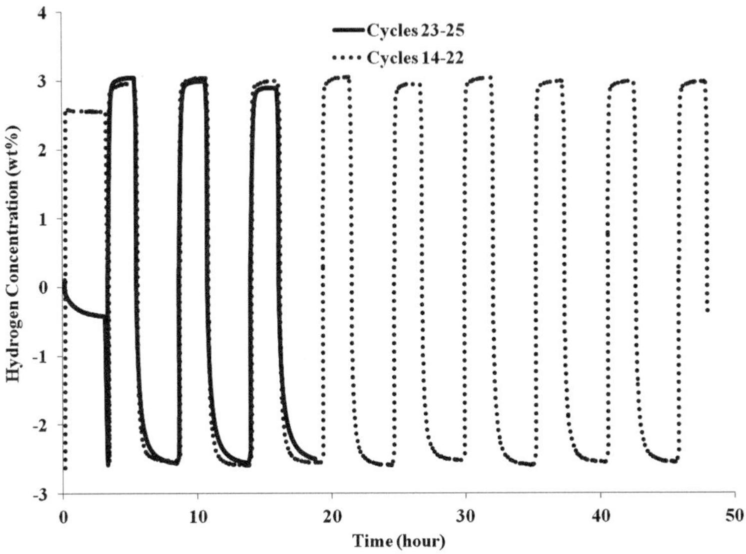

Fig. 8.22 Hydrogen sorption kinetics at room temperature from 14th cycle to 25th cycle showing little degradation in kinetics and no degradation in the storage capacity within the measured cycles

sorbed in the nanofibers without any plateau region, it is clear that the remaining 3.5 wt% of hydrogen is stored via physisorption. The reversibility of hydrogen absorption and desorption of 3–4 wt% at room temperature was unambiguously demonstrated by Fig. 8.21a, b. It is also noticeable from this figure that by increasing the sorption temperature from room temperature to 50 °C in the 5th cycle, the plateau pressure region disappears, which indicates that these nanofibers react with hydrogen more effectively at room temperature than that at a moderate temperature of 50 °C.

Furthermore, it is clear that the main driving force behind hydrogen sorption in these nanofibers is not temperature, but instead pressure. This means that a pressure of approximately 80 bar is required to keep the hydrogen stored in the material. If the pressure is reduced from 80 bar to near vacuum, the hydrogen is very quickly released, as is evident in Fig. 8.22 – the kinetic behavior of the nanofibers. Conversely, if the pressure is increased instantaneously to 80 bar, the hydrogen is quickly reabsorbed. After several hydrogen sorption cycles, the absorption kinetics at the 13th cycle was recorded and compared with the 1st cycle. It was observed that the reaction kinetics in the 13th cycle remain the same as for the 1st cycle as shown in Fig. 8.20. Hydrogen absorption and desorption from the 14th to 25th cycle are shown in Fig. 8.22. It is clearly seen from this figure that there was no deterioration in the hydrogen storage capacity of 3–4 wt% during these reversible cycles. Furthermore, it can be seen that it only takes approximately 6 min to absorb the hydrogen. With increasing cycles, however, there is a slight reduction in the kinetics of the material, most notably in the hydrogen desorption.

Fig. 8.23 SEM image of PANI nanofibers after hydrogen cycling

While the absorption kinetics stay virtually unchanged, the desorption kinetics slow from 6 min to about 30 min for full hydrogen release with consecutive hydrogen absorption/desorption cycles.

8.4.1.4 Hydrogen Cycling Effects on PANI Nanofibers

This decrease in hydrogen kinetics can be explained when looking at the microstructure of the PANI nanofibers after hydrogen cycling, as seen in Fig. 8.23. It is evident that the nanofiber structure that was initially present is no longer intact. The nanofibers are in fact wholly missing. The interaction of the hydrogen with the nanofibers, through repeated pressure changes at close to room temperature, first caused chemical bonding to the PANI which resulted in termination of bonds. Additionally, the force caused by the pressure variations of the hydrogen essentially resulted in the nanofibers being repeatedly compressed and decompressed, thereby forcing the microstructural changes. As can be seen in the SEM image, however, the PANI surface morphology is still characterized by a porous nature, created by the diffusion of hydrogen in and out of the structure. Since the microstructure changed so significantly, a surface area analysis was performed on the PANI nanofibers before and after hydrogen cycling to confirm the initial findings of unchanged surface area. The results of the BET surface area analysis are presented in Fig. 8.24.

It can be seen that the surface area of the PANI nanofibers before hydrogen cycling was approximately 3.26 m^2/g and after hydrogen cycling actually increased slightly to 3.56 m^2/g. This surface area is significantly lower than most other porous polymers reported in literature, which often have areas larger than 100 m^2/g. The intrinsic nature of polyaniline, however, allows the hydrogen to bond with many different sites both through chemisorption and physisorption, as previously shown.

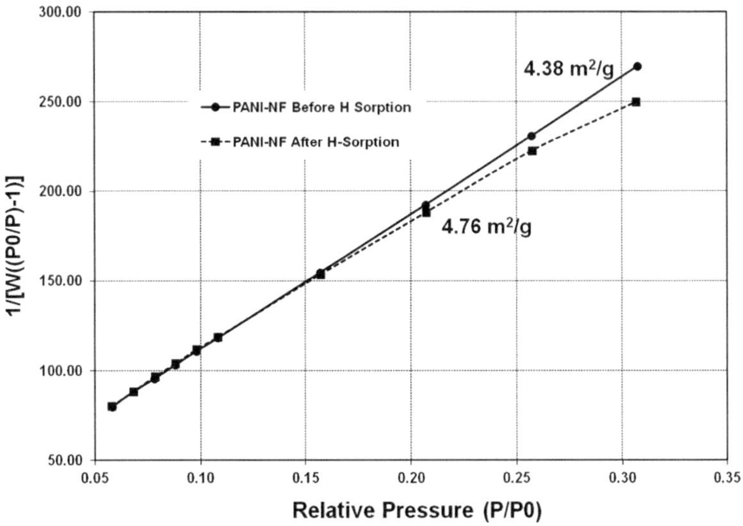

Fig. 8.24 BET surface area analysis of PANI nanofibers (CM) before and after hydrogen cycling

8.4.2 PANI Nanofibers (CM) Summary

In summary, polyaniline nanofibers (PANI-NF), synthesized by chemical templating technique using various surfactants as dopants and ammonium persulfate as the oxidant, were successfully shown to reversibly store hydrogen. The characteristics of PANI-NF are understood regarding their structure, microstructure, bonding, and thermal stability behaviors. The signature bands of benzenoid and quinoid transitions in the frequency of 1,500 and 1,600 cm^{-1}, respectively, as evidenced from FTIR, confirm the formation of the polyaniline nanostructure irrespective of the surfactants used in the precursor preparation. The rate of hydrogen sorption during the initial cycle is very rapid (95 % hydrogen storage capacity (3–4 wt%) absorbed in less than 6 min). Moreover, these PANI nanofibers demonstrate excellent reversibility (up to 25 cycles measured) at room temperature. This behavior has previously not been reported in literature. Another important feature discernible from the PCT isotherms was that during the second hydrogen absorption run, the plateau pressure occurred around 30 bar, and it diminished in subsequent cycles. This means that hydrogen first bonded to vacancy sites and unterminated bonds created during the chemical synthesis. The reduction in the plateau pressure as well as the reduction in plateau capacity with subsequent cycles indicates that the chemisorption sites disappear as hydrogen is chemically bonded. This chemisorption is further proven by the lack of plateau regions in desorption cycles, as the low temperature (around room temperature) does not provide enough energy for the hydrogen to break the stronger chemical bonds. Nevertheless, the reversible capacity of 3–4 wt% was maintained throughout 25 cycles through physisorption with pressure changes as the main driving force in hydrogen absorption and desorption. Furthermore, it was shown that the nanofibrallar nature of the PANI nanofibers disappears through hydrogen cycling, most likely due to repeated compression of the structure from 0 to 80 bar of hydrogen pressure. The surface area, however, remains virtually unchanged and even increases slightly. This means that the binding sites for the hydrogen remain intact, even though the morphology is changed significantly. While a capacity of 3–4 wt% is not enough to meet the DOE targets for a practical hydrogen storage medium, the rapid kinetics and low temperature required for hydrogen cycling make these chemically synthesized nanofibers an attractive hydrogen storage medium for smaller applications.

8.5 Electrospun PANI Nanofibers

In this section, the synthesis and characterization of polyaniline nanofibers by electrospinning process of polymer solution on a collection substrate are investigated. These as-deposited nanostructures were characterized using Fourier transform infrared (FTIR) spectroscopy for determination of stretching modes and vibrations. An important aspect of hydrogen adsorption and desorption behavior in these nanofibers was estimated by high pressure hydrogen sorption measurements. Microstructural changes due to hydrogen sorption were observed by scanning electron microscopy (SEM) in the imaging mode.

8.5.1 Synthesis and Characterizations of Electrospun Polyaniline Nanofibers

As with the chemically grown nanofibers, the polyaniline nanofibers (PANI-NF-ES) were grown by oxidative polymerization of aniline monomer at 0 °C in an ice bath using ammonium persulfate as the oxidant in the presence of surfactant, though with slightly different conditions. Aniline, ammonium persulfate, dodecyl benzene sulfonic acid, acrylmethylpropyl sulfonic acid, and camphorsulfonic acid are used as received from Sigma-Aldrich. Sulfonic acid-based surfactants as the dopant and ammonium persulfate as the oxidant were used in the present synthesis of polyaniline nanofibers.

Calculated quantities of aniline monomer (0.01 mol) were mixed with 50 mL of distilled water and stirred using magnetic stirrer for 10 min. Meanwhile, calculated quantities of surfactant (0.75 mol) and oxidant (0.01 mol) were dissolved separately in distilled water and stirred for 10 min in an ice bath. The surfactant solution was first added into the aniline monomer aqueous solution, and then the previously cooled oxidant solution was added dropwise after which the mixture was allowed to react for 15 h in an ice bath (Fig. 8.25).

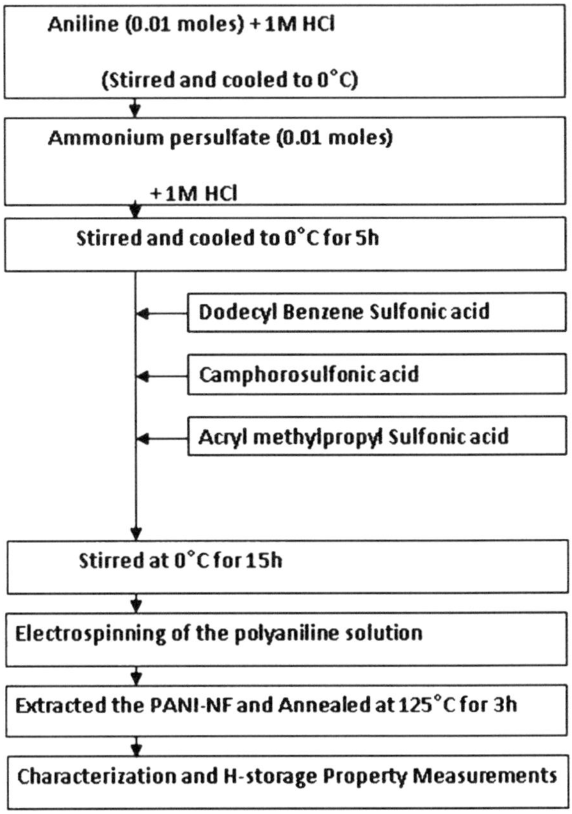

Fig. 8.25 Flow chart for the synthesis of polyaniline nanofibers prepared using electrospinning of polyaniline solution

The obtained polymer solution was then sprayed by the electrospun method schematically represented in Fig. 8.26. In the electrospinning process, a high electric field is applied to the viscous polymer solution, held in a capillary tube, inducing a charge density on the liquid surface. The distance between the tip of the needle and the substrate was optimized to 20 cm with a voltage of 15 kV.

Mutual charge repulsion causes a force directly opposite to the surface tension. When the electric field is sufficiently high, the surface of the solution in proximity of the tip of the capillary tube elongates and forms a cone, named Taylor cone, which in turn allows for the formation of randomly arranged nanofibers. These polyaniline nanofibers are collected and then dried in vacuum at room temperature for 24 h. Later, the vacuum-dried sample is annealed at 125 °C for 3 h. Finally, the dried electrospun nanofibers are characterized and tested for the hydrogen uptake and release measurements.

8.5.1.1 Scanning Electron Microscopy Results

The microstructure of the as-synthesized PANI nanofibers was studied by Hitachi S800 scanning electron microscope (SEM). A fixed working distance of 5 mm and

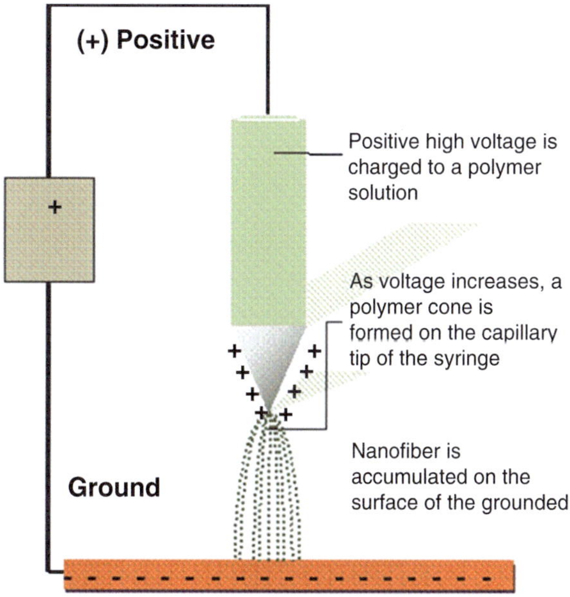

Fig. 8.26 Schematic diagram of electrospinning method to produce polyaniline nanofibers

a voltage of 5–25 kV were used. Sample preparation for the SEM measurement was carried out inside the glove box by covering the sample holder with Parafilm® for minimal exposure to oxygen while transferring it to the secondary emission chamber. EDAX Genesis software was used to analyze the SEM images. The electrospun nanofibers were successfully grown and can be seen in Fig. 8.27. It is important to note that the nanofibers are more or less of the same dimensions. The average diameter is approximately 1.5 μm, thereby not technically nanofibers. The fibers have very large aspect ratios, with lengths of several μm. Additionally, the fibers are very smooth in appearance, which is typical of electrospun grown fibers. No bulk PANI is visible from the SEM images, meaning that there is essentially a 100 % yield of nanofibers.

8.5.1.2 FTIR Characterization Results

Fourier transform infrared (FTIR) spectra of electrospun polyaniline nanofibers (PANI-NF-ES) before hydrogen sorption are shown in Fig. 8.28. The major bonding environment remains unchanged for both the pristine and hydrogenated PANI-nanofibrous structures. The presence of two bands in the vicinity of 1,500 and 1,600 cm^{-1} is assigned to the nonsymmetric C6 ring stretching modes. The higher frequency vibration at 1,600 cm^{-1} is for the quinoid rings, while the lower frequency mode at 1,500 cm^{-1} depicts the presence of benzenoid ring units. Furthermore, the

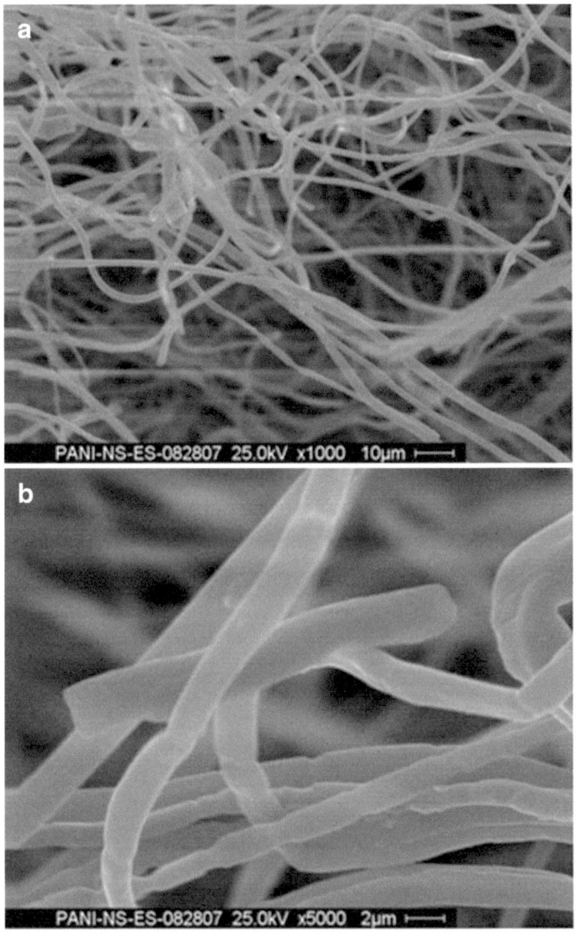

Fig. 8.27 SEM image of electrospun PANI nanofibers at a magnification of (**a**) 1,000 and (**b**) 5,000

peaks at 1,250 cm^{-1} and at 800 cm^{-1} are assigned to vibrations associated with the C–N stretching vibration of aromatic amine out of plane deformation of C–H of 1,4 disubstituted rings. The aromatic C–H bending in the plane (1,167 cm^{-1}) and out of the plane (831 cm^{-1}) for a 1,4 disubstituted aromatic ring indicates a linear structure.

It is interesting to note that the quinoid ring stretch around 1,650 cm^{-1} is significantly more pronounced relative to the other peaks and also shifted by 50 cm^{-1} as compared to the bulk polyaniline. Additionally, the benzenoid ring stretch around 1,450 cm^{-1} is not as broad, but instead composed of several individual smaller peaks. Overall, the FTIR spectrum confirms that polyaniline in its emeraldine form is present, as all the major characteristic peaks are observed. However, there is a definite difference in the FTIR spectra of the electrospun nanofibers as compared to even the chemically grown nanofibers, which essentially had an identical spectrum to the bulk chemically grown polyaniline.

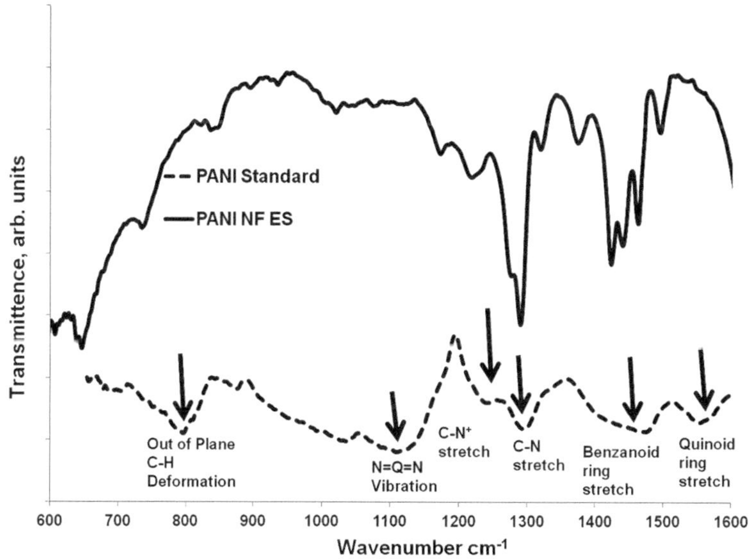

Fig. 8.28 FTIR spectra of bulk PANI and electrospun PANI nanofibers

8.5.1.3 Thermogravimetric Results

The thermal stability of PANI nanofibers (both CM and ES methods) and the standard samples were characterized using TA Instruments' simultaneous DSC and TGA (SDT-Q600) tool. A pre-weighed sample was loaded into the ceramic pan and covered with the ceramic lid inside the glove box to prevent moisture from getting into the sample during transfer. The ramp rate of 5 °C/min was used for all the measurements. TA's Universal Analysis 2000 software program was used to analyze the TGA and DSC profiles.

This TGA comparison is shown in Fig. 8.29. The initial weight loss up to 15 % is due to moisture that was absorbed during regular storage, since the polyaniline samples were all kept in sealed, but non-inert containers. After the release of the stored moisture, it can be seen that the PANI nanofibers (ES) are stable up to 150 °C. This is to be expected, as the sample preparation includes annealing of the PANI at 125 °C. A temperature of higher than 150 °C is not required anyway as this would be too high a temperature for practical hydrogen storage. It should be noted, however, that the electrospun nanofibers are less thermally stable than the bulk or chemically grown counterparts. This is most likely due to the slight chemical differences that were observed during FTIR analysis, as previously mentioned.

8.5.1.4 Hydrogen Sorption Results

The electrospun PANI nanofibers were subjected to both kinetic and PCT hydrogen measurements. Figure 8.30 demonstrates the hydrogen sorption life cycle kinetic

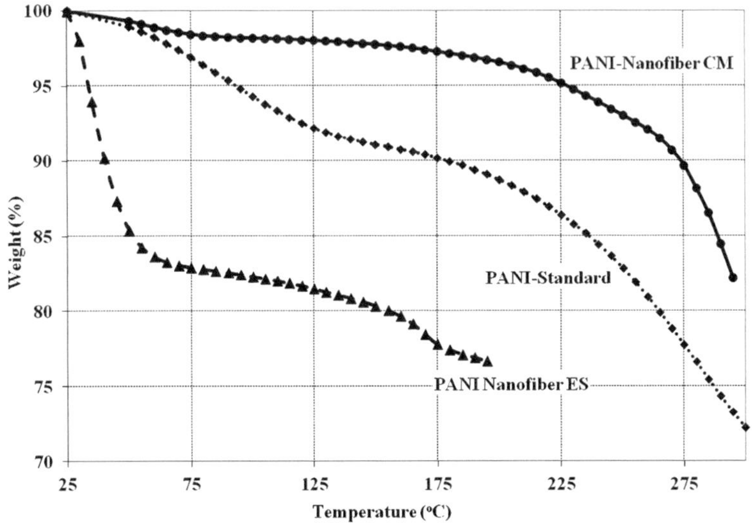

Fig. 8.29 Thermogravimetric weight loss analysis of PANI samples in both bulk and nanofiber form

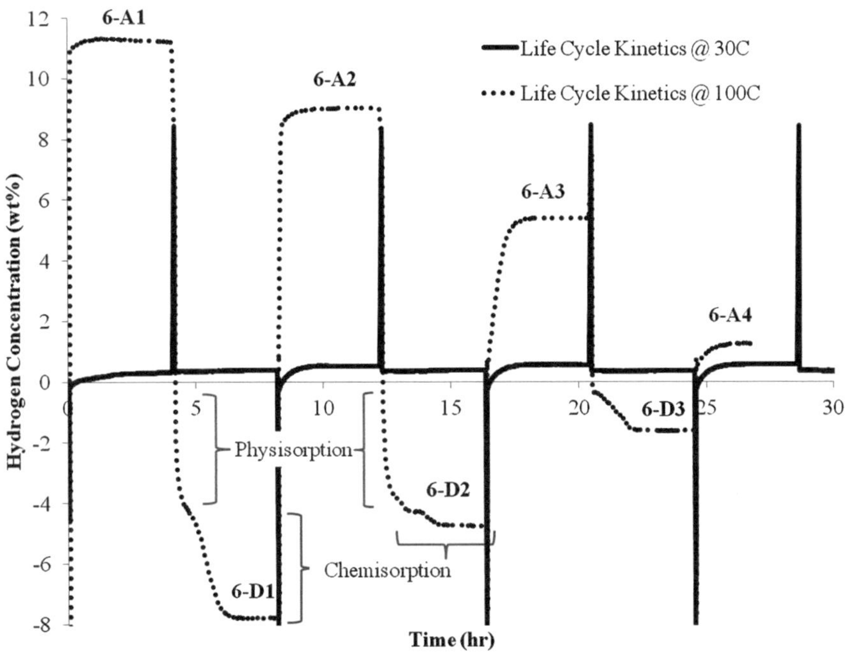

Fig. 8.30 Hydrogen sorption kinetics curves for the PANI-NF-ES at 30 and 100 °C

curves of PANI-NF at two different temperatures, 30 and 100 °C. As with previous samples, the first hydrogen sorption measurement was performed at 30 °C. However, when the nanofibers were subjected to a high hydrogen pressure of 80 bar, the hydrogen uptake was found to be less than 0.75 wt%, which is reminiscent of the bulk PANI samples. Additionally, there was no real desorption observed at this low temperature. It was therefore decided to perform the hydrogen measurements at various temperatures, and finally, at 100 °C (see Fig. 8.30), hydrogen was absorbed. Surprisingly, a hydrogen uptake (A1 in Fig. 8.30) of close to 11 wt% with rapid adsorption kinetics (approximately 95 % hydrogen adsorption within 5–10 min) was observed at 100 °C. The desorption cycle of the same sample at 100 °C then exhibited a hydrogen release of close to 8 wt%. This means that the other 3 wt% of hydrogen that was previously absorbed remained in the structure. Interestingly, the desorption (labeled as D1 in the figure) was composed of both a rapid hydrogen release step in which approximately 4 wt% was released within a few minutes and a slower hydrogen release step, accounting for another 4 wt% in approximately 45 min. When the desorption equilibrium pressure was reached, hydrogen was reabsorbed into the sample. This time (A2) the full 8 wt% that was previously released was reabsorbed as rapidly as had occurred in the first hydrogen absorption. After this, only about 5 wt% hydrogen was released, again in a two-step process. Only this time the slower release step released only about 1 wt%, while the fast release regime remained at a capacity of 4 wt%. The full 5 wt% was then reabsorbed, though this time the kinetics of the hydrogen uptake decreased from about 5 to 25 min.

After this cycle, the hydrogen capacity and kinetics dropped very quickly until finally virtually no hydrogen could be sorbed. The two-step release observed for the electrospun nanofibers is clearly evident that both physisorption and chemisorption properties are taking place simultaneously. The physisorption, again driven mainly by pressure changes, allowed for very rapid hydrogen release and uptake, while the slower second step during release was due to chemisorption, as more energy or more time in lack of a higher temperature is required to release the hydrogen. The decrease in the hydrogen absorption capacity from cycle to cycle is clearly due to chemisorption of the hydrogen to the nanofibers. Once the strong primary bond between the structure and the hydrogen is formed, it is virtually impossible to get the hydrogen back out without destroying the polymeric structure, as a temperature of more than 150 °C has been shown to start the decomposition process.

Since only kinetic measurements were performed on the nanofibers and also in order to confirm the initial hydrogen uptake of 10 wt% that resulted from these measurements, a new batch of electrospun nanofibers was loaded into the PCT reactor. The hydrogen PCT absorption and desorption behavior of this batch is shown in Figs. 8.31 and 8.32, respectively. It is important to note that only PCT measurements were made on this batch and no kinetic measurements.

It was observed that virtually no hydrogen was absorbed or desorbed until a measurement temperature of approximately 50 °C. After the initial hydrogen uptake of about 2.5 wt% at 50 °C, the hydrogen capacity was found to increase with

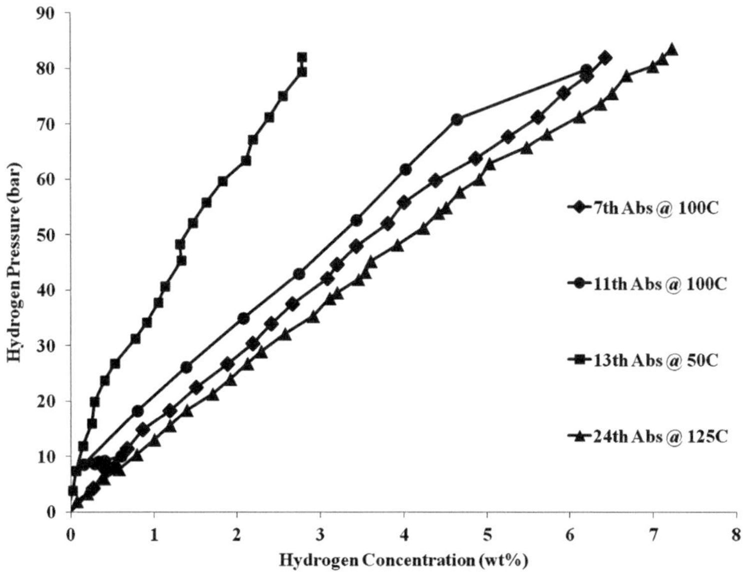

Fig. 8.31 Hydrogen adsorption PCT curves for the PANI-NF-ES at different temperatures

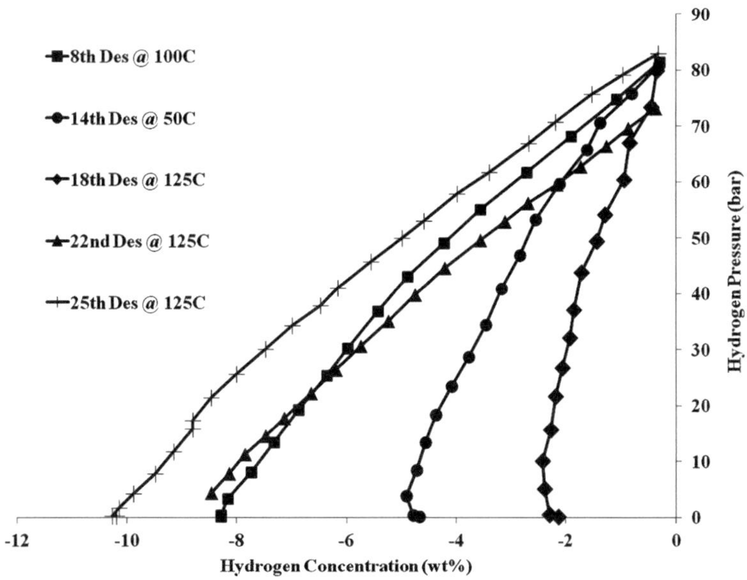

Fig. 8.32 Hydrogen desorption PCT curves for the PANI-NF-ES at different temperatures

increasing temperature up to 125 °C in various cycles as seen in Fig. 8.31. At approximately 100–125 °C, a twofold increase of capacity (6–8 wt%) was invariably obtained at various hydrogenation cycles. At the end of each adsorption PCT, desorption experiments were performed and the results are depicted in Fig. 8.32. A hydrogen storage capacity of 2–10 wt% was obtained at temperature range of 50–125 °C.

While it might appear that the amount of hydrogen that is released from the nanofibers is larger than the amount that is absorbed, this is in fact not the case. The data presented in Fig. 8.32 is actually the amount of hydrogen released without taking the hydrogen itself into account which means that the numbers are slightly higher than they should be. It should be regarded as raw data. When one actually takes into account the weight of the material plus the weight of hydrogen stored in it, the maximum amount of hydrogen released changes from 8.5 wt% to approximately 7.5 wt%, thereby corresponding to the amount of hydrogen previously absorbed.

The nanofibers that were tested for their capacity using PCT only are from the same synthesis batch and have absolutely no difference in terms of their chemical or physical composition to the batch of sample that was subjected only to kinetic measurements. However, the type of measurement that was employed made a large difference in capacity. As previously mentioned, hydrogen through physisorption or chemisorption is often based on diffusion. When one performs a kinetic measurement, the sample is subjected to either a high (80 bar) or a low (0 bar) hydrogen pressure. A PCT measurement, however, allows the sample to experience gradual pressure increase as opposed to this sudden pressure change. The hydrogen, therefore, has enough time to diffuse through the material as the pressure is incrementally increased or decreased for absorption or desorption, respectively. While this has not affected any previous samples, the electrospun nanofibers were much more sensitive to the type of measurement. Most likely, this is due to the difference in chemical structure, as was shown by the FTIR spectrum in Fig. 8.28.

8.5.1.5 Hydrogen Cycling Effects on Electrospun PANI Nanofibers

In order to confirm the hypothesis of chemisorbed hydrogen from the kinetic measurements, scanning electron microscopy was performed on the cycled nanofibers. The SEM image of the electrospun PANI nanofibers before hydrogen sorption, shown in Fig. 8.33a, is compared to the SEM image of the nanofibers after hydrogen cycling, as shown in Fig. 8.33b. It becomes evident that the density of the nanofibers was not altered significantly. The SEM micrographs of the hydrogenated and dehydrogenation samples show nanofibrallar swelling, breakage of fiber length, and precipitations, which is most likely the cause for hydrogen loading saturation and poor cyclic reversibility. In addition to the swelling of the nanofibers due to the chemical bonding of hydrogen to the PANI, it is clear that the nanofibers also break into smaller pieces, thereby losing the high aspect ratio and returning more to the bulk form of polyaniline previously described.

Fig. 8.33 SEM micrographs of PANI-NF (**a**) before and (**b**) after hydrogen sorption cycling

After hydrogen cycling, FTIR was performed to observe any possible changes in the structure of the nanofibers. As can be seen from Fig. 8.34, there is absolutely no change in the nature of the chemical bonds. The quinoid and benzenoid ring vibrations and stretches are all intact. Due to the nature of sample preparation, though, the intensity of the transmittance does vary between the measurements. The relative intensities of the individual peaks remain constant, though, indicating no chemical change after hydrogen interaction.

8.5.2 Electrospun PANI Nanofibers Summary

Conducting polyaniline nanofibers were synthesized using chemical templating method followed by an electrospinning process. The FTIR spectra of these PANI nanofibers reveal the presence of two bands in the vicinity of 1,500 and 1,600 cm^{-1},

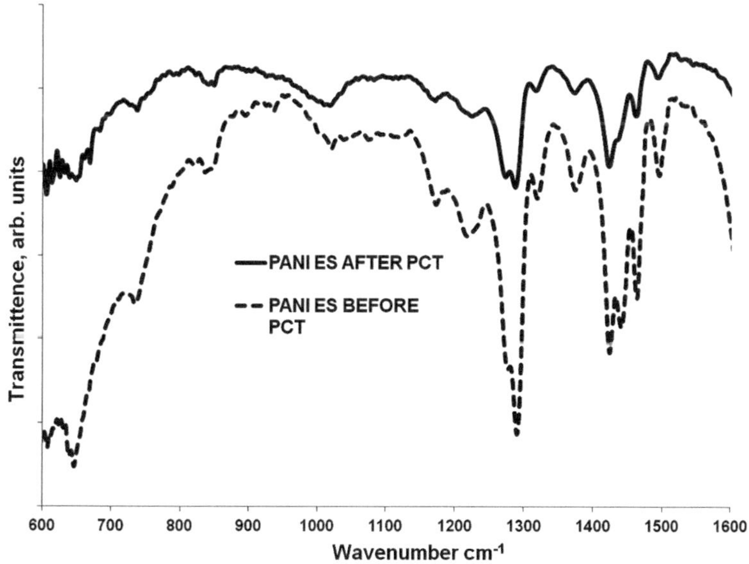

Fig. 8.34 FTIR spectra of electrospun PANI nanofibers before and after hydrogen cycling

which are assigned to the nonsymmetric C6 ring stretching modes. The lower frequency vibration at 1,500 cm^{-1} is for the benzenoid rings, while the higher frequency mode at 1,600 cm^{-1} corresponds to benzenoid to quinoid transition indicating that the synthesized polymer was polyaniline in oxidized form. These nanofibers have been compared with their standard bulk counterpart and found to be stable up to 150 °C, slightly lower than their bulk chemical counterparts. It was found that electrospun polyaniline nanofibers have a high hydrogen uptake of 10 wt% at around 100 °C in the first absorption run. However, in the consecutive hydrogenation and dehydrogenation cycles, the capacity diminishes. This is due to a combination of chemisorption and physisorption during kinetic measurements where the sample is exposed to varying extreme pressures. Eventually, the electrospun PANI nanofibers become saturated with hydrogen during kinetic loading. When the nanofibers are exposed to PCT measurements, though, where the pressure is either gradually increased or decreased, the nanofibers exhibit a reversible hydrogen storage capacity of 3–8 wt% at different temperatures. This is due to the fact that the hydrogen is given enough time to diffuse and reach local equilibrium at varying pressures. Unfortunately, this also means that a larger time is required to either absorb or desorb the hydrogen. The surface morphologies before and after hydrogen sorption on these PANI nanofibers encompass significant changes in the microstructure such as nanofibrallar swelling and slight deterioration in the length of the fibers.

It was confirmed that the hydrogen data is reproducible with high hydrogen storage capacity of 3–8 wt% at different temperatures, as summarized in Table 8.1. From this table, it is discernible that the hydrogen capacity increases with an

Table 8.1 Hydrogen sorption capacity of electrospun nanofibers at various temperatures and sorption cycles

Cycle number	Adsorption/ desorption	Temperature (°C)	Hydrogen concentration (wt%)
7	Adsorption	100	6.5
8	Desorption	100	6.5
13	Adsorption	50	3
14	Desorption	50	3
24	Adsorption	125	8.5
25	Desorption	125	8.5

increase of temperature. However, in order to attain this higher capacity, the nanofibers need to be exposed to gradual pressure changes rather than be exposed to large pressure changes, as would be typical with kinetic measurements.

8.6 Polyaniline Hydrogen Storage Summary

Polyaniline, in its emeraldine form, was successfully synthesized through chemical means. The bulk, or as-synthesized form, and three different nanostructures of PANI were characterized for physical, chemical, and hydrogen sorption characteristics. It was found that the inclusion of additives to the bulk form PANI had virtually no effect on its hydrogen sorption characteristics. Bulk PANI was found to have a hydrogen capacity of less than 0.5 wt% at a temperature of 100 °C. It was observed that the capacity of hydrogen in this standard sample increased slightly with increasing temperature. This increase with temperature is due to the expansion of PANI, which in turn allows for the hydrogen to diffuse readily into the polymer. While others have claimed that standard PANI can absorb as much as 8 wt%, this was refuted by later reports [2, 16].

The nanostructured PANI nanospheres, with an average diameter of 100 nm, were synthesized by slightly modifying the chemical synthesis technique used to produce the standard PANI. FTIR analysis, however, shows that the nanospheres do possess the same chemical composition as the standard PANI. While an initial hydrogen uptake of 5.5 wt% of hydrogen was observed at 30 °C, the capacity of each consecutive desorption cycle was reduced by roughly 1 wt%, though the desorbed hydrogen was reabsorbed. Since the kinetics of hydrogen uptake were on the order of several hours, with slower kinetics for each consecutive cycle, it was found that the main driving force behind the hydrogen sorption of the nanospheres was chemisorption, most likely in the form of hydrogen reacting with unterminated bonds that were a result of the synthesis. Finally, a capacity of only 0.6 wt%, virtually identical to that of bulk PANI, was observed at a temperature of 90 °C. SEM analysis showed that the repeated hydrogen cycling caused the nanospheres to swell after hydrogen reaction and also to agglomerate until it finally had an appearance reminiscent of bulk polyaniline. The interaction of hydrogen with the sample also led to the formation of microcracks, which are caused by the diffusion of hydrogen through the material.

By again slightly modifying the chemical synthesis technique, chemically grown PANI nanofibers were created. FTIR investigation again proved that the nanofibers were of the emeraldine form. The fibers had an average diameter of 250 nm with varying lengths. The nanofibers were also characterized by a rough surface. Hydrogen cycling of the nanofibers showed that an initial capacity of 3–4 wt% was absorbed in less than 10 min, while approximately 1–2 wt% hydrogen was released at room temperature. PCT measurements performed on the nanofibers showed that roughly 3.5 wt% of hydrogen was reversibly stored in the nanostructures. The initial hydrogen absorption showed a hydrogen plateau pressure around 30 bar, indicative of chemisorption. Through repeated cycling, however, the plateau pressure region disappeared, though the capacity remained unchanged at 3.5 wt% for many cycles. SEM analysis of the chemically grown nanofibers after hydrogen cycling showed no evidence of the initial nanofibrous morphology. Instead, the fibers had virtually disappeared and a rather porous material remained. Surface area analysis revealed that the surface area remained unchanged after hydrogen cycling. Unlike the nanospheres, whose structure also disappeared, the nanofibrous nature of the PANI allowed for it to agglomerate into a porous material, thereby losing the initial physical characteristics but still maintaining the surface area. This is an important factor for hydrogen storage materials, as was previously discussed.

Finally, electrospun nanofibers were investigated for their hydrogen performance. By using a well-known electrospinning technique, the nanofibers were synthesized with average diameters of 1.5 μm and lengths of several microns. The nanofibers were, unlike the chemically grown nanofibers, rather smooth. While FTIR again confirmed the presence of the typical benzenoid and quinoid rings of emeraldine base PANI, it was observed that the quinoid peak was much larger in relation to the other FTIR peaks. The quinoid peak was also shifted by approximately 50 cm^{-1}, making it unique among all the polyaniline structures investigated. Hydrogen kinetic measurements on the electrospun nanofibers at 30 °C showed no hydrogen sorption, but absorption at 100 °C revealed a very promising initial uptake of over 10 wt% in less than 10 min. The desorption that followed showed an almost instantaneous release of 4 wt% hydrogen and another 4 wt% with much slower kinetics. Each consecutive absorption cycle reabsorbed the previously released hydrogen with slightly lower kinetics, but still in a matter of minutes rather than hours. Desorption capacity decreased by roughly 2 wt% for each cycle, but the fast hydrogen desorption remained until, finally, after approximately three absorption/desorption cycles; virtually no hydrogen was sorbed. While the fast kinetics of the absorption measurements clearly indicate that physisorption is taking place, the desorption cycles are clearly a combination of both chemisorption (slower kinetics) and physisorption (faster kinetics). The combination of physisorption and chemisorption is also evidenced by the breakage and swelling of the nanofibers, though overall they do remain intact. Hydrogen PCT measurements on a new batch of this sample, however, revealed that approximately 8 wt% hydrogen could be

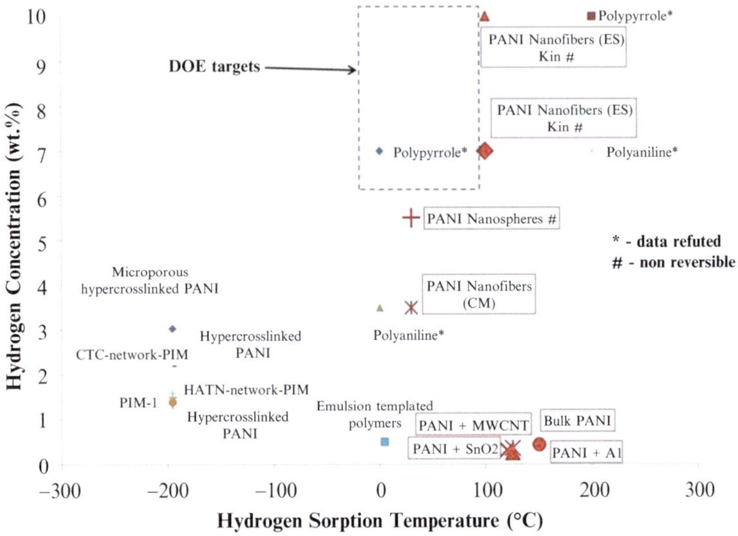

Fig. 8.35 Comparison of novel PANI results with previous polymer hydrogen storage research

reversibly stored in the electrospun nanofibers as long as the pressure was gradually increased or decreased. This slow exposure to hydrogen is essential in that the nanofibers are not exposed to pressure extremes, which can result in the destruction of the nanofiber morphology. Additionally, the hydrogen is allowed enough time to diffuse into the material and bond. While this method of storage requires a higher amount of time, it is still the most preferential method of storing hydrogen in polyaniline. The capacity of 8 wt% along with the relatively low temperature of 100 °C makes the electrospun polyaniline an ideal candidate for hydrogen storage.

It appears that the main factor in storing hydrogen in polyaniline is due to their surface area, as was previously thought for polymers, but in fact lies more in the exact chemical composition. Unterminated bonds, which provide chemisorption sites for hydrogen, are an excellent initial hydrogen storage site but require an amount of energy that is simply too high for practical purposes, therefore requiring high temperatures that would in fact destroy the chemical structure of the polyaniline. The relative intensity of the quinoid ring vibration with respect to the other FTIR peaks appears to play a vital role in the hydrogen storage behavior of polyaniline. Also, the form of hydrogen storage, namely exposure of the material to hydrogen, is extremely important when dealing with relatively weak materials such as PANI. Though it might require more time to store and release the hydrogen, a gradual pressure change is essential in maintaining the hydrogen characteristics of the material.

The work on these polymers is the first to investigate various nanostructures of polymers at temperatures above 77 K. A comparison of the results obtained from this research with other work that was previously conducted on hydrogen storage in polymers is shown in Fig. 8.35. The significant results described within this chapter are highlighted in red boxes, showing a closing in on the DOE goals with the PANI nanofibers produced via electrospinning method almost meeting the DOE targets.

Acknowledgments The author wish to thank his post doctoral advisors, colleagues, his graduate students and collaborators for the successful execution of the projects related to polymeric materials for hydrogen storage. The author also thank the editors, Professors B. Ding and J. Yu for their critical comments and suggestions to complete this book chapter project. The financial support from federal agencies, US Department of Energy and National Science Foundation are gratefully acknowledged.

References

1. Cho SJ, Song KS (2002) Hydrogen sorption in HCl-treated polyaniline and polypyrrole: new potential hydrogen storage media. ACS Div Fuel Chem 47(2):790–791
2. Panella B, Kossykh L (2005) Volumetric measurement of hydrogen storage in HCl-treated polyaniline and polypyrrole. Synth Met 151(3):208–210
3. McKeown NB, Budd PM (2006) Polymers of intrinsic microporosity (PIMs): organic materials for membrane separations, heterogeneous catalysis and hydrogen storage. Chem Soc Rev 35(8):675–683
4. Banyay GA, Shaltout MM (2007) Polymer and composite foam for hydrogen storage application. J Mater Process Technol 191(1–3):102–105
5. Lee J-Y, Bradshaw WCD, Cooper AI (2006) Hydrogen adsorption in microporous hypercrosslinked polymers. Chem Commun (Camb) 5:2670–2672
6. Germain J, Hradil J (2006) High surface area nanoporous polymers for reversible hydrogen storage. Chem Mater 18(18):4430–4435
7. McKeown NB, Gahnem B (2006) Towards polymer-based hydrogen storage materials: engineering ultramicroporous cavities within polymers of intrinsic microporosity. Angew Chem Int Ed Engl 45(11):1804–1807
8. Stejskal J (2002) Preparation of a conducting polymer (IUPAC technical report). Pure Appl Chem 74(5):857–867
9. MacDiarmid AG, Epstein AJ (1989) Polyanilines: a novel class of conducting polymers. Faraday Discuss Chem Soc 88:317–332
10. Trivedi DC (1997) Polyanilines. In: Nalwa HS (ed) Handbook of organic conductive molecules and polymers, 2nd edn. Wiley, Chichester, pp 505–572
11. Fusalba F, Villers GP, Belanger D (2001) Electrochemical characterization of polyaniline in nonaqueous electrolyte and its evaluation as electrode material for electrochemical supercapacitors. J Electrochem Soc 148:A1–A6
12. Roßberg K, Dunsch L, Ludwig S (1998) The influence of porosity and the nature of the charge storage capacitance on the impedance behaviour of electropolymerized polyaniline films. J Electroanal Chem 443(1):49–62
13. Virji S, Kaner RB, Weiller BH (2006) Hydrogen sensors based on conductivity changes in polyaniline nanofibers. J Phys Chem B 110(44):22266–22270
14. Cho SJ, Choo K et al (2007) H_2 sorption in HCl-treated polyaniline and polypyrrole. Catal Today 120(3–4):336–340
15. Germain J, Frechet JMJ, Svec F (2007) Hypercrosslinked polyanilines with nanoporous structure and high surface area: potential adsorbents for hydrogen storage. J Mater Chem 17:4989–4997
16. Jurczyk MU, Kumar A et al (2007) Polyaniline-based nanocomposite materials for hydrogen storage. Int J Hydrogen Energy 32(8):1010–1015

17. McKeown NB, Budd PM, Book D (2007) Microporous polymers as potential hydrogen storage materials. Macromol Rapid Commun 28:995–1002
18. Nikitin A, Li X, Zhang Z, Ogasawara H, Dai H, Nilsson A (2008) Hydrogen storage in carbon nanotubes through the formation of stable C–H bonds. Nano Lett 8:682–691
19. Niemann MU, Srinivasan SS, Phani AR, Kumar A, Goswami DY, Stefanakos EK (2008) Nanomaterials for hydrogen storage applications: a review. J Nanomater 2008:9. doi:10.1155/2008/950967
20. Wang Y, Jing X (2008) Formation of polyaniline nanofibers: a morphological study. J Phys Chem B 112(4):1157–1162
21. Sadek AZ, Trinchi A, Wlodarski W, Kalantar-zadeh K, Galatsis K, Baker C, Kaner RB (2005) A room temperature polyaniline nanofiber hydrogen gas sensor. In: SENSORS 2005: proceedings of the 2005 IEEE sensors conference. IEEE, Piscataway, pp 207–210
22. Athawale AA, Bhagwat SV (2003) Synthesis and characterization of novel copper/polyaniline nanocomposite and application as a catalyst in the Wacker oxidation reaction. J Appl Polym Sci 89:2412–2417
23. Yang Y, Ouyang J, Ma L, Tseng RJ-H, Chu C-W (2006) Electrical switching and bistability in organic/polymeric thin films and memory devices. Adv Funct Mater 16:1001–1014
24. Deng J, Ding X, Zhang W, Peng Y, Wang J, Long X, Li P, Chan ASC (2002) Carbon nanotube-polyaniline hybrid materials. Eur Polym J 38(12):2497–2501(5)
25. Deshpande R, Dillon AC, Mahan AH, Alleman J, Mitra S (2006) Hydrogen adsorption in single-walled and multi-walled carbon nanotubes grown in a hot-wire CVD (Cat-CVD) reactor. Thin Solid Films 501(1–2):224–226
26. Li GJ, Zang XH, Kawi S (1999) Relationships between sensitivity, catalytic activity, and surface areas of SnO_2 gas sensors. Sens Actuators B 60:64–70
27. Trchová M, Sedenkova I, Tobolková E, Stejskal J (2004) FTIR spectroscopic and conductivity study of the thermal degradation of polyaniline films. Polym Degrad Stab 86:179–185
28. Quillard S, Louarn G, Lefrant S, Macdiarmid AG (1994) Vibrational analysis of polyaniline: a comparative study of leucoemeraldine, emeraldine, and pernigraniline bases. Phys Rev B 50(17):12496–12508
29. Fan Y-Y, Kaufmann A, Mukasyan A, Varma A (2006) Single- and multi-wall carbon nanotubes produced using the floating catalyst method: synthesis, purification and hydrogen up-take. Carbon 44:2160–2170
30. Huczko A (2000) Template-based synthesis of nanomaterials. Appl Phys A 70(4):365–376
31. Rahy A, Yang DJ (2008) Synthesis of highly conductive polyaniline nanofibers. Mater Lett 62:3849–3851
32. Zhang D, Wang Y (2006) Synthesis and applications of one-dimensional nano-structured polyaniline: an overview. Mater Sci Eng B 134(1):9–19

Chapter 9
Electrospinning of Phase-Change Materials for Thermal Energy Storage

Linge Wang, Changzhong Chen, and Yong Huang

Abstract Thermal energy storage using phase-change materials (PCMs) is one of the effective methods of conserving energy. PCMs can absorb, retain, and release large quantity of thermal energy during phase transitions over a certain temperature range. Electrospinning the combination of conventional PCMs in a polymer matrix for fabricating novel composite fibers that have form-stable and thermoregulating properties has been developed in the past decade. In this chapter, we give an overview of the chemical composition of the novel form-stable PCM composite fibers, the factors affecting and tuning their morphologies and thermal properties, and the different techniques or fabrication methods developed for producing those thermal-storage materials. In the novel form-stable PCM composite fibers, organic PCMs such as paraffin waxes, fatty acids and their blends, polyethylene glycol (PEG), etc., were widely used and encapsulated within the polymer (cellulose acetate (CA), polyethylene terephthalate (PET), polyvinylpyrrolidone (PVP), etc.) fibrous matrix, which served as structure-supporting component. The type of the PCMs and the ratio of PCM/polymer (or PCM content) dramatically affected the morphology and the thermal properties (such as phase transition temperatures

L. Wang (✉)
Department of Polymer Science and Engineering, School of Materials Science and Engineering, South China University of Technology, 381 Wushan Road, Guangzhou 510640, China
e-mail: lingewang@scut.edu.cn

C. Chen
School of Chemistry and Pharmacy Engineering, Nanyang Normal University, Nanyang 473061, China
e-mail: czcheng@126.com

Y. Huang (✉)
Technical Institute of Physics and Chemistry, Chinese Academy of Sciences, 29# Zhongguancun East Road, Beijing 100190, China
e-mail: yhuang@mail.ipc.ac.cn

B. Ding and J. Yu (eds.), *Electrospun Nanofibers for Energy and Environmental Applications*, Nanostructure Science and Technology,
DOI 10.1007/978-3-642-54160-5_9, © Springer-Verlag Berlin Heidelberg 2014

and latent of heats) of the fibers. Different approaches (such as melt coaxial electrospinning) are introduced to fabricate the composite fibers with various morphologies and structures. The further directions of developments are discussed.

Abbreviations

CA	Cellulose acetate
CBT mixture	CVL, bisphenol A and 1-dodecanol mixture
CVL	Crystal violet lactone
DADOEs	Diacid dioctadecyl esters
LA	Lauric acid
MA	Myristic acid
PA	Palmitic acid
PA6	Polyamide 6
PAN	Polyacrylonitrile
PCMs	Phase-change materials
PEG	Polyethylene glycol
PET	Polyethylene terephthalate
PMMA	Poly(methyl methacrylate)
PS	Polystyrene
PU	Polyurethane
PVDF	Polyvinylidene fluoride
PVP	Polyvinylpyrrolidone
SA	Stearic acid
SS	Stearyl stearate
T_c	Crystallizing temperature
T_m	Melting temperature
TSMs	Thermal-storage materials
ΔH_c	Latent heat of crystallization or enthalpy of crystallization
ΔH_f	Latent heat of fusion or enthalpy of melting

9.1 Introduction

With the growing energy crisis, there are more and more interests in the research of thermal energy storage (TES) and relative thermal-storage materials (TSMs) nowadays. Generally, thermal energy can be stored in three kinds of storage forms: sensible heat storage, latent heat storage, and reversible chemical reaction heat storage. Among those forms, latent heat storage which is carried out by using phase-change materials (PCMs) [1, 2] is the most attractive one, because of its high-energy storage density and small temperature variation from storage to retrieval [1]. In that case, thermal energy is stored or released by PCMs during the phase-change process. As renewable and clean energy storage materials, PCMs, including various inorganic materials, organic materials, and their blends [1–3], have been studied for several decades and applied in many different areas such as solar energy storage [4], buildings [5, 6], textiles [7], and thermal protection [1].

Since PCMs melt and crystallize repeatedly during use, there is a risk of PCM leakage. Therefore, form-stable PCMs or shape-stabilized PCMs, a kind of compound material of PCM and supporting material (commonly using polymer matrix), were developed for their attractive advantages such as direct use without additional encapsulation, cost-effective, and easily prepared with desirable dimensions [8–11]. With the development of nanoscience and technology, a novel kind of form-stable or shape-stabilized PCMs, ultrafine fibers of PCM/polymer composite [12–15], has been developed via electrospinning (which is a simple, convenient, and versatile technique for generating ultrafine fibers with diameters on both the micro- and nanoscales [16–19]). In general, paraffin waxes, fatty acids, and their blends, polyethylene glycol (PEG), are widely used as latent heat storage component in developing form-stable PCMs, while other electrospinable polymers serve as structure-supporting component. McCann et al. [12] firstly prepared phase-change nanofibers consisting of long-chain hydrocarbon cores and TiO_2–polyvinylpyrrolidone (PVP) sheaths by melt coaxial electrospinning in 2006. Later, Chen and co-workers [13–15] reported the preparation of PEG/cellulose acetate (CA) ultrafine composite fibers [13] and a series of fatty acids/polyethylene terephthalate (PET) [14, 15] ultrafine composite fibers as form-stable PCMs by conventional electrospinning method, respectively. All those composite fibers showed good thermal stability and reliability with the supporting effect of the polymer matrix, which are promising in serving as novel thermal-storage materials for thermal energy storage.

9.2 Fabrication, Morphology, and Structure

PCMs have particularly attracted much attention among the research of thermal energy storage due to their phase-transformation reversibility and thermal-storage and thermoregulation properties. In the past several decades, a large amount of studies have been carried out on the use and the application of PCMs, and several encapsulating and storage methods have been developed for reducing the reactivity of PCMs with outside environment, decreasing the evaporation and diffusion rates, and promoting the ease of handling. Three major encapsulation methods are as follows: microencapsulation of PCMs, impregnation of PCMs into polyurethane (PU) foam, and shape stabilizing by embedding PCMs into a matrix of another material [20]. Among them, shape-stabilized (or form-stable) PCMs have been more focused due to their several remarkable features such as direct use without additional encapsulation, cost-effective, shape stable, and easily prepared with desirable dimensions [8, 21–24].

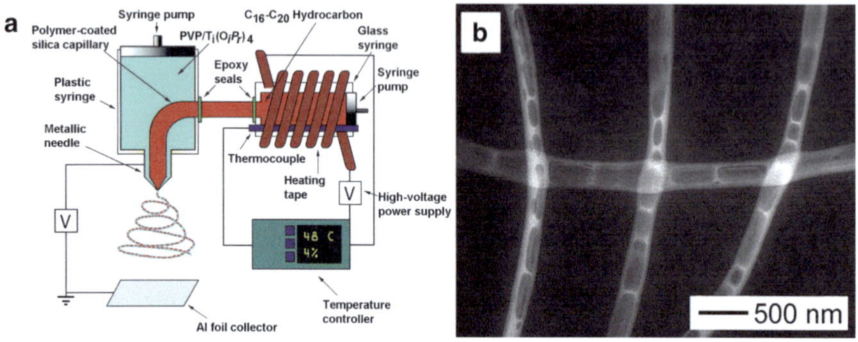

Fig. 9.1 (**a**) Schematic of the melt coaxial electrospinning setup used for fabricating octadecane, hexadecane, and eicosane @TiO_2−PVP nanofibers, (**b**) TEM image of these fibers after the core (octadecane) had been removed by soaking in hexane for 24 h (Reprinted with the permission from McCann et al. [12]. Copyright 2006 American Chemical Society)

Electrospinning is a remarkably simple, convenient, and versatile technique for fabricating ultrafine fibers from a wide variety of polymers, polymer blends, and nanoparticle-impregnated polymers [16–19]. Great variation in electrospun fiber's morphologies and assemblies can be effected through the design and construction of the fabrication methods (or electrospinning apparatus and setups), including the spinneret, the manipulation of the electric field, and the collector [25]. There are two main categories of the spinneret design, one spinneret (conventional electrospinning) and coaxial (or dual-capillary) spinneret design (coaxial electrospinning). In some cases, multiple spinnerets design was also used as a high-yield fabrication method for producing fibers. Coaxial spinneret design has been utilized by various researchers either to protect or to exhibit functionalizing agents or to electrospun materials that cannot otherwise be electrospun directly perhaps owing to high surface tension or low molecular weight [26–28].

In 2006, McCann et al. [12] successfully developed a method, which combined melt electrospinning with a coaxial spinneret, for the encapsulation of organic PCMs such as paraffin waxes (cores) in a composite or polymer matrix (sheaths) to generate nanofiber-based PCMs in one step. The schema of the combined melt coaxial electrospinning setup was shown in Fig. 9.1a; a coaxial spinneret is constructed by inserting an inner capillary into a plastic syringe for loading the polymer solution. The inner capillary is connected to a glass syringe that for loading the core solution is controlled by a temperature controller. With this spinneret, materials that are solid at room temperature can be injected concentrically into a spinning jet together with a carrier (polymer) solution [12]. During the spinning process, cooling of the jet due to solvent evaporation causes the inner liquid to quickly solidify, leading to its encapsulation in the nanofibers. The core–sheath structure of the nanofibers was characterized by transmission electron microscopy (TEM), as shown in Fig. 9.1b. The segmented morphologies indicated that the PCM (octadecane) formed elongated, compartmentalized domains along the long axis of

Table 9.1 Fabrication method of electrospun PCM/polymer composite fibers

PCMs			
Classification	Detail	Polymer	Fabrication method
Paraffin waxes	Octadecane [12], hexadecane [12, 29], eicosane [12], dodecane [30]	PVP [12], PAN [29], Zein [30]	Melt coaxial electrospinning, melting [12], conventional electrospinning [29, 30], coaxial electrospinning [30]
PEG		CA [13, 31, 32], nylon 6,6 [33], PVDF [34–36]	Conventional electrospinning [13, 31–33, 36], coaxial electrospinning [34, 35]
Fatty acids and their derivatives	LA [14, 15, 37–40], MA [14], PA [14], SA [14], SS [41], capric acid (CA) [42], binary fatty acid eutectics [38, 43]	PET [14, 15, 37, 38, 41–43], PA6 [39, 40]	Conventional electrospinning [14, 15, 37–43]
CBT mixture	CVL, bisphenol A and 1-dodecanol mixture	PMMA [44, 45]	Melt coaxial electrospinning [44], coaxial electrospinning [45]
Bi- or multi-component PCM	Hexadecane and eicosane [46], hexadecane (HD) and PEG [47], binary fatty acid eutectics [38, 43]	PVP [46], PS [47], PET [38, 43]	Melt coaxial electrospinning [46], conventional electrospinning [38, 43, 47]
DADOEs		PET [48]	Conventional electrospinning [48]
Bio-PCMs	Soybean wax [49]	PU [49]	Coaxial electrospinning [49]

the fibers. Meanwhile, the results of differential scanning calorimetry (DSC) clearly demonstrated that there was no thermal evaporation in the heating process and the capability of these core–sheath PCM nanofibers to stabilize temperature.

The melt coaxial electrospinning method is quite useful for the encapsulation of active solids in fibrous matrices. As listed in Table 9.1, various core–sheath-structured PCMs/polymer composite nanofibers have been successfully fabricated via this method [12, 44, 46]. In some cases, different variations of the setup have been developed in accordance with some specific requests. The variations include removing the heating part and replacing the core PCM melt by PCM solution [30, 34, 35, 45, 49] or applying multiple core spinnerets (or multifluidic compound-jet electrospinning) for encapsulating multicomponent PCM composite nanofibers [46]. As shown in Fig. 9.2, the side view scanning electron microscopy (SEM) images revealed there are individual channels formed inside the composite fibers, which indicated the multicomponent PCM microfibers have been successfully fabricated via the multifluidic compound-jet electrospinning. The multiresponsive effect benefits from a special multichannel tubular microstructure that could controllably encapsulate different phase-change materials into the channels independently. Aside from the fabrication of multicomponent phase-change microfibers, the melt multifluidic compound-jet electrospinning is promising for applications related to microencapsulation and multifunctional material fields [46].

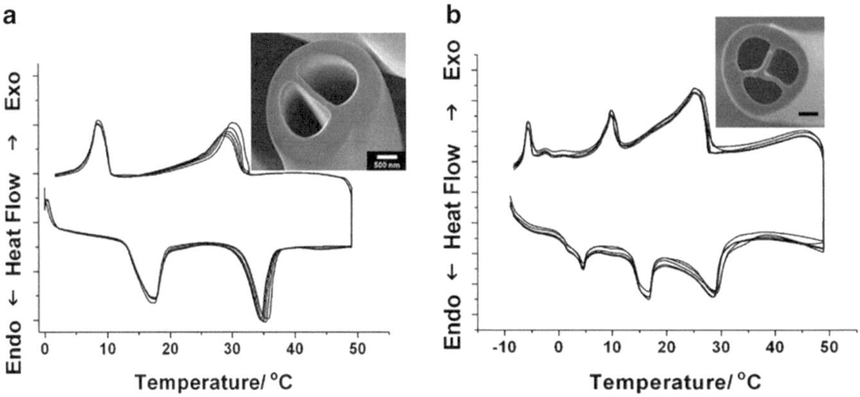

Fig. 9.2 DSC measurements and side view SEM images of (**a**) the two-component and (**b**) three-component phase-change fibers after inner PCMs have been removed (Reproduced from Wang et al. [46] by permission of John Wiley & Sons Ltd)

In brief, the novel one-step (melt) coaxial electrospinning technique allows for the fabrication of form-stable PCM fibers with new core–sheath structure. The inner PCMs were encapsulated by the outside polymer matrix. The sheath gives a maximum protection to avoid the leakage of the core PCMs after melting. Compared with the conventional electrospinning, however, the coaxial spinneret and the temperature controlling setup are more complicated, which brings some inconveniences for setting apparatus. Thus, researchers began to fabricate PCM/polymer composite fibers from the mixture solution via the conventional electrospinning, which just uses one spinneret.

Chen et al. [13] first successfully prepared PEG/CA composite ultrafine fibers directly from the as-prepared PEG/CA mixture solution using the conventional electrospinning approach. In the composite fibers, PEG acted as model PCM, while CA acted as the supporting matrix. The SEM observation (Fig. 9.3a, b) from the electrospun PEG/CA composite fibers revealed that the fibers were cylindrical in shape and had a smooth external surface. Since PEG is a water-soluble polymer and CA is not, the morphology of the washed composite fibers provided information about the distribution of PEG. From Fig. 9.3b, the water-treated electrospun PEG/CA composite fibers still remained in a cylindrical form; however, a number of grooves were found on the washed fibers' surface, demonstrating that there was PEG distributed on the surface of the electrospun fibers. From the TEM images (Fig. 9.3c, d), there were cavities within the washed fibers, which were not present in the original composite fibers (Fig. 9.3c). Therefore, the information from the SEM and TEM images indicated that PEG was not only distributed on the surface but also within the core of the composite fibers themselves.

In order to improve the water-resistant ability and thermal stability of the electrospun PEG/CA composite fibers for potential thermal-storage applications, Chen and co-workers cross-linked the electrospun PEG/CA fibers by using toluene-2,4-diisocyanate (TDI) [50]. From the SEM images (Fig. 9.4), the cross-linked

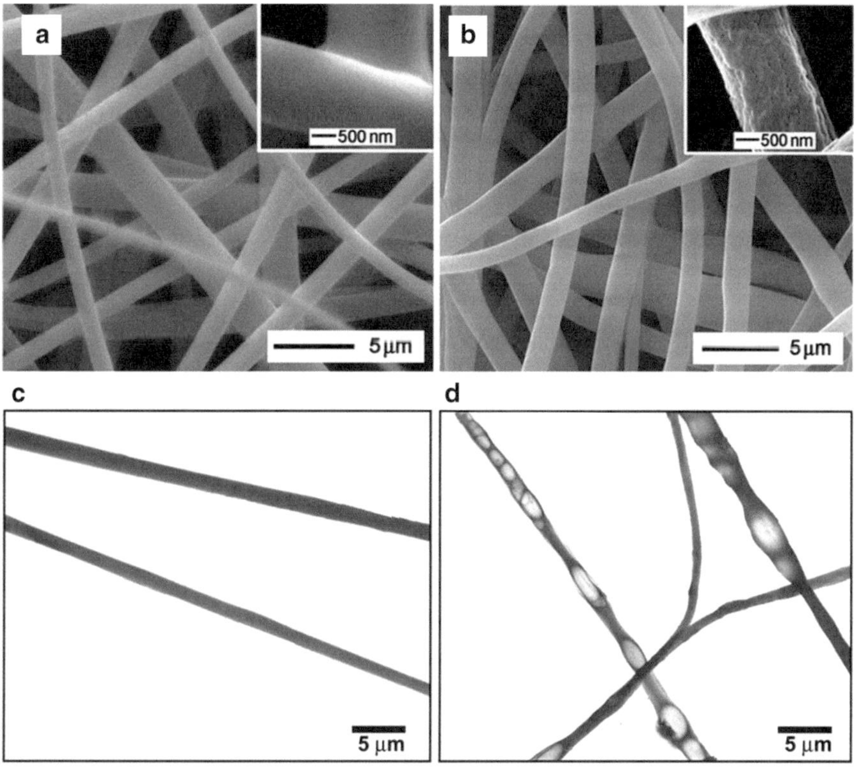

Fig. 9.3 SEM micrographs of (**a**) the electrospun PEG/CA composite fibers and (**b**) the washed electrospun PEG/CA composite fibers. TEM micrographs of (**c**) the original and (**d**) the washed electrospun PEG/CA composite fibers (Reprinted from Chen et al. [13], Copyright 2007, with permission from Elsevier)

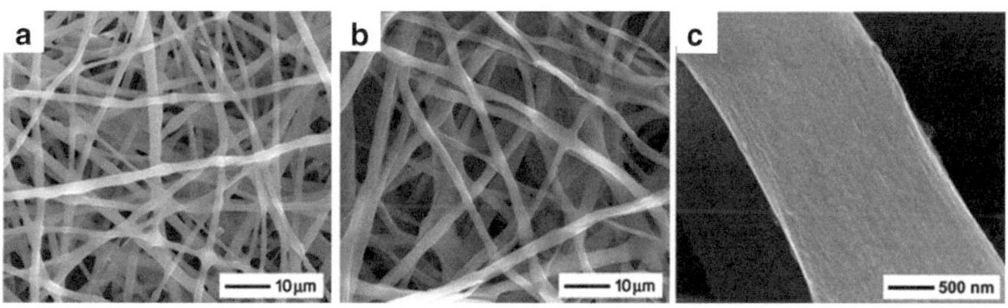

Fig. 9.4 SEM micrographs of the cross-linked electrospun PEG/CA composite fibers (**a**) and immersed in deionized water for 24 h (**b**) and (**c**) (Reprinted from Chen et al. [50], Copyright 2008, with permission from Elsevier)

PEG/CA fibers showed cylindrical shape with a smooth surface which was similar with the morphology of the original composite fibers (Fig. 9.3a). Meanwhile, some "cross-links" or cohesion occurred between the fiber junctions, as shown in Fig. 9.4a, b, demonstrating that the fibers have formed a dense network membrane by the cross-linking. This result was also confirmed by the morphological change of the fibers after water-resistant tests. The morphology of the cross-linked PEG/CA fibers was well preserved after the soaking treatment (Fig. 9.4c); the fibers still showed smooth surface, and no obvious grooves appeared on the surface compared with the one without cross-linking (Fig. 9.3b, d). It suggested that the water-resistant ability of the electrospun PEG/CA was enhanced after the cross-linking.

Besides the effect of cross-linking, Chen and co-workers also studied the effects of PEG molecular weight [31] and PEG content [32] on the morphology of the electrospun PEG/CA composite fibers. For the study on effect of PEG molecular weight [31], five PEG samples (2, 4, 6, 10, and 20k) were individually added into the CA solution with equal weight proportion of CA (PEG/CA = 1/1 by weight) in all composite fibers. All the electrospun PEG/CA fibers were cylindrical in shape with smooth surfaces, and the fiber diameter and its distribution were different apparently for the five types of electrospun fibers. The average fiber diameter of the PEG/CA composite fibers increased as the molecular weight of PEG was increased, ranging between 1,029 and 1,752 nm. The increase of the fiber diameter after PEG adding into the original CA solution was attributed to the variations of the solutions such as a significant increase of viscosity of the PEG/CA mixed solution [31]. For the study on effect of PEG content [32], different contents of PEG (0–70 wt%) were mixed into the CA solution and electrospun. Normally, very homogeneous and smooth PEG/CA fibers without beads were electrospun from the mixed solutions with lower PEG content (<50 wt%). With the increase of PEG content in the composite fibers, PEG/CA fibers became nonuniform along the fiber axial and markedly inferior. Moreover, the average fiber diameter increased with the increase of PEG content, which is also due to the increase of viscosity of the solution. Similar results were also reported in other electrospun PCM/polymer composite fibers such as fatty acid (lauric acid, LA)/PET fibers [14], fatty acid (LA)/polyamide 6 (PA6) fibers [39], PEG/nylon 6,6 fibers [33], and diacid dioctadecyl esters (DADOEs)/PET fibers [48].

9.3 Thermal Properties

For the potential application in thermal energy storage, good thermal performances such as high latent heats and appropriate phase transition temperatures are the most desirable characteristics for PCMs and TSMs. For instance, in the study by McCann et al. [12], the long-chain hydrocarbons were chosen as core PCMs due to their melting points near room temperature (octadecane) and body temperature (eicosane) as these temperature ranges are most valuable in practice [12]. Via DSC measurement, the thermal properties (melting temperature (T_m), crystallizing temperature (T_c), latent heat of fusion or enthalpy of melting (ΔH_f), and latent heat of crystallization (ΔH_c) of the samples) and the thermal behavior of the fibers were measured and analyzed. The T_m and ΔH_f of 45 wt% octadecane were measured to be 27 °C and 114 J/g, respectively, while those of 36 wt% eicosane were 37 °C and 88 J/g, respectively [12]. It is indicated that the thermal properties (phase transition

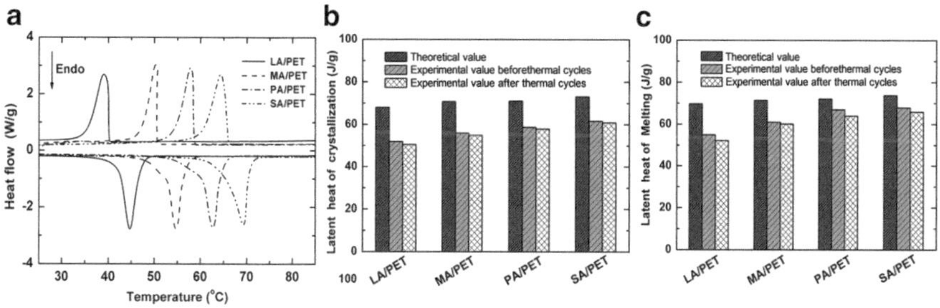

Fig. 9.5 DSC curves of electrospun fatty acid/PET (70/100) composite fibers in heating and cooling process (**a**). Theoretical and experimental values of ΔH_c (**b**) and ΔH_f (**c**) of the four types of electrospun fatty acid/PET (70/100) composite fibers before and after 100 heating–cooling cycles (Reprinted from Chen et al. [14], Copyright 2008, with permission from Elsevier)

temperatures and latent heats) of the form-stable PCMs can be tunable by changing the type of the PCMs, complementing similar results reported in the studies of other form-stable PCM fibers [14, 31, 38, 43, 48], which were fabricated via both coaxial and conventional electrospinning methods.

Chen and co-workers first electrospun ultrafine fibers based on the composites of PET and a series of fatty acids, LA, myristic acid (MA), palmitic acid (PA), and stearic acid (SA) as form-stable PCMs [14]. It was found that the thermal performances of the four samples were different from each other, and both the latent heats and the phase transition temperatures have a tendency to increase with the increase of alkyl chain of the fatty acid in the composite fibers. The T_m was about 45, 55, 63, and 69 °C for LA/PET, MA/PET, PA/PET, and SA/PET composite fibers, respectively. The ΔH_f was about 54.91, 60.99, 66.91, and 67.88 J/g for LA/PET, MA/PET, PA/PET, and SA/PET composite fibers, respectively, while the ΔH_c was about 51.93, 55.93, 58.65, and 61.62 J/g, respectively. Therefore, it is obvious that both of the phase transition temperatures and the latent heats of the PCM/polymer composite fibers were dominantly related (or be tunable) with the type of PCMs, as shown in Fig. 9.5. It is worth noting that the latent heats of the composite fibers were lower than those of pure PCM due to the introduction of polymer in the PCM/polymer system, so the efficiency of enthalpies (the ratio between the experimental value and the theoretical value) and the thermal reliability of the composite fibers, which are important for their potential applications, were less than 1. Theoretically, the enthalpy values of the fatty acid/PET composite fiber were obtained by multiplying the latent heat of pure fatty acid and the mass percentage of fatty acid in the composite fibers. Figure 9.5b, c illustrated the theoretical values and experimental values of the latent heats for the four composite fibers with same PCM/PET mass ratio before and after the 100 heating–cooling thermal cycles. From the chart in Fig. 9.5, the experimental values (both before and after the thermal cycles) of latent heat for all the fatty acid/PET composite fibers were obviously lower than the corresponding theoretical values, and the efficiency of enthalpy (melting/crystallization) before the thermal cycle treatment for the electrospun

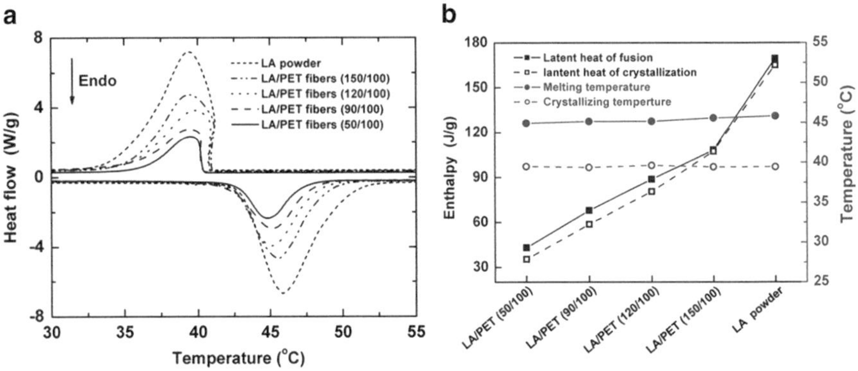

Fig. 9.6 DSC curves of electrospun LA/PET composite fibers and LA powder in heating and cooling process (**a**) and the latent heat and phase transition temperature of LA powder and LA/PET composite fibers with different mass ratios (**b**) (Reprinted from Chen et al. [14], Copyright 2008, with permission from Elsevier)

LA/PET, MA/PET, PA/PET, and SA/PET composite fibers was 78.72 %/76.38 %, 85.42 %/79.06 %, 92.85 %/82.55 %, and 92.09 %/84.31 %, respectively. It is believed that the deviation (reduction) of the thermal properties from the theoretical values was mainly caused by a retardation of the crystallization process of PCM in the composite fibers, which was hindered by the quench process of electrospinning [13, 51, 52] and the dilution effect of PET in the composite fibers. Moreover, all the enthalpies of the composite fibers after 100 heating–cooling thermal cycles, namely, the experimental values after thermal cycles, slightly decreased compared with those of the corresponding experimental values before thermal cycles (e.g., the ratio of the ΔH_f after thermal cycles to that before thermal cycles was 95.10, 98.64, 95.53, and 97.14 % for the LA/PET, MA/PET, PA/PET, and SA/PET composite fibers, respectively), which demonstrated good stable and reliable thermal properties of those composite fibers.

Meanwhile, Chen and co-workers also investigated the effect of PCM content on the thermal properties of the electrospun PCM/polymer fibers [14]. Figure 9.6 showed the DSC curves of LA powder and electrospun LA/PET composite fibers with different LA/PET mass ratios and the corresponding data of thermal properties. Clearly, the latent heats (ΔH_f and ΔH_c) of the four LA/PET composite fibers were lower than that of LA powder (169.41 J/g and 165.12 J/g, respectively) because PET in the composite fibers has few contribution to the latent heat at this temperature range, and the latent heats of the fibers increased with the increase of LA/PET mass ratio. However, the phase transition temperatures (T_m and T_c) of all the electrospun LA/PET composite fibers have no obvious variations compared with those of LA (less than 1 °C). It indicated that the PCM/polymer mass ratio plays an important role on the latent heat of the form-stable PCM/polymer composite fibers but has less effect on their phase-change temperature.

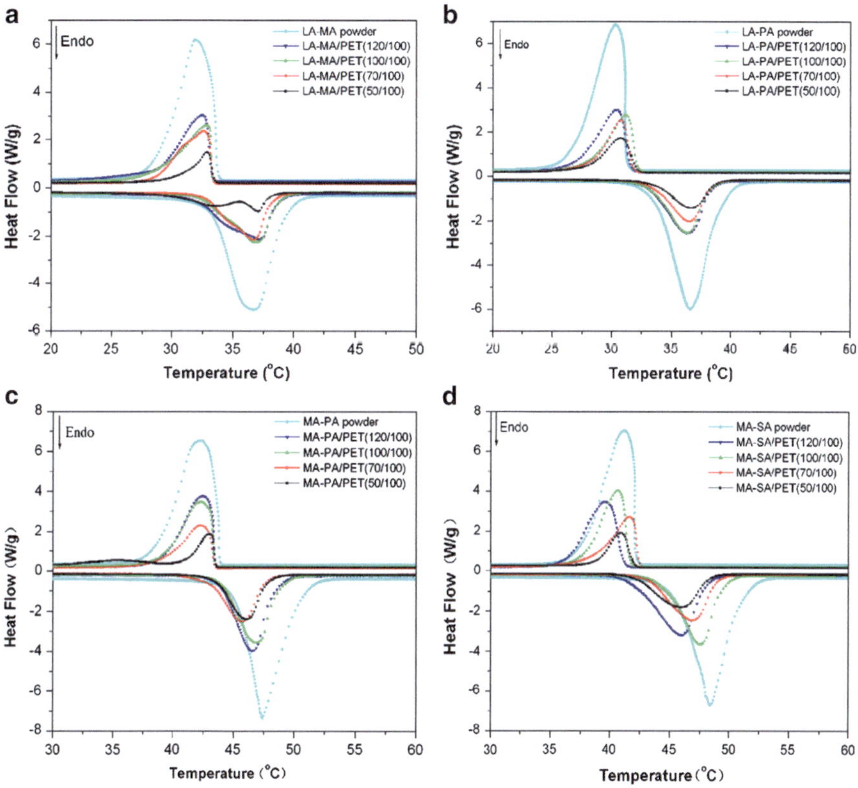

Fig. 9.7 DSC curves of electrospun (**a**) LA–MA/PET, (**b**) LA–PA/PET, (**c**) MA–PA/PET, and (**d**) MA–SA/PET composite with different mass ratios (Reprinted from Cai et al. [38], Copyright 2012, with permission from Elsevier)

Based on the above reports, Cai et al. [38] and Ke et al. [43] fabricated form-stable PCM fibers based on the composites of PET and a series of binary fatty acid eutectics, respectively. In Cai's study [38], five binary fatty acid eutectics including LA–MA, LA–PA, MA–PA, MA–SA, and PA–SA were prepared according to Schrader equation and then were selected as an innovative type of solid–liquid PCMs. The T_m of binary fatty acid eutectics was found to be 36.75 °C for LA–MA in the combination ratio of 61.27:38.73 wt%, 36.52 °C for LA–PA in the ratio of 77.51:22.49 wt%, 47.39 °C for MA–PA in the ratio of 66.92:33.08 wt%, 48.37 °C for MA–SA in the ratio of 77.42:22.58 wt%, and 56.16 °C for PA–SA in the ratio of 62.99:37.01 wt%, respectively, while the ΔH_f of them was 173.6 J/g, 169.6 J/g, 183.1 J/g, and 180.6 J/g, respectively. The DSC curves of the LA–MA/PET, LA–PA/PET, MA–PA/PET, MA–SA/PET, and PA–SA/PET ultrafine composite fibers with varied mass ratios of 50/100, 70/100, 100/100, and 120/100 are presented in Fig. 9.7, respectively. The intensity changes of DSC curves indicated that the endothermic peaks and exothermic peaks of all composite fibers increased with the increase of binary fatty acid eutectic content. Thus, the binary fatty acid eutectic mass percentage in the composite fibers became the dominating factor for the variations of enthalpies of the fibers. DSC measurements suggested that the ΔH_c and ΔH_f of the ultrafine composite fibers increased gradually with increasing fatty acid eutectic amounts, ranging from approximately 33.91 to 94.29 J/g (ΔH_f). Their phase transition temperatures had almost no obvious variation as relative to the corresponding fatty acid eutectics. Meanwhile, the characteristic temperatures and heat enthalpies of the ultrafine composite fibers varied with the different types of binary fatty acid eutectics. Those results and conclusions were coincided with that found in the electrospun fatty acid/PET composite fibers [14].

In Ke's work [43], other four binary fatty acid eutectics (capric acid–LA, capric acid–MA, capric acid–PA, and capric acid–SA) were prepared according to Schrader equation and then were selected as an innovative type of solid–liquid PCMs. And the capric acid-based binary fatty acid eutectic/PET ultrafine composite fibers with varied mass ratios of 50/100, 100/100, and 150/100 were electrospun successfully. The T_m of binary fatty acid eutectics was found to be 22.70 °C for capric acid–LA in the combination ratio of 66.75:33.25 wt%, 26.02 °C for capric acid–MA in the ratio of 78.39:21.61 wt%, 28.71 °C for capric acid–PA in the ratio of 89:11 wt%, and 31.17 °C for capric acid–SA in the ratio of 94.47:5.53 wt%, respectively, while the ΔH_f of them was 127.2 J/g, 155.2 J/g, 141.4 J/g, and 156.8 J/g, respectively. Clearly, both the phase transition temperatures and the latent heats of the capric acid-based binary fatty acid eutectics were lower than that of LA- and MA-based binary fatty acid eutectics [38]. The DSC results indicated that the heat enthalpies (ΔH_c and ΔH_f) of the composite nanofibers increased gradually with increasing fatty acid eutectic amounts, ranging from approximately 34.18 to 95.24 J/g (ΔH_f), while they had appropriate phase transition temperatures based upon climatic requirement. The characteristic temperatures and enthalpies of the composite nanofibers varied with the type of binary fatty acid eutectics. In addition, DSC thermal cycling test suggested that thermal properties of electrospun composite nanofibers were well retained after thermal cycling, which indicated that the composite nanofibers had good thermal reliability. Again, the conclusions were quite similar to that mentioned above.

Besides the fatty acids and their derivatives/PET composite fibers, PEG/CA composite fibers [31, 32] and DADOE/PET [48] composite fibers were also fabricated and studied by Chen and co-workers, respectively. In the research of electrospun PEG/CA composite fibers, the effects of PEG molecular weight (like different types of PCMs) [31] and PEG content [32] on the thermal properties of the electrospun PEG/CA composite fibers were also investigated. For the study on effect of PEG molecular weight (M_n) [31], five PEG samples (2, 4, 6, 10, and 20k), which have different thermal properties, were individually added into the CA solution with equal weight proportion of CA (PEG/CA = 1/1 by weight) in all composite fibers. Figure 9.8 showed the DSC curves of the five types of electrospun PEG/CA composite fibers. Obviously, all the PEG/CA composite fibers had balanced thermal-storage and released properties at heating and cooling process.

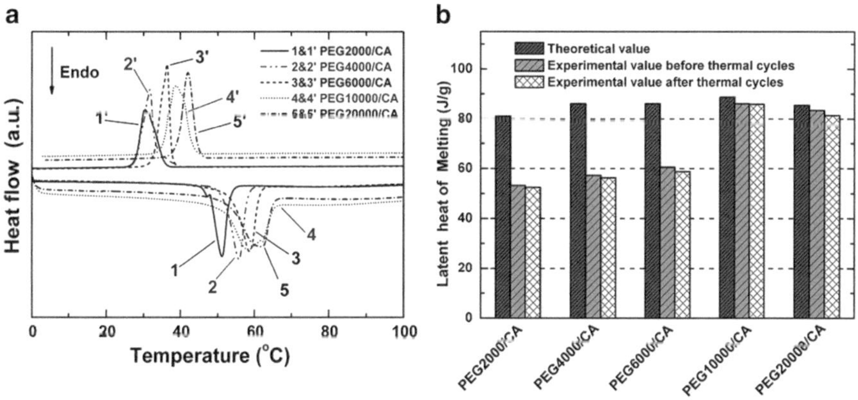

Fig. 9.8 (**a**) DSC curves and (**b**) theoretical and experimental values of melting enthalpy of the five types of electrospun PEG/CA composite fibers (Reproduced from Chen et al. [31] by permission of John Wiley & Sons Ltd.)

It could be seen that the T_m (ranging from 51.14 to 62.27 °C), ΔH_f (ranging from 53.23 to 86.03 J/g), T_c (ranging from 30.45 to 41.89 °C), and ΔH_c (ranging from 47.93 to 73.48 J/g) of the PEG/CA composite fibers increased generally with the increase of the molecular weight of PEG in the composite fibers, which could be interpreted by the crystalline behaviors of the composite fibers. The crystalline behaviors of the CA fibers and the PEG/CA composite fibers were studied by their wide-angle X-ray diffraction (WAXD) patterns. All the PEG/CA composite fibers had similar diffraction patterns and diffraction angles at about 19.5° and 24°, which were the same as the corresponding pure PEG powders. It was easily inferred that the diffraction peaks of the PEG/CA composite fibers were aroused by the component of PEG, and CA acted merely as a diluent in the composite fibers. The main difference between the WAXD curves of the five types of PEG/CA fibers was the diffraction peaks' shape, and the diffraction peaks became sharper with the increase of the molecular weight of PEG in the composite fibers, which meant that the degree of crystallization of the composite fibers increased, that is, the larger the crystallization region and more regular the structure of the PEG molecular chains in the composite fibers. Consequently, the enthalpies for the phase change of the PEG/CA composite fibers in general increased. As shown in Fig. 9.8, before and after 100 heating–cooling cycles, the experimental values of melting enthalpy of the PEG/CA composite fibers were lower than the corresponding theoretical values, and the efficiency of enthalpy before the thermal cycles for low M_n PEG/CA electrospun fibers was less than 70.3 %. However, it increased to above 97.0 % for the PEG/CA composite fibers with a large M_n PEG (above 10k). It was believed that the reduction of the enthalpies was mainly caused by a retardation of the crystallization process of PEG in the composite fibers, which is not only hindered by the quench process of electrospinning but also influenced by the hydrogen bonding interactions between the PEG and CA chains [13]. Since the solvent is removed from the polymer in

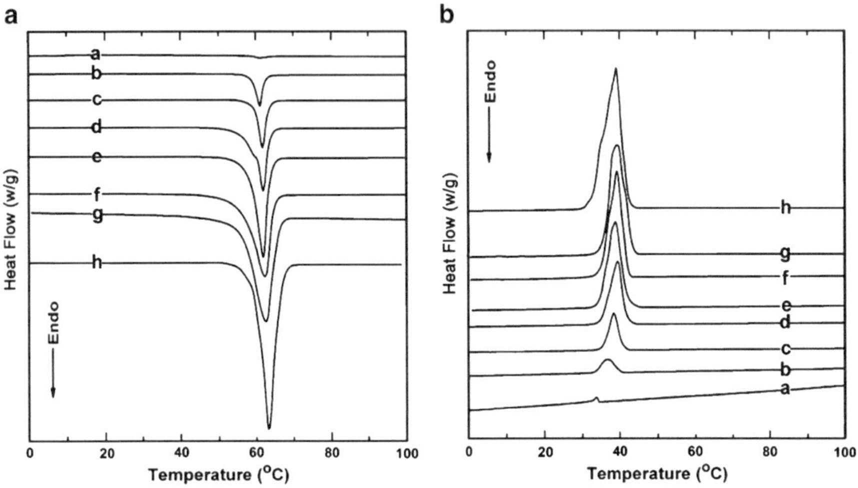

Fig. 9.9 DSC curves of (**a**) heating and (**b**) cooling curves of PEG/CA fibers with different PEG contents: (*a*) 10 wt%, (*b*) 20 wt%, (*c*) 30 wt%, (*d*) 40 wt%, (*e*) 50 wt%, (*f*) 60 wt%, (*g*) 70 wt%, and (*h*) pristine PEG (Reprinted from Chen et al. [32], Copyright 2011, with permission from Elsevier)

sub-second time scales during the electrospinning process, the molecular chains of PEG have no time to form a well-defined, microscale equilibrium structure in the composite fibers. Meanwhile, PEG was found to be distributed on both the surface and within the core of the composite fibers; the crystalline areas of PEG were very tiny within ultrafine fiber with a few microns and surrounded by hydrogen bonding between the OH group of PEG and the carbonyl group of CA. Therefore, the short chain length of the lower molecular weight PEG led to lower crystalline areas of PEG in the composite fibers and a larger deviation of the enthalpies from the theoretical values. In addition, thermal cycling test indicated that the prepared composites had excellent thermal stability and reliability even they were subjected to 100 heating–cooling thermal cycles (Fig. 9.8). The ratio of the enthalpies after thermal cycles to that before thermal cycles was above 97 % for all the PEG/CA composite fibers.

For the study on effect of PCM content [32], different contents of PEG (0–70 wt%) were mixed into the CA solution and electrospun. Figure 9.9 showed the DSC curves of pristine PEG and electrospun PEG/CA fibers with different PEG contents. From the DSC measurement, the phase transition temperatures of all the electrospun PEG/CA composite fibers were lower than those of PEG powder because of the introduction of CA. The enthalpy of melting (ΔH_f) and enthalpy of crystallization (ΔH_c) of PEG were about 177.38 J/g and 167.75 J/g, respectively. Endothermic peaks and exothermic peaks of the composite fibers increased with the increase of PEG content, which means ΔH_f and ΔH_c of the composite fibers have increased obviously with increasing PEG content from the intensity changes of DSC curves. ΔH_f (and ΔH_c) of the composite fibers from 10 to 70 wt% PEG

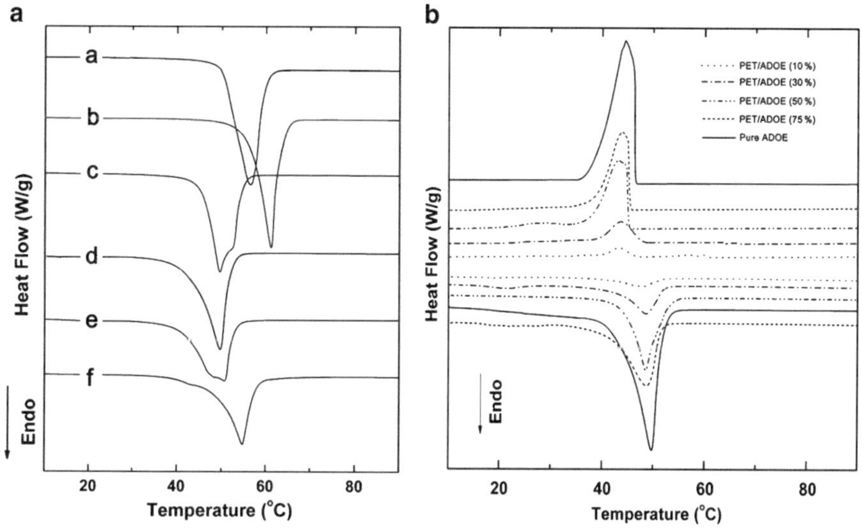

Fig. 9.10 (**a**) DSC heating curves of *(a)* OD, *(b)* EDOE, *(c)* BDOE, *(d)* ADOE, *(e)* ODOE, and *(f)* SDOE. (**b**) DSC curves of ADOE/PET composite fibers with various ADOE contents (Reprinted from Chen et al. [48], Copyright 2012, with permission from Elsevier)

were 3.83 J/g (1.61 J/g), 16.35 J/g (10.87 J/g), 30.18 J/g (21.74 J/g), 57.48 J/g (43.86 J/g), 86.03 J/g (65.75 J/g), 101.79 J/g (79.83 J/g), and 120.18 J/g (104.40 J/g), respectively. Again, the PEG mass percentage in the composite fibers becomes the dominating factor for the variations of enthalpies of the fibers.

Octadecanol (OD) is a common solid–liquid PCM for the potential application in thermal energy storage, but it has poor thermal stability for its sublimation. In order to improve the thermal stability of OD, recently, in an attempt of Chen's study, a series of DADOEs (ethane diacid dioctadecyl ester (EDOE), butane diacid dioctadecyl ester (BDOE), adipic acid dioctadecyl ester (ADOE), octane diacid dioctadecyl ester (ODOE), and sebacic acid dioctadecyl ester (SDOE)) were synthesized as novel solid–liquid PCMs through the esterification of OD and ethane diacid, butane diacid, adipic acid, octane diacid, and sebacic acid, and then DADOE/PET ultrafine composite fibers with different DADOE contents were prepared by electrospinning as form-stable PCMs for TES [48]. Figure 9.10a showed the DSC curves of pure OD and the five types of synthesized DADOE. Apparently, phase transition enthalpies of DADOEs were lower than those of OD. Moreover, the thermal performances of the five samples were different from each other, and the phase transition enthalpies decrease with the increase of alkyl chain between the two ester groups in DADOE molecules, and the enthalpy of melting gradually decreases from 176.8 to 133.1 J/g. Figure 9.10b showed the DSC curves of the ADOE/PET composite fibers with different ADOE contents. The intensities of the DSC peaks for all ADOE/PET composite fibers are lower than those of pure ADOE, which indicated that ΔH_f and ΔH_c of ADOE/PET composite fibers with

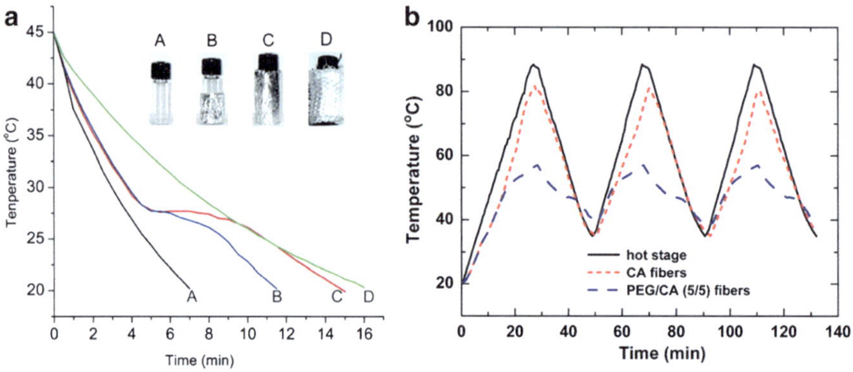

Fig. 9.11 Demonstration of thermal insulation capability of octadecane@TiO_2–PVP nanofibers (**a**), where 1 cm^3 of water at 60 °C was allowed to cool in a 4 °C environment in glass vials covered with different insulation jackets. Sample *A* had no insulation, sample *B* was half covered with the PCM nanofiber jacket, and sample *C* was fully covered by the PCM nanofiber jacket. Sample *D* was covered by a jacket of conventional fiberglass (Reprinted with the permission from McCann et al. [12]. Copyright 2006 American Chemical Society) and (**b**) temperature-regulating capability of PEG/CA fibers, the thermoregulating curves of the hot stage, the electrospun CA fibers, and the PEG/CA fibers with 50 wt% PEG (Reprinted from Chen et al. [32], copyright 2011, with permission from Elsevier)

different ADOE contents were lower than those of pure ADOE because PET in the composite fibers had no contribution to the latent heats in this temperature range (0–100 °C), and both ΔH_f and ΔH_c increased with the increase of ADOE content. Moreover, the phase transition temperatures of all the electrospun ADOE/PET composite fibers had no obvious variations compared with those of ADOE.

In summary, the thermal properties of the electrospun form-stable PCM/polymer composite fibers can be affected dramatically according to different types of PCMs [14, 31, 38, 43, 48] and various PCM/polymer mass ratios or PCM contents [14, 33, 37–39, 41, 43, 48, 49]. The former factor is in charge of the intrinsic thermal properties of the novel thermal-storage materials, because it fixes the phase transition temperatures and limits the maximum value of the latent heats of the composite fibers. The latter one plays a key role in the latent heats of the form-stable PCM/polymer composite fibers but has less effect on their phase transition temperatures. All those novel composite fibers showed good thermal stability and reliability no matter what electrospinning methods were applied.

In addition, the novel form-stable PCM/polymer composite fibers not only show excellent performances in thermal energy storage but also have the capability of temperature regulation. In the testament of the core–sheath octadecane@TiO_2–PVP nanofibers, supercooling was not observed in the temperature history curve for the PCM nanofibers [12], as shown in Fig. 9.11a. The insulating capacity was reduced by the removal of PCM insulation, with the sample reaching 20 °C after 7 min for the uninsulated sample A, 12 min for the sample B with half insulation, and 16 min for the sample C that was fully covered by the PCM fibers. The fiberglass fibers were as effective as the PCM nanofibers in insulating the vial although the former jacket was approximately four times thicker than the PCM fiber-based one. More importantly, the jacket based on PCM fibers allowed one to stabilize the temperature (close to the melting point of octadecane) in the vial for 5 min. In contrast, the temperature continuously dropped with time for all other types of insulating jackets.

The thermoregulating capability test of the PEG/CA fibers was performed with a hot stage by self-designed direct method [32], in which the hot stage simulated an exterior environment temperature swing. Figure 9.11b shows the temperature vs time relationship curves of the hot stage, electrospun CA fibers, and PEG/CA composite fibers with 50 wt% PEG, respectively. As to PEG/CA composite fibers, at the beginning of the heating process, the fibers' temperature increased as the temperature of the hot stage increased. And then, the fibers' temperature responded less to the hot stage (exterior environment) temperature changes, and it only raised to about 58 °C (close to the T_m of PEG/CA fibers), while the hot stage temperature was close to 90 °C. This phenomenon was because of PEG absorbing heats isothermally during the melting process. In the cooling process, the temperature slowly decreased and stabilized at around 47 °C due to the releasing of latent heats in the crystallization process of PEG. In contrast, the temperature vs time curve of CA fibers followed closely the variation of the hot stage. This result demonstrated that PEG/CA fibers have a good capability to regulate their interior temperature as the ambient temperature alters.

9.4 Future Trends

In the past decade, the novel electrospun form-stable PCM/polymer composite fibers have attracted a large amount of interest in both academia and industry, owing to its promising future for using as thermal-storage materials. The first attempt is made by McCann and the co-workers [12], which uses a melt coaxial electrospinning. The approach has several distinct advantages over its solution-based counterpart, most notably environment friendliness and low cost due to the absence of solvent. In contrast, the conventional electrospinning method has the advantages of easy setup and wide choice of various polymer solutions. Recently, a new type of multicomponent PCM microfibers has been successfully fabricated by a novel one-step temperature control multifluidic electrospinning technique [46]. Different kinds of PCMs could be encapsulated independently in fibers and display their own thermal properties. By this method, multicomponent phase-change fibers would be fabricated with controllable phase transition point. Besides, the enthalpy of the fibers could be facilely regulated just by tuning the flow rate of the inner fluid in a certain range. It is a simple and powerful method for the independent encapsulation of different functional materials in fibers. Furthermore, new composite materials with different functional components independently encapsulated could be fabricated by just replacing inner fluids with other functional materials.

Since the type of PCMs plays a dominant role on the morphology and the thermal properties of the electrospun form-stable PCM/polymer composite fibers, many innovative types of PCMs were used to fabricate novel form-stable PCM fibers. Those PCMs include CBT mixture, bi- or multicomponent PCMs, and bio-PCMs. CBT mixture [44, 45] is a mixture of crystal violet lactone (CVL), bisphenol A, and 1-tetradecanol, which has a good optical transmission and a thermo-responsive property. Via melt coaxial electrospinning, thermochromic phase-change nanofibers with a transparent PMMA sheath and a PCM (CBT) core were successfully fabricated. With body-temperature PCMs and 1-tetradecanol as a filling agent, these nanofibers could have obvious body-temperature-responsive properties. The fibers have obvious thermal energy management and thermochromic properties, good reversibility, and fluorescence [44, 45]. With the increasing of more complex practical needs of PCMs, besides body-temperature PCMs, it is of particular significance to fabricate new PCMs with different phase transition points, that is, bi- or multicomponent PCMs. It could be directly assumed that these kinds of multicomponent phase-change fibers could be obtained by doping several kinds of microcapsules with different phase transition points, such as two inner PCM cores (hexadecane and eicosane) or multiple inner PCM cores (tetradecane, hexadecane, and octadecane) fabricated via the multifluidic compound-jet electrospinning technique [46]. It could be a simple mixture of two PCMs, for example, nanocapsules containing hexadecane as core material and polystyrene (PS) as sheath were electrospun with PEG as a matrix material into the fiber webs [47]. It also could be the binary fatty acid eutectics [38, 43] prepared according to a careful calculation (Schrader equation) and come tuning the thermal properties as wish.

Besides, electrospinning PCM/polymer composite fibers from bio-PCM or biopolymer is also a potential direction of the TSM development. Recently, Hu and Xun successfully prepared a bio-PCM (soy wax)/polyurethane (PU) ultrafine fiber via coaxial electrospinning technique [49]. Natural soy wax was used as the core PCM for thermal storage, and PU was used as the sheath material for encapsulation. The results indicated a uniform fiber morphology with a core–sheath structure and a homogeneous wax distribution throughout the core of the fibers. Thermal analysis results showed that the enthalpy of melting increased (from 9.24 to 36.47 J/g) with wax content (10–60 wt%). The fibrous structures exhibited balanced thermal-storage and released properties for thermoregulating functions. The thermal properties were unaltered after 100 heating–cooling test cycles, demonstrating that the composite fibers have good thermal stability and reliability. Meanwhile, in early 2013, Pérez-Masiá and co-workers successfully fabricated micro-, submicro-, and nanoencapsulation structures based on zein (a maize protein) and dodecane (a paraffin which has a transition temperature at −10 °C) with various dodecane/zein mass ratio via both conventional and coaxial electrospinning methods. The new thermal-storage material has good thermal properties and a reliable thermal stability, which is promising in the potential application in food biopackaging such as the smart packaging materials to preserve the cold chain [30].

References

1. Zalba B, Marin JM, Cabeza LF, Mehling H (2003) Review on thermal energy storage with phase change: materials, heat transfer analysis and applications. Appl Therm Eng 23(3):251–283. doi:10.1016/s1359-4311(02)00192-8
2. Farid MM, Khudhair AM, Razack SAK, Al-Hallaj S (2004) A review on phase change energy storage: materials and applications. Energy Convers Manage 45(9–10):1597–1615. doi:10.1016/j.enconman.2003.09.015
3. Verma P, Varun P, Singal SK (2008) Review of mathematical modeling on latent heat thermal energy storage systems using phase-change material. Renew Sust Energy Rev 12(4):999–1031. doi:10.1016/j.rser.2006.11.002
4. Kaygusuz K (2003) Phase change energy storage for solar heating systems. Energy Source 25(8):791–807. doi:10.1080/00908310390207837
5. Tyagi VV, Buddhi D (2007) PCM thermal storage in buildings: a state of art. Renew Sust Energy Rev 11(6):1146–1166. doi:10.1016/j.rser.2005.10.002
6. Khudhair AM, Farid MM (2004) A review on energy conservation in building applications with thermal storage by latent heat using phase change materials. Energy Convers Manage 45(2):263–275. doi:10.1016/s0196-8904(03)00131-6
7. Mondal S (2008) Phase change materials for smart textiles – an overview. Appl Therm Eng 28(11–12):1536–1550. doi:10.1016/j.applthermaleng.2007.08.009
8. Alkan C, Sari A, Uzun O (2006) Poly(ethylene glycol)/acrylic polymer blends for latent heat thermal energy storage. Aiche J 52(9):3310–3314. doi:10.1002/aic.10928
9. Pielichowski K, Flejtuch K (2003) Differential scanning calorimetry study of blends of poly(ethylene glycol) with selected fatty acids. Macromol Mater Eng 288(3):259–264. doi:10.1002/mame.200390022
10. Lin KP, Zhang YP, Xu X, Di HD, Yang R, Qin PH (2004) Modeling and simulation of underfloor electric heating system with shape-stabilized PCM plates. Build Environ 39(12):1427–1434. doi:10.1016/j.buildenv.2004.04.005
11. Sari A, Alkan C, Kolemen U, Uzun O, Eudragit S (2006) (Methyl methacrylate methacrylic acid copolymer)/fatty acid blends as form-stable phase change material for latent heat thermal energy storage. J Appl Polym Sci 101(3):1402–1406. doi:10.1002/app.23478
12. McCann JT, Marquez M, Xia YN (2006) Melt coaxial electrospinning: a versatile method for the encapsulation of solid materials and fabrication of phase change nanofibers. Nano Lett 6(12):2868–2872. doi:10.1021/nl0620839
13. Chen CZ, Wang L, Huang Y (2007) Electrospinning of thermo-regulating ultrafine fibers based on polyethylene glycol/cellulose acetate composite. Polymer 48(18):5202–5207. doi:10.1016/j.polymer.2007.06.069
14. Chen CZ, Wang LG, Huang Y (2008) Morphology and thermal properties of electrospun fatty acids/polyethylene terephthalate composite fibers as novel form-stable phase change materials. Sol Energy Mat Sol C 92(11):1382–1387. doi:10.1016/j.solmat.2008.05.013
15. Chen CZ, Wang LG, Huang Y (2008) A novel shape-stabilized PCM: electrospun ultrafine fibers based on lauric acid/polyethylene terephthalate composite. Mater Lett 62(20):3515–3517. doi:10.1016/j.matlet.2008.03.034
16. Reneker DH, Chun I (1996) Nanometre diameter fibres of polymer, produced by electrospinning. Nanotechnology 7(3):216–223. doi:10.1088/0957-4484/7/3/009
17. Fang X, Reneker DH (1997) DNA fibers by electrospinning. J Macromol Sci Phys B36(2):169–173. doi:10.1080/00222349708220422
18. Jaeger R, Bergshoef MM, Batlle CMI, Schonherr H, Vancso GJ (1998) Electrospinning of ultra-thin polymer fibers. Macromol Symp 127:141–150. doi:10.1002/masy.19981270119
19. Bergshoef MM, Vancso GJ (1999) Transparent nanocomposites with ultrathin, electrospun nylon-4,6 fiber reinforcement. Adv Mater 11(16):1362–1365. doi:10.1002/(SICI)1521-4095(199911)11:16<1362::AID-ADMA1362>3.0.CO;2-X

20. Sarier N, Onder E (2012) Organic phase change materials and their textile applications: an overview. Thermochim Acta 540:7–60. doi:10.1016/j.tca.2012.04.013
21. Hong Y, Ge XS (2000) Preparation of polyethylene-paraffin compound as a form-stable solid–liquid phase change material. Sol Energy Mat Sol C 64(1):37–44. doi:10.1016/s0927-0248(00)00041-6
22. Sari A (2004) Form-stable paraffin/high density polyethylene composites as solid–liquid phase change material for thermal energy storage: preparation and thermal properties. Energy Convers Manage 45(13–14):2033–2042. doi:10.1016/j.enconman.2003.10.022
23. Lin KP, Zhang YP, Xu X, Di HF, Yang R, Qin PH (2005) Experimental study of under-floor electric heating system with shape-stabilized PCM plates. Energy Build 37(3):215–220. doi:10.1016/j.enbuild.2004.06.017
24. Zhang YP, Lin KP, Yang R, Di HF, Jiang Y (2006) Preparation, thermal performance and application of shape-stabilized PCM in energy efficient buildings. Energy Build 38(10):1262–1269. doi:10.1016/j.enbuild.2006.02.009
25. Wang L, Ryan AJ (2011) Chapter 1: Introduction to electrospinning. In: Electrospinning for tissue regeneration. Woodhead Publishing Ltd, Cambridge
26. Larsen G, Velarde-Ortiz R, Minchow K, Barrero A, Loscertales IG (2003) A method for making inorganic and hybrid (organic/inorganic) fibers and vesicles with diameters in the submicrometer and micrometer range via sol–gel chemistry and electrically forced liquid jets. J Am Chem Soc 125(5):1154–1155. doi:10.1021/ja028983i
27. Li D, Xia YN (2004) Direct fabrication of composite and ceramic hollow nanofibers by electrospinning. Nano Lett 4(5):933–938. doi:10.1021/nl049590f
28. Sun ZC, Zussman E, Yarin AL, Wendorff JH, Greiner A (2003) Compound core-shell polymer nanofibers by co-electrospinning. Adv Mater 15(22):1929–1932. doi:10.1002/adma.200305136
29. Alay S, Gode F, Alkan C (2010) Preparation and characterization of poly(methylmethacrylate-co-glycidyl methacrylate)/n-hexadecane nanocapsules as a fiber additive for thermal energy storage. Fiber Polym 11(8):1089–1093. doi:10.1007/s12221-010-1089-2
30. Perez-Masia R, Lopez-Rubio A, Lagaron JM (2013) Development of zein-based heat-management structures for smart food packaging. Food Hydrocolloid 30(1):182–191. doi:10.1016/j.foodhyd.2012.05.010
31. Chen CZ, Wang LG, Huang Y (2009) Role of M(n) of PEG in the morphology and properties of electrospun PEG/CA composite fibers for thermal energy storage. Aiche J 55(3):820–827. doi:10.1002/aic.11708
32. Chen CZ, Wang LG, Huang Y (2011) Electrospun phase change fibers based on polyethylene glycol/cellulose acetate blends. Appl Energy 88(9):3133–3139. doi:10.1016/j.apenergy.2011.02.026
33. Seifpoor M, Nouri M, Mokhtari J (2011) Thermo-regulating nanofibers based on nylon 6,6/polyethylene glycol blend. Fiber Polym 12(6):706–714. doi:10.1007/s12221-011-0706-z
34. Thuy TTN, Lee JG, Park JS (2011) Fabrication and characterization of coaxial electrospun polyethylene glycol/polyvinylidene fluoride (core/sheath) composite non-woven mats. Macromol Res 19(4):370–378. doi:10.1007/s13233-011-0409-8
35. Do CV, Thuy TTN, Park JS (2012) Fabrication of polyethylene glycol/polyvinylidene fluoride core/shell nanofibers via melt electrospinning and their characteristics. Sol Energy Mat Sol C 104:131–139. doi:10.1016/j.solmat.2012.04.029
36. Thuy TTN, Park JS (2011) Fabrication of electrospun nonwoven mats of polyvinylidene fluoride/polyethylene glycol/fumed silica for use as energy storage materials. J Appl Polym Sci 121(6):3596–3603. doi:10.1002/app.34148
37. Cai YB, Ke HZ, Dong J, Wei QF, Lin JL, Zhao Y, Song L, Hu YA, Huang FL, Gao WD, Fong H (2011) Effects of nano-SiO_2 on morphology, thermal energy storage, thermal stability, and combustion properties of electrospun lauric acid/PET ultrafine composite fibers as form-stable phase change materials. Appl Energy 88(6):2106–2112. doi:10.1016/j.apenergy.2010.12.071

38. Cai YB, Ke HZ, Lin L, Fei XZ, Wei QF, Song L, Hu Y, Fong H (2012) Preparation, morphology and thermal properties of electrospun fatty acid eutectics/polyethylene terephthalate form-stable phase change ultrafine composite fibers for thermal energy storage. Energy Convers Manage 64:245–255. doi:10.1016/j.enconman.2012.04.018
39. Cai YB, Gao CT, Xu XL, Fu Z, Fei XZ, Zhao Y, Chen Q, Liu XZ, Wei QF, He GF, Fong H (2012) Electrospun ultrafine composite fibers consisting of lauric acid and polyamide 6 as form-stable phase change materials for storage and retrieval of solar thermal energy. Sol Energy Mat Sol C 103:53–61. doi:10.1016/j.solmat.2012.04.031
40. Cai YB, Xu XL, Gao CT, Wang L, Wei QF, Song L, Hu Y, Qiao H, Zhao Y, Chen Q, Fong H (2012) Effects of carbon nanotubes on morphological structure, thermal and flammability properties of electrospun composite fibers consisting of lauric acid and polyamide 6 as thermal energy storage materials. Fiber Polym 13(7):837–845. doi:10.1007/s12221-012-0837-x
41. Chen CZ, Wang LG, Huang Y (2009) Ultrafine electrospun fibers based on stearyl stearate/polyethylene terephthalate composite as form stable phase change materials. Chem Eng J 150(1):269–274. doi:10.1016/j.cej.2009.03.007
42. Ke HZ, Li DW, Zhang HD, Cai YB, Huang FL, Wei QF (2012) Electrospun capric acid/polyethylene terephthalate composite nanofibres for storage and retrieval of thermal energy. Mater Res Innov 16(6):429–437. doi:10.1179/1433075x12y.0000000042
43. Ke HZ, Li DW, Zhang HD, Wang XL, Cai YB, Huang FL, Wei QF (2013) Electrospun form-stable phase change composite nanofibers consisting of capric acid-based binary fatty acid eutectics and polyethylene terephthalate. Fiber Polym 14(1):89–99. doi:10.1007/s12221-013-0089-4
44. Li FY, Zhao Y, Wang S, Han D, Jiang L, Song YL (2009) Thermochromic core-shell nanofibers fabricated by melt coaxial electrospinning. J Appl Polym Sci 112(1):269–274. doi:10.1002/app.29384
45. Malherbe I, Sanderson RD, Smit E (2010) Reversibly thermochromic micro-fibres by coaxial electrospinning. Polymer 51(22):5037–5043. doi:10.1016/j.polymer.2010.09.018
46. Wang N, Chen HY, Lin L, Zhao Y, Cao XY, Song YL, Jiang L (2010) Multicomponent phase change microfibers prepared by temperature control multifluidic electrospinning. Macromol Rapid Commun 31(18):1622–1627. doi:10.1002/marc.201000185
47. Faridi-Majidi R, Madani M, Sharifi-Sanjani N, Khoee S, Fotouhi A (2012) Multi-phase composite nanofibers via electrospinning of latex containing nanocapsules with core-shell morphology. Polym-Plast Technol 51(4):364–368. doi:10.1080/03602559.2011.639326
48. Chen CZ, Liu SS, Liu WM, Zhao YY, Lu YZ (2012) Synthesis of novel solid–liquid phase change materials and electrospinning of ultrafine phase change fibers. Sol Energy Mat Sol C 96(1):202–209. doi:10.1016/j.solmat.2011.09.057
49. Hu W, Yu X (2012) Encapsulation of bio-based PCM with coaxial electrospun ultrafine fibers. RSC Adv 2(13):5580–5584. doi:10.1039/c2ra20532g
50. Chen CZ, Wang LG, Huang Y (2009) Crosslinking of the electrospun polyethylene glycol/cellulose acetate composite fibers as shape-stabilized phase change materials. Mater Lett 63(5):569–571. doi:10.1016/j.matlet.2008.11.033
51. Zong XH, Kim K, Fang DF, Ran SF, Hsiao BS, Chu B (2002) Structure and process relationship of electrospun bioabsorbable nanofiber membranes. Polymer 43(16):4403–4412. doi:10.1016/S0032-3861(02)00275-6
52. Li Y, Huang ZM, Lu YD (2006) Electrospinning of nylon-6,66,1010 terpolymer. Eur Polym J 42(7):1696–1704. doi:10.1016/j.eurpolymj.2006.02.002

Chapter 10
Electrospun Nanofibrous Sorbents and Membranes for Carbon Dioxide Capture

Xianfeng Wang and Bingyun Li

Abstract CO_2 emission resulting from combustion of fossil fuel is a major anthropogenic factor for global warming. Current commercial CO_2 capture approaches using aqueous amine scrubbing require high energy consumption for regeneration which leads to significantly decreased efficiency. Therefore, novel CO_2 capture materials and technologies for economical CO_2 capture are of the utmost importance. Nanofibers, mainly fabricated by electrospinning, have unique porous structure, high surface area, and good mechanical properties thereby exhibit potential as advanced sorbents for CO_2 capture and storage. More significantly, nanofiber-based sorbents are expected to have extremely low resistance for gas transport and extremely fast kinetics due to the unique structure of nanofibers. In this chapter, we summarize recent progress in the development of electrospun nanofibrous sorbents or membranes (e.g., nanofiber-supported metal-organic frameworks, carbon nanofibers, ionic liquid-based nanofibrous membranes, metal oxide nanofibers, etc.) for CO_2 capture, describe the types of nanofibrous materials that have been developed, and discuss their fabrication variables and CO_2 adsorption performance in detail. This chapter may pave the way for developing advanced nanofibrous sorbents for CO_2 capture from power plants and even the atmosphere.

X. Wang • B. Li (✉)
National Energy Technology Laboratory, U.S. Department of Energy, Morgantown, WV 26505, USA

Department of Orthopaedics, School of Medicine, West Virginia University, Morgantown, WV 26506, USA
e-mail: bili@hsc.wvu.edu

B. Ding and J. Yu (eds.), *Electrospun Nanofibers for Energy and Environmental Applications*, Nanostructure Science and Technology,
DOI 10.1007/978-3-642-54160-5_10,

10.1 Introduction

The global climate change phenomenon, which is mainly caused by CO_2 emission into the atmosphere, has attracted more and more attention [1–3]. It is reported that the concentration of CO_2 in the atmosphere has increased from about 310 to 394 ppm during the last half century and it is projected to continue to rise if anthropogenic sources remain unchecked [4, 5]. Technologies for CO_2 capture and storage (CCS) present some of the most promising and effective options for large-scale reduction in CO_2 emissions [1]. The CCS strategies embody a group of technologies for the capture of CO_2 from power plants, followed by compression, transport, and permanent storage. Capturing CO_2 from the mixed-gas streams produced during power generation is a first and critical step for CCS. Three strategies for incorporating capture into power generation scenarios are of primary focus today: precombustion, post-combustion, and oxyfuel processes.

To reduce the amount of CO_2 released into the atmosphere, four main methods are being developed for CCS: solution absorption, adsorption, membrane diffusion, and cryogenic distillation [6, 7]. It is evident that the achievement of most of these technologies depends heavily on the development of materials. Among them, adsorption or sorbent is of great interest because of its low energy consumption, low equipment cost, and ease of application. A number of materials have been used for CO_2 adsorption, and materials with large surface area (e.g., zeolites, activated carbon) have been widely investigated. Nanomaterials have shown potential in CO_2 capture due to their high surface area and adjustable properties and characteristics. Higher surface areas provide more sites of reaction and hence boost the adsorption or absorption capacity [8]. For instance, CaO nanopods with higher surface area (16.92 m^2/g) showed higher CO_2 capture capacity [17.5 mmol CO_2/(g sorbent)] than that [12.1 mmol CO_2/(g sorbent)] of commercial CaO sorbents with surface area of 0.40 m^2/g [9].

Nanomaterials also offer an advantage in that their structure can be tailor-made to improve the properties and characteristics of a compound or element. For example, carbon nanotubes are able to be produced in different dimensions and diameters. Nanofibers, as one of the most important one-dimensional (1D) nanomaterials, are expected to have extremely low resistance for gas transport and extremely fast kinetics; therefore, they have great potential for CO_2 capture and storage. As a nanofabrication technique, electrospinning has been established as a robust and versatile method for fabricating fibers with diameters down to the nanometer by applying a high voltage to a polymer solution [10, 11]. A variety of materials such as polymers, ceramics, and even metals have been electrospun into fibers with well-controlled sizes, compositions, and structures [12]. It has been shown that the outstanding properties of such nanofibers are highly attractive to numerous applications including biotechnology, textiles, filters, sorbents, sensors, and so on [13]. Benefitting from the unique properties (e.g., high porosity, high surface area, and good mechanical properties) of nanofibrous structures [14], nanofibrous adsorbent has emerged as a new class of adsorption materials, and considerable effort has been applied to their adsorbent design and optimization [15, 16]. In this chapter, currently emerging new materials for CO_2 capture are briefly introduced and then the recent advances of nanofibrous sorbents used in CO_2 capture and separation are highlighted. We conclude with a summary of current and future research efforts and opportunities in the development of electrospun nanofibers for CO_2 capture applications.

10.2 New Materials for CO_2 Capture

A number of promising new materials exist for CO_2 capture from precombustion, post-combustion, and oxyfuel processes [17]. Examples of new materials include ionic liquids (ILs), metal-organic frameworks (MOFs), membranes, and fibrous and nanofibrous sorbents (Fig. 10.1).

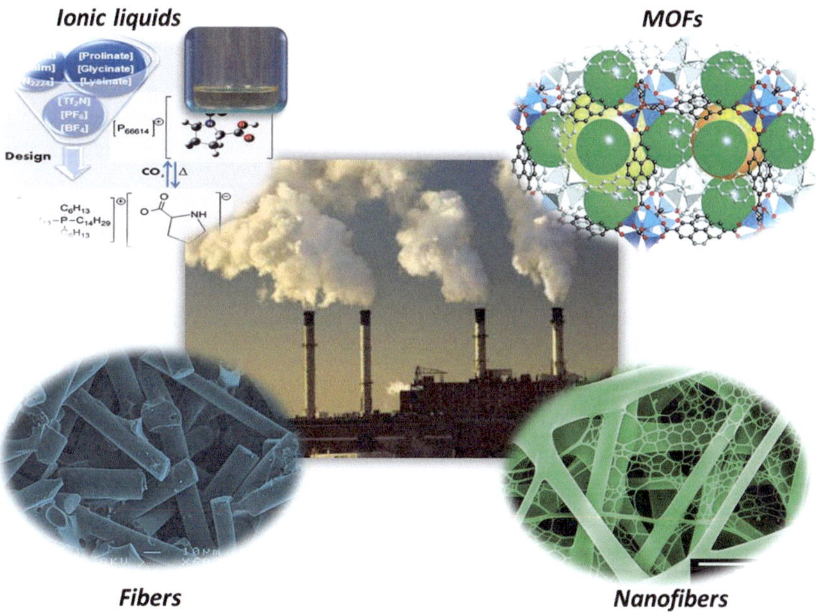

Fig. 10.1 Schematic of new materials for CO_2 capture. Permissions for individual pictures were obtained (Reprinted with the permission from Brennecke and Gurkan [18]. Copyright 2010 American Chemical Society. Reprinted with the permission from Rowsell and Yaghi [22]. Copyright 2006 American Chemical Society. Reprinted from Thiruvenkatachari et al. [27]. Copyright 2009, with permission from Elsevier. Reprinted with permission from Wang et al. [30]. Copyright 2010 IOP Publishing Ltd.)

10.2.1 Ionic Liquids (ILs)

ILs are low-melting salts that are attractive for a number of applications as they are relatively nonvolatile, nonflammable, environmentally benign, and exceptionally thermally stable [18]. In addition, there are numerous combinations of cations and anions that can be used to produce ILs, and thus chemical and physical properties of ILs can be tuned, which is needed to design an energy-efficient liquid absorbent for CO_2 capture. The mechanism for CO_2 capture in ILs is often based on physisorption and involves a weak association between the IL and CO_2 molecules [19]. Once the CO_2 has been removed from the gas mixture, it can be released from the ILs (which would be reused) by either a decrease in pressure or an increase in temperature [18]. While the viscosity of ILs minimizes solvent loss from the gas stream, this attribute also limits mass transfers, and they often suffer from low rates of absorption. To overcome these shortcomings and increase the capacity of simple ILs, amine-functionalized ILs have been developed, which allow higher rates of sorption to be achieved at pressures relevant to flue streams [19, 20]. A number of reports have also demonstrated high CO_2/N_2 selectivity in polymerized ILs, which exhibit enhanced CO_2 solubility relative to the monomeric ILs [21].

10.2.2 Metal-Organic Frameworks (MOFs)

MOFs are novel materials constructed by coordinate bonds between multidentate ligands and metal atoms or small metal-containing clusters [4], which have recently attracted intense research interest because of their permanent porous structures, large surface areas, and potential applications as novel adsorbents [22]. Most of the MOF materials have three-dimensional (3D) structures with uniform pores and a network of channels. The integrity of these pores and channels can be retained after careful removal of the guest species. The remaining voids within the 3D structures can then adsorb other guest molecules. Several reviews have summarized the research efforts in gas adsorption applications for MOFs, such as hydrogen and methane storage, and CO_2 capture [4, 19]. Recently, Liu and co-workers contributed a review on the progress and challenges in using MOFs for adsorption-based CO_2 capture including both experimental and simulation studies [4].

10.2.3 Membranes

A gas separation membrane typically consists of multiple layers with different functions. A dense ultrathin selective layer ($\sim$100 nm in thickness) performs the molecular separation, while a microporous support structure provides mechanical strength and minimal transport resistance [23]. The constituent materials of membranes are typically classified as inorganic (e.g., ceramic, metal oxide, metallic, molecular sieves, and MOFs) or organic (e.g., cellulose acetate, polysulfone, polyamide, and polyimide) [19].

A membrane separates species by selectively permeating certain components of a mixture faster than others through a thin barrier in response to an external driving force, such as a concentration, partial pressure, or, more generally, a chemical potential gradient. A variety of mechanisms influence the separation of gases by a membrane, with the most important being solution diffusion and molecular sieving. Membranes offer a number of inherent advantages over other technologies for separating gases, including (1) simple, passive operation with no moving parts; (2) environmentally benign separation without the use of hazardous chemicals; (3) small footprint, which is critically important in some applications (e.g., aboard aircraft, spacecraft, or on offshore natural gas platforms); and (4) lower energy use because they can separate species without a phase change.

Clearly, membranes represent a promising technology for gas separation; however, they suffer a number of drawbacks, particularly with regard to CO_2 capture from flue gas. In this case, the low CO_2 partial pressure provides a minimal driving force for gas separation, which creates an energy penalty due to the need for compression of the feed gas. Membrane materials also suffer from a decrease in permeability over time due to particulate deposition on the surface [19].

10.2.4 Fibrous Sorbents

Solid amine adsorbents using a fibrous structure instead of particles as the matrix are expected to offer amazing benefits (e.g., fast kinetics and high CO_2 capture capacity) for the adsorption of CO_2 because of their high external surface area and porosity, low pressure drops, and flexibility of the matrix fibers. Recently, Li et al. [24] developed a novel fibrous adsorbent for CO_2 capture by coating polyethylenimine (PEI) on a glass fiber matrix using epoxy resin as a cross-linking agent. They found that a maximum adsorption capacity of 6.3 mmol CO_2/(g of PEI) was obtained at a PEI/epoxy resin ratio of 10:1. Then, they replaced epoxy resin ($M_W = 370$) with a lower molecular weight cross-linking agent (i.e., epichlorohydrin, $M_W = 92.5$) and developed adsorbents of PEI modified glass fibers [23]. The resultant fibrous sorbent improved adsorption performance (e.g., higher CO_2 capacity, faster kinetics, and better regenerability) [25]. The maximum CO_2 adsorption capacity of 13.08 mmol CO_2/(g of PEI) was achieved at a coating weight of 45 wt%. Polymer supports for amines have attracted considerable attention since they are light in weight, flexible, and easy to handle [6]. Yang et al. [26] demonstrated that solid amine-containing fibrous adsorbent could be prepared by pre-irradiation grafting copolymerization of allylamine onto polyacrylonitrile fiber (PAN-AF). The higher grafting degree resulted in higher CO_2 uptake, and the adsorption capacity of PAN-AF reached 6.22 mmol CO_2/(g of PAN-AF) at the grafting degree of 60 wt%. They attributed the good performance of this fibrous adsorbent to the fibrous structure, which might have reduced the resistance to gas flow, thereby maximizing sorbent performance while minimizing energy consumption, and the highly stable interface (i.e., grafting copolymerization of amines onto fibers) between amines and the support [26].

Apart from the amine-modified fibrous adsorbents, porous carbon materials are well-known adsorbents for CO_2 adsorption due to their highly developed porosity, extended surface area, flexible surface chemistry, and high thermal stability. Recently, some new classes of carbon materials (e.g., carbon fiber, carbon molecular sieves) have emerged as adsorbents for gas separation and storage. Carbon fiber refers to fibers which are at least 90 wt% carbon in composition obtained by the controlled pyrolysis of an appropriate precursor material (e.g., pitch, PAN, rayon, nonheterocyclic aromatic polymers) [27]. Much research has been devoted to the preparation of PAN-based activated carbon fiber (PAN-ACF) from modified PAN, and some attention has been paid to the use of PAN-ACFs for CO_2 removal. Hierarchical porous PAN-ACFs with a large Brunauer-Emmett-Teller (BET) surface area were made from PAN fibers through pre-oxidation and chemical activation. This type of material contains a large number of nitrogen-containing groups (N content > 8.1 wt%) and consequently basic sites, leading to a faster adsorption rate and a higher CO_2 adsorption capacity (2.4 mmol/g). Moreover, PAN-ACF adsorbents had stable CO_2 adsorption/desorption performance under multiple cycling conditions [28].

10.3 Why Electrospun Nanofibers?

Due to the unique structure and mechanical properties of nanofibers, nanofiber-based sorbents are expected to have extremely low resistance for gas transport and thereby extremely fast kinetics. Due to the high surface area and low density of nanofibers, the developed nanofibrous sorbents are also expected to have high CO_2 capture capacity. Various nanofibrous sorbents (e.g., nanofibrillated cellulose and graphite nanofibers or GNFs) have been developed for CO_2 capture. For instance, Meng et al. [16] found that porous GNFs could be fabricated using a KOH etching method at temperatures in the 700–1,000 °C range (Fig. 10.2a). The CO_2 adsorption isotherms revealed that GNFs treated under 900 °C had the highest BET surface area (567 m^2/g) and the best CO_2 adsorption capacity of 59.2 mg/g (Fig. 10.2b).

Electrospinning is a simple and versatile technique to produce continuous nanofibers with a diameter down to nanometers. Due to the combination of a high fiber production rate and the simplicity of the setup, the electrospinning approach has the unique ability to produce nanofibers with most materials (organic, inorganic, or hybrid), easily providing various fibers with the desired composition and surface properties [15]. Their high surface-to-volume ratio, large porosity (up to over 80 %), and adjustable functionality make electrospun fibrous membranes useful for numerous applications in particulate gas separation [29, 30].

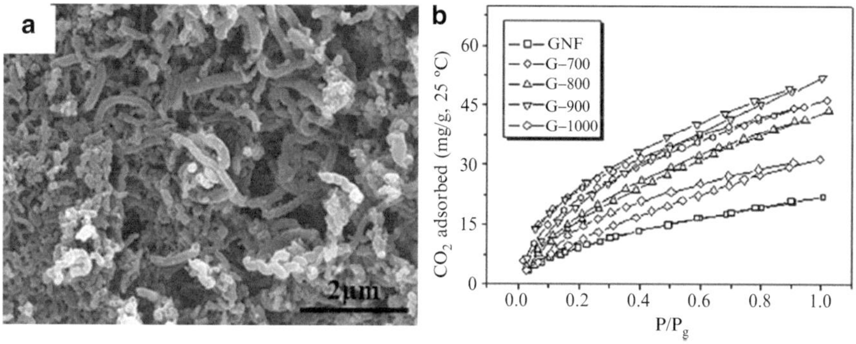

Fig. 10.2 (**a**) SEM image of chemical-treated GNFs. (**b**) CO_2 isotherms of the pristine and chemical-treated GNFs (Reprinted from Meng and Park [16]. Copyright 2010, with permission from Elsevier)

10.4 Electrospun Nanofibers for CO_2 Capture

10.4.1 Electrospun Nanofiber-Supported MOF for CO_2 Capture

Great attention has been paid to preparing MOF particles with novel structures and desired properties. Recently, new focus has been placed on the fabrication of supported MOF thin films and membranes [31], particularly given their utility in engineering-related applications such as membrane-based molecular separators, reactors, and chemical sensors [15]. The use of compact substrates and surface modification with organic functional groups are the most commonly used approaches for fabricating MOF thin films [32]. More recently, several attempts have been made to generate more useful MOF membranes by growing continuous MOF crystals on porous organic polymer substrates [33, 34]. Nevertheless, well-intergrown, free-standing, and high MOF-loading films were difficult to achieve. Therefore, a new class of porous substrates with easily tunable structural parameters (e.g., composition, porosity, thickness, and size), in particular, surface properties, or with seed crystals tightly embedded on the surface, is highly desirable.

Nanofibrous membranes produced by electrospinning are ideal porous substrates for developing chemical systems due to their high specific surface area, large porosity, and enormous structural and chemical tenability [15]. Recently, Ostermann et al. [35] reported the fabrication of MOF-containing polymer nanofibers via electrospinning. However, since the MOF particles were embedded in the electrospun polymer nanofibers, an additional diffusion barrier formed around the MOF particles and thus resulted in reduced gas uptake and access into MOFs. In order to overcome this drawback, Wu et al. [15] reported a new strategy to produce free-standing MOF membranes using electrospun nanofibrous membranes as skeletons. A two-step procedure was developed, namely, the preparation of MOF nanocrystal (e.g., zeolite-imidazolate framework-8 or ZIF-8) doped electrospun fibers followed by a second growth. By using a second-growth approach with MOF crystal embedded electrospun fibrous mats as the seeding layer and skeleton, an MOF membrane mostly consisting of MOF crystals was obtained. Compared with the MOF crystal embedded electrospun fibers, the second-growth MOF membranes exhibit a much higher gas adsorption capacity. Gas separation experiments of the prepared free-standing ZIF-8 membrane were tested in custom-made analytical equipment. In the case of the 1:1 mixture of N_2/CO_2, due to the preferred affinity of ZIF-8 for CO_2 and its high specific surface area, the membrane exhibited capture of the CO_2 component from the mixture. After the "capture" by the ZIF membranes, the proportion of CO_2 (50 %) in the effluent was reduced to less than 30 %, and for N_2, it increased to more than 70 %. The separation factor of N_2/CO_2 reached 2.4 on average.

10.4.2 Electrospun Carbon Nanofibers for CO_2 Capture

Porous carbon materials have been preferred for CO_2 adsorption due to their highly developed porosity, extended surface area, surface chemistry, and thermal stability. Various carbon materials such as activated carbons, activated carbon fibers (ACF), carbon molecular sieves, and carbon nanotubes have been used as adsorbents for CO_2 [36, 37]. Recently, Bai et al. [38] reported the fabrication of oxyfluorinated activated electrospun carbon nanofibers (OFACFs) for CO_2 capture. Electrospun CFs were prepared from a PAN/*N*, *N*-dimethylformamide (DMF) solution via electrospinning and heat treatment. The electrospun CFs were chemically activated in order to generate the pore structure, and then oxyfluorination was used to modify the surface. Through the N_2 adsorption isotherm, the specific surface area and pore volume decreased slightly as a result of oxyfluorination treatment. Nevertheless, the CO_2 uptake of OFACFs increased up to 16.2 wt% due to the highly developed microporous structure and semi-ionic interaction effect of oxyfluorination. The mechanism of the improved effects of oxyfluorinated activated carbon nanofibers was proposed. The nonpolar CO_2 molecules near the carbon pores are affected by the semi-ionic interaction of oxygen groups, which has the lone-pair electron, causing the electron attraction in the CO_2 molecules. This reaction might play an important role as a guide for enhancing the CO_2 storage capacity. Eventually, the CO_2 gas can be stored in the silt pores of carbon. Some of the residual CO_2 molecules in the carbon silt pores can also be affected by oxygen group effects, such as the grabbing effects due to semi-ionic interaction, resulting in the high efficiency of CO_2 storage [38].

Recently, Katepalli et al. [39] reported the synthesis of a carbon polymer micro-/nanocomposite by electrospinning PAN nanofibers directly onto ACF substrates to prepare hierarchical fabric structures (ACF-PANS) (Fig. 10.3a). The composite materials were subsequently pyrolyzed, followed by steam activation, to yield the hierarchical activated carbon fabrics consisting of PAN-derived carbon nanofiber (CNF) on ACF (ACF-PANC). These multi-scale fabrics (ACF-PANS

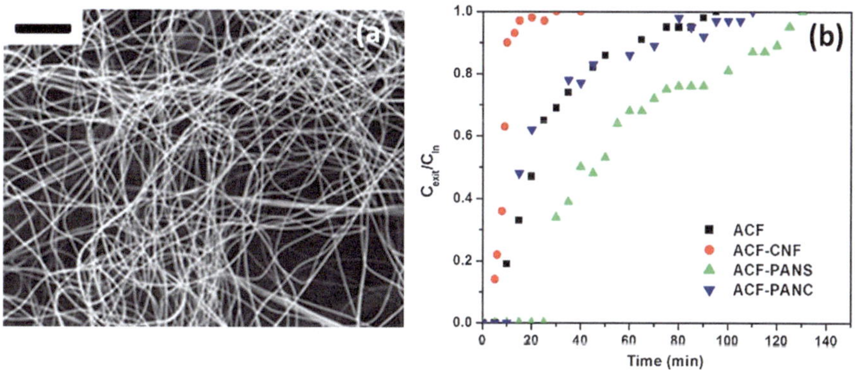

Fig. 10.3 (**a**) SEM image of ACF-PANS (scale bar equals to 20 μm) and (**b**) breakthrough profiles of SO_2 over different materials (ACF, ACF-CNF, ACFPAN, ACF-PANS, and ACF-PANC) (Reprinted from Katepalli et al. [39]. Copyright 2011, with permission from Elsevier)

and ACF-PANC) were tested for their adsorption properties toward common atmospheric pollutants (e.g., SO_2, CO_2, NO, and toluene), and the performance was compared to ACF and another hierarchical carbon fabric fabricated by growing carbon nanofibers on metal-impregnated ACF (ACF-CNF) by chemical vapor deposition. Interestingly, the performance of the ACF-PANS was found to be superior to that of ACF, ACF-PANC, and ACF-CNF fabrics (Fig. 10.3b). The superior adsorptive performance of ACF-PANS may be attributed to the large number of nitrogen-based surface functional groups, which favored the adsorption and catalytic oxidation of SO_2 or CO_2.

10.4.3 Ionic Liquid-Based Nanofibrous Membranes for CO_2 Separation

To date, many kinds of CO_2 separation membranes have been reported, including polymeric membranes, composite membranes, and facilitated transport membranes. Further improvements in membrane performance depend on effective CO_2 separation materials, and one candidate is ILs. It has been reported that ILs have good CO_2 selectivity, suggesting that they may be a possibility for the development of new CO_2 separation materials. Since ILs are liquid at room temperature, it is necessary to affix ILs to appropriate support materials. Supported IL membranes have been prepared by impregnation of commercial porous polymer films with 1-*n*-hexyl-3-methylimidazolium bis(trifluoromethanesulfonyl)imide ([hmim][Tf_2N]) and have obtained good CO_2/He separation properties [40]. Recently, electrospun Nafion/polyethylene oxide (PEO)-supported IL membranes were fabricated for CO_2 separation [41]. In this composite membrane, the electrospun Nafion/PEO material acted as a gutter layer for ILs and PEO was added to form clean nanofibrous

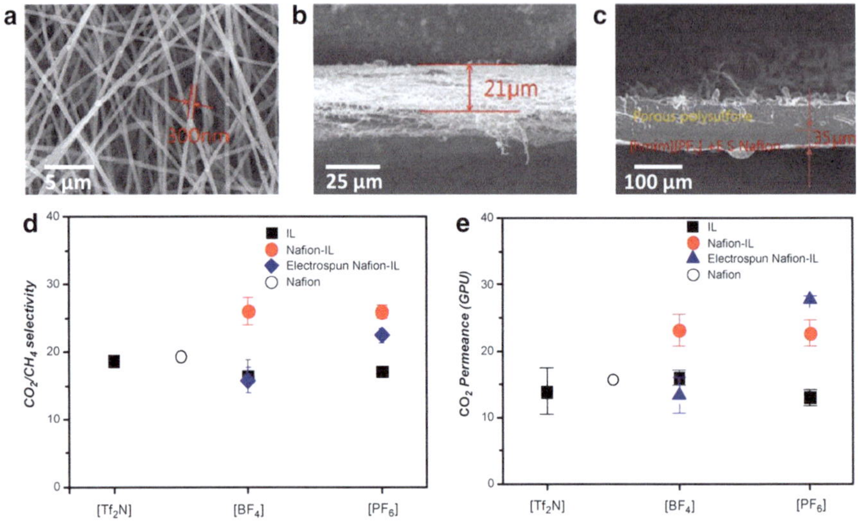

Fig. 10.4 SEM images of the (**a**) surface and (**b**) cross-section of an electrospun Nafion/PEO composite membrane. (**c**) Cross-sectional image of an electrospun Nafion/PEO composite membrane containing [hmim][PF_6]. (**d**) Ideal selectivity and (**e**) CO_2 permeance of the membrane prepared with IL, electrospun Nafion-IL, and Nafion-IL membranes (Reprinted from Yoo et al. [41]. Copyright 2010, with permission from Elsevier)

membranes via electrospinning. Figure 10.4 shows the representative scanning electron microscopy (SEM) images of the Nafion/PEO electrospun structures. It can be seen that dry membranes are composed of numerous randomly oriented fibers with diameters of about 300 nm (Fig. 10.4a). The SEM image of IL-swollen Nafion membranes showed that almost all the voids among fibers were filled (Fig. 10.4c). Figure 10.4d and e presents the CO_2 separation parameters of IL, Nafion-IL dense membranes, and electrospun Nafion/PEO supported IL membranes. The gas permeance and selectivity of electrospun Nafion/PEO/[hmim][BF_4] maintained the characteristics of supported IL membranes containing [hmim][BF_4], while the CO_2 permeance through the electrospun Nafion/PEO/[hmim][PF_6] increased from 20 to 28 GPU. The selectivity of CO_2/CH_4 slightly increased from 17.1 to 22.4, implying that the high adsorption of [hmim][PF_6] into Nafion made a definite contribution in CO_2 separation.

Recently, Tang et al. [21] reported that poly(ionic liquid)s showed significantly enhanced and fast CO_2 absorption compared to ILs. Especially, the polymers of tetraalkylammonium-based ILs have CO_2 sorption capacities 6.0–7.6 times of those of room temperature ILs. Electrospinning of the poly(ionic liquid) solution was first demonstrated by Chen and co-workers [42]. Figure 10.5 shows the electrospun fibers from various concentrations of poly(1-[(2-methacryloyloxy)ethyl]-3-butylimidazolium tetrafluoroborate or MEBIm-BF_4) in cosolvent 3/1 acetonitrile/DMF. At 2.5 wt%, a concentration below the entanglement concentration, only droplets were obtained (Fig. 10.5a). Beaded fibers were observed at 5 and

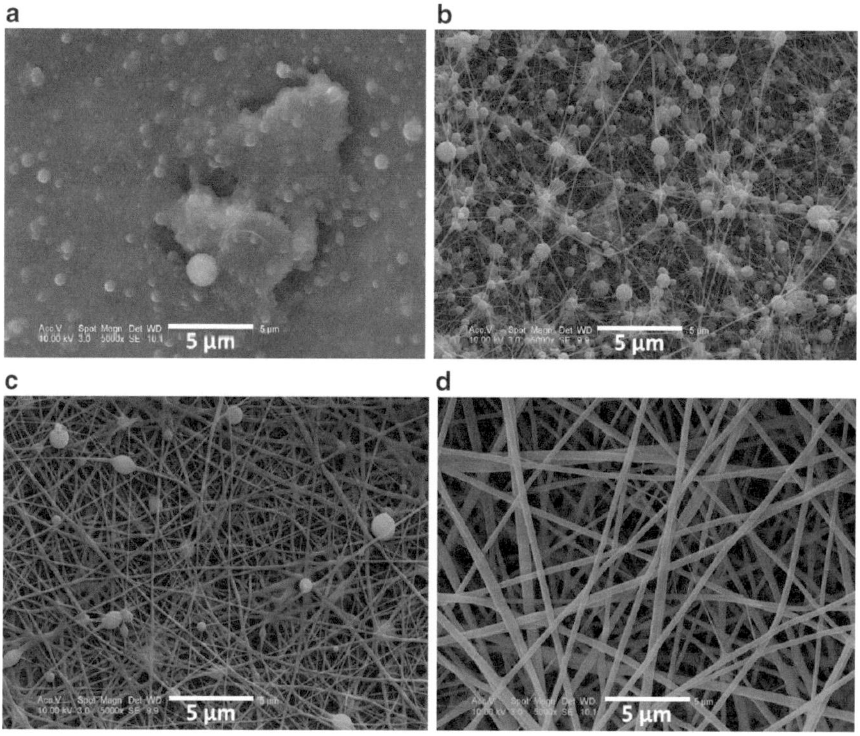

Fig. 10.5 Concentration effect on the electrospinning of poly(MEBIm-BF_4) in 3/1 acetonitrile/DMF. Polymer concentrations: (**a**) 2.5, (**b**) 5, (**c**) 6, and (**d**) 8 wt% (Reprinted with the permission from Chen and Elabd [42]. Copyright 2009 American Chemical Society)

6 wt% (Fig. 10.5b, c) and defect-free fibers were formed at 8 wt% (Fig. 10.5d). Due to high solution conductivities, electrospinning produces fibers approximately an order of magnitude smaller than neutral polymers at equivalent normalized solution concentrations. Although the CO_2 capture performance of poly(ionic liquid) nanofibers has not been reported, we can expect that these novel nanofibrous materials will be very prospective as sorbent and membrane for CO_2 separation.

10.4.4 CO_2 Sequestered in Electrospun Metal Oxide Nanofibers

Most of the electrospinning procedures that yield metal oxide nanofibers involve pyrolysis of a guide polymer as a final processing step. Bender and co-workers demonstrated, using diffuse reflectance Fourier transform infrared spectroscopy (DRIFTS), that CO_2 is sequestered in electrospun metal oxide nanofibers upon

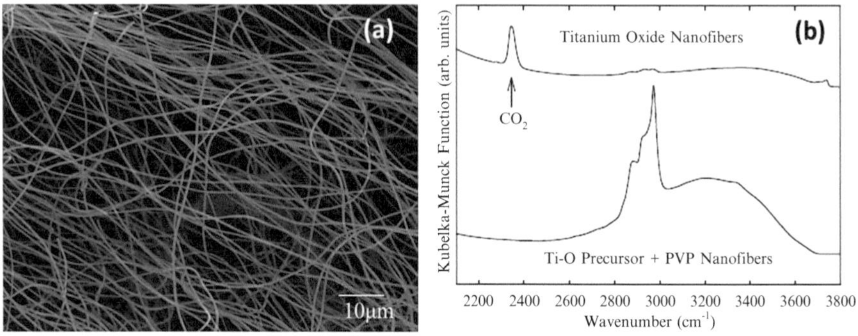

Fig. 10.6 (**a**) SEM image of TiO_2 nanofibers and (**b**) DRIFTS spectra of TiO_2 nanofibers before and after pyrolysis (Reprinted from Bender et al. [43]. Copyright 2006, with permission from Elsevier)

removal of the guide polymers by pyrolysis [43]. CO_2 is not present before pyrolysis and is formed during combustion of the polymer. Figure 10.6 shows a representative SEM image of titania (TiO_2) nanofibers after 700 °C pyrolysis. When the composite nanofibers are heated to 700 °C, it shows a DRIFTS feature around 2,340–2,350 cm^{-1} from CO_2 (Fig. 10.6b). They propose that CO_2 is trapped within the fibers as opposed to being adsorbed on the surface and the nature of the metal oxide nanofiber structure is responsible for the CO_2 entrapment.

10.5 Concluding Remarks and Outlook

CO_2 removal from post-combustion flue gas at large point sources has been spotlighted in recent years as a potential way to reduce greenhouse gas emissions. Among a range of separation technologies, adsorption with nanomaterial-based sorbents is emerging to be one of the most promising CO_2 capture strategies. In this chapter, we reviewed recent progress in developing electrospun nanofibrous sorbents with potential applications for CO_2 removal. Clearly, the selection of capture materials is essential for any technologies in CO_2 removal. In general, nanofibrous materials have advantages such as ease of design and synthesis, high porosity, tailored pore properties, high surface area, and good mechanical properties, which make them highly attractive for the development of sorbents with high CO_2 capture capacity and fast kinetics.

Despite recent advances toward the development of nanofibrous sorbents for CO_2 capture applications, several challenges still remain. (1) Some nanofibrous materials do not have CO_2 adsorption ability until they are modified with amine groups or ILs as described in this chapter. However, the undesired blocking of fiber surfaces/interior with amines or ILs will reduce CO_2 capacity and kinetics of the adsorbents. Further research should focus on the achievement of effective surface modification strategies. For example, surface modifying nanofibrous materials using layer-by-layer nano-assembly technology [44, 45] by depositing a CO_2-adsorbing amine polymer will be a potential method to solve this problem. (2) The gas flows treated in practical CO_2 capture always involve water, and it is not economically feasible to dry the flue gas by an additional process before separation. Therefore, adsorption materials require a high tolerance to water or superhydrophobicity. However, quite few investigations of the effect of water on the capture performances of nanofibrous adsorbents have been reported. Addressing this issue should include research on both the physical co-adsorption of water in the nanofibrous materials and the material structure and functional design. (3) In parallel with experimental studies, computational modeling methods must be further developed as a tool to predict the performance of nanofibrous sorbents which are proposed for a given separation process. Despite the numerous challenges surrounding CO_2 removal, further understanding of fibrous sorbent structure-property relationships and the subsequent improvement of sorbent performance under realistic operating conditions will likely allow for the realization of nanofibrous sorbents in practical CO_2 capture processes in the near future.

Acknowledgments This technical effort was performed with support of the US Department of Energy, National Energy Technology Laboratory's ongoing research in carbon management under RES contract DE-FE0004000. Support from WV NASA EPSCoR was also acknowledged. Any opinions, findings, conclusions, or recommendations expressed in this material are those of the author(s) and do not necessarily reflect the views of the funding agencies or their institutions. The authors thank Suzanne Danley for proofreading.

References

1. Du N, Park HB, Dal-Cin MM, Guiver MD (2012) Advances in high permeability polymeric membrane materials for CO_2 separations. Energy Environ Sci 5(6):7306–7322. doi:10.1039/c1ee02668b
2. Xing W, Liu C, Zhou Z, Zhang L, Zhou J, Zhuo S, Yan Z, Gao H, Wang G, Qiao SZ (2012) Superior CO_2 uptake of N-doped activated carbon through hydrogen-bonding interaction. Energy Environ Sci 5(6):7323–7327. doi:10.1039/c2ee21653a
3. Banerjee R, Phan A, Wang B, Knobler C, Furukawa H, O'Keeffe M, Yaghi OM (2008) High-throughput synthesis of zeolitic imidazolate frameworks and application to CO_2 capture. Science 319(5865):939–943. doi:10.1126/science.1152516
4. Liu J, Thallapally PK, McGrail BP, Brown DR (2012) Progress in adsorption-based CO_2 capture by metal-organic frameworks. Chem Soc Rev 41(6):2308–2322. doi:10.1039/c1cs15221a
5. Jiang B, Wang X, Gray ML, Duan Y, Luebke D, Li B (2013) Development of amino acid and amino acid-complex based solid sorbents for CO_2 capture. Appl Energy 109:112–118, doi: http://dx.doi.org/10.1016/j.apenergy.2013.03.070
6. Choi S, Drese JH, Jones CW (2009) Adsorbent materials for carbon dioxide capture from large anthropogenic point sources. ChemSusChem 2(9):796–854
7. Li B, Duan Y, Luebke D, Morreale B (2013) Advances in CO_2 capture technology: a patent review. Appl Energy 102:1439–1447, doi: http://dx.doi.org/10.1016/j.apenergy.2012.09.009
8. Lee ZH, Lee KT, Bhatia S, Mohamed AR (2012) Post-combustion carbon dioxide capture: evolution towards utilization of nanomaterials. Renew Sust Energy Rev 16(5):2599–2609. doi:10.1016/j.rser.2012.01.077

9. Yang ZH, Zhao M, Florin NH, Harris AT (2009) Synthesis and characterization of CaO nanopods for high temperature CO_2 capture. Ind Eng Chem Res 48(24):10765–10770. doi:10.1021/Ie901137s
10. Ding B, Wang MR, Wang XF, Yu JY, Sun G (2010) Electrospun nanomaterials for ultrasensitive sensors. Mater Today 13(11):16–27, doi: http://dx.doi.org/10.1016/S1369-7021(10)70200-5
11. Wang XF, Ding B, Yu JY, Wang MR (2011) Highly sensitive humidity sensors based on electrospinning/netting a polyamide 6 nano-fiber/net modified by polyethyleneimine. J Mater Chem 21(40):16231–16238. doi:10.1039/C1jm13037d
12. Wang XF, Ding B, Yu JY, Wang MR (2011) Engineering biomimetic superhydrophobic surfaces of electrospun nanomaterials. Nano Today 6(5):510–530. doi:10.1016/j.nantod.2011.08.004
13. Ramakrishna S, Fujihara K, Teo WE, Yong T, Ma ZW, Ramaseshan R (2006) Electrospun nanofibers: solving global issues. Mater Today 9(3):40–50. doi:10.1016/S1369-7021(06)71389-X
14. Li D, Xia Y (2004) Electrospinning of nanofibers: reinventing the wheel? Adv Mater 16(14):1151–1170. doi:10.1002/adma.200400719
15. Wu YN, Li FT, Liu HM, Zhu W, Teng MM, Jiang Y, Li WN, Xu D, He DH, Hannam P, Li GT (2012) Electrospun fibrous mats as skeletons to produce free-standing MOF membranes. J Mater Chem 22(33):16971–16978. doi:10.1039/C2jm32570e
16. Meng LY, Park SJ (2010) Effect of heat treatment on CO_2 adsorption of KOH-activated graphite nanofibers. J Colloid Interface Sci 352(2):498–503. doi:10.1016/j.jcis.2010.08.048
17. Han KK, Ma L, Zhao HM, Li X, Chun Y, Zhu JH (2012) In situ synthesis of SBA-3/cotton fiber composite materials: a hybrid device for CO_2 capture. Microporous Mesoporous Mater 151:157–162. doi:10.1016/j.micromeso.2011.10.043
18. Brennecke JE, Gurkan BE (2010) Ionic liquids for CO_2 capture and emission reduction. J Phys Chem Lett 1(24):3459–3464. doi:10.1021/Jz1014828
19. D'Alessandro DM, Smit B, Long JR (2010) Carbon dioxide capture: prospects for new materials. Angew Chem Int Ed 49(35):6058–6082. doi:10.1002/anie.201000431
20. Wang X, Akhmedov NG, Duan Y, Luebke D, Li B (2013) Immobilization of amino acid ionic liquids into nanoporous microspheres as robust sorbents for CO_2 capture. J Mater Chem A 1(9):2978–2982. doi:10.1039/C3TA00768E
21. Tang JB, Tang HD, Sun WL, Plancher H, Radosz M, Shen YQ (2005) Poly(ionic liquid)s: a new material with enhanced and fast CO_2 absorption. Chem Commun 26:3325–3327. doi:10.1039/B501940k
22. Rowsell JLC, Yaghi OM (2006) Effects of functionalization, catenation, and variation of the metal oxide and organic linking units on the low-pressure hydrogen adsorption properties of metal-organic frameworks. J Am Chem Soc 128(4):1304–1315. doi:10.1021/Ja056639q
23. Pandey P, Chauhan R (2001) Membranes for gas separation. Prog Polym Sci 26(6):853–893, doi: http://dx.doi.org/10.1016/0376-7388(93)80013-N
24. Li P, Zhang S, Chen S, Zhang Q, Pan J, Ge B (2008) Preparation and adsorption properties of polyethylenimine containing fibrous adsorbent for carbon dioxide capture. J Appl Polym Sci 108(6):3851–3858. doi:10.1002/app.27937
25. Li P, Ge B, Zhang S, Chen S, Zhang Q, Zhao Y (2008) CO_2 capture by polyethylenimine-modified fibrous adsorbent. Langmuir 24(13):6567–6574. doi:10.1021/la800791s
26. Yang Y, Li H, Chen S, Zhao Y, Li Q (2010) Preparation and characterization of a solid amine adsorbent for capturing CO_2 by grafting allylamine onto PAN fiber. Langmuir 26(17):13897–13902. doi:10.1021/la101281v
27. Thiruvenkatachari R, Su S, An H, Yu XX (2009) Post combustion CO_2 capture by carbon fibre monolithic adsorbents. Prog Energy Combust Sci 35(5):438–455. doi:10.1016/j.pecs.2009.05.003
28. Shen W, Zhang S, He Y, Li J, Fan W (2011) Hierarchical porous polyacrylonitrile-based activated carbon fibers for CO_2 capture. J Mater Chem 21(36):14036–14040. doi:10.1039/c1jm12585k

29. Yoon K, Hsiao BS, Chu B (2008) Functional nanofibers for environmental applications. J Mater Chem 18(44):5326–5334. doi:10.1039/B804128h
30. Wang X, Ding B, Yu J, Wang M, Pan F (2010) A highly sensitive humidity sensor based on a nanofibrous membrane coated quartz crystal microbalance. Nanotechnology 21(5). doi:10.1088/0957-4484/21/5/055502
31. Zacher D, Shekhah O, Woll C, Fischer RA (2009) Thin films of metal-organic frameworks. Chem Soc Rev 38(5):1418–1429. doi:10.1039/B805038b
32. Schoedel A, Scherb C, Bein T (2010) Oriented nanoscale films of metal-organic frameworks by room-temperature gel-layer synthesis. Angew Chem Int Ed 49(40):7225–7228. doi:10.1002/anie.201001684
33. Li YS, Bux H, Feldhoff A, Li GL, Yang WS, Caro J (2010) Controllable synthesis of metal-organic frameworks: from MOF nanorods to oriented MOF membranes. Adv Mater 22(30):3322–3324. doi:10.1002/adma.201000857
34. Venna SR, Jasinski JB, Carreon MA (2010) Structural evolution of zeolitic imidazolate framework-8. J Am Chem Soc 132(51):18030–18033. doi:10.1021/Ja109268m
35. Ostermann R, Cravillon J, Weidmann C, Wiebcke M, Smarsly BM (2011) Metal-organic framework nanofibers via electrospinning. Chem Commun 47(1):442–444. doi:10.1039/C0cc02271c
36. Tenney C, Lastoskie C (2006) Molecular simulation of carbon dioxide adsorption in chemically and structurally heterogeneous porous carbons. Environ Prog 25(4):343–354
37. Moon S-H, Shim J-W (2006) A novel process for CO_2/CH_4 gas separation on activated carbon fibers-electric swing adsorption. J Colloid Interface Sci 298(2):523–528, doi: http://dx.doi.org/10.1016/j.jcis.2005.12.052
38. Bai BC, Kim JG, Im JS, Jung S-C, Lee Y-S (2011) Influence of oxyfluorination on activated carbon nanofibers for CO_2 storage. Carbon Lett 12:236–242, doi: http://carbonlett.org/10.5714/CL.2011.12.4.236
39. Katepalli H, Bikshapathi M, Sharma CS, Verma N, Sharma A (2011) Synthesis of hierarchical fabrics by electrospinning of PAN nanofibers on activated carbon microfibers for environmental remediation applications. Chem Eng J 171(3):1194–1200. doi:10.1016/j.cej.2011.05.025
40. Ilconich J, Myers C, Pennline H, Luebke D (2007) Experimental investigation of the permeability and selectivity of supported ionic liquid membranes for CO_2/He separation at temperatures up to 125 degrees C. J Membr Sci 298(1–2):41–47. doi:10.1016/j.memsci.2007.03.056
41. Yoo S, Won J, Kang SW, Kang YS, Nagase S (2010) CO_2 separation membranes using ionic liquids in a nafion matrix. J Membr Sci 363(1–2):72–79. doi:10.1016/j.memsci.2010.07.013
42. Chen H, Elabd YA (2009) Polymerized ionic liquids: solution properties and electrospinning. Macromolecules 42(9):3368–3373. doi:10.1021/Ma802347t
43. Bender ET, Katta P, Lotus A, Park SJ, Chase GG, Ramsier RD (2006) Identification of CO_2 sequestered in electrospun metal oxide nanofibers. Chem Phys Lett 423(4–6):302–305. doi:10.1016/j.cplett.2006.03.092
44. Jiang B, Kish V, Fauth DJ, Gray ML, Pennline HW, Li B (2011) Performance of amine-multilayered solid sorbents for CO_2 removal: effect of fabrication variables. Int J Greenhouse Gas Control 5(5):1170–1175. doi:10.1016/j.ijggc.2011.05.035
45. Li B, Jiang B, Fauth DJ, Gray MML, Pennline HW, Richards GA (2011) Innovative nanolayered solid sorbents for CO_2 capture. Chem Commun 47(6):1719–1721

Part III
Electrospun Nanofibers for Environmental Applications

Chapter 11
Electrospun Nanofiber-Based Sensors

Xueqin Wang, Yan Li, and Bin Ding

Abstract Monitoring of usable, explosive, and hazardous materials (gases, heavy metal ions, and organic contaminants) is one of the most important concerns that has attracted growing attention in the past few decades. Numerous sensors have been developed to meet the demands and the pursuit of highly sensitive, selective, reversible, fast response and recovery, cost-effective, and portable sensors which have gained great interest. Electrospun nanofibrous membranes that possess large surface area, high porosity, good interconnected porous structure, and flexibility of surface functionalization are expected to be ideal candidates of substrates to improve the performance of the sensors. More interestingly, the novel two-dimensional nanofiber/net membranes fabricated by electrospinning/netting are demonstrated to have extremely large surface area and unique pore structure, which could further enhance the sensitivity, stability, and response speed of the sensors. In this chapter, we summarize recent progress in the development of electrospun nanofiber-based sensors, describe the design of different types of sensors, and discuss their sensing performances in detail. This chapter might bring further development and evolution of sensors based on electrospun nanofibers for the detection of various analytes.

X. Wang
State Key Laboratory for Modification of Chemical Fibers and Polymer Materials,
College of Materials Science and Engineering, Donghua University,
Shanghai 201620, China

Y. Li
Nanomaterials Research Center, Modern Textile Institute, Donghua University, Shanghai 200051, China

B. Ding (✉)
State Key Laboratory for Modification of Chemical Fibers and Polymer Materials,
College of Materials Science and Engineering, Donghua University,
Shanghai 201620, China

Nanomaterials Research Center, Modern Textile Institute, Donghua University,
Shanghai 200051, China
e-mail: binding@dhu.edu.cn

B. Ding and J. Yu (eds.), *Electrospun Nanofibers for Energy and Environmental Applications*, Nanostructure Science and Technology,
DOI 10.1007/978-3-642-54160-5_11,

11.1 Introduction

Environmental pollution caused by industry and agriculture activities, which include toxic gases (formaldehyde, HCl, NH_3, NO_x, etc.), organic contaminants (phenol, H_2O_2, nitrobenzene, dyes, etc.), organophosphorus compounds (pesticides, insecticides), and heavy metal ions (Pb(II), Hg(II), As(III), Cd(VI), etc.), poses a significant risk to environmental security and human health [1–3]. For instance, heavy metal ion contamination has attracted widespread concern since it can cause systemic disorders and accumulate in the human body, eventually leading to enormous damage to human health. Therefore, it is demanding to develop accurate, sensitive, selective, rapid response, and cost-effective detection methods to measure these toxic substances [4, 5].

To date, numerous techniques have been developed to monitor the environmental pollutants, such as inductively coupled plasma mass spectrometry [6, 7], high-performance liquid chromatography [8, 9], gas chromatography–mass spectrometry [10, 11], and atomic absorption spectroscopy [12]. Although these strategies provide relatively accurate detection, most of them require sophisticated equipment, tedious sample pretreatment, and skilled personnel to operate, which are unsuitable for routine pollutant detection. In the past few decades, large numbers of different sensors, involving electrochemical sensors [13], optical sensors [14], and piezoelectric sensors [15], have been used to monitor environmental contaminants owing to their easy-to-use, cost-effective high sensitivity and good selectivity. However, utilizing flat films, quartz, sol–gel matrix, or even indium tin oxide glass as substrate makes the sensitivity, stability, and response speed of the sensors relatively low, by virtue of the small surface area of these substrates. Thus, it is urgent to explore materials possessing high surface area to alternate those substrates.

In recent years, the rapid growth of nanotechnology has gained tremendous interest in environmental applications. Nanomaterials, involving nanoparticles, nanotubes, nanowires, and nanofibers which possess large specific surface area, high surface reactivity, and strong adsorption capacity, could become ideal candidates for detecting environmental contamination [16]. Due to the excellent properties of nanomaterials, the nanomaterial-based sensors possessed smaller size, faster response, better stability, and higher sensitivity and selectivity compared with traditional sensors [17, 18]. It is worth noting that the nanofibrous membranes can have interconnected open pore structures which can potentially allow the analytes to diffuse to the surface of sensing materials fast and lead to fast response accordingly. Furthermore, unlike nanoparticles, nanofibrous membrane-based sensors are easy to recycle, which is important for practical use and in protecting the environment [19].

There are a number of processing techniques such as drawing, template synthesis, phase separation, and self-assembly which have been used to prepare polymer nanofibers in the past decade [20, 21]. Considering the shortages of those techniques, such as material limitation, time-consuming, and discontinuity, another technique called electrospinning has become an internationally highly recognized method for the preparation of polymer nanofibers with diameters down to a few nanometers among those methods [19, 22]. Actually, the development history of electrospinning is not smooth; after the principle of electrospinning was studied first in the 1930s, it has been suffering ignorance for a long time, then regained considerable attention in the 1990s [23].

A typical electrospinning process is described as follows: a high voltage is applied to form the electrical charged jet of polymer solution, and before the jet reaches the collector, the solvent of the solution would evaporate and the fibers would be stretched; finally the nanofibers are collected on the collector. If the appropriate parameters are chosen, a broad range of complex architectures of core-shell, porous, multichannel, tubular, multi-core, cable-like nanofibrous structures could be obtained [24]. Moreover, for the pursuit of ever higher specific surface areas and porosity, the scientists have made every endeavor to achieve that goal. Recently, the novel two-dimensional (2D) nanowebs were generated in 3D fibrous materials newly discovered nanoarchitecture, named nanofiber/net (NFN) membranes were fabricated by optimizing processing parameter during process [25, 26]. The nanonets can create enhanced interconnectivity and additional specific surface area; thus, NFN membranes also became attractive candidates as sensing substrates to load with receptors for fabrication of ultrasensitive sensors [27, 28].

In comparison with other nanomaterials, electrospun nanofibers are featured with good continuity, extremely large length-to-diameter ratio, high surface-area-to-volume ratio, and ease of surface functionalization. Meanwhile, multiple extra functions can be incorporated into electrospun nanofibers to broaden their significances in applications. Because of this, we have found in the open literature that more than 50 polymers have been successfully fabricated into sensors, involving polyamide-6 (PA-6), polyamide-6,6 (PA-6,6), polyacrylonitrile (PAN), polyurethane (PU), poly(acrylic acid) (PAA), TiO_2, ZnO, cellulose acetate (CA), polyethylene oxide (PEO), polystyrene (PS), poly(vinylidene fluoride) (PVdF), etc. (Fig. 11.1).

11.2 Sensors Based on Electrospun Nanofibers

In recent years, considerable efforts have been made to develop varied sensors for the assay of contaminants. Benefiting from the large surface-area-to-volume ratio, tailored pore structures, large stacking density, and ease of surface modification of electrospun nanofibrous membranes, the nanofiber-based sensors achieve high sensitivity, fast response and recovery, and good reversibility. In the following parts, we will mainly introduce five types of nanofiber-based sensors: acoustic wave sensor, resistive sensors, optical sensors, optoelectronic sensors, and amperometric sensors, classified by the detecting mechanisms.

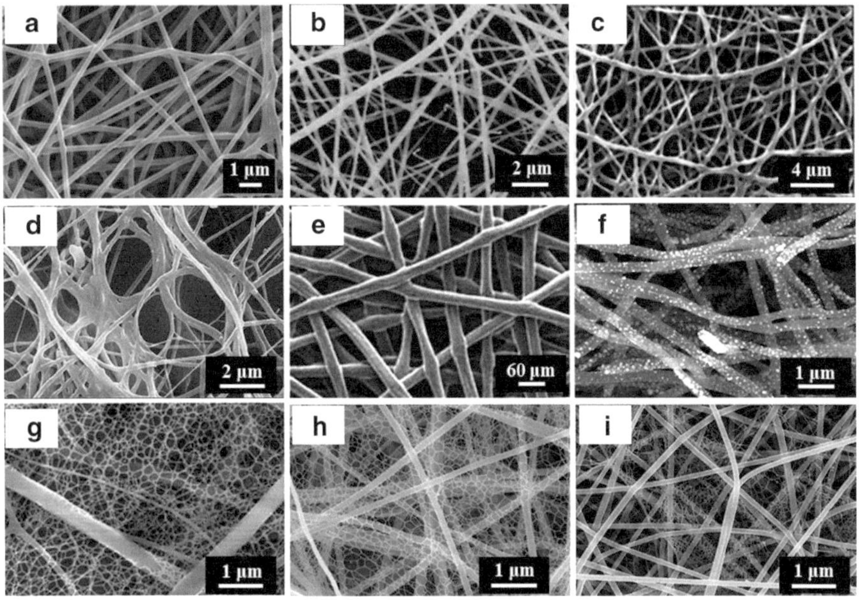

Fig. 11.1 Several typical morphologies of electrospun sensing membranes based on various systems. (**a**) PA-6,6/1-butyl-3-methyldazolium hexafluorophosphate ($BMIMPF_6$), (**b**) ZnO, (**c**) PU, (**d**) PVdF/poly(aminophenylboronic acid) (PAPBA), (**e**) PS/polydiacetylenes (PDAs), (**f**) carbon nanofiber/NiO, (**g**) PAA, (**h**) PA-6/polyaniline (PANI), (**i**) PA-6 (**a**: Reprinted with permission from Kang et al. [29]. © 2012 Springer; **b**: Reprinted with permission from Wang et al. [30]. ©2008 Wiley Periodicals, Inc.; **c**: Reprinted with permission from Huang et al. [31]. © 2013 Royal Society of Chemistry. **d**: Reprinted with permission from Ding et al. [32]. © 2007 Elsevier; **e**: Reprinted with permission from Jung et al. [33]. © 2011 Royal Society of Chemistry; **f**: Reprinted with permission from Manesh et al. [34]. © 2009 Elsevier; **g**: Reprinted with permission from Wang et al. [35]. © 2011 Royal Society of Chemistry; **h**: Reprinted with permission from Ding et al. [36]. © 2011 Royal Society of Chemistry; **i**: Reprinted with permission from Bin et al. [37]. © 2011 Royal Society of Chemistry)

11.2.1 The Acoustic Wave Sensor

11.2.1.1 Surface Acoustic Wave (SAW) Sensor

Since Wohltjen [38] applied SAW sensors in the detection of chemicals in 1979, SAW sensors have attracted worldwide attention for its high sensitivity, fast response, easy preparation, small size, and low cost. A SAW sensor can be obtained by casting a sensing layer onto the active area of the SAW device. The fundamental sensing mechanism of the SAW sensors relied on monitoring the change in the velocity of an acoustic wave on a piezoelectric substrate surface caused by the adsorption of analytes. The change in electrical conductivity or mass of the sensing layer perturbs the propagation velocity of acoustic wave due to the mechanical and piezoelectric effects; accordingly, the resonant frequency of the SAW sensor is

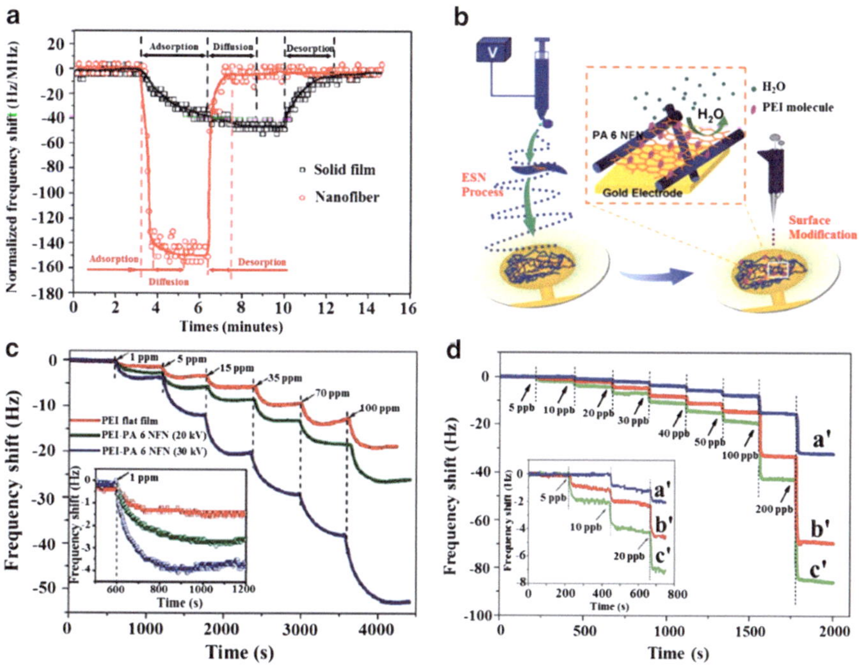

Fig. 11.2 (**a**) Comparison of frequency shift of SAW sensors based on nanofibrous membranes and solid thin film (*solid lines* are average fitting lines). (**b**) A schematic diagram illustrating the preparation of sensing coatings on QCM. (**c**) Dynamic response of QCM sensors coated with three different sensing structures upon exposure to increasing formaldehyde concentrations. (**d**) Responses of QCM sensors based on nanofibrous membranes of different surface area (a') 2.5, (b') 16, and (c') 43 $m^2\ g^{-1}$ upon exposure to Cr^{3+} (**a**: Reprinted with permission from Liu et al. [39]. © 2011 Elsevier; **b**: Reprinted with permission from Wang et al. [40]. 2011 Royal Society of Chemistry; **c**: Reprinted with permission from Bin et al. [37]. 2011 Royal Society of Chemistry; **d**: Reprinted with permission from Sun et al. [41]. 2012 Royal Society of Chemistry)

shifted. Through measuring the changes in resonant frequency of the SAW device, the concentration of the analytes can be assayed.

Intensive research effort has been focused on using nanoscale structures to significantly improve the performances (sensitivity, selectivity, response and recovery speed) of the SAW sensors. Liu et al. [39] reported that dynamic chemical vapor SAW sensors have been developed by depositing electrospun polyethylene oxide nanofibrous membranes on the surface of ST-cut quartz; the response time of the sensors was 5 min (Fig. 11.2a). Lin [42] constructed a SAW humidity sensor by casting polyaniline/poly(vinyl butyral) nanofibers on SAW resonator; the as-prepared sensor could detect 0.5 % relative humidity in 1 s. Electrospun poly(vinyl pyrrolidone) (PVP)/$LiTaO_3$ composite nanofibrous membranes were coated onto the surface of SAW electrode to assay H_2; the results indicated that the sensors' sensitivity, response time, and recovery time could be remarkably improved and reached 1 %, 120, and 200 s by regulating the electrospinning parameters [43].

11.2.1.2 Quartz Crystal Microbalance (QCM) Sensor

QCM is one of the acoustic wave techniques which can sensitively respond to mass changes on a nanogram scale. The QCM comprises a thin quartz crystal sandwiched between two metal electrodes that establish an alternating electrical field across the crystal, causing vibrational motion of the crystal at its resonant frequency [44]. As a sensor, the resonance frequency is measured and related to the mass change induced by the analyte binding to the recognition layer immobilized on the crystal surface. For instance, Yoo et al. [45] demonstrated a sensor for water vapor (gas phase) detection prepared by deposing poly(methyl methacrylate) (PMMA) on QCM. Growing attempts have been made to develop nanostructured coatings on QCM to improve the sensitivity and responsiveness of sensors [46].

Taking advantage of the large specific surface area, high porosity, and good interconnectivity of the electrospun fibrous assemblies, trimethylamine [35], formaldehyde [47, 48], HCl [49], and moisture [40] have been successfully detected by nanofibrous membrane-based QCM devices; a typical method of fabricating QCM sensors is shown in Fig. 11.2b. It is reported that the polyethyleneimine (PEI)/poly(vinyl alcohol) (PVA) nanofibrous membrane-coated QCM sensors exhibited ultrahigh sensitivity towards formaldehyde at low concentration of 10 ppm, which is three times higher than that of corresponding flat membrane-coated sensors [50]. In order to further improve the sensitivity of the sensors, Ding et al. [37] employed polyamide 6 NFN membranes as a unique platform for PEI molecule modification. Benefiting from the extremely large specific surface area, high porosity, large stacking density, and strongly tight adhesive force to the device of NFN, the as-prepared QCM sensors achieved an exceedingly low detection limit of 50 ppb (Fig. 11.2c). Sabri et al. [51] prepared a QCM-based Hg vapor sensor modified with electrospun polypyrrole (PPy) nanofibers supported Pd(II) acetate or Pd(II) nitrate. The developed sensors were observed to operate with a detection limit of 0.12 ppm, a response time of 45 min, and a stability of over 50 h, revealed a good choice for room temperature Hg vapor sensing applications.

Additionally, nanofiber-based QCM sensors could also be used to detect environmental contamination in liquid. Sun et al. [41] selected PEI-modified electrospun PS membranes as a highly sensitive coating to construct it on a QCM electrode for Cr(III) assaying; the detection limit was down to 5 ppb and they also found that the sensitivity of the sensor is increased when the surface area increases (Fig. 11.2d). Pei et al. [52] utilized ZnO-modified MnO_2 nanofiber-coated QCM sensor to detect the tiny dimethyl methylphosphonate which is a significant environmental and food chain pollutant. The experimental data revealed that the sensor could detect 35 ppb of DMMP in 300 s and possessed good selectivity and reversibility.

These researches have demonstrated the possibility of combining acoustic wave technique and electrospinning technique to develop acoustic wave sensor for contamination detection. Through regulating the structure of the nanofibrous membranes and modifying with different probes, the sensors could obtain higher surface activity and analyte sensitivity and stability.

Table 11.1 Semiconducting metal oxide-based resistive sensors

Nanofibers	Category	Analyte	Detection limit (ppm)	Response time (s)	Recovery time (s)	Operating temperature (°C)	Refs.
ZnO	n-type	NO_2	0.03	96	36	210	[61]
ZnO	n-type	Ethanol	2.8	3	5	270	[62]
WO_3	n-type	NH_3	10	600	600	300	[63]
SnO_2	n-type	Toluene	10	1	5	350	[64]
SnO_2	n-type	Ethanol	0.01	13	14	330	[57]
Co_3O_4	p-type	Ethanol	5	2.4	7.9	300	[55]
Fe_2O_3	p-type	Ethanol	100	3	5	300	[65]
$LaFeO_3$	p-type	Ethanol	50	11	19	200	[66]
In_2O_3/SnO_2	n-type	Methanol	2	8	15	280	[67]
ZnO/SnO_2	n-type	Ethanol	1	5	6	300	[68]
ZnO/SnO_2	n-type	Toluene	1	6	12	190	[69]
NiO/SnO_2	p-n junction	H_2	5	3	3	320	[70]
NiO/SnO_2	p-n junction	H_2S	50	50	124	215	[71]
Co_2O_3/SnO_2	p-n junction	H_2	100	2	3	330	[72]

11.2.2 Resistive Sensors

Resistive sensors relied on changes in electrical resistance upon surface reactions with reactive analytical samples. The amounts of samples could be obtained by measuring the resistivity or electrical conductivity. So far, many attempts were carried out to prepare ultrasensitive sensors to detect NH_3, H_2S, CO, NO_2, O_2, CO_2, moisture, and volatile organic compounds (VOCs) [53–56].

11.2.2.1 Inorganic Nanofibers

One-dimensional (1D) nanostructured metal oxide fibers have attracted much attention as chemical sensor materials due to their large space-to-volume ratio, porous structures, excellent surface activities, and high space-charge modulation depth. Specially, 1D semiconducting metal oxides such as ZnO, SnO_2, TiO_2, NiO, and $LaFeO_3$ have been widely applied in the creation of highly sensitive gas sensors [57–60] (Table 11.1). In recent years, the electrospun metal oxide nanofiber-based sensors have attracted a great deal of attention due to its ease of design and use, high sensitivity, and fast response and recovery.

First, the sensing mechanism could be clarified, expressed as follows (using NH_3 and NO_2 as models) and schematically shown in Fig. 11.3a, b. The detection response is defined as the relative ratio between the initial resistance in air, R_{air}, and the target gas sensing resistance, R_{gas}, Response $= R_{air}/R_{gas}$. When the semiconductor-based sensors are surrounded by air, oxygen molecules will adsorb on the surface of nanofibers and then transform to negatively charged oxygen

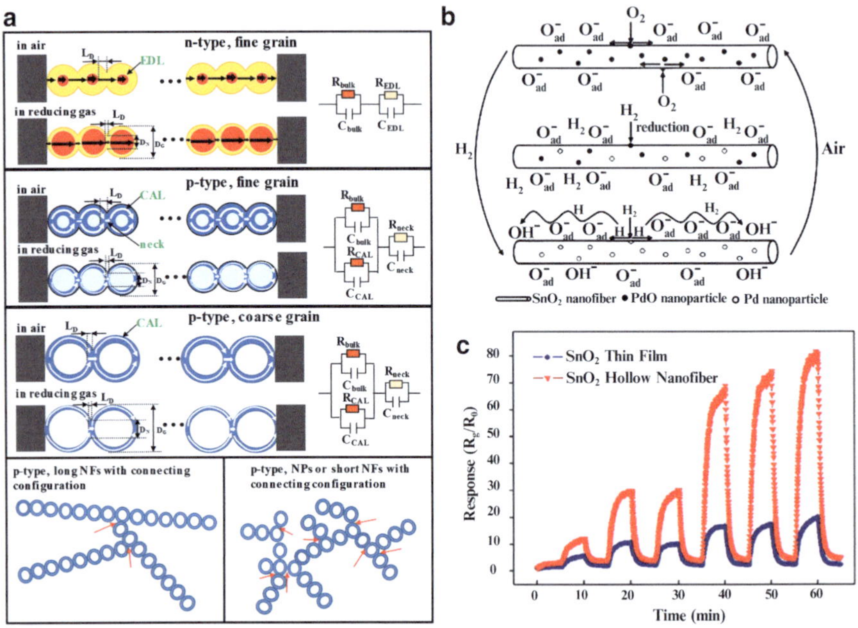

Fig. 11.3 (**a**) Gas sensing mechanisms of n-type and p-type semiconductors and the different connecting configuration between long nanofibers and nanoparticles or short nanofibers. (**b**) Schematic model of the three steps for the sensitization mechanism of the as-prepared nanofiber sensor. (**c**) Response of SnO_2 hollow nanofibers versus thin films serving as a reference (**a**: Reprinted with permission from Yoon et al. [55]. © 2012 Elsevier; **b**: Reprinted with permission from Zhang et al. [73]. © 2010 Elsevier; **c**: Reprinted with permission from Cho et al. [74]. © 2011 Elsevier)

species (e.g., O^{2-}, O^-, O^{2-}) by capturing free electrons from the conduction band of semiconductor. This electron-capture process resulted in the formation of a negative surface potential and a depletion region around the grain boundary of metal oxide crystals, which results in a decrease in conductivity of electron carrier along the nanofiber [63, 75]. When the sensors are exposed to measuring gases, gas molecules can chemisorb on the surface sites of the semiconductor.

Meanwhile, gas molecules can undergo several competitive reactions with oxygen adsorbate on the surface of the nanofiber. Several main reactions have been proposed for NH_3 oxidation being accompanied with the generation of a free electron (for oxidizing gases like NO_2, it could deplete the electron): $2NH_3 + 3O^-(ad) \rightarrow N_2 + 3H_2O + 3e^-$, $2NH_3 + 5O^-(ad) \rightarrow 2NO + 3H_2O + 5e^-$, $2NH_3 + 4O^-(ad) \rightarrow N_2O + 3H_2O + 4e^-$ ($NO_2 + e^- \rightarrow NO_2^-(ad)$, $NO + e^- \rightarrow NO^-(ad)$). These surface reactions at the solid-gas interface lead to an increase (or a decrease) of free electrons in the conduction band, thus a decrease (or an increase) in inter-grain barrier potential. The barrier energy decrease (or increase) upon exposure to the analysis gases can explain the observed response, for example, a conductivity increase (or decrease) in the nanofiber-based sensor systems [64, 76].

A great deal of n-type semiconductors such as ZnO, SnO_2, TiO_2, In_2O_3, WO_3, and ZnO/SnO_2 have proven to be excellent gas sensing materials for the detection of both reducing and oxidizing gases, including H_2, ethanol, NH_3, acetone, and toluene. It is reported that ZnO nanofibers with an average diameter of 150 nm exhibit excellent sensing properties against ethanol at an operating temperature of 300 °C, including a rapid response (about 3 s), a short recovery time (about 8 s), and a high selectivity [30]. By combining the electrospinning and oxygen plasma etching together, the highly porous SnO_2 nanofibers were prepared to serve as an ethanol sensor which exhibited fast response (7 s), wide linear response range, and low detection limit (<1 ppb) [77]. Benefiting from the enhanced surface activity such as extremely large surface-area-to-volume ratio (outer and inner layer), fast diffusion, high surface permeability, and great space-charge modulation depth, the SnO_2 hollow nanofiber-based NO_2 gas sensor achieved a low detection limit of 0.5 ppm, which was four times lower than that of film-based ones, and a rapid response (55 s) [74] (Fig. 11.3c). On the other hand, doping is an effective method to improve the sensing properties of the sensors. Zhang et al. [78] prepared a high ethanol sensor based on the ZnO/In_2O_3 composite nanofibers with a double-layer structure. Compared with the ZnO nanofibers, and ZnO/In_2O_3/ZnO nanofibers, the obtained ZnO/In_2O_3 nanofiber-based sensor exhibits improved and excellent sensing properties (detection limit of 1 ppm, short response and recovery time of 2 and 1 s).

Apart from n-type semiconductors, p-type semiconductors involving NiO, Cr_2O_3, $LaFeO_3$, CuO, LaOCl/NiO, etc., have also been used to fabricate vapor sensors. As a p-type semiconductor, the major carrier holes transmit inside the materials in the influence of external electrical field, and the holes transmission plays an important role in response and recovery properties of the sensor [79]. Fan and co-workers [66] prepared the electrospun $LaFeO_3$ nanofiber-based ethanol sensors with good reversibility and selectivity and fast response and recovery. Moreover, the as-prepared sensor showed a better sensing performance than $LaFeO_3$ nanobelts and nanoparticles. Similarly, the sensing characteristics of hierarchical structure p-type semiconductor-based sensors have been investigated in depth. The conductive electrospun LaOCl/NiO composite nanofibers are demonstrated to have superior performance in ethanol sensing against CO, H_2, NH_3, and NO_2. The enhanced gas sensing characteristics of LaOCl/NiO composite fiber mats can be attributed to the porous structure and high surface area which facilitates fast gas transport and effective surface reaction. In addition, NiO in the composite fibers is considered to have the catalytic function for promotion of gas sensing reaction [80].

More importantly, the p-type metal oxide semiconductors could combine with n-type metal oxide semiconductors by forming p-n junction, resulting in band bending in the depletion layers on either side of the physical interface to accommodate the equalization of the Fermi levels, thereby increasing the electrical resistance of the sensor (Fig. 11.4a). Through this combination, the sensing characteristics including sensitivity, operating temperature, and response and recovery time were remarkably improved [70, 71]. The past decade has witnessed the development of the resistance sensor by selecting different metal oxides to generate p-n heterojunction aiming

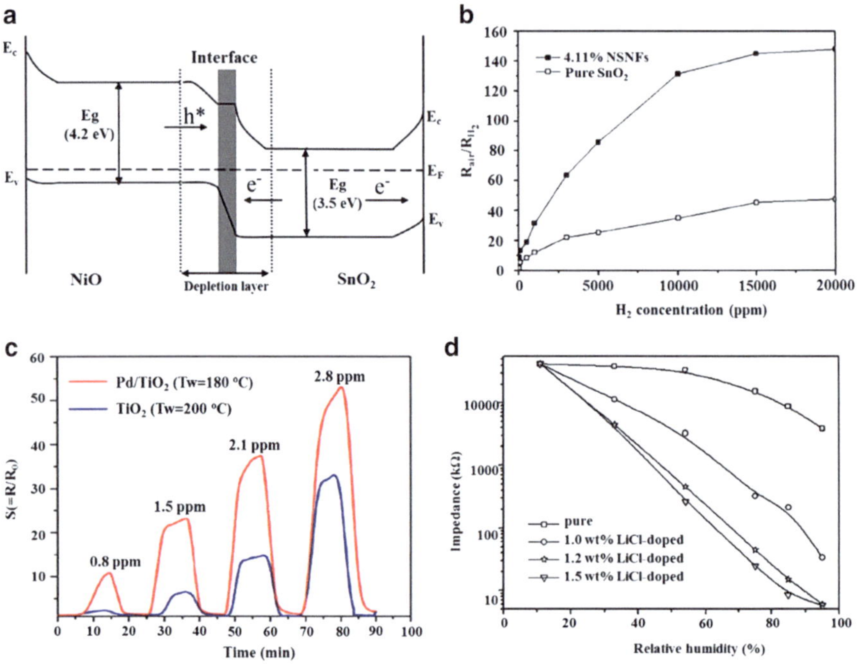

Fig. 11.4 (**a**) Proposed ideal band structure for a p-NiO/n-SnO_2 heterojunction (E_C, lower level of conduction band; E_F, Fermi level; E_V, upper level of valence band). (**b**) Responses to H_2 concentration in the range from 5 to 20,000 ppm for pure SnO_2 and 4.11 mol% NiO/SnO_2 nanofiber sensors. (**c**) The response comparison between Pd-doped TiO_2 and pure TiO_2 nanofiber-based sensors as a function of NO_2 concentrations. (**d**) The dependence of impedance on the RH for ZnO nanofibers containing different contents of LiCl (**a** and **b**: Reprinted with permission from Wang et al. [70]. ©2010 American Chemical Society; **c**: Reprinted with permission from Moon et al. [81]. © 2010 Elsevier; **d**: Reprinted with permission from Wang et al. [82]. ©2009 Elsevier)

at higher sensitivities and lower operating temperatures (Fig. 11.4b). For example, Wang et al. [83] prepared a C_2H_2 gas sensor based on NiO-doped ZnO electrospun nanofibers, and the sensitivity of the sensor is 6.5 times larger than that of pure ZnO nanofibers. Liu et al. [72] reported that the Co_2O_3-doped SnO_2 nanofibers achieved 8 times larger response than that of SnO_2 nanofibers and the sensor possesses low detection limit of 100 ppm, good selectivity, rapid response (2 s), and fast recovery (3 s).

Significantly, in order to improve the sensitivity and response and recovery speed of the prepared sensors, several methods are introduced to the design of resistance sensors: (1) attaching nanoparticles which have catalytic function (e.g., Pd, Ag, Pt) onto the surface of nanofibers [73, 81] and (2) choosing metal salts such as KCl, LiCl, NaCl, and $MgCl_2$ as the dopant to add into the nanofibers, especially in humidity sensors; the influences of the nanoparticles or metal salts are illustrated in Fig. 11.4c, d [82, 84]. For example, Xu et al. [85] fabricated Ag nanoparticle-coated ZnO–SnO_2 nanofibers for the detection of formaldehyde. Due to the "spillover effect" afforded by the metal nanoparticles, the sensitivity of the sensor was enhanced dramatically, and the sensor achieved a very low detection limit of 9 ppb. Carbon nanofibrous mat attached with Pd nanoparticles was prepared for the application of H_2 sensor [86]. The experimental data indicated that the carbon nanofibers with Pd nanoparticles had substantially higher sensitivity on H_2, due to the electronic interactions among H_2 molecules, Pd nanoparticles, and carbon nanofibers; the adsorbed H_2 molecules would react with Pd nanoparticles to form palladium hydride when the sensor was exposed to H_2. Since palladium hydride possessed lower work function than Pd, which would facilitate the transfer of electrons to carbon nanofibers, the large decrease of resistance for the carbon nanofibers was detected. Li et al. [87] reported a highly sensitive and stable humidity nanosensor based on LiCl-doped TiO_2 electrospun nanofibers. LiCl could be dissolved by the adsorbed water molecule, and then release Li(I) ions that could transfer freely among fibers, thus leading to a dramatical decrease of the impedance. Compared with the initial impedance, the impedance of sensors under humid conditions was approximately four time smaller than initial one [88]. Accordingly, the as-prepared sensor exhibited excellent sensing characteristics, including fast response time (3 s) and recovery time (7 s) for measuring relative humidity in a wide range of 11–95 % at room temperature.

11.2.2.2 Organic Nanofibers

During the last decade, organic nanofibers have also been investigated for the detection of environmental contaminants. As alternatives to inorganic semiconductors, organic polymers, mainly conducting polymers (CPs), provide great potential to produce low-cost, large-scale, lightweight, and flexible device due to their attractive features such as mechanical flexibility, ease of processing, and modifiable electrical conductivity [89–91]. Up to now, an enormous amount of research has been devoted to the development of nanosensors based on CPs such as PANI, polythiophene, PPy, and their derivatives. However, the application of CPs is hampered by the poor solubility in common solvents. In order to overcome this drawback, CPs are incorporated into other polymeric systems (e.g., PS, PEO, CA) or synthesized in other conducting forms (oxidized, reduced) that could dissolve in common organic solvents [92–95]. Electrospun nanofibers possessing large surface area, high porosity, and large stacking density which could provide large amount of activated sites for CP immobilization have attracted tremendous attention in recent years.

For CPs where charge transport is dominated via hopping mechanism, the change in the polymer resistance in the presence of a sensing gas occurs mainly due to the chemical changes (doping/de-doping), conformational change, or polymer swelling (increase in interchain separation). It is demonstrated that the electrospun isolated nanofibers of poly(3,4-ethylenedioxythiophene) doped with poly(styrene sulfonic acid) (PEDOT-PSSA) were used to sense vapors (aliphatic alcohols, NH_3, HCl, NO_2) [96]. When the as-prepared sensor was exposed to different vapors, the resistance changes because of different reasons: for aliphatic alcohols, the increase in resistance is attributed to the polymer swelling, and the sensitivity and response rate becomes weaker as the alcohol molecules become larger; for NH_3, HCl, and NO_2, the resistance changes are due to de-doping, acid doping, and oxidizing doping. Aussawasathien et al. [97] prepared camphorsulfonic-doped poly(*o*-anisidine)-PS composite nanofibers for the determination of water and ethanol. Owing to the hydrogen bonds between ethanol and water molecules and amine nitrogen in poly(*o*-anisidine) structure and the polymer swelling, the as-prepared sensor showed high sensitivity and fast response and recovery to water and ethanol. Chen et al. [98] fabricated a single PANI nanofiber field-effect transistor NH_3 gas sensor, and the sensor reached a detection limit under 1 ppm with a response time of less than 10 s. Bai et al. [99] found that the sensitivity of the gas sensor based on PPy/PMMA composite nanofibers exhibited greatly improved performance compared with those of devices based on PPy flat films due to the large surface area and the blend structures of the prepared composite fibers.

Besides CPs, nanofibers containing ionic liquids are also utilized to detect vapors. Kang et al. [29] prepared PA-6,6 nanofibers containing an ionic liquid of $BMIMPF_6$ via electrospinning and used as a chemiresistor to detect organic gases for the first trial. In the resultant sensor, the $BMIM^+$ and PF_6^- acted as charge carrier to make the resistance decrease. The results showed that the sensors exhibited sensitivity to all the organic solvents (methanol, ethanol, and acetone) with good reproducibility. Moving forward, the sensitivity, selectivity, and response and recovery rate should be further improved.

11.2.2.3 Hybrid Nanofibers

In order to overcome the drawbacks of inorganic nanofiber-based sensors (need high operating temperature) and organic nanofiber-based sensors (low sensitivity caused by the low conductivity and small surface-area-to-volume ratio of organic nanofibers), the hybrid nanofiber-based sensing devices have been developed to combine the advantages of inorganic nanofiber-based and organic nanofiber-based sensors. Scattered researches have investigated the performances of sensors based on hybrid nanofibers.

Wang and co-workers [100] synthesized an NH_3 resistance transducer based on highly porous PPy-coated TiO_2/ZnO nanofibrous membranes. The gas sensor exhibited a fast response over a wide dynamic range and high sensitivity with a much lower detection limit of 60 ppb compared with reported pristine PPy film. The dramatic enhanced sensitivity is attributed to the porous structure of composite nanofibers and ultrathin PPy layer which could offer a large surface area for the free access of NH_3 to nanofibers and a decrease in the gas diffusion resistance. Choi et al. [101] fabricated a chemiresistor for low molecular weight alcohol vapor sensing utilizing mutiwalled carbon nanotubes (MWCNT) doped PA-6,6 nanofibers. The hydrogen bond among the alcohol vapors, hydroxyl group on MWCNT-OH, and amide groups in PA-6,6 led to the changes of the electrical resistance; consequently, the as-prepared sensors demonstrated fast response, good reversibility, and high sensitivity to alcohol in low molecular weight. By combining the in situ polymerization of aniline in the presence of PA-6 nanofibers and a sputtering-deposition process with TiO_2 sputtering target, the TiO_2-PANI/PA-6 composite nanofibers were fabricated for NH_3 gas sensing. The adsorption of NH_3 would not only make PANI deprotonated but also change the resistance of p-n junction formed between p-type PANI and n-type TiO_2, thereby resulting in an increase in the sensitivity of the nanofiber sensors at low range of TiO_2 content. Compared with PANI/PA-6 nanofiber-based sensors, the as-prepared sensor exhibited better sensitivity, faster response and recovery, and good selectivity. Moving forward, the work should be devoted to improving the sensitivity, stability, and reusability of the sensors to variety of analytes.

11.2.3 Optical Sensors

11.2.3.1 Colorimetric Sensors

During the last decade, monitoring of contaminants has gained enormous attention due to environmental and public safety concerns posed by exposure to these substances. Therefore, a great deal of researches have been carried out to develop sensitive, selective, easy-to-use, fast responding sensors for analyte detection. Among output signals, the colorimetric mode is considered as the most convenient sensing platform for developing a simple naked eye detector because it minimizes the need for extensive signal transduction hardware [102, 103]. Nanofibrous or NFN membranes which could afford a great number of activity sites for immobilization of sensing materials and reactions have attached tremendous attention. Generally, the nanofibrous or NFN membrane-based colorimetric sensors are prepared by immobilizing the probe which could specifically react with analytes on nanofibrous membranes through various methods, including in situ polymerization, dip coating, or direct electrospinning [33, 104]. Due to the outstanding properties of colorimetric sensors, such as cost-effectiveness, portability, easy to use (detected by the color change that could be observed by the naked eye), and semiquantitative determination, many researchers have focused on the development of colorimetric sensors towards various analytes.

CPs are a class of polymers which could undergo distinct color changes under various external stimuli such as heat, oxidation or reduction, doping and de-doping, and ligand-receptor interactions, which makes them the most promising sensing materials for pollutant assay [105, 106]. Interestingly, a series of PDA-embedded electrospun nanofibers are prepared by two steps: first, electrospinning the blend which contains PEO, tetraethyl orthosilicate, and several commercially available or easily obtainable diacetylene (DA) monomers and, second, polymerization of the molecular-assembled DAs to form polymer nanofibers containing embedded PDAs [107, 108] (Fig. 11.5a). The as-prepared sensors show different colorimetric response upon exposure to different organic solvents (Fig. 11.5b). Lee et al. [33]

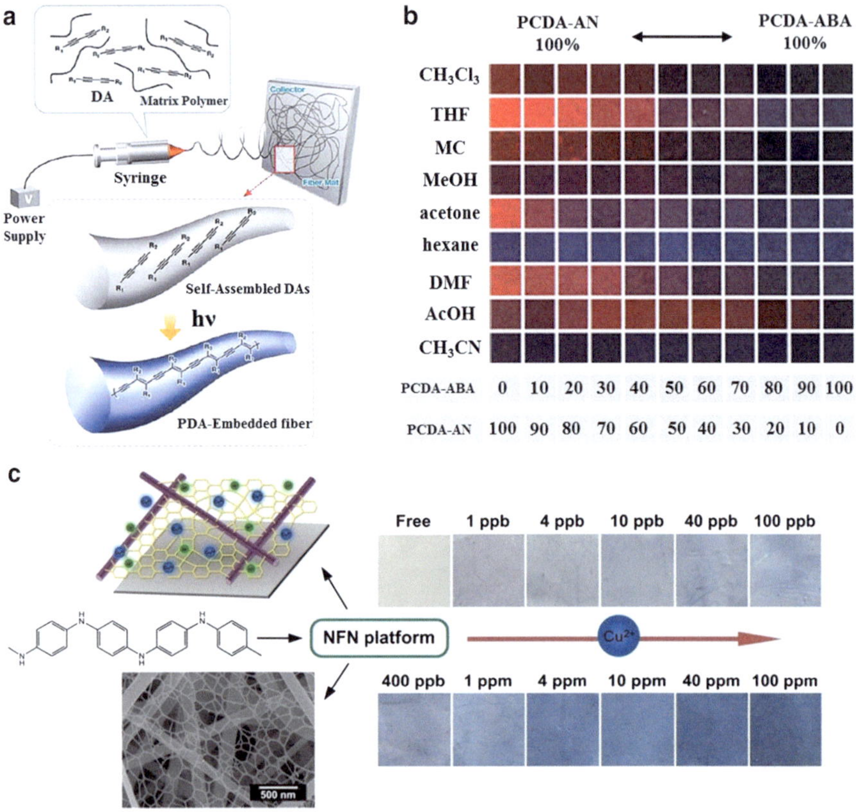

Fig. 11.5 (**a**) Representation of fabrication of a conjugated PDA-embedded electrospun organic microfiber. (**b**) Photographs of the polymerized PDA-embedded electrospun fiber mats after exposure to organic solvents at 25 °C for 30 s. (**c**) PA-6/PANI NFN sensing membranes for the colorimetric detection of Cu^{2+} in concentration of 1–100 ppm (**a**: Reprinted with permission from Jung et al. [33]. ©2011 Royal Society of Chemistry; **b**: Reprinted with permission from Yoon et al. [108]. ©2009 Wiley Periodicals, Inc.; **c**: Reprinted with permission from Ding et al. [36]. ©2011 Royal Society of Chemistry)

applied PDA-embedded electrospun nanofibers to detect gasoline quality. The distinct color transition from blue to red can be monitored with the naked eye when adulterated gasoline is applied to the conjugated polymer-encapsulated nanofibers. The results demonstrate that the components of the adulterated gasoline can be selectively detected by changing the structure of the nanofibers. Taking advantage of the extremely large surface area, high porosity, and unique structure of the 2D NFN membranes, the sensitivity, response rate, and reversibility of the colorimetric sensors would be improved dramatically. Ding et al. [36] successfully prepared a novel, ultrasensitive, selective, and flexible colorimetric strip based on

PANI/PA-6 NFN membranes for naked eye detection of Cu(II) in water. Based on the sensing mechanism which involves the reversible transformations between different oxidation and doping forms of PANI, the sensors achieved an extremely low detection limit of 1 ppb, excellent selectivity (specific reaction), and good reproducibility (Fig. 11.5c).

Apart from CPs, other sensing materials also have been used for colorimetric assay of analytes. Poltue and co-workers [109] served the electrospun dimethylglyoxime/poly(caprolactone) (DMG/PCL) blend nanofibers as an optical sensor for the Ni(II) detection based on the formation of red $Ni(DMG)_2$ complex. It is reported that the as-spun sensors could be used to detect Ni(II) in the concentration of 1 ppm with a good linear response between 1 and 10 ppm. Li et al. [110] prepared a sensor strip utilized electrospun PA-6/nitrocellulose NFN membranes assemble bovine serum albumin decorated Au nanoparticle (BAu probe) for naked eye colorimetric assay of Pb(II). Pb(II) could accelerate the specific leaching reaction between BAu probe and $S_2O_3^{2-}$ ions in the leaching liquor, thus leading to the color change from pink to white (gradually faded with the increasing concentration of Pb(II)) which could be detected by the naked eye. Compared with the commercial nitrocellulose films, the as-prepared colorimetric strips reached a much lower detection limit of 0.2 μM with excellent selectivity, fast response, and good stability and reversibility attributed to the extremely large specific surface area, unique structure, and high porosity of NFN membranes. The incorporation of pH-sensitive dyes in nanofibrous structures offers potential for producing halochromic textile structures and has been investigated by researchers. Van der Schueren et al. [111, 112] investigated the performances of pH indicators prepared by electrospinning the blend solution of various dyes and polymers. Due to the interactions between fibers and dyes, the as-spun nanofibers showed a clear color change with a change in pH, and the halochromic behavior is different from the dyes in solution.

11.2.3.2 Fluorescence Sensor

Classically, the design of a fluorescent indicator includes two moieties, a receptor responsible for the molecular recognition of the analyte and a fluorophore responsible of signal recognition [113, 114]. There are three main strategies for the design of fluorescent indicators for chemical sensing: first, intrinsic fluorescent probes (fluorescent molecules), in which the mechanism for signal transduction involves interaction of the analyte with a ligand which is part of the p-system of fluorophore; second, extrinsic fluorescent probes, in which the receptor moiety and the fluorophore are covalently linked but are electronically independent; and, third, chemosensing ensemble, based on a competitive assay in which a receptor-fluorophore ensemble is selectively dissociated by the addition of an appropriate competitive analyte able to interact efficiently with the receptor resulting in a detectable response of the fluorophore [115].

In fluorescence techniques, suitable indicators which are sensitive to analyte concentrations and exhibit changes in fluorescence intensity are used as molecular recognition materials. Adewuyi et al. [116] developed a fluorescent nanofiber probe for the detection of Ni(II) via electrospinning of a covalently functionalized pyridylazo-2-naphthol-poly(acrylic acid) polymer. Importantly, the sensing system demonstrated high sensitivity and good selectivity in its fluorescence "turn-off" response towards Ni(II) against other metal ions. Moreover, the reversibility process could be realized by breaking of an indicator-quencher complexation and a fluorescence "turn-on." Wang et al. [117] developed a fluorescence sensor based on naphalimide functionalized PMMA fibrous membranes. Here, naphalimide was chose as an indicator due to its large molar extinction coeffiecient and high fluorescence quantum yield. The results revealed that the sensors possessed a high sensitivity (the detection limit for Cu(II) is 20 μM) due to the large surface-area-to-volume ratio of nanofibrous membranes and a good selectivity to Cu(II) against other interfering heavy metal ions. In another study [118], rhodamine derivatives were selected as a new fluorophore moiety and developed a sensing system in which the surface of the PMMA is modified with rhodamine derivatives to form a "turn-on" fluorescence sensor poly(MMA-co-AHPA) for Cu(II) detection. The utilized surface modification method avoids the aggregation and self-quenching of molecular fluorophores efficiently and exhibits high fluorescence sensitivity, short response and recovery time, and good reversibility and reproducibility (Fig. 11.6a, b). Davis et al. [119] reported a protein sensor by encapsulating the anionic fluorescent dendrimer in electrospun CA fibers, which were deacetylated to cellulose to generate secondary porous structure that is desirable for enhancing molecular interactions. The results revealed that the sensor could be used to detect cytochrome-*c* of 0.1 μM with good reversibility (Fig. 11.6b). CPs have been proven to be high-performance sensing probes due to their excellent molar absorptivity, fluorescence quantum yield, and amplified sensory response compared with small fluorescent molecules for the analysis of chemical and biomolecules. Electrospun PS nanofibers doped with a synthesized fluorescent CP is developed as a sensory device for assay of explosive 2,4-dinitrotoluene (DNT) [120]. The electrospun nanofibrous membranes adequately maintained the fluorescence properties of the as-synthesized CP and showed higher sensitivity towards DNT compared to the conventional spin-casting films owing to the large surface area of nanofibers, indicating that this approach could effectively avoid the aggregation of CP through π-π stacking interaction. Wang et al. [121] fabricated a conjugated fluorescent probe by an electrostatically assembled hydrolyzed poly[2-(3-thienyl) ethanol butyl carbonyl-methyl urethane] onto the surface of CA electrospun nanofibrous membranes through electrostatic layer-by-layer adsorption technique for the detection of methyl viologen and cytochrome-*c* in aqueous solutions. Thanks to the high surface-area-to-volume ratio of the electrospun nanofibers and interaction between fluorescent conjugated polymer and analytes, the as-prepared sensor could be applied for the detection of cytochrome-*c* in 0.12 μM.

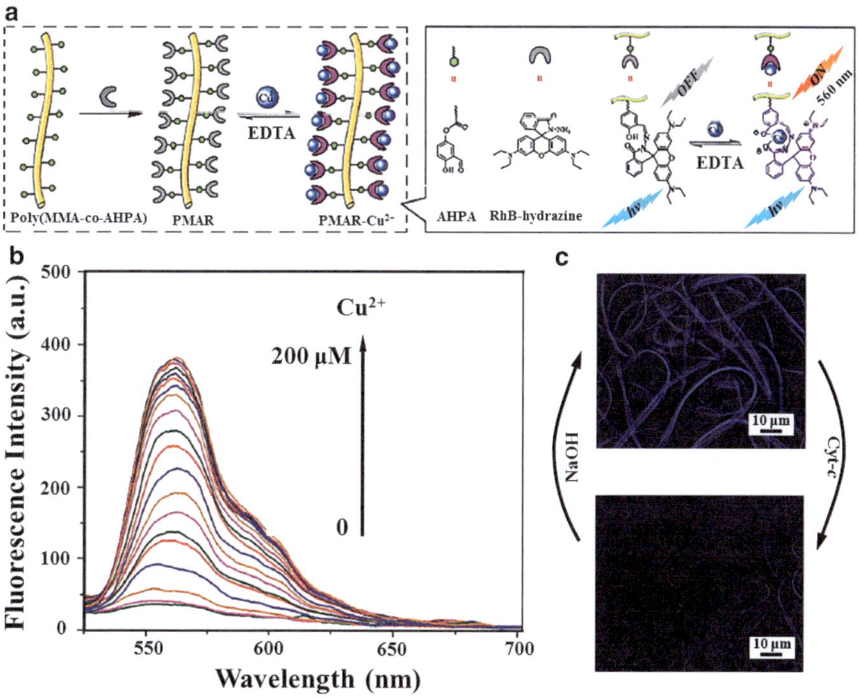

Fig. 11.6 (**a**) Chemical and schematic illustration of the preparation of nanofibrous film fluorescent sensors for Cu^{2+} ions. (**b**) Fluorescent spectra of the PMAR nanofibrous film in the absence and presence of Cu^{2+} (1–200 μM). (**c**) Confocal fluorescence images of the electrospun nanofibers before and after incubation in cyt-*c* or NaOH solution (**a** and **b**: Reprinted with permission from Wang et al. [118]. © 2013 Royal Society of Chemistry; **c**: Reprinted with permission from Davis et al. [119]. © 2010 American Chemical Society)

11.2.3.3 Luminescence Sensors

Of the many strategies for optical detection, some of the most powerful are based on luminescence, which could provide exquisite sensitivity and selectivity. A useful type of luminescent probe is based on the quenching of a reporter molecule by the analyte [122–124]. The presence of a quencher in the system results in more rapid depletion of the excited state population which is detected either as a concomitant decrease in steady-state luminescence intensity or as a shorter emission decay time. Change in either intensity or decay time could be used to quantitate the amount of analyte present [125].

In recent years, the performances of nanofibrous membrane-based luminescence sensors have been extensively studied. The synthesis of luminescent transition metal complexes receives much attention due to their favorable optical and electrical properties, which make them good candidates for application in electro-optical devices. For example, glucose oxidase-functionalized PS electrospun optical fibrous membranes were fabricated for quick and highly sensitive optical detection of glucose [126]. Because of the doped iridium complex, the fibrous membranes emitted yellow luminescence when excited at 405 nm, and the luminescence was significantly enhanced with the presence of glucose. The prepared optical sensor achieved high sensitivity (detection limit was 0.1 nM), wide linear range (0.3 nM to 130 mM), and fast response (1 s) attributed to the high surface-area-to-volume of the porous fibrous membranes and the fast electron or energy transfer between dissolved oxygen and optical membranes. Wang et al. [127] integrated Cu(I) complex with PS to form electrospun composite nanofibers for oxygen determination. The emissive state of Cu(I) complexes is quenched by the oxygen molecule, leading to the emission quenching, thus forming strong interaction between the excited state electron localizing on LUMO of Cu(I) complexes and ground state molecular oxygen ($^3O_2$). In addition, the $[Cu(P\text{-}P)(N\text{-}N)]^+$ complexes with a large coplanar conjugation diamine ligand may offer higher sensitivities. Therefore, the optical oxygen sensors achieved high sensitivity and fast response and recovery time (10 and 20 s) attributed to the above reasons and the large surface-area-to-volume ratio of electrospun PS nanofibers. It is reported that luminescent electrospun fibers for pH-tunable colorimetric sensors were prepared from binary blends of poly(phenylquinoline)-*b*-PS/PS with a single-capillary spinneret [128]. The as-prepared nanofibers revealed a higher pH-tunable sensing ability than the spin-coated film, and the fibers from CH_2Cl_2 exhibited a dramatic shifting of the emission maximum as the pH value increased from 7 to 1 because of the smaller poly(phenylquinoline) aggregation and the promoted efficient protonation caused by the porous structure. A conjugated rod-coil-coil triblock copolymer of polyfluorene-*b*-poly(*N*-propylacrylamide)-*b*-poly(*N*-methylolacrylaimde) was successfully synthesized and electrospun into nanofibers which showed outstanding dimension stability and a reversible on/off transition on photoluminescence as the temperature varied [129]. Furthermore, the as-prepared sensors exhibited much enhanced temperature sensitivity and response speed compared to those of the drop-cast film owing to the high surface-area-to-volume ratio.

11.2.4 Optoelectronic Sensors

Since Yang's discovery of UV photodetectors based on single ZnO nanowire, vigorous researches in correlative field have been carried out worldwide [130, 131]. Owing to the good chemical stability, wide bandgap of 3.37 eV, and high exciton-binding energy of 60 meV at room temperature, ZnO has been investigated by researchers as an active component in a range of devices including ultraviolet photodetectors, field-effect transistors, and chemical/biological sensors. The mechanism of photoconductance of ZnO can be explained as follows: under dark conditions, oxygen molecules adsorbed onto nanofiber surface and captured free electrons from the conduction band, thus forming negatively charged ions on the surface and creating a low conductivity depletion layer near nanofiber surface. The photogeneration and recombination of electron–hole pair can be held

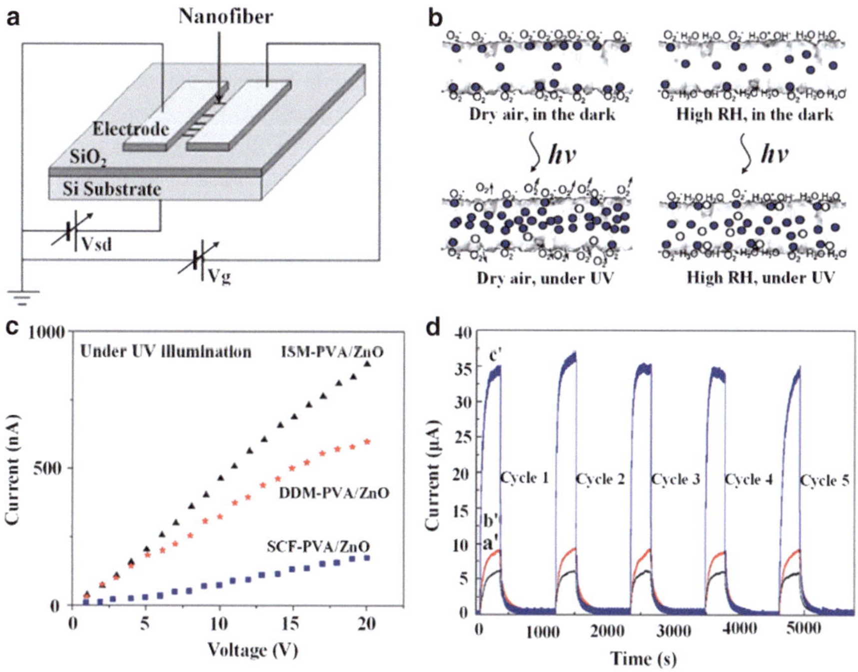

Fig. 11.7 (**a**) Schematic view of the nanofiber field-effect transistor configuration. (**b**) Schematic of adsorption/desorption of surface species and charge carrier trapping and transport processes in electrospun ZnO nanofibers in dry air and air with high RH. The *blue dots* represent electrons, while the *white dots* represent holes. (**c**) Comparative photocurrent measurement of three different types of PVA/ZnO composites under UV illumination. (**d**) The reversible response to 365 nm UV light of the nanofibers of (a') ZnO; (b') 0.05 mol% Au/ZnO; (c') 0.20 mol% Au/ZnO (**a**: Reprinted with permission from Wu et al. [132]. © 2009 Wiley Periodicals, Inc.; **b**: Reprinted with permission from Lai et al. [133]. © 2013 Royal Society of Chemistry; **c**: Reprinted with permission from Anitha et al. [134]. © 2013 Elsevier; **d**: Reprinted with permission from Li et al. [135]. © 2012 Elsevier)

responsible for the photoresponse of ZnO. When the UV light falls on the sample, electron–hole pairs are created. These photogenerated holes migrate to the surface along the potential gradient produced by the band bending and get neutralized by the negatively charged adsorbed oxygen ions. Consequently, oxygen is desorbed from the surface, thus increasing the free carrier concentration and thereby leading to a decrease in the depletion layer. The remaining unpaired electrons are the major contributors to the current; the device and mechanism are displayed in Fig. 11.7a, b [133].

A UV nanosensor based on uniaxially aligned electrospun ZnO nanofibers was demonstrated to display superior sensitivity and reversibility and could be operated at room temperature. Further researches indicated that the photosensitivity of the device was enhanced twofold by modifying the surface of electrospun ZnO nanofibers with PVP [136]. On this basis, Anitha and co-workers [134] compared the photoresponse of three different types of PVA/ZnO composites, including solution-cast film (SCF-PVA/ZnO), electrospun PVA/ZnO fibrous membranes by direct dispersion method (DDM-PVA/ZnO), and in situ sol–gel method (ISM-PVA/ZnO) (Fig. 11.7c). The comparative results displayed that the photocurrent in ISM-PVA/ZnO was six times higher than in SCF-PVA/ZnO and 1.4 times higher than DDM-PVA/ZnO, which implied that the large surface area of the electrospun nanofibers can significantly enhance the sensitivity of the optoelectronic sensors.

In order to improve the sensitivity of the photodetectors, the metal nanoparticles (Ag, Au, etc.) were loaded onto the ZnO nanofibers to enhance the photoresponse. The origin for the enhancement is related to the enhanced separation of photogenerated electron–hole pairs and the resulted efficient discharge of surface states. For example, the photoresponsive properties of Au/ZnO nanofibers were dramatically enhanced compared with ZnO nanofiber by regulating the concentration of the nanoparticle content ones (Fig. 11.7d). Furthermore, ZnO nanofibers could also be used to detect organic gases under UV illumination, such as ethanol, acetone, toluene, and benzene; the Au nanoparticle decoration would also remarkably enhance the sensitivity of the sensors as expected [135].

In addition, because of the wide direct bandgap of 3.39 eV and ability to operate at extreme conditions (temperature and voltage bias), photoconductive GaN detectors were intensively studied [137] during the initial development of UV detectors. Wu et al. [132] reported a strategy to synthesize extremely long GaN nanofibers on a large scale based on electrospinning and applied it as high-performance UV photodetector. Attributing to the rough surface, high surface-area-to-volume ratio of GaN nanofibers, and polycrystalline structure, the GaN nanofibers showed much higher sensitivity to UV compared with smooth GaN nanowire ones. Besides high sensitivity, the electrospun GaN nanofiber based UV sensor also showed advantages in response speed and reversibility.

11.2.5 Amperometric Sensors

In recent years, intensive research activities have been devoted to the development of amperometric biosensors; among them, glucose sensors have been investigated because of its importance in diagnosis and treatment of diabetes mellitus as well as in monitoring of food processes [138]. Due to the high sensitivity and selectivity to glucose, glucose oxidase (GOD) has been widely used to construct various amperometric biosensors for glucose detection; the general mechanism is presented in Fig. 11.8a [139, 141]. Scampicchio et al. [140] illustrated a new nanofibrous membrane-based glucose sensor; the sensing unit is based on a carbon electrode coated with PA-6 fibrous membranes in which GOD is covalently tethered. Porous activated carbon nanofibers with large specific surface area were used to immobilize GOD for determination of glucose with a detection limit of 1 mM in 7 s [142] (Fig. 11.8b). Ren et al. [143] prepared a glucose sensor by electrospinning PVA/GOD biocomposite nanofibrous membranes on Au electrode

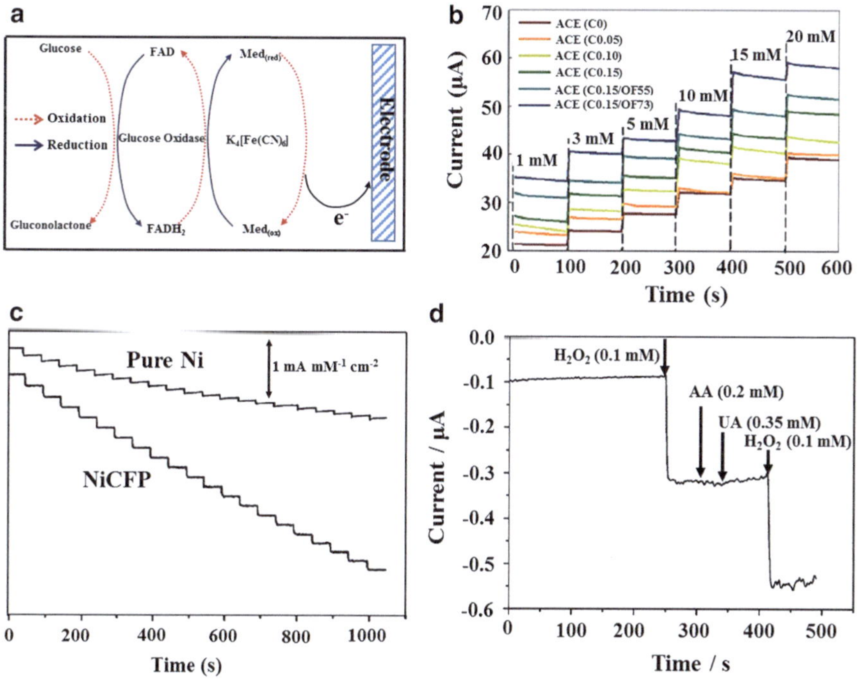

Fig. 11.8 (**a**) The sensing mechanism of GOD-based sensors. (**b**) Amperometric response of the glucose sensor at various glucose concentrations. (**c**) Successive amperometric responses of the Ni bulk electrode and the NiCFP electrode. (**d**) Current responses of the Pd/CNF-CPE to the sequential additions of 0.1 mM H_2O_2, 0.2 mM AA, 0.35 mM UA, and 0.1 mM H_2O_2 into N_2-saturated phosphate buffer solution (**a**: Reprinted with permission from Im et al. [139]. © 2011 Elsevier; **b**: Reprinted with permission from Scampicchio et al. [140]. © 2012 Elsevier; **c**: Reprinted with permission from Manesh et al. [34]. © 2009 Elsevier; **d**: Reprinted with permission from Huang et al. [31]. © 2008 Wiley Periodicals, Inc.)

and then immobilizing GOD through cross-linking the PVA/GOD membranes, and the as-prepared sensor could detect 0.05 mM of glucose in 1 s.

However, the drawbacks of the GOD-based sensors such as the insufficient stability, low output current density, and the gradual deterioration of the enzyme activity have limited their practical application [32, 144]. Therefore, nonenzymatic glucose detection has been intensively developed in the past few decades. Manesh et al. [34] used electrospun PVdF/PAPBA nanofibrous membranes as electrode to prepare a glucose sensor. The sensor displayed an excellent linear response to glucose for the concentration range of 1–15 mM with a response time of less than 6 s; in addition, the sensor electrode retained 90 % of original activity after 50 days (Fig. 11.8c). Liu and co-workers [145] developed a nonenzymatic glucose sensor based on the renewable Ni nanoparticle-loaded carbon nanofiber paste (NiCFP) electrode. Taking advantage of the large specific area of carbon nanofibers, the sensor reached a lower detection limit of 1 μM in 5 s without being poisoned by chloride ions.

In addition, the analytical assay of H_2O_2 has also attracted growing interest, ascribable to the fact that it is the end product of enzymatic reaction and the essential of environmental analysis [31, 146]. The detection of H_2O_2 could also be classified into two categories: enzyme-based sensing and nonenzymatic detection. Much research has focused on the immobilization of enzyme on materials such as nanoparticles, nanowires, and nanofibers. For example, scientists [147] immobilized a direct electron transfer-type enzyme, horseradish peroxidase, onto the electrospun carbon nanofibers' webs to detect the concentration of H_2O_2. The response to the concentration of H_2O_2 was linear over the range from 1 to 10 μM, and the as-prepared H_2O_2 sensor also displayed fast response (within 2 s), satisfactory selectivity, and good reproducibility and stability (Fig. 11.8d). But the enzymatic detection of H_2O_2 has suffered from the same problem with the enzymatic sensing of glucose. Therefore, nonenzymatic biosensors have aroused numerous attentions. Ouyang et al. [148] reported a nonenzymatic amperometric biosensor to detect H_2O_2 by directly electrospinning PU/carbon nanotube/Ag nanoparticle nanofibers on the surface of glassy carbon electrode. The as-fabricated sensors achieved a relatively low detection limit of 18.6 μM and could store for 15 days.

Besides glucose and H_2O_2, amperometric method has been utilized to assay fructose, methanol, NO_2, hydrazine, and ethanol either by enzyme sensing or by nonenzymatic detection. Although the stability, output current density, and the reactivity have been enhanced, the selectivity of the nonenzymatic sensors was decreased; further researches should be done to keep both in balance.

11.3 Summary and Perspectives

In summary, large numbers of sensors based on electrospun nanofibers have been used to detect various contaminants in the environment. By virtue of the large surface-area-to-volume ratio, high length-to-diameter ratio, tailored pore structure, and large stacking density, nanofiber-based sensors achieved high sensitivity, fast response and recovery, and excellent stability. In further study, the following aspects should be taken into consideration: first, the surface area and pore structures of the nanofibrous membranes should be further improved to promote the sensitivity and response and recovery speed; second, the approaches for modification of probes onto the surface of nanofibers should be renovated, so that the probe could be easy to modify and stable and interact with analytes sufficiently; third, the reversibility of the sensors should be improved to decrease the loss of sensitivity after probe regeneration; finally, the performance of the as-prepared sensors under the realistic conditions should be studied to realize the real-time and on-site determination of practical contaminants in the future. Moving forward, it is reasonable to believe that the electrospun nanofiber-based sensors may throw a light on the development of sensors.

Acknowledgement This work is supported by the National Natural Science Foundation of China (No. 51173022 and 51273038), the Shanghai Nano Special Projects (11nm0502900), the Shanghai Committee of Science and Technology (No. 12JC1400101), the Huo Yingdong Foundation (131070), the Program for New Century Talents of the University in China, and the Fundamental Research Funds for the Central Universities.

References

1. Noyes PD, McElwee MK, Miller HD, Clark BW, Van Tiem LA, Walcott KC, Erwin KN, Levin ED (2009) The toxicology of climate change: environmental contaminants in a warming world. Environ Int 35(6):971–986. doi:10.1016/j.envint.2009.02.006
2. Nriagu JO, Pacyna JM (1988) Quantitative assessment of worldwide contamination of air, water and soils by trace-metals. Nature 333(6169):134–139. doi:10.1038/333134a0
3. Jarup L (2004) Health and environment information systems for exposure and disease mapping, and risk assessment. Environ Health Perspect 112(9):995–997. doi:10.1289/ehp.6736
4. Onyido I, Norris AR, Buncel E (2004) Biomolecule-mercury interactions: modalities of DNA base-mercury binding mechanisms. Remediation strategies. Chem Rev 104(12):5911–5929. doi:10.1021/cr030443w
5. Maserti BE, Ferrara R (1991) Mercury in plants, soil and atmosphere near a chlor-alkali complex. Water Air Soil Pollut 56(1):15–20. doi:10.1007/bf00342257
6. Komarek M, Ettler V, Chrastny V, Mihaljevic M (2008) Lead isotopes in environmental sciences: a review. Environ Int 34(4):562–577. doi:10.1016/j.envint.2007.10.005
7. Yuan CG, Shi JB, He B, Liu JF, Liang LN, Jiang GB (2004) Speciation of heavy metals in marine sediments from the East China Sea by ICP-MS with sequential extraction. Environ Int 30(6):769–783. doi:10.1016/j.envint.2004.01.001
8. Mateos R, Espartero JL, Trujillo M, Rios JJ, Leon M, Alcudia F, Cert A (2001) Determination of phenols, flavones, and lignans in virgin olive oils by solid-phase extraction and high-performance liquid chromatography with diode array ultraviolet detection. J Agric Food Chem 49(5):2185–2192. doi:10.1021/jf0013205
9. Carabias-Martinez R, Rodriguez E, Fernandez ME, Calvo L, Sanchez FJ (2003) Evolution over time of the agricultural pollution of waters in an area of Salamanca and Zamora (Spain). Water Res 37(4):928–938. doi:10.1016/s0043-1354(02)00366-4
10. Galloway TS, Sanger RC, Smith KL, Fillmann G, Readman JW, Ford TE, Depledge MH (2002) Rapid assessment of marine pollution using multiple biomarkers and chemical immunoassays. Environ Sci Technol 36(10):2219–2226. doi:10.1021/es010300w
11. Basheer C, Balasubramanian R, Lee HK (2003) Determination of organic micropollutants in rainwater using hollow fiber membrane/liquid-phase microextraction combined with gas chromatography–mass spectrometry. J Chromatogr A 1016(1):11–20. doi:10.1016/s0021-9673(03)01295-0
12. Sturgeon RE, Willie SN, Berman SS (1985) Preconcentration of selenium and antimony from seawater for determination by graphite-furnace atomic-absorption spectrometry. Anal Chem 57(1):6–9. doi:10.1021/ac00279a006
13. Kalcher K, Kauffmann JM, Wang J, Svancara I, Vytras K, Neuhold C, Yang Z (1995) Sensors based on carbon-paste in electrochemical analysis – a review with particular emphasis on the period 1990–1993. Electroanalysis 7(1):5–22. doi:10.1002/elan.1140070103
14. Karton N, Segal E, Omer L, Portnoy M, Satchi R, Shabat D (2011) A unique paradigm for a turn-on near-infrared cyanine-based probe: noninvasive intravital optical imaging of hydrogen peroxide. J Am Chem Soc 133(28):10960–10965. doi:10.1021/ja203145v

15. Janshoff A, Galla HJ, Steinem C (2000) Piezoelectric mass-sensing devices as biosensors-an alternative to optical biosensors? Angew Chem-Int Ed 39(22):4004–4032. doi:10.1002/1521-3773(20001117)39:22<4004::aid-anie4004>3.0.co;2-2
16. Khin MM, Nair AS, Babu VJ, Murugan R, Ramakrishna S (2012) A review on nanomaterials for environmental remediation. Energy Environ Sci 5(8):8075–8109. doi:10.1039/c2ee21818f
17. Zhang LD, Fang M (2010) Nanomaterials in pollution trace detection and environmental improvement. Nano Today 5(2):128–142. doi:10.1016/j.nantod.2010.03.002
18. Su S, Wu W, Gao J, Lu J, Fan C (2012) Nanomaterials-based sensors for applications in environmental monitoring. J Mater Chem 22(35):18101–18110. doi:10.1039/c2jm33284a
19. Huang ZM, Zhang YZ, Kotaki M, Ramakrishna S (2003) A review on polymer nanofibers by electrospinning and their applications in nanocomposites. Compos Sci Technol 63(15):2223–2253. doi:10.1016/s0266-3538(03)00178-7
20. Zhang SG (2003) Fabrication of novel biomaterials through molecular self-assembly. Nat Biotechnol 21(10):1171–1178. doi:10.1038/nbt874
21. Barnes CP, Sell SA, Boland ED, Simpson DG, Bowlin GL (2007) Nanofiber technology: designing the next generation of tissue engineering scaffolds. Adv Drug Deliv Rev 59(14):1413–1433. doi:10.1016/j.addr.2007.04.022
22. Teo WE, Ramakrishna S (2006) A review on electrospinning design and nanofibre assemblies. Nanotechnology 17(14):R89–R106. doi:10.1088/0957-4484/17/14/r01
23. Subbiah T, Bhat GS, Tock RW, Pararneswaran S, Ramkumar SS (2005) Electrospinning of nanofibers. J Appl Polym Sci 96(2):557–569. doi:10.1002/app.21481
24. Lin J, Wang X, Ding B, Yu J, Sun G, Wang M (2012) Biomimicry via electrospinning. Crit Rev Solid State 37(2):94–114. doi:10.1080/10408436.2011.627096
25. Ding B, Li C, Miyauchi Y, Kuwaki O, Shiratori S (2006) Formation of novel 2D polymer nanowebs via electrospinning. Nanotechnology 17(15):3685–3691. doi:10.1088/0957-4484/17/15/011
26. Wang X, Ding B, Sun G, Wang M, Yu J (2013) Electro-spinning/netting: a strategy for the fabrication of three-dimensional polymer nano-fiber/nets. Prog Mater Sci 58(8):1173–1243, doi:http://dx.doi.org/10.1016/j.pmatsci.2013.05.001
27. Ding B, Wang M, Yu J, Sun G (2009) Gas sensors based on electrospun nanofibers. Sensors 9(3):1609–1624. doi:10.3390/s90301609
28. Ding B, Wang M, Wang X, Yu J, Sun G (2010) Electrospun nanomaterials for ultrasensitive sensors. Mater Today 13(11):16–27. doi:10.1016/S1369-7021(10)70200-5
29. Kang E, Kim M, Oh JS, Park DW, Shim SE (2012) Electrospun $BMIMPF_6$/nylon 6,6 nanofiber chemiresistors as organic vapour sensors. Macromol Res 20(4):372–378. doi:10.1007/s13233-012-0043-0
30. Wang W, Huang H, Li Z, Zhang H, Wang Y, Zheng W, Wang C (2008) Zinc oxide nanofiber gas sensors via electrospinning. J Am Ceram Soc 91(11):3817–3819. doi:10.1111/j.1551-2916.2008.02765.x
31. Huang JS, Wang DW, Hou HQ, You TY (2008) Electrospun palladium nanoparticle-loaded carbon nanofibers and their electrocatalytic activities towards hydrogen peroxide and NADH. Adv Funct Mater 18(3):441–448. doi:10.1002/adfm.200700729
32. Ding Y, Liu Y, Parisi J, Zhang L, Lei Y (2011) A novel NiO-Au hybrid nanobelts based sensor for sensitive and selective glucose detection. Biosens Bioelectron 28(1):393–398. doi:10.1016/j.bios.2011.07.054
33. Jung L, Balakrishnan S, Jaehwa C, Seong J, Jong K (2011) Detection of adulterated gasoline using colorimetric organic microfibers. J Mater Chem 21(8):2648–2655. doi:10.1039/c0jm02287j
34. Manesh KM, Santhosh P, Gopalan A, Lee KP (2007) Electrospun poly(vinylidene fluoride)/poly(aminophenylboronic acid) composite nanofibrous membrane as a novel glucose sensor. Anal Biochem 360(2):189–195. doi:10.1016/j.ab.2006.09.021
35. Wang X, Ding B, Yu J, Si Y, Yang S, Sun G (2011) Electro-netting: fabrication of two-dimensional nano-nets for highly sensitive trimethylamine sensing. Nanoscale 3(3):911–915. doi:10.1039/c0nr00783h

36. Ding B, Si Y, Wang X, Yu J, Feng L, Sun G (2011) Label-free ultrasensitive colorimetric detection of copper(II) ions utilizing polyaniline/polyamide-6 nano-fiber/net sensor strips. J Mater Chem 21(35):13345–13353. doi:10.1039/c1jm11851j
37. Bin D, Xianfeng W, Jianyong Y, Moran W (2011) Polyamide 6 composite nano-fiber/net functionalized by polyethyleneimine on quartz crystal microbalance for highly sensitive formaldehyde sensors. J Mater Chem 21(34):12784–12792. doi:10.1039/c1jm11847a
38. Henry Wohltjen RD (1979) Surface acoustic wave probe for chemical analysis. Anal Chem 51(9):1458–1464. doi:10.1021/ac50045a024
39. Liu S, Sun H, Nagarajan R, Kumar J, Gu Z, Cho J, Kurup P (2011) Dynamic chemical vapor sensing with nanofibrous film based surface acoustic wave sensors. Sens Actuator A Phys 167(1):8–13. doi:10.1016/j.sna.2011.02.007
40. Wang X, Ding B, Yu J, Wang M (2011) Highly sensitive humidity sensors based on electro-spinning/netting a polyamide 6 nano-fiber/net modified by polyethyleneimine. J Mater Chem 21(40):16231–16238. doi:10.1039/c1jm13037d
41. Sun M, Ding B, Yu J (2012) Sensitive metal ion sensors based on fibrous polystyrene membranes modified by polyethyleneimine. RSC Adv 2(4):1373–1378. doi:10.1039/c1ra00673h
42. Lin Q, Li Y, Yang M (2012) Highly sensitive and ultrafast response surface acoustic wave humidity sensor based on electrospun polyaniline/poly(vinyl butyral) nanofibers. Anal Chim Acta 748:73–80. doi:10.1016/j.aca.2012.08.041
43. He XL, Arsat R, Sadek AZ, Wlodarski W, Kalantar K, Li JP (2010) Electrospun PVP fibers and gas sensing properties of PVP/36 deg YX $LiTaO_3$ SAW device. Sens Actuator B Chem 145(2):674–679. doi:10.1016/j.snb.2010.01.013
44. Buttry DA, Ward MD (1992) Measurement of interfacial processes at electrode surfaces with the electrochemical quartz crystal microbalance. Chem Rev 92(6):1355–1379. doi:10.1021/cr00014a006
45. Yoo HY, Bruckenstein S (2013) A novel quartz crystal microbalance gas sensor based on porous film coatings. A high sensitivity porous poly(methylmethacrylate) water vapor sensor. Anal Chim Acta 785:98–103. doi:10.1016/j.aca.2013.04.052
46. Vashist SK, Vashist P (2011) Recent advances in quartz crystal microbalance-based sensors. J Sens 2011:571405–571418. doi:10.1155/2011/571405
47. Wang X, Cui F, Lin J, Ding B, Yu J, Al SS (2012) Functionalized nanoporous TiO_2 fibers on quartz crystal microbalance platform for formaldehyde sensor. Sens Actuator B Chem 171–172:658–665. doi:10.1016/j.snb.2012.05.050
48. Hu W, Chen S, Liu L, Ding B, Wang H (2011) Formaldehyde sensors based on nanofibrous polyethyleneimine/bacterial cellulose membranes coated quartz crystal microbalance. Sens Actuator B Chem 157(2):554–559. doi:10.1016/j.snb.2011.05.021
49. Wang X, Wang J, Si Y, Ding B, Yu J, Sun G, Luo W, Zheng G (2012) Nanofiber-net-binary structured membranes for highly sensitive detection of trace HCl gas. Nanoscale 4(23):7585–7592. doi:10.1039/c2nr32730a
50. Wang X, Ding B, Sun M, Yu J, Sun G (2010) Nanofibrous polyethyleneimine membranes as sensitive coatings for quartz crystal microbalance-based formaldehyde sensors. Sens Actuator B Chem 144(1):11–17. doi:10.1016/j.snb.2009.08.023
51. Sabri YM, Kojima R, Ippolito SJ, Wlodarski W, Kalantar K, Kaner RB, Bhargava SK (2011) QCM based mercury vapor sensor modified with polypyrrole supported palladium. Sens Actuator B Chem 160(1):616–622. doi:10.1016/j.snb.2011.08.038
52. Pei Z, Ma X, Ding P, Zhang W, Luo Z, Li G (2010) Study of a QCM dimethyl methylphosphonate sensor based on a ZnO-modified nanowire-structured manganese dioxide film. Sensors 10(9):8275–8290. doi:10.3390/s100908275
53. Kim ID, Rothschild A, Lee BH, Kim DY, Jo SM, Tuller HL (2006) Ultrasensitive chemiresistors based on electrospun TiO_2 nanofibers. Nano Lett 6(9):2009–2013. doi:10.1021/nl061197h
54. Choi SH, Hwang IS, Lee JH, Oh SG, Kim ID (2011) Microstructural control and selective C_2H_5OH sensing properties of Zn_2SnO_4 nanofibers prepared by electrospinning. Chem Commun 47(33):9315–9317. doi:10.1039/c1cc10707k

55. Yoon JW, Choi JK, Lee JH (2012) Design of a highly sensitive and selective C_2H_5OH sensor using p-type Co_3O_4 nanofibers. Sens Actuator B Chem 161(1):570–577. doi:10.1016/j.snb.2011.11.002
56. Wang G, Ji Y, Huang X, Yang X, Gouma PI, Dudley M (2006) Fabrication and characterization of polycrystalline WO_3 nanofibers and their application for ammonia sensing. J Phys Chem B 110(47):23777–23782. doi:10.1021/jp0635819
57. Zhang Y, He X, Li J, Miao Z, Huang F (2008) Fabrication and ethanol-sensing properties of micro gas sensor based on electrospun SnO_2 nanofibers. Sens Actuator B Chem 132(1): 67–73. doi:10.1016/j.snb.2008.01.006
58. Leng JY, Xu XJ, Lv N, Fan HT, Zhang T (2011) Synthesis and gas-sensing characteristics of WO_3 nanofibers via electrospinning. J Colloid Interface Sci 356(1):54–57. doi:10.1016/j.jcis.2010.11.079
59. Ruggieri F, Di Camillo D, Lozzi L, Santucci S, De A, Ferri G, Giancaterini L, Cantalini C (2013) Preparation of nitrogen doped TiO_2 nanofibers by near field electrospinning (NFES) technique for NO_2 sensing. Sens Actuator B Chem 179:107–113. doi:10.1016/j.snb.2012.10.094
60. Choi SW, Park JY, Kim SS (2011) Dependence of gas sensing properties in ZnO nanofibers on size and crystallinity of nanograins. J Mater Res 26(14):1662–1665. doi:10.1557/jmr.2011.209
61. Lee HU, Ahn K, Lee SJ, Kim JP, Kim HG, Jeong SY, Cho CR (2011) ZnO nanobarbed fibers: fabrication, sensing NO_2 gas, and their sensing mechanism. Appl Phys Lett 98(19). doi:10.1063/1.3590202
62. Zhang Z, Li X, Wang C, Wei L, Liu Y, Shao C (2009) ZnO hollow nanofibers: fabrication from facile single capillary electrospinning and applications in gas sensors. J Phys Chem C 113(45):19397–19403. doi:10.1021/jp9070373
63. Tuan N, Park S, Kim JB, Kim TK, Seong GH, Choo J, Kim YS (2011) Polycrystalline tungsten oxide nanofibers for gas-sensing applications. Sens Actuator B Chem 160(1): 549–554. doi:10.1016/j.snb.2011.08.028
64. Qi Q, Zhang T, Liu L, Zheng X (2009) Synthesis and toluene sensing properties of SnO_2 nanofibers. Sens Actuator B Chem 137(2):471–475. doi:10.1016/j.snb.2008.11.042
65. Zheng W, Li Z, Zhang H, Wang W, Wang Y, Wang C (2009) Electrospinning route for α-Fe_2O_3 ceramic nanofibers and their gas sensing properties. Mater Res Bull 44(6):1432–1436. doi:10.1016/j.materresbull.2008.12.013
66. Fan HT, Xu XJ, Ma XK, Zhang T (2011) Preparation of $LaFeO_3$ nanofibers by electrospinning for gas sensors with fast response and recovery. Nanotechnology 22(11):115502–115509. doi:10.1088/0957-4484/22/11/115502
67. Zheng W, Lu X, Wang W, Dong B, Zhang H, Wang Z, Xu X, Wang C (2010) A rapidly responding sensor for methanol based on electrospun In_2O_3-SnO_2 nanofibers. J Am Ceram Soc 93(1):15–17. doi:10.1111/j.1551-2916.2009.03354.x
68. Song X, Liu L (2009) Characterization of electrospun ZnO-SnO_2 nanofibers for ethanol sensor. Sens Actuator A Phys 154(1):175–179. doi:10.1016/j.sna.2009.06.010
69. Wei S, Zhang Y, Zhou M (2011) Toluene sensing properties of SnO_2-ZnO hollow nanofibers fabricated from single capillary electrospinning. Solid State Commun 151(12):895–899. doi:10.1016/j.ssc.2011.03.031
70. Wang Z, Li Z, Sun J, Zhang H, Wang W, Zheng W, Wang C (2010) Improved hydrogen monitoring properties based on p-NiO/n-SnO_2 heterojunction composite nanofibers. J Phys Chem C 114(13):6100–6105. doi:10.1021/jp9100202
71. Xu L, Zheng R, Liu S, Song J, Chen J, Dong B, Song H (2012) NiO@ZnO heterostructured nanotubes: coelectrospinning fabrication, characterization, and highly enhanced gas sensing properties. Inorg Chem 51(14):7733–7740. doi:10.1021/ic300749a
72. Liu L, Guo C, Li S, Wang L, Dong Q, Li W (2010) Improved H_2 sensing properties of Co-doped SnO_2 nanofibers. Sens Actuator B Chem 150(2):806–810. doi:10.1016/j.snb.2010.07.022

73. Zhang HN, Li ZY, Liu L, Xu XR, Wang ZJ, Wang W, Zheng W, Dong B, Wang C (2010) Enhancement of hydrogen monitoring properties based on Pd–SnO_2 composite nanofibers. Sensor Actuator B Chem 147(1):111–115. doi:10.1016/j.snb.2010.01.056
74. Cho NG, Yang DJ, Jin MJ, Kim HG, Tuller HL, Kim ID (2011) Highly sensitive SnO_2 hollow nanofiber-based NO_2 gas sensors. Sensor Actuator B Chem 160(1):1468 1472. doi:10.1016/j.snb.2011.07.035
75. Cho S, Kim DH, Lee BS, Jung J, Yu WR, Hong SH, Lee S (2012) Ethanol sensors based on ZnO nanotubes with controllable wall thickness via atomic layer deposition an O_2 plasma process and an annealing process. Sens Actuator B Chem 162(1):300–306. doi:10.1016/j.snb.2011.12.081
76. Modafferi V, Panzera G, Donato A, Antonucci PL, Cannilla C, Donato N, Spadaro D, Neri G (2012) Highly sensitive ammonia resistive sensor based on electrospun V_2O_5 fibers. Sens Actuator B Chem 163(1):61–68. doi:10.1016/j.snb.2012.01.007
77. Zhang Y, Li J, An G, He X (2010) Highly porous SnO_2 fibers by electrospinning and oxygen plasma etching and its ethanol-sensing properties. Sens Actuator B Chem 144(1):43–48. doi:10.1016/j.snb.2009.10.012
78. Zhang XJ, Qiao GJ (2012) High performance ethanol sensing films fabricated from ZnO and In_2O_3 nanofibers with a double-layer structure. Appl Surf Sci 258(17):6643–6647, doi:http://dx.doi.org/10.1016/j.apsusc.2012.03.098
79. Choi SH, Choi SJ, Min BK, Lee WY, Park JS, Kim ID (2013) Facile synthesis of p-type perovskite $SrTi_{0.65}Fe_{0.35}O_3$ nanofibers prepared by electrospinning and their oxygen-sensing properties. Macromol Mater Eng 298(5):521–527. doi:10.1002/mame.201200375
80. Hwang DK, Kim S, Lee JH, Hwang IS, Kim ID (2011) Phase evolution of perovskite $LaNiO_3$ nanofibers for supercapacitor application and p-type gas sensing properties of LaOCl-NiO composite nanofibers. J Mater Chem 21(6):1959–1965. doi:10.1039/c0jm02256j
81. Moon J, Park J, Lee S, Zyung T, Kim I (2010) Pd-doped TiO_2 nanofiber networks for gas sensor applications. Sensor Actuator B Chem 149(1):301–305. doi:10.1016/j.snb.2010.06.033
82. Wang W, Li Z, Liu L, Zhang H, Zheng W, Wang Y, Huang H, Wang Z, Wang C (2009) Humidity sensor based on LiCl-doped ZnO electrospun nanofibers. Sens Actuator B Chem 141(2):404–409. doi:10.1016/j.snb.2009.06.029
83. Wang X, Zhao M, Liu F, Jia J, Li X, Cao L (2013) C_2H_2 gas sensor based on Ni-doped ZnO electrospun nanofibers. Ceram Int 39(3):2883–2887. doi:10.1016/j.ceramint.2012.09.062
84. Zhang H, Li Z, Liu L, Wang C, Wei Y, MacDiarmid AG (2009) Mg^{2+}/Na^{+}-doped rutile TiO_2 nanofiber mats for high-speed and anti-fogged humidity sensors. Talanta 79(3):953–958. doi:10.1016/j.talanta.2009.05.035
85. Xu L, Xing R, Song J, Xu W, Song H (2013) ZnO-SnO_2 nanotubes surface engineered by Ag nanoparticles: synthesis, characterization, and highly enhanced HCHO gas sensing properties. J Mater Chem C 1(11):2174–2182. doi:10.1039/c3tc00689a
86. Zhang L, Wang X, Zhao Y, Zhu Z, Fong H (2012) Electrospun carbon nano-felt surface-attached with Pd nanoparticles for hydrogen sensing application. Mater Lett 68:133–136. doi:10.1016/j.matlet.2011.10.064
87. Li Z, Zhang H, Zheng W, Wang W, Huang H, Wang C, MacDiarmid AG, Wei Y (2008) Highly sensitive and stable humidity nanosensors based on LiCl doped TiO_2 electrospun nanofibers. J Am Chem Soc 130(15):5036–5037. doi:10.1021/ja800176s
88. Casalbore G, Yang MJ, Camaioni N, Mari CM, Li Y, Sun H, Ling M (2000) Investigations on the ion transport mechanism in conducting polymer films. Solid State Ion 131(3–4):311–321. doi:10.1016/s0167-2738(00)00688-3
89. Lin Q, Li Y, Yang M (2012) Polyaniline nanofiber humidity sensor prepared by electrospinning. Sens Actuator B Chem 161(1):967–972. doi:10.1016/j.snb.2011.11.074
90. Liu HQ, Kameoka J, Czaplewski DA, Craighead HG (2004) Polymeric nanowire chemical sensor. Nano Lett 4(4):671–675. doi:10.1021/nl049826f
91. Li P, Li Y, Ying B, Yang M (2009) Electrospun nanofibers of polymer composite as a promising humidity sensitive material. Sens Actuator B Chem 141(2):390–395. doi:10.1016/j.snb.2009.07.006

92. Wang S, Chao D, Berda EB, Jia X, Yang R, Wang X, Jiang T, Wang C (2013) Fabrication of electroactive oligoaniline functionalized poly(amic acid) nanofibers for application as an ammonia sensor. RSC Adv 3(12):4059–4065. doi:10.1039/c3ra00056g
93. Aussawasathien D, Sahasithiwat S, Menbangpung L (2008) Electrospun camphorsulfonic acid doped poly(o-toluidine)-polystyrene composite fibers: chemical vapor sensing. Synthetic Met 158(7):259–263. doi:10.1016/j.synthmet.2008.01.007
94. Haynes AS, Gouma PI (2008) Electrospun conducting polymer-based sensors for advanced pathogen detection. IEEE Sens J 8(5–6):701–705. doi:10.1109/jsen.2008.923039
95. Rojas R, Pinto NJ (2008) Using electrospinning for the fabrication of rapid response gas sensors based on conducting polymer nanowires. IEEE Sens J 8(5–6):951–953. doi:10.1109/jsen.2008.923932
96. Pinto NJ, Rivera D, Melendez A, Ramos I, Lim JH, Johnson ATC (2011) Electrical response of electrospun PEDOT-PSSA nanofibers to organic and inorganic gases. Sens Actuator B Chem 156(2):849–853. doi:10.1016/j.snb.2011.02.053
97. Aussawasathien D, Sahasithiwat S, Menbangpung L, Teerawattananon C (2011) Poly(o-anisidine)-polystyrene composite fibers via electrospinning process: surface morphology and chemical vapor sensing. Sens Actuator B Chem 151(2):341–350. doi:10.1016/j.snb.2010.07.048
98. Chen D, Lei S, Chen Y (2011) A single polyaniline nanofiber field effect transistor and its gas sensing mechanisms. Sensors 11(7):6509–6516. doi:10.3390/s110706509
99. Bai H, Zhao L, Lu C, Li C, Shi G (2009) Composite nanofibers of conducting polymers and hydrophobic insulating polymers: preparation and sensing applications. Polymer 50(14):3292–3301. doi:10.1016/j.polymer.2009.04.066
100. Wang Y, Jia W, Strout T, Schempf A, Zhang H, Li B, Cui J, Lei Y (2009) Ammonia gas sensor using polypyrrole-coated TiO_2/ZnO nanofibers. Electroanalysis 21(12):1432–1438. doi:10.1002/elan.200904584
101. Jaewon C, Eun Joo P, Dong Wha P, Sang Eun S (2010) MWCNT-OH adsorbed electrospun nylon 6,6 nanofibers chemiresistor and their application in low molecular weight alcohol vapours sensing. Synthetic Met 160(23–24):2664–2669. doi:10.1016/j.synthmet.2010.10.022
102. Eaidkong T, Mungkarndee R, Phollookin C, Tumcharern G, Sukwattanasinitt M, Wacharasindhu S (2012) Polydiacetylene paper-based colorimetric sensor array for vapor phase detection and identification of volatile organic compounds. J Mater Chem 22(13):5970–5977. doi:10.1039/c2jm16273c
103. Chae SK, Park H, Yoon J, Lee CH, Ahn DJ, Kim JM (2007) Polydiacetylene supramolecules in electrospun microfibers: fabrication, micropatterning, and sensor applications. Adv Mater 19(4):521–524. doi:10.1002/adma.200602012
104. Wang X, Si Y, Wang J, Ding B, Yu J, Al SS (2012) A facile and highly sensitive colorimetric sensor for the detection of formaldehyde based on electro-spinning/netting nano-fiber/nets. Sens Actuator B Chem 163(1):186–193. doi:10.1016/j.snb.2012.01.033
105. Yarimaga O, Jaworski J, Yoon B, Kim JM (2012) Polydiacetylenes: supramolecular smart materials with a structural hierarchy for sensing, imaging and display applications. Chem Commun 48(19):2469–2485. doi:10.1039/c2cc17441c
106. Pumtang S, Siripornnoppakhun W, Sukwattanasinitt M, Ajavakom A (2011) Solvent colorimetric paper-based polydiacetylene sensors from diacetylene lipids. J Colloid Interface Sci 364(2):366–372. doi:10.1016/j.jcis.2011.08.074
107. Yoon J, Chae SK, Kim JM (2007) Colorimetric sensors for volatile organic compounds (VOCs) based on conjugated polymer-embedded electrospun fibers. J Am Chem Soc 129(11):3038–3039. doi:10.1021/ja067856+
108. Yoon J, Jung YS, Kim JM (2009) A combinatorial approach for colorimetric differentiation of organic solvents based on conjugated polymer-embedded electrospun fibers. Adv Funct Mater 19(2):209–214. doi:10.1002/adfm.200800963
109. Poltue T, Rangkupan R, Dubas ST, Dubas L (2011) Nickel(II) ions sensing properties of dimethylglyoxime/poly(caprolactone) electrospun fibers. Mater Lett 65(14):2231–2234. doi:10.1016/j.matlet.2011.04.012

110. Li Y, Si Y, Wang X, Ding B, Sun G, Zheng G, Luo W, Yu J (2013) Colorimetric sensor strips for lead (II) assay utilizing nanogold probes immobilized polyamide-6/nitrocellulose nano-fibers/nets. Biosens Bioelectron 48:244–250. doi:10.1016/j.bios.2013.03.085
111. Van der Schueren L, Mollet T, Ceylan Ö, De Clerck K (2010) The development of polyamide 6.6 nanofibres with a pH-sensitive function by electrospinning. Eur Polym J 46(12):2229–2239, doi:http://dx.doi.org/10.1016/j.eurpolymj.2010.09.016
112. Van der Schueren L, De Meyer T, Steyaert I, Ceylan O, Hemelsoet K, Van Speybroeck V, De Clerck K (2013) Polycaprolactone and polycaprolactone/chitosan nanofibres functionalised with the pH-sensitive dye Nitrazine Yellow. Carbohyd Polym 91(1):284–293. doi:10.1016/j.carbpol.2012.08.003
113. Valeur B, Leray I (2000) Design principles of fluorescent molecular sensors for cation recognition. Coord Chem Rev 205(1):3–40. doi:10.1016/s0010-8545(00)00246-0
114. Wang M, Meng G, Huang Q, Qian Y (2012) Electrospun 1,4-DHAQ-doped cellulose nanofiber films for reusable fluorescence detection of trace Cu^{2+} and further for Cr^{3+}. Environ Sci Technol 46(1):367–373. doi:10.1021/es202137c
115. Basabe L, Reinhoudt DN, Crego M (2007) Design of fluorescent materials for chemical sensing. Chem Soc Rev 36(6):993–1017. doi:10.1039/b609548h
116. Adewuyi S, Ondigo DA, Zugle R, Tshentu Z, Nyokong T, Torto N (2012) A highly selective and sensitive pyridylazo-2-naphthol-poly(acrylic acid) functionalized electrospun nanofiber fluorescence "turn-off" chemosensory system for Ni^{2+}. Anal Method 4(6):1729–1735. doi:10.1039/c2ay25182e
117. Wang W, Yang Q, Sun L, Wang H, Zhang C, Fei X, Sun M, Li Y (2011) Preparation of fluorescent nanofibrous film as a sensing material and adsorbent for Cu^{2+} in aqueous solution via copolymerization and electrospinning. J Hazard Mater 194:185–192. doi:10.1016/j.jhazmat.2011.07.083
118. Wang W, Wang X, Yang Q, Fei X, Sun M, Song Y (2013) A reusable nanofibrous film chemosensor for highly selective and sensitive optical signaling of Cu^{2+} in aqueous media. Chem Commun 49(42):4833–4835. doi:10.1039/c3cc41317a
119. Davis BW, Niamnont N, Hare CD, Sukwattanasinitt M, Cheng Q (2010) Nanofibers doped with dendritic fluorophores for protein detection. ACS Appl Mater Interface 2(7):1798–1803. doi:10.1021/am100345g
120. Long Y, Chen H, Yang Y, Wang H, Yang Y, Li N, Li K, Pei J, Liu F (2009) Electrospun nanofibrous film doped with a conjugated polymer for DNT fluorescence sensor. Macromolecules 42(17):6501–6509. doi:10.1021/ma900756w
121. Wang XY, Kim YG, Drew C, Ku BC, Kumar J, Samuelson LA (2004) Electrostatic assembly of conjugated polymer thin layers on electrospun nanofibrous membranes for biosensors. Nano Lett 4(2):331–334. doi:10.1021/nl034885z
122. Silva AP, Fox DB, Huxley AJM, Moody TS (2000) Combining luminescence, coordination and electron transfer for signalling purposes. Coord Chem Rev 205(1):41–57. doi:10.1016/s0010-8545(00)00238-1
123. Prodi L, Bolletta F, Montalti M, Zaccheroni N (2000) Luminescent chemosensors for transition metal ions. Coord Chem Rev 205(1):59–83. doi:10.1016/s0010-8545(00)00242-3
124. Keefe MH, Benkstein KD, Hupp JT (2000) Luminescent sensor molecules based on coordinated metals: a review of recent developments. Coord Chem Rev 205(1):201–228. doi:10.1016/s0010-8545(00)00240-x
125. Demas JN, DeGraff BA, Coleman PB (1999) Oxygen sensors based on luminescence quenching. Anal Chem 71(23):793A–800A. doi:10.1021/ac9908546
126. Zhou C, Shi Y, Ding X, Li M, Luo J, Lu Z, Xiao D (2013) Development of a fast and sensitive glucose biosensor using iridium complex-doped electrospun optical fibrous membrane. Anal Chem 85(2):1171–1176. doi:10.1021/ac303107d
127. Wang LY, Xu Y, Lin Z, Zhao N, Xu YH (2011) Electrospinning fabrication and oxygen sensing properties of Cu(I) complex-polystyrene composite microfibrous membranes. J Lumin 131(7):1277–1282. doi:10.1016/j.jlumin.2011.03.017

128. Kuo CC, Tung YC, Chen WC (2010) Morphology and pH sensing characteristics of new luminescent electrospun fibers prepared from poly(phenylquinoline)-block-polystyrene/polystyrene blends. Macromol Rapid Commun 31(1):65–70. doi:10.1002/marc.200900566
129. Liu CL, Lin CH, Kuo CC, Lin ST, Chen WC (2011) Conjugated rod-coil block copolymers: synthesis, morphology, photophysical properties, and stimuli-responsive applications. Prog Polym Sci 36(5):603–637. doi:10.1016/j.progpolymsci.2010.07.008
130. Kind H, Yan HQ, Messer B, Law M, Yang PD (2002) Nanowire ultraviolet photodetectors and optical switches. Adv Mater 14(2):158–160. doi:10.1002/1521-4095(20020116)14:2<158::aid-adma158>3.0.co;2-w
131. Lin D, Wu H, Zhang W, Li H, Pan W (2009) Enhanced UV photoresponse from heterostructured Ag-ZnO nanowires. Appl Phys Lett 94(17). doi:10.1063/1.3126045
132. Wu H, Sun Y, Lin D, Zhong R, Zhang C, Pan W (2009) GaN nanofibers based on electrospinning: facile synthesis, controlled assembly, precise doping, and application as high performance UV photodetector. Adv Mater 21(2):227–231. doi:10.1002/adma.200800529
133. Lai C, Wang X, Zhao Y, Fong H, Zhu Z (2013) Effects of humidity on the ultraviolet nanosensors of aligned electrospun ZnO nanofibers. RSC Adv 3(18):6640–6645. doi:10.1039/c3ra23420g
134. Anitha S, Brabu B, Rajesh KP, Natarajan TS (2013) Fabrication of UV sensor based on electrospun composite fibers. Mater Lett 92:417–420. doi:10.1016/j.matlet.2012.11.005
135. Li Y, Gong J, He G, Deng Y (2012) Enhancement of photoresponse and UV-assisted gas sensing with Au decorated ZnO nanofibers. Mater Chem Phys 134(2–3):1172–1178. doi:10.1016/j.matchemphys.2012.04.014
136. Zhu ZT, Zhang LF, Howe JY, Liao YL, Speidel JT, Smith S, Fong H (2009) Aligned electrospun ZnO nanofibers for simple and sensitive ultraviolet nanosensors. Chem Commun 18:2568–2570. doi:10.1039/b901426h
137. Garrido JA, Monroy E, Izpura I, Munoz E (1998) Photoconductive gain modelling of GaN photoconductors. Semicond Sci Technol 13(6):563–568. doi:10.1088/0268-1242/13/6/005
138. Avnir D, Braun S, Lev O, Ottolenghi M (1994) Enzymes and other proteins entrapped in sol–gel materials. Chem Mater 6(10):1605–1614. doi:10.1021/cm00046a008
139. Im JS, Kim JG, Bae TS, Yu HR, Lee YS (2011) Surface modification of electrospun spherical activated carbon for a high-performance biosensor electrode. Sens Actuator B Chem 158(1):151–158. doi:10.1016/j.snb.2011.05.058
140. Scampicchio M, Arecchi A, Lawrence NS, Mannino S (2010) Nylon nanofibrous membrane for mediated glucose biosensing. Sens Actuator B Chem 145(1):394–397. doi:10.1016/j.snb.2009.12.042
141. Wilson R, Turner APF (1992) Glucose-oxidase-an ideal enzyme. Biosens Bioelectron 7(3):165–185. doi:10.1016/0956-5663(92)87013-f
142. Im JS, Yun J, Kim JG, Bae TS, Lee YS (2012) The effects of carbon nanotube addition and oxyfluorination on the glucose-sensing capabilities of glucose oxidase-coated carbon fiber electrodes. Appl Surf Sci 258(7):2219–2225. doi:10.1016/j.apsusc.2011.08.017
143. Ren G, Xu X, Liu Q, Cheng J, Yuan X, Wu L, Wan Y (2006) Electrospun poly(vinyl alcohol)/glucose oxidase biocomposite membranes for biosensor applications. React Funct Polym 66(12):1559–1564. doi:10.1016/j.reactfunctpolym.2006.05.005
144. Liu G, Zheng B, Jiang Y, Cai Y, Du J, Yuan H, Xiao D (2012) Improvement of sensitive CuO NFs-ITO nonenzymatic glucose sensor based on in situ electrospun fiber. Talanta 101:24–31. doi:10.1016/j.talanta.2012.08.040
145. Liu Y, Teng H, Hou H, You T (2009) Nonenzymatic glucose sensor based on renewable electrospun Ni nanoparticle-loaded carbon nanofiber paste electrode. Biosens Bioelectron 24(11):3329–3334. doi:10.1016/j.bios.2009.04.032
146. Xiao Y, Ju HX, Chen HY (1999) Hydrogen peroxide sensor based on horseradish peroxidase-labeled Au colloids immobilized on gold electrode surface by cysteamine monolayer. Anal Chim Acta 391(1):73–82. doi:10.1016/s0003-2670(99)00196-8

147. Mao X, Simeon F, Rutledge GC, Hatton TA (2013) Electrospun carbon nanofiber webs with controlled density of states for sensor applications. Adv Mater 25(9):1309–1314. doi:10.1002/adma.201203045
148. Ouyang Z, Li J, Wang J, Li Q, Ni T, Zhang X, Wang H, Li Q, Su Z, Wei G (2013) Fabrication, characterization and sensor application of electrospun polyurethane nanofibers filled with carbon nanotubes and silver nanoparticles. J Mater Chem B 1(18):2415–2424. doi:10.1039/c3tb20316f

Chapter 12
Electrospun Nanofibers for Air Filtration

Na Wang, Xue Mao, Shichao Zhang, Jianyong Yu, and Bin Ding

Abstract Fine particle pollution resulting from the rapid urbanization and industrialization is one of the most serious sources of air pollution. In response to the recognized threats of fine particles to the environment and human bodies, the pursuit of novel, high-performance, energy saving, and environmentally friendly filter medium has gained great interest in recent years. Electrospun nanofibrous membranes, as one of the utmost promising and versatile filter media for fine particle filtration, possess several fascinating features such as remarkable specific surface area, high open porosity, and interconnected porous structure. More significantly, electrospun nanofiber-based filter media are expected to have extremely high filtration efficiency for fine particle and relatively low pressure drop due to the unique structure of electrospun nanofibers. In this chapter, we summarize the recent progress in the development of electrospun nanofibrous membranes (e.g., organic, hybrid, inorganic) for fine particle filtration, describe the types of nanofibrous materials that have been developed, and discuss their structure variables and particle filtration performance in detail. This chapter may trigger the development of advanced nanofibrous filter media for fine particle emissions from anthropogenic polluted atmosphere.

N. Wang • X. Mao • S. Zhang
State Key Laboratory for Modification of Chemical Fibers and Polymer Materials,
College of Materials Science and Engineering, Donghua University,
Shanghai 201620, China

J. Yu • B. Ding (✉)
State Key Laboratory for Modification of Chemical Fibers and Polymer Materials,
College of Materials Science and Engineering, Donghua University,
Shanghai 201620, China

Nanomaterials Research Center, Modern Textile Institute, Donghua University,
Shanghai 200051, China
e-mail: binding@dhu.edu.cn

B. Ding and J. Yu (eds.), *Electrospun Nanofibers for Energy and Environmental Applications*, Nanostructure Science and Technology,
DOI 10.1007/978-3-642-54160-5_12,

12.1 Introduction

In the past several decades, the quality of the environment has seriously deteriorated due to the rapid urbanization and industrialization. Particularly, fine particles (particulate matter <2.5 μm aerodynamic diameter, PM2.5, promulgated by the US Environmental Protection Agency), as one of the most serious sources of air pollution, turned out to be a major cause of adverse health effects [1, 2]. It has both acute and chronic effects on human health, ranging from minor upper respiratory irritation to chronic respiratory and lung cancer, acute respiratory infections in children, and chronic bronchitis in adults, aggravating preexisting heart and lung diseases or asthmatic attacks [3]. Furthermore, stricter emission limits have also contributed to the urgent need for better quality of air. At a global scale, the emissions of fine particles are mainly from anthropogenic sources (e.g., industrial activities, intensive road transport, biomass burning, secondary nitrate and secondary sulfates), which can be classified as normal and medium-high-temperature particles in terms of the conditions that the particles produced [4].

Given the recognized threats of fine particles to the environment and human bodies, many prevention methods are being developed for addressing air pollution problems: source control, development of clean energy, filtration technologies, and so on [5]. Among them, air filtration technology is of great interest because of its low equipment cost, low energy consumption, high performance, and wide application range. Nowadays, nonwoven fibrous membrane based on fibers (e.g., glass, polyethylene (PE), polypropylene (PP), polyester, and aramid fibers), as a kind of porous media, has been widely used in different filtration applications, including disposable respirators, industrial gas cleaning equipment, cleanroom air purification systems, automotive cabin air filters, and indoor air purifiers [6]. However, conventional fibrous media usually suffer from many structural and performance disadvantages, such as large fiber diameter, nonuniform fiber diameter and pore size, relatively low filtration efficiency, high basis weight, and poor high-temperature resistance [7]. It is well documented that the smaller the fiber diameter, the better the filtration performances. Nanofibers boost the filtration efficiency by providing an enhanced "slip effect" that could cause less resistance against airflow, consequently leading to a smaller pressure drop across the fibrous media [8–10]. Additionally, nanofibrous filter media also show improved filter life and more contaminant holding capacity in laboratory tests and actual operating environments [11].

Up to now, a range of techniques such as template synthesis, electrospinning, drawing, self-assembly, solvothermal synthesis, phase separation, etc., have been adopted toward the design of nanofibrous membranes [12–15]. Among these techniques, electrospinning is regarded as a simple and versatile top-down approach for fabrication of uniform nanofibers in a continuous process and at a long length scale [16–20]. Electrospun nanofibrous membranes possess several fascinating features that make them very attractive in an air filtration area, such as remarkable specific surface area, high open porosity, and interconnected porous structure [6, 20–23]. Taking advantage of these intriguing characters, electrospun nanofibrous membranes are of huge interest for use in air filters. Additionally, the economic benefit of such high-performance filtration membranes applied in environmental applications is prominent day by day. There is no doubt that electrospun nanofibers have risen as a shining star in the horizon on the path of the scientists' searching for new materials for future environmental applications. In this chapter, currently emerging electrospun nanofibrous membranes are introduced according to the temperature of particle source. Meanwhile, we highlighted the recent advances of nanofibrous membranes used in air filtration. Finally, we conclude with a summary of current and future research efforts and opportunities in the development of electrospun nanofibers for air filtration applications and the prospects of future research directions.

12.2 Normal-Temperature Filter

Normal-temperature filtration media, which can be conveniently used under 150 °C for a long time, are usually regarded as the core filtration part of vehicle cabin filters, building filters, and personal respiratory. Therefore, the materials used in these filtration applications are usually natural and synthetic polymers, including biocompatible and biodegradable polymers, polymer blends, and polymers incorporated with functional materials. A comprehensive summary of materials which have been successfully fabricated into filtration membranes to date is listed in Table 12.1 for a quick and cursory review. Figure 12.1 shows the typical morphologies of electrospun filtration membranes based on different polymer systems, such as polyamide-6 (PA-6), polyacrylonitrile (PAN), poly(vinyl alcohol) (PVA),

Table 12.1 Filtration performances of various electrospun nanofibrous membranes

Materials	Test particle size (μm)	Basis weight (g/m^2)	Face velocity (cm/s)	Filtration efficiency (%)	Pressure drop (Pa)	Ref.
PA-6	0.3	10.75	5	99.993	1,470	[31]
PAN	0.3	3.895	5.3	99.999	356	[40]
PU	0.08	0.893	5.7	99.66	190	[45]
PVA	0.6	1.9	8.33	100	/	[53]
Polyethylene oxide (PEO)	0.6–180	95.4	400	99.43	17	[61]
Polycarbonate (PC)	0.3–1.8	/	2–10	>99.995	132	[64]
PAN/polymethyl methacrylate (PMMA)	0.02–0.3	3.5	5.3	99.9	30	[39]
Polyvinyl chloride (PVC)/PU	0.3	20.72	5.3	99.5	144	[47]
PA-6/boehmite	0.3	/	5.3	96.8	48	[24]
PAN/TiO_2	0.1–2.0	5.37	80	>98	/	[67]

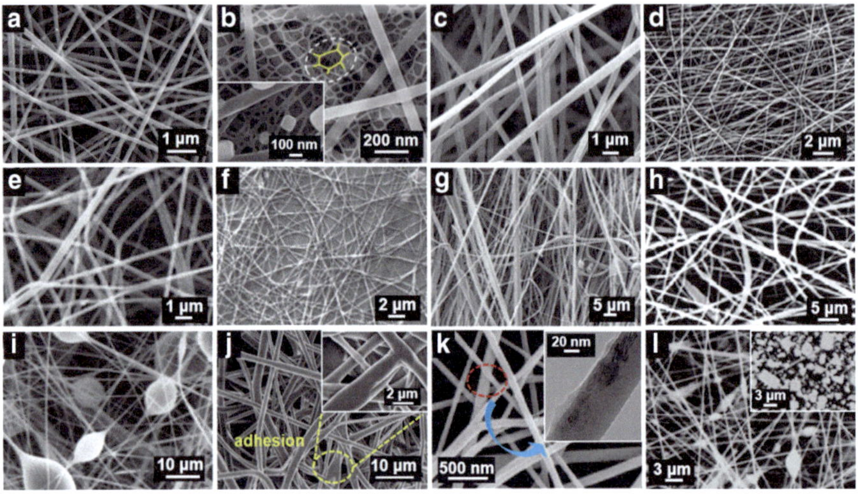

Fig. 12.1 Several typical morphologies of electrospun filtration membranes based on various systems. (**a**) PA-6, (**b**) PA-66, (**c**) PAN, (**d**) PU, (**e**) PVA, (**f**) PEO, (**g**) PC, (**h**) silk, (**i**) PAN/PMMA, (**j**) PVC/PU, (**k**) PA-6/boehmite, (**l**) PAN/TiO_2 ((**a**) Reprinted with permission from [31]. © 2008 Springer. (**b**) Reprinted with permission from [11]. © 2012 Royal Society of Chemistry. (**c**) Reprinted with permission from [40]. © 2012 Wiley Periodicals, Inc. (**d**) Reprinted with permission from [46]. © 2011 Elsevier. (**e**) Reprinted with permission from [53]. © 2008 Wiley Periodicals, Inc. (**f**) Reprinted with permission from [61]. © 2010 Elsevier. (**g**) Reprinted with permission from [64]. © 2011 Wiley Periodicals, Inc. (**h**) Reprinted with permission from [63]. © 2010 Springer. (**i**) Reprinted with permission from [39]. © 2010 Elsevier. (**j**) Reprinted with permission from [47]. © 2013 Elsevier. (**k**) Reprinted with permission from [24]. © 2010 Springer. (**l**) Reprinted with permission from [67]. © 2013 Elsevier)

polyurethane (PU), etc. In this part, we present some typical recently developed electrospun nanofibrous membranes based on various polymers and their filtration property characterizations. Particularly, three-dimensional (3D) filtration modeling and particle capture theoretical predictions based on electrospun nanofiber structure and filtration performances are proposed. Additionally, researches have also made successful attempts for functional material-polymer composite filtration membranes [24–26]. Their functionalization can offer exceptional properties in composite filter media. This section will also review these hybrid electrospun filtration membranes, illustrating their fabrication and discussing their intrinsic characteristics.

12.2.1 Organic Fibrous Filtration Membranes

12.2.1.1 Polyamide (PA)

PA is a kind of polymorphic, biocompatible, and synthetic polymeric material which presents good mechanical and physical properties. PA nanofibers can be easily fabricated in a controlled manner, and these nanofibrous membranes exhibit

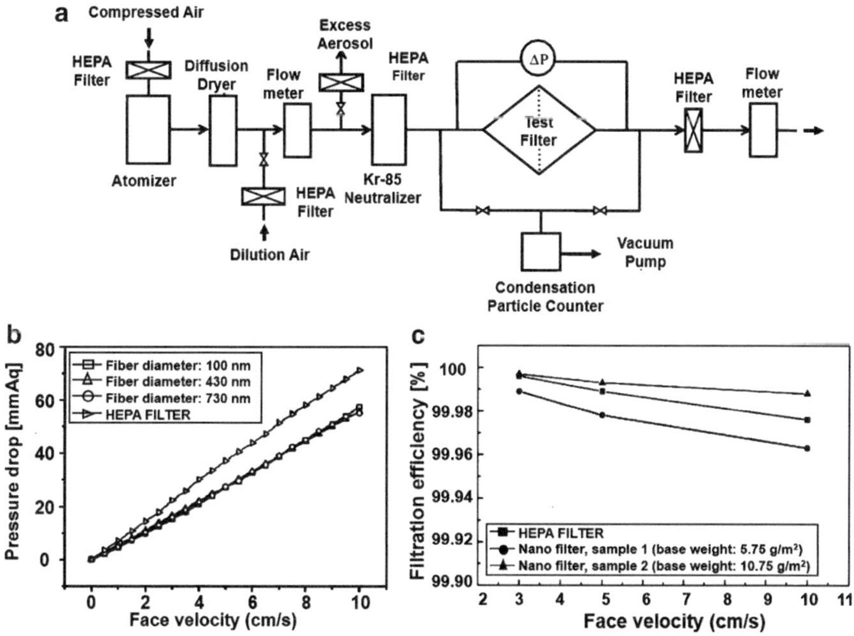

Fig. 12.2 (**a**) Schematic diagram for the filtration performance evaluation of the electrospun fibrous membranes. (**b**) Pressure drop of the PA-6 fibrous membranes and the HEPA filter as a function of face velocity for various fiber diameters. (**c**) Filtration efficiency of the PA-6 fibrous membranes and the HEPA filter as a function of particle size for various fiber diameters ((**a**) and (**b**) Reprinted with permission from [31]. © 2008 Springer. (**c**) Reprinted with permission from [28]. © 2006 Elsevier)

many interesting morphological characteristics, such as high surface area to volume ratio, high porosity, and vast possibilities for surface functionalization. These advantages make PA electrospun membranes good candidates for air filtration media [27–29]. Gibson et al. [30] first demonstrated the fabrication of PA-66 filtration membranes and revealed that electrospun nanofiber membranes were extremely efficient at trapping airborne particles (0.5–200 μm). Kim et al. [31] investigated the filtration characteristics of PA-6 nanofibrous membranes as a function of the fiber diameter ranging from 100 to 730 nm and evaluated the filtration performances of filter membranes by the filter testing apparatus, as shown in Fig. 12.2a. The PA-6 electrospun filter membrane shows a much lower pressure drop (27 mmAq) compared with the commercial high-efficiency particulate air (HEPA) filter media (37.1 mmAq), while their filtration efficiency is over 99.98 % (Fig. 12.2b). Ahn et al. [28] reported that PA-6 electrospun membranes (basis weight 10.75 g/m^2, average pore size 0.24 μm) presented higher filtration efficiency (99.993 %) than a commercial HEPA filter (99.97 %, basis weight 78.2 g/m^2, average pore size 1.7 μm) (Fig. 12.2c), implying a high filtration performance to fine particles and can be considered as an excellent candidate for air filtration applications.

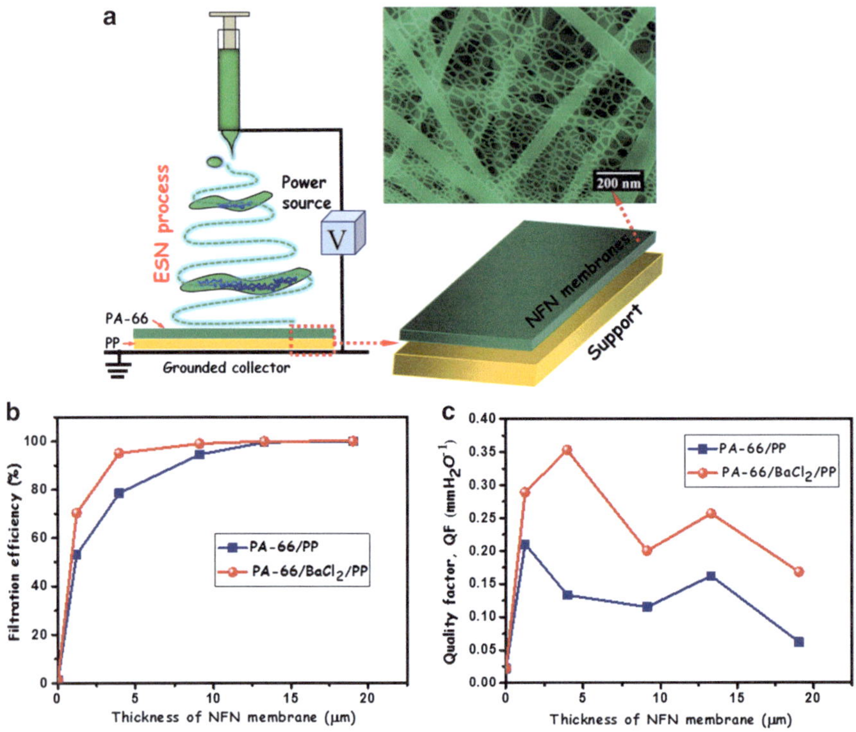

Fig. 12.3 (**a**) Schematic diagram illustrating the fabrication of PA-66 nanofiber/net membranes on nonwoven PP scaffold. (**b**) and (**c**) Filtration properties of composite membranes variation versus thickness of NFN membranes ((**a–c**) Reprinted with permission from [11]. © 2012 Royal Society of Chemistry)

Very recently, Wang et al. [11] first fabricated PA-66 nanofiber/net (NFN) membranes that comprise common electrospun nanofibers and two-dimensional (2D) spiderweb-like nano-nets via electrospinning/netting (ESN) technique (Fig. 12.3a). As the "second doorkeeper," the average pore size of nano-nets (118 nm) is smaller than the size of test NaCl particles (300 nm), so the nano-nets could filter the nanoparticles mainly through interception mechanism. The filtration efficiency of NFN/PP composite filter was 99.9 % (Fig. 12.3b, c), which can be used as the core filter medium in the air filtration field, such as respiratory protection equipment, HEPA filters, and ultralow penetration air (ULPA) filters.

Additionally, the adhesion of the nanofiber layer and support materials is another problem that needed to be solved to meet the requirement of post-processing of air filter media. Thus, a hot-press lamination process using a thermoplastic resin as glue was applied to improve the adhesion of the nanofiber layer onto the textile support. After 1,500 cycles of repeated compression and torsion, the PA-6 nanofiber layer was still firmly attached to the support, while mechanical damage is visible in some areas [32]. Recently, Guibo and co-workers [33] reported the preparation of high-efficiency PA-6 filter membrane and investigated the mechanical properties of the resultant membranes. The stress–strain measurements revealed that the heat treatment had significantly affected the mechanical properties of electrospun membranes. And the resultant nanofibrous membranes demonstrated excellent dimensional stability; the breaking tenacity and elongation reached 4.71 ± 1.66 MPa and 69.97 ± 6.56 %, respectively. Meanwhile, the permeability of PA-6 membranes with 72.9 ± 1.04 μm thickness still maintained 516 L/m^2, and the actual filtration efficiency of 0.3 μm particles reached 99.98 %, which shows the potential applications in HEPA and ULPA. In particular, Hung et al. [8] presented continuous efforts toward the aim of deeply understanding the filtration process of fibrous membranes. They proposed a semiempirical correlation on single fiber efficiency due to diffusion as follows: $h = 1.502Pe^{-0.62}$, where Pe is the Peclet number. This improved correlation offers a good agreement with experimental filtration data under large slip flow for nano-aerosols dominated by diffusion. Furthermore, they demonstrated that the filtration efficiency was decreased with the increased face velocity, and the reduction becomes larger at smaller aerosol sizes. This phenomenon could be due to the larger airflow that reduces the retention time of aerosol within the nanofiber network, thus lowering the chance for aerosol to deposit on fibers through diffusion mechanism.

12.2.1.2 Polyacrylonitrile (PAN)

PAN is used on a large scale in textile industry and one of the most suitable and widely used polymers for making high-performance functional fibers. Fitzer [34] and Harrison [35] believed that the close association of operational and physical properties is expected to play an important role in the filtration performances and permeability, both of which are critical parameters for the filter media. Additionally, the investigation of the construction and attaining uniformity of fibers in membranes provides deep understanding of the filtration mechanism of nanofiber media. Therefore, it is of fundamental interest to understand the inherent correlation of membrane structure and filtration performances. Growing attempts were reported to regulate the structure of PAN fibrous membrane to achieve the designed filtration functions. Barhate et al. [36] first reported that the pore size, packing density and alignment of fibers, and the extent of fiber crossing can be optimized by coordination of the drawing and collection rates (Fig. 12.4a–c) and concluded that the improved structural (pore size distribution, pore interconnectivity, and porosity) and transport properties of fibrous filter media can be attained. It should be noted that the differential pressure versus airflow curves of membranes collected at rotational speeds of 1.8 and 2.7 m/s show no difference, but the fibrous membranes obtained at higher rotational speed (2.7 m/s) showed lower pressure drop per thickness due to the better pore interconnectivity or networking in the vertical direction (Fig. 12.4d). Additionally, in order to investigate the effect of columbic force on the particle collection efficiency, Yun and co-workers [37] investigated the penetration of charged, uncharged, and charge-neutralized particles through

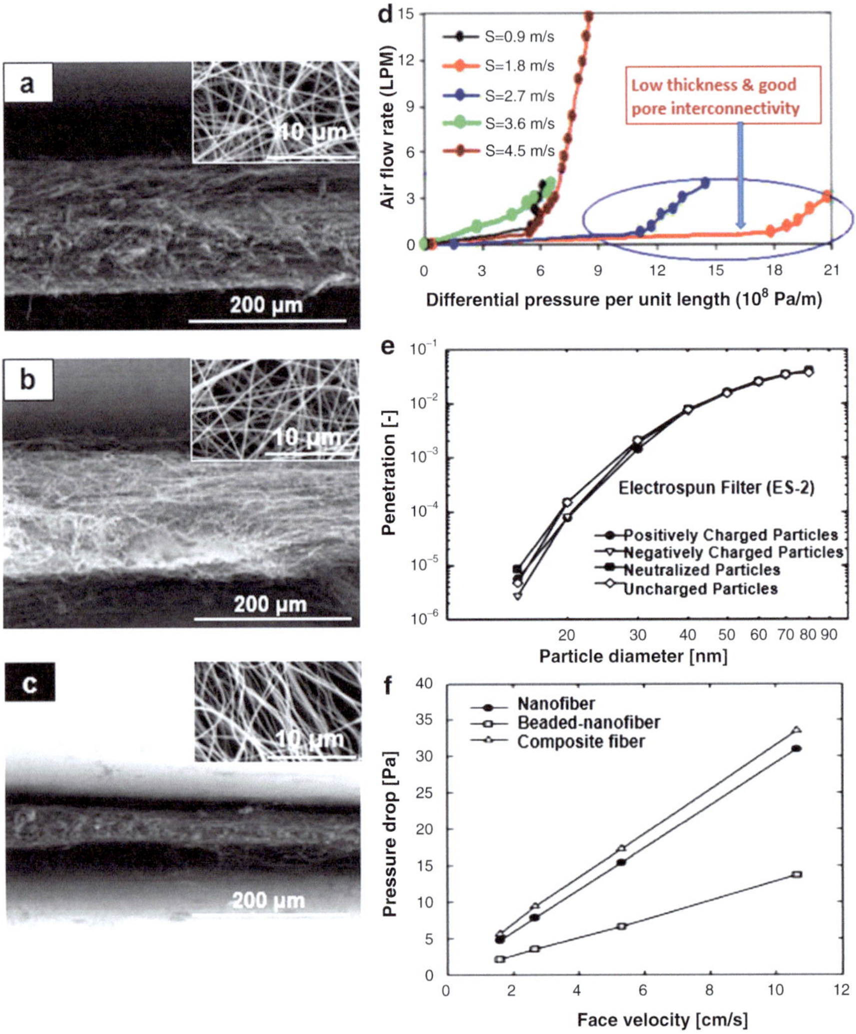

Fig. 12.4 (**a**–**c**) Vertical section of nanofiber membranes with increasing rational speed of collector. (**d**) Effect of collection speed on the wet curve of nanofiber membranes. (**e**) Penetration of nanoparticles of various charge states through electrospun fibrous membranes. (**f**) Pressure drop versus velocity plot for three kinds of tested fibrous filters versus face velocity ((**a**–**d**) Reprinted with permission from [36]. © 2006 Elsevier. (**e**) Reprinted with permission from [37]. © 2007 Elsevier. (**f**) Reprinted with permission from [39]. © 2010 Elsevier)

the electrospun PAN filter. From the filtration test, it was observed that particle penetration was independent of particle charge state (Fig. 12.4e), which implies that columbic forces present between the particles and the fiber did not lead to enhancement of particle collection. Besides, the research that involves structure-activity relationship between fibrous membrane structure and filtration performance also has been developed.

Recent reports have revealed that the morphology or structure optimization of electrospun membranes was an effective method for the improvement of filtration performances [38]. For example, Yun et al. [39] reported the fabrication of the composite PAN/polymethyl methacrylate (PMMA) membranes (beaded nanofiber and particle/nanofiber) and used the quality factor (QF, defined as

$$QF = -\frac{\ln(1-\eta)}{\Delta P}, \qquad (12.1)$$

where η and ΔP are the filtration efficiency and pressure drop across the filter, respectively) to evaluate the filtration performances of different membranes. The presence of the PMMA beads or particles in the composite membranes has a small effect on the pressure drop and nanoparticle penetration through the PAN/PMMA filters, as show in Fig. 12.4f. Furthermore, the QF analysis of composite membranes showed that the nanofiber filter without PMMA polymer beads or particles in the membranes was the lowest. They concluded that the increased QF of beaded-nanofibrous membranes could be due to the improvement of the volume fraction by the PMMA beads and particles in the nanofibers. Following this approach, Mei et al. [40] also reported that electrospun PAN nanofibrous membranes with bimodal fiber size distributions exhibited better filtration performance than those membranes constructed by unimodal nanofibers. Compared with the construction method of nanofibers, regulation of the structure of nanofibrous filter membranes is another versatile and effective way to obtain filter media with high filtration performances. Zhang et al. [41] recently examined the properties of filter media made of multiple thin nanofiber layers. It is found that the multiple filter media showed a much better filtration performance in terms of the pressure drop compared with single thick-layer nanofiber membranes. By stacking three layers of the short-term electrospun PAN nanofiber membranes, the corresponding pressure drop of relevant multiple membranes decreased dramatically (from 98 Pa to 41 Pa), while the filtration efficiency was almost unchanged (from 99.98 % to 99.92 %). Nevertheless, the filtration mechanisms of multiple filter media for fine particles have not been revealed; thus, further study not only should focus on the controllable fabrication of high-performance nanofiber media but also should give more attention to the capture mechanisms of multiple filter media.

12.2.1.3 Polyurethane (PU)

PU is a versatile class of thermoplastic polymer that possesses a range of excellent properties, such as excellent elastomeric property, resistance to abrasion, and high durability [42–44]. Sambaer et al. [45] proposed a new approach to create a simple 3D nanofiber membrane model based on the real structure features of PU membranes. The slip/transition/free molecular flow regime, particle-fiber interactions, air/particle slip, sieve, and homogenous flow field have been considered in this full 3D particle filtration model (Fig. 12.5a). They suggested that in the force

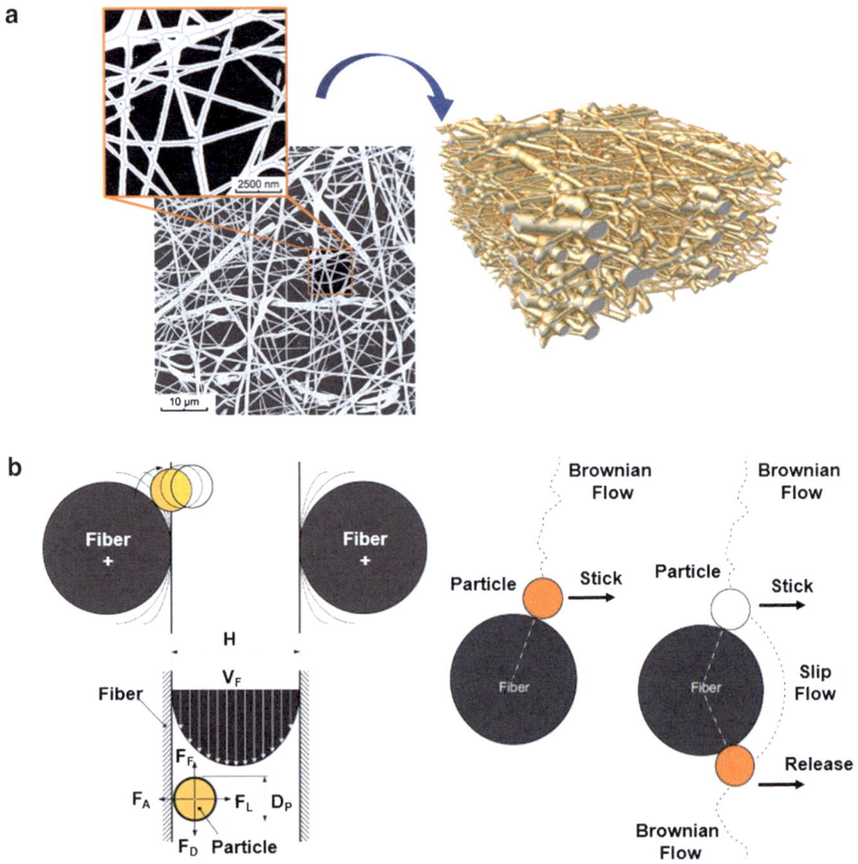

Fig. 12.5 (**a**) Full 3D model for PU nanofiber membranes based on SEM image. (**b**) Particle-fiber force balance visualization inside the tube pore and two possible particle-fiber configurations ((**a**) and (**b**) Reprinted with permission from [45]. © 2011 Elsevier)

balance model, four different forces should be considered for capturing particle-fiber interactions: the drag force (F_D), the lift force (F_L), the adhesion (van der Waals) force (F_A), and the friction force (F_F), which are defined in Fig. 12.5b. Additionally, the theoretical predictions of the effect of air velocity, viscosity, temperature, pressure, and friction coefficient between the particle and fiber were discussed in detail. They concluded that the particle-fiber friction coefficient has the highest effects on the filtration efficiency; thus, the particle-fiber surface modification could provide useful guidelines for the optimization and enhancement of filtration performances. Furthermore, 3D modeling of filtration process via multilayered electrospun PU nanofiber-based nonwoven filters was also proposed [46]. For the model validation purposes, two PU nanofiber layers (having identical average fiber diameter, 120 nm, but different average pore size) prepared by electrospinning were combined to form different multilayer filter media with comparable mass area and

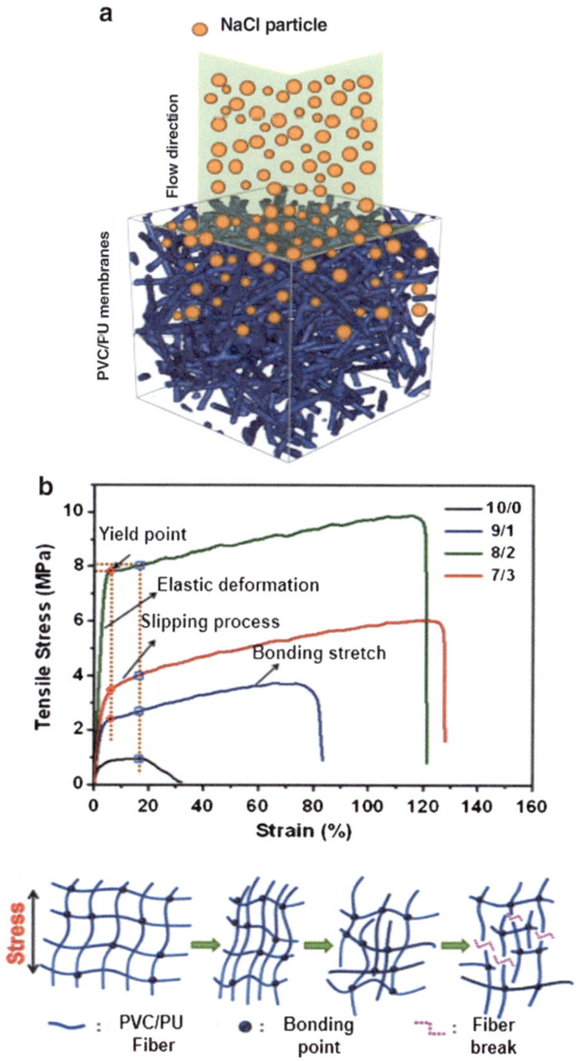

Fig. 12.6 (**a**) The filtration process of PVC/PU membranes for 300–500 nm NaCl aerosol particles. (**b**) Stress–strain curves of PVC/PU fibrous membranes and the proposed three-step break mechanism of tensile fracture process upon external stress ((**a**) and (**b**) Reprinted with permission from [47]. © 2013 Elsevier)

thickness. It has been found that the proposed 3D model was able to reasonably predict the measured filtration efficiency, pressure drop, and quality factor for PU-based nanofiber filter media. It is believed that the proposed 3D modeling tools in these studies will help to evaluate the filtration efficiency, pressure drop, and quality factor of different nanofiber nonwoven filters.

In view of the excellent mechanical properties of PU, considerable efforts have been devoted toward the design of PU or PU-blended filter media [43, 47, 48]. Wang et al. [47] reported the tortuously structured PVC/PU fibrous membranes (Fig. 12.6a) which showed high efficiency (99.5 %) and relatively low pressure drop (144 Pa) with the test particles of 0.3 μm in diameter. Furthermore, the tensile strength of PVC/PU membranes showed the maximum tensile strength of 9.9 MPa when the PU content reached 20 wt% relative to polymer, which derived from the most available bonding structure formed between PU and PVC elastomer (Fig. 12.6b). Meanwhile, the abrasion resistance of composite membranes remarkably increased to 237 cycles with the incorporation of PU, but the air permeability decreased regularly with the increasing PU content, both of which can be ascribed to the enhanced adhesion structure among adjacent fibers. Thus, in order to satisfy the real application of air filter, they demonstrated that the weight ratio of PU and PVC in the solutions needed reasonable adjustment; research and observations are under way.

12.2.1.4 Poly(vinyl alcohol) (PVA)

PVA is the largest volume synthetic resin produced in the world as a kind of water-soluble polyhydroxy polymer. The excellent chemical resistance, physical properties, and complete biodegradability of PVA resins have led to their broad practical applications [25, 49, 50]. Thus, engineering PVA into nanomaterials has attracted increasing attentions during the past several years. As typical electrospun fibrous membranes, the PVA nanofibers were randomly oriented as a porous membrane with a relatively narrow fiber diameter distribution. Qin et al. [51] reported the fabrication of the PVA filtration membranes supported by PP spunbonded or PP meltblown substrate. The filtration performances of relevant PVA membranes for the test particles with the mean size 0.6 μm showed significant differences according to the kind of substrate. The filtration efficiency of PVA membranes (2.4 g/m^2) deposited on the PP meltblown substrate was 100 %, while the PVA (2.9 g/m^2)/PP spunbonded complex filter media showed a much lower value (95 %). However, the press drop of nanofibers that were electrospun on the meltblown sub-layer increases sharply than the same basis weight of nanofibers deposited on the spunbonded sub-layer. Thus, they concluded that there is an optimum region for maximum filtration efficiency with a minimum pressure drop and that this balance comes out at a lower basis weight for meltblown nanofibers than for spunbonded nanofibers. Despite the fact that PVA has good filtration properties in the dry state, its applications were limited by its high hydrophilicity [52]. Therefore, PVA fibers can be readily cross-linked to improve mechanical properties and antiwater solubility. Qin and co-workers [53] did continuous studies on PVA-based filtration membranes. The PVA fibers were efficiently cross-linked by maleic acid with the vitriolic acid used as catalyst activator. The filtration results proved that cross-linked PVA nanofibers can improve the filtration efficiency.

As we all know, the needle electrospinning has a very low production rate, and the spinneret was frequently clogged during the electrospinning process [54, 55]. Recently, needleless electrospinning technique appeared with the aim of producing nanofibers on a large scale [56, 57]. Li et al. [58] prepared PVA nanofibers with an average fiber diameter of 100 nm using needleless electrospinning technique. The filtration performance of PVA/PP composite filter was evaluated by measuring the

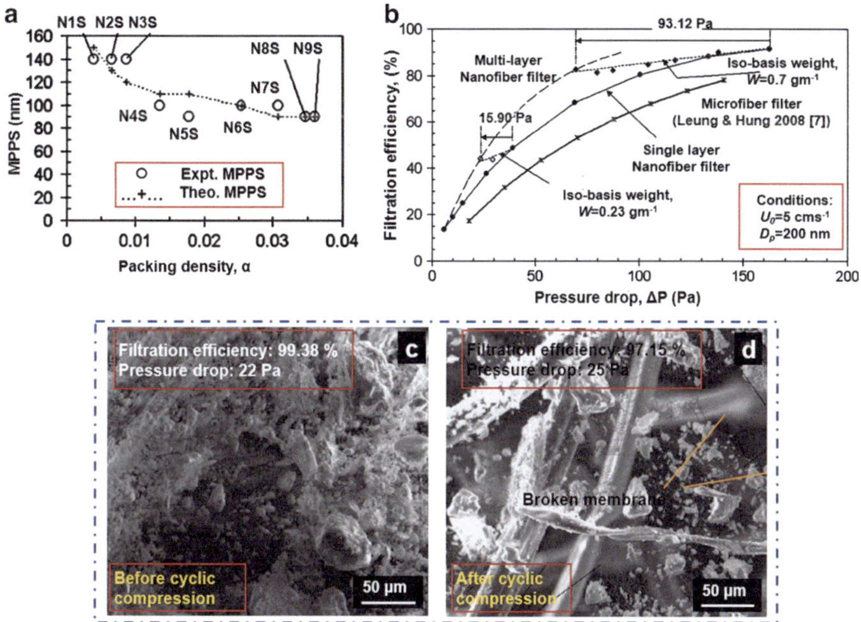

Fig. 12.7 (**a**) MPPS as a function of nanofiber packing density. (**b**) Operation curves ($U_0 = 5$ cm s^{-1}, $D_p = 200$ nm) of multilayer filters formed by stacking up constituent units from N1S to N9S. (**c**) and (**d**) SEM images of the structural changes in the nanofiber membranes deposited over nonwoven filter media before and after cyclic compression ((**a**) and (**b**) Reprinted with permission from [60]. © 2010 Elsevier. (**c**) and (**d**) Reprinted with permission from [61]. © 2010 Elsevier)

filtration of NaCl particles (75 ± 20 nm) through the filters. They found that the filtration efficiency of the composite filter media for NaCl nanoparticles increased along with the thickness of nanofiber membranes. Additionally, the multilayer arrangement for fabrication of the composite filter can achieve high filtration efficiency (up to 99.95 %) which was comparable to commercial HEPA.

12.2.1.5 Polyethylene Oxide (PEO)

PEO is another type of water-soluble synthetic polymer, which has been extensively studied both from the fundamental standpoint of understanding the behavior of polymer solutions and for its wide range of applications [59]. Leung and co-workers [60] prepared different kinds of PEO nanofibers and evaluated the effect of face velocity, fiber packing density and membrane thickness on filtration efficiency, and pressure drop of filters with nanofibers coated on a nonwoven microfiber substrate. The test aerosol particle is NaCl with size ranging from 50 to 480 nm. The results show that the most penetrating particle size (MPPS) decreases from 140 to 90 nm when nanofiber packing density increases from 3.9 to 36×10^{-3}, as shown in Fig. 12.7a. The filtration efficiency generally decreased when face velocity increased from 5 to 10 cm s^{-1}, and the reduction becomes larger at smaller particle sizes especially for those below 100 nm. They also suggested that the basis weight of nanofibers has less prominent effect on MPPS compared with fiber packing density. The QF value of PEO-based filter media decreased by increasing the basis weight of nanofibers and was subjected to a lower decrease rate especially for those larger than 0.2×10^{-2} g m^{-2}. Therefore, in order to obtain better filtration performance, it is recommended to adopt the lowest possible face velocity in real filter applications. Additionally, two and three layers of PEO nanofibers were stacked up in order to construct multilayer filter media with the same packing density but increased thickness as compared to a single layer of PEO fibers. Given the filtration efficiency and pressure drop across a filter media as η and ΔP, respectively, it can be derived from Eq. 12.1 that the QF of a composite filter media formed by stacking up k identical layers of the filter remains unchanged and described as

$$QF_k = -\frac{\ln (1-\eta)^k}{k\,(\Delta P)} = -\frac{\ln (1-\eta)}{\Delta P} = QF \qquad (12.2)$$

The experimental QF of different layers of PEO membranes, which are close to each other as expected from Eq. 12.2. The pressure drop savings through "multilayering" show more significance at higher fiber basis weight (Fig. 12.7b). Hence, "multilayering" can be regarded as an effective method to yield high filtration efficiency and low pressure drop under a fixed nanofiber basis weight. Patanaik et al. [61] prepared sandwich-structured nonwoven/PEO/nonwoven filter media and investigated the filtration performances before and after cyclic compression. The composite filtration membranes displayed good filtration performances: on one side, the filtration efficiency for 0.6–180 μm particles could reach 99.38 % with a low pressure drop (22 Pa); on the other side, the filtration properties of composite media showed no significant changes after 50 cyclic compression (Fig. 12.7c, d). Thus, sandwich-structured composite media consisting of nanofibrous membrane are a promising media for long-term air filtration applications.

12.2.1.6 Others

Grafe et al. [62] compared the filtration performances of a composite cellulose/nanofiber filter and a standard (wet-laid cellulose) filter and discussed the performance of nanofiber media in a mining vehicle cabin air filter. These two kinds of filters were tested independently on a Caterpillar 992G wheel loader, mounted with two particle sampling devices, one inside the cabin and the other outside. With the standard cellulose filter, the filtration efficiencies of submicron and respirable (>1 μm) dust were 68 % and 86 %, respectively. However, with the cellulose/nanofiber composite filter, a higher reduction of approximately 92 % has been observed for both the submicron and respirable dusts. Besides, the results showed that submicron salt crystal preferentially collected on nanofibers (compared to large spunbonded fiber in the upper left) due to smaller fiber diameter. Further measurements revealed that the nanofiber filter life and initial pressure drop were comparable with the standard filters [20].

Silk is a well-described natural fiber obtained from silkworms and spiders and has been used in textile industries for thousands of years. Recently, silk nanofiber-based filter media that contain fibers of diameter ranging from 100 to 1,000 nm have been conveniently constructed by electrospinning technique. Improved multiple structures have been developed in this study in which a controlled amount of fine fiber was placed on both sides of the media to result in a substantial improvement in filtration efficiency and lifetime [63].

Hsiao et al. [64] prepared PC nanofibers (average diameter of about 0.17 μm) using the air blowing-assisted electrospinning process. The air blowing pressure and fiber diameter have great influence on the Frazier air permeability of blowing-assisted electrospun fibers obtained from 16 % PC solutions. As can be seen, an increase in the air blowing pressure narrows the distribution of the fiber diameter of the PC nanofibers, which leads to a decrease of Frazier air permeability. This study indicated that the air blowing pressure has a negative effect on Frazier air permeability. Additionally, the air blowing pressure also narrowed the distribution of PC fiber pore size and resulted in smaller mean flow pore size. Furthermore, the filtration performances of PC fibers were evaluated under various face velocities and different mean flow pore size. As shown in Fig. 12.8a, the filtration efficiency of the PC/PP filter media decreased from 98 % to 86 % at a 2.67 cm/s face velocity. With an increase in the face velocity, more particles penetrated through the PC/PP filter media, resulting in an obvious reduction of filtration efficiency. The dependence of the pressure drop on the mean flow pore size was similar to efficiency. Meanwhile, with increasing surface velocity, the reduction of pressure drop became larger, reaching a maximum of 5.0 mmH_2O at 13.33 cm/s, which translates into more resistance to airflow through the filter media (Fig. 12.8b).

Besides, Morozov et al. [65] developed electrospinning-neutralization technology to make water-soluble nanofilters from polyvinylpyrrolidone (PVP) solutions to be used in the collection of biological micro- and nano-aerosols for analysis of biohazards. Aerosol particle penetration and resistance to airflow were characterized as a function of filter thickness, relative air humidity, and storage stability. As can be seen, the filtration efficiency of particle capturing at different flow rates shows little, if any, dependence at all the rates studied. Additionally, PVP filter media were also characterized for their ability to collect aerosol particles from the air at different humidity. And they concluded that the PVP filters can retain their performance at low and moderate humidity only. At high humidity (exceeding 80 %), a rapid deterioration of the filtration performance occurs which is accompanied by increased resistance to airflow, as illustrated in Fig. 12.8c. Filters of different thickness and filtration ability were attached to substrates and kept in ambient air with a humidity of 30–50 %. After a certain storage time, no notable changes in their collection ability were detected over more than a month of storage time for all the filters (Fig. 12.8d).

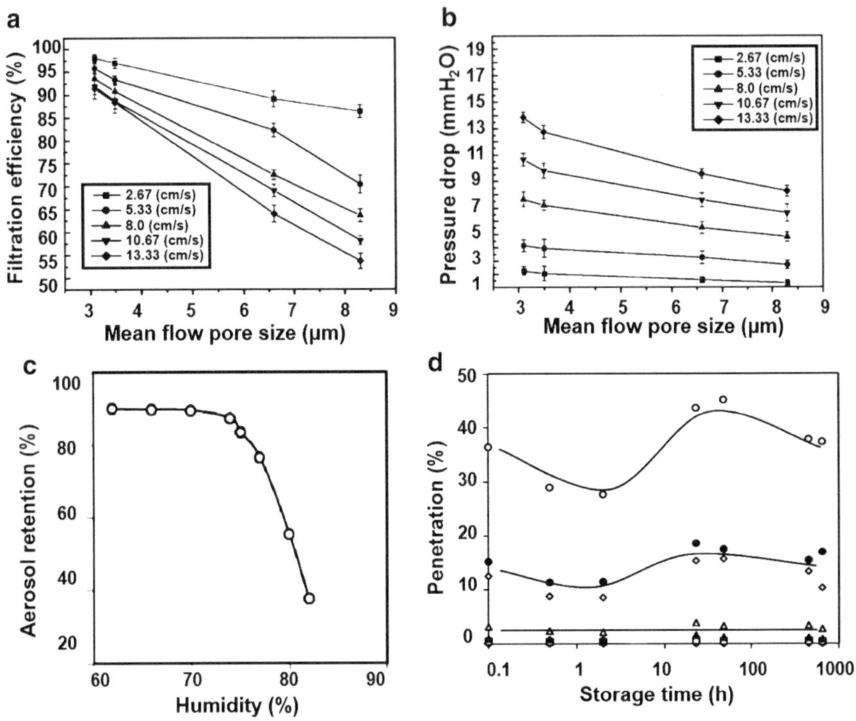

Fig. 12.8 (**a**) Effect of mean flow pore size on filtration efficiency of PC filter media. (**b**) Effect of mean flow pore size on pressure drop of PC filter media. (**c**) Effect of humidity on retention of aerosol particles larger than 0.3 μm by a PVP filter media. (**d**) Effect of storage on the performance of PVP fibrous filter media with fiber density ((**a**) and (**b**) Reprinted with permission from [64]. © 2011 Wiley Periodicals, Inc.. (**c**) and (**d**) Reprinted with permission from [65]. © 2012 Elsevier)

12.2.2 Hybrid Fibrous Membranes

As mentioned above, interception, inertial impaction, and diffusion were the dominant capturing mechanisms for the fine particles during filtration process of electrospun filter media, while the gravitational settling and electrostatic attraction should always be negligible. Thus, extensive researches have been performed on developing more proficient strategies to improve the filtration performances of nanofibrous filter media [14, 32, 66]. Among them, the incorporation of electrostatic charging agent into organic fibers in order to enhance the electrostatic capture ability of electrospun fibers for fine particles has recently become a subject of special interest.

Yeom et al. [24] reported the fabrication of PA-6 nanofiber membranes incorporated with boehmite nanoparticles as an electrostatic charging agent. The morphology analysis showed that the average fiber diameter of PA-6 nanofibers was 73 nm and the boehmite nanoparticles had little effect on the fiber diameter (Fig. 12.9a). They investigated the filtration performances of discharged,

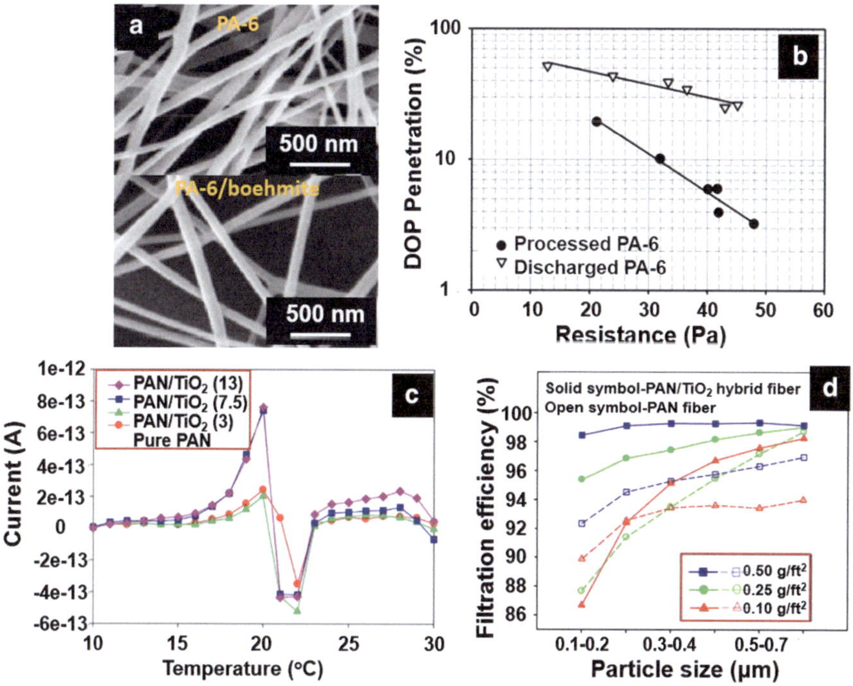

Fig. 12.9 (**a**) SEM images of pure PA-6 and PA-6/boehmite nanofibers. (**b**) DOP penetration of processed and discharged filter media. (**c**) Thermally stimulated current spectra of as-spun hybrid nanofibers near room temperature. (**d**) Filtration efficiency of the filter media with different basis weight as a function of particle size ((**a**) and (**b**) Reprinted with permission from [24]. © 2010 The Polymer Society of Korea and Springer Netherlands. (**c**) and (**d**) Reprinted with permission from [67]. © 2013 Elsevier)

PA-6/boehmite (processed), and corona-charged PA-6 nanofiber membranes. As shown in Fig. 12.9b, the processed PA-6 nanofiber membranes show higher efficiency (96.8 %) and pressure drop (48 Pa) than discharged PA-6 membranes. The QF of processed PA-6 membranes was 0.662 mmH_2O^{-1} with 0.959 of a correlation factor, r^2, while that of discharged membranes was 0.219 mmH_2O^{-1} with $r^2 = 0.917$. However, the filtration performance of charged nanofiber membranes did not follow the same trend as those for processed and discharged membranes. Additionally, the filtration index analysis showed the same trend and reached a maximum value at 3 % content of boehmite. It implied that high-efficiency fibrous filter media can be produced with the addition of electrostatic charging agent without sacrificing the pressure drop. Cho et al. [67] also improved the air filtration efficiency of cellulose filters coated with PAN nanofibers via inclusion of electrostatically active TiO_2 nanoparticles. The thermally stimulated current measurement spectra showed that by increasing the TiO_2 nanoparticle content in the composite fibers, the measured currents became higher (Fig. 12.9c). The filtration efficiency of the PAN/TiO_2 filter was found to be greater (4–6 %) than that of those made with just pure PAN fibers, especially for smaller particles in the range 0.1–0.5 μm (Fig. 12.9d). In addition, the pressure drop of the composite system showed much lower than the pure PAN-based filter media. Therefore, the filter media constructed by PAN/TiO_2 present attractive filtration features of high filtration efficiency and low pressure drop; thus, the operating life of the filter media is eventually expected to be longer.

12.3 Medium-High-Temperature Filter

Most of the industrial dedusting processes such as filtration units for municipal waste incinerators, cement plants, and coal-fired boilers operate within a range between 150 °C and 260 °C. The normal-temperature filtration media such as PA-6, PAN, PVA, and PU nanofibrous membranes that have been mentioned above could not fulfill the requirements of medium-temperature filtration. The high-performance polymer nanofibrous membranes with rigid chain segments such as polybenzoxazole [68], polyethersulfone (PES) [69], polyether imide [70], polyimide [71], and polymeta-phenylene isophthalamide nanofibrous membranes [72] have excellent thermal stability in the temperature range of 200–350 °C and can be used in the abovementioned industrial dedusting processes.

Recently, PES nanofibrous membranes with thermal resistance up to 220 °C were reported as the high-heat-resistant air filter [69]. To investigate the filtration performance of the PES nanofibrous membranes, they prepared the filtration membranes with various fiber diameters and pore sizes by regulating PES concentrations and the process parameters. The results revealed that the fiber diameter and pore size were clearly affected by the PES concentration, feeding rate, and tip-to-collector distance. Meanwhile, the filtration efficiency and pressure drop were steeply decreased over the pore size of 3.0 μm. The PES membranes with average pore size of 3.2 μm showed the best filtration performance with high filtration efficiency (99.9998 %) and low pressure drop (215 Pa) for 300 nm particles, which can fit the HEPA requirement.

To date, most studies have focused on polymer membranes; however, the relative low thermal stability of normal polymer and even heat-resistant polymer membranes prevents them from being directly used as filtration media of the high-temperature dust over 1,000 °C such as metallurgical processes. Currently, due to low thermal stability of polymer filtration media, the high-temperature waste gases from process and power industries need to be cooled down to 250 °C and even lower before filtration. However, the cooling process typically consumes a lot of energy, and the cooled and cleaned gases have low enthalpy for recovering heat from these gases; therefore, the cooling process becomes economically ineffective.

Ceramic fibrous membranes with excellent resistance against corrosion, chemical erosion, and thermal stability are of particular interest in many separation processes. Electrospun ceramic nanofibrous membranes which combine the unique thermal stability and nanometer fiber diameter can be used as the high-temperature filtration media in the range of 1,000 °C without expensive pretreatments for hot waste gases.

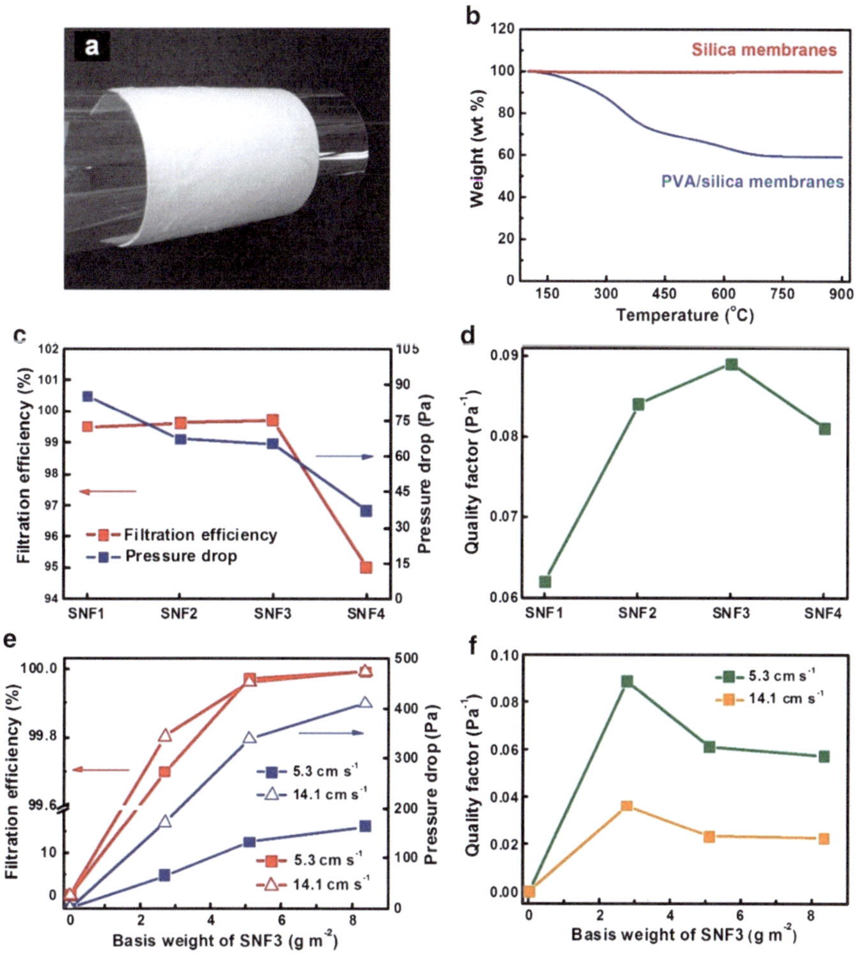

Fig. 12.10 (**a**) Optical image of flexible silica fibrous membranes when bent with PET film. (**b**) Thermogravimetric analysis of silica nanofibrous membranes (*red line*) and PVA/silica hybrid membranes (*blue line*) was conducted from 100 °C to 900 °C in air. (**c**) Filtration efficiency and pressure drop and (**d**) QF values of various silica nanofiber membranes at the face velocity of 5.3 cm s^{-1}. (**e**) Filtration efficiency and pressure drop and (**f**) QF value variation versus basis weight of SNF3 membranes under the face velocity of 5.3 and 14.1 cm s^{-1} ((**a**) Reprinted with permission from [76]. © 2010 American Chemical Society. (**b–f**) Reprinted with permission from [79]. © 2012 The Royal Society of Chemistry)

Over the past decade, considerable effort has been devoted to fabricating ceramic nanofibers, including SiO_2 [73], Al_2O_3 [74], ZrO_2 [75], and so on; however, the brittleness of current ceramic membranes significantly limits their practical applications. Recently, silica nanofibrous (SNF) membranes with remarkable flexibility and thermal stability were prepared by a facile combination of electrospinning and solgel methods [76–79]. Figure 12.10a showed the optical image of flexible silica fibrous membranes when bent with poly(ethylene terephthalate) (PET) film, and no crack appeared during the bending [76]. Additionally, no weight loss of the silica membranes from 100 °C to 900 °C is observed in Fig. 12.10b, which confirms the remarkable thermal stability of the membranes at least up to 900 °C [79].

The filtration performance of silica membranes with various fiber diameters for 300–500 nm sodium chloride aerosols was also discussed in Fig. 12.10c [79]. Figure 12.10d compares the filtration efficiency and pressure drop of silica membranes with different fiber diameters from 160 (SNF1), 205 (SNF2), and 303 (SNF3) to 448 (SNF4) nm, and the SNF3 membranes with a fiber diameter of 303 nm show the best filtration efficiency with the relative higher filtration efficiency (99.70 %) and relative lower pressure drop (65 Pa) at low face velocity of 5.3 cm s^{-1}. Meanwhile, the QF value indicated that the best filtration media were the SNF3 membranes, compared to SNF1, SNF2, and SNF4 membranes with a similar basis weight of 2.60 g m^{-2} (Fig. 12.10d). Additionally, the influence of face velocity and basis weight on the filtration performance was also discussed in Fig. 12.10e. When the face velocity increased from 5.3 to 14.1 cm s^{-1}, the pressure drop rose rapidly with the increased basis weight. However, the filtration efficiency remained unchanged and the pressure drop was quite linearly dependent on the basis weight, which can also be seen in Fig. 12.10e, f. Significantly, the SNF3 membranes with excellent flexibility (0.0156 gf cm), tensile strength (5.5 MPa), and thermal stability (up to 1,000 °C) showed a high filtration efficiency (99.99 %) and a low pressure drop (163 Pa) for 300–500 nm sodium chloride aerosols. The development of flexible electrospun silica nanofibrous membranes suggests a promising candidate for high-temperature filtration filter.

12.4 Concluding Remarks and Outlook

In the past 10 years, electrospun nanofibrous membranes have been spotlighted as an effective filter media to capture fine particles. In addition to the basic studies of electrospinning process to better understand the membrane construction process, researchers from all over the world focus on the study of the relationships between the structure characteristics of electrospun nanofibrous membranes (fiber diameter, pore size, porosity, surface area, etc.) and filtration performances (filtration efficiency, pressure, air permeability, etc.). In this chapter, recent advances in fabricating nanofibrous filter media via electrospinning process have been reviewed. In particular, filtration performances and relevant mechanical properties are discussed in detail. It is interesting that the electrospun nanofibrous membranes have been able to outperform conventional nonwoven membranes fabricated essentially by using the meltblown or spunbonded process.

Despite recent advances toward the development of electrospun nanofibrous media for particle filtration, challenges in this area have also been met, and many practical problems have to be solved. First, the control of the morphology of electrospun nanofibrous membranes, such as fiber diameter and its distribution, packing density, and porosity, needs more fundamental and experimental studies. Second, the nanofibers are usually deposited on nonwoven substrate to construct a composite filter medium and cannot be used separately due to their low mechanical strength. Addressing this issue should include research on both the mechanical properties and the filtration performances. Third, in parallel with experimental studies, relevant industrial equipment must be further developed to change the low production rate of current electrospun nanofibers. It is expected that the continuous efforts on exploration of electrospun nanofibrous filter media inspired by their significant properties will address the current challenges and push forward the rapid development of air filter media.

Acknowledgments This work is supported by the National Key Technology Research and Development Program of the Ministry of Science and Technology of China (2013BAC01B02), National Basic Research Program of China (973 Program, 2011CB606103), the National Natural Science Foundation of China (No. 51173022 and U1232116), the Huo Yingdong Foundation (131070), and the Program for New Century Talents of the University in China.

References

1. Rodríguez S, Querol X, Alastuey A, Viana MM, Alarcón M, Mantilla E, Ruiz CR (2004) Comparative p M10–PM2.5 source contribution study at rural, urban and industrial sites during PM episodes in Eastern Spain. Sci Total Environ 328(1–3):95–113, http://dx.doi.org/10.1016/S0048-9697(03)00411-X
2. Querol X, Alastuey A, Rodriguez S, Plana F, Mantilla E, Ruiz CR (2001) Monitoring of PM10 and PM2.5 around primary particulate anthropogenic emission sources. Atmos Environ 35(5):845–858, http://dx.doi.org/10.1016/S1352-2310(00)00387-3
3. Kampa M, Castanas E (2008) Human health effects of air pollution. Environ Pollut 151(2):362–367, http://dx.doi.org/10.1016/j.envpol.2007.06.012
4. Peukert W (1998) High temperature filtration in the process industry. Filtr Separat 35(5): 461–464, http://dx.doi.org/10.1016/S0015-1882(98)80015-3
5. Song Y, Zhang Y, Xie S, Zeng L, Zheng M, Salmon LG, Shao M, Slanina S (2006) Source apportionment of PM2.5 in Beijing by positive matrix factorization. Atmos Environ 40(8):1526–1537, http://dx.doi.org/10.1016/j.atmosenv.2005.10.039
6. Wang CS, Otani Y (2012) Removal of nanoparticles from gas streams by fibrous filters: a review. Ind Eng Chem Res. doi:10.1021/ie300574m
7. Adiletta JG (1999) Fibrous nonwoven web. US Patent 5,954,962, 21 Sept 1999
8. Hung CH, Leung WWF (2011) Filtration of nano-aerosol using nanofiber filter under low Peclet number and transitional flow regime. Sep Purif Technol 79(1):34–42, http://dx.doi.org/10.1016/j.seppur.2011.03.008
9. Yoon K, Hsiao BS, Chu B (2008) Functional nanofibers for environmental applications. J Mater Chem 18(44):5326–5334. doi:10.1039/b804128h
10. Ma H, Hsiao BS, Chu B (2011) Thin-film nanofibrous composite membranes containing cellulose or chitin barrier layers fabricated by ionic liquids. Polymer 52(12):2594–2599. doi:10.1016/j.polymer.2011.03.051
11. Wang N, Wang X, Ding B, Yu J, Sun G (2012) Tunable fabrication of three-dimensional polyamide-66 nano-fiber/nets for high efficiency fine particulate filtration. J Mater Chem 22(4):1445–1452. doi:10.1039/c1jm14299b

12. Tanaka S, Doi A, Nakatani N, Katayama Y, Miyake Y (2009) Synthesis of ordered mesoporous carbon films, powders, and fibers by direct triblock-copolymer-templating method using an ethanol/water system. Carbon 47(11):2688–2698, http://dx.doi.org/10.1016/j.carbon.2009.05.024
13. Wang D, Sun G, Chiou BS (2007) A high-throughput, controllable, and environmentally benign fabrication process of thermoplastic nanofibers. Macromol Mater Eng 292(4):407–414. doi:10.1002/mame.200600460
14. Qiu P, Mao C (2010) Biomimetic branched hollow fibers templated by self-assembled fibrous polyvinylpyrrolidone structures in aqueous solution. ACS Nano 4(3):1573–1579. doi:10.1021/nn9009196
15. Wang X, Li Y (2002) Selected-control hydrothermal synthesis of α- and β-MnO_2 single crystal nanowires. J Am Chem Soc 124(12):2880–2881. doi:10.1021/ja0177105
16. Fong H, Chun I, Reneker DH (1999) Beaded nanofibers formed during electrospinning. Polymer 40(16):4585–4592, http://dx.doi.org/10.1016/S0032-3861(99)00068-3
17. Ding B, Kim HY, Lee SC, Lee DR, Choi KJ (2002) Preparation and characterization of nanoscaled poly(vinyl alcohol) fibers via electrospinning. Fiber Polym 3(2):73–79. doi:10.1007/bf02875403
18. Ding B, Kimura E, Sato T, Fujita S, Shiratori S (2004) Fabrication of blend biodegradable nanofibrous nonwoven mats via multi-jet electrospinning. Polymer 45(6):1895–1902. doi:10.1016/j.polymer.2004.01.026
19. Wu J, Wang N, Zhao Y, Jiang L (2013) Electrospinning of multilevel structured functional micro-/nanofibers and their applications. J Mater Chem A 1(25):7290–7305. doi:10.1039/c3ta10451f
20. Thavasi V, Singh G, Ramakrishna S (2008) Electrospun nanofibers in energy and environmental applications. Energ Environ Sci 1(2):205–221. doi:10.1039/b809074m
21. Sahay R, Kumar PS, Sridhar R, Sundaramurthy J, Venugopal J, Mhaisalkar SG, Ramakrishna S (2012) Electrospun composite nanofibers and their multifaceted applications. J Mater Chem 22(26):12953–12971. doi:10.1039/c2jm30966a
22. Zhou C, Chu R, Wu R, Wu Q (2011) Electrospun polyethylene oxide/cellulose nanocrystal composite nanofibrous mats with homogeneous and heterogeneous microstructures. Biomacromolecules 12:2617–2625. doi:10.1021/bm200401p
23. Si Y, Ren T, Li Y, Ding B, Yu J (2012) Fabrication of magnetic polybenzoxazine-based carbon nanofibers with Fe_3O_4 inclusions with a hierarchical porous structure for water treatment. Carbon 50(14):5176–5185. doi:10.1016/j.carbon.2012.06.059
24. Yeom B, Shim E, Pourdeyhimi B (2010) Boehmite nanoparticles incorporated electrospun nylon-6 nanofiber web for new electret filter media. Macromol Res 18(9):884–890. doi:10.1007/s13233-010-0910-5
25. Gule NP, de Kwaadsteniet M, Cloete TE, Klumperman B (2012) Electrospun poly(vinyl alcohol) nanofibres with biocidal additives for application in filter media, 1-properties affecting fibre morphology and characterisation. Macromol Mater Eng 297(7):609–617. doi:10.1002/mame.201100275
26. Kim SJ, Nam YS, Rhee DM, Park HS, Park WH (2007) Preparation and characterization of antimicrobial polycarbonate nanofibrous membrane. Eur Polym J 43(8):3146–3152, http://dx.doi.org/10.1016/j.eurpolymj.2007.04.046
27. Zhang S, Shim WS, Kim J (2009) Design of ultra-fine nonwovens via electrospinning of Nylon 6: spinning parameters and filtration efficiency. Mater Design 30(9):3659–3666, http://dx.doi.org/10.1016/j.matdes.2009.02.017
28. Ahn YC, Park SK, Kim GT, Hwang YJ, Lee CG, Shin HS, Lee JK (2006) Development of high efficiency nanofilters made of nanofibers. Current Applied Physics 6(6):1030–1035, http://dx.doi.org/10.1016/j.cap.2005.07.013
29. Heikkilä P, Taipale A, Lehtimäki M, Harlin A (2008) Electrospinning of polyamides with different chain compositions for filtration application. Polym Eng Sci 48(6):1168–1176. doi:10.1002/pen.21070

30. Gibson P, Schreuder GH, Rivin D (2001) Transport properties of porous membranes based on electrospun nanofibers. Colloid Surf A: Physicochem Eng Asp 187–188:469–481, http://dx.doi.org/10.1016/S0927-7757(01)00616-1
31. Kim G, Ahn Y, Lee J (2008) Characteristics of nylon 6 nanofilter for removing ultra fine particles. Korean J Chem Eng 25(2):368–372. doi:10.1007/s11814-008-0061-y
32. Faccini M, Vaquero C, Amantia D (2012) Development of protective clothing against nanoparticle based on electrospun nanofibers. J Nanomater 2012:1–9. doi:10.1155/2012/892894
33. Guibo Y, Qing Z, Yahong Z, Yin Y, Yumin Y (2013) The electrospun polyamide 6 nanofiber membranes used as high efficiency filter materials: filtration potential, thermal treatment, and their continuous production. J Appl Polym Sci 128(2):1061–1069. doi:10.1002/app.38211
34. Fitzer E (1989) Pan-based carbon fibers-present state and trend of the technology from the viewpoint of possibilities and limits to influence and to control the fiber properties by the process parameters. Carbon 27(5):621–645, http://dx.doi.org/10.1016/0008-6223(89)90197-8
35. Chen JC, Harrison IR (2002) Modification of polyacrylonitrile (PAN) carbon fiber precursor via post-spinning plasticization and stretching in dimethyl formamide (DMF). Carbon 40(1):25–45, http://dx.doi.org/10.1016/S0008-6223(01)00050-1
36. Barhate RS, Loong CK, Ramakrishna S (2006) Preparation and characterization of nanofibrous filtering media. J Membrane Sci 283(1–2):209–218, http://dx.doi.org/10.1016/j.memsci.2006.06.030
37. Yun KM, Hogan CJ Jr, Matsubayashi Y, Kawabe M, Iskandar F, Okuyama K (2007) Nanoparticle filtration by electrospun polymer fibers. Chem Eng Sci 62(17):4751–4759, http://dx.doi.org/10.1016/j.ces.2007.06.007
38. Kim K, Lee C, Kim I, Kim J (2009) Performance modification of a melt-blown filter medium via an additional nano-web layer prepared by electrospinning. Fiber Polym 10(1):60–64. doi:10.1007/s12221-009-0060-6
39. Yun KM, Suryamas AB, Iskandar F, Bao L, Niinuma H, Okuyama K (2010) Morphology optimization of polymer nanofiber for applications in aerosol particle filtration. Sep Purif Technol 75(3):340–345. doi:10.1016/j.seppur.2010.09.002
40. Mei Y, Wang Z, Li X (2013) Improving filtration performance of electrospun nanofiber mats by a bimodal method. J Appl Polym Sci 128(2):1089–1094. doi:10.1002/app.38296
41. Zhang Q, Welch J, Park H, Wu CY, Sigmund W, Marijnissen JCM (2010) Improvement in nanofiber filtration by multiple thin layers of nanofiber mats. J Aerosol Sci 41(2):230–236, http://dx.doi.org/10.1016/j.jaerosci.2009.10.001
42. Mao X, Chen Y, Si Y, Li Y, Wan H, Yu J, Sun G, Ding B (2013) Novel fluorinated polyurethane decorated electrospun silica nanofibrous membranes exhibiting robust waterproof and breathable performances. RSC Adv 3(20):7562–7569. doi:10.1039/c3ra23326j
43. Ge J, Si Y, Fu F, Wang J, Yang J, Cui L, Ding B, Yu J, Sun G (2013) Amphiphobic fluorinated polyurethane composite microfibrous membranes with robust waterproof and breathable performances. RSC Adv 3(7):2248–2255. doi:10.1039/c2ra22111j
44. Osman MA, Mittal V, Morbidelli M, Suter UW (2003) Polyurethane adhesive nanocomposites as gas permeation barrier. Macromolecules 36(26):9851–9858. doi:10.1021/ma035077x
45. Sambaer W, Zatloukal M, Kimmer D (2011) 3D modeling of filtration process via polyurethane nanofiber based nonwoven filters prepared by electrospinning process. Chem Eng Sci 66(4):613–623, http://dx.doi.org/10.1016/j.ces.2010.10.035
46. Sambaer W, Zatloukal M, Kimmer D (2012) 3D air filtration modeling for nanofiber based filters in the ultrafine particle size range. Chem Eng Sci 82:299–311, http://dx.doi.org/10.1016/j.ces.2012.07.031
47. Wang N, Raza A, Si Y, Yu J, Sun G, Ding B (2013) Tortuously structured polyvinyl chloride/polyurethane fibrous membranes for high-efficiency fine particulate filtration. J Colloid Interface Sci 398:240–246, http://dx.doi.org/10.1016/j.jcis.2013.02.019
48. Wang J, Raza A, Si Y, Cui L, Ge J, Ding B, Yu J (2012) Synthesis of superamphiphobic breathable membranes utilizing SiO_2 nanoparticles decorated fluorinated polyurethane nanofibers. Nanoscale 4(23):7549–7556. doi:10.1039/c2nr32883f

49. Stejskal J, Kratochvíl P, Helmstedt M (1996) Polyaniline dispersions. 5. Poly(vinyl alcohol) and poly(*N*-vinylpyrrolidone) as steric stabilizers. Langmuir 12(14):3389–3392. doi:10.1021/la9506483
50. Ding B, Kim HY, Lee SC, Shao CL, Lee DR, Park SJ, Kwag GB, Choi KJ (2002) Preparation and characterization of a nanoscale poly(vinyl alcohol) fiber aggregate produced by an electrospinning method. J Polym Sci Polym Phy 40(13):1261–1268. doi:10.1002/polb.10191
51. Qin XH, Wang SY (2006) Filtration properties of electrospinning nanofibers. J Appl Polym Sci 102(2):1285–1290. doi:10.1002/app.24361
52. Deitzel JM, Kleinmeyer JD, Hirvonen JK, Beck Tan NC (2001) Controlled deposition of electrospun poly(ethylene oxide) fibers. Polymer 42(19):8163–8170, http://dx.doi.org/10.1016/S0032-3861(01)00336-6
53. Qin XH, Wang SY (2008) Electrospun nanofibers from crosslinked poly(vinyl alcohol) and its filtration efficiency. J Appl Polym Sci 109(2):951–956. doi:10.1002/app.28003
54. Yu DG, Li XY, Wang X, Chian W, Liao YZ, Li Y (2013) Zero-order drug release cellulose acetate nanofibers prepared using coaxial electrospinning. Cellulose 20(1):379–389. doi:10.1007/s10570-012-9824-z
55. Yoo HS, Kim TG, Park TG (2009) Surface-functionalized electrospun nanofibers for tissue engineering and drug delivery. Adv Drug Delivery Rev 61(12):1033–1042, http://dx.doi.org/10.1016/j.addr.2009.07.007
56. Jirsak O, Sanetrnik F, Lukas D, Kotek V, Martinova L, Chaloupek J (2009) A method of nanofibres production from a polymer solution using electrostatic spinning and a device for carrying out the method. EP Patent 1,673,493, 8 Aug 2009
57. Niu H, Lin T, Wang X (2009) Needleless electrospinning. I. A comparison of cylinder and disk nozzles. J Appl Polym Sci 114(6):3524–3530. doi:10.1002/app.30891
58. Li J, Gao F, Liu L, Zhang Z (2013) Needleless electro-spun nanofibers used for filtration of small particles. Express Polym Lett 7(8):683–689. doi:10.3144/expresspolymlett.2013.65
59. Devanand K, Selser JC (1990) Polyethylene oxide does not necessarily aggregate in water. Nature 343(6260):739–741. doi:10.1038/343739a0
60. Leung WWF, Hung CH, Yuen PT (2010) Effect of face velocity, nanofiber packing density and thickness on filtration performance of filters with nanofibers coated on a substrate. Sep Purif Technol 71(1):30–37. doi:10.1016/j.seppur.2009.10.017
61. Patanaik A, Jacobs V, Anandjiwala RD (2010) Performance evaluation of electrospun nanofibrous membrane. J Membrane Sci 352(1–2):136–142, http://dx.doi.org/10.1016/j.memsci.2010.02.009
62. Grafe TH, Graham KM (2003) Nanofiber webs from electrospinning. In: Proceedings of the 5th international conference on nonwovens in filtration, Stuttgart, p 1
63. Kanafchian M, Valizadeh M, Haghi A (2011) Electrospun nanofibers with application in nanocomposites. Korean J Chem Eng 28(2):428–439. doi:10.1007/s11814-010-0376-3
64. Hsiao HY, Huang CM, Liu YY, Kuo YC, Chen H (2012) Effect of air blowing on the morphology and nanofiber properties of blowing-assisted electrospun polycarbonates. J Appl Polym Sci 124(6):4904–4914. doi:10.1002/app.35599
65. Morozov VN, Mikheev AY (2012) Water-soluble polyvinylpyrrolidone nanofilters manufactured by electrospray-neutralization technique. J Membrane Sci 403–404:110–120, http://dx.doi.org/10.1016/j.memsci.2012.02.028
66. Uyar T, Balan A, Toppare L, Besenbacher F (2009) Electrospinning of cyclodextrin functionalized poly(methyl methacrylate) (PMMA) nanofibers. Polymer 50(2):475–480, http://dx.doi.org/10.1016/j.polymer.2008.11.021
67. Cho D, Naydich A, Frey MW, Joo YL (2013) Further improvement of air filtration efficiency of cellulose filters coated with nanofibers via inclusion of electrostatically active nanoparticles. Polymer 54(9):2364–2372, http://dx.doi.org/10.1016/j.polymer.2013.02.034
68. Duan G, Jiang S, Chen S, Hou H (2010) Heat and solvent resistant electrospun polybenzoxazole nanofibers from methoxy-containing polyaramide. J Nanomater 2010:1–5. doi:10.1155/2010/219562

69. Nakata K, Hun Kim S, Ohkoshi Y, Gotoh Y, Nagura M (2007) Electrospinning of poly (ether sulfone) and evaluation of the filtration efficiency. Sen'i Gakkaishi 63(12):307–312, http://jlc.jst.go.jp/JST.JSTAGE/fiber/63.307
70. Moon S, Choi J, Farris RJ (2008) Preparation of aligned polyetherimide fiber by electrospinning. J Appl Polym Sci 109(2):691–694. doi:10.1002/app.27172
71. Fukushima S, Karube Y, Kawakami H (2010) Preparation of ultrafine uniform electrospun polyimide nanofiber. Polym J 42(6):514–518. doi:10.1038/pj.2010.33
72. Huang C, Chen S, Reneker DH, Lai C, Hou H (2006) High-strength mats from electrospun poly(p-phenylene biphenyltetracarboximide) nanofibers. Adv Mater 18(5):668–671. doi:10.1002/adma.200501806
73. Shao C, Kim H, Gong J, Lee D (2002) A novel method for making silica nanofibres by using electrospun fibres of polyvinylalcohol/silica composite as precursor. Nanotechnology 13(5):635. doi:10.1088/0957-4484/13/5/319
74. Yu H, Guo J, Zhu S, Li Y, Zhang Q, Zhu M (2012) Preparation of continuous alumina nanofibers via electrospinning of PAN/DMF solution. Mater Lett 74:247–249, http://dx.doi.org/10.1016/j.matlet.2012.01.077
75. Zhao Y, Tang Y, Guo Y, Bao X (2010) Studies of electrospinning process of zirconia nanofibers. Fiber Polym 11(8):1119–1122. doi:10.1007/s12221-010-1119-0
76. Guo M, Ding B, Li X, Wang X, Yu J, Wang M (2009) Amphiphobic nanofibrous silica mats with flexible and high-heat-resistant properties. J Phys Chem C 114(2):916–921. doi:10.1021/jp909672r
77. Zhao F, Wang X, Ding B, Lin J, Hu J, Si Y, Yu J, Sun G (2011) Nanoparticle decorated fibrous silica membranes exhibiting biomimetic superhydrophobicity and highly flexible properties. RSC Adv 1(8):1482–1488. doi:10.1039/c1ra00605c
78. Yang L, Raza A, Si Y, Mao X, Shang Y, Ding B, Yu J, Deyab SS (2012) Synthesis of superhydrophobic silica nanofibrous membranes with robust thermal stability and flexibility via in situ polymerization. Nanoscale 4(20):6581–6587. doi:10.1039/c2nr32095a
79. Mao X, Si Y, Chen Y, Yang L, Zhao F, Ding B, Yu J (2012) Silica nanofibrous membranes with robust flexibility and thermal stability for high-efficiency fine particulate filtration. RSC Adv 2(32):12216. doi:10.1039/c2ra22086e

Chapter 13
Electrospun Nanofibrous Membranes for Liquid Filtration

Ran Wang, Benjamin S. Hsiao, and Benjamin Chu

Abstract Liquids are essential for most of the lives and activities on earth, where liquid filtration is an important tool to remove undesired suspended solids, emulsified particles, dissolved molecules, and ions from liquids by separating the permeate from the unwanted impurities through filters. Electrospun nanofibrous membranes (ENMs) have been developed as filtration media having unique advantages over traditional membranes, due to high porosity, interconnected pore structures, large surface-to-volume ratio with capability for chemical/physical functionalization, and cost-effectiveness. The new fibrous membrane format has been used in many applications, including filtration, (membrane) distillation, and pervaporation.

Abbreviations

DCMD	Direct contact membrane distillation system
DMF	Dimethylformamide
ENM	Electrospun nanofibrous membrane
FO	Forward osmosis
HA	Hyaluronic acid
HPC	Hydroxypropyl cellulose
LRV	Log reduction value
MD	Membrane distillation
MF	Microfiltration
NF	Nanofiltration
PBI	Polybenzimidazole
PC	Polycarbonate
PCL	Polycaprolactone

R. Wang • B.S. Hsiao (✉) • B. Chu (✉)
Department of Chemistry, Stony Brook University, Stony Brook, NY 11794-3400, USA
e-mail: bhsiao@notes.cc.sunysb.edu; bchu@notes.cc.sunysb.edu

B. Ding and J. Yu (eds.), *Electrospun Nanofibers for Energy and Environmental Applications*, Nanostructure Science and Technology,
DOI 10.1007/978-3-642-54160-5_13,

PEG	Polyethylene glycol
PEI	Polyethylenimine
PEO	Poly(ethylene oxide)
PES	Poly(ethersulfone)
PGA	Polyglycolic acid
PHB	Polyhydroxybutyrate
PI	Polyimide
PMMA	Poly(methyl methacrylate)
PTT	Polytrimethylene terephthalate
PU	polyurethane
PVA	Polyvinyl alcohol
PVC	Polyvinyl chloride
PVDF	Polyvinylidene fluoride
PVP	Polyvinylpyrrolidone
RO	Reverse osmosis
THF	Tetrahydrofuran
UF	Ultrafiltration

13.1 Introduction

13.1.1 Liquid Filtration

Liquid filtration is a mechanical or physical process that separates undesired components, such as solid particles, bacteria, viruses, and metal ions, from the liquid host by passing the mixture through a filter. The input liquid with the undesired components in the separation process is known as the feed solution. The objectives of performing a liquid filtration process usually involve the following: (1) the purification of the liquid and (2) the recovery of unwanted components. However, a more challenging process may also include the retention of desired components in the liquid after the filtration. Selective removal or retention is often more difficult to accomplish. Filtration is one of the most efficient methods for the separation of mixtures. Other commonly used methods include sieving, decantation, adsorption, centrifugation, and distillation. In sieving, particles that are too big to pass through the holes on the top surface of the sieve are retained, while small particles can be retained due to its inability to follow the tortuous "molecular" channels of the filter, leading to a more refined separation. Decantation can remove the stable top layer of liquid, when a precipitate has been fully settled, but it requires a relatively long time period and more cautious operation. Adsorption uses mainly charge interactions to remove unwanted species, instead of by means of size exclusion. Centrifugation forces the suspended particles with higher density than the suspending medium to precipitate to the bottom of the container to achieve the desired separation. Distillation separates liquid mixtures by differentiating the volatility of components. However, centrifugation and distillation usually consume more energy and become less cost-effective when compared with filtration. It should be noted that there is no universal filtration method for the removal of all impurities in a liquid because filters have limited effective range and capacity. For example, a variety of membrane distillation, adsorption, and micro-reactor devices contain filters, although filtration is not their only function. The details will be discussed later in this chapter.

The range of industrial applications to liquid filtration is very broad. Water purification is one of the most important applications in liquid filtration. The processes of water purification include microfiltration (MF), ultrafiltration (UF), nanofiltration (NF), and reverse osmosis (RO), depending on the size of the undesired particles/molecules to be removed. Some examples of industrial applications including fuel filters, water filters, carbon filter, and wet/dry filters are being widely used in the production of textile, cosmetics, food and beverage, electronic components, papers, ship builders, and oil and gas wells.

13.1.2 Liquid Filtration Classification

Filtration systems can be classified according to the following criteria.

(a) Flow direction of liquid with respect to filter media: cross-flow versus dead end
(b) Type of filter medium: screen versus depth filter
(c) Operation of filtration: continuous versus single use
(d) Pore size: MF, UF, NF, and RO
(e) Driving forces generated across filters: differences of pressure, concentration, temperature, and electrical potential

Among the above categories, the pore size classification is most widely used because it is directly related to the properties of impurities to be removed. In Fig. 13.1 [1], the filtration spectrum presents the relationships among pore size, impurity materials, and filtration technology in varying filter media. In Sect. 13.3.1, a description of those filtration membranes made of electrospun nanofibers is summarized.

13.1.3 Liquid Filter Media

The filter medium is the core of the whole filtration process. A premium filter with a desired selectivity should have a defined pore size and a narrow pore size distribution. In addition, filters with high porosity, smooth surface, and liquid affinity should be beneficial toward a higher permeation rate and lower fouling, as the feed liquid is being passed from the filter surface into the porous pathway inside the filter.

There are two major types of filter media: surface filter and depth filter. In the former, like sieving, solid particles are blocked at the surface of the filter (e.g., Buchner funnel, cross-flow filter, and screen filter). In a depth filter, a relatively thick and granular medium of varying porosity and density retains the solid particles

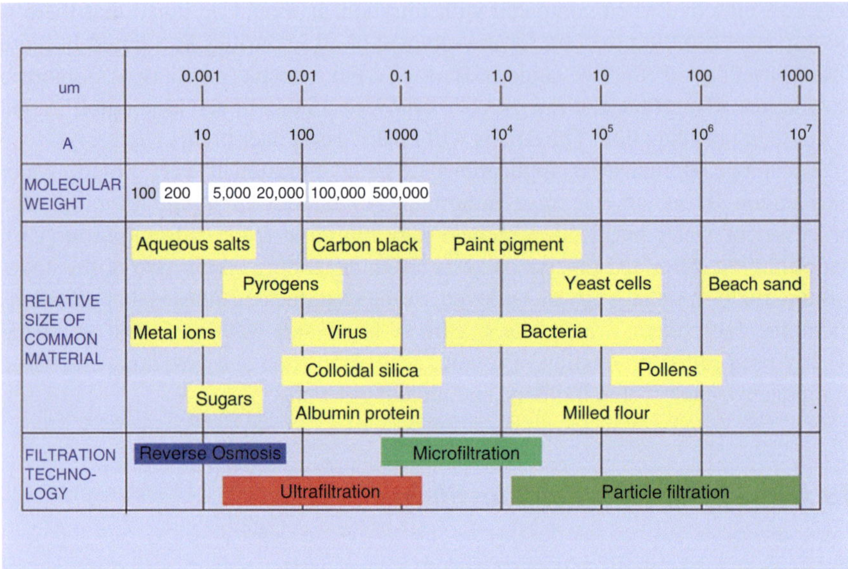

Fig. 13.1 The filtration spectrum [1] (Reprinted with permission from Ref. [1]. Copyright 2007, Elsevier)

inside as they pass through the filter (e.g., sand filter). Surface filters allow the solid particles to be collected on the filter surface, usually forming a caked layer; depth filters sort the particles by density and pore diameter, with smaller particles floating to the top and larger particles to the bottom. Both kinds of filters suffer the same problem of fouling, which is a phenomenon that oversized or charged particles may aggregate and form a caked layer blocking the porous structure and preventing the liquid from penetrating through the filter. The antifouling ability of the filter medium can be improved by using a variety of methods, such as precoagulation, backwashing, or/and increasing the hydrophilic nature of the filter.

Depending on the particular filtration devices and applications, various filter media have been used, such as sand [2], charcoal [3], cotton [4], wool [5], zeolite [6], earthenware [7], ceramic [8], synthetic nanofibrous membranes [9], and more sophisticated laboratory biological membranes [10]. In this chapter, we focus on filters based on electrospun nanofibers.

13.1.4 Electrospun Filters

Electrospinning technology has gained attention to create nanofibrous membranes. Specifically, ENMs possess higher porosity (typically around 80 %), lower base weight, larger effective surface area (up to 40 m^2/g depending on the fiber diameter and porosity), and continuously interconnected pores, when compared with conventional polymeric or ceramic membranes [11, 12]. These attributes can

Table 13.1 Major companies with electrospun nanofiber filters

Name	Country	Website
Donaldson	USA	www.donaldson.com
Esfil Tehno	Estonia	www.esfiltehno.ee
Electrospinning Company	UK	www.electrospinning.co.uk
Electrospunra	Singapore	www.electrospunra.com
Elmarco	Czech	www.elmarco.com
eSpin	USA	www.espintechnologies.com
Finetex	Korea	www.ftene.com
Toray	Japan	www.toray.com
Sorbent	Russia	www.sorbent.su

directly improve the flux performance without sacrificing the contaminant rejection ratio. The unique structures in nanofibrous membranes can benefit not only water filtration but also separations in liquids in general.

Advantages of electrospun filters have attracted industrial companies for further development and production. Table 13.1 lists major industrial players in the electrospun nanofibrous filters. However, major challenges on scale-up productions from laboratory practice remain.

13.2 Preparation and Characterization of Electrospun Filters

13.2.1 Preparation of Electrospun Filters

13.2.1.1 Historical Background

In 1934, Formals first patented an experimental setup (US Patent 1975504), termed electrospinning, to produce cellulose acetate polymer filaments from acetone using electrostatic forces. In 1936, I. V. Petryanov-Sokolov first produced fine electrospun fibers, known as filters of Petryanov [13]. In the 1960s, Taylor first reported the jet formation from the droplet of polymer solution. They noted that when the solution surface tension balanced the electric strength, a stable conical shape of the droplet was formed. This jet was called the Taylor cone [14]. In the 1970s, Baumgarten investigated the control on the morphology of electrospun nanofibers by changing the polymer solution viscosity. Fibers obtained from more concentrated solution were observed to be thicker [15]. From the 1990s, electrospinning has been further developed as a crucial method to fabricate nanofibers [16–18]. By controlling the electrospinning conditions and polymer/solution mixtures, nanofibers with different diameters (10–2,000 nm) and shapes could be produced. The simple deposition of electrospun nanofibers on the collector could form an interconnected nonwoven structure. Potential applications based on such nanofibrous membranes specifically as liquid filtration membranes were subsequently realized [19].

Table 13.2 Polymers and solvents used to prepare electrospun membranes

Polymer	Solvent	References
Nylon-6	Formic acid	[20, 21]
PAN	DMF	[22, 23]
PVA	Water	[25]
PET	Formic acid	[24]
PS	DMF, chloroform	[26, 27]
Polyimide	DMAc	[31, 32]
PES	DMF	[30]
PEO	DMF, chloroform, ethanol, and water	[33]
PU	DMF	[34]
PVDF	Acetone, DMAc	[28]
PC	Chloroform, THF, DMF	[35]
PCL	TFE	[36]
PBI	DMAc	[37, 38]
PVP	DMF	[39]
PTT	TFA/MC	[40]
PVC	THF	[41]
PMMA	Acetone	[42]
HPC	Ethanol, 2-propanol	[43]
PHB	Chloroform	[45]
PGA	Water	[44]
CA	Acetone, acetic acid, DMAc	[46, 47]
Chitosan	TFA	[48]
Chitin	HFIP	[49]
Collagen	HFP	[50]
Gelatin	Water	[51]
HA	Water	[52]
DNA	Water	[54]

13.2.1.2 Materials in Electrospinning

Various polymers have been successfully electrospun from solution, sol–gel suspension, or melt into ultrafine nanofibers. For example, as listed in Table 13.2, these polymers included nylon-6 [20, 21], polyacrylonitrile (PAN) [22, 23], polyethylene terephthalate (PET) [24], polyvinyl alcohol (PVA) [25], polystyrene (PS) [26, 27], polyvinylidene fluoride (PVDF) [28, 29], polyethersulfone (PES) [30], polyimides (PI) [31, 32], polyethylene oxide (PEO) [33], polyurethanes (PU) [34], polycarbonates (PC) [35], polycaprolactone (PCL) [36], polybenzimidazole (PBI) [37, 38], polyvinylpyrrolidone (PVP) [39], polytrimethylene terephthalate (PTT) [40], polyvinyl chloride (PVC) [41], polymethylmethacrylate (PMMA) [42], hydroxypropyl cellulose (HPC) [43], polyglycolic acid (PGA) [44], polyhydroxybutyrate (PHB) [45], cellulose acetate (CA) [46, 47] and many more. To be used as filtration membranes, nanofibers made of water-soluble polymers have to be further crosslinked after the electrospinning process. As polymers having different physical and

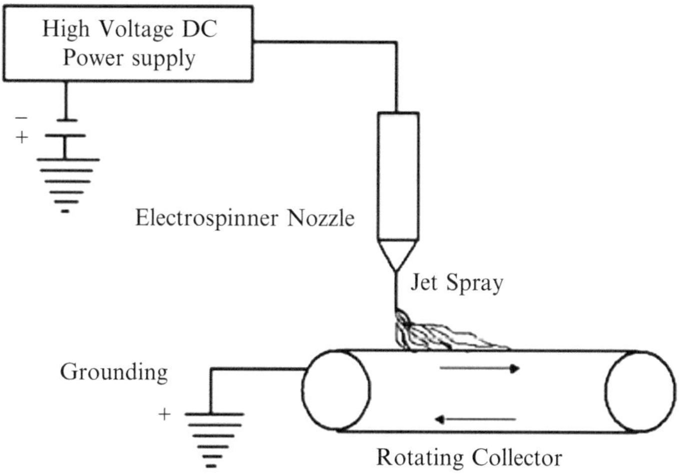

Fig. 13.2 Schematic diagram of electrospinning [71] (Reprinted with permission from Ref. [60]. Copyright 2009, John Wiley and Sons)

chemical properties, copolymers and block polymers have been utilized to achieve desired membrane properties in the nanofiber form.

Natural biomacromolecules have also been used for electrospinning into nanofibers, These polymers include chitosan [48], chitin [49], collagen [50], gelatin [51], hyaluronic acid [52, 53], lipase, silk, and DNA [54].

Another class of polymers used in electrospinning is nanocomposites, containing the matrix polymer and nanoparticles such as Ag [55], Pd [56], ZnO [57], CuO [58], SnO_2 [59], MgO [60], TiO_2 [61], Al_2O_3 [62], NiO [63], CeO_2 [64], Mn_3O_4 [65], MoO_3 [66], Y_2O_3 [67], Gd_2O_3 [68], PbS [68], CdS [69], Co_3O_4 [70], and other oxides. The objective of using these materials is to develop a new class of bimetallic hybrid nanocomposites in the nanofiber format, which are of great interest because the resulting product combines the advantages of large specific surface areas of nanoparticles and anisotropic properties of highly porous nanofibrous membranes.

13.2.1.3 Preparation Techniques

Electrospinning is a fabrication method that can create a nonwoven nanofibrous structure. In the typical electrospinning process (Fig. 13.2) a high electric potential is used to charge the droplet of polymer solution or melt and to stretch it to form a jet stream before it is solidified to form a polymer fiber on the collector [71]. Usually, an electrode is placed near the spinning solution and another by the collector. As the electric field strength is increased, the hemispherical surface of the solution at the tip of the spinneret elongates to form a conical shape. With further increase in the electric field strength, the Taylor cone is formed at a critical value and a charged jet of solution is ejected from the tip of the Taylor cone. The discharged polymer solution jet is further stretched, especially in a whipping process. In the meantime, the solvent is also being evaporated, leaving behind a solidified polymer fiber, which lays itself on a grounded collector.

Modification of the apparatus and solution can enhance and create unique structures and properties of electrospun nanofibers. For examples, a coaxial electrospinning setup can generate core-shell nanofibers from immiscible fluids and porous nanofibers from miscible fluids [36, 72]. Emulsion electrospinning is similar to the normal electrospinning, but the starting fluid is an emulsion with a water phase and an immiscible solvent phase. After electrospinning, the resulting nanofibers can possess the core-shell or composite fibers [73]. To deal with the environmental concerns of solvent evaporation in great quantity during typical electrospinning, the use of molten polymers, such as PET and PE, has been demonstrated [74]. However, the diameters of melt-electrospun nanofibers are usually much larger than those obtained from solution electrospinning.

Another innovation in the electrospinning process is to increase the production of stable nanofibers by using multiple-jet format containing secondary electrodes [40]. Also in 2003, a novel method to produce nanofibers was patented, called Nanospider, by Elmarco, a Czechic company, and its academic partner, Technical University of Liberec. The new technology is a needle-free, high-voltage, free liquid surface process. The Taylor cones and solution flow are generated from the tip of a capillary and a thin film of a polymer solution.

13.2.2 *Characterization of Electrospun Filters*

13.2.2.1 Electrospun Filter Morphology

Fundamental studies on the electrospinning process have revealed pathways to control the nanofiber morphology, structure, surface functionality, and appropriate processing conditions for electrospinning of various polymers. In order to prepare an effective filter, efforts have been made to produce nanofibrous membrane having a relatively uniform thickness and narrow fiber diameter size distributions. The morphological characteristics depend mainly on solution properties and operating conditions. The important solution characteristics, such as solvent vapor pressure, viscosity, surface tension, and conductivity, depend on the nature of polymer and of solvent, polymer molecular weight, and molecular weight distribution, while the important features of operating conditions depend on the applied electric field strength, distance between spinneret and collector, temperature, and humidity. Figure 13.3 shows SEM images of PAN nanofibers, with diameters from 80 nm to hundreds of nanometers. At higher polymer concentrations, polymer chains could be entangled to a greater extent, thereby favoring the formation of thicker fibers [22]. Xia and co-workers demonstrated that increasing the conductivity of the electrospinning solution, by adding a small amount of salt, will yield

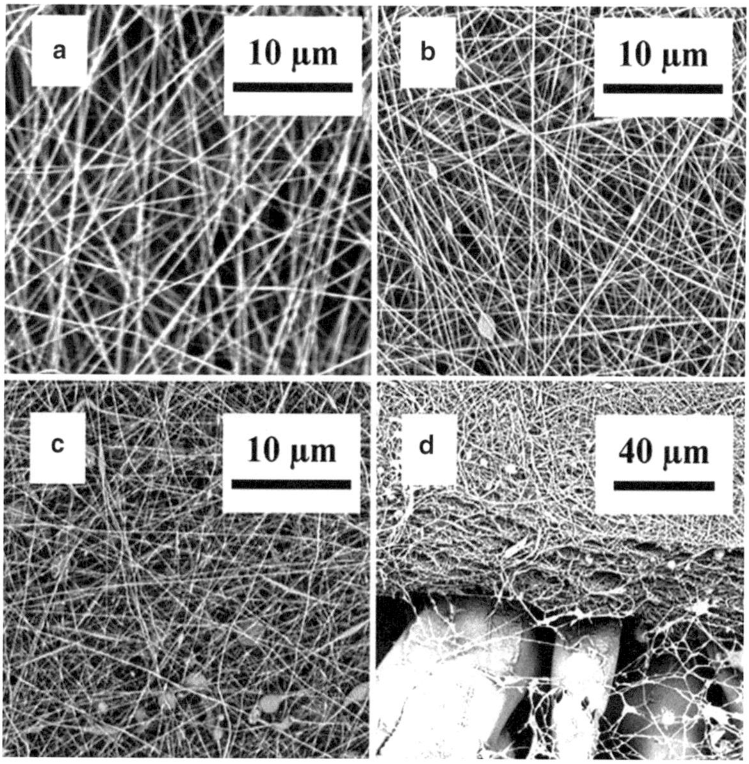

Fig. 13.3 SEM images of surfaces of PAN nanofibrous membranes electrospun from PAN solutions at 8 wt% (**a**), 6 wt% (**b**), and 4 wt% (**c**), respectively, and the cross section of a membrane electrospun from the 6 wt% PAN solution (**d**) [22] (Reprinted with permission from Ref. [18]. Copyright 2012, Elsevier)

bead-free nanofibers with smaller diameters [75]. In general, more concentrated solution, lower conductivity, smaller electric field strength, and smaller feeding solution rate will result in a thicker nanofiber. The semiempirical findings are in reasonable agreement with theoretical modeling carried out by Rutledge and co-workers [17]. A diameter reduction in nanofibers by means of electrospinning can be related to stretching forces exerted on the entangled polymer chains in the jet stream. Depending on the electrospinning conditions, the nanofiber diameter could vary from 10 to 2,000 nm. Another concern of nanofiber morphology is the presence of defects, such as beads or spindles. Reneker and co-workers summarized the formation of beaded nanofibers being from an unbalance of surface tension, viscosity, and electrical forces at a certain area of the jet tip. Beads can be eliminated by decreasing the surface tension and/or increasing the viscosity and electrical forces.

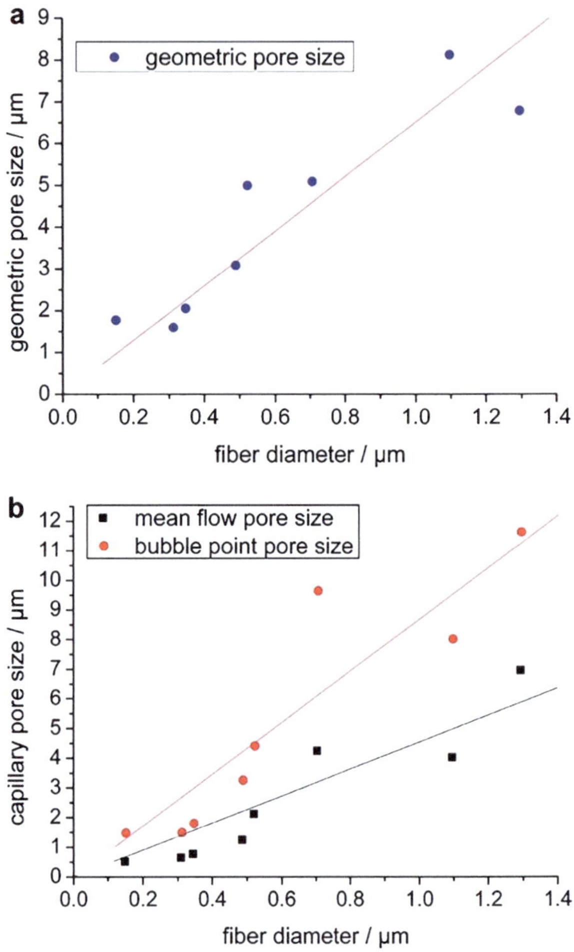

Fig. 13.4 Dependence of pore diameters on fiber diameters for PAN-based nonwovens: (**a**) geometric pore size; (**b**) capillary pore sizes, largest pore sizes (*bubble-point pores*) and average pore sizes (*mean flow pores*) [77] (Reprinted with permission from Ref. [66]. Copyright 2010, Elsevier)

13.2.2.2 Electrospun Filter Porometry

Pore size is directly related to the filtration performance. Unlike the sieve, electrospun nanofibrous filters do not have regular pores. The effective maximum pore size of electrospun filters is reflected as the maximum size of particles which can pass through the interconnected tortuous path of the membrane. In practice, the maximum pore size and mean flow pore size are measured by means of a capillary porometer [76]. Hussain and co-workers correlated the mean fiber diameter and pore size of ENMs [77]. In Fig. 13.4, they found that both values of bubble-point pore size and mean flow pore size could be positively correlated with fiber diameters. The average correlation ratio of the geometric pore size (diameter) to the fiber diameter amounted to about 6. The number became about 5 for the mean capillary pore size (mean flow pore size) and about 9 for the largest capillary pore

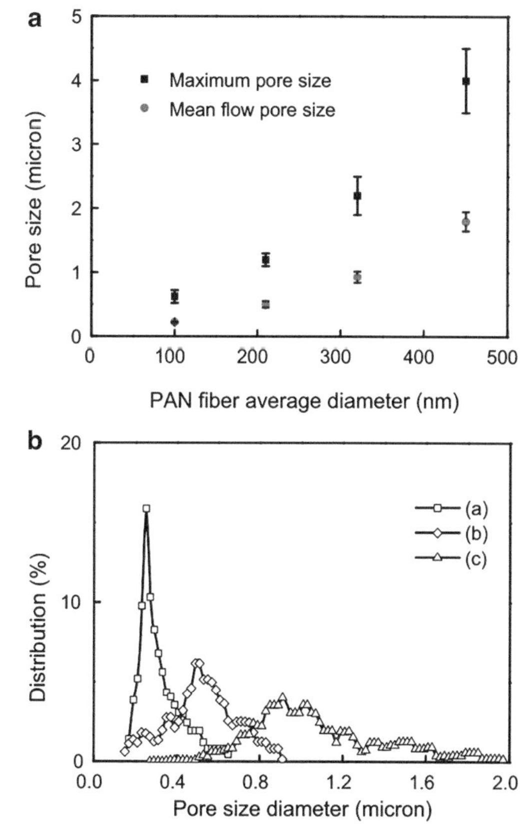

Fig. 13.5 Effects of the nanofiber diameter on the maximum pore size and mean flow pore size (**a**) and pore size distribution (**b**) of the electrospun PAN membrane. Membranes a, b, and c have average fiber diameters of 100, 210, and 320 nm, respectively [22] (Reprinted with permission from Ref. [18]. Copyright 2012, Elsevier)

size (bubble-point pore size). In Fig. 13.5, Chu and co-workers confirmed such a correlation. They also found out that membranes with thinner fiber diameters not only have a smaller pore size but also a narrower pore size distribution [22]. In a planar random deposition with the same amount of materials, thinner fibers could lead to more fiber crossings in the network, which in turn could reduce the effective pore size. Besides experimental measurements on porosity, the correlation between porosity and fiber morphology was also investigated by computational simulations [78]. In Fig. 13.6, Tomadakis and co-workers correlated the average fiber diameter, the average fiber radius, and the porosity of nonwovens. For a given porosity, the average pore size decreased with fiber radius, and there was a simulated correlation ratio, corresponding to the porosity.

Another advantage, based on nanometer-sized diameters, is the high surface-to-volume ratio. Li and co-workers produced ENMs made of poly(D,L-lactide-*co*-glycolide). The result from a mercury porosimeter test showed that the specific surface area was 23.54 m^2/g, when the porosity was 91.63 % and the total pore volume was 9.69 mL/g [79]. Tan and co-workers fabricated ENMs by using a PEO/water solution. The specific surface area, obtained from Brunauer-Emmett-Teller (BET)

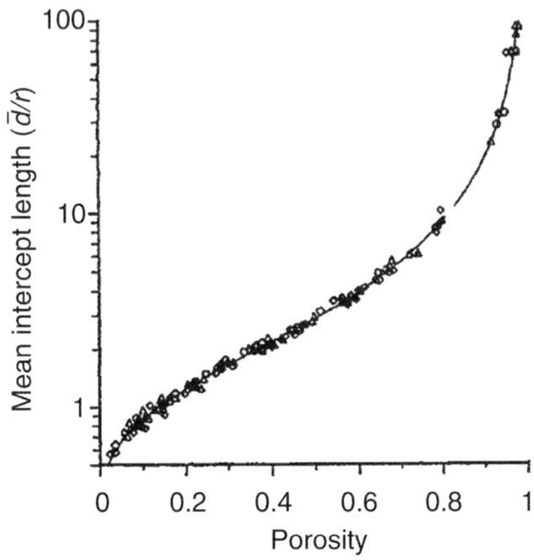

Fig. 13.6 Correlation of the average pore diameter (*d*), the average fiber radius (*r*), and porosity of nonwovens, as determined by Monte Carlo simulations [78] (Reprinted with permission from Ref. [67]. Copyright 1991, John Wiley and Sons)

measurements, ranged from 10 to 20 m^2/g [80]. Those values are several orders of magnitude higher than the number of ordinary textile fibers and even comparable to some mesoporous materials. In addition to the interconnected porous structure, the entire surface of electrospun filters is accessible.

13.3 Applications of Electrospun Nanofibrous Filters for Liquid Filtration

13.3.1 Membrane Filtration

13.3.1.1 MF Filter

The unique physical and chemical properties of electrospun nanofibrous filters can be used to improve the flux performance without sacrificing the contaminant rejection ratio. Stony Brook University and the National University of Singapore are two main players in utilizing ENMs in microfiltration.

In 2006, Gopal and co-workers first explored the viability of using ENM for liquid filtration and demonstrated the separation performance on particulate removal [81]. A complete self-supporting membrane was electrospun with PVDF in DMF/acetone solution, with heating as a posttreatment. The membrane had a thickness of 300 μm and a fiber diameter range of 380 ± 106 nm, with an effective pore size range of 4.0–10.6 μm, which were close to the properties of conventional microfiltration membranes. The filtration performance was examined by separation of 1-, 5-, and 10-μm polystyrene particles. More than 90 % of the microparticles were successfully rejected, as shown in Fig. 13.7. However, the fouling on the top layer of the membrane surface due to particle deposition compromised the

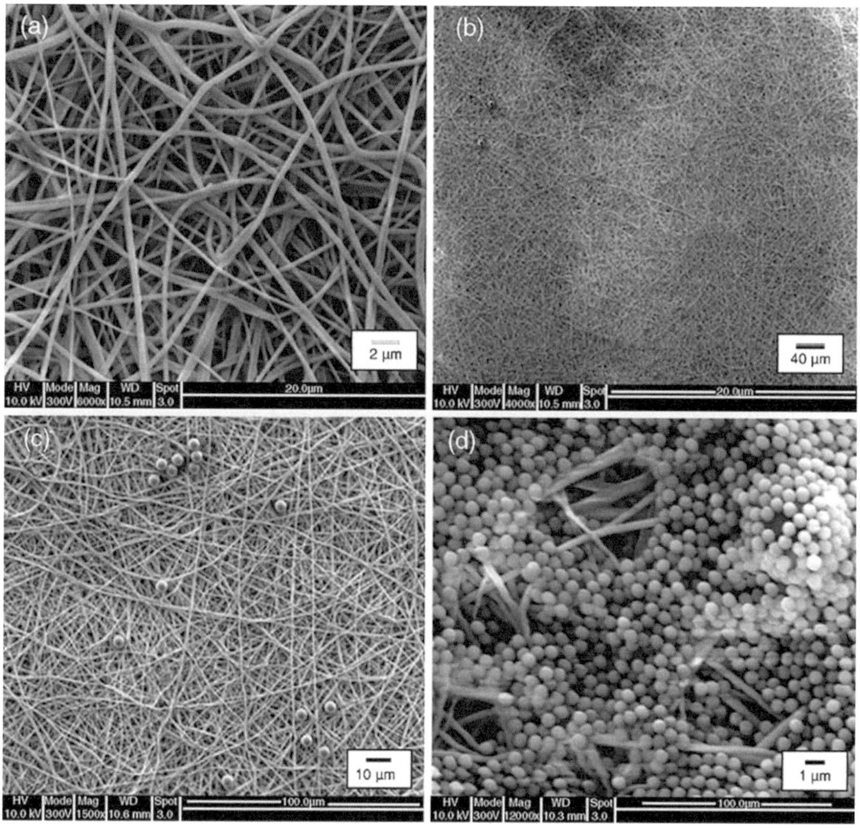

Fig. 13.7 FESEM micrographs of PVDF membrane: (**a**) before separation, (**b**) after 10-μm, (**c**) after 5-μm, and (**d**) after 1-μm separation [81] (Reprinted with permission from Ref. [70]. Copyright 2006, Elsevier)

permeation rate. With proper design on antifouling, there is reasonable potential using those membrane filters for microfiltration.

Daels and co-workers studied the use of a functionalized nanofibrous microfiltration membrane as an antibacterial water filter [82]. Polyamide nanofibrous membrane was electrospun with several biocides that are soluble in the electrospinning solvent, and biocides were incorporated partially on the nanofiber surface. Examples of biocides were silver nanoparticles, poly[(dimethylimino)(2-hydroxy-1.3-propanedily) chloride] (WSCP or Busan 77), 2,2-dibromo-3-nitrilopropionamide (DBNPA), thiocyanic acid, (2-benzothiazolylthio)methyl ester (Busan 72), and bronopol. The results from Fig. 13.8 showed that bacteria were not likely to pass the functionalized ENMs, with a log reduction ratio of 3–5.2, depending on the bacteria and biocides. The increased removal was contributed to by two factors: PA ENM had (1) a high surface area and (2) hydrophilicity that facilitated the interaction with aqueous bacteria, where functionalizing agents could then kill the bacteria

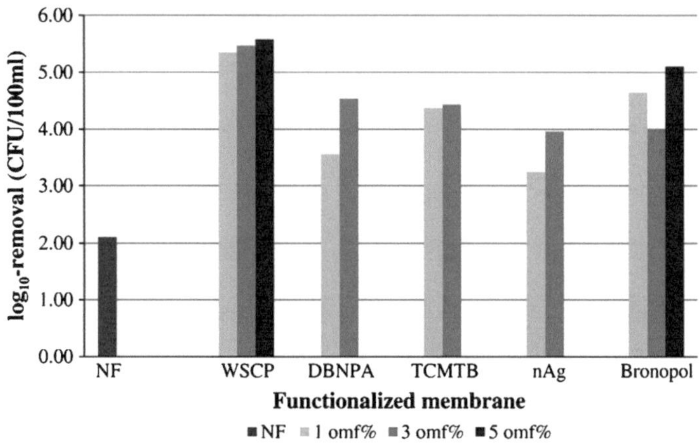

Fig. 13.8 Pathogen removal by nanofiber membrane filtration. Plate counts were performed on the influent (hospital wastewater) and filtrated through a nonfunctionalized (NF) membrane and functionalized nanofiber membranes with 1, 3, and 5 omf% WSCP, 1 and 3 omf% DBNPA, 1 and 3 omf% TCMTB, 1 and 3 % nAg, and 1, 3, and 5 omf% bronopol [82] (Reprinted with permission from Ref. [71]. Copyright 2011, Elsevier)

during the intimate contact. Kim and co-workers investigated similar antimicrobial ability of electrospun filters, with an additive of quaternary ammonium salt (benzyl triethylammonium chloride, BTEAC) [83].

Wang and co-workers studied the relation between the structural parameters of electrospun nanofibrous membranes (i.e., average fiber diameter, fiber diameter size distribution, membrane thickness, and porosity) and the microfiltration performance [22]. Based on the study, a high-flux nanofibrous membrane, containing a fairly small pore size (i.e., a maximum pore size of 0.62 ± 0.03 μm and a mean flow pore size of 0.22 ± 0.01 μm) based on the electrospun PAN nanofibrous scaffold and a nonwoven PET substrate, was fabricated. This membrane could successfully remove micron-sized particles and *E. coli* from contaminated water down to 1 ppm. Furthermore, the same membrane could achieve several times higher permeability than that of the commercial MF membrane (GSWP, Millipore) with a nominal 0.22-μm pore size. The Millipore filter had been widely used for bacteria removal and other water filtration applications. Electrospun membranes will be useful for drinking water purification as well as prefilters for UF, NF, or RO to minimize the possible fouling and contamination from microorganisms or microparticles.

Kaur and co-workers developed a method to fabricate hydrophilic ENM by using a blending technique [84]. In direct liquid penetration, hydrophilic membrane favors a high permeation rate and low fouling. PVDF, as a major solute of the electrospinning solution, was blended with hydrophilic surface modifying macromolecules, based on polyurethane prepared from the synthesis of bis(p-phenyl isocyanate) with PEGs and PPGs. PVDF membranes blended with MDI-PPG3500-PEG1000 had a static contact angle of 0°, compared with a value of 131.5° of the original PVDF membranes. The pure water flux of blended PVDF membranes was 20 % higher than that of original PVDF membranes. Also, the hydrophilicity of the modified membrane was influenced by the hydrophilicity of the macromolecules and their orientation incorporated in the nanofibers.

Homaeigohar and co-workers enhanced the mechanical properties of polyethersulfone ENMs by inducing interfiber adhesion through thermal treatment [85]. The thermal treatment was simple and cost-efficient. After electrospinning, the membranes were heated to extract residual solvent from the nanofibers, and the residual solvent could partially dissolve and stick the nanofibers to each other at the crossing points. This interfiber adhesion increased the mechanical properties of the nanofibrous membrane without a significant change in the porous structure.

Liu and co-workers prepared high-flux microfiltration filters made of PVA electrospun nanofibers [25]. The filters were fabricated by electrospinning of PVA onto a nonwoven support and followed by chemical cross-linking with glutaraldehyde in acetone. Due to the high porosity and hydrophilicity, the filters outperformed Millipore GSWP 0.22-μm filters with 3–7 times pure water flux and had an ability to reject 98 % of 0.2-μm-sized polycarboxylate microsphere particles. The PVA ENMs should be a good candidate as microfiltration filters for water purification.

Veleirinho and Lopes-da-Silva produced self-supporting nanofibrous PET membranes with good mechanical strength and applied them in the apple juice clarification process [86]. The experimental findings demonstrated that the ENM had a higher throughput, resulting in more economical operations than traditional processes. The same approach could be transferred to other applications in the food industry.

ENM filters with antimicrobial functionality have also attracted growing attention. Zhang and co-workers prepared PAN ENMs with fiber diameters of ~450 nm and treated the ENMs in hydroxylamine aqueous solution to introduce –C(NH2) = N–OH groups, which were used for coordination with Ag^+ ions to form silver nanoparticles. Then, the functionalized membranes were challenged by antimicrobial tests against *Staphylococcus aureus* and *Escherichia coli*. The results indicated that the new membranes became antimicrobial without distinguishable morphological variations [87].

13.3.1.2 UF, NF, and RO Filters

Halaui and co-workers first developed microscale hollow fibrous ultrafiltration membranes by using asymmetric microtubes produced by the co-electrospinning technique. PCL microtubes with an average inner diameter of 4 μm and a wall thickness of 0.9 μm and PVDF microtubes with an average inner diameter of 1.4 μm and a wall thickness of 0.9 μm were electrospun and then cut to the desired size. In Fig. 13.9, the gaps between the parallel fibers in a bundle were clogged by glue in order to prevent streaming in, and all the liquid flowed through inside the nanofibers. Microscale UF membranes exhibited relatively high rejection values and considerable flux [88].

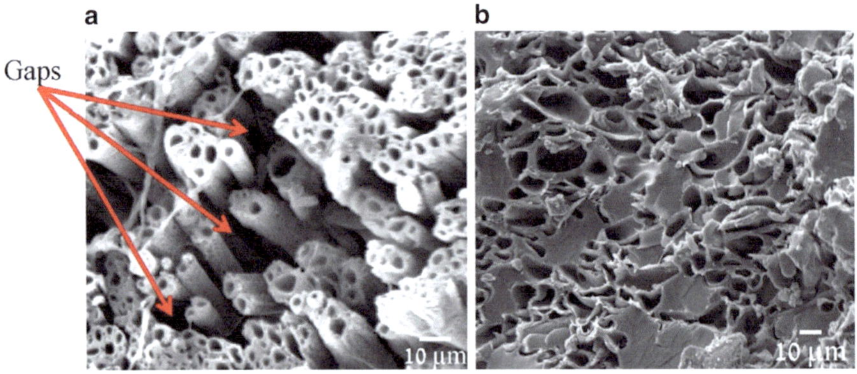

Fig. 13.9 (**a**) Hollow fibers before the gluing process. (**b**) Hollow fibers after the gluing process [88] (Reprinted with permission from Ref. [77]. Copyright 2011, Elsevier)

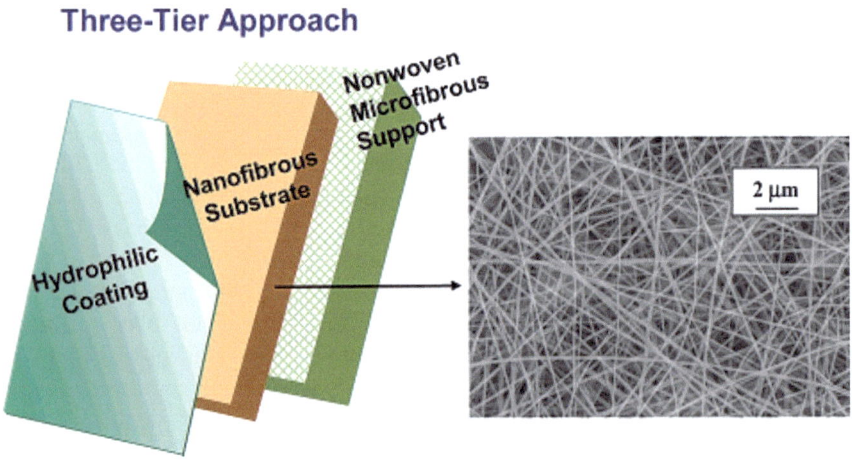

Fig. 13.10 Schematic diagram of the three-tier composite membrane (*left*) and representative SEM image of electrospun PVA substrate (*right*) [12] (Reprinted with permission from Ref. [11]. Copyright 2005, American Chemical Society)

Besides UF filters made of electrospun hollow fibers, there are more and more multiple-layer UF, NF, and RO filters made of ENMs. Traditionally, those filters were manufactured by phase immersion methods, with limitations of lower flux and higher fouling. Research at Stony Brook University showed that ENMs with high porosity and open connected pore structures could overcome those disadvantages [89].

Wang and co-workers constructed a high-flux UF medium for oil/water emulsion separation [12]. As shown in Fig. 13.10, the medium consisted of a three-tier composite structure, i.e., a nonporous hydrophilic nanocomposite top layer, an electrospun PVA nanofibrous substrate middle layer, and a conventional nonwoven

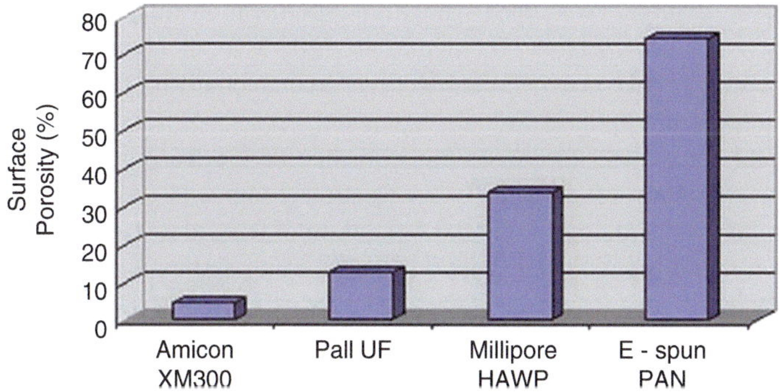

Fig. 13.11 Surface porosity of conventional ultrafiltration membranes and nanofibrous support (the porosity was measured by electron microscopy images) [11] (Reprinted with permission from Ref. [11]. Copyright 2006, Elsevier)

microfibrous support. The electrospun nanofibrous layer provided good mechanical strength, and an extremely lightweight and interconnected porous structure with a large specific surface area, making them excellent candidates as filtration supporting scaffolds. Combined with the top barrier layer, the oil/water emulsion tests showed a high flux rate (up to 330 L/m^2h at the feed pressure of 100 psi) and an excellent total organic solute rejection ratio (99.8 %).

Yoon and co-workers demonstrated that the torturous porosity of ENMs benefited the UF/NF filters [11]. They prepared a high-flux UF/NF composite mat based on an electrospun PAN nanofibrous scaffold coupled with a thin top layer of hydrophilic, water-resistant, but water-permeable chitosan coating. The porosity of the ENM support was larger than 70 %, which was much higher than that of conventional asymmetric porous membranes, often with porosity in the range of 30–40 % (Fig. 13.11). The filtration results exhibited that at the same rejection efficiency (>99.9 %) for oily wastewater filtration, the flux rate of the new composite filter was an order magnitude higher than those of commercial NF membranes over 24 h. of operation.

Tang and co-workers prepared a novel class of high-flux thin-film nanofibrous composite (TFNC) membranes, containing a thin hydrophilic top-layer coating, a PES nanofibrous mid-layer scaffold, and a PET nonwoven microfibrous support. Also, the layer-to-layer adhesion was evaluated, since weak adhesion will lead to delamination during handling or operation. The adhesion was improved by adjusting various processing parameters, such as using mixed solvent, increasing PES concentration and flow rate, and adding high molecular weight PEOs [90].

Yung and co-workers demonstrated a TFNC membrane as NF filters [91]. The investigated TFNC membranes contained an electrospun PES nanofibrous scaffold as the mid-layer, a PET nonwoven as the mechanical support, and an ultrathin highly cross-linked amine polymer layer, formed by interfacial polymerization, as the barrier layer. By optimizing the interfacial polymerization of top barrier layer, TFNC membranes showed comparable permeation flux and rejection ratio with those of a commercial NF membrane (NF-270). It should be anticipated that substantial further improvement in performance could be feasible with better processing tools and optimization of parameters.

Kaur and co-workers discussed the influence of electrospun fiber size on the separation efficiency of thin-film nanofiltration composite membrane [92]. The experimental results indicated that the separation of salts increased with decreasing electrospun nanofiber diameter, with a corresponding decrease in the flux. So, the best combination was to decrease the thickness of the ENM layer, together with smaller fiber diameter; the desired result of an increase in the flux together with a high salt rejection should be feasible. For practical water treatment applications, the pore size and thickness of the ENM layer could be adjusted to achieve the desired combination of rejection and flux.

Kim and co-workers demonstrated a novel RO filter, made of a sulfonated poly(arylene ether) top barrier layer and a middle layer of end-group cross-linked poly(arylene ether) electrospun nanofibers [93]. The salt rejection was measured at 1,000 psi in a dead-end filtration system. The order of salt rejection was as follows: Na_2SO_4 (88.4 %), NaCl (79.3 %), $MgSO_4$ (70.2 %), and $MgCl_2$ (62.3 %). Also, the NaCl rejection and permeability of the composite membranes were controlled as the different sulfonation group ratio of poly(arylene ether).

13.3.1.3 FO Filter

Forward osmosis (FO) is an osmotic process for water purification, with very low energy input. Unlike MF, UF, NF, and RO, FO filters purify the water by taking advantage of the osmotic pressure gradient across the membrane. Water diffuses naturally from one side of the membrane with the higher osmotic pressure gradient, usually due to higher salt concentration, to the side with the low osmotic pressure gradient. This low-energy process can be used in desalination, wastewater treatment, and energy generation.

Conventional FO membranes have a similar structure with UF, NF, and RO, consisting of a top thin barrier layer and a thick support layer. The drawback is due to the phenomenon of internal concentration polarization (ICP), caused by the tortuous and dense support layer hindering the compensate diffusion passing through the support layer. ICP leads to a lower water flux, and it gets worse with solute concentration increase [94]. Most of the conventional FO membranes have water flux rate of less than 25 L/m^2h. Loeb and co-workers described that the appropriate support layer for FO should have low tortuosity, high porosity, and a thin structure [95].

ENMs have recently been demonstrated as a novel supporting layer for FO, and the structural advantages of ENMs guarantee direct paths for salt and water diffusion, leading to an improvement in the flux rate. In Fig. 13.12a, SEM images of the cross section of the RO membranes presented the differences. Compared with phase inversion FO (PI-FO), nanocomposite FO (NC-FO) has unique structural

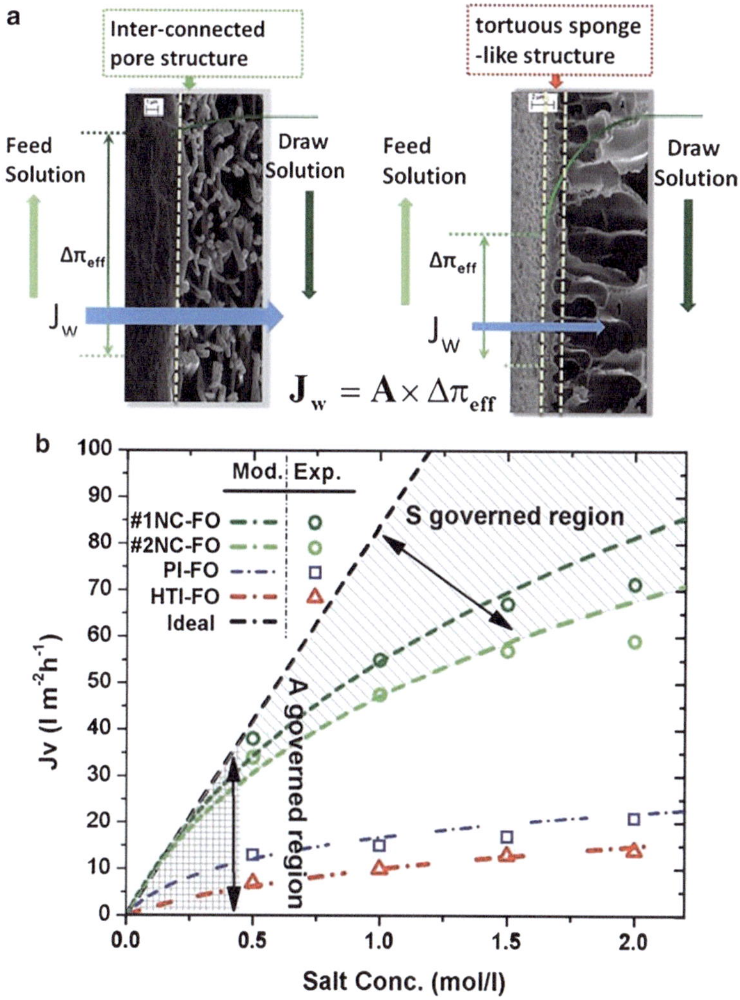

Fig. 13.12 (**a**) Comparison of the salt concentration profiles for the NC-FO (*left*) and PI-FO (*right*) membranes; (**b**) the curves show water fluxes of different FO membranes as a function of salt concentrations, where the feed solution is deionized (DI) water and draw solute is NaCl [96] (Reprinted with permission from Ref. [85]. Copyright 2011, John Wiley and Sons)

advantages, such as the interconnected pore structure and high porosity, favoring faster salt diffusion and higher osmotic driving force. In Fig. 13.12b, the authors combined the modeling and experimental data of the osmosis results to compare the two NC-FO, the PI-FO, and the Hydration Technology Innovations FO (HTI-FO). The results confirmed the fact that NC-FO had a higher water permeation value and a smaller value of the structure parameter [96].

Bui and co-workers prepared a thin-film composite (TFC) membrane for potential FO application. The TFC membrane consisted of an ENM support layer and a polyamide skin layer formed by in situ polymerization. The best TFC membranes exhibited 2–5 times higher flux with up to 100 times lower salt flux than conventional FO membranes [97].

13.3.2 Membrane Distillation

13.3.2.1 Water Desalination

Membrane distillation (MD) is a new and promising technology for water desalination because of the high salt rejection capability, no extensive pretreatment, and low energy consumption. The driving force in MD is the thermal difference, which leads vapor molecules migrate through the membrane pores from the higher vapor pressure side to the lower side. Other water desalination technologies, such as reverse osmosis and distillation, usually have a higher flux rate than conventional MD membranes, but new MD membranes made of electrospun nanofibers have attracted more and more attention.

In 2008, Feng and co-workers first demonstrated a PVDF ENM could be used in MD to produce drinking water (NaCl concentration < 280 ppm) from a saline water of NaCl concentration of 6 % by using an air-gap membrane distillation process. The flux rate was comparable to those of commercial MF membranes (5–28 kg/m^{2}h) at temperature differences ranging from 25 °C to 83 °C [98].

Essalhi and Khayet demonstrated the feasibility of ENMs for desalination by using a direct contact membrane distillation system (DCMD) and analyzed the PVDF ENMs thickness effect on the DCMD performance [99]. The PVDF ENMs in DCMD had a permeation flux of 15.2×10^{-3} kg/m^{2}s and a salt (NaCl) rejection ratio of 99.39 %, with a feed temperature of 80 °C. After 25 h, the PVDF ENMs stayed dry and the distilled water permeate flux lost less than 5 %. The membrane thickness was analyzed and optimized for a better MD performance. As shown in Fig. 13.13, the permeate flux decreased with the END thickness, because of the energy loss by heat conduction through the ENMs. So, the design of ENM thickness should be balanced with the void volume fraction, sizes of interfiber space, and the mechanism of mass transport through ENMs. In a theoretical study, Essalhi and Khayet applied the kinetic theory of gases through porous media to water vapor transport through ENMs and analyzed the polarization effects and thermal efficiency of self-sustained ENMs for DCMD. The model showed a correlation with the permeate flux of the PVDF ENMs over a wide range of feed temperature and salt concentration in DCMD and could predict the permeate value of other ENMs. Also, temperature polarization effect was found more dominate than concentration polarization effect [100].

Su and co-workers employed two kinds of ENM made of PVDF and poly(vinylidene fluoride-co-hexafluoropropylene) (PVDF-co-HFP) in DCMD. The authors optimized the feed solution flow rate and temperature differences for better permeate flux and salt rejection. Furthermore, the effect of hydrophobicity of

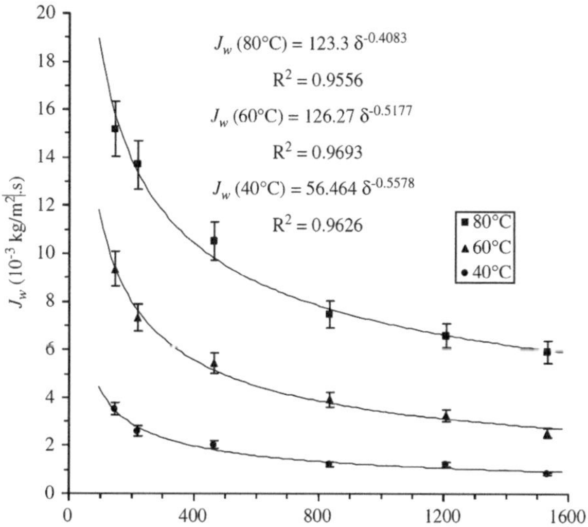

Fig. 13.13 Effect of PVDF ENMs thickness (δ) on the water permeate flux (J_w) at different feed temperatures ($T_{b,\delta}$). The stirring rate of the feed and permeate liquid solutions is 500 rpm and the permeate temperature ($T_{b,p}$) is 20° [99] (Reprinted with permission from Ref. [88]. Copyright 2013, Elsevier)

nanofibrous membrane on the distillation was examined. From a 12-h experiment, more hydrophobic PVDF-HFP membrane (water contact angle of 128°) had a higher NaCl rejection ratio (99.9901 %) than that of PVDF membrane (water contact angle of 125° and rejection ratio of 99.9888 %) [101].

Researchers from the Middle East spent a great deal of efforts on MD with ENMs. Lalia and co-workers prepared PVDF-HFP membranes and pressed them under high temperature to control the membranes dimensional stability and mechanical strength [102]. The data demonstrated that hot pressing could improve the characteristics of ENM for MD applications (i.e., smaller mean pore size, narrower pore distribution, good mechanical strength while keeping high permeability and porosity). Francis and co-workers fabricated ENMs of Matrimid (a soluble thermoplastic polyimide), with a high water contact angle of 130°, while a conventional phase inversion membranes of Matrimid had 85° as the contact angle. The improvement increased the water vapor flux of ENM to 56 L/m^{2}h in the DCMD test, which outperformed most of the other MD membranes fabricated by more conventional methods, including casting [103].

13.3.2.2 Other Membrane Distillations

The porous structures in electrospun nanofibrous membranes could result in much less resistance to mass transport than structures developed by other membranes. This advantage has been transferred to other materials applications.

Feng and co-workers introduced PVDF ENMs to remove volatile organic compounds (VOC) from water by gas strip membrane distillation (GSMD) [104]. PVDF ENMs with high surface hydrophobicity and appropriate pore sizes were tested. The overall mass transfer coefficient of chloroform was 2.40×10^{-5} m/s.

Table 13.3 Ethanol concentration in the permeate versus feed ethanol concentration and its comparison with vapor/liquid equilibrium [105] (Reprinted with permission from Ref. [94]. Copyright 2013, Elsevier)

Feed ethanol concentration (wt%)	Permeate ethanol concentration (wt%)	Vapor/liquid equilibrium (wt% alcohol)
20	38	67.0
40	37	77.3
60	80	81.6
80	85	87.1

That value was higher than the value of a hollow fiber membrane in GSMD. The reasons could be the surface hydrophobicity or the lower barrier layer resistance.

Feng and co-workers also applied PVDF ENMs to separate ethanol/water mixtures by vacuum membrane distillation (VMD) [105]. Table 13.3 summarized the comparison of ethanol concentration in feed solution and that in the permeate solution. The permeate flux could be positively correlated with the ethanol concentration in the feed solution and the operating temperature. The effective separation factors were changed from 1.42 to 2.45.

13.3.3 Membrane Adsorption

ENMs have unique structural properties for adsorption, including high specific surface area, interconnected pores, and the potential of introducing new functional groups as adsorption sites. In general, there are two methods to incorporate functional groups on the ENMs: blending of the active adsorbent in the electrospinning solution and modification of ENMs after electrospinning. The first method is simple and easy to operate but could sacrifice a substantial portion of the adsorbents, i.e., they might be embedded inside the ENMs. The second method guarantees that the functional sites are located on the surface of ENMs, but the linkage could be weak and more complex chemistry could result.

Ki and co-workers prepared ENMs of wool keratose/silk fibroin blend for heavy metal ion adsorption [106]. The ENMs have fibers of 200–400 nm, thickness of $\sim$ 126 μm, and Cu^{2+} adsorption value of 2.88 μg/mg. The ability to adsorb heavy metal ions could be attributed to the numerous polar groups which could bind positively charged metal ions, especially many amino acids from wool keratose. Furthermore, the ENMs could be reused by desorption. Lin and co-workers prepared PAN/$FeCl_2$ composite porous membranes based on blend electrospinning [107]. The composite membranes had an excellent adsorption of $Cr_2O_7{}^{2-}$, due to the existence of PAN Fe(II) bonds on the ENM surface, which were hydrolyzed during soaking in water and adsorbed the negative $Cr_2O_7{}^{2-}$. The Cr removal capacity was more than 110 mg Cr/g $FeCl_2$, but the hydrolysis and adsorption depend highly on the pH value.

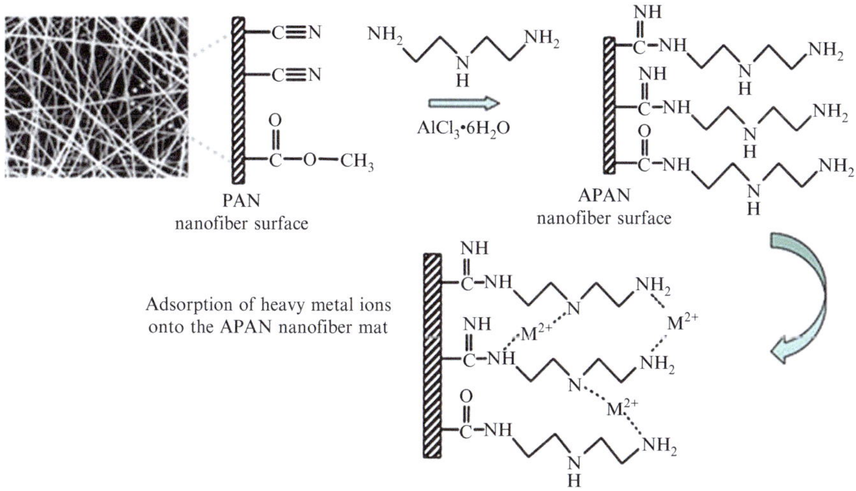

Fig. 13.14 Scheme of chemical reaction between PAN and DETA [109] (Reprinted with permission from Ref. [98]. Copyright 2010, American Chemical Society)

Park and co-workers prepared PAN functionalized ENMs, with chemical modification of amidoxime groups after electrospinning. The modification did not only increase the strength of the membranes but also gave the functional ENMs the saturation adsorption capacities for Cu(II) and Pb(II) of 52.70 and 263.45 mg/g, respectively [108]. As shown in Fig. 13.14, Kampalanonwat and Supaphol prepared ENMs with a combination of electrospinning and modified amidino diethylenediamine chelating groups in the polymer. The ability of functional ENMs to remove metal ions was tested, with four types of metal ions, including Cu(II), Ag(I), Fe(II), and Pb(II) ions. After the optimization of pH values and contact time, the maximal adsorption capacities of metal ions were 150.6, 155.5, 116.5, and 60.6 mg/g for Cu(II), Ag(I), Fe(II), and Pb(II) ions, respectively [109].

13.3.4 Membrane Bioreactor

Membrane bioreactor (MBR) is another promising process for water treatment. It combines conventional biological treatment processes with membrane filtration to provide an advanced level of organic and suspended solids removal. In an MBR system (Fig. 13.15), the membranes are submerged in an aerated biological reactor, and fouling must be controlled for cost and operating effectiveness. The collaboration of University College West Flanders and Ghent University from Belgium represents the main players in this field [110].

Daels and co-workers evaluated the use of ENMs in three different MBR setups, including an active sludge MBR (AS-MBR), an AS-MBR with flux performing

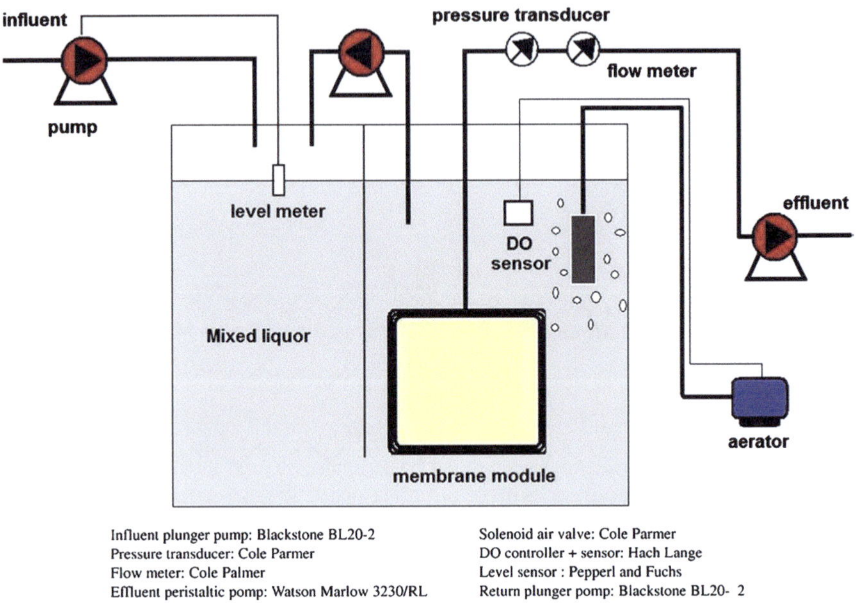

Fig. 13.15 Schematic presentation of MBR [110] (Reprinted with permission from Ref. [99]. Copyright 2009, Elsevier)

enhancer MPE50, and a trickling filter MBR (TF-MBR). The best performance was from TF-MBR, because the trickling filter and module configuration reduced membrane fouling. Furthermore, the ENMs remained effective with proper NaClO/HCl cleaning [82]. Bilad and co-workers optimized the preparation of ENMs for MBR applications by adding a nonwoven support and introducing heat treatment on the sheet surface at the feed side to prevent fouling. The results showed ENMs could be used as a substitute for conventional MBR membranes made by phase inversion [111].

13.3.5 Membrane Pervaporation

In 2012, Yeh and co-workers first applied ENMs in water/ethanol pervaporation [112]. As shown in Fig. 13.16, the separation membrane system consists of a cross-linked PVA hydrophilic top layer, a cellulose nanofibrous buffer layer, an ENM layer with high porosity and fully interconnected pore structure, and a PET support layer. In water/ethanol pervaporation test, the separation factor of the new membrane could reach 80.8, and flux was 765 kg/m^{2}h, which were significantly higher than conventional membranes (e.g., Sulzer 1210). Using ENMs promised the novel pervaporation membrane a higher permeate flux than conventional ones, and using the cellulose nanofibrous layer prevented the penetration of the barrier layer.

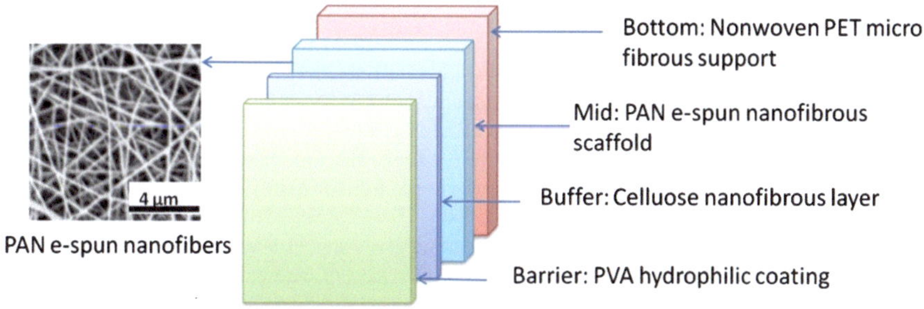

Fig. 13.16 Schematic representation of a high-flux pervaporation membrane with a four-layer structure [112] (Reprinted with permission from Ref. [101]. Copyright 2013, American Institute of Physics)

The combined features could lead to a new class of pervaporation membranes. The pervaporation performance of ENMs can be further improved by logical selection and thickness reduction of the barrier layer.

13.4 Summary and Outlook

In the last 20 years, electrospun nanofibrous membranes began to play an important role as a new filter medium in liquid filtration. In addition to the advance of fundamental understanding of electrospinning through physics, engineering, instrumentation, and theoretical modeling to improve the fabrication techniques, researchers from all over the world have also demonstrated the unique structural advantages of ENMs, including high porosity, interconnected pore size, and great potentials on functionalization. In almost every liquid filtration application, ENMs have been able to outperform conventional membranes fabricated essentially by using the phase inversion method. The performance of ENMS can be further improved. On the one hand, the control of the morphology of ENMs, such as fiber diameter (sub 10 nm), its distribution and orientation can be improved by gaining more fundamental and experimental data in processing. On the other hand, innovative blending of different polymers and more effective incorporation of nanocomposites in the matrix polymer can lead to even broader applications.

References

1. Pearce G (2007) Introduction to membranes: filtration for water and wastewater treatment. Filtr Separat 44:24–27
2. Benjamin MM, Sletten RS, Bailey RP, Bennett T (1996) Sorption and filtration of metals using iron-oxide-coated sand. Water Res 30:2609–2620

3. Khan AH, Rasul SB, Munir AKM, Habibuddowla M, Alauddin M, Newaz SS, Hussam A (2000) Appraisal of a simple arsenic removal method for groundwater of Bangladesh. J Environ Sci Health A Toxic/Hazard Subst Environ Eng 35:1021–1041
4. Badawy SM, Sokker HH, Othman SH, Hashem A (2005) Cloth filter for recovery of uranium from radioactive waste. Radiat Phys Chem 73:125–130
5. Lambertini E, Spencer SK, Bertz PD, Loge FJ, Kieke BA, Borchardt MA (2008) Concentration of enteroviruses, adenoviruses, and noroviruses from drinking water by use of glass wool filters. Appl Environ Microbiol 74:2990–2996
6. Tavolaro A, Drioli E (1999) Zeolite membranes. Adv Mater 11:975–996
7. Jones TC, Olveira MMG, Rocha H (1974) Effectiveness of earthenware household filters for water-purification. Lancet 2:275–277
8. Oyanedel-Craver VA, Smith JA (2008) Sustainable colloidal-silver-impregnated ceramic filter for point-of-use water treatment. Environ Sci Technol 42:927–933
9. Gopal R, Kaur S, Feng CY, Chan C, Ramakrishna S, Tabe S, Matsuura T (2007) Electrospun nanofibrous polysulfone membranes as pre-filters: particulate removal. J Membr Sci 289:210–219
10. Li XY, Chu HP (2003) Membrane bioreactor for the drinking water treatment of polluted surface water supplies. Water Res 37:4781–4791
11. Yoon K, Kim K, Wang XF, Fang DF, Hsiao BS, Chu B (2006) High flux ultrafiltration membranes based on electrospun nanofibrous PAN scaffolds and chitosan coating. Polymer 47:2434–2441
12. Wang XF, Chen XM, Yoon K, Fang DF, Hsiao BS, Chu B (2005) High flux filtration medium based on nanofibrous substrate with hydrophilic nanocomposite coating. Environ Sci Technol 39:7684–7691
13. Lushnikov A (1997) Igor' Vasilievich Petryanov-Sokolov (1907–1996) – Obituary. J Aerosol Sci 28:545–546
14. Melcher JR, Taylor GI (1969) Electrohydrodynamics – a review of role of interfacial shear stresses. Annu Rev Fluid Mech 1:111–146
15. Baumgart PK (1971) Electrostatic spinning of acrylic microfibers. J Colloid Interf Sci 36:71–79
16. Reneker DH, Chun I (1996) Nanometre diameter fibres of polymer, produced by electrospinning. Nanotechnology 7:216–223
17. Shin YM, Hohman MM, Brenner MP, Rutledge GC (2001) Experimental characterization of electrospinning: the electrically forced jet and instabilities. Polymer 42:9955–9967
18. Huang ZM, Zhang YZ, Kotaki M, Ramakrishna S (2003) A review on polymer nanofibers by electrospinning and their applications in nanocomposites. Compos Sci Technol 63:2223–2253
19. Bognitzki M, Czado W, Frese T, Schaper A, Hellwig M, Steinhart M, Greiner A, Wendorff JH (2001) Nanostructured fibers via electrospinning. Adv Mater 13:70–72
20. Ahn YC, Park SK, Kim GT, Hwang YJ, Lee CG, Shin HS, Lee JK (2006) Development of high efficiency nanofilters made of nanofibers. Curr Appl Phys 6:1030–1035
21. Ryu YJ, Kim HY, Lee KH, Park HC, Lee DR (2003) Transport properties of electrospun nylon 6 nonwoven mats. Eur Polym J 39:1883–1889
22. Wang R, Liu Y, Li B, Hsiao BS, Chu B (2012) Electrospun nanofibrous membranes for high flux microfiltration. J Membr Sci 392:167–174
23. Sato A, Wang R, Ma HY, Hsiao BS, Chu B (2011) Novel nanofibrous scaffolds for water filtration with bacteria and virus removal capability. J Electron Microsc (Tokyo) 60:201–209
24. Hong KH, Kang TJ (2006) Hydraulic permeabilities of PET and nylon 6 electrospun fiber webs. J Appl Polym Sci 100:167–177
25. Liu Y, Wang R, Ma HY, Hsiao BS, Chu B (2013) High-flux microfiltration filters based on electrospun polyvinylalcohol nanofibrous membranes. Polymer 54:548–556
26. Uyar T, Besenbacher F (2008) Electrospinning of uniform polystyrene fibers: the effect of solvent conductivity. Polymer 49:5336–5343
27. Wang C, Hsu CH, Lin JH (2006) Scaling laws in electrospinning of polystyrene solutions. Macromolecules 39:7662–7672

28. Kim JR, Choi SW, Jo SM, Lee WS, Kim BC (2004) Electrospun PVdF-based fibrous polymer electrolytes for lithium ion polymer batteries. Electrochim Acta 50:69–75
29. Seoul C, Kim YT, Baek CK (2003) Electrospinning of poly(vinylidene fluoride)/dimethylformamide solutions with carbon nanotubes. J Polym Sci B Polym Phys 41:1572–1577
30. Lin Y, Yao YY, Yang XZ, Wei N, Li XQ, Gong P, Li RX, Wu DC (2008) Preparation of poly(ether sulfone) nanofibers by gas-jet/electrospinning. J Appl Polym Sci 107:909–917
31. Chen D, Liu TX, Zhou XP, Tjiu WC, Hou HQ (2009) Electrospinning fabrication of high strength and toughness polyimide nanofiber membranes containing multiwalled carbon nanotubes. J Phys Chem B 113:9741–9748
32. Chung GS, Jo SM, Kim BC (2005) Properties of carbon nanofibers prepared from electrospun polyimide. J Appl Polym Sci 97:165–170
33. Son WK, Youk JH, Lee TS, Park WH (2004) The effects of solution properties and polyelectrolyte on electrospinning of ultrafine poly(ethylene oxide) fibers. Polymer 45:2959–2966
34. Kimmer D, Slobodian P, Petras D, Zatloukal M, Olejnik R, Saha P (2009) Polyurethane/multiwalled carbon nanotube nanowebs prepared by an electrospinning process. J Appl Polym Sci 111:2711–2714
35. Krishnappa RVN, Desai K, Sung CM (2003) Morphological study of electrospun polycarbonates as a function of the solvent and processing voltage. J Mater Sci 38:2357–2365
36. Zhang YZ, Huang ZM, Xu XJ, Lim CT, Ramakrishna S (2004) Preparation of core-shell structured PCL-r-gelatin bi-component nanofibers by coaxial electrospinning. Chem Mater 16:3406–3409
37. Weber J (2010) Nanostructured poly(benzimidazole): from mesoporous networks to nanofibers. Chemsuschem 3:181–187
38. Kim C, Park S–H, Lee W-J, Yang K-S (2004) Characteristics of supercapacitor electrodes of PBI-based carbon nanofiber web prepared by electrospinning. Electrochim Acta 50:877–881
39. Jin WJ, Lee HK, Jeong EH, Park WH, Youk JH (2005) Preparation of polymer nanofibers containing silver nanoparticles by using poly (*N*-vinylpyrrolidone). Macromol Rapid Commun 26:1903–1907
40. Fang DF, Chu B, Hsiao B (2003) Multiple-jet electrospinning of non-woven nanofiber articles. Abstr Pap Am Chem Soc 226:U424–U424
41. Lee KH, Kim HY, La YM, Lee DR, Sung NH (2002) Influence of a mixing solvent with tetrahydrofuran and *N*, *N*-dimethylformamide on electrospun poly(vinyl chloride) nonwoven mats. J Polym Sci B Polym Phys 40:2259–2268
42. Piperno S, Lozzi L, Rastelli R, Passacantando M, Santucci S (2006) PMMA nanofibers production by electrospinning. Appl Surf Sci 252:5583–5586
43. Shukla S, Brinley E, Cho HJ, Seal S (2005) Electrospinning of hydroxypropyl cellulose fibers and their application in synthesis of nano and submicron tin oxide fibers. Polymer 46:12130–12145
44. Boland ED, Wnek GE, Simpson DG, Pawlowski KJ, Bowlin GL (2001) Tailoring tissue engineering scaffolds using electrostatic processing techniques: a study of poly(glycolic acid) electrospinning. J Macromol Sci Pure 38:1231–1243
45. Guan DH, Chen ZQ, Huang CP, Lin YH (2008) Attachment, proliferation and differentiation of BMSCs on gas-jet/electrospun nHAP/PHB fibrous scaffolds. Appl Surf Sci 255:324–327
46. Liu HQ, Hsieh YL (2002) Ultrafine fibrous cellulose membranes from electrospinning of cellulose acetate. J Polym Sci B Polym Phys 40:2119–2129
47. Ma ZW, Kotaki M, Ramakrishna S (2005) Electrospun cellulose nanofiber as affinity membrane. J Membr Sci 265:115–123
48. Ohkawa K, Cha DI, Kim H, Nishida A, Yamamoto H (2004) Electrospinning of chitosan. Macromol Rapid Commun 25:1600–1605
49. Min BM, Lee SW, Lim JN, You Y, Lee TS, Kang PH, Park WH (2004) Chitin and chitosan nanofibers: electrospinning of chitin and deacetylation of chitin nanofibers. Polymer 45:7137–7142

50. Matthews JA, Wnek GE, Simpson DG, Bowlin GL (2002) Electrospinning of collagen nanofibers. Biomacromolecules 3:232–238
51. Huang ZM, Zhang YZ, Ramakrishna S, Lim CT (2004) Electrospinning and mechanical characterization of gelatin nanofibers. Polymer 45:5361–5368
52. Li JX, He AH, Zheng JF, Han CC (2006) Gelatin and gelatin-hyaluronic acid nanofibrous membranes produced by electrospinning of their aqueous solutions. Biomacromolecules 7:2243–2247
53. Ji Y, Ghosh K, Shu XZ, Li BQ, Sokolov JC, Prestwich GD, Clark RAF, Rafailovich MH (2006) Electrospun three-dimensional hyaluronic acid nanofibrous scaffolds. Biomaterials 27:3782–3792
54. Fang X, Reneker DH (1997) DNA fibers by electrospinning. J Macromol Sci Phys B36: 169–173
55. Zhao HT, Lu BG, Xu J, Xie EQ, Wang TH, Xu Z (2013) Electrospinning-thermal treatment synthesis: a general strategy to decorate highly porous nanotubes on both internal and external side-walls with metal oxide/noble metal nanoparticles. Nanoscale 5:2835–2839
56. Huang JS, Liu Y, Hou HQ, You TY (2008) Simultaneous electrochemical determination of dopamine, uric acid and ascorbic acid using palladium nanoparticle-loaded carbon nanofibers modified electrode. Biosens Bioelectron 24:632–637
57. Hassan MS, Amna T, Sheikh FA, Al-Deyab SS, Choi KE, Hwang IH, Khil MS (2013) Bimetallic Zn/Ag doped polyurethane spider net composite nanofibers: a novel multipurpose electrospun mat. Ceramics Int 39:2503–2510
58. Wang W, Feng ZY, Jiang W, Zhan JH (2013) Electrospun porous CuO-Ag nanofibers for quantitative sensitive SERS detection. Crystengcomm 15:1339–1344
59. Xu L, Xing RQ, Song J, Xu W, Song HW (2013) ZnO-SnO_2 nanotubes surface engineered by Ag nanoparticles: synthesis, characterization, and highly enhanced HCHO gas sensing properties. J Mater Chem C 1:2174–2182
60. Aykut Y (2013) Electrospun MgO-loaded carbon nanofibers: enhanced field electron emission from the fibers in vacuum. J Phys Chem Solids 74:328–337
61. Chuangchote S, Sagawa T, Yoshikawa S (2008) Efficient dye-sensitized solar cells using electrospun $TiO_{(2)}$ nanofibers as a light harvesting layer. Appl Phys Lett 93:033310
62. Dadvar S, Tavanai H, Morshed M (2011) UV-protection properties of electrospun polyacrylonitrile nanofibrous mats embedded with MgO and Al_2O_3 nanoparticles. J Nanopart Res 13:5163–5169
63. Guan HY, Shao CL, Wen SB, Chen B, Gong J, Yang XH (2003) Preparation and characterization of NiO nanofibres via an electrospinning technique. Inorg Chem Commun 6:1302–1303
64. Yang XG, Shao CL, Liu YC, Mu RX, Guan HY (2005) Nanofibers of CeO_2 via an electrospinning technique. Thin Solid Films 478:228–231
65. Shao CL, Guan HY, Liu YC, Li XL, Yang XH (2004) Preparation of Mn_2O_3 and Mn_3O_4 nanofibers via an electrospinning technique. J Solid State Chem 177:2628–2631
66. Li W, Zhao S, Qi B, Du Y, Wang XH, Huo MX (2009) Fast catalytic degradation of organic dye with air and MoO_3:Ce nanofibers under room condition. Appl Catal B Environ 92:333–340
67. Dong XT, Liu L, Wang JX, Liu GX (2010) Fabrication and characterization of Y_2O_3: Yb^{3+}, Er^{3+} upconversion nanofibers by electrospinning. Chem J Chin Univ 31:20–25
68. Azad AM, Matthews T, Swary J (2005) Processing and characterization of electrospun Y_2O_3-stabilized ZrO_2 (YSZ) and Gd_2O_3-doped CeO_2 (GDC) nanofibers. Mater Sci Eng B Solid State Mater Adv Technol 123:252–258
69. Bashouti M, Salalha W, Brumer M, Zussman E, Lifshitz E (2006) Alignment of colloidal CdS nanowires embedded in polymer nanofibers by electrospinning. Chemphyschem 7:102–106
70. Guan HY, Shao CL, Wen SB, Chen B, Gong J, Yang XH (2003) A novel method for preparing Co_3O_4 nanofibers by using electrospun PVA/cobalt acetate composite fibers as precursor. Mater Chem Phys 82:1002–1006

71. Feng C, Khulbe KC, Matsuura T (2010) Recent progress in the preparation, characterization, and applications of nanofibers and nanofiber membranes via electrospinning/interfacial polymerization. J Appl Polym Sci 115:756–776
72. Dong H, Nyame V, Macdiarmid AG, Jones WE (2004) Polyaniline/poly(methyl methacrylate) coaxial fibers: the fabrication and effects of the solution properties on the morphology of electrospun core fibers. J Polym Sci B Polym Phys 42:3934–3942
73. Qi HX, Hu P, Xu J, Wang AJ (2006) Encapsulation of drug reservoirs in fibers by emulsion electrospinning: morphology characterization and preliminary release assessment. Biomacromolecules 7:2327–2330
74. Lyons J, Li C, Ko F (2004) Melt-electrospinning part I: processing parameters and geometric properties. Polymer 45:7597–7603
75. Li D, Xia YN (2004) Electrospinning of nanofibers: reinventing the wheel? Adv Mater 16:1151–1170
76. Jena A, Gupta K (2010) Advances in pore structure evaluation by porometry. Chem Eng Technol 33:1241–1250
77. Hussain D, Loyal F, Greiner A, Wendorff JH (2010) Structure property correlations for electrospun nanofiber nonwovens. Polymer 51:3989–3997
78. Tomadakis MM, Sotirchos SV (1991) Effective Knudsen diffusivities in structures of randomly overlapping fibers. Aiche J 37:74–86
79. Li WJ, Laurencin CT, Caterson EJ, Tuan RS, Ko FK (2002) Electrospun nanofibrous structure: a novel scaffold for tissue engineering. J Biomed Mater Res 60:613–621
80. Deitzel JM, Kleinmeyer J, Harris D, Tan NCB (2001) The effect of processing variables on the morphology of electrospun nanofibers and textiles. Polymer 42:261–272
81. Gopal R, Kaur S, Ma ZW, Chan C, Ramakrishna S, Matsuura T (2006) Electrospun nanofibrous filtration membrane. J Membr Sci 281:581–586
82. Daels N, De Vrieze S, Sampers I, Decostere B, Westbroek P, Dumoulin A, Dejans P, De Clerck K, Van Hulle SWH (2011) Potential of a functionalised nanofibre microfiltration membrane as an antibacterial water filter. Desalination 275:285–290
83. Kim SJ, Nam YS, Rhee DM, Park HS, Park WH (2007) Preparation and characterization of antimicrobial polycarbonate nanofibrous membrane. Eur Polym J 43:3146–3152
84. Kaur S, Rana D, Matsuura T, Sundarrajan S, Ramakrishna S (2012) Preparation and characterization of surface modified electrospun membranes for higher filtration flux. J Membr Sci 390–391:235–242
85. Homaeigohar S, Koll J, Lilleodden ET, Elbahri M (2012) The solvent induced interfiber adhesion and its influence on the mechanical and filtration properties of polyethersulfone electrospun nanofibrous microfiltration membranes. Sep Purif Technol 98:456–463
86. Veleirinho B, Lopes-da-Silva JA (2009) Application of electrospun poly(ethylene terephthalate) nanofiber mat to apple juice clarification. Process Biochem 44:353–356
87. Zhang L, Luo J, Menkhaus TJ, Varadaraju H, Sun Y, Fong H (2011) Antimicrobial nanofibrous membranes developed from electrospun polyacrylonitrile nanofibers. J Membr Sci 369:499–505
88. Halaui R, Moldavsky A, Cohen Y, Semiat R, Zussman E (2011) Development of micro-scale hollow fiber ultrafiltration membranes. J Membr Sci 379:370–377
89. Burger C, Hsiao BS, Chu B (2006) Nanofibrous materials and their applications. In: Annual review of materials research. Annual Reviews, Palo Alto, pp 333–368
90. Tang ZH, Qiu CQ, McCutcheon JR, Yoon K, Ma HY, Fang DF, Lee E, Kopp C, Hsiao BS, Chu B (2009) Design and fabrication of electrospun polyethersulfone nanofibrous scaffold for high-flux nanofiltration membranes. J Polym Sci B Polym Phys 47:2288–2300
91. Yung L, Ma HY, Wang X, Yoon K, Wang R, Hsiao BS, Chu B (2010) Fabrication of thin-film nanofibrous composite membranes by interfacial polymerization using ionic liquids as additives. J Membr Sci 365:52–58
92. Kaur S, Sundarrajan S, Rana D, Matsuura T, Ramakrishna S (2012) Influence of electrospun fiber size on the separation efficiency of thin film nanofiltration composite membrane. J Membr Sci 392:101–111

93. Kim Y-J, Shim H-S, Lee J-S (2010) Desalination properties of a novel composite membrane prepared by electrospinning method. Desalin Water Treat 15:84–91
94. Gray GT, McCutcheon JR, Elimelech M (2006) Internal concentration polarization in forward osmosis: role of membrane orientation. Desalination 197:1–8
95. Loeb S, Titelman L, Korngold E, Freiman J (1997) Effect of porous support fabric on osmosis through a Loeb-Sourirajan type asymmetric membrane. J Membr Sci 129:243–249
96. Song XX, Liu ZY, Sun D (2011) Nano gives the answer: breaking the bottleneck of internal concentration polarization with a nanofiber composite forward osmosis membrane for a high water production rate. Adv Mater 23:3256–3260
97. Nhu-Ngoc B, McCutcheon JR (2013) Hydrophilic nanofibers as new supports for thin film composite membranes for engineered osmosis. Environ Sci Technol 47:1761–1769
98. Feng C, Khulbe KC, Matsuura T, Gopal R, Kaur S, Ramakrishna S, Khayet M (2008) Production of drinking water from saline water by air-gap membrane distillation using polyvinylidene fluoride nanofiber membrane. J Membr Sci 311:1–6
99. Essalhi M, Khayet M (2013) Self-sustained webs of polyvinylidene fluoride electrospun nanofibers at different electrospinning times: 1. Desalination by direct contact membrane distillation. J Membr Sci 433:167–179
100. Essalhi M, Khayet M (2013) Self-sustained webs of polyvinylidene fluoride electrospun nanofibers at different electrospinning times: 2. Theoretical analysis, polarization effects and thermal efficiency. J Membr Sci 433:180–191
101. Su CI, Shih JH, Huang MS, Wang CM, Shih WC, Liu YS (2012) A study of hydrophobic electrospun membrane applied in seawater desalination by membrane distillation. Fiber Polym 13:698–702
102. Lalia BS, Guillen-Burrieza E, Arafat HA, Hashaikeh R (2013) Fabrication and characterization of polyvinylidenefluoride-co-hexafluoropropylene (PVDF-HFP) electrospun membranes for direct contact membrane distillation. J Membr Sci 428:104–115
103. Francis L, Maab H, AlSaadi A, Nunes S, Ghaffour N, Amy GL (2013) Fabrication of electrospun nanofibrous membranes for membrane distillation application. Desalin Water Treat 51:1337–1343
104. Feng CY, Khulbe KC, Tabe S (2012) Volatile organic compound removal by membrane gas stripping using electro-spun nanofiber membrane. Desalination 287:98–102
105. Feng C, Khulbe KC, Matsuura T, Tabe S, Ismail AF (2013) Preparation and characterization of electro-spun nanofiber membranes and their possible applications in water treatment. Sep Purif Technol 102:118–135
106. Ki CS, Gang EH, Um NC, Park YH (2007) Nanofibrous membrane of wool keratose/silk fibroin blend for heavy metal ion adsorption. J Membr Sci 302:20–26
107. Lin YX, Cai WP, Tian XY, Liu XL, Wang GZ, Liang CH (2011) Polyacrylonitrile/ferrous chloride composite porous nanofibers and their strong Cr-removal performance. J Mater Chem 21:991–997
108. Saeed K, Haider S, Oh T-J, Park S-Y (2008) Preparation of amidoxime-modified polyacrylonitrile (PAN-oxime) nanofibers and their applications to metal ions adsorption. J Membr Sci 322:400–405
109. Kampalanonwat P, Supaphol P (2010) Preparation and adsorption behavior of aminated electrospun polyacrylonitrile nanofiber mats for heavy metal Ion removal. Acs Appl Mater Inter 2:3619–3627
110. Bjorge D, Daels N, De Vrieze S, Dejans P, Van Camp T, Audenaert W, Hogie J, Westbroek P, De Clerck K, Van Hulle SWH (2009) Performance assessment of electrospun nanofibers for filter applications. Desalination 249:942–948
111. Bilad MR, Westbroek P, Vankelecom IFJ (2011) Assessment and optimization of electrospun nanofiber-membranes in a membrane bioreactor (MBR). J Membr Sci 380:181–191
112. Yeh T-M, Yang L, Wang X, Mahajan D, Hsiao BS, Chu B (2012) Polymeric nanofibrous composite membranes for energy efficient ethanol dehydration. J Renew Sustain Ener 4:041406

Chapter 14
Protective Clothing Based on Electrospun Nanofibrous Membranes

Aikifa Raza, Yang Li, Junlu Sheng, Jianyong Yu, and Bin Ding

Abstract Recently, researchers have focused on the development of specialized protective clothing used to protect military personnel and emergency responders when they are at risk of exposure to chemical or biological threats such as toxic industrial chemicals or chemical or biological warfare agents. These breathable chemical and biological detoxifying protective fabrics with comparable comfort level containing peculiar membranes provide full barrier protection by absorbing or blocking toxic agents. Nanofibrous membranes, mainly fabricated by electrospinning, have exhibited great potential for many emerging applications including smart protective clothing. They can be considered as one of the safest nanomaterials due to their extremely long length (can be up to hundreds of kilometers) and their ability to be embedded within other media. The new generation of protective fabrics based on electrospun nanofibrous membranes is envisioned not only to absorb or block toxic chemical and biological agents but also to detoxify them to reduce the risk of secondary contamination. The aim of this chapter is to examine the use

A. Raza • Y. Li
State Key Laboratory for Modification of Chemical Fibers and Polymer Materials, College of Materials Science and Engineering, Donghua University, Shanghai 201620, China

J. Sheng
Nanomaterials Research Center, Modern Textile Institute, Donghua University, Shanghai 200051, China

J. Yu • B. Ding (✉)
State Key Laboratory for Modification of Chemical Fibers and Polymer Materials, College of Materials Science and Engineering, Donghua University, Shanghai 201620, China

Nanomaterials Research Center, Modern Textile Institute, Donghua University, Shanghai 200051, China
e-mail: binding@dhu.edu.cn

B. Ding and J. Yu (eds.), *Electrospun Nanofibers for Energy and Environmental Applications*, Nanostructure Science and Technology,
DOI 10.1007/978-3-642-54160-5_14, © Springer-Verlag Berlin Heidelberg 2014

of electrospun nanofibers for developing protective clothing materials that could provide both good protection performance from various environmental hazards and thermal comfort for the users.

14.1 Introduction

The increasing apprehension for professional protection of health care, agriculture production, military personnel, emergency responses, and over exposure to biological and chemical hazards has stimulated the developments of novel protective clothing materials [1]. The presently used protective clothing materials are typically impermeable barriers to block penetration of any contaminants or containing scavengers to absorb any penetrated toxins [2]. However, the achievement of such functionalities alters the wearing and comfort performance of clothing, including bulky materials, heavy weight, and high thermal burden to poor water vapor transmission (WVT). Various protective clothing materials are used to reduce the skin exposure of workers to variety of biological and chemical hazards, ranging from single-use nonwoven personal protective equipment to impermeable polymeric suits [3, 4].

Use of nonwovens for single-use protective clothing continues to grow due to relatively inexpensive, lightweight, and effective protection. A porous spunbond nonwoven with high air permeability exhibited low barrier performance, whereas a compact polyethylene spunbond nonwoven provided high level of protection but low level of thermal transport [5]. Microporous membranes and laminated fabrics, which are used as barrier material for certain personal protective equipment applications, offer higher level of protection but lower air permeability. Monolithic polymeric films are used for impermeable polymeric suits for highly toxic chemicals and provide the highest level of protection at the cost of low level of comfort in a hot, humid environment [6]. The development of nanotechnologies has also inspired a variety of innovations of nanoscaled materials for protective clothing applications [7]. Electrospinning, an efficient and simple method for generating polymeric fibers with diameters both on the micro- and nanoscale, has gained increasing attention as it can not only control the fiber diameter easily but can also manipulate the individual fiber structures via variable tuning [8–11].

The development of nanofibrous membranes for protective clothing materials provides a promising pathway to circumvent the conventional problems due to their advantageous properties, such as lightweight, ultrafine fiber and pore size, highly open porous structure, and extraordinary high surface area to volume ratio. One peculiar feature of electrospun webs for protective clothing use could be the direct application of electrospun webs to the garment systems [13]. Fibers may be electrospun directly onto three-dimensional forms on cloth surface, so that the thickness of the electrospun fibrous web could be varied at numerous locations as needed, making "zoned" materials in protective clothing. This direct application of electrospun nanofibers to the clothing systems would reduce the costly manufacturing steps and resolve seam-sealing problems [13, 14].

14.2 Electrospun Fibers Used in Protective Clothing

Recent research has demonstrated that electrospun membranes possessed excellent barrier performance against toxins in aerosol form and maintained high WVT simultaneously. Electrospun nanofiber-coated nonwoven fabrics remarkably blocked liquid pesticide penetration through the materials which make them practically more desirable as compared to the conventional fabrics and porous coatings (Fig. 14.1). The aim of this chapter is then to give a detailed overview of recent advances in the application of breathable and waterproof, chemical and antimicrobial, and radiation protective materials fabricated from electrospinning.

14.2.1 Breathable and Waterproof Clothing

Waterproof-breathable fabric has been of interest in the area of protective clothing and sportswear industry. For these materials, the protection level against water is known to have an inverse relationship with the vapor permeability or breathability. The electrospun nanofibers can be used to construct unique functional nanostructures [15–17]. The nano- or micropores of the electrospun web provide good moisture and vapor release properties, while it maintains water and wind resistance so that it can be well applicable to waterproof-breathable fabrics. High thermal insulation resulting from large volume of air in microscopic pores would be another advantage [18, 19]. Polyurethanes (PU) have been used effectively in textile coatings for garments, such as raincoats and industrial safety clothing against various hazards, and they are notable as being comfortable to wear and easy to care for. Thermoplastic polyether-based PU are elastomeric and resistant to various microorganisms and bear exceptional hydrolytic stability [20, 21]. For breathable and waterproof clothing, the blend of high barrier performance with thermal comfort is provided by the electrospun PU nanofibers which is not attainable with available conventional protective clothing materials [5].

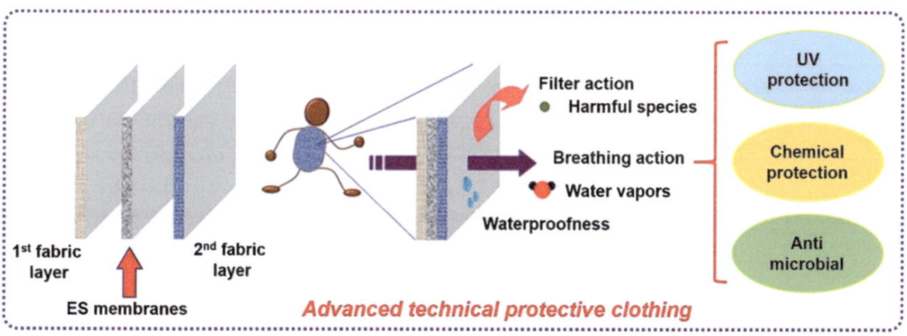

Fig. 14.1 Schematics of the use of electrospun fibrous membranes in protective clothing (Reprinted with permission from Gugliuzza and Drioli [12]. Copyright © 2013, Elsevier)

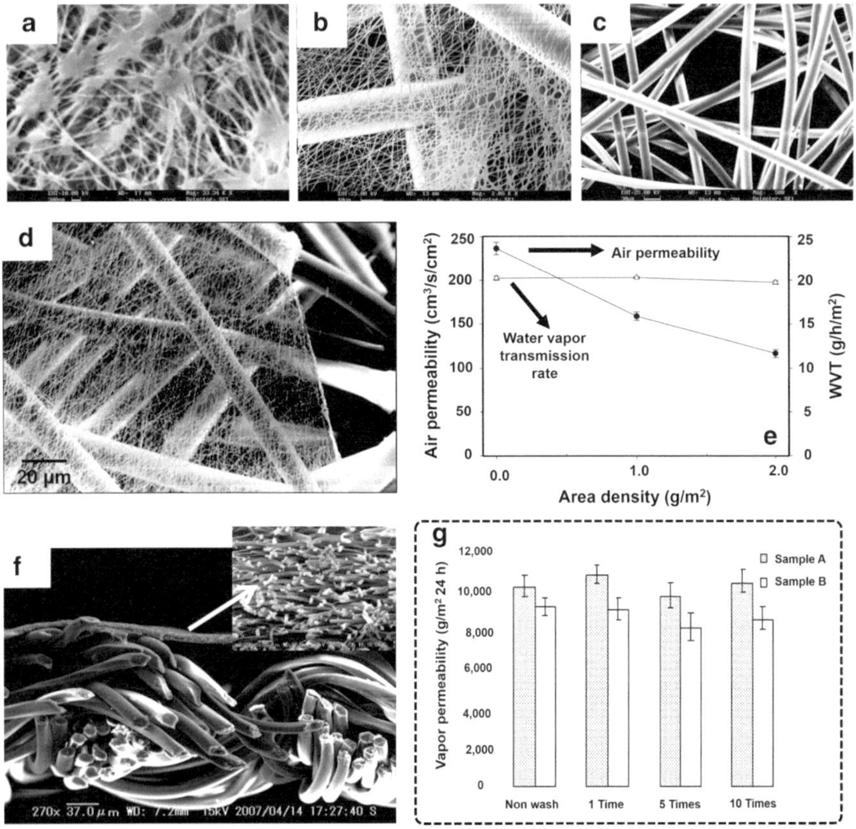

Fig. 14.2 Comparison of SEM micrographs: (**a**) microporous membrane, (**b**) layered fabric system with electrospun PU nanofiber web, and (**c**) spunbond nonwoven fabric. (**d**) SEM micrographs of electrospun PU nanofiber web. (**e**) Effects of area density of electrospun polyurethane web on air/moisture vapor transport properties of layered fabric systems. (**f**) Cross section of laminated nanofiber nonwoven web and (**g**) vapor permeability before/after wash (**a**–**c**: Reprinted with permission from Lee and Obendorf [23]. Copyright © 2007, The Korean Fiber Society. **d**, **e**: Reprinted with permission from Lee and Obendorf [24]. Copyright © 2007, SAGE Publications. **f**, **g**: Reprinted with permission from Kimura et al. [25]. Copyright © 2010, SAGE Publications)

Kang et al. first used direct electrospinning of PU onto the substrate fabrics (polyester/nylon blended fabric, which contained stainless steel yarn in weft) to prepare the waterproof-breathable fabric. PU was solvent electrospun onto the substrate fabric to develop the waterproof-breathable fabric with lightweight and thermal insulation [22]. The reported air permeability and WVT rate were 0.5 $cm^3/cm^2/sec$ and 375.80 $g/h/m^2$, respectively, for electrospun PU coated fabric, which were far better than the PU coated fabrics. Lee et al. developed electrospun PU layered fabric systems in which nanofibrous web was layered on spunbonded nonwoven to check the possibility of developing protective clothing materials as barriers to liquid diffusion (Fig. 14.2a–c). Layered clothing systems with electrospun nanofibrous web showed air permeability, WVT, and barrier performance in the range between microporous materials and nonwovens used for protective clothing (Fig. 14.2d, e) [23, 24].

Later, they also explored the possibility of using mass-produced laminated nanofiber web in breathable, waterproof cloth by measuring several functional properties including waterproofness, vapor permeability, warm/cool feeling, and thermal conductivity. They suggested that the hot-melt laminating method is more suitable for nanofiber web lamination and the maintenance of nanofiber web morphology despite repeated laundering allows the retention of the water and thermal transfer properties of the material (Fig. 14.2f, g) [25, 27]. Yoon et al. developed the waterproof-breathable materials by fabricating layered fabric systems with varying composite structures. Specifically, layered fabric structures based on electrospun nanofiber webs with different levels of nanofiber web density, as well as different substrates and layer structures. The layered structures based on electrospun nanofiber webs provided a higher level of resistance to water penetration than densely woven fabrics and a higher degree of moisture vapor and air permeability than microporous membrane laminates and coated fabrics, with a proper selection of layer structure, substrate fabric, and lamination process [26].

Bagherzadeh et al. used polyacrylonitrile (PAN) multilayered electrospun nanowebs as a middle layer in the fabric and compared them with a well-known commercial protective multilayered fabric, Gortex™. It could be concluded that using the nanofiber membranes, instead of coating material (such as polytetrafluoroethylene), not only presented high windproof properties and acceptable water repellent properties but also improved water vapor permeability behavior. By changing the area density (time of electrospinning) of electrospun mats in layered fabric systems, different barrier and level of comfort protection could be developed according to the desired application [27].

Electrospun PU membranes were proven to be the appropriate choice for the preparation of waterproof and breathable clothing by the researchers due to the desirable properties such as elastomericity, resistance to abrasion, and excellent hydrolytic stability [28, 29]. However, in PU, the hydrophilic carbamate groups and ether groups in the main chain of polymer lead to a relatively high surface energy (40 mJ m^{-2}); thus, the waterproof property of relevant electrospun membranes was usually lower than 5 kPa, which limits the practical use of these membranes. To overcome these limitations, Ge et al. fabricated amphiphobic, PU composite membranes with robust breathable and waterproof performances by electrospinning of novel fluorinated PU (FPU) containing terminal perfluoroalkane segment. By employing the FPU incorporation, the hydrophilicity of PU membranes was transformed to superhydrophobicity with water contact angle of 156° and comparable oleophobicity with oil contact angle of 145°. The as-prepared membranes have shown good air permeability (8.46 L m^{-2} s^{-1}), high water resistance (39.3 kPa), and water vapor transmittance (0.384 kg m^{-2} h^{-1}) with comparable tensile strength (10 MPa), as shown in Fig. 14.3a–c [30]. Later the same group fabricated the SiO_2 nanoparticles (SNP) incorporated FPU membranes which have shown superhydrophobic characteristics with a water contact angle of 165° and

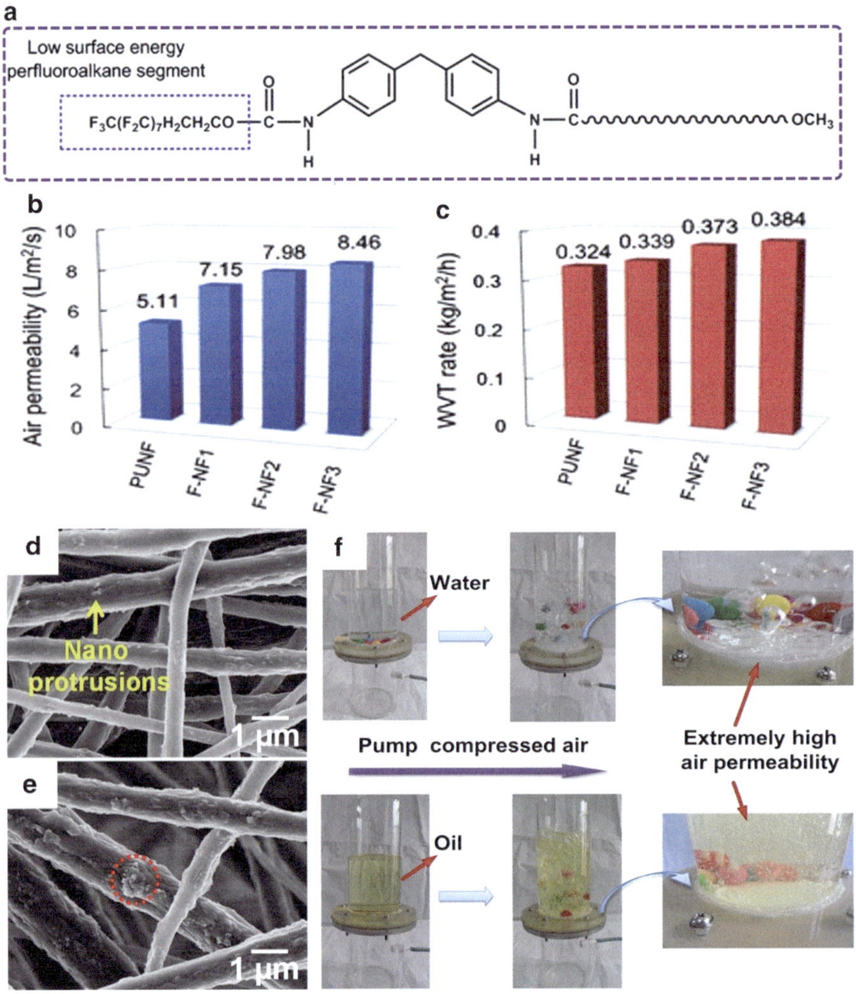

Fig. 14.3 (**a**) Chemical structure of FPU. (**b**) Air permeability. (**c**) WVT rate of relevant FPU membranes. FE-SEM images of (**d**) FPU-18/SNP-0.1 and (**e**) FPU-18/SNP-1 fibrous membranes. (**f**) Robust water/oil proof and breathable performances of FPU-18/SNP-1 membranes (**a–c**: Reprinted with permission from Ge et al. [30]. Copyright © 2012, Royal Society of Chemistry. **d–f**: Reprinted with permission from Wang et al. [31]. Copyright © 2012, Royal Society of Chemistry)

peculiar superoleophobicity (oil contact angle of 151°). Furthermore, a designed concept test confirmed that the as-prepared membranes could bear a load 1.5 kg water or oil at an extremely high air permeability of 2 L min^{-1}, suggesting their usage as promising materials for diverse potential applications including protective clothing (Fig. 14.3d–f) [31].

Recently, Han et al. used shape memory PU nanofibrous membranes which allowed the material's shape to be retained and recovered through heating.

They evaluated the shape memory performance and air and water vapor permeability, and all samples showed shape recoveries of at least 99 % and shape retentions of at least 94 %. The stretched and fixed nanowebs showed greater differences of water vapor permeability at 10 °C with 90 % relative humidity (RH) and 15 °C with 90 % RH, than the original nanowebs did, because stretched shapes of the samples could be retained by more than 70 % at 10 °C or 15 °C. The 40 mm shape memory PU nanoweb showed the best performance, maintaining high shape memory and the largest difference in water vapor permeability between when stretched and not stretched at 15 °C with 90 % RH [32].

14.2.2 Chemical Protective Clothing

The increased concern over exposure to the hazards of chemical and biological toxins, along with the growing chemical and biological threats due to intentional or accidental release of toxic agents, motivates the development of countermeasures to provide effective protection for military personnel and emergency responders. Currently, the protective clothing systems such as "hazardous material suits" or "joint service lightweight integrated suit technology" are widely used to achieve full protection. These protective systems are based upon either full barrier protection through blocking contaminant permeation or on air-permeable adsorptive protective over garments in which all the toxins are adsorbed on contact [2]. High thermal loads from poor water vapor permeability and excess insulation as well as weight and bulkiness of these protective fabrics and suits impair a wearer's performance. Recently, protective fabrics have been introduced based on selectively permeable membranes, which allow for permeation of water vapor while remaining resistant to the permeation of organic molecules [33]. The new generation of protective fabrics is envisioned not only to absorb or block toxic chemical and biological agents but also to detoxify them to reduce the risk of secondary contamination [2].

Gibson and co-workers showed that the lightweight electrospun fabrics exhibit higher breathability than other barrier materials while displaying better airflow resistance and enhanced aerosol particle retention compared to current commercially available membranes [14, 34]. Obendorf et al. showed that laminated fabrics with electrospun polypropylene fiber layers significantly limit the penetration of liquid pesticides while still maintaining better water vapor permeability. The combination of high breathability and efficient barrier properties of electrospun fabrics makes them promising candidates for the next generation of protective clothing. Moreover, the high specific surface areas of electrospun fiber mats allow attachment of functional compounds to obtain chemical or biological detoxifying protective clothing [35]. Ramakrishna et al. successfully electrospun fibers with a reactive compound, (3-carboxy-4-iodosobenzyl) oxy-β-cyclodextrin, and showed that these reactive fabrics can decompose paraoxon, an organophosphate pesticide [36]. In another study, electrospun zinc titanate nanofibers were tested as reactive sorbents capable of detoxifying nerve and mustard agent simulants [37].

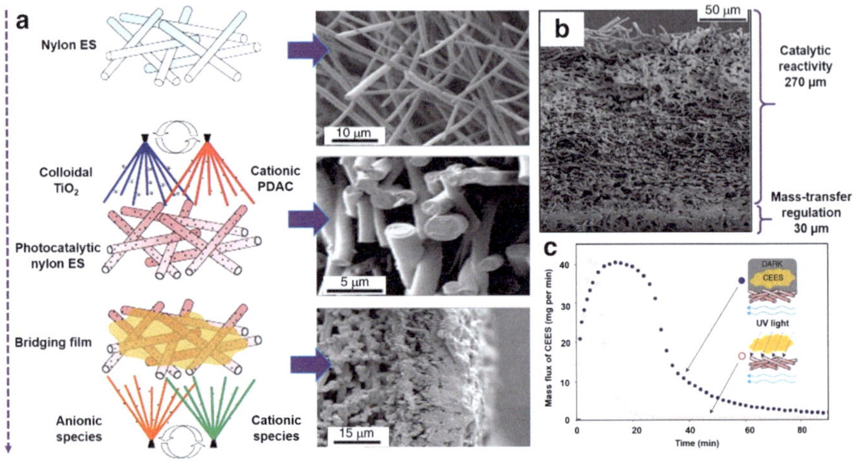

Fig. 14.4 (**a**) Schematic representation and respective SEM images of conformal spray coating of electrospun nanofibrous membranes to create (poly(dimethyldiallylammonium chloride) (PDDC)/TiO_2)$_{25}$ and (PDDC/sodium 4-styrenesulfonate)$_{50}$ conformal coatings. (**b**) An electrospun nylon sample that has been treated with (PDDC/TiO_2)$_{25}$ in the presence of a pressure gradient to create a high-surface-area photocatalytic region, followed by (poly(amidoamine)/poly(acrylic acid))$_{50}$ treatment in the absence of a gradient to create a (chloroethyl ethyl sulfide) CEES transport barrier. (**c**) On exposure to 3 μL of CEES, the mass flux of CEES across the membrane and into the sweep gas is observed in the presence of ultraviolet light as well as in the dark (**a–c**: Reprinted with permission from Krogman et al. [42]. Copyright © 2009, Rights Managed by Nature Publishing Group)

Lev et al. have studied chemical methods to counteract nerve agents and remediate organophosphate (OP) contamination by means of nanoparticles, polymers, and nanofibers functionalized by α-nucleophilic agents [38, 39]. Fibrous mats functionalized with α-nucleophilic oxime moieties were prepared by either electrospinning blends of polyacrylamidoxime and PAN or surface oximation of prefabricated PAN fibrous mats and demonstrated to possess a pronounced capability to hydrolyze chemical nerve agent simulants in the presence of moisture [39]. The layer-by-layer electrostatic assembly technique offers another strategy for electrospun fiber surface functionalization [40]. For the application in protective fabrics, Lee et al. demonstrated that electrospun fibers used as the substrate for titanium dioxide nanoparticle coatings resulted in increased specific substrate surface areas of about 10^4 that of the flat film, which enhanced photocatalytic decomposition of toxic industrial chemicals (Fig. 14.4a–c) [41, 42].

Faccini et al. described the development of effective protective clothing which could filter nanoparticulate aerosols by depositing nanofibrous webs of polyamide 6 on the surface of nonwoven viscose substrate. A hot-press lamination procedure was applied to strongly adhere the nanofibrous layer onto the textile support with the help of thermoplastic resin as a glue. The NaCl particle penetration with diameter ranging from 15 to 300 nm through the as-prepared electrospun textiles was found to be majorly dependent on nanofibrous layer thickness. A really thin nanofiber coating provides up to 80 % retention of 20 nm size particles and over 50 % retention of 200 nm size nanoparticles. By increasing the thickness of nanofibrous mat, the filtration efficiency was increased by 99 % along the whole nanoparticle range [43].

14.2.3 Antimicrobial Clothing

Nanofibers which are functionalized to serve as potential antimicrobial filters in air and liquid filtration are also desired to be used in protective clothing and wound dressing applications. By selecting appropriate polyelectrolytes of opposite charges and varying functionalities, one can generate multifunctional electrospun, fiber-based protection fabrics for both chemical and biological protection. In this study, a nucleophilic and chemically reactive polyanion, polyhydroxamic acid, and an antimicrobial polycation, poly(*N*-vinylguanidine), were synthesized and assembled onto prefabricated PAN fibrous mats. The performance of the functionalized mats in OP decomposition was tested with diisopropyl fluorophosphate, a widely used simulant for G-type nerve agents [44, 45]. The antibacterial properties of these functionalized mats were examined with *Escherichia coli* and *Staphylococcus epidermidis*, gram-negative and gram-positive microorganisms, respectively.

Various biocides such as silver nanoparticles, quaternary ammonium salts or their derivatives, compounds with biguanide groups, and *N*-halamine have been incorporated into electrospun fiber membranes to serve as antimicrobial filters or to create biological protective clothing [46–48]. Silver ions and silver compounds have long been recognized as potent antimicrobial agents. The incorporation of silver nanoparticles into nanofibers offers synergies of two nanotechnologies and which could be incorporated onto the electrospun nanofibrous membranes by several methods (Fig. 14.5) [46].

Antibacterial polymers have also been incorporated into electrospun fibers to enhance their bactericidal properties for protective clothing. Commonly, quaternary ammonium salts or their derivatives or polymers with biguanide groups have been blended with polymers to provide antibacterial properties. Fan et al. have tested the bactericidal properties of alginate and carboxymethyl chitosan blended nanofibers, treated with *N*-(2-hydroxy)-propyl-3-trimethylammonium chitosan chloride (HTCC) and silver nitrate [49]. They have found that silver nitrate-treated nanofibers had the best reduction rate (>99.99 %) of *Staphylococcus aureus*, when compared with HTCC (86–90 %). However, this was still much higher than the untreated nanofibers which had a bacterial reduction rate of 5–10 %. In another work, quaternary ammonium salts generated on electrospun diblock copolymers by *N*-alkylation of tertiary amine groups were tested for their antibacterial efficiency [47]. The nanofibers were found to exhibit high antibacterial characteristics with 95 % of *E. coli* and 97 % *S. aureus* killed within 10 min of contact with 50 mg of fibers.

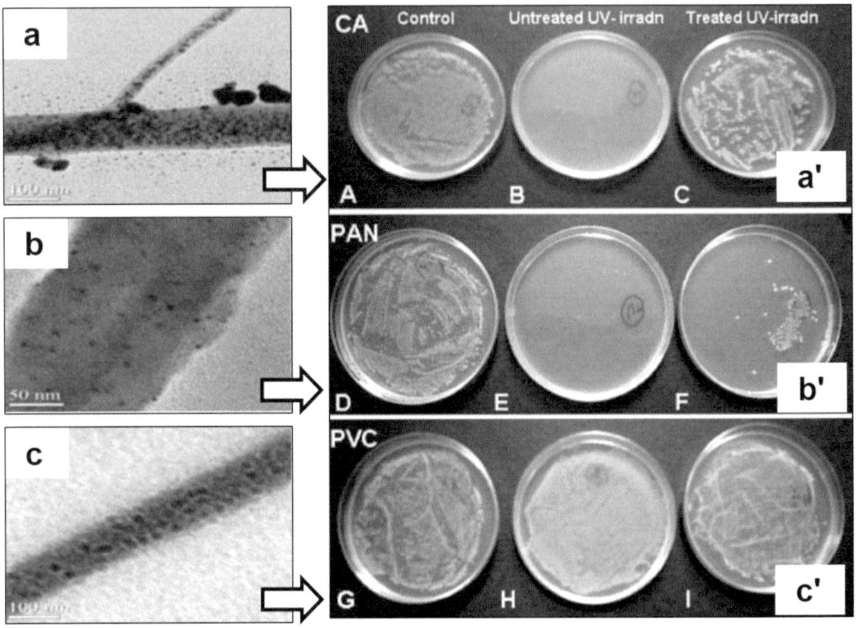

Fig. 14.5 TEM images of (**a**) cellulose acetate with 50 wt% of $AgNO_3$ nanofibers, (**b**) PAN with 5 wt% of $AgNO_3$ nanofibers, (**c**) polyvinylchloride with 5 wt% of $AgNO_3$ nanofibers after UV-irradiation. (**a′**–**c′**) Test results of antibacterial on *E. coli* after 12 h incubated on respective nanofibrous mats (**a**–**c**: Reprinted with permission from Lala et al. [46]. Copyright © 2007 Wiley Periodicals, Inc.)

14.2.4 Ultraviolet (UV) Radiations Protective Clothing

In recent years, increasing concern over excessive terrestrial UV radiation as a kind of environmental pollution has pushed the occurrence of extensive studies of UV-blocking textiles [50, 51]. The prolonged exposure to the sun increases the risk of permanent skin damage from ultraviolet (UV) radiations. Exposure to UVA (315–400 nm) produces the signs of aging, whereas exposure to UVB (280–315 nm) causes erythema and is believed to be responsible for the development of skin cancers [52]. In order to reduce the risk of skin injury associated with ultraviolet exposure, appropriate textiles are considered to be one of the best protection measures for sun protection and skin cancer prevention. This is crucial especially for outdoor workers such as agricultural workers or construction workers since they are at high risk of getting skin cancer due to their high exposure to UV radiation over extended periods of time. Lightweight fabrics commonly worn in summer are known to provide insufficient protection against UV radiation. The formation of UV-protective coating surfaces on textiles, particularly those based on dip–pad–cure and dip-coating processes and solgel and hydrothermal methods, has been studied extensively by many researchers [53–55]. Some inorganic oxides, such as

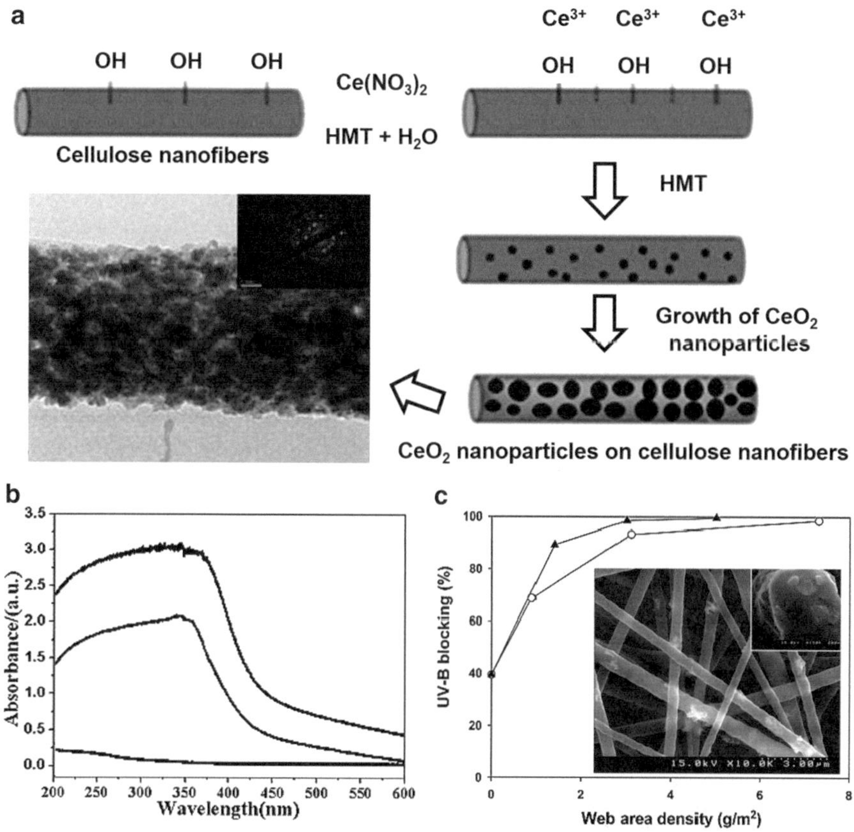

Fig. 14.6 (**a**) Scheme of the formation process of CeO_2 nanoparticles on the cellulose nanofiber and TEM image of cellulose/CeO_2 nanocomposites. (**b**) UV-visible absorption spectra of CeO_2 unmodified and CeO_2 modified cellulose nanofibers. (**c**) Effects of ZnO concentration and web area density on blocking percentages of layered fabric systems for UVB. *Inset* is the SEM image of electrospun PU/zinc oxide nanocomposite fibers (**a**, **b**: Reprinted with permission from Seungsin [56]. Copyright © 2013 Wiley Periodicals, Inc. **c**: Reprinted with permission from Li et al. [57]. Copyright © 2009, The Korean Fiber Society and Springer-Verlag Berlin Heidelberg GmbH)

ZnO, TiO_2, and cerium dioxide (CeO_2), have been widely used in the field of UV blocking; this has included applications on the surface of textile fabrics to provide UV protection.

Lee et al. explored the application of zinc oxide nanoparticles to polypropylene nonwoven fabrics via electrospinning as one approach to impart UV-protective properties. In addition, moisture and air vapor transport characteristics of layered clothing systems with various levels of electrospun mat area density are assessed to observe the consequence of electrospun web layers on thermal comfort level of the layered structures (Fig. 14.6c) [56]. Ultrafine cellulose fibers from natural cotton lines, whose degree of polymerization was above 10,000, were electrospun. The cellulose nanofibers treated with the hydrothermal incorporation of CeO_2 nanoparticles into the nanofiber substrate surface showed good UV-shielding properties compared to the natural cotton cellulose nanofibers. CeO_2 nanoparticles were attached to the surface of the nanofiber substrate because of strong interfacial and electrostatic interactions between the active groups of the nanofiber surface and the CeO_2 nanoparticles as shown in Fig. 14.6a, b [57].

14.3 Concluding Remarks and Future Perspectives

Electrospun nanofibrous membranes have exhibited great advantages over conventional materials for developing efficient protective clothing against various hazards; there remain a large number of challenges. Apart from other avenues, one-dimensional nanostructured materials have significantly contributed to meeting some of these challenges as the structural control of nanomaterials can be as important as the composition of the materials themselves. However, the inherent weakness of electrospun nanofibrous mats and the need of deposition on mechanically strong substrates such as porous woven fabrics that can provide good structural support without altering the nanofiber characteristics are still challenging. New approaches in structure manipulation, such as controlled fibrous morphology, porous nanofibers, interfiber adhesion, and modification of nanofiber surfaces using other multiple techniques, have been rapidly demonstrated. In particular, the processes of electrospinning the polymeric solutions with multifunctional characteristics to produce smart fabrics could be particularly important to be used as protective clothing against over exposure to biological and chemical hazards. These new approaches and new chemistries may all be incorporated in the fabrication of better materials (lightweight electrospun fabrics, better airflow resistance, and enhanced aerosol particle retention) for protective clothing. In conclusion, the convenient adaptability of the nonwoven nanofibrous mats to the conventional fabric materials allows functional nanofibers to become technically one of the most critical elements for the development of smart protective clothing against various hazards.

Acknowledgments This work is supported by the National Basic Research Program of China (973 Program, 2011CB606103, and 2012CB525005), the National Natural Science Foundation of China (No. 51322304, 51173022, U1232116, and 51273038), the Shanghai Nano Special Projects (11 nm0502900), the Shanghai Committee of Science and Technology (No. 12JC1400101), the Huo Yingdong Foundation (131070), and the Program for New Century Talents of the University in China.

References

1. Zhu J, Bahramian Q, Gibson P, Schreuder-Gibson H, Sun G (2012) Chemical and biological decontamination functions of nanofibrous membranes. J Mater Chem 22(17):8532–8540. doi:10.1039/c2jm16605d
2. Schreuder-Gibson HL, Truong Q, Walker JE, Owens JR, Wander JD, Jones WE (2003) Chemical and biological protection and detection in fabrics for protective clothing. MRS Bull 28(8):574–578. doi:10.1557/mrs2003.168

3. Yoon B, Lee S (2011) Designing waterproof breathable materials based on electrospun nanofibers and assessing the performance characteristics. Fiber Polym 12(1):57–64. doi:10.1007/s12221-011-0057-9
4. Ahn HW, Park CH, Chung SE (2011) Waterproof and breathable properties of nanoweb applied clothing. Text Res J 81(14):1438–1447. doi:10.1177/0040517510392462
5. Lee S, Obendorf SK (2007) Barrier effectiveness and thermal comfort of protective clothing materials. J Text Inst 98(2):87–97. doi:10.1533/joti.2005.0143
6. Lin DJ, Chang HH, Chen TC, Lee YC, Cheng LP (2006) Formation of porous poly(vinylidene fluoride) membranes with symmetric or asymmetric morphology by immersion precipitation in the water/TEP/PVDF system. Eur Polym J 42(7):1581–1594. doi:10.1016/j.eurpolymj.2006.01.027
7. Lin T, Wang X (2009) Nano related research in fibres and textiles. Int J Nanotechnol 6(5–6):579–598. doi:10.1504/IJNT.2009.024648
8. Wang X, Ding B, Yu J, Wang M (2011) Engineering biomimetic superhydrophobic surfaces of electrospun nanomaterials. Nano Today 6(5):510–530. doi:10.1016/j.nantod.2011.08.004
9. Wang X, Ding B, Sun G, Wang M, Yu J (2013) Electro-spinning/netting: a strategy for the fabrication of three-dimensional polymer nano-fiber/nets. Prog Mater Sci 58(8):1173–1243, http://dx.doi.org/10.1016/j.pmatsci.2013.05.001
10. Lin J, Wang X, Ding B, Yu J, Sun G, Wang M (2012) Biomimicry via electrospinning. Crit Rev Solid State 37(2):94–114. doi:10.1080/10408436.2011.627096
11. Ding B, Wang M, Wang X, Yu J, Sun G (2010) Electrospun nanomaterials for ultrasensitive sensors. Mater Today 13(11):16–27, http://dx.doi.org/10.1016/S1369-7021(10)70200-5
12. Gugliuzza A, Drioli E (2013) A review on membrane engineering for innovation in wearable fabrics and protective textiles. J Membr Sci 446:350–375, http://dx.doi.org/10.1016/j.memsci.2013.07.014
13. Mukhopadhyay A, Midha VK (2008) A review on designing the waterproof breathable fabrics part i: Fundamental principles and designing aspects of breathable fabrics. J Ind Text 37(3):225–262. doi:10.1177/1528083707082164
14. Gibson P, Schreuder-Gibson H, Rivin D (2001) Transport properties of porous membranes based on electrospun nanofibers. Colloid Surf A 187:469–481. doi:10.1016/s0927-7757(01)00616-1
15. Frenot A, Chronakis IS (2003) Polymer nanofibers assembled by electrospinning. Curr Opin Colloid Interface 8(1):64–75. doi:10.1016/s1359-0294(03)00004-9
16. Shang TC, Yang F, Wang C (2007) Array of polymeric nanofibers via electrospinning. In: Bai C, Xie S, Zhu X (eds) Nanoscience and technology, Pts 1 and 2, vol 121–123, Solid state phenomena., Trans Tech Publications, Stafa-Zurich, Switzerland, pp 583–586. doi:10.4028/www.scientific.net/SSP.121-123.583
17. Greiner A, Wendorff JH (2008) Functional self-assembled nanofibers by electrospinning. In: Shimizu T (ed) Self-assembled nanomaterials I: Nanofibers, vol 219, Advances in polymer science., Springer Berlin Heidelberg, pp 107–171. doi:10.1007/12_2008_146
18. Ding B, Kim CK, Kim HY, Se MK, Park SJ (2004) Titanium dioxide nanofibers prepared by using electrospinning method. Fiber Poly 5(2):105–109. doi:10.1007/bf02902922
19. Huang ZM, Zhang YZ, Kotaki M, Ramakrishna S (2003) A review on polymer nanofibers by electrospinning and their applications in nanocomposites. Compos Sci Technol 63(15): 2223–2253. doi:10.1016/s0266-3538(03)00178-7
20. Cha DI, Kim KW, Chu GH, Kim HY, Lee KH, Bhattarai N (2006) Mechanical behaviors and characterization of electrospun polysulfone/polyurethane blend nonwovens. Macromol Res 14(3):331–337. doi:10.1007/BF03219090
21. Pedicini A, Farris RJ (2003) Mechanical behavior of electrospun polyurethane. Polymer 44(22):6857–6862. doi:10.1016/j.polymer.2003.08.040
22. Yun KK, Chung HP, Jooyoun K, Tae JK (2007) Application of electrospun polyurethane web to breathable water-proof fabrics. Fiber Polym 8(5):564–570. doi:10.1007/bf02875881
23. Lee S, Obendorf SK (2007) Transport properties of layered fabric systems based on electrospun nanofibers. Fiber Polym 8(5):501–506. doi:10.1007/bf02875872

24. Lee S, Obendorf SK (2007) Use of electrospun nanofiber web for protective textile materials as barriers to liquid penetration. Text Res J 77(9):696–702. doi:10.1177/0040517507080284
25. Kimura LSD, Keun Hyung L, Park JC, Ick Soo K (2010) The effect of laundering on the thermal and water transfer properties of mass-produced laminated nanofiber web for use in wear. Text Res J 80(2):99–105. doi:10.1177/0040517508102308
26. Boram Y, Seungsin L (2011) Designing waterproof breathable materials based on electrospun nanofibers and assessing the performance characteristics. Fiber Poly 12(1):57–64. doi:10.1007/s12221-011-0057-9
27. Bagherzadeh R, Latifi M, Najar SS, Tehran MA, Gorji M, Kong L (2012) Transport properties of multi-layer fabric based on electrospun nanofiber mats as a breathable barrier textile material. Text Res J 82(1):70–76. doi:10.1177/0040517511420766
28. Yao C, Li X, Neoh KG, Shi Z, Kang ET (2008) Surface modification and antibacterial activity of electrospun polyurethane fibrous membranes with quaternary ammonium moieties. J Membr Sci 320(1–2):259–267. doi:10.1016/j.memsci.2008.04.012
29. Wu J, Wang N, Wang L, Dong H, Zhao Y, Jiang L (2012) Unidirectional water-penetration composite fibrous film via electrospinning. Soft Matter 8(22):5996–5999. doi:10.1039/c2sm25514f
30. Ge J, Si Y, Fu F, Wang J, Yang J, Cui L, Ding B, Yu J, Sun G (2013) Amphiphobic fluorinated polyurethane composite microfibrous membranes with robust waterproof and breathable performances. RSC Adv 3(7):2248–2255. doi:10.1039/c2ra22111j
31. Wang J, Raza A, Si Y, Cui L, Ge J, Ding B, Yu J (2012) Synthesis of superamphiphobic breathable membranes utilizing SiO_2 nanoparticles decorated fluorinated polyurethane nanofibers. Nanoscale 4(23):7549–7556
32. Han HR, Chung SE, Park CH (2013) Shape memory and breathable waterproof properties of polyurethane nanowebs. Text Res J 83(1):76–82. doi:10.1177/0040517512450757
33. Wilusz E, Truong OT, Rivin D, Kendrick CE (1997) Development of selectively permeable membranes for chemical protective clothing. Abstr Pap Am Chem S 214:341, PMSE
34. Gibson PW, Schreuder-Gibson HL, Rivin D (1999) Electrospun fiber mats: transport properties. Aiche J 45(1):190–195. doi:10.1002/aic.690450116
35. Seungsin L, Kay Obendorf S (2006) Developing protective textile materials as barriers to liquid penetration using melt-electrospinning. J Appl Polym Sci 102(4):3430–3437. doi:10.1002/app.24258
36. Ramaseshan R, Sundarrajan S, Liu Y, Barhate RS, Lala NL, Ramakrishna S (2006) Functionalized polymer nanofibre membranes for protection from chemical warfare stimulants. Nanotechnology 17(12):2947–2953. doi:10.1088/0957-4484/17/12/021
37. Ramaseshan R, Ramakrishna S (2007) Zinc titanate nanofibers for the detoxification of chemical warfare simulants. J Am Ceram Soc 90(6):1836–1842. doi:10.1111/j.1551-2916.2007.01633.x
38. Bromberg L, Schreuder-Gibson H, Creasy WR, McGarvey DJ, Fry RA, Hatton TA (2009) Degradation of chemical warfare agents by reactive polymers. Ind Eng Chem Res 48(3): 1650–1659. doi:10.1021/ie801150y
39. Chen L, Bromberg L, Schreuder-Gibson H, Walker J, Hatton TA, Rutledge GC (2009) Chemical protection fabrics via surface oximation of electrospun polyacrylonitrile fiber mats. J Mater Chem 19(16):2432–2438. doi:10.1039/b818639a
40. Ogawa T, Ding B, Sone Y, Shiratori S (2007) Super-hydrophobic surfaces of layer-by-layer structured film-coated electrospun nanofibrous membranes. Nanotechnology 18(16). doi:10.1088/0957-4484/18/16/165607
41. Lee JA, Krogman KC, Ma M, Hill RM, Hammond PT, Rutledge GC (2009) Highly reactive multilayer-assembled TiO_2 coating on electrospun polymer nanofibers. Adv Mater 21(12):1252. doi:10.1002/adma.200802458
42. Krogman KC, Lowery JL, Zacharia NS, Rutledge GC, Hammond PT (2009) Spraying asymmetry into functional membranes layer-by-layer. Nat Mater 8(6):512–518. doi:10.1038/nmat2430
43. Faccini M, Vaquero C, Amantia D (2012) Development of protective clothing against nanoparticle based on electrospun nanofibers. J Nanomater 2012:1–9. doi:10.1155/2012/892894

44. Yang YC, Baker JA, Ward JR (1992) Decontamination of chemical warfare agents. Chem Rev 92(8):1729–1743. doi:10.1021/cr00016a003
45. Chen L, Bromberg L, Lee JA, Zhang H, Schreuder-Gibson H, Gibson P, Walker J, Hammond PT, Hatton TA, Rutledge GC (2010) Multifunctional electrospun fabrics via layer-by-layer electrostatic assembly for chemical and biological protection. Chem Mater 22(4):1429–1436. doi:10.1021/cm902834a
46. Lala NL, Ramaseshan R, Li B, Sundarrajan S, Barhate RS, Liu YJ, Ramakrishna S (2007) Fabrication of nanofibers with antimicrobial functionality used as filters: protection against bacterial contaminants. Biotechnol Bioeng 97(6):1357–1365. doi:10.1002/bit.21351
47. Fu GD, Yao F, Li Z, Li X (2008) Solvent-resistant antibacterial microfibers of self-quaternized block copolymers from atom transfer radical polymerization and electrospinning. J Mater Chem 18(8):859–867. doi:10.1039/b716127a
48. Chen L, Bromberg L, Hatton TA, Rutledge GC (2008) Electrospun cellulose acetate fibers containing chlorhexidine as a bactericide. Polymer 49(5):1266–1275. doi:10.1016/j.polymer.2008.01.003
49. Fan L, Du Y, Zhang B, Yang J, Zhou J, Kennedy JF (2006) Preparation and properties of alginate/carboxyrnethyl chitosan blend fibers. Carbohyd Polym 65(4):447–452. doi:10.1016/j.carbpol.2006.01.031
50. Lu HF, Fei B, Xin JH, Wang RH, Li L (2006) Fabrication of UV-blocking nanohybrid coating via miniemulsion polymerization. J Colloid Interf Sci 300(1):111–116. doi:10.1016/j.jcis.2006.03.059
51. Tragoonwichian S, O'Rear EA, Yanumet N (2008) Broad ultraviolet protection by copolymerization of 2-3-(2H-benzotriazol-2-yl)-4-hydroxyphenyl ethyl methacrylate and 2-hydroxy-4-acryloyloxybenzophenone on cotton via admicellar polymerization. J Appl Polym Sci 108(6):4004–4013. doi:10.1002/app.28049
52. Broasca G, Borcia G, Dumitrascu N, Vrinceanu N (2013) Characterization of ZnO coated polyester fabrics for UV protection. Appl Surf Sci 279:272–278. doi:10.1016/j.apsusc.2013.04.084
53. Wang RH, Xin JH, Tao XM (2005) UV-Blocking property of dumbbell-shaped ZnO crystallites on cotton fabrics. Inorg Chem 44(11):3926–3930. doi:10.1021/ic0503176
54. Wu DY, Long MC, Zhou JY, Cai WM, Zhu XH, Chen C, Wu YH (2009) Synthesis and characterization of self-cleaning cotton fabrics modified by TiO_2 through a facile approach. Surf Coat Technol 203(24):3728–3733. doi:10.1016/j.surfcoat.2009.06.008
55. Wu ZB, Gu ZL, Zhao WR, Wang HQ (2007) Photocatalytic oxidation of gaseous benzene over nanosized TiO_2 prepared by solvothermal method. Chin Sci Bull 52(22):3061–3067. doi:10.1007/s11434-007-0456-x
56. Seungsin L (2009) Developing UV-protective textiles based on electrospun zinc oxide nanocomposite fibers. Fiber Polym 10(3):295–301. doi:10.1007/s12221-009-0295-2
57. Li C, Shu S, Chen R, Chen B, Dong W (2013) Functionalization of electrospun nanofibers of natural cotton cellulose by cerium dioxide nanoparticles for ultraviolet protection. J Appl Polym Sci 130(3):1524–1529. doi:10.1002/app.39264

Chapter 15
Electrospun Nanofiber-Based Photocatalysts

Xiaofeng Lu and Ce Wang

Abstract Electrospinning technique combining with the calcination process affords a simple and versatile method to generate inorganic nanostructures with well-defined compositions and structures. Additionally, the electrospun polymeric nanofibers with large surface area and porous structures provide a platform to support inorganic semiconductor nanostructures. Various kinds of inorganic nanostructured semiconductors such as titanium dioxide (TiO_2), zinc oxide (ZnO), stannic oxide (SnO_2), tungsten trioxide (WO_3), cadmium sulfide (CdS), etc., have been proved to be good photocatalysts for many kinds of applications including decomposition of organic pollutants, hydrogen production from water splitting, photocatalytic reduction of CO_2, dye-sensitized solar cells, fabricating self-cleaning surface, and antibacterial or anticancer activity. The electrospun nanofiber-based photocatalysts take the advantages of the optional design of the nanostructures (such as solid, porous, hollow, core-sheath, hierarchical fibers, etc.), thus exhibiting a high photocatalytic property. In this chapter, we present the applications of electrospun inorganic semiconductor nanofibers and the electrospun polymeric nanofibers containing semiconductor nanoparticles for photocatalysis. Within these nanostructures, the effect of the chemical composition, the size, the crystal phase, and the doping state of the semiconductors on the photocatalytic performance is widely discussed.

X. Lu • C. Wang (✉)
Alan G. MacDiarmid Institute, College of Chemistry, Jilin University, Changchun 130012, China
e-mail: cwang@jlu.edu.cn

B. Ding and J. Yu (eds.), *Electrospun Nanofibers for Energy and Environmental Applications*, Nanostructure Science and Technology,
DOI 10.1007/978-3-642-54160-5_15,

15.1 Fundamental Principles of Photocatalysis

In 1972, the semiconductor of titanium dioxide (TiO_2) photocatalysis under light irradiation was first demonstrated by Fujishima and Honda [1]. Since then, a large number of studies in this field have emerged and showed many kinds of potential applications in environmental purification, hydrogen production, dye-sensitized solar cells, etc. [2–5]. Up to now, many kinds of semiconductors such as TiO_2, zinc oxide (ZnO), stannic oxide (SnO_2), tungsten trioxide (WO_3), cerium dioxide (CeO_2), iron oxide (Fe_2O_3), gallium nitride (GaN), cadmium sulfide (CdS), zinc sulfide (ZnS), bismuthous sulfide (Bi_2S_3), etc., have been used as efficient photocatalysts. During the photocatalytic process, the semiconductors will generate photoexcited electrons and positively charged holes when the light energy of the incident photons overcomes the band gap energy. Thus, the electrons will initiate or accelerate the photocatalytic reduction reaction, while the positively charged holes are often able to take part in the photocatalytic oxidation process.

In general, the mechanism of the photocatalytic reaction in the presence of semiconductors mainly involves three steps (Fig. 15.1): first, electrons will be promoted from the valence band (VB) to the conduction band (CB) under the light with energy higher than the band gap energy of semiconductors, leaving the equal number of positively charged holes in the VB; second, the electrons and holes diffuse or migrate to the surface of semiconductors; third, the electrons and holes react with the electron acceptors and electron donors in the environment [3]. In fact, the recombination of electron-hole pairs will occur in a very short time on the surface or in the bulk of the photocatalysts during the second step and release energy in the form of heat or photons, which significantly decreases the photocatalytic performance of the semiconductors. To prevent the recombination between the electrons and holes when they move to the surface of the semiconductors, the photocatalysts are usually combined with noble metals or forming heterojunction coupling with other kinds of semiconductors. Therefore, an internal electric field

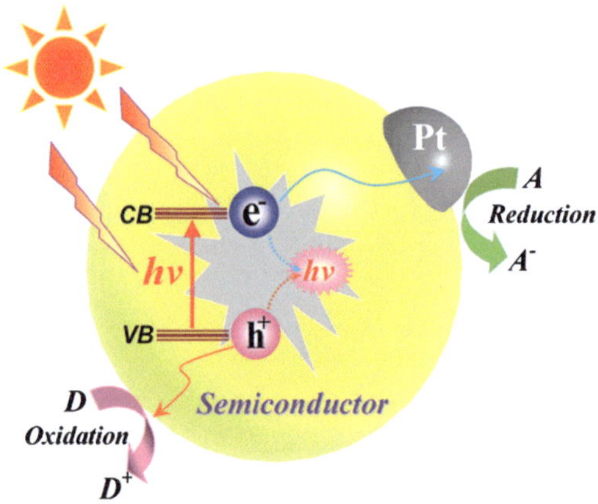

Fig. 15.1 Schematic illustration of basic mechanism of a semiconductor. Light (hυ) excites electrons from VB to CB (Reproduced from Tong et al. [3] by permission of John Wiley and Sons Ltd.)

between the noble metals or the added semiconductors and the host semiconductors will be formed, which facilitates the separation of the photogenerated electrons and holes.

15.2 Electrospun Nanofibers for Photocatalysis

One-dimensional (1D) nanomaterials have been extensively studied in the past few years due to their small dimension, large surface area, and high aspect ratio, which could be used to transport electrical carriers along one controllable direction efficiently. The unique properties of the 1D nanomaterials provided their intriguing application in the area of photocatalysis [6]. In the past two decades, many kinds of advanced techniques have been brought forward to fabricate 1D nanomaterials with well-controlled morphology and structures. Among those techniques, electrospinning has been proven to be a simple and versatile approach capable of producing 1D nanomaterials [7–10]. Up to now, many types of 1D nanostructures such as solid nanofibers, core-sheath fibers, nanotubes, nanobelts, porous nanofibers, multi-channel fibers, anisotropic fibers, hierarchical fibers, and others have been prepared by the electrospinning technique. In the early days, electrospinning technique was mainly used to produce polymeric nanofibers. In 2002, Kim and co-workers firstly fabricated silica nanofibers by using the electrospinning technique [11]. Afterwards, a large number of inorganic nanofibers have been prepared by the electrospinning way and the calcination process [12, 13]. Especially, the inorganic nanofibers of some semiconductors such as TiO_2, ZnO, SnO_2, WO_3, CeO_2, Fe_2O_3, etc., have been successfully applied in the field of photocatalysis. On the other hand, the electrospun polymeric nanofibers have also been used as the supports for the semiconductor photocatalysts because they possess large surface area and porous structures. In this chapter, we will demonstrate the photocatalysts of both electrospun inorganic nanofibers and the electrospun polymeric nanofibers as supports for semiconductors.

15.2.1 Electrospun Inorganic Semiconductor Nanofibers for Photocatalysis

15.2.1.1 TiO_2

TiO_2 is broadly used as a standard photocatalyst because of its high photocatalytic activity, good thermal stability, low cost, and low toxicity [14]. TiO_2 is an n-type semiconductor and usually exhibits three crystal forms, i.e., anatase, rutile, and brookite. Among the three TiO_2 crystals, anatase was most widely studied and usually showed a higher photocatalytic activity. The band gap energy of the anatase is located at 3.2 eV, which corresponds to an optical absorption edge of 387 nm. Therefore, the anatase can only be activated by the photons of light mainly in

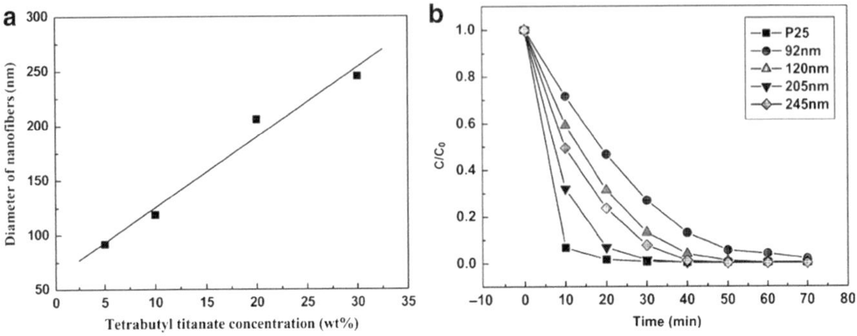

Fig. 15.2 (**a**) The diameters of TiO_2 nanofibers that obtained with different tetrabutyl titanate contents in the precursor solution. (**b**) The relationship between the photocatalytic activity and the diameters of TiO_2 nanofibers for degradation of RB under UV light irradiation (Reproduced from Li et al. [17] by permission of John Wiley and Sons Ltd.)

the ultraviolet (UV) domain with the wavelength smaller than 387 nm [15]. It is well known that the solar spectrum only contains 4–5 % UV light while covers about 50 % visible light. The inability to utilize the visible light limits its practical efficiency for solar applications.

Many kinds of strategies have been developed to improve the photocatalytic efficiency of TiO_2, including the control of the morphologies and chemical composition, increasing the surface area and porosity, etc. Through the electrospinning technique, different morphologies and structures of TiO_2 could be fabricated, which could be used as good candidates for photocatalysis. The photocatalytic performance of the electrospun TiO_2 nanofibers was significantly relying on the morphologies, sizes, structures, crystal forms, and chemical composition. Li and co-workers have studied the relationship of the morphologies of electrospun TiO_2 with its photocatalytic performance [16]. By adjusting the content of polyvinylpyrrolidone (PVP) polymer in the electrospinning precursor and followed by the calcination process, irregular bead-like, ultrafine, and thick nanofibers of TiO_2 could be obtained. It was found that the ultrafine TiO_2 nanofibers exhibited the highest photocatalytic activity toward the degradation of methylene blue (MB) under the UV light irradiation.

The diameter of the electrospun TiO_2 nanofibers has also been found to play an important role on their photocatalytic activity. Pan and co-workers prepared pure anatase TiO_2 nanofibers with different diameters (92, 120, 205, 245 nm) via an electrospinning process and followed the calcination treatment only by adjusting the content of tetrabutyl titanate in the precursor solution [17]. The photocatalytic property studies showed that the activity increased with the increasing of the fiber diameter up to about 200 nm and then it decreased (Fig. 15.2). The photocatalysis mechanism might be due to dynamics of the recombination of electron/hole on the surface sites.

The crystalline structure of the electrospun TiO_2 nanofibers is also significantly related to their photocatalytic properties. Alves and co-workers prepared TiO_2

nanofibers with different diameters of 540, 450, and 200 nm after heat treatment at 500, 750, and 1,000 °C, respectively [18]. It is well known that the calcined temperature highly influences the crystalline structure of TiO_2 nanofibers. XRD analysis showed that anatase structure was obtained at calcination temperature of 500 °C, while rutile structure was presented at 1,000 °C. Both anatase and rutile structures were observed when the calcination temperature was set 750 °C. The electrospun nanofibers at the calcination temperature of 500 °C exhibited better photocatalytic activity, which could be due to the increasing anatase content and the surface areas. The photocatalytic activity for hydrogen evolution by TiO_2 nanofibers under different calcination temperatures was also studied by Yoshikawa and co-workers [19]. The results showed that the TiO_2 nanofibers calcined at 450 °C with anatase structure possessed the highest activity, which is also higher than commercial TiO_2 nanoparticles and the samples prepared by the hydrothermal method. These results indicated that the crystalline forms much influenced the photocatalytic performance of the semiconductors.

It is well known that high surface area is of great significance to enhance the photocatalytic efficiency of the catalyst. To increase the surface-to-volume ratio of the electrospun nanofibers, an electrospinning technique combining with the alkali-dissolution process has been demonstrated [20]. Leaching out the silica from the electrospun TiO_2/SiO_2 composite nanofibers could generate porous TiO_2 nanofibers. The porous TiO_2 nanofibers produced from TiO_2/SiO_2 composite nanofibers containing 20 wt% silica have a BET surface area of about 97.90 m^2/g and a pore size of lower than 50 nm. Moreover, the porous TiO_2 nanofibers exhibited an enhanced photocatalytic activity compared to the solid TiO_2 nanofibers. The porous TiO_2 nanofibers have also been prepared by the calcination of the as-spun PVP/titanium tetraisopropoxide/carbon nanospheres composite nanofibers [21]. The thermal decomposition of carbon nanospheres in the composite nanofibers produced the pores in the TiO_2 nanofibers. The porous TiO_2 nanofibers showed a remarkably higher photocatalytic activity than the nonporous TiO_2 nanofibers.

In comparison to the solid nanofibers, the fabrication of the hollow nanotubes is another way to increase the surface area of the electrospun TiO_2 nanofibers. For example, Nagamine and co-workers first prepared polyvinyl alcohol/TiO_2 core-sheath nanofibers through introducing the threadlike droplets into the electrospun titanium tetraisopropoxide/hexane precursor [22]. Then, the hollow TiO_2 nanofibers could be synthesized by a series of alkaline treatment, acid treatment, and calcination process. The increased external surface area and the high crystallinity of the hollow TiO_2 nanotubes enhanced their photocatalytic activity toward the oxidative degradation of MB. The hollow TiO_2 nanotubes could also be prepared by using the electrospinning approach together with the layer-by-layer technique [23]. After the removal of the template, hollow TiO_2 multilayer nanofibers have been obtained. The diameter and the shell thickness of the hollow multilayer fiber are about 700 and 140 nm, respectively. The photocatalytic activity of the hollow TiO_2 multilayer nanofibers was much higher than that of TiO_2 film toward the degradation of MB because of their unique hollow structure.

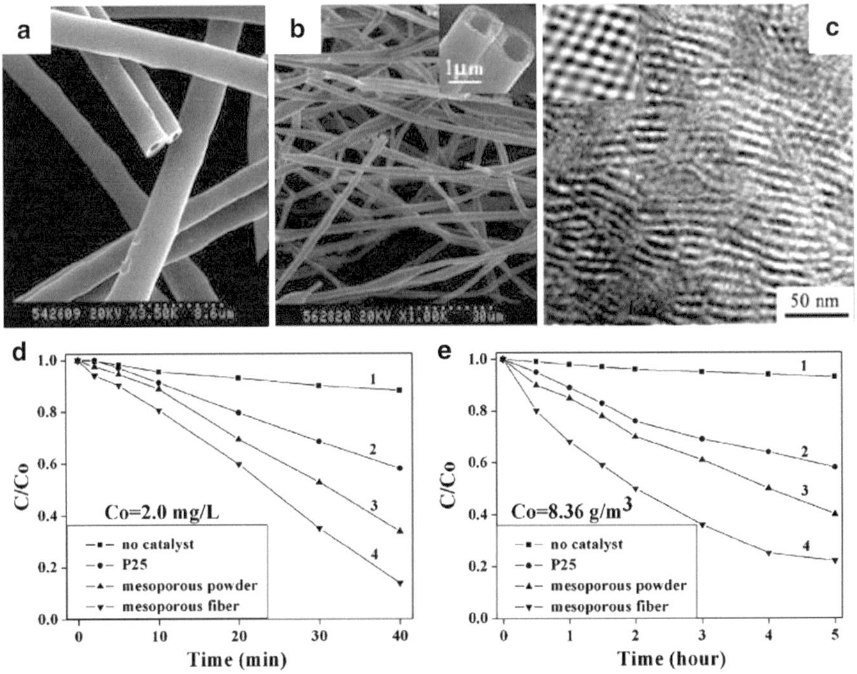

Fig. 15.3 SEM images of (**a**) xerogel and (**b**) TiO_2 hollow fibers. (**c**) HRTEM image of cross section (perpendicular to the axis) of TiO_2 hollow fibers calcined at 400 °C. Time courses for UV photodegradation of (**d**) MB and (**e**) gaseous formaldehyde (Reprinted with the permission from Zhan et al. [24], Copyright 2006 American Chemical Society)

In addition, the hollow TiO_2 nanotubes with mesoporous walls have also been prepared by the coaxial electrospinning technique through introducing the pore-directing agent (Pluronic, P123, $H(C_2H_5O)_{20}(C_3H_7O)_{70}(C_2H_5O)_{20}OH$) (Fig. 15.3) or two immiscible polymers in the electrospun precursors [24, 25]. And the hollow TiO_2 nanotubes with porous structure could also be fabricated by single-needle emulsion electrospinning technique [26]. All of these hollow porous TiO_2 nanotubes exhibited improved photocatalytic efficiency because of their large surface area and the effective separation of the electrons and holes at the mixed interface.

Interestingly, TiO_2 hollow fibers with controllable multichannel and nanoporous wall structures have been fabricated, which results in an effective increasement of photocatalytic activity toward the degradation of gaseous acetaldehyde because this novel structure exhibited a cooperative effect of trapping more gaseous acetaldehyde molecules and reflection of light inside the channels [27].

15.2.1.2 ZnO

The band gap energy of the ZnO is found to be 3.2 eV, thus the photocatalytic activity of ZnO is expected to be similar with that of TiO_2. However, the chemical stability of ZnO is not as good as the TiO_2 because ZnO was found to be stable only in a small pH range. Liu and co-workers fabricated ZnO nanofibers by using electrospinning technique and studied their photocatalytic properties for the degradation of Rhodamine B (RB) and acid fuchsin [28]. The diameter of the as prepared ZnO nanofibers is smaller than 100 nm. The obtained ZnO nanofibers showed higher photocatalytic activity than ZnO nanoparticles under visible-light irradiation. The high photocatalytic performance of the ZnO nanofibers could be due to their porous structure and the prevention of their aggregation compared to ZnO nanoparticles.

Recently, hierarchical nanostructured ZnO has been fabricated through electrospinning technique to enhance the photocatalytic performance. For example, Kanjwal and co-workers prepared ZnO nanobranches by electrospinning of the colloidal solution containing zinc nanoparticles, zinc acetate dehydrate, and PVA, followed by the calcination process and the hydrothermal treatment [29]. The photocatalytic activity of the hierarchical structure of ZnO is much higher than that of traditional electrospun ZnO nanofibers for the degradation of MB under UV light irradiation.

The ZnO hierarchical structure has also been fabricated via a coprecipitation reaction approach [30]. Through an annealing process of the HCOOH-treated electrospun PVA/zinc acetate nanofibers, a three-dimensional ZnO hierarchical structure consisting of primary ZnO fibers and secondary ZnO nanowires could be simply prepared. The hierarchical structured ZnO exhibited an increased photocatalytic activity compared with ZnO nanoparticles or nanofibers toward the photochemical degradation of acetaldehyde, which could be due to the easy diffusion and mass transportation of the gaseous acetaldehyde molecules. ZnO nanofibers with mesoporous structure have also been prepared to increase their surface area by the electrospinning technique [31]. The as-prepared electrospun ZnO nanofibers with a diameter of 60 nm exhibited a high photocatalytic degradation of the polycyclic aromatic hydrocarbon dyes.

Interestingly, the electrospinning technique not only provided a simple method to fabricate ZnO nanofibers but also exhibited an ability to produce ZnO nanospheres. Lu, Xie, and co-workers prepared hollow ZnO nanospheres by annealing the electrospun PVP/ZnO nanofibers from PVP and zinc acetate dehydrate solution in water and ethanol solvent [32]. The results showed that the hollow ZnO nanospheres exhibited a high photocatalytic activity to decompose RB dye under UV light irradiation, which should be attributed to the efficient capture of large number of UV light photons in the hollow structure.

15.2.1.3 SnO_2

Similar with TiO_2 and ZnO, SnO_2 is also an n-type semiconductor. The band gap energy of SnO_2 is about 3.8 eV, which is higher than that of TiO_2 and ZnO, indicating that it could be only capable of being excited by UV light below 330 nm. Moreover, the CB position of SnO_2 is lower than that of the standard potential of H_2/H_2O redox couples, indicating that SnO_2 could not reduce the water molecules for producing H_2. However, by adjusting the morphology, SnO_2 nanomaterials could exhibit a narrower band gap. For example, Wang and co-workers prepared an interesting nanosized V-shaped bipod, which showed a broad photoresponse in the visible-light range of 310–600 nm [33]. This should be caused by the presence of oxygen vacancies. On the other hand, coupled with other kinds of semiconductors, the photocatalytic performance of SnO_2 could also be enhanced due to the efficient charge separation [34].

Recently, the electrospun SnO_2 nanofibers have also been studied for the photocatalysis. Wei, Zhang, and co-workers have prepared two types of SnO_2 nanofibers by electrospinning technique and calcination process [35]. The results showed that the obtained SnO_2 nanofibers by using polyvinyl acetate (PVAc) polymer solution as precursor were formed by large aggregated SnO_2 nanoparticles, while those by using PVP polymer solution as precursor kept an intact and uniform fibrous morphology. The photocatalytic performance indicated that the intact SnO_2 nanofibers showed a higher photocatalytic activity due to their efficient adsorption of a large amount of dyes.

SnO_2 hollow tubes have also been prepared by electrospinning technique followed with calcination process. The SnO_2 hollow tubes exhibited a good photocatalytic activity and stable repeatability toward the degradation of RB under UV light irradiation [36].

15.2.1.4 Fe_2O_3 and WO_3

Fe_2O_3 and WO_3 both possess relatively narrow band gaps compared to TiO_2 and ZnO, which enable them to absorb visible light to produce electrons and holes. The band gap energy of Fe_2O_3 is only 2.1 eV, which is about 1.1 eV narrower than TiO_2. Therefore, Fe_2O_3 could harness more visible light from the solar energy. Yu, Zhan, and co-workers have prepared hierarchical Fe_2O_3 nanofibers by an electrospinning technique [37]. However, the photocatalytic property toward the degradation of MB by the hierarchical Fe_2O_3 nanofibers was evaluated under UV light irradiation. The Fe_2O_3 nanofibers showed a higher photocatalytic activity than commercial α-Fe_2O_3 powder and nanoparticles.

Recently, photocatalytic properties of the electrospun Fe_2O_3 nanofibers under the visible light have also been demonstrated [38]. Through the electrospinning technique, a novel 1D nanobraid and nanoporous α-Fe_2O_3 nanofibers have been prepared. Under the visible-light irradiation, about 91.2 % Congo red dye has been photodegraded by the α-Fe_2O_3 nanofibers within 140 min, indicating their excellent photocatalytic performance. The superior photocatalytic performance of the α-Fe_2O_3 nanofibers was attributed to the high crystallinity, the small crystallite size, and the porous structure of the nanofibers.

The advantage of WO_3 for photocatalysis is its narrow band gap of about 2.6 eV, which can be excited by visible light [39]. In addition, the chemical stability of WO_3 is much better than ZnO in acidic aqueous solution, which makes it applicable for wastewater treatment in a broad pH range. Zhou and co-workers prepared ultrafine

WO_3 nanofibers by electrospinning the PVA/$H_3PW_{12}O_{40}$ precursor and followed by a calcination treatment [40]. The diameter of the obtained WO_3 nanofibers was in the range of 200–600 nm. The band gap energy of the as-spun WO_3 nanofibers was 2.63 eV that investigated by the UV-vis diffuse reflectance spectrum. The narrow band gap energy and small size of the WO_3 nanofibers enable them as excellent photocatalysts for the degradation of RB under UV light irradiation. The photocatalytic properties of the electrospun WO_3 nanofibers under visible-light irradiation were also studied by Intasanta and co-workers [41]. It was found that the as-prepared WO_3 nanofibers showed excellent photocatalytic activity for the degradation of MB under visible-light irradiation.

15.2.1.5 Bi-Based Photocatalysts

In the past few years, Bi-based photocatalysts such as Bi_2O_3, BiOCl, Bi_2WO_6, Bi_2MoO_6, $Bi_4Ti_3O_{12}$, $Bi_2Fe_4O_9$, etc., have attracted much attention because they could effectively photodegrade organic compounds under UV and visible-light irradiation. Bi_2O_3 mainly has five kinds of phases, including the alpha, beta, gamma, delta, and omega. Among these polymorphs, β-Bi_2O_3 has been proved to be a good candidate for heterogeneous photocatalyst. Liu and co-workers have prepared Bi_2O_3 nanofibers with diameters in the range of 70–200 nm by an electrospinning technique [42]. The XRD analysis showed that β-Bi_2O_3 phase was obtained under the calcinations at 500 °C. After being calcined at 550–600 °C, a little content of α-Bi_2O_3 phase was observed. The band gap energy of the as-prepared electrospun Bi_2O_3 nanofibers is determined to be in the range of 2.55–2.97 eV, which was dependent on the calcined temperatures. The obtained Bi_2O_3 nanofibers that were obtained at a calcination temperature of 500 °C exhibited a good photocatalytic activity for the decomposition of RB under UV light irradiation. By using the similar electrospinning technique, BiOCl nanofibers could also be prepared with diameters in the range of 80–140 nm [43]. The as-spun BiOCl nanofibers showed the absorption edge of about 390 nm. The band gap energy was determined from UV-vis diffuse reflectance spectrum and was found to be about 3.2 eV. The photocatalytic performance of the obtained BiOCl nanofibers was much higher than that of Bi_2O_3 nanofibers toward the decomposition of RB under UV light irradiation.

Bi_2WO_6 is a typical Aurivillius oxide with novel physical and chemical properties, showing potential applications in the field of photocatalysis. Liu and co-workers prepared Bi_2WO_6 nanosheet combined with nanoparticles by the electrospinning and calcination process [44]. The as-prepared Bi_2WO_6 nanomaterials showed an excellent photocatalytic property for the photodegradation of MO under visible-light irradiation, which could be applicable for the wastewater treatment by utilizing the solar energy. With the similar chemical structure, Bi_2MoO_6 is also a good photocatalyst in the visible region. By using the electrospun polyacrylonitrile (PAN) nanofibers as templates, Bi_2MoO_6 microtubes were prepared by a calcination process [45]. Because of the large surface area and the hollow multichannelled

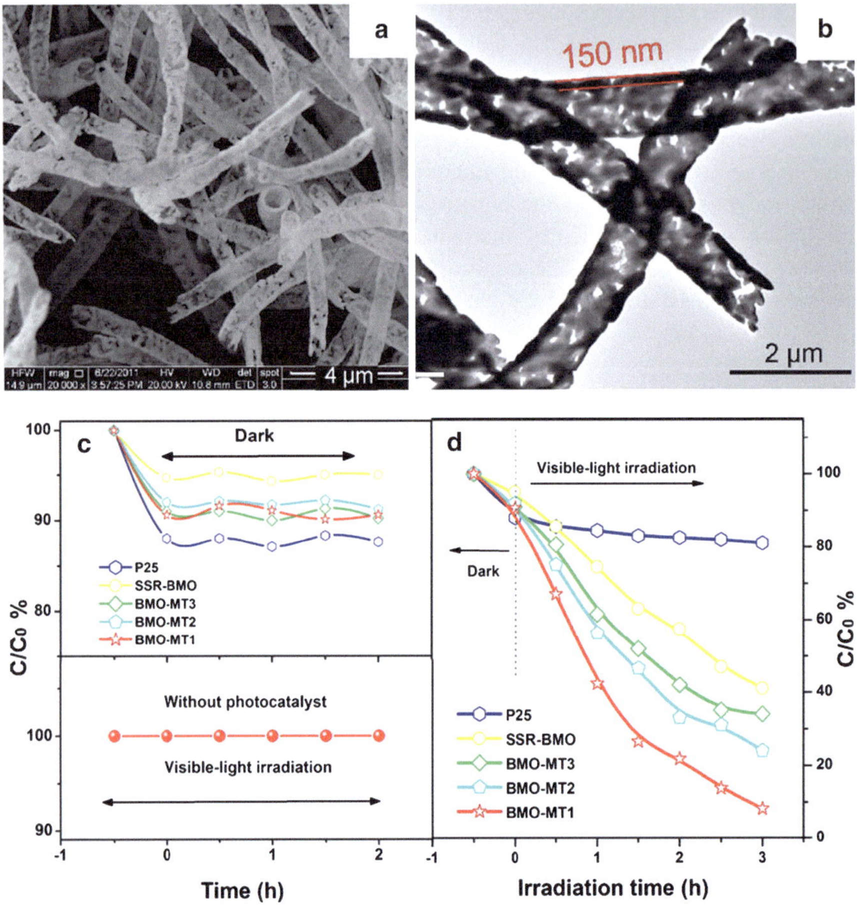

Fig. 15.4 (**a**) SEM and (**b**) TEM images of Bi_2MoO_6 nanotubes (BMO-MT1). (**c**) Degradation profiles of RB in the presence of different photocatalysts in the dark (*up*) and with visible-light irradiation in the absence of the nanofiber photocatalysts (*down*). (**d**) Degradation curves of RB over different catalysts under visible light (Reprinted from Zhang et al. [45], Copyright 2012 with the permission from Elsevier)

structure, the as-prepared Bi_2MoO_6 microtubes exhibited a higher photocatalytic activity than that of the standard photocatalyst P25 and Bi_2MoO_6 prepared by the solid-state reaction under visible-light irradiation (Fig. 15.4).

$Bi_2Fe_4O_9$ is also a visible-light-driven photocatalyst with a band gap around 2.1 eV, which has a good ability to oxidative decompose different kinds of organic contaminants. $Bi_2Fe_4O_9$ nanofibers have been prepared by an electrospinning technique combined with a sol-gel and calcination process [46]. The diameters of the obtained $Bi_2Fe_4O_9$ nanofibers were in the range of 150–200 nm, which was dependent on the calcined temperatures. The as-prepared $Bi_2Fe_4O_9$ nanofibers showed a good photocatalytic efficiency of 45 % for the photodegradation of methyl orange (MO) under visible-light irradiation for the sample calcined at 600 °C. With the aid of H_2O_2, the efficiency could increase up to 70 %, indicating its potential application in the visible-light-driven photocatalyst.

Bismuth titanate ($Bi_4Ti_3O_{12}$), as one kind of layered perovskite compounds, has been extensively studied for their applications in optical displays, piezoelectrics, optoelectronics, and photocatalysis in the past few years. Recently, Yu, Hu, and co-workers firstly prepared $Bi_4Ti_3O_{12}$ nanofibers with diameters in the range of 50–100 nm through an electrospinning approach [47]. The as-synthesized $Bi_4Ti_3O_{12}$ nanofibers are porous, which are composed of aggregated nanoparticles with diameters of 30–50 nm. The porous nanostructure of the $Bi_4Ti_3O_{12}$ nanofibers enables them to exhibit an excellent photocatalytic performance toward photodecomposition of RB under visible-light irradiation, and the photodegradation efficiency could reach 67 % by the electrospun $Bi_4Ti_3O_{12}$ nanofibers within 240 min. The good photocatalytic performance should be attributed to the nanofibrous morphology, the grain shape, and the intrinsic structure of oxygen vacancies.

15.2.1.6 Other Single Component Semiconductor Photocatalysts

In addition to the above common photocatalysts, many kinds of other single semiconductor nanofibers, such as sodium tantalite ($NaTaO_3$), niobium oxide (Nb_2O_5), $InVO_4$, $LaCrO_4$, $LaCrO_3$, and $ZnMn_2O_4$, etc., have also been prepared by the electrospinning technique and applicable in the photocatalysis [48–52]. It is expected that by controlling the chemical composition, crystal form, crystal facet, shape and morphology, and surface energy, the electrospun inorganic nanofibers could be used as efficient photocatalysts for the decomposition of organic pollutants, photolysis of water to yield hydrogen, etc.

15.2.1.7 Metal/Semiconductor Composite

As we discussed earlier, the fast recombination of electron-hole pairs significantly decreases the photocatalytic performance of the semiconductors. To prevent the fast electrons/holes recombination, noble metal nanoparticles are usually incorporated into the semiconductor matrix. Combining noble metal nanoparticles with semiconductor could produce a high Schottky barrier, resulting in a rectified charge carrier transfer [53]. When the noble metal/semiconductor composites are illuminated by UV or visible light with photon energy higher than the band gap energy of semiconductor, electrons in the VB of semiconductor will be excited to the CB, leaving the same amount of positively charged holes in the VB. Then, the deflexed energy band drives the electrons from semiconductor to noble metal nanoparticles, which make the Fermi level of the metal nanoparticles to more negative value. The electrons accumulated at the noble metal nanoparticles or the conduction band of semiconductor could be transferred to the surface-adsorbed species to participate the reduction reaction. On the other hand, the photoinduced holes in the VB will transfer to the surface to take part in the oxidation reaction. Several kinds of nanosized noble metal including Ag, Au, and Pt have been reported to combine with semiconductors such as TiO_2, ZnO, etc., which have indeed shown an enhanced photocatalytic performance.

Among various kinds of metal/semiconductor composite, Ag/TiO_2 was widely studied for their applications in photocatalytic degradation. Kanjwal and co-workers prepared Ag/TiO_2 composite nanofibers by electrospinning the precursor of silver nitrate/titanium isopropoxide/poly(vinyl acetate) sol-gel and then followed by a calcination process [54]. The authors studied the effects of silver content on the photocatalytic performance for the decomposition of MB and RB under UV light irradiation. It was found that with the increasing the content of the silver, the photocatalytic activity increased accordingly. Moreover, the photocatalytic activity of the Ag/TiO_2 composite nanofibers was much higher than that of Ag/TiO_2 composite nanoparticles. These results indicated that the morphology and the noble metal loading could affect the photocatalytic performance of the semiconductors.

The Ag/TiO_2 composite nanofibers have also been fabricated by a two-step approach [55]. TiO_2 nanofiber was firstly prepared by an electrospinning technique, and then Ag nanoparticles could be deposited on the surface of TiO_2 nanofibers via photoreduction of silver ions. The silver nanoparticles significantly influence the photocatalytic performance of the composite nanofibers for the degradation of MB under UV light irradiation. The results showed that 2 % nanosilver-decorated TiO_2 nanofibers showed the highest photocatalytic activity. This should be attributed to the efficient reduction of the recombination rate of electron/hole pairs in the 2 % nanosilver-decorated TiO_2 nanofibers.

Most of the studies have been focused on the photocatalytic performance of the Ag/TiO_2 composite nanofibers under UV light irradiation [56–58]. Recently, the visible-light-driven Ag/TiO_2 composite nanofiber photocatalysts have been prepared combining an electrospinning technique and a solvothermal process [59]. The size of the Ag nanostructures could be tailored by controlling the reaction time and the solvent during the solvothermal process. The as-prepared Ag/TiO_2 composite nanofibers exhibited a higher photocatalytic activity than the pure TiO_2 nanofibers toward the degradation of RB under visible-light irradiation. On the one hand, the introducing silver into the TiO_2 nanofibers improved the separation of the photogenerated electrons and holes. In addition, the silver nanoparticles extended the response of TiO_2 to visible light because the silver nanoparticles exhibited a good light absorption capability. Moreover, the Ag/TiO_2 composite nanofibers could be easily recovered from the reaction solution because of their long 1D structure. In a word, the fabrication of the electrospun Ag/TiO_2 composite nanofibers affords a novel kind of visible-light-driven photocatalysts.

In addition to Ag/TiO_2 nanofibers, Au/TiO_2 composite nanofibers have also been fabricated by a combination of electrospinning technique and sputtering technology [60]. The density of the Au islands on the surface of TiO_2 nanofibers increased with the sputtering time. The Au/TiO_2 composite nanofibers showed an excellent photocatalytic activity for the decomposition of acetaldehyde under UV light irradiation. The photocatalytic activity increased with the increasing content

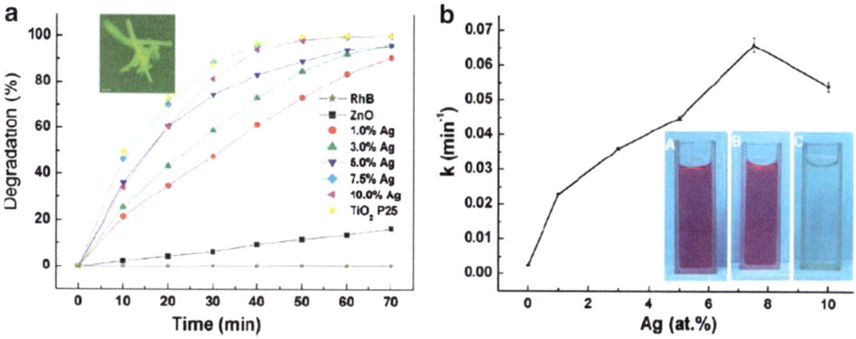

Fig. 15.5 (**a**) Kinetics of the photodegradation of an aqueous solution of RB by Ag-ZnO heterojunction nanofibers. The inset is the typical morphology of the photocatalyst dispersed in the dye solution (scale bar: 5 μm). (**b**) Degradation rate constants for Ag-ZnO nanofibers obtained in the presence of different Ag contents (the *inset* illustrates photos for comparison of the RB solutions photodegraded with various Ag-ZnO nanofibers for 70 min; *A* without catalyst; *B* pure ZnO; and *C* 7.5 atom% Ag-ZnO) (Reprinted with the permission from Lin et al. [61], Copyright 2009 American Chemical Society)

of Au islands. The photocatalytic removal rate could reach above 94 % over the Au/TiO_2 composite nanofibers with high density (2 min deposition) for 70 min, which is much higher than that of pure TiO_2 nanofibers.

Recently, Ag/ZnO composite nanofibers with a dimer-type heterostructure and a diameter in the range of 80–150 nm have also been prepared by a simple electrospinning and thermal treatment [61]. The size of the Ag nanoparticles in the ZnO nanofibers was from several to 15 nm. The introduction of silver nanoparticles in the ZnO nanofibers promoted the charge separation, which enhanced the photocatalytic efficiency for the decomposition of RB under UV light irradiation. Compared to the pure ZnO nanofibers, the photocatalytic activity of the optimal Ag/ZnO composite nanofibers was enhanced by a factor of more than 25 with the silver content of 7.5 atom% (Fig. 15.5).

15.2.1.8 Coupled Semiconductor Photocatalyst

In the past few years, TiO_2/metal oxide composite nanofibers have also been extensively studied for their excellent photocatalytic properties. Typical examples of the TiO_2/metal oxide composite nanofibers are incorporation of SnO_2 component into the TiO_2 matrix [62–64]. It is well known that SnO_2 have a band gap energy of 3.8 eV and a low CB level than TiO_2, thus the photogenerated electrons will flow from TiO_2 to SnO_2, while the holes will move from SnO_2 to TiO_2, resulting in an efficient separation of electron/hole pairs and a high quantum efficiency (Fig. 15.6). In order to make the TiO_2 and SnO_2 fully exposed to the surface and allow both the electrons and holes to react with various reactants in the system, Liu, Sun, and co-workers designed the fabrication of TiO_2/SnO_2 composite nanofibers through a

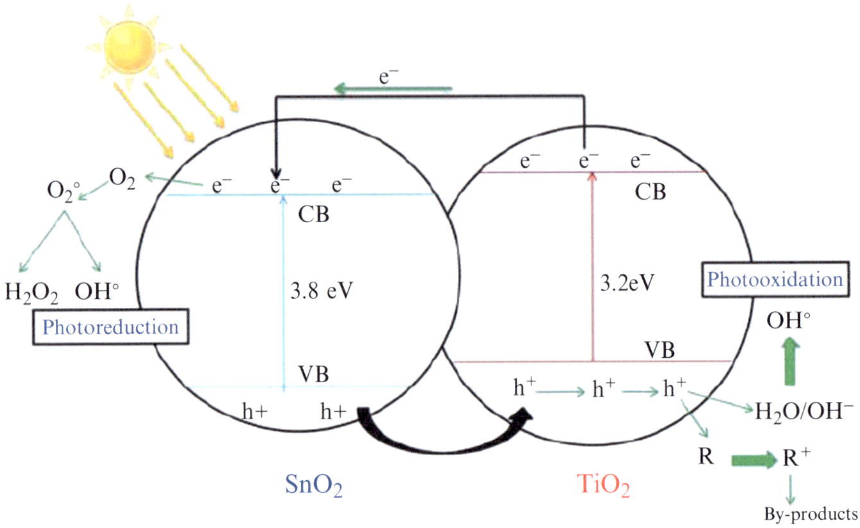

Fig. 15.6 TiO_2/SnO_2 photocatalyst configuration (Reprinted with the permission from Daghrir et al. [5], Copyright 2013 American Chemical Society)

side-by-side dual spinneret electrospinning technique, which increased the charge separation of the photogenerated electrons and allowed both the electrons and holes to take part in the photocatalytic reaction [65]. The as-prepared bicomponent TiO_2/SnO_2 nanofibers showed a much higher photocatalytic activity than pure TiO_2 nanofibers (Fig. 15.7). The electrospun TiO_2/SnO_2 composite nanofibers could also be prepared by decoration of SnO_2 nanoparticles on the surface of TiO_2 nanofibers via a hydrothermal reaction [66]. By controlling the fabrication parameters, SnO_2 nanoparticles or nanorods could grow uniformly on the surface of TiO_2 nanofibers. The photocatalytic performance of the obtained SnO_2/TiO_2 composite nanofibers was superior to the bare TiO_2 nanofibers under UV light irradiation. Recently, the coaxial electrospinning technique was also demonstrated to fabricate core-shell SnO_2/TiO_2 nanofibers [67]. By adjusting the precursor concentration, three different interior morphologies have been prepared and the tubular SnO_2/TiO_2 nanofibers showed much higher photocatalytic activity than the solid TiO_2 nanofibers and commercial P25 nanoparticles alone under UV light irradiation.

Combining the electrospinning technique and the hydrothermal reaction, CeO_2/TiO_2 heterostructured nanofibers have been fabricated [68, 69]. The obtained CeO_2 nanowalls were uniformly distributed on the surface of TiO_2 nanofibers. The BET surface area of the CeO_2/TiO_2 heterostructured nanofibers was about 27.82 m^2/g, which was higher than that of pure TiO_2 nanofibers [69]. On the other hand, the heterostructured nanofibers of CeO_2/TiO_2 promoted the separation of electron/hole pairs. Both of these characteristics enhanced the photocatalytic properties of the CeO_2/TiO_2 heterostructured nanofibers for the degradation of RB under UV light irradiation.

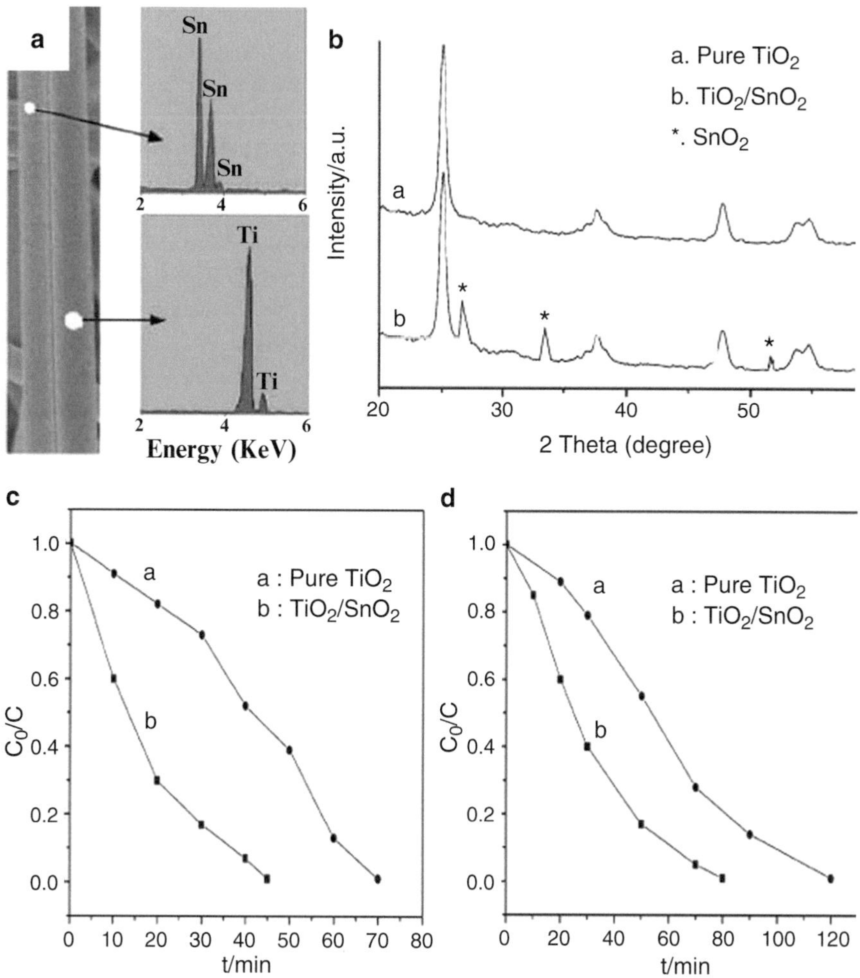

Fig. 15.7 (**a**) Typical EDS microanalysis on selected areas of a single nanofiber. (**b**) Typical XRD diffraction patterns of the electrospun nanofibers that were calcined at 500 °C for 2 h. (**c**) Concentration changes of RB dye and (**d**) TOC changes as a function of irradiation time under UV light with different nanofiber photocatalysts (Reprinted with the permission from Liu et al. [65], Copyright 2007 American Chemical Society)

TiO_2/ZnO nanocomposites have also been widely studied for photocatalysis due to their efficient charge separation of electrons and holes [70, 71]. The TiO_2/ZnO composite nanofibers could be prepared by the electrospinning technique and calcination process [71]. A proper blending of ZnO component into the TiO_2 nanofibers improved the UV light absorption. The photocatalytic ability showed that the as-prepared TiO_2/ZnO composite nanofibers had much higher photocatalytic efficiency than pure TiO_2 nanofibers for the decomposition of RB and phenol molecules in water under UV light irradiation. By using the similar procedure,

CdO/TiO_2 composite nanofibers with a BET surface area of about 65.3 m^2/g have also been prepared [72]. The obtained CdO/TiO_2 composite nanofibers could degrade the MO dye completely within about 75 min, which was superior to that of pristine TiO_2 nanofibers. The magnetic $CoFe_2O_4/TiO_2$ composite nanofibers have also been prepared by the electrospinning technique and atomic layer deposition process [73]. The advantage of the $CoFe_2O_4/TiO_2$ composite nanofibers in the photocatalytic process is their good reusability, which could be easily separated from the photocatalytic reaction system after the reaction.

In addition to TiO_2-based composite nanofibers, other kinds of coupled semiconductor photocatalyst have also been fabricated for photocatalysis, such as ZnO-SnO_2, ZnO-NiO, ZnO-CdO, ZnO-CeO_2, ZnO-$BaTiO_3$, etc. [74–82]. Take the electrospun ZnO-SnO_2 composite nanofibers as an example, they could be prepared by electrospinning the ZnO and SnO_2 precursors and then followed by a calcination process [75]. The diameter of the composite nanofibers was in the range of 100–150 nm. Compared with the electrospun ZnO and SnO_2 nanofibers, the photocatalytic activity of the obtained ZnO-SnO_2 composite nanofibers was much higher under UV light irradiation, which could be attributed to the high specific surface area and the formation of the ZnO-SnO_2 heterojunction.

In the past few years, electrospun nanofibers with multicomponents have also been fabricated for electrocatalytic application [83–86]. For example, Shao and co-workers prepared TiO_2/ZnO nanofibers with well-dispersed Au nanoparticles on their surface, which exhibited an enhanced photocatalytic activity via a three-way synergistic heterostructure [85]. Liu, Sun, and co-workers designed a novel forest-like nanostructure composed of TiO_2 nanofibers as trunks, ZnO nanorods as branches, and CuO nanoparticles as the leaves [86]. The TiO_2/ZnO/CuO composite nanofibers displayed excellent photocatalytic activity toward the hydrogen generation under UV light irradiation, which is much higher than most of the newly developed photocatalysts such as Pt/TiO_2.

As we discussed earlier, some important photocatalysts including TiO_2 and ZnO have broad band gap energies, which enable them only to be excited under UV light irradiation. Coupling TiO_2 and ZnO with other semiconductors with narrower band gap energy could promote the separation of the photoexcited electron/hole pairs, as well as extend the absorption wavelength range into the visible-light region. For example, when TiO_2 is coupled with a narrow band gap semiconductor such as CdS under visible-light irradiation, only CdS is excited and the photogenerated electrons will flow from CB of CdS to that of TiO_2 because CdS possesses more negative CB level. On the other hand, as the VB level of CdS is more cathodic than that of TiO_2, the holes photogenerated will leave in the VB of CdS, which increases the separation of the electron/hole pairs and then enhances the photocatalytic performance of the semiconductors.

By using the electrospinning technique combining with a hydrothermal process, Shao and co-workers prepared the TiO_2/CdS composite nanofibers with a uniform diameter in the range of 200–300 nm [87]. The obtained hexagonal wurtzite CdS nanoparticles with a size of about 6–40 nm were uniformly immobilized on the surface of electrospun anatase TiO_2 nanofibers. By introducing the CdS

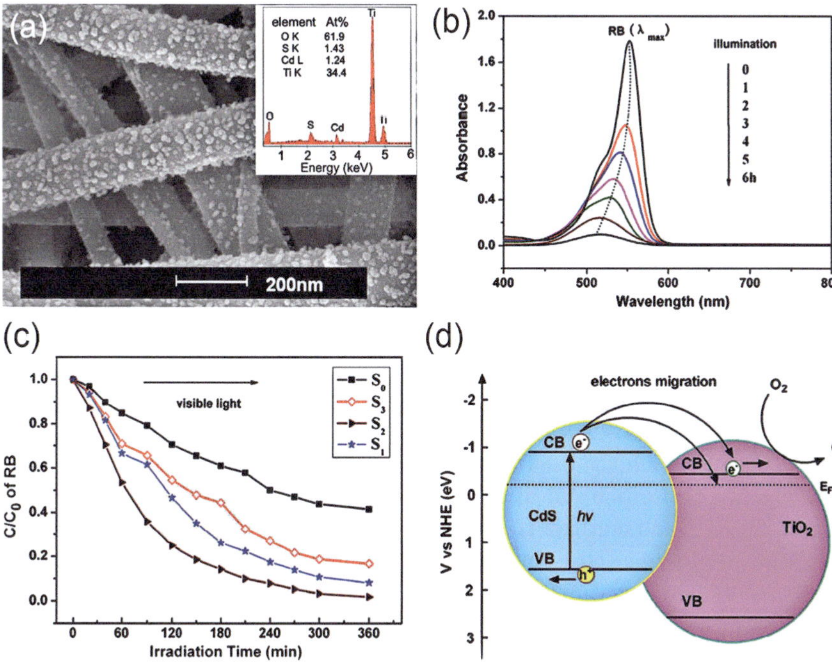

Fig. 15.8 (**a**) FESEM images of TiO_2/CdS heteroarchitectures (S2) nanofibers. *Inset* part in figure is the EDX spectrum of the TiO_2/CdS heteroarchitectures nanofibers. (**b**) The time-dependent absorption spectra of RB solution in the presence of the TiO_2/CdS heteroarchitectures (S2) nanofibers. (**c**) Comparison of photocatalytic degradation of RB over different catalysts under visible light. (**d**) The schematic profile depicting the energy band structure and occurrence of vectorial electron transfer in the TiO_2/CdS heteroarchitectures (Reprinted from Su et al. [87], Copyright 2011 with the permission from Elsevier)

nanocrystals, the composite nanofibers exhibited an absorption in the visible region with the scope of 390–500 nm. The investigation of photocatalytic performance showed that the density of CdS nanoparticles on the surface of TiO_2 nanofibers much influenced the photocatalytic activity for the decomposition of RB under visible-light irradiation. The as-prepared TiO_2/CdS composite nanofibers containing a proper content of CdS nanoparticles displayed an enhanced visible photocatalytic activity for the degradation of RB dye, which should be due to the increased visible light absorbing and the efficient separation of photogenerated electrons and holes (Fig. 15.8).

In addition to the CdS nanocrystals, ultrathin hexagonal SnS_2 nanosheets have also been decorated on the surface of TiO_2 nanofibers via the hydrothermal reaction [88]. The lateral size and the thickness of the SnS_2 nanosheets on the surface of TiO_2 nanofibers were 40–70 nm and 4–8 nm, respectively. The photocatalytic ability of the SnS_2/TiO_2 (AR) is higher than that of SnS_2/TiO_2 (A) and TiO_2 (AR) nanofibers, which is ascribed to the high separation efficiency by the three-way photosynergistic effect among SnS_2, TiO_2 (A), and TiO_2 (R) in the SnS_2/TiO_2 hierarchical nanofibers.

Similar with CdS nanoparticles, modification of metal oxide or other semiconductors such as CuO, In_2O_3, V_2O_5, and WO_3 and some Bi-based semiconductors on the surface of TiO_2 nanofibers could also produce visible-light-driven photocatalysts [89–96]. For example, Chen, Shao, and co-workers prepared In_2O_3/TiO_2 heteroarchitectures by using an electrospinning technique and followed by a solvothermal process [90]. The as-prepared In_2O_3/TiO_2 heteroarchitectures showed an increased photocatalytic activity under visible-light irradiation to degrade RB over the pure TiO_2 nanofibers. By using the similar procedure, 3D hierarchical heterostructure of Bi_2WO_6/TiO_2 was also fabricated. The Bi_2WO_6 nanoplates which composed of nanoparticles were grown on the surface of TiO_2 nanofibers [93]. Compared with the TiO_2 nanofibers, Bi_2WO_6 nanoparticles, and Bi_2WO_6/TiO_2 powder, the as-prepared Bi_2WO_6/TiO_2 hierarchical heterostructure showed an improved photocatalytic activity, which is due to its high surface area, small grain size, and the hierarchical heterostructure. Metallophthalocyanine is also a typical organic semiconductor, which has been widely investigated in dye photosensitization due to its intensive absorption in the visible-light region. Shao and co-workers fabricated 2,9,16,23-tetranitrophthalocyanine copper(II) (TNCuPc)/TiO_2 hierarchical nanostructures by an electrospinning technique and solvothermal reaction [96]. By modification of the experimental parameters, TNCuPc nanoflowers could be grown on the surface of TiO_2 nanofibers. Under visible-light irradiation, the obtained TNCuPc/TiO_2 hierarchical nanostructures exhibited an enhanced photocatalytic activity in the decomposition of RB over TiO_2 nanofibers and TNCuPc alone.

15.2.1.9 Doped Semiconductor Photocatalyst

It is well known that semiconductors with wide band gap energy, such as TiO_2, ZnO, etc., have been long used as active photocatalysts. But the limitation of the utilization of solar energy hinders their widely applications in the field of photocatalysis. Up to now, there are several approaches to change the absorption edge into the visible-light region to efficiently utilize the solar energy. Among these methods, doping with ionic species or anionic nonmetals are meaningful objects for the fabrication of visible-light-driven photocatalysts because the light absorption range could be widen by doing such species and the recombination of the photogenerated electrons and holes could also be inhibited. On the other hand, the content of dopant and the photocorrosion process will also influence the chemical and physical properties at the surface of semiconductors.

In the past few years, ionic species such as W^{6+}, Bi^{3+}, Fe^{3+}, and V^{5+} and rare earth ions have been doped in semiconductors to extend the absorption range and increase the carrier lifetime of semiconductor [97–103]. For example, Wang and co-workers have prepared mesoporous W^{6+}-doped TiO_2 nanofibers by electrospinning technique, which exhibited a BET surface area of 191 m^2/g and a pore size of about 5.2 nm [97]. The photocatalytic activity of the W^{6+}-doped TiO_2 nanofibers is strongly dependent on the content of W^{6+}. 3 wt% W^{6+}-doped TiO_2 nanofibers showed higher photocatalytic activity for the decomposition of MB

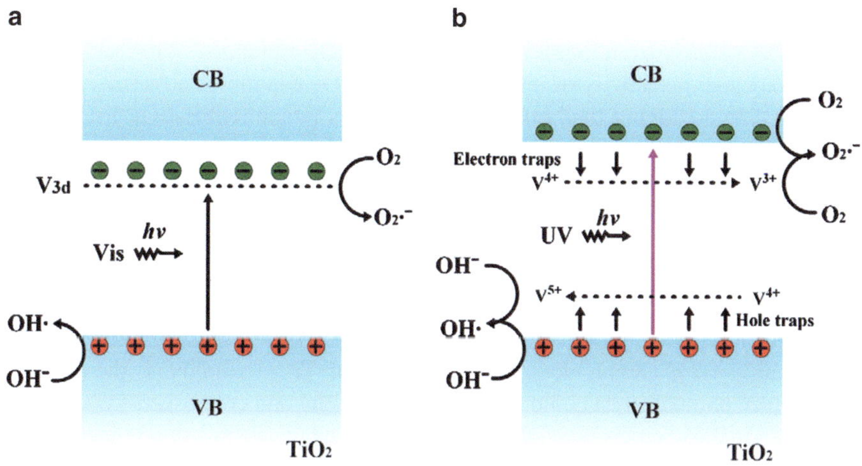

Fig. 15.9 The energy band diagram and photocatalytic mechanism of V-doped TiO_2 nanofibers (Reprinted from Zhang et al. [100], Copyright 2010 with the permission from Elsevier)

under visible-light irradiation than other samples and commercial P25 nanoparticles. The V (V^{4+} or V^{5+})-doped TiO_2 nanofibers have also exhibited the similar photocatalytic results [100]. However, under UV light irradiation, 1.0 wt.% V-doped TiO_2 nanofiber showed the highest photocatalytic activity. While under visible-light irradiation, 5.0 wt.% V-doped TiO_2 nanofiber showed higher photocatalytic activity than other samples. This should be due to the different photocatalytic mechanism (Fig. 15.9). Under visible-light irradiation, the electrons are excited from VB of TiO_2 to the t2g level of V3d orbital because the VB of TiO_2 is a little higher. Then, the separated electrons and holes could take part in the reduction and oxidation process. Under UV light irradiation, the electrons in the VB of TiO_2 will be excited to the CB, leaving the same amount of holes in the VB. On the other hand, by trapping photogenerated electrons and holes, V^{4+} could be changed into V^{3+} and V^{5+} ions. The V^{3+} and V^{5+} ions could participate in the reduction and oxidation process, which enhanced the photocatalytic activity of the TiO_2 nanofibers. However, V^{4+} ions could also serve as the recombination centers of electrons and holes photogenerated, so high concentration of V^{4+} will decrease the photocatalytic activity of TiO_2 nanofibers.

In addition to ionic species, anionic nonmetals have also been extensively studied to be doped in semiconductors to increase the photocatalytic performance. Among such systems, N-doped TiO_2 nanofibers attracted much attention because of the effects of band gap narrowing, oxygen vacancies creation, or intra-band gap surface states generation. Lozzi and co-workers prepared N-doped TiO_2 nanofibers by using ethylenediamine as the N source by an electrospinning technique and heat treatment [104]. The existence of the nitrogen does not influence the crystalline phase of the TiO_2 nanofibers. These nanofibers showed an improved photocatalytic activity for the degradation of MB under visible-light irradiation compared to the pure TiO_2

nanofibers. The N-doped TiO_2 nanofibers could also be prepared by a surface nitridation process by treating the electrospun TiO_2 nanofibers with NH_3 [105]. The photocatalytic activity of optimal N-doped TiO_2 nanofibers possesses a factor of more than 12 higher than pure TiO_2 nanofibers under visible-light irradiation. These examples fully indicated that the N-doping have significantly enhanced the visible-light photocatalytic activity of the semiconductors.

Carbon/TiO_2 composite nanotubes have been fabricated via the electrospinning technique by using PVA as the carbon source [106]. The crystalline phase of the obtained TiO_2 nanotubes is anatase. Compared with commercial P25 nanoparticles, the obtained carbon/TiO_2 composite nanotubes that calcined at 600 °C exhibited a superior photocatalytic performance under UV light irradiation. By sensitization with carbon dots on the surface of TiO_2 nanofibers, the visible-light-driven photocatalyst could also be prepared. Zhu and co-workers synthesized TiO_2/carbon dots composite nanofibers by combining the electrospinning technique and reflux method [107]. The obtained composite nanofibers exhibited enhanced photocatalytic efficiency for the decomposition of RB under visible-light irradiation over commercial P25 nanoparticles and pure TiO_2 nanofibers. The mechanism of the carbon dots sensitized TiO_2 nanofibers could be attributed to the photogenerated electrons and holes by carbon dots under visible-light irradiation. Afterward, the electrons photogenerated will flow to the CB of TiO_2 nanofibers, taking part in the oxidation reaction.

15.2.2 Electrospun Polymeric Nanofibers Containing Semiconductors for Photocatalysis

The electrospun polymeric nanofibers have large surface area and porous structure as non-woven fabrics, which are excellent candidates to support semiconductor nanoparticles for photocatalysis. The loading of semiconductor nanoparticles on the surface of polymeric nanofibers could prevent their application. Therefore, there are many reports for the fabrication of polymeric nanofibers containing semiconductors by electrospinning technique in the past two decades.

15.2.2.1 Polymer/TiO_2 Composite Nanofibers

Among those polymer/semiconductor composite nanofibers, polymer/TiO_2 was most extensively studied because TiO_2 has a high photocatalytic activity under UV light irradiation. Many kinds of polymers such as PAN, poly(methyl methacrylate) (PMMA), polyvinyl alcohol (PVA), nylon 6, poly(dimethyl siloxane) (PDMS), polyaniline (PANI), polycarbosilane (PCS), poly(vinylidene difluoride) (PVDF) mixing methacrylic acid (MAA)-trifluoroethyl acrylate (TFA) copolymer (fluoropolymer), cellulose acetate (CA), etc., have been used to support TiO_2 nanoparticles for photocatalysis [108–116]. For example, Xu and co-workers have prepared

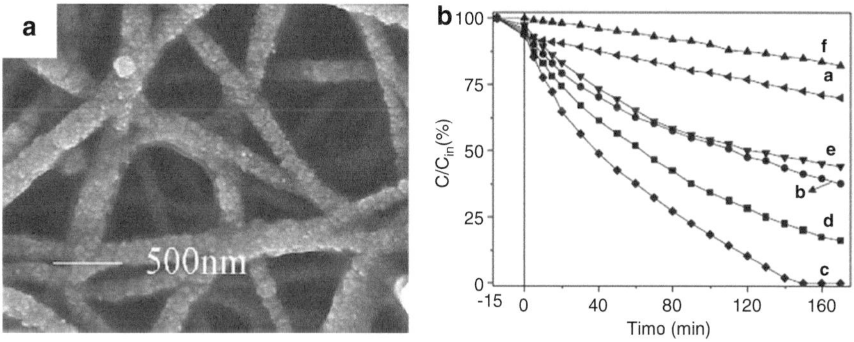

Fig. 15.10 (**a**) SEM images of TiO_2-fluoropolymer fiber nanocomposites prepared at 150 °C for 6 h. (**b**) Photocatalytic degradation of methylene blue by (*a*) poly(MAA-co-TFA)/PVDF electrospun fiber mats; TiO_2-fluoropolymer fiber nanocomposites prepared at 150 °C for (*b*) 3 h, (*c*) 6 h, (*d*) 12 h; (*e*) Degussa P25; and (*f*) blank sample (Reprinted from He et al. [115], Copyright 2009 with the permission from Elsevier)

TiO_2/fluoropolymer composite nanofibers by combining the electrospinning technique and hydrothermal complex-precipitation [115]. The obtained TiO_2 nanoparticles were well dispersed on the surface of fluoropolymer nanofibers. The photocatalytic activity of the as-prepared TiO_2/fluoropolymer composite nanofibers is higher than that of P25 nanoparticles (Fig. 15.10). Moreover, the good recycling and stability of the composite nanofibers enable them to be applicable in the wastewater treatment.

To increase the photocatalytic activity of polymer/TiO_2 composite nanofibers, metal component and TiO_2 nanoparticles have been simultaneously introduced into the polymer nanofibers. Kim and co-workers prepared Ag/TiO_2/nylon 6 composite nanofibers via an electrospinning technique and followed by photoreduction process [117]. The obtained Ag/TiO_2/nylon 6 composite nanofiber exhibited a higher photocatalytic activity than that of TiO_2/nylon 6 composite nanofibers and TiO_2 powders. And the as-prepared Ag/TiO_2/nylon 6 composite nanofibers also showed an excellent recycling property because of the stability of the nylon 6 nanofiber support.

15.2.2.2 Polymer/ZnO Composite Nanofibers

By using the electrospinning technique and hydrothermal process, "firecracker-shaped" ZnO/polyimide (PI) composite nanofibers have been prepared [118]. The hexagonal-shaped ZnO nanorods with diameters in the range of 100–300 nm and average length of about 700 nm were densely deposited on the surface of PI nanofibers. The as-prepared ZnO/PI composite nanofibers showed excellent photocatalytic performance, which could completely photodegrade RB dyes in 2 h under UV light irradiation.

ZnO/nylon 6 core-shell nanofibers could also be produced by combining the electrospinning technique and atomic layer deposition [119]. The ZnO shell exhibited a polycrystalline nature with hexagonal wurtzite structure and the thickness of the ZnO layer on the surface of nylon 6 nanofibers is about 90 nm. The ZnO/nylon 6 composite nanofibers with the thinner diameter showed better photocatalytic activity due to their higher surface area. Furthermore, the stability of the ZnO/nylon 6 composite nanofibers exhibited a good recycling property for the second cycle photocatalytic test.

15.2.2.3 Polymer/Sulfide Composite Nanofibers

It is well known that the band gap energy of CdS is about 2.4 eV, which could be used as an efficient visible-light-driven photocatalyst. However, it is difficult to synthesize pure CdS nanofibers by electrospinning technique. Therefore, the electrospun polymeric nanofibers containing CdS component were extensively studied for photocatalysis under visible-light irradiation [120–123]. The CdS-OH/PAN composite nanofibers have been fabricated by a one-step co-electrospinning technique [122]. The CdS-OH nanoparticles with a uniform size of 5 nm which distributed in the PAN nanofibers exhibited a cubic zinc blended crystallographic structure. The photocatalytic hydrogen production efficiency of the CdS-OH/PAN composite nanofibers showed a rate of 13.5 μmol/h g. And the composite nanofibers could also be recycled easily from the reaction system. To enhance the photocatalytic performance of CdS/polymer composite nanofibers, PdS has been introduced into the nanocomposites as an oxidation cocatalyst. Kim and co-workers have prepared core-shell CdPdS/PVAc nanofibers and evaluated their photocatalytic performance toward Reactive black 5 and Reactive Orange 16 degradation [123]. It was found that the core-shell CdPdS/PVAc nanofibers exhibited an excellent photocatalytic activity and recycling property.

ZnS nanoparticles were also immobilized on the surface of polymeric nanofibers for photocatalysis [124, 125]. For example, Xu and co-workers have prepared ZnS/fluoropolymer composite nanofibers by an electrospinning technique. The photocatalytic activity for the decomposition of MB by ZnS/fluoropolymer composite nanofibers is much higher than that of ZnS powders. And the photocatalytic activity changes indistinctively after ten times recycling tests, indicating their good recycling property.

15.2.2.4 Polymer/Phthalocyanine Composite Nanofibers

Recently, metal phthalocyanines, as one kind of porphyrin derivatives, have attracted much attention for their applications in visible-light-driven photocatalysis. However, the metal phthalocyanine photocatalyst in the homogeneous system is difficult to recover from the reaction solution. Therefore, it is a meaningful object to immobilize them onto some low-cost supports. In the past few years, there are

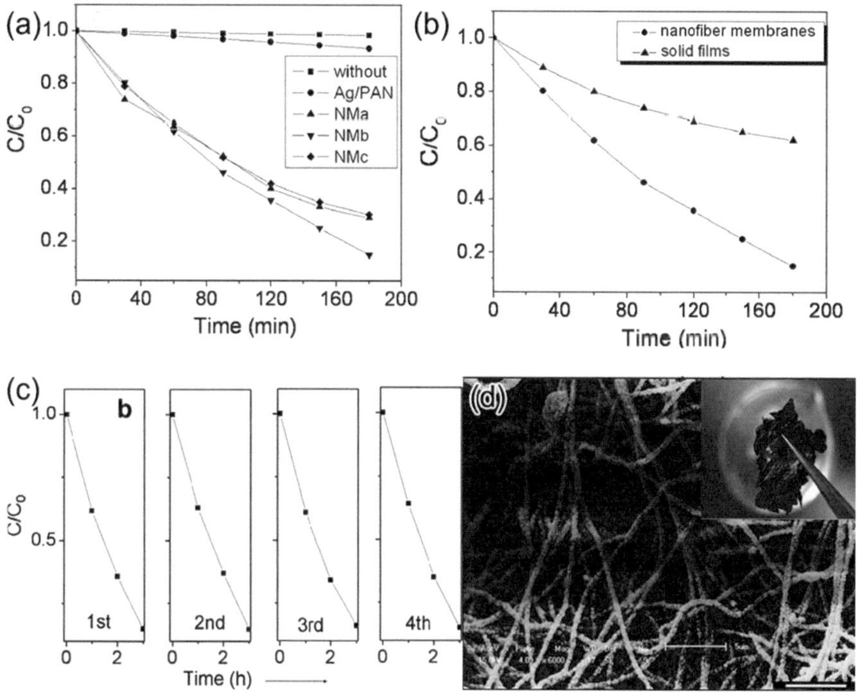

Fig. 15.11 (**a**) The photocatalytic degradation of MO dye in solution over the as-prepared photocatalysts under visible-light irradiation as a function of time. (**b**) The comparison of the photocatalytic activities between the Ag/AgCl-coated PAN nanofiber membranes and solid films. (**c**) Recycling photocatalytic test over Ag/AgCl-coated PAN nanofiber membrane (NMb). (**d**) SEM image of Ag/AgCl-coated PAN nanofiber membranes after four times degradation of MO solution under visible-light irradiation for 3 h each. The scale bar: 5 μm. *Inset* in the figure is a picture of Ag/AgCl-coated PAN nanofiber photocatalysts after reaction (Reprinted from Lei et al. [131], Copyright 2011 with the permission from Elsevier)

a few reports on the fabrication of metalloporphyrins or metal phthalocyanines on the surface of electrospun nanofibers for photocatalysis [126–129]. Take cadmium phthalocyanine (CdPc) supported on the surface of PAN nanofibers as an example [129]. The photocatalytic activity to degrade organic pollutions by the CdPc/PAN composite nanofibers is much higher than that of CdPc nanoparticles and commercial P25 nanoparticles. In addition, the CdPc/PAN composite nanofibers could be separated from reaction solution easily and reused by sedimentation. This indicated that the obtained CdPc/PAN composite nanofibers have potential applications to remove the organic pollutants from wastewater.

15.2.2.5 Polymer/Ag/AgCl Composite Nanofibers

Recently, silver/silver halide nanocomposites have been used as a novel and efficient visible-light-driven photocatalyst by utilizing the plasmon resonance of silver nanoparticles, which was first reported by Huang and co-workers [130]. However, the small silver/silver halide nanocomposites are easily aggregated during the photocatalytic process, which decreased the photocatalytic activity. Wang and co-workers have prepared Ag/AgCl-coated electrospun PAN nanofibers by an electrospinning technique and followed by an electroless plating method [131]. The as-prepared Ag/AgCl/PAN composite nanofibers exhibited an excellent photocatalytic activity under visible-light irradiation. The composite nanofibers also showed a long-term stability and recycling property, which could be used as a novel photocatalyst for practical application (Fig. 15.11).

15.3 Conclusions

Electrospinning technique is a simple, versatile, and environmentally friendly approach for fabrication of inorganic and polymeric nanofibers. As described in this chapter, some recent achievements in the field of electrospun nanofibers for photocatalysis have been selectively discussed. Especially, the electrospun inorganic semiconductor nanofibers show a great advantage as photocatalysts such as large surface area and high aspect ratio. By combining with noble metal nanoparticles, coupling with other semiconductors, doping with ionic species and anionic nonmetals will enhance the photocatalytic performance of the electrospun nanofibers. On the other hand, the electrospun nanofibers containing semiconductors for photocatalysis application have also been discussed because they afford an excellent recycling property. It is expected that the research on the electrospun nanofibers as photocatalysts will become more and more extensive in the near future and larger number of new and efficient photocatalysts will emerge, which could be applicable in the photocatalytic wastewater treatment, hydrogen production from water splitting, photocatalytic reduction of CO_2, etc.

References

1. Fujishima A, Honda K (1972) Electrochemical photolysis of water at a semiconductor electrode. Nature 238(5358):37–38. doi:10.1038/238037a0
2. Kubacka A, Fernández-García M, Colón G (2012) Advanced nanoarchitectures for solar photocatalytic applications. Chem Rev 112(3):1555–1614. doi:10.1021/cr100454n
3. Tong H, Ouyang SX, Bi YP, Umezawa N, Oshikiri M, Ye JH (2012) Nano-photocatalytic materials: possibilities and challenges. Adv Mater 224(2):229–251. doi: 10.1002/adma.201102752
4. Zhang HJ, Chen GH, Bahnemann DW (2009) Photoelectrocatalytic materials for environmental applications. J Mater Chem 19(29):5089–5121. doi:10.1039/b821991e
5. Daghrir R, Drogui P, Robert D (2013) Modified TiO_2 for environmental photocatalytic applications: a review. Ind Eng Chem Res 52(10):3581–3599. doi:10.1021/ie303468t
6. Li CJ, Xu GR (2011) Electrospinning technique in the nano-photocatalyst research. E-Polymers No. 012

7. Huang ZM, Zhang YZ, Kotaki M, Ramakrishna S (2003) A review on polymer nanofibers by electrospinning and their applications in nanocomposites. Compos Sci Technol 63(15):2223–2253. doi:10.1016/S0266-3538(03)00178-7
8. Li D, Xia YN (2004) Electrospinning of nanofibers: reinventing the wheel? Adv Mater 16(14):1151–1170. doi:10.1002/adma.200400719
9. Greiner A, Wendorff JH (2007) Electrospinning: a fascinating method for the preparation of ultrathin fibres. Angew Chem Int Ed 46(30):5670–5703. doi:10.1002/anie.200604646
10. Lu XF, Wang C, Wei Y (2009) One-dimensional composite nanomaterials: synthesis by electrospinning and their applications. Small 5(21):2349–2370. doi:10.1002/smll.200900445
11. Shao CL, Kim H, Gong J, Lee D (2002) A novel method for making silica nanofibres by using electrospun fibres of polyvinylalcohol/silica composite as precursor. Nanotechnology 13(5):635–637. doi:10.1088/0957-4484/13/5/319
12. Li D, McCann JT, Xia YN (2006) Electrospinning: a simple and versatile technique for producing ceramic nanofibers and nanotubes. J Am Ceram Soc 89(6):1861–1869. doi:10.1111/j.1551-2916.2006.00989.x
13. Dai YQ, Liu WY, Formo F, Sun YM, Xia YN (2011) Ceramic nanofibers fabricated by electrospinning and their applications in catalysis, environmental science, and energy technology. Polym Adv Technol 22(3):326–338. doi:10.1002/pat.1839
14. Morales-Torres S, Pastrana-Martinez LM, Figueiredo JL, Faria JL (2012) Design of graphene-based TiO_2 photocatalysts—a review. Environ Sci Pollut Res 19(9):3676–3687. doi:10.1007/s11356-012-0939-4
15. Pelae M, Nolan NT, Pillai SC, Seery MK, Falaras P, Kontos AG, Dunlop PSM, Hamilton WJ, Byrne JA, Shea KO, Entezari MH, Dionysiou DD (2012) A review on the visible light active titanium dioxide photocatalysts for environmental applications. Appl Catal B 125:331–349. doi:10.1016/j.apcatb.2012.05.036
16. Li SD, Dong YH, Guo MR (2012) Photocatalytic activity of different morphologies TiO_2 nanofibers. Appl Surf Sci 258(20):8015–8019. doi:10.1016/j.apsusc.2012.04.160
17. Li HP, Zhang W, Li B, Pan W (2010) Diameter-dependent photocatalytic activity of electrospun TiO_2 nanofiber. J Am Ceram Soc 93(9):2503–2506. doi:10.1111/j.1551-2916.2010.03841.x
18. Alves AK, Berutti FA, Clemens FJ, Graule T, Bergmann CP (2009) Photocatalytic activity of titania fibers obtained by electrospinning. Mater Res Bull 44(2):312–317. doi:10.1016/j.materresbull.2008.06.001
19. Chuangchote S, Jitputti J, Sagawa T, Yoshikawa S (2009) Photocatalytic activity for hydrogen evolution of electrospun TiO_2 nanofibers. ACS Appl Mater Interface 1(5):1140–1143. doi:10.1021/am9001474
20. Li QM, Sun DX, Kim H (2011) Fabrication of porous TiO_2 nanofiber and its photocatalytic activity. Mater Res Bull 46(11):2094–2099. doi:10.1016/j.materresbull.2011.06.034
21. Liu SH, Liu BS, Nakata K, Ochiai T, Murakami T, Fujishima A (2012) Electrospinning preparation and photocatalytic activity of porous TiO2 nanofibers. J Nanomater 2012:491927(1–5) doi:10.1155/2012/491927
22. Nagamine S, Ochi S, Ohshima M (2011) Fabrication of TiO_2 hollow fibers with surface nanostructure. Mater Res Bull 46(12):2328–2332. doi:10.1016/j.materresbull.2011.08.034
23. Zhang TZ, Ge LQ, Wang X, Gu ZZ (2008) Hollow TiO_2 containing multilayer nanofibers with enhanced photocatalytic activity. Polymer 49(12):2898–2902. doi:10.1016/j.polymer.2008.04.035
24. Zhan SH, Chen DR, Jiao XL, Tao CH (2006) Long TiO_2 hollow fibers with mesoporous walls: sol-gel combined electrospun fabrication and photocatalytic properties. J Phys Chem B 110(23):11199–11204. doi:10.1021/jp057372k
25. Zhang X, Thavasi V, Mhaisalkar SG, Ramakrishna S (2012) Novel hollow mesoporous 1D TiO_2 nanofibers as photovoltaic and photocatalytic materials. Nanoscale 4(5):1707–1716. doi:10.1039/c2nr11251e

26. Lu BG, Zhu CQ, Zhang ZX, Lan W, Xie EQ (2012) Preparation of highly porous TiO_2 nanotubes and their catalytic applications. J Mater Chem 22(4):1375–1379. doi:10.1039/c1jm15242d
27. Zhao TY, Liu ZY, Nakata K, Nishimoto S, Murakami T, Zhao Y, Jiang L, Fujishima A (2010) Multichannel TiO_2 hollow fibers with enhanced photocatalytic activity. J Mater Chem 20(24):5095–5099. doi:10.1039/c0jm00484g
28. Liu HQ, Yang JX, Liang JH, Huang YX, Tang CY (2008) ZnO nanofiber and nanoparticle synthesized through electrospinning and their photocatalytic activity under visible light. J Am Ceram Soc 91(4):1287–1291. doi:10.1111/j.1551-2916.2008.02299.x
29. Kanjwal MA, Sheikh FA, Barakat NAM, Li XQ, Kim HY, Chronakis IS (2012) Zinc oxide's hierarchical nanostructure and its photocatalytic properties. Appl Surf Sci 258(8):3695–3702. doi:10.1016/j.apsusc.2011.12.008
30. Pan C, Dong L, Qiu JS (2012) Fabrication and photocatalytic property of three-dimensional ZnO hierarchical structure. Chem J Chin Univ 33(5):1031–1035. doi:10.3969/j.issn.0251-0790.2012.05.030
31. Singh P, Mondal K, Sharma A (2013) Reusable electrospun mesoporous ZnO nanofiber mats for photocatalytic degradation of polycyclic aromatic hydrocarbon dyes in wastewater. J Colloid Interface Sci 394:208–215. doi:10.1016/j.jcis.2012.12.006
32. Zhu CQ, Lu BG, Su Q, Xie EQ, Lan W (2012) A simple method for the preparation of hollow ZnO nanospheres for use as a high performance photocatalyst. Nanoscale 4(10):3060–3064. doi:10.1039/c2nr12010k
33. Wang G, Lu W, Li JH, Choi J, Jeong Y, Choe SY, Park JB, Ryu MK, Lee K (2005) V-shaped tin oxide nanostructures featuring a broad photocurrent signal: an effective visible-light-driven photocatalyst. Small 2(12):1436–1439. doi:10.1002/smll.200600216
34. Wang C, Zhao JC, Wang XM, Mai BX, Sheng GY, Peng PA, Fu JM (2002) Preparation, characterization and photocatalytic activity of nano-sized ZnO/SnO_2 coupled photocatalysts. Appl Catal B 39(3):269–279. doi:10.1016/S0926-3373(02)00115-7
35. Xia X, Li SL, Wang X, Liu JX, Wei QF, Zhang XW (2013) Structures and properties of SnO_2 nanofibers derived from two different polymer intermediates. J Mater Sci 48(9):3378–3385. doi:10.1007/s10853-012-7122-7
36. Wang X, Fan HQ, Ren PR (2013) Electrospinning derived hollow SnO_2 microtubes with highly photocatalytic property. Catal Commun 31:37–41. doi:10.1016/j.catcom.2012.11.009
37. Liu Y, Yu HB, Zhan SH, Li Y, Lv ZN, Yang XQ, Yu YJ (2011) Fast degradation of methylene blue with electrospun hierarchical α-Fe_2O_3 nanostructured fibers. J Sol-gel Sci Technol 58(3):716–723. doi:10.1007/s10971-011-2451-6
38. Sundaramurthy J, Kumar PS, Kalaivani M, Thavasi V, Mhaisalkar SG, Ramakrishna S (2012) Superior photocatalytic behaviour of novel 1D nanobraid and nanoporous α-Fe_2O_3 structures. RSC Adv 2(21):8201–8208. doi:10.1039/c2ra20608k
39. Santato C, Odziemkowski M, Ulmann M, Augustynski J (2001) Crystallographically oriented mesoporous WO_3 films: synthesis, characterization, and applications. J Am Chem Soc 123(43):10639–10649. doi:10.1021/ja011315x
40. Sui CH, Gong J, Cheng TX, Zhou GD, Dong SF (2011) Fabrication of tungsten oxide microfibers with photocatalytic activity by electrospinning from $PVA/H_3PW_{12}O_{40}$ gel. Appl Surf Sci 257(20):8600–8604. doi:10.1016/j.apsusc.2011.05.028
41. Srisitthiratkul C, Yaipimai W, Intasanta V (2012) Environmental remediation and superhydrophilicity of ultrafine antibacterial tungsten oxide-based nanofibers under visible light source. Appl Surf Sci 259:349–355. doi:10.1016/j.apsusc.2012.07.050
42. Wang CH, Shao CL, Wang LJ, Zhang LN, Li XH, Liu YC (2009) Electrospinning preparation, characterization and photocatalytic properties of Bi_2O_3 nanofibers. J Colloid Interface Sci 333(1):242–248. doi:10.1016/j.jcis.2008.12.077
43. Wang CH, Shao CL, Liu YC, Zhang LN (2009) Photocatalytic properties BiOCl and Bi_2O_3 nanofibers prepared by electrospinning. Scrip Mater 59(3):332–335. doi:10.1016/j.scriptamat.2008.03.038

44. Zhao G, Liu SW, Lu QF, Shi M, Song LJ (2011) Preparation of Bi_2WO_6 by electrospinning: researching their synthesis mechanism and photocatalytic activity. J Clust Sci 22(4):621–631. doi:10.1007/s10876-011-0403-5
45. Zhang MY, Shao CL, Zhang P, Su CY, Zhang X, Liang PP, Sun YY, Liu YC (2012) Bi_2MoO_6 microtubes: controlled fabrication by using electrospun polyacrylonitrile microfibers as template and their enhanced visible light photocatalytic activity. J Hazard Mater 225:155–163. doi:10.1016/j.jhazmat.2012.05.006
46. Qi SH, Zuo RZ, Wang Y, Wong HL, Chan LW (2013) Synthesis and photocatalytic performance of the electrospun $Bi_2Fe_4O_9$ nanofibers. J Mater Sci 48(11):4143–4150. doi:10.1007/s10853-013-7227-7
47. Hou DF, Luo W, Huang YH, Yu JC, Hu XL (2013) Synthesis of porous $Bi_4Ti_3O_{12}$ nanofibers by electrospinning and their enhanced visible-light-driven photocatalytic properties. Nanoscale 5(5):2028–2035. doi:10.1039/c2nr33750a
48. Yi X, Li JL (2010) Synthesis and optical property of $NaTaO_3$ nanofibers prepared by electrospinning. J Sol-gel Sci Technol 53(2):480–484. doi:10.1007/s10971-009-2110-3
49. Qi SS, Zuo RZ, Liu Y, Wang Y (2013) Synthesis and photocatalytic activity of electrospun niobium oxide nanofibers. Mater Res Bull 48(3):1213–1217. doi: 10.1016/j.materresbull.2012.11.074
50. Song LJ, Liu SW, Lu QF, Zhao G (2012) Fabrication and characterization of electrospun orthorhombic $InVO_4$ nanofibers. Appl Surf Sci 258(8):3789–3794. doi: 10.1016/j.apsusc.2011.12.029
51. Sun X, Dong XT, Wang JX, Liu GX (2011) Electrospinning fabrication and photocatalytic properties of $LaCrO_4$ and $LaCrO_3$ nanobelts. Chem J Chin Univ 32(10):2262–2267
52. Li CJ, Xu GR (2011) Zn–Mn–O heterostructures: study on preparation, magnetic and photocatalytic properties. Mater Sci Eng B 176(7):552–558. doi:10.1016/j.mseb.2011.01.011
53. Linsebigler AL, Lu GQ, Yates JT (1995) Photocatalysis on TiO_2 surfaces: principles, mechanisms, and selected results. Chem Rev 95(3):735–758. doi:10.1021/cr00035a013
54. Kanjwal MA, Barakat NAM, Sheikh FA, Baek W, Khil MS, Kim HY (2010) Effects of silver content and morphology on the catalytic activity of silver-grafted titanium oxide nanostructure. Fiber Polym 11(5):700–709. doi:10.1007/s12221-010-0700-x
55. Srisitthiratkul C, Pongsorrarith V, Intasanta N (2011) The potential use of nanosilver-decorated titanium dioxide nanofibers for toxin. Appl Surf Sci 257(21):8850–8856. doi:10.1016/j.apsusc.2011.04.083
56. Park JY, Hwang KJ, Lee JW, Lee IH (2011) Fabrication and characterization of electrospun Ag doped TiO_2 nanofibers for photocatalytic reaction. J Mater Sci 46(22):7240–7246. doi:10.1007/s10853-011-5683-5
57. Mishra S, Ahrenkiel SP (2012) Synthesis and characterization of electrospun nanocomposite TiO_2 nanofibers with Ag nanoparticles for photocatalysis applications. J Nanomater 2012(90291):1–6. doi:10.1155/2012/902491
58. Barakat NAM, Kanjwal MA, Chronakisc IS, Kim HY (2012) Influence of temperature on the photodegradation process using Ag-doped TiO_2 nanostructures: negative impact with the nanofibers. J Mol Catal A 366:333–340. doi:10.1016/j.molcata.2012.10.012
59. Su CY, Liu L, Zhang MY, Zhang Y, Shao CL (2012) Fabrication of Ag/TiO_2 nanoheterostructures with visible light photocatalytic function via a solvothermal approach. CrystEngComm 14(11):3989–3999. doi:10.1039/c2ce25161b
60. Pan C, Dong L (2009) Fabrication of gold-doped titanium dioxide (TiO_2:Au) nanofibers photocatalyst by vacuum ion sputter coating. J Macromol Sci Part B 48(5):919–926. doi:10.1080/00222340903028662
61. Lin DD, Wu H, Zhang R, Pan W (2009) Enhanced photocatalysis of electrospun Ag-ZnO heterostructured nanofibers. Chem Mater 21(15):3479–3484. doi:10.1021/cm900225p
62. Lee SS, Bai HW, Liu ZY, Sun DD (2012) Electrospun TiO_2/SnO_2 nanofibers with innovative structure and chemical properties for highly efficient photocatalytic H_2 generation. Int J Hydrogen Energy 37(14):10575–10584. doi:10.1016/j.ijhydene.2012.04.098

63. Hwang SH, Kim C, Jang J (2011) SnO_2 nanoparticle embedded TiO_2 nanofibers-highly efficient photocatalyst for the degradation of rhodamine B. Catal Commun 12(11): 1037–1041. doi:10.1016/j.catcom.2011.02.024
64. Zhang R, Wu H, Lin DD, Pan W (2009) Preparation of necklace-structured TiO_2/SnO_2 hybrid nanofibers and their photocatalytic activity. J Am Ceram Soc 92(10):2463–2466. doi:10.1111/j.1551-2916.2009.03223.x
65. Liu ZY, Sun DD, Guo P, Leckie JO (2007) An efficient bicomponent TiO_2/SnO_2 nanofiber photocatalyst fabricated by electrospinning with a side-by-side dual spinneret method. Nano Lett 7(4):1081–1085. doi:10.1021/nl061898e
66. Wang CH, Shao CL, Zhang XT, Liu YC (2009) SnO_2 nanostructures-TiO_2 nanofibers heterostructures: controlled fabrication and high photocatalytic properties. Inorg Chem 48(15):7261–7268. doi:10.1021/ic9005983
67. Peng XH, Santulli AC, Sutter E, Wong SS (2012) Fabrication and enhanced photocatalytic activity of inorganic core-shell nanofibers produced by coaxial electrospinning. Chem Sci 3(4):1262–1272. doi:10.1039/c2sc00436d
68. Cao TP, Li YJ, Wang CH, Wei LM, Shao CL, Liu YC (2010) Fabrication, structure, and enhanced photocatalytic properties of hierarchical CeO_2 nanostructures/TiO_2 nanofibers heterostructures. Mater Res Bull 45(10):1406–1412. doi:10.1016/j.materresbull.2010.06.043
69. Cao TP, Li YJ, Wang CH, Wei LM, Shao CL, Liu YC (2010) Three-dimensional hierarchical CeO_2 nanowalls/TiO_2 nanofibers heterostructure and its high photocatalytic performance. J Sol-gel Technol 55(1):105–110. doi:10.1007/s10971-010-2221-x
70. Wang HY, Yang Y, Li X, Li LJ, Wang C (2010) Preparation and characterization of porous TiO_2/ZnO composite nanofibers via electrospinning. Chin Chem Lett 21(9):1119–1123. doi:10.1016/j.cclet.2010.03.009
71. Liu RL, Ye HY, Xiong XP, Liu HQ (2010) Fabrication of TiO_2/ZnO composite nanofibers by electrospinning and their photocatalytic property. Mater Chem Phys 121(3):432–439. doi:10.1016/j.matchemphys.2010.02.002
72. Kanjwal MA, Barakat NA, Sheikh FA, Kim HY (2010) Electronic characterization and photocatalytic properties of TiO_2/CdO electrospun nanofibers. J Mater Sci 45(5):1272–1279. doi:10.1007/s10853-009-4078-3
73. Santala E, Kemell M, Leskela M, Ritala M (2009) The preparation of reusable magnetic and photocatalytic composite nanofibers by electrospinning and atomic layer deposition. Nanotechnology 20(3):035602. doi:10.1088/0957-4484/20/3/035602
74. Wang ZJ, Li ZY, Zhang HN, Wang C (2009) Improved photocatalytic activity of mesoporous ZnO-SnO_2 coupled nanofibers. Catal Commun 11(4):257–260. doi:10.1016/j.catcom.2009.10.006
75. Zhang ZY, Shao CL, Li XH, Zhang L, Xue HM, Wang CH, Liu YC (2010) Electrospun nanofibers of ZnO-SnO_2 heterojunction with high photocatalytic activity. J Phys Chem C 114(17):7920–7925. doi:10.1021/jp100262q
76. Liu RL, Huang YX, Xia AH, Liu HQ (2010) Preparation and photocatalytic property of mesoporous ZnO/SnO_2 composite nanofibers. J Alloys Compd 503(1):103–110. doi:10.1016/j.jallcom.2010.04.211
77. Asokan K, Park JY, Choi SW, Kim SS (2010) Nanocomposite ZnO-SnO_2 nanofibers synthesized by electrospinning method. Nanoscale Res Lett 5(4):747–752. doi:10.1007/s11671-010-9552-y
78. Zhang ZY, Shao CL, Li XH, Wang CH, Zhang MY, Liu YC (2010) Electrospun nanofibers of p-type NiO/n-type ZnO heterojunctions with enhanced photocatalytic activity. ACS Appl Mater Interface 2(10):2915–2923. doi:10.1021/am100618h
79. Yousef A, Barakat NAM, Amna T, Unnithan AR, Al-Deyab SS, Kim HY (2012) Influence of CdO-doping on the photoluminescence properties of ZnO nanofibers: effective visible light photocatalyst for waste water treatment. J Luminescen 132(7):1668–1677. doi:10.1016/j.jlumin.2012.02.031

80. Li CR, Chen R, Zhang XQ, Shu SX, Xiong J, Zheng YY, Dong WJ (2011) Electrospinning of CeO_2-ZnO composite nanofibers and their photocatalytic property. Mater Lett 65(9):1327–1330. doi:10.1016/j.matlet.2011.01.075
81. Li CR, Zhang XQ, Dong WJ, Liu YS (2012) High photocatalytic activity material based on high-porosity ZnO/CeO_2 nanofibers. Mater Lett 80:145–147. doi:10.1016/j.matlet.2012.04.105
82. Ren PR, Fan HQ, Wang X (2012) Electrospun nanofibers of ZnO/$BaTiO_3$ heterostructures with enhanced photocatalytic activity. Catal Commun 25:32–35. doi:10.1016/j.catcom.2012.04.003
83. Wang RY, Guo J, Chen D, Miao YE, Pan JS, Tjiu WW, Liu TX (2011) "Tube brush" like ZnO/SiO_2 hybrid to construct a flexible membrane with enhanced photocatalytic properties and recycling ability. J Mater Chem 21(48):19375–19380. doi:10.1039/c1jm13979g
84. Fragala ME, Cacciotti I, Aleeva Y, Nigro RL, Bianco A, Malandrino G, Spinella C, Pezzotti G, Gusmano G (2010) Core-shell Zn-doped TiO_2-ZnO nanofibers fabricated via a combination of electrospinning and metal–organic chemical vapour deposition. CrystEngComm 12(11):3858–3865. doi:10.1039/c004157b
85. Zhang P, Shao CL, Li XH, Zhang MY, Zhang X, Sun YY, Liu YC (2012) In situ assembly of well-dispersed Au nanoparticles on TiO_2/ZnO nanofibers: a three-way synergistic heterostructure with enhanced photocatalytic activity. J Hazard Mater 237–238:331–338. doi:10.1016/j.jhazmat.2012.08.054
86. Bai HW, Liu ZY, Sun DD (2012) The design of a hierarchical photocatalyst inspired by natural forest and its usage on hydrogen generation. Int J Hydrogen Energy 37(19):13998–14008. doi:10.1016/j.ijhydene.2012.07.041
87. Su CY, Shao CL, Liu YC (2011) Electrospun nanofibers of TiO_2/CdS heteroarchitectures with enhanced photocatalytic activity by visible light. J Colloid Interface Sci 359(1):220–227. doi:10.1016/j.jcis.2011.03.044
88. Zhang ZY, Shao CL, Li XH, Sun YY, Zhang MY, Mu JB, Zhang P, Cuo ZC, Liu YC (2013) Hierarchical assembly of ultrathin hexagonal SnS_2 nanosheets onto electrospun TiO_2 nanofibers: enhanced photocatalytic activity based on photoinduced interfacial charge transfer. Nanoscale 5(2):606–618. doi:10.1039/c2nr32301j
89. Yousef A, Barakat NAM, Amna T, Al-Deyab SS, Hassan MS, Abdel-hay A, Kim HY (2012) Inactivation of pathogenic *Klebsiella pneumoniae* by CuO/TiO_2 nanofibers: a multifunctional nanomaterial via one-step electrospinning. Ceram Int 38(6):4525–4532. doi:10.1016/j.ceramint.2012.02.029
90. Mu JB, Chen B, Zhang MY, Guo ZC, Zhang P, Zhang ZY, Sun YY, Shao CL, Liu YC (2012) Enhancement of the visible-light photocatalytic activity of In_2O_3-TiO_2 nanofiber heteroarchitectures. ACS Appl Mater Interfaces 4(1):424–430. doi:10.1021/am201499r
91. Wang Y, Zhang JW, Liu LX, Zhu CQ, Liu XQ, Su Q (2012) Visible light photocatalysis of V_2O_5/TiO_2 nanoheterostructures prepared via electrospinning. Mater Lett 75:95–98. doi:10.1016/j.matlet.2012.01.074
92. Lu BG, Li XD, Wang TH, Xie EQ, Xu Z (2013) WO_3 nanoparticles decorated on both sidewalls of highly porous TiO_2 nanotubes to improve UV and visible-light photocatalysis. J Mater Chem A 1(12):3900–3906. doi:10.1039/c3ta01444d
93. Shang M, Wang WZ, Zhang L, Sun SM, Wang L, Zhou L (2009) 3D Bi_2WO_6/TiO_2 hierarchical heterostructure: controllable synthesis and enhanced visible photocatalytic degradation performances. J Phys Chem C 113(33):14727–14731. doi:10.1021/jp9045808
94. Zhang YP, Fei LF, Jiang XD, Pan CX, Wang Y (2011) Engineering nanostructured Bi_2WO_6-TiO_2 toward effective utilization of natural light in photocatalysis. J Am Ceram Soc 94(12):4157–4161. doi:10.1111/j.1551-2916.2011.04905.x
95. Cao TP, Li YJ, Wang CH, Zhang ZY, Zhang MY, Shao CL, Liu YC (2011) $Bi_4Ti_3O_{12}$ nanosheets/TiO_2 submicron fibers heterostructures: in situ fabrication and high visible light photocatalytic activity. J Mater Chem 21(19):6922–6927. doi:10.1039/c1jm10343a

96. Zhang MY, Shao CL, Guo ZC, Zhang ZY, Mu JB, Cao TP, Liu YC (2011) Hierarchical nanostructures of copper(II) phthalocyanine on electrospun TiO_2 nanofibers: controllable solvothermal-fabrication and enhanced visible photocatalytic properties. ACS Appl Mater Interfaces 3(2):369–377. doi:10.1021/am100989a
97. Yang Y, Wang HY, Li X, Wang C (2009) Electrospun mesoporous W^{6+}-doped TiO_2 thin films for efficient visible-light photocatalysis. Mater Lett 63(2):331–333. doi:10.1016/j.matlet.2008.10.037
98. Xu J, Wang WZ, Shang M, Gao EP, Zhang ZJ, Ren J (2011) Electrospun nanofibers of Bi-doped TiO_2 with high photocatalytic activity under visible light irradiation. J Hazard Mater 196:426–430. doi:10.1016/j.jhazmat.2011.09.010
99. Park JY, Lee JH, Choi DY, Hwang CH, Lee JW (2012) Influence of Fe doping on phase transformation and crystallite growth of electrospun TiO_2 nanofibers for photocatalytic reaction. Mater Lett 88:156–159. doi:10.1016/j.matlet.2012.07.080
100. Zhang ZY, Shao CL, Zhang LN, Li XH, Liu YC (2010) Electrospun nanofibers of V-doped TiO_2 with high photocatalytic activity. J Colloid Interface Sci 351(1):57–62. doi:10.1016/j.jcis.2010.05.067
101. Lee DY, Kim BY, Cho NI, Oh YJ (2011) Electrospun Er^{3+}-TiO_2 nanofibrous films as visible light induced photocatalysts. Current Appl Phys 11(3):S324–327. doi: 10.1016/j.cap.2011.03.022
102. Lee DY, Lee MH, Cho NI (2012) Preparation and photocatalytic degradation of erbium doped titanium dioxide nanorods. Current Appl Phys 12(4):1229–1233. doi: 10.1016/j.cap.2012.03.007
103. Cacciotti I, Bianco A, Pezzotti G, Gusmano G (2011) Synthesis, thermal behaviour and luminescence properties of rare earth-doped titania nanofibers. Chem Eng J 166(2):751–764. doi:10.1016/j.cej.2010.07.008
104. Camillo DD, Ruggieri F, Santucci S, Lozzi L (2012) N-doped TiO_2 nanofibers deposited by electrospinning. J Phys Chem C 116(34):18427–18431. doi:10.1021/jp302499n
105. Li HP, Zhang W, Huang SY, Pan W (2012) Enhanced visible-light-driven photocatalysis of surface nitrided electrospun TiO_2 nanofibers. Nanoscale 4(3):801–806. doi:10.1039/c1nr11346a
106. Im JH, Yang SJ, Yun CH, Park CR (2012) Simple fabrication of carbon/TiO_2 composite nanotubes showing dual functions with adsorption and photocatalytic decomposition of Rhodamine B. Nanotechnology 23(3):035604. doi:10.1088/0957-4484/23/3/035604
107. Li HP, Zhu YH, Cao HM, Yang XL, Li CZ (2013) Preparation and characterization of photocatalytic carbon dots-sensitized electrospun titania nanostructured fibers. Mater Res Bull 48(2):232–237. doi:10.1016/j.materresbull.2012.10.052
108. Im JS, Kim MI, Lee YS (2008) Preparation of PAN-based electrospun nanofiber webs containing TiO_2 for photocatalytic degradation. Mater Lett 62(21–22):3652–3655. doi:10.1016/j.matlet.2008.04.019
109. Wang QQ, Wang X, Li XJ, Cai YB, Wei QF (2011) Surface modification of PMMA/O-MMT composite microfibers by TiO_2 coating. Appl Surf Sci 258(1):98–102. doi: 10.1016/j.apsusc.2011.08.013
110. Costa RGF, Ribeiro C, Mattoso HC (2013) Study of the effect of rutile/anatase TiO_2 nanoparticles synthesized by hydrothermal route in electrospun PVA/TiO_2 nanocomposites. J Appl Polym Sci 127(6):4463–4469. doi:10.1002/app.38031
111. Lombardi M, Palmero P, Sangermano M, Varesano A (2011) Electrospun polyamide-6 membranes containing titanium dioxide as photocatalyst. Polym Int 60(2):234–239. doi:10.1002/pi.2932
112. Kim YB, Cho D, Park WH (2010) Fabrication and characterization of TiO_2/poly(dimethyl siloxane) composite fibers with thermal and mechanical stability. J Appl Polym Sci 116(1):449–454. doi:10.1002/app.31480
113. Neubert S, Pliszka D, Thavasi V, Wintermantel E, Ramakrishna S (2011) Conductive electrospun PANi-PEO/TiO_2 fibrous membrane for photo catalysis. Mater Sci Eng B 176(8):640–646. doi:10.1016/j.mseb.2011.02.007

114. Seo DK, Jeun JP, Kang PH (2011) Preparation and characterization of the electrospun PCS/TiO_2 fiber mat by electron beam irradiation. J Ind Eng Chem 17(3):803–807. doi:10.1016/j.jiec.2011.05.003
115. He TS, Zhou ZF, Xu WB, Ren FM, Ma HH, Wang J (2009) Preparation and photocatalysis of TiO_2-fluoropolymer electrospun fiber nanocomposites. Polymer 50(13):3031–3036. doi:10.1016/j.polymer.2009.04.015
116. Bedford NM, Steckl AJ (2010) Photocatalytic self cleaning textile fibers by coaxial electrospinning. ACS Appl Mater Interface 2(8):2448–2455. doi:10.1021/am1005089
117. Pant HR, Rajpandeya D, Nam KT, Baek W, Hong ST, Kim HY (2011) Photocatalytic and antibacterial properties of a TiO_2/nylon-6 electrospun nanocomposite mat containing silver nanoparticles. J Hazard Mater 189(1–2):465–471. doi:10.1016/j.jhazmat.2011.02.062
118. Chang ZJ (2011) "Firecracker-shaped" ZnO/polyimide hybrid nanofibers via electrospinning and hydrothermal process. Chem Commun 47(15):4427–4429. doi:10.1039/c0cc05634k
119. Kayaci F, Ozgit-Akgun C, Donmez I, Biyikli N, Uyar T (2012) Polymer–inorganic core–shell nanofibers by electrospinning and atomic layer deposition: Flexible nylon–ZnO core–shell nanofiber mats and their photocatalytic activity. ACS Appl Mater Interfaces 4(11):6185–6194. doi:10.1021/am3017976
120. Afeesh R, Barakat NAM, Al-Deyab SS, Yousef A, Kim HY (2012) Nematic shaped cadmium sulfide doped electrospun nanofiber mat: highly efficient, reusable, solar light photocatalyst. Colloids Surf A 409:21–29. doi:10.1016/j.colsurfa.2012.05.021
121. Ardona HAM, Paredes FU, Arellano IHJ, Arco SD (2013) Electrospun PET supported-ionic liquid-stabilized CdS catalyst for the photodegradation of Rhodamine B under visible light. Mater Lett 91:96–99. doi:10.1016/j.matlet.2012.09.069
122. Wang QY, Chen YJ, Liu RL, Liu HQ, Li ZH (2012) Fabrication and characterization of electrospun CdS-OH/polyacrylonitrile hybrid nanofibers. Compos Part A 43(11):1869–1876. doi:10.1016/j.compositesa.2012.07.023
123. Unnithan AR, Barakat NAM, Abadir MF, Yousef A, Kim HY (2012) Novel CdPdS/PVAc core-shell nanofibers as an effective photocatalyst for organic pollutants degradation. J Mol Catal A 363–364:186–194. doi:10.1016/j.molcata.2012.06.008
124. Zhou ZF, Feng Y, Xu WB, Ren FM, Ma HH (2009) Preparation and photocatalytic activity of zinc sulfide/polymer nanocomposites. J Appl Polym Sci 113(2):1264–1269. doi:10.1002/app.30139
125. He TS, Ma HH, Zhou ZF, Xu WB, Ren FM, Shi ZF, Wang J (2009) Preparation of ZnS-fluoropolymer nanocomposites and its photocatalytic degradation of methylene blue. Polym Degrad Stab 94(12):2251–2256. doi:10.1016/j.polymdegradstab.2009.08.012
126. Shao LJ, Xing GY, Lv WX, Yu HJ, Qiu MX, Zhang XM, Qi CZ (2013) Photodegradation of azo-dyes in aqueous solution by polyacrylonitrile nanofiber mat-supported metalloporphyrins. Polym Int 62(2):289–294. doi:10.1002/pi.4298
127. Guo ZC, Shao CL, Mu JB, Zhang MY, Zhang ZY, Zhang P, Chen B, Liu YC (2011) Controllable fabrication of cadmium phthalocyanine nanostructures immobilized on electrospun polyacrylonitrile nanofibers with high photocatalytic properties under visible light. Catal Commun 12(10):880–885. doi:10.1016/j.catcom.2011.02.004
128. Zugle R, Nyokong T (2013) Comparative phototransformation of environmental pollutants using metallophthalocyanines supported on electrospun polymer fibers. J Appl Polym Sci 128(2):1131–1142. doi:10.1002/app.38381
129. Zugle R, Antunes E, Khene S, Nyokong T (2012) Photooxidation of 4-chlorophenol sensitized by lutetium tetraphenoxy phthalocyanine anchored on electrospun polystyrene polymer fiber. Polyhedron 33(1):74–81. doi:10.1016/j.poly.2011.11.005
130. Wang P, Huang BB, Qin XY, Zhuang XY, Dai Y, Wei JY, Whangbo MH (2008) Ag@AgCl: a highly efficient and stable photocatalyst active under visible light. Angew Chem Int Ed 47(41):7931. doi:10.1002/anie.200802483
131. Lei JY, Wang W, Song MX, Dong B, Li ZY, Wang C, Li LJ (2011) Ag/AgCl coated polyacrylonitrile nanofiber membranes: synthesis and photocatalytic properties. React Funct Polym 71(11):1071–1076. doi:10.1016/j.reactfunctpolym.2011.08.002

Chapter 16
Multicomponent Nanofibers via Electrospinning of Polymers and Colloidal Dispersions for Environmental and Optical Applications

Keqin Zhang, Wei Yuan, Ning Zhou, and Chaojie Wu

Abstract Electrospinning of solutions, melts, and coaxial configurations is a well-known conventional method for the fabrication of nano- and microfibers. Electrospinning of colloids, or colloid electrospinning, is a relatively new technique that offers the possibility of producing customized nanomaterials with multiple components and hierarchical nanostructures. Nevertheless, the physical and chemical properties of colloids and their compatibilities with polymer solutions significantly increase the complexity of the system's theoretical predictions. Additionally, these fundamental properties also impact control of the composite fibers' fine structures through changing the process parameters. This chapter summarizes recent and significant progresses in colloid electrospinning in terms of their novel structures and the prospective applications in the fields of the environmental protection and optical devices. Additionally, applications of electrospun colloidal fibers in energy conversion have been intensively reviewed in other chapters and are thus omitted here to avoid redundancy.

16.1 Introduction

One-dimensional (1D) nanostructures have attracted significant attention due to their unique properties and useful applications in many areas [1, 2]. A large number of fabrication methods for 1D nanostructures have already been demonstrated; these methods have successfully generated nanostructures in the forms of fibers, wires, rods, belts, tubes, and rings from various materials [1]. Among these methods, electrospinning is currently the only technique that allows the fabrication of continuous fibers with diameters down to a few nanometers [3–6]. The porous

K. Zhang (✉) • W. Yuan • N. Zhou • C. Wu
National Engineering Laboratory for Modern Silk, College of Textile and Clothing Engineering, Soochow University, Suzhou 215123, China
e-mail: kqzhang@suda.edu.cn

B. Ding and J. Yu (eds.), *Electrospun Nanofibers for Energy and Environmental Applications*, Nanostructure Science and Technology,
DOI 10.1007/978-3-642-54160-5_16,

three-dimensional (3D) membranes assembled from electrospun fibers show large specific surface area, high porosity, and good interconnectivity [7–9]. The qualities displayed by electrospun nanomaterials are highly desirable for applications such as filtration, sensors, drug delivery platforms, and tissue engineering [10–12].

Compared to electrospinning of monophasic solutions and melts, colloid electrospinning [13–15] is a significant variant of the electrospinning technique that has only recently emerged in popularity. Colloid electrospinning is performed by electrospinning the polymer solution when it has been blended with organic/inorganic colloidal nanoparticles [16–29]. The incorporation of colloidal particles in polymer solution during the electrospinning process serves the purpose of embedding functional nanoparticles into the fibrous materials without posttreatment. The incorporation of colloidal particles also endows the polymeric nanofibers with multiple functionalities, including superhydrophobic nonwoven fabrics [23], magnetic fibers [27], and ultrafiltration [28].

16.2 Colloid Electrospinning

16.2.1 Basic Principles of the Colloid Electrospinning

The basic laboratory setup for colloid electrospinning is similar to the environment for solution electrospinning. Figure 16.1 shows the three major components required for colloid electrospinning: a high-voltage power supply, a microinjection pump consisting of a syringe and metallic needle, and an electrically conductive collector placed opposite to the metal needle. To form fibers from colloidal dispersions containing inorganic or organic colloids, a fiber-forming polymer is usually needed to act as a template. The template solution containing the dispersed colloids is fed through the syringe at a constant flow rate, forming a droplet at the tip of the metal needle. When a strong electrical field is applied to the liquid, the droplet, suspended in equilibrium at the tip of a capillary, is distorted into a conical shape [30]. When a sufficiently strong electric field is applied, charges on the droplet surface overcome the surface tension, inducing the formation of a liquid jet that accelerates toward a grounded collector. Due to solvent evaporation and electrostatic repulsion forces, this jet undergoes a strong thinning and whipping process on its way toward the collector [31], resulting in nonwovens composed of fibers with diameters ranging from several micrometers to nanometers.

16.2.2 Possible Structures Obtained via Electrospinning of Colloidal Dispersions

The morphology of the hybrid fibers produced by colloid electrospinning is governed by colloid size, compatibility between colloids and polymer solutions, and the microstructure of phase separation between the colloids and the polymer

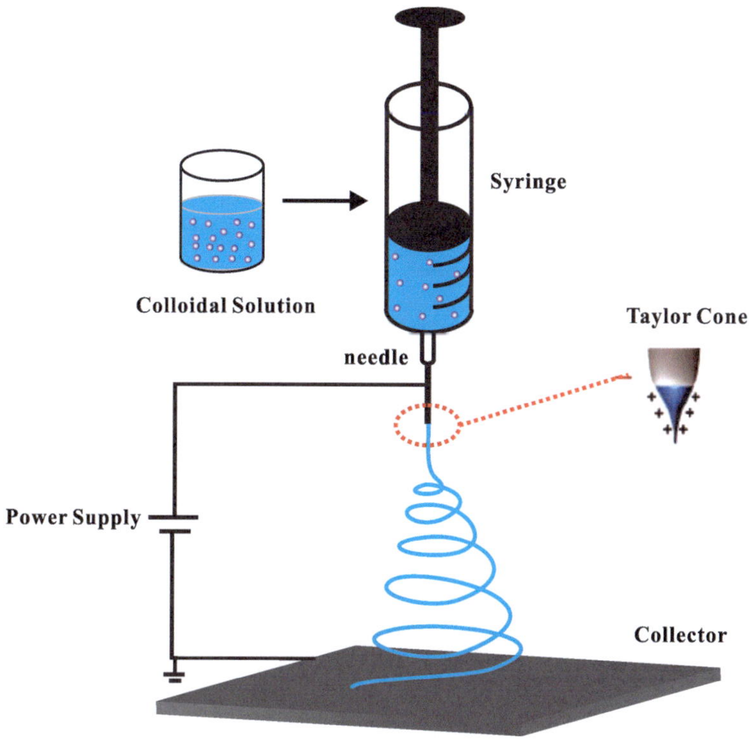

Fig. 16.1 Basic laboratory colloid-electrospinning setup

solution due to fast solvent evaporation [26]. If the size of the colloids is under 20 nm, the nanoparticles usually embed inside the fibers, as shown in Table 16.1a. In Kim et al.'s [32] study of Au nanoparticle dispersion in electrospun poly(ethylene oxide) (PEO)/Au composite nanofibers, it was observed that the particles linearly packed and formed chain-like structures along the length of the fiber (Fig. 16.2). The good compatibility between Au surfaces and polymer solution aided in embedding the nanoparticles inside the fibers; the small sizes of the Au particles led to an even distribution along the thickness of the fibers.

When the size of the colloids was increased to several hundred nanometers, the morphology of the hybrid fibers obtained via colloidal electrospinning varied with different experimental conditions as shown in Table 16.1b. Yuan et al. [15] systemically studied the spin ability of a polymer solution (PVA) containing colloidal nanoparticles (PSNs) based on PVA:PSN weight ratio, the blended solution's viscosity, and PSNs sizes. It was found that PVA nanofibers acted as a confining template to restrain PSNs into string-on-bead and necklace-like nanofibers. The diameter of the PVA-fiber-string PSNs could be adjusted via the viscosity of the blend solution in a PVA-dominant blend solution. In contrast, PSNs were wrapped by the PVA to form a core-shell structure in a PSN-dominant blend solution. Under such conditions, PVA worked as an adhesive, causing PSNs to adhere in structures

Table 16.1 Possible morphologies obtained via electrospinning of colloids and polymer hybrid solutions

Colloid Size	Fiber Morphology
(a) ○ < 20nm	
(b) 50nm < ○ < 500nm	More Polymer   More Colloids 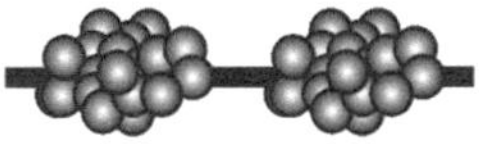 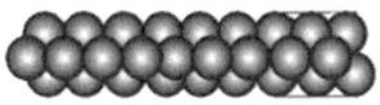
(c) 800nm < ○	

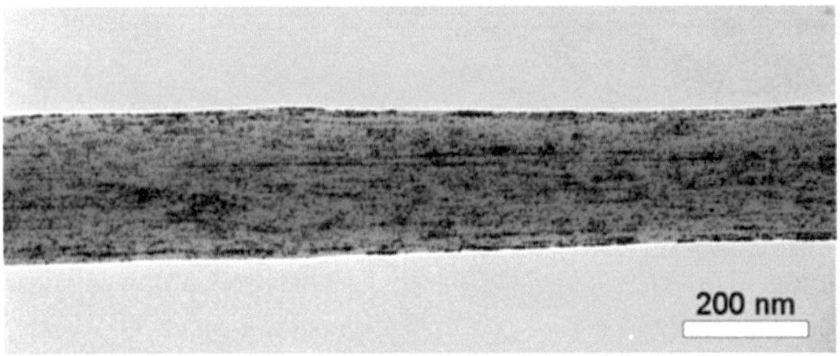

Fig. 16.2 TEM micrograph of 1D chain-like arrays of Au nanoparticles in PEO nanofiber (Reprinted with the permission from Ref. [32]. Copyright 2005 American Chemical Society)

such as blackberry-like aggregates, uniform colloidal fibers, and colloidal balls. The structural changes of electrospun hybrid fibers were controlled by the viscosity of the blend solution (Fig. 16.3).

Lim et al. [24] observed that hydrophilic polyacrylamide (PAM) nanofibers effectively wrapped the hydrophilic silica colloids, while hydrophobic polyacrylonitrile (PAN) nanofibers embedded large amounts of hydrophobic silica colloids. Moreover, the addition of a small amount of hexanol to the PEO solution successfully confined all silica colloids inside PEO fibers. Hexanol acted as a surfactant, decreasing the interfacial tension between the silica colloids and PEO (Fig. 16.4). Based on the observations, it was postulated that encapsulation of silica colloids was mainly governed by the wettability of the polymers on the particles.

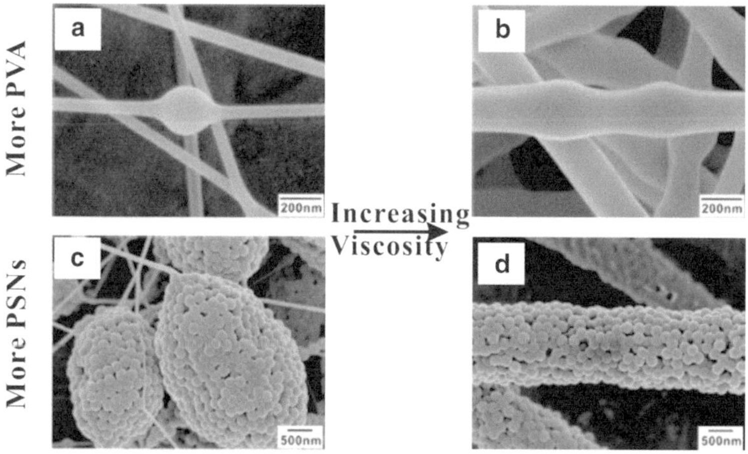

Fig. 16.3 Structures of electrospun PVA/PSN fibers with PVA concentrations of (**a**, **b**) 7 wt% and (**c**, **d**) 13 wt%. (**a**, **c**) Weight ratio was fixed at 4:1 and (**c**, **d**) 1:4 (Reprinted with the permission from Ref. [15]. Copyright 2012 American Chemical Society)

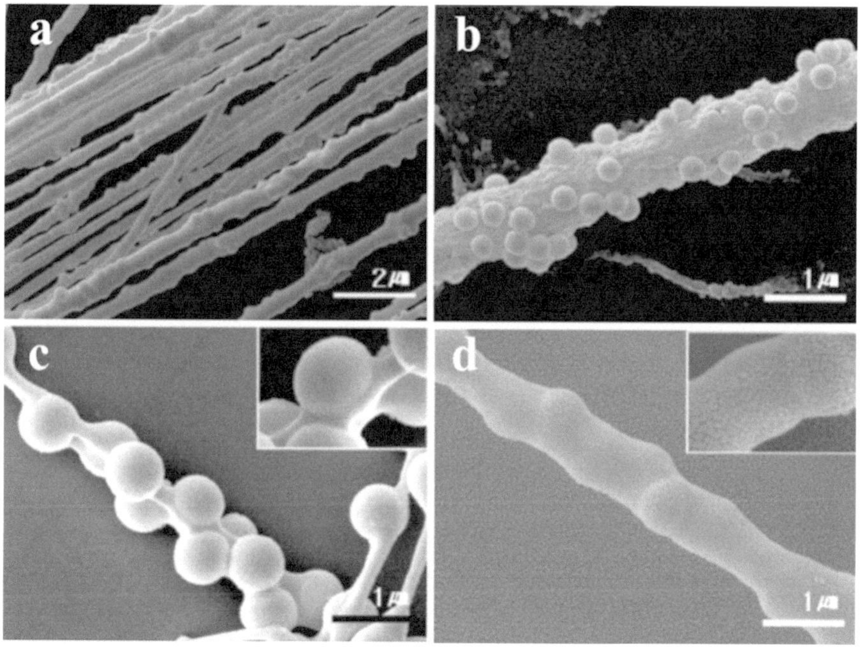

Fig. 16.4 SEM images of silica particle assemblies with PAM (**a**) and PAN (**b**) for 300 nm silica particles. SEM images of PEO nanofiber with 700 nm silica particles (**c**) without and (**d**) with hexanol (Reprinted with the permission from Ref. [24]. Copyright 2006 American Chemical Society)

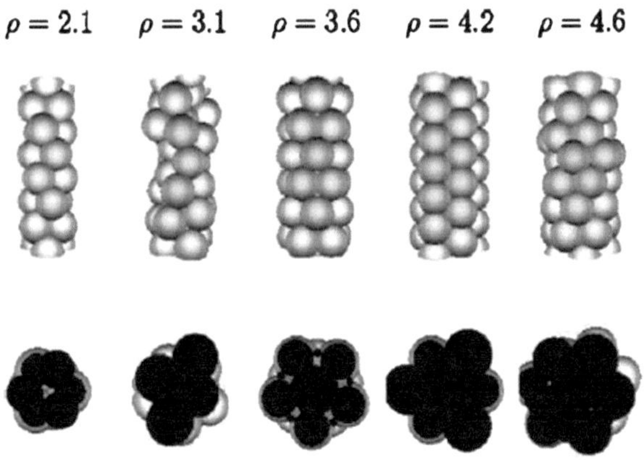

Fig. 16.5 Calculated fiber shapes for different densities. The *bottom* row shows a cross section of the fiber, with cut spheres marked in *black* (Reproduced from Ref. [22] by permission of John Wiley & Sons Ltd)

F. Grossmann [22, 32] reported that a contracting flow that occurred during the electrospinning process pulled the particles toward the center of the fiber, forming regular and irregular chains. The formation of the chains can be modeled with a phenomenological contracting potential and an approximate inter-particle potential. Mechanical relaxation of random initial conditions resulted in configurations that were similar to experimental findings. The key parameter is ρ, the number of latex spheres multiplied by their diameter. The minimum value for a connected string of spheres was found to be $\rho = 1$. Examples of fibers with higher densities are shown in Fig. 16.5. Values of $\rho = 2$, 3.5, and 4 formed regular arrangements of spheres with several strands of spheres. For intermediate values, the arrangements were typically irregular, as the particles that did not fit and either attached to smaller regular chains or caused gaps in the bigger regular chains.

Furthermore, under certain conditions, colloids with larger sizes were more feasible in the formation of beads-on-string and necklace-like structures, as shown in Table 16.1c. Yang et al. [18] demonstrated that silica particles either closely packed with each other or linearly linked as a string in electrospun fibers; the formation depended on the dimension ratio between the fiber thickness and silica diameter, as shown in Fig. 16.6.

16.2.3 Electrospinning of Inorganic Nanoparticles

Inorganic nanoparticles are an extremely important group of nanomaterials in modern technology. Due to their simple, inexpensive synthesis and mass production, they are readily integrated into colloidal electrospinning. Compared to organic colloids, inorganic nanoparticles possess higher electronic density than polymer templates and can thus be localized into fibers during electrospinning. Inorganic nanoparticles commonly hail from metal, metalloid, and metal and metalloid oxide groups, as well as minerals such as carbonates, aluminosilicates, hydroxyapatite, and non-oxide ceramics.

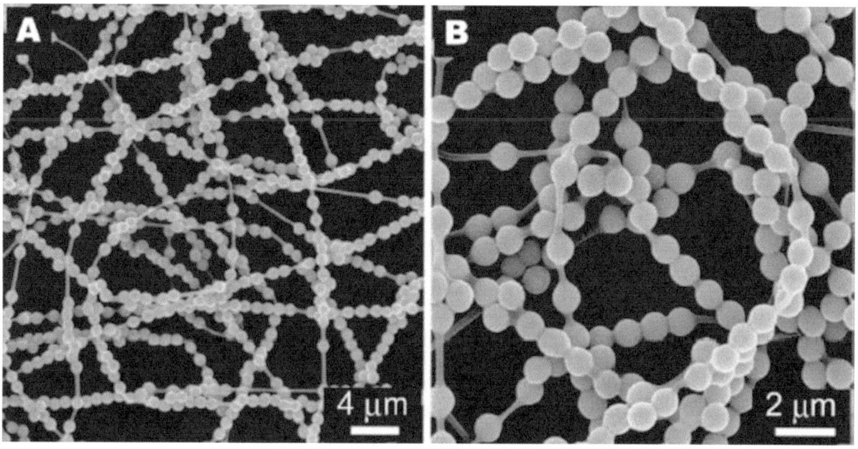

Fig. 16.6 Necklace-like structures of PVA/SiO_2 (910 nm) ES mats by removing some of the water in the solutions. PVA:SiO_2 = 300:700, PVA 12 %, work distance 10 cm, and voltage 15 kV (Reprinted with the permission from Ref. [18]. Copyright 2010 American Chemical Society)

16.2.3.1 Metal Nanoparticles

Based on reports, the most commonly used metal nanoparticles in colloidal electrospinning are silver and gold nanoparticles. As discovered in previous studies, various sizes of silver nanoparticles could be electrospun with synthetic polymers such as polyacrylonitrile (PAN) [33], poly(ethylene oxide) (PEO) [34], poly(N-vinylpyrrolidone) (PVP) [35], poly(vinyl alcohol) (PVA) [36], polyamide 6 [37], poly(lactic acid) (PLA) [38], and poly(phenylene vinylene) (PPV) [39] or with renewable resource-based polymers such as chitosan/gelatin blends [40] and N-carboxyethyl-chitosan [41]. For example, Wang et al. [42] prepared PAN nanofiber-coating silver nanoparticles via electrospinning. The conductivity of silver nanoparticles/PAN fiber rose from 10^{-14} $S \cdot cm^{-1}$of pure PAN fiber to 10^{-7} $S \cdot cm^{-1}$. Khan et al. [43] employed PEO at electrospinnable molecular weights of 600, 1,000, and 2,000 kDa for use as electrospinning polymers and reactants in the reduction of silver salt to zerovalent silver nanoparticles. This synthesis was found to be particularly convenient and free of organic solvents. The silver nanoparticles inside the fibers reduced fiber diameters and enhanced fiber quality due to increased electrical conductivity (Fig. 16.7). Youk et al. [44] utilized two different methods for preparation of polymer nanofibers containing silver nanoparticles. The first method involved direct electrospinning of PVP nanofibers containing silver nanoparticles; this method utilized N,N-dimethylformamide as both a solvent for the PVP and a reducing agent for the silver ions in the PVP solution. In the second method, PVA aqueous solution was electrospun with 5 wt% PVP-containing silver nanoparticles. PVP-containing silver nanoparticles were used to introduce silver nanoparticles to other PVP-miscible polymer nanofibers, resulting with even distribution of silver nanoparticles in PVA nanofibers. Cheng et al. [45]

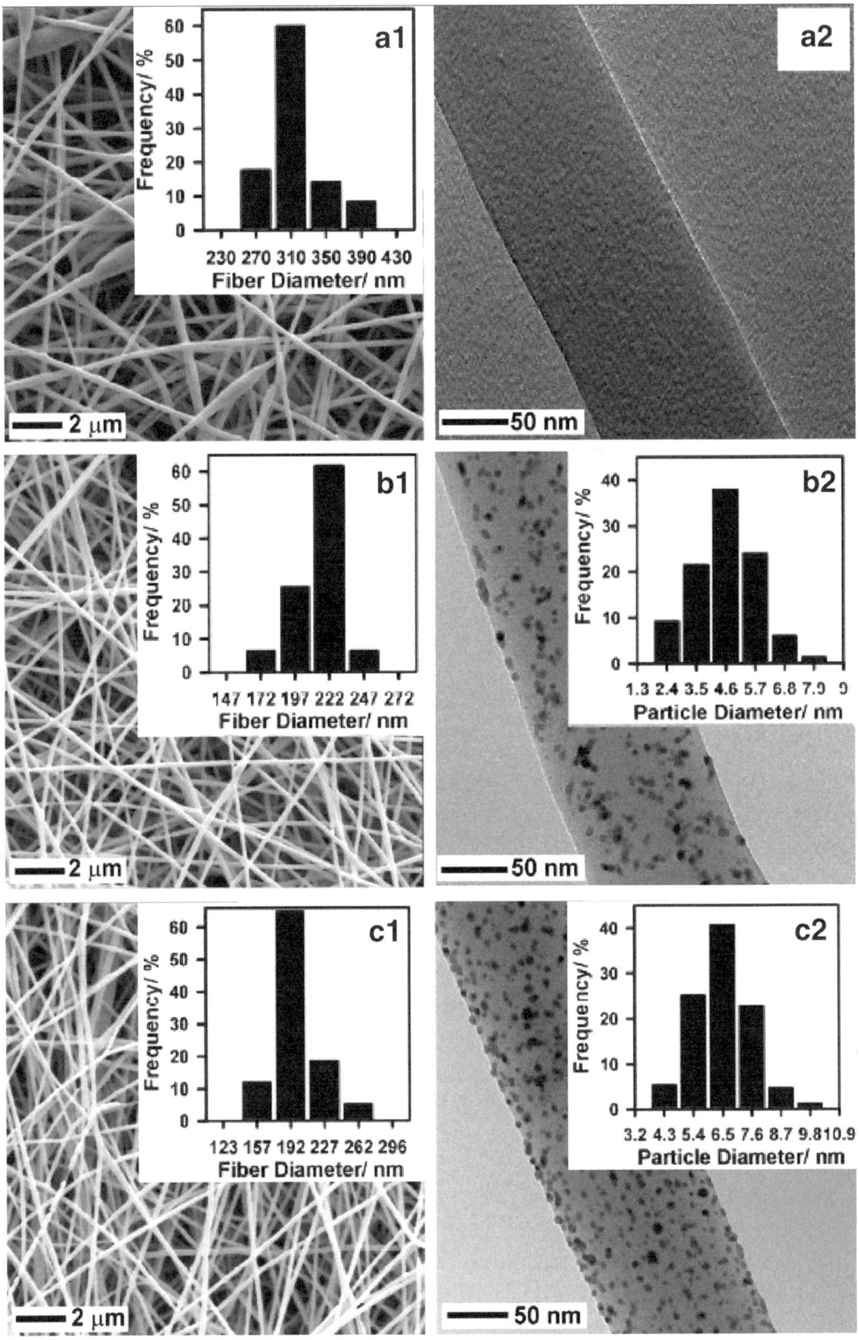

Fig. 16.7 SEM (a1, b1, c1) and TEM (a2, b2, c2) images of the silver nanoparticles/PEO nanofiber mats fabricated from aqueous solution of 4 wt% 600 kDa PEO and $AgNO_3$ at concentrations of (**a**) 0, (**b**) 0.17, and (**c**) 0.26 wt% (Reproduced from Ref. [43] by permission of John Wiley & Sons Ltd)

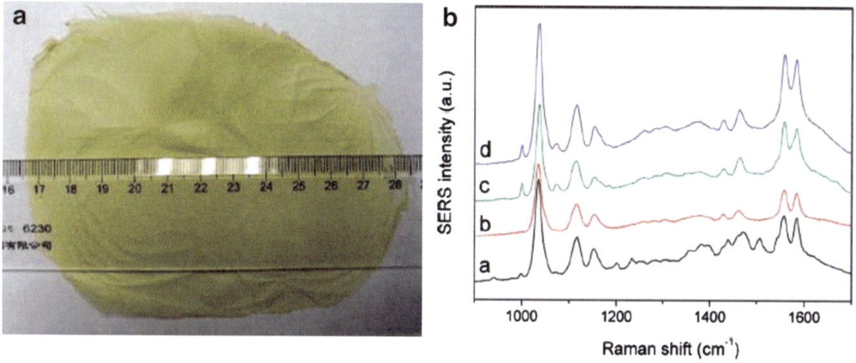

Fig. 16.8 (**a**) Photograph of silver nanoparticle/PVA nanofiber mat with a PVA/silver nanoparticles molar ratio of 530:3. (**b**) SERS spectra of 4-mercaptobenzoic acid collected on a set of silver nanoparticles/PVA nanofiber mats with different molar ratios (Reprinted with the permission from Ref. [47]. Copyright 2009 American Chemical Society)

reported electrospinning as a novel method for the fabrication of silver nanoparticle-containing chitosan/gelatin. Using chitosan as a reducing agent and stabilizer, silver nanoparticles ranging from 1 nm to 5 nm were synthesized at room temperature. The silver nanoparticles/chitosan composites were dissolved in gelatin containing acetic acid solution, and the prepared solution was electrospun into chitosan/gelatin nanofibers containing silver nanoparticles. In some cases, silver nanoparticles/polymer nanofibers displayed possibility for antibacterial applications as a result of antimicrobial properties in silver nanoparticles. Hota et al. [46] successfully synthesized PAN nanofibers impregnated with silver nanoparticles via electrospinning at room temperature. It was found that silver nanoparticles/PAN composite nanofibers showed strong antibacterial activity toward both gram-positive and gram-negative bacteria. In recent years, the optical application of silver nanoparticles/polymer nanofibers has increasingly attracted attention from researchers. Via electrospinning, Yu et al. [47] successfully developed silver nanoparticles/PVA nanofibers with chain-like arrays; these nanoparticles could be used as surface-enhanced Raman scattering (SERS) substrates. The aggregation state of the obtained Ag nanoparticles significantly contributed to the high sensitivity of SERS to 4-mercaptobenzoic acid with an enhancement factor of 10^9 (Fig. 16.8).

Gold nanoparticles were also electrospun, mostly in polymers such as PEO [48], PVA [49], PS [50], PVP [51], or PLA [52]. Kim et al. [31] first fabricated one-dimensional arrays of gold nanoparticles within nanofibers, employing PEO as a template for the controlled nanoscale organization of nanoparticles. Greiner et al. [53] prepared gold nanowires by electrospinning highly concentrated aqueous dispersions of gold nanoparticles in the presence of PVA. The obtained product was subsequently annealed at high temperatures, resulting in continuous wires of sintered gold nanoparticles. It was found that the anticancer drug daunorubicin could efficiently self-assemble on surfaces of the gold nanoparticles/PLA nanofibers

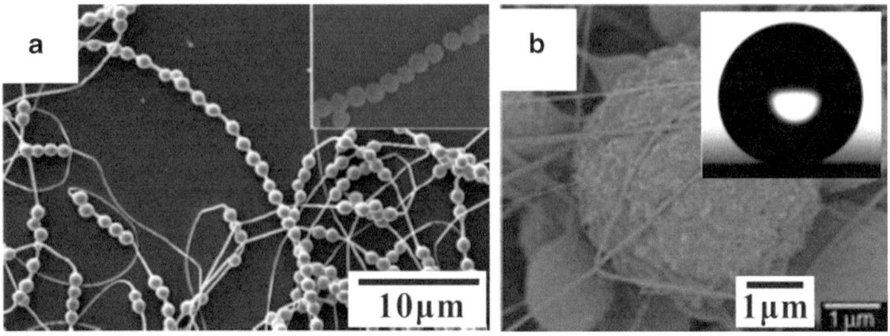

Fig. 16.9 (**a**) SEM micrographs of silica nanoparticles in polyacrylamide (PAM) fibers (Reprinted with the permission from Ref. [24]. Copyright 2006 American Chemical Society). (**b**) Surface nanostructure of a single porous microsphere produced via electrospraying. The insert picture is the profile of a water droplet on the corresponding membrane (Reproduced from Ref. [65] by permission of John Wiley & Sons Ltd)

nanocomposites, making this structure relevant for biomedical applications [54]. It was reported that the presence of the gold nanoparticles in PEO fibers strongly reduced the orientation of the polymer crystals without significantly affecting the degree of polymer crystallinity [55]. Furthermore, Shi et al. [56] reported a facile approach to immobilizing gold nanoparticles in electrospun polyethyleneimine (PEI)/PVA composite nanofibers for catalytic applications. The results showed that gold nanoparticles containing nanofibers displayed excellent catalytic activity and reusability for the transformation of 4-nitrophenol to 4-aminophenol. Recently, Fan et al. [57] successfully fabricated surface-enhanced Raman scattering (SERS) substrates of gold-silver composite nanoparticles/PVA nanofibers via a simple electrospinning approach. The Raman spectra displayed an apparent enhancement in the signal of 4-mercaptobenzoic acid, pyridine, and thiophenol molecules.

Additionally, the zerovalent nanoparticles used for colloidal electrospinning extended to silicon [58] and transition metals such as zinc [59], cobalt [60], titanium [61], and copper [62]. Zhang et al. [63] developed an innovative energy-storage material by embedding silica nanoparticles in electrospun carbon nanofibers. Anodes made from these carbon/silicon composite nanofibers combine the advantages of carbon (long cycle life) and silicon (high lithium-storage capacity) and are thus promising candidates for energy storage in high-performance batteries.

16.2.3.2 Metal Oxide Nanoparticles

Many types of metal oxide nanoparticles can be successfully electrospun; examples include magnesium, aluminum, silicon, titanium, iron, zinc, and mixed metal oxide nanoparticles. Silica nanoparticles electrospun with polystyrene dispersions are extensively utilized in the production of superhydrophobic meshes (Fig. 16.9a) [23, 24]. Nanoporous silica nanofibers with ultrahigh surface area have been fabricated via electrospinning blend solutions of PVA and colloidal silica nanoparticles, followed by selective removal of PVA components. Porous silica membranes modified with fluoroalkylsilane showed superhydrophobicity due to their porous structures [64]. The diameters of silica nanoparticles, voltage, and ingredient of solutions significantly influenced the morphology of electrospun mats. Jiang et al. [65] showed that electrospinning and spraying a PS solution in dimethylformamide (DMF) yielded a superhydrophobic film composed of porous microparticles and nanofibers (Fig. 16.9b). Yang et al. [18] systemically investigated the influence of silica nanoparticle diameter, PVA: silica weight ratio, voltage, and relative content of PVA/silica/water on the morphology of electrospun fibers. The results showed that silica nanoparticles with diameters of 143 nm tended to aggregate into bunches in the fibers, while 265 nm and 910 nm silica nanoparticles aligned along the fibers one by one to form a necklace-like structure.

Titanium dioxide is a popular metal oxide nanoparticle due to its photocatalytic activity. The electrospinning of titanium dioxide in hydrophilic polymers such as PEO [66] and PVP [67] or hydrophobic polymers such as PVA [68], PAN [69], and PSU has been reported [70]. Xia et al. [71] described an electrospinning-based procedure for generating composite nanofibers made of PVP and amorphous TiO_2. By performing calcination in air at 500 °C, it was possible to convert these nanofibers into anatase without changing their morphology. Through variation of parameters, the average diameter of these ceramic nanofibers could be controlled in the range of 20–200 nm. Lee et al. [72] successfully synthesized PVA/TiO_2 composite polymer membranes at different TiO_2 concentrations (20 %, 30 %, 40 %, and 50 %) using electrospinning. TiO_2 nanoparticles were embedded in the PVA fiber and dispersed linearly along the fiber. The average diameters of obtained fibers were dependent on the total loadings used. Wang et al. [73] prepared highly photoluminescent TiO_2/poly(phenylene vinylene) (PPV) hybrid nanoparticle-polymer fibers by electrospinning a PPV precursor added to a TiO_2 sol–gel solution. The diameters of the hybrid fibers were in the range of 100–300 nm and the average size of TiO_2 nanoparticles within the fibers was 10–60 nm. Electrospinning TiO_2 in the presence of polymer materials has potential applications as photoelectrodes of dye-sensitized solar cells. Kim et al. [74] reported a new application of electrospun TiO_2 fibers as an electrode for dye-sensitized solar cells (DSSCs). TiO_2 fiber electrodes were directly electrospun onto a conducting glass substrate from a mixture of titanium (IV) propoxide and poly(vinyl acetate) (PVAc) in dimethylformamide. Using an electrospinning technique, Nuraje et al. [75] fabricated polystyrene and polyvinyl chloride (PVC) fibers incorporated into TiO_2 nanoparticles and graphene nanoflakes. The nanocomposite fibers can be utilized in designing photoelectrodes of DSSCs as self-cleaning and anti-icing materials for long-term cell efficiency.

Some metal oxide nanoparticles possess unique, significant applications when electrospun with certain polymers. Lee [76] successfully developed layered fabric systems with electrospun polyurethane/zinc oxide nanocomposite fiber webs layered on cotton substrates, which imparted UV protection and antibacterial functions. Okuyama et al. [77] used mixed solutions of indium tin oxide (ITO) nanoparticle sol and PEO as precursors of the nanofibers to prepare ITO nanofibers via

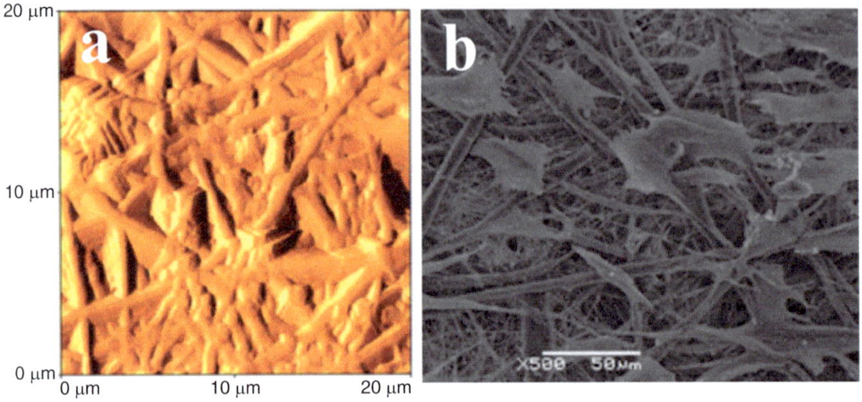

Fig. 16.10 (**a**) AFM images of electrospun 30/70 hydroxyapatite/collagen fibers (Reproduced from Ref. [80] by permission of John Wiley & Sons Ltd). (**b**) SEM images of PLLA/MWNTs/HAp membranes (Reprinted with the permission from Ref. [82]. Copyright 2007 American Chemical Society)

electrospinning, followed by annealing at 400 °C. The ITO nanofibers could be used in high-efficiency dye-sensitized solar cells, optical applications, electromagnetic field filters, and high-performance sensing applications. Shao et al. [78] prepared PVA/chromium nitrate (Cr_2O_3)/aluminum nitrate (Al_2O_3) composite nanofibers using sol–gel processing and electrospinning techniques. Cr_2O_3/Al_2O_3 composite nanofibers, which could be used as a catalyst due to their high surface area, were successfully obtained after high-temperature calcinations. Via an electrospinning technique similar to the previously described approach for gold and TiO_2 nanoparticle fabrication, Andrew et al. [79] fabricated polyvinylidene difluoride (PVDF) fibers with continuously dispersed coupling agent ($Ni_{0.5}Zn_{0.5}Fe_2O_4$) nanoparticles from dimethylformamide solutions. The presence of the nanoparticles significantly promoted the formation of the ferroelectric phases of PVDF.

16.2.3.3 Nanoparticles from Other Minerals

Minerals such as hydroxyapatite are widely used in colloidal electrospinning, especially for bone tissue engineering. Nanoparticles of hydroxyapatite are usually electrospun with biocompatible or biodegradable polymers. Both natural and synthetic polymers are used as supporting templates, depending on preferences of individual researchers. Stanishevsky et al. [80] prepared hydroxyapatite (HAp) nanoparticle/collagen composite materials with various HAp/collagen weight ratios from HAp/collagen dispersions using solution deposition and electrospinning with static or rotating collectors (Fig. 16.10a). Wang et al. [81] fabricated poly(lactide-co-glycolide) (PLGA)/HA composite scaffolds with varying HAp contents (0 %, 5 %, and 10 %) via electrospinning. Yang et al. [82] developed a novel type of guided tissue regeneration (GTR) membrane by electrospinning a suspension consisting of poly(l-lactic acid) (PLLA), multiwall carbon nanotubes, and hydroxyapatite (PLLA/MWNTs/HAp). MWNTs/HAp nanoparticles were uniformly dispersed in the membranes, and the degradation characteristics were greatly improved (Fig. 16.10b). Lim et al. [83] reported the preparation of innovative biomimetic nanocomposite nanofibers of hydroxyapatite/chitosan (HAp/CTS) via combining an in situ coprecipitation synthesis approach with an electrospinning process. The electrospun nanocomposite nanofibers of HAp/CTS, with compositional and structural features close to their natural mineralized nanofibril counterparts, hold potential in bone tissue engineering applications.

In addition to HAp/polymer composite nanofibers, examples of hybrid fibers produced by electrospinning inorganic nanoparticles include boehmite in polyamide 6 [84] and calcium carbonate nanoparticles in poly(ε-caprolactone) [85].

In conclusion, a wide variety of inorganic nanoparticles were electrospun. Metal nanoparticles are usually embedded as separated nanoparticles in the fibers due to their low amount in dispersion. Metal oxide nanoparticles are commonly dispersed in higher concentrations in the electrospinning solutions, therefore often forming hierarchical structures.

16.2.4 Electrospinning of Polymeric Nanoparticles

Unlike the results of electrospinning with inorganic nanoparticles, described above, the use of organic nanoparticles in electrospinning is seldom reported. Carbon black nanoparticles are the most widely used organic nanoparticles and have been electrospun in various polymer fibers. Farris et al. [86] added carbon black nanoparticles to three different polymer solutions, producing color changes by electrospinning a black solution with white or off-white fiber mats for all three polymer/solvent systems. The electrospun mats reverted to a black color when heated, indicating potential for use in applications such as temperature sensors. Ghosh et al. [87] successfully fabricated porous conducting composite fiber webs containing carbon black nanoparticles by using the unique approach of incorporating nanoparticles into PU polymer via electrospinning. Supaphol et al. [88] prepared PVA nanofiber mats with incorporation of carbon black nanoparticles via electrospinning. Incorporation of nanoparticles did not affect the morphology and average diameter of the obtained fibers, but affected both the mechanical and the electro-rheological properties of the PVA/carbon black fiber mats, causing the mats to become more rigid with increasing nanoparticle concentrations. Furthermore, carbon nanotubes were electrospun in various polymer fibers. The electrospinning process was used to successfully fabricate nanofibers of PEO with embedded multiwall carbon nanotubes (MWCNT). The MWCNT were found to be oriented along the fiber axis [89]. Highly oriented, large-area, continuous composite nanofiber sheets made from surface-oxidized multiwall carbon nanotubes and PAN polymer were successfully developed using electrospinning [90]. Polystyrene can be considered another widely utilized type

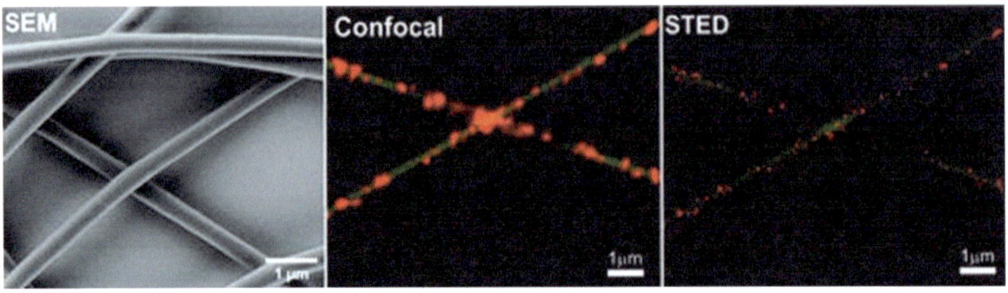

Fig. 16.11 SEM, laser confocal, and two-color STED microscopy of PS/PVA composite nanofibers (Reprinted with the permission from Ref. [14]. Copyright 2011 American Chemical Society)

of nanoparticle in colloidal electrospinning. A model system for multicomponent nanofibers was fabricated via colloid electrospinning. The obtained nanomaterials consisted of fluorescent PS nanoparticles that were synthesized in miniemulsion and embedded in fluorescently labeled PVA nanofibers (Fig. 16.11) [14]. PS nanoparticles with sizes from 100 nm to over 300 nm were mixed with a PVA solution and electrospun. The particles were in close contact during the thinning of the jet and became packed into the resulting fibers. The obtained structure was less ordered than self-assembled structures in a block crystal with similar polymer nanoparticles [16]. Moreover, PS/PVA composite nanofibers were prepared via electrospinning of dispersions of styrene-acrylate copolymers in an aqueous environment. The synthetic versatility of emulsion polymerization techniques allowed incorporation of cross-linking strategies to improve the thermal and mechanical stability of fibers [91]. Finally, the colloidal electrospinning of PU [92], poly(lactide-co-glycolide) [93], and poly(hexamethylene adipate-block-ethylene oxide) [94] nanoparticles was also reported.

16.3 Electrospun Colloidal Nanofibers for Environmental Applications

The presence of one or several colloids endows the obtained electrospun colloidal nanofibers with more complicated structures than conventional electrospun nanofibers, making them promising candidates for multiple emerging environmental functionalities.

16.3.1 Filtrations

Due to industrialization and rapid urbanization in the past decades, many environmental problems have emerged. Filtration is considered a vital solution for many of these issues and possesses a vital role in environmental protection. Fig. 16.12 illustrates common filtration processes, with membranes functioning as screen

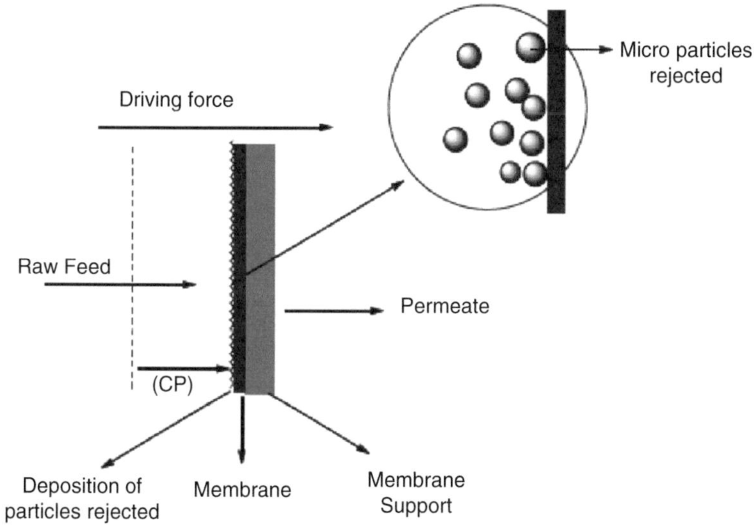

Fig. 16.12 Schematic illustration of membrane filtration process depicting various fouling methods: concentration polarization (*CP*), deposition of particles rejected on the feed side, and fouling of the membrane on the skin void space (Reprinted from Ref. [95], with kind permission from Springer Science + Business Media)

filters [95]. Alternatively, molecules and compounds can also be rejected by being adsorbed or entrapped within the porous structure of the membrane; membranes with these functions are referred to as depth filters [96]. The development of more effective, energy-efficient membranes is essential to the future of membrane technology. Filtration systems made from electrospun colloidal nanofibers possess dimensions up to 800 times smaller than conventional filtration media. This provides better efficiency due to more inertial impaction and interception, thus justifying the increased pressure drop compared to larger fibers [97].

The electrospinning of silica particles dispersions in precursor silica-PVP solution followed by calcination can fabricate hierarchical SiO_2 nanofibers with fiber diameters of ~500 nm and particle sizes of tens of nanometers [98]. Such electrospun SiO_2 nanofibers have important applications in adsorption and filtration. In order to decrease the weight loss and shrinkage of nanofiber membranes during sintering, Guo et al. cut the precursor electrospun composite membrane into many pieces with a dimension of 30 mm by 6 mm. These were layered into a mold and pressed together at a pressure of 100 Mpa before sintering [99]. Approximately 92 % of both submicron-sized and respirable dust can be effectively reduced by using nanofiber composite filters. Further test results showed that the nanofiber filter's life and initial pressure drop were better than standard filters [100]. Gopal et al. have successfully utilized nanofiber membranes made from polysulfone to remove micron-sized particles. It was possible to alter colloidal electrospinning conditions to adjust membrane pore sizes [101].

16.3.2 Catalysis

Industrialization creates large amounts of environmentally damaging pollution. Catalysis plays an important role in dealing with these contaminants by significantly accelerating the contamination's degradation rate. Electrospun colloidal nanofibers are promising for catalysis application due to the enhanced surface-to-volume ratio and one-dimensional hierarchical structure [69]. Hierarchically structured metal oxide/silica nanofibers were successfully obtained by colloid electrospinning, showing improved mechanical integrity compared to neat metal oxide fibers [102]. Two different metal oxides were used for the creation of fibers; these metal oxides, CeO_2 and $LiCoO_2$, were used as models for simple and binary oxides, respectively. The higher catalysis activity is explained by the fact that CeO_2 was well distributed along the fiber surface, therefore accessible for the degrading of the molecules [102]. The fabrication of other metal or metalloid oxide fibers has been recently reviewed [13]. Mesoporous anatase nanofibers were prepared via electrospinning of preformed TiO_2 nanoparticles in an adequate solvent mixture without the use of any surfactant [103]. This approach is more flexible than the electrospinning of sol–gel precursors and is a versatile method for the fabrication of catalytic substrates.

Energy obtained from fossil fuels generates large amounts of harmful greenhouse gas. The development of clean energy devices is necessary to protect the environment. The dimensionality, directionality, and compositional flexibility of electrospun colloidal nanofibers are also investigated intensively as the electrocatalyst materials of clean energy devices, such as fuel cells, solar cells, lithium ion batteries, and supercapacitors. The synthesis of catalytic electrospun nanowires (NWs) of Pt-Rh and Pt-Ru with diameters of 50 nm was reported by Kim et al. [104]. Catalytic supporting materials should be dispersed uniformly to improve the anode catalytic performance. Colloidal electrospinning technique can uniformly disperse the nanoparticles onto nanofibers (Fig. 16.13) without the involvement of a surfactant or template. The dispersion of catalyst particles, with less time for Pt particle deposition, enabled large surface areas in nanofibers (Fig. 16.14) [105]. More strategies are currently in development, including photodeposition [106] and in situ electrospinning-based methods [107].

16.3.3 Smart Membranes

A number of useful properties can be obtained by incorporating nanosized fillers into electrospun nanofibers via tuning the parameters to control structures. These smart membranes have a wide range of applications due to their unique properties, including environmental friendly applications [108]. Some examples are investigated in this section. Ding et al. introduced silica nanoparticles to porous polystyrene nanofibers and controlled the distributions of particles via tuning the solvent compositions in electrospinning. The membranes displayed superhydrophobicity because the silica nanoparticles on the fiber surfaces enhanced

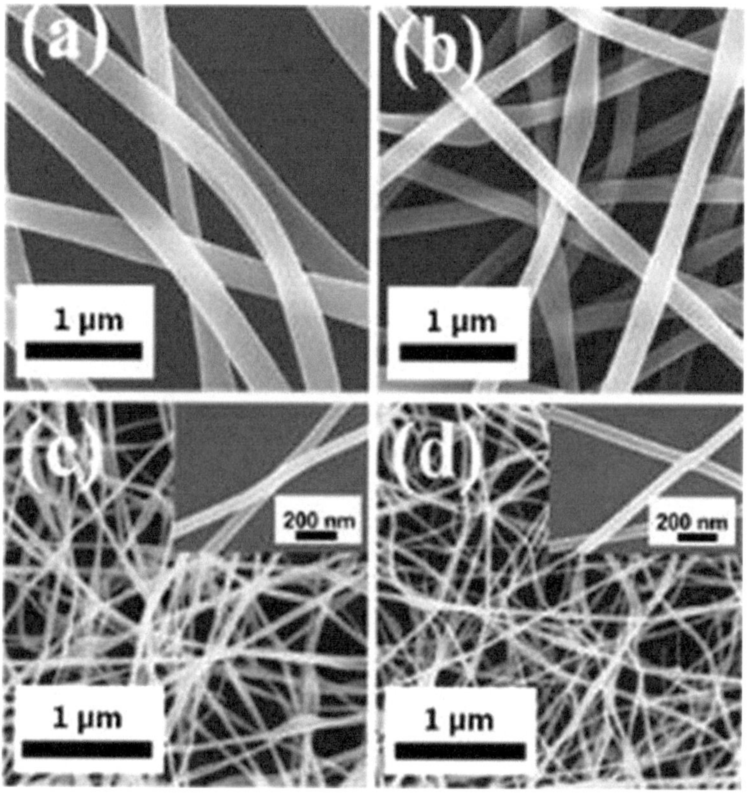

Fig. 16.13 SEM images of as-spun (**a**) Pt-Rh precursor/PVP and (**b**) Pt-Ru precursor/PVP and bimetallic nanowires of (**c**) Pt1Rh1 and (**d**) Pt1Ru1; insets of (**c**) and (**d**) show the diameter of bimetallic NWs (Reprinted from Ref. [104]. Copyright 2008, with permission from Elsevier)

the surface roughness with numerous papillae [109]. By creating pores on the superhydrophobic-oleophilic polystyrene nanofibers and controlling the diameter, high oil sorption capacity and oil/water selective adsorption can be obtained [110]. The ability to separate oil from water endows the materials with potential environmental applications. Biodegradable polymer nanoparticles are eco-friendly and hold great promise for controlled and targeted drug delivery process [111]. Colloidal electrospinning can localize polymeric colloids in the core region of electrospun nanofibers, allowing membranes to display a significantly retarded release [26].

16.3.4 Sensors

Nanofibers fabricated via colloidal electrospinning have specific surfaces approximately one to two orders of magnitude larger than conventional flat films, making

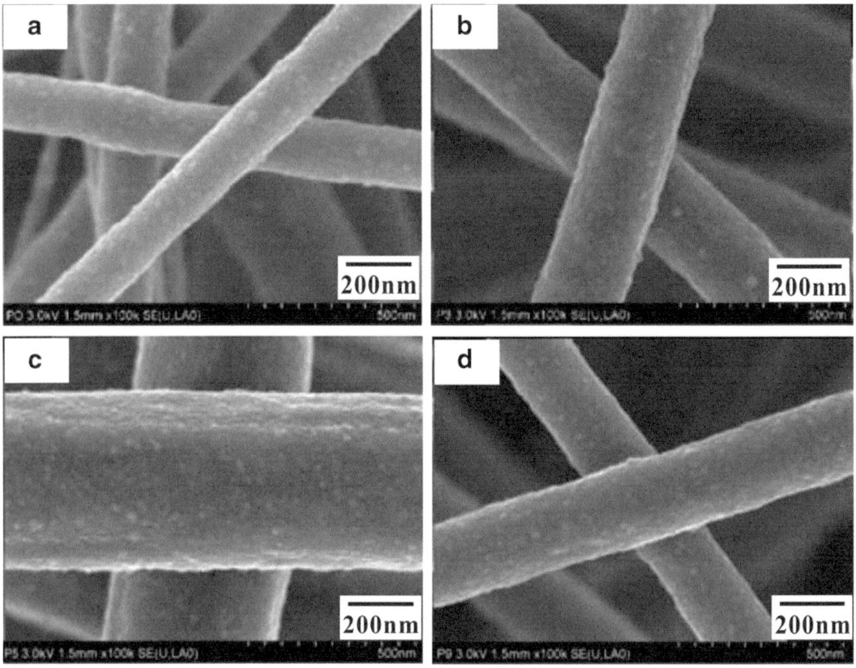

Fig. 16.14 SEM images of Pt/TiO_2-CNF composites according to the UV irradiation time: (**a**) 0 min, (**b**) 3 min, (**c**) 5 min, and (**d**) 10 min (Reproduced from Ref. [105] by permission of John Wiley & Sons Ltd)

them excellent for sensor applications. The presence of fillers in the nanofiber gives them unique sensing properties. Zhang et al. reported colloid-electrospinning fabrication of Au nanorods with Ag nanowires within polymer nanofiber matrix for enhanced surface-enhanced Raman scattering (SERS), with which both Au nanorods and Ag nanowires were stabilized and arranged along the axis of polymer nanofibers [112]. Porous CuO-Ag nanofibers have been synthesized via electrospinning a combination of the polymer and metal/metal oxide particles followed by the removal of polymer. The obtained porous structure nanofibers have a quantitative sensitive SERS detection application [113]. Nanoporous films consisting of weak polyelectrolytes and TiO_2 nanoparticles fabricated via layer-by-layer (LBL) self-assembly methods were used as ammonia gas sensors by setting the membranes on the electrode of quartz crystal microbalances [114]. The fact that various sensors comprising electrospun nanofibers of conducting polymer composites, polyelectrolytes, and semiconducting metal oxides were successfully fabricated was discussed in many comprehensive reviews [115]. It is expected that electrospun colloidal nanofiber sensors will play an important role in environmental detection and protection.

16.4 Electrospun Colloidal Nanofibers for Optical Applications

Nanofibers obtained by colloidal electrospinning have special optical properties which have attracted increasing amounts of attention in recent years. For example, Ramakrishna et al. [116] fabricated anatase TiO_2 nanofibers of average diameters 60, 100, and 150 nm by controlled electrospinning of a polymeric solution and subsequent sintering of the as-spun fibers. They found that the absorption spectra of the fibers had a red shift with an increase in the fiber diameter, which is attributed to an increase in the surface stress with a decrease in the fiber diameter. Greiner et al. [53] prepared gold nanowires via the electrospinning of highly concentrated aqueous dispersions of gold nanoparticles (AuNPs) in the presence of poly(vinyl alcohol) (PVA) and subsequent annealing at higher temperatures. They inspected the gold nanowires by optical microscopy, and the optical microscopy image showed long wires at least several millimeters in length; this was not unexpected, as electrospun fibers are typically long. Therefore, based on these special properties, researchers have developed several optical applications such as photocatalytic, waveguide, sensor, and optoelectronic devices.

16.4.1 Optoelectronic and Photocatalytic Devices

Li et al. [117] prepared novel sodium tantalite nanofibers by electrospinning a sol–gel precursor solution of poly(vinylpyrrolidone)/sodium tantalite, followed by careful sintering of the as-electrospun composite fibers at 550 °C for 3 h. UV–Vis spectroscopic analysis showed the $NaTaO_3$ nanofibers possessed a novel optical absorption property of red shift, giving them potential as good candidates for photocatalytic applications. Kim et al. [118] successfully prepared PVA/zinc acetate organic/inorganic hybrid nanofibers by electrospinning with poly(vinyl alcohol) (PVA) and zinc acetate; pure zinc oxide fibers were obtained by high-temperature calcination of the hybrid fibers in air. The photoluminescence spectra under excitation at 325 nm showed an ultraviolet emission at 3.13 eV and a green emission at 2.21 eV. These nanofibers could be used as light-emitting devices in nanoscale optoelectronic applications. Barakat et al. [119] produced silver nanofibers by electrospinning a sol–gel consisting of poly(vinyl alcohol) (PVA) and silver nitrate. The resulting nanofibers possessed a distinct plasmon resonance compared to the reported silver nanoparticles, which had a blue-shifted plasmon resonance at 330 nm. Moreover, the optical properties studied indicated that the synthesized nanofibers have two band gap energies of 0.75 and 2.34 eV. These advantageous physical features greatly increase the applications of the prepared nanofibers in various fields, especially optical.

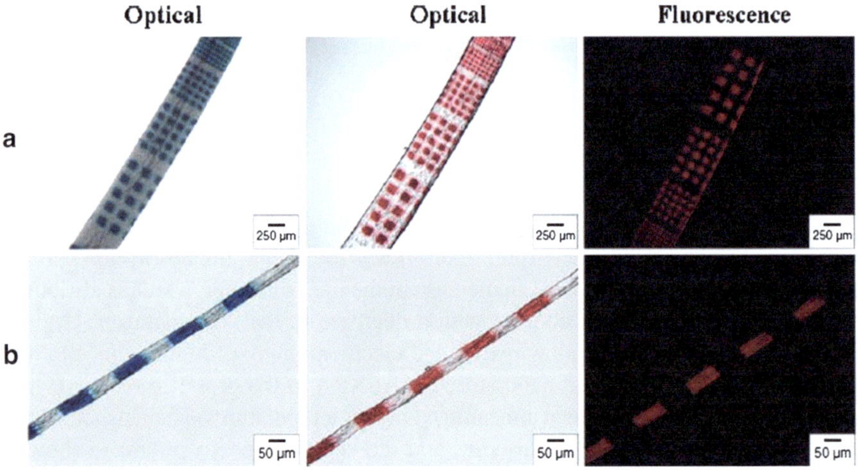

Fig. 16.15 Patterned images of a single PCDA-embedded PMMA fiber, obtained by electrospinning of a CH_2Cl_2 solution containing (**a**) 30 and (**b**) 20 wt% PMMA (Mw: 120,000 g mol^{-1}). Optical images (*left*) obtained after UV irradiation of the electrospun fiber (254 nm, 1 $mWcm^{-2}$, 3 min). Optical (*middle*) and fluorescence (*right*) microscopy images obtained after heat treatment of the blue-patterned fiber (110 °C, 1 min) (Reproduced from Ref. [121] by permission of John Wiley & Sons Ltd)

16.4.2 Colorimetric Sensors

Samuelson et al. [120] used a fluorescent polymer, poly(acrylic acid)-poly(pyrene methanol) (PAA-PM), as a sensing material and successfully developed nanofibrous thin-film optical chemical sensors for metal ion (Fe^{3+} and Hg^{2+}) detection using the electrospinning technique. Kim et al. [121] prepared polymer microfibers encapsulating polydiacetylene (PDA) supramolecules via electrospinning. Randomly distributed diacetylene monomers were found to self-assemble during fiber formation and enable the generation of blue-colored PDA-containing polymer fibers upon UV irradiation. Additionally, the feasibility of PDA-embedded electrospun microfibers as potential sensor materials was demonstrated by fluorescence generation upon specific ligand-receptor interactions (Fig. 16.15). Chen et al. [122] successfully prepared light-emitting electrospun (ES) nanofibers with diameters of 250–750 nm through the binary blends of polyfluorene derivative/poly(methyl methacrylate) (PMMA) using a single-capillary spinneret. Uniform ES fibers produced from the binary blends of PFO/PMMA, PFQ/PMMA, PFBT/PMMA, and PFTP/PMMA exhibited the following luminescence characteristics (peak maximum (nm), color): 443, blue; 483, green; 539, yellow; and 628, red. Such fibers could be potentially used for sensory device applications (Fig. 16.16).

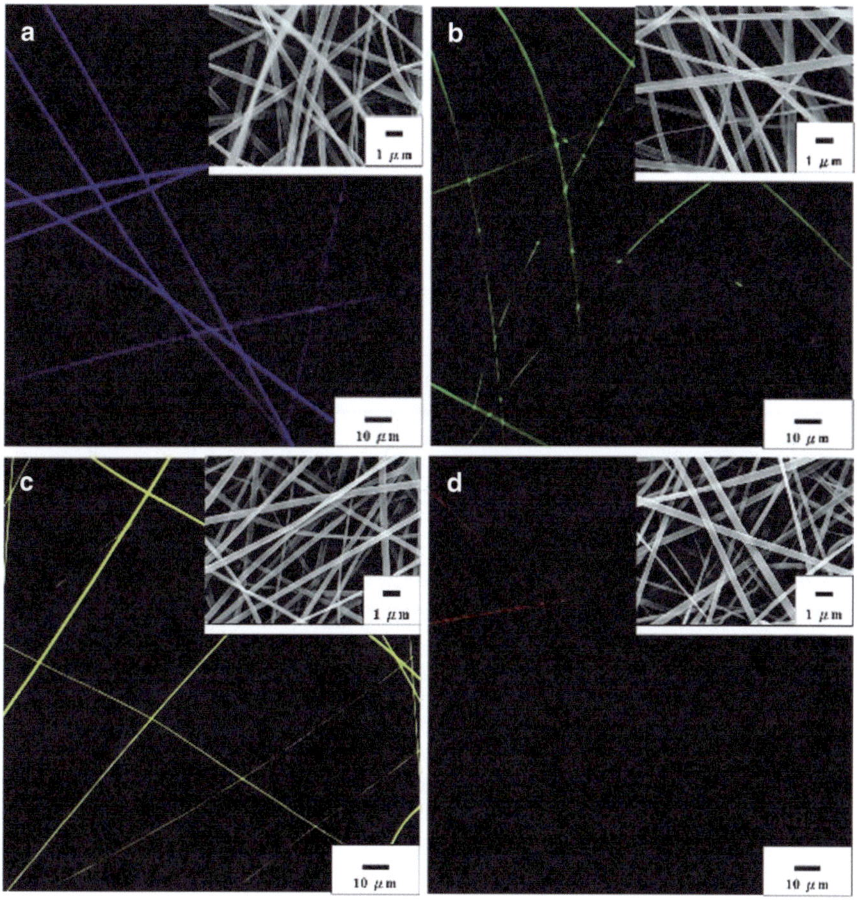

Fig. 16.16 Laser confocal microscopy images of 5 wt% polyfluorene derivatives/PMMA blend ES nanofibers with (**a**) PFO, (**b**) PFQ, (**c**) PFBT, and (**d**) PFTP (Reprinted with the permission from Ref. [122]. Copyright 2007 American Chemical Society)

16.4.3 Other Optical Applications

Craighead et al. [123] fabricated a suspended inorganic-organic (quantum dots/SU8) heterostructure cylindrical waveguide by a one-step electrospinning approach and demonstrated that the subwavelength-sized SU8 nanofiber acted as a waveguiding structure with a length of at least several micrometers. Hou et al. [124] prepared one-dimensional $CaMoO_4{:}Ln^{3+}$ (Ln = Eu, Tb, Dy) nanofibers via a combination of sol–gel and electrospinning process. The nanofibers showed strong characteristic

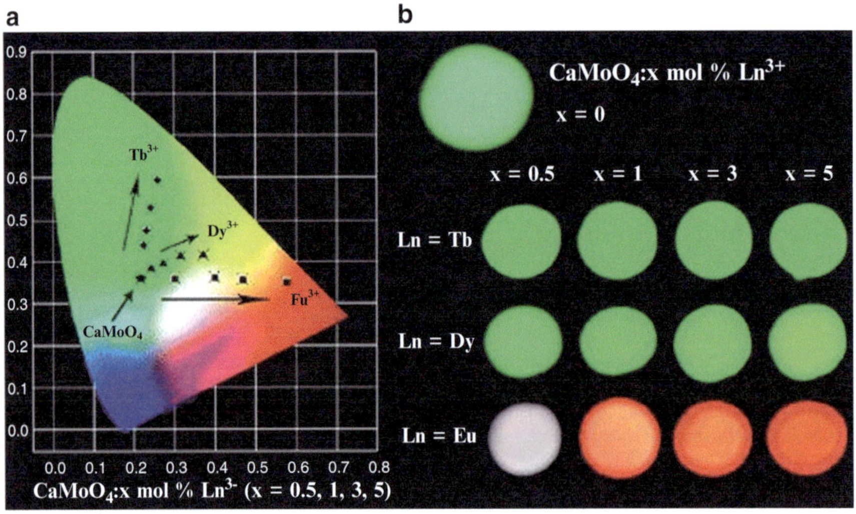

Fig. 16.17 CIE chromaticity diagram (**a**) and luminescent photographs for $CaMoO_4$ and $CaMoO_4$: × mol% Ln^{3+} (**b**). The pictures were taken in a *dark room* by using a 254 nm ultraviolet lamp as excitation source (Reprinted with the permission from Ref. [124]. Copyright 2009 American Chemical Society)

emission under ultraviolet excitation and low-voltage electron beam excitation. Furthermore, the emission colors of $CaMoO_4$:Ln^{3+} nanofibers were easily changeable from blue green to green, yellow, and orange red by adjusting the doping concentrations of Ln^{3+} ions, making the materials potentially applicable for fluorescent lamps and field emission displays (Fig. 16.17).

16.5 Conclusions and Perspectives

The electrospinning of colloidal dispersions allows the creation of materials with hierarchical levels of nanostructures and multiple components; this is accomplished by adjusting the physical and chemical properties of particles and the fiber morphology via control of process parameters. Inorganic nanoparticles, such as metals, metal oxide, and other mineral nanoparticles, are not notably deformed or affected during the electrospinning process, and their functionalities remain in the final composite fibers. Meanwhile, organic/polymeric nanoparticles are also incorporated into the fiber matrices via electrospinning. Theoretically, functional colloids with complex structures, such as microporous organic particles or nanocrystalline domains for specific optical properties, can be easily embedded in nanofiber matrixes. The polymeric colloids form ordered and interconnected structures in the fiber, which can be selectively removed to obtain highly porous nano- or microfibers. Moreover, more functional colloids such as fluorescence and surface plasmon resonance can be embedded into the fiber matrices via electrospinning. More interestingly, the incorporation of living cells and viruses into the fibers via electrospinning poses the possibility as a new route in preparation of biofunctional fibers for tissue engineering.

The functional nano- or microfibers obtained through the colloid electrospinning have the more significant applications compared to the monocomponent fibers via conventional electrospinning. The electrospun colloidal fibers' advantages of ultrahigh surface area, conductivity, energy conversion ability, and high sensitivity render their possibility in playing a critical role in areas of energy conversion and environmental protection, such as in the applications for fuel cells, dye-sensitized solar cells, supercapacitors, filters, and gas sensors. However, such fibers are generally weak. The fragile fibers can be difficult to utilize in practical usages. Therefore, stronger fibers should be developed by changing matrix polymer during colloid electrospinning.

Further research is required in the field of colloid electrospinning. One particular area is modeling the relationship between the properties of colloidal nanoparticles and fiber morphology. Furthermore, the possibility of isolating nanoparticles in fiber matrixes via low amounts of dispersed phase in the electrospinning solution is fundamentally significant for all scientific fields. The electrospinning technique possesses a possibility as a suitable preparation method for investigating the effects of nanoconfinement on physical and chemical properties of polymers. Finally, the incorporation of the living cells and organs into the fibers is an emerging field of research, with a possibility as a promising method to develop high-quality biomaterials if combined with three-dimensional electrospinning or printing.

Acknowledgments The authors acknowledge the National Scientific Foundation of China under Grants 51073113, 51373110, and 91027039 and the Natural Science Foundation of the Jiangsu Higher Education Institutions of China under Grant 10KJA540046. The authors also acknowledge support from the Priority Academic Program Development of Jiangsu Higher Education Institutions (PAPD) and Qing Lan Project for Excellent Scientific and Technological Innovation Team of Jiangsu Province (2012 and 2013).

References

1. Xia YN, Yang PD, Sun YG, Wu YY, Mayers B, Gates B, Yin YD, Kim F, Yan HQ (2003) One-dimensional nanostructures: synthesis, characterization, and applications. Adv Mater 15: 353–389
2. Hu JT, Odom TW, Lieber CM (1999) Chemistry and physics in one dimension: synthesis and properties of nanowires and nanotubes. Acc Chem Res 32:435–445
3. Li D, Xia YN (2004) Electrospinning of nanofibers: reinventing the wheel? Adv Mater 16:1151–1170
4. Greiner A, Wendorff JH (2007) Electrospinning: a fascinating method for the preparation of ultrathin fibers. Angew Chem Int Ed 46:5670–5703
5. Teo WE, Ramakrishna S (2006) A review on electrospinning design and nanofibre assemblies. Nanotechnology 17:R89–R106

6. Huang ZM, Zhang YZ, Kotaki M, Ramakrishna S (2003) A review on polymer nanofibers by electrospinning and their applications in nanocomposites. Compos Sci Technol 63:2223–2253
7. Reneker DH, Yarin AL (2008) Electrospinning jets and polymer nanofibers. Polymer 49:2387–2425
8. Sill TJ, Recum HA (2008) Electrospinning: applications in drug delivery and tissue engineering. Biomaterials 29:1989e2006
9. Frenot A, Chronakis IS (2003) Polymer nanofibers assembled by electrospinning. Curr Opin Colloid Interface Sci 8:64–75
10. Ding B, Wang MR, Wang XF, Yu JY, Sun G (2010) Electrospun nanomaterials for ultrasensitive sensors. Mater Today 11:16–27
11. Bhardwaj N, Kundu SC (2010) Electrospinning: a fascinating fiber fabrication technique. Biotechnol Adv 28:325–347
12. Rutledge GC, Fridrikh SV (2007) Formation of fibers by electrospinning. Adv Drug Deliv Rev 59:1384–1391
13. Crespy D, Friedemann K, Popa AM (2012) Colloid-electrospinning: fabrication of multicompartment nanofibers by the electrospinning of organic or/and inorganic dispersions and emulsions. Macromol Rapid Commun 33:1978–1995
14. Friedemann K, Turshatov A, Landfester K, Crespy D (2011) Characterization via two-color STED microscopy of nanostructured materials synthesized by colloid electrospinning. Langmuir 27:7132–7139
15. Yuan W, Zhang KQ (2012) Structural evolution of electrospun composite fibers from the blend of polyvinyl alcohol and polymer nanoparticles. Langmuir 28:15418–15424
16. Stoiljkovic A, Ishaque M, Justus U, Hamel L, Klimov E, Heckmann W, Eckhardt B, Wendorff JH, Greiner A (2007) Preparation of water-stable submicron fibers from aqueous latex dispersion of water-insoluble polymers by electrospinning. Polymer 48:3974–3981
17. Stoiljkovic A, Venkatesh R, Klimov E, Raman V, Wendorff JH, Greiner A (2009) Poly(styrene-co-n-butyl acrylate) nanofibers with excellent stability against water by electrospinning from aqueous colloidal dispersions. Macromolecules 42:6147–6151
18. Jin Y, Yang DY, Kang DY, Jiang XY (2009) Fabrication of necklace-like structures via electrospinning. Langmuir 26:1186–1190
19. Herrmann C, Turshatov A, Crespy D (2012) Fabrication of polymer ellipsoids by the electrospinning of swollen nanoparticles. ACS Macro Lett 1:907–909
20. Friedemann K, Corrales T, Kappl M, Landfester K, Crespy D (2012) Facile and large-scale fabrication of anisometric particles from fibers synthesized by colloid-electrospinning. Small 8:144–153
21. Xue MQ, Li FW, Wang Y, Cai XJ, Pan F, Chen JT (2013) Ultralow-limit gas detection in nano-dumbbell polymer sensor via electrospinning. Nanoscale 5:1803–1805
22. Agarwal S, Eckhardt B, Grossmann F, Greiner A, Göring P, Wehrspohn RB, Wendorff J (2010) Gradient nanowires and nanotubes. Phys Status Solid B 247:2451–2457
23. Lim JM, Yi GR, Moon JH, Heo CJ, Yang SM (2007) Superhydrophobic films of electrospun fibers with multiple-scale surface morphology. Langmuir 23:7981–7989
24. Lim JM, Moon JH, Yi GR, Heo CJ, Yang SM (2006) Fabrication of one-dimensional colloidal assemblies from electrospun nanofibers. Langmuir 22:3445–3449
25. Kanehata M, Ding B, Shiratori S (2007) Nanoporous ultra-high specific surface inorganic fibres. Nanotechnology 18:315602
26. Jo E, Lee S, Kim KT, Won YS, Kim HS, Cho EC, Jeong U (2009) Core-sheath nanofibers containing colloidal arrays in the core for programmable multi-agent delivery. Adv Mater 21:968–972
27. Yang DY, Lu B, Zhao Y, Jiang XY (2007) Fabrication of aligned fibrous arrays by magnetic electrospinning. Adv Mater 19:3702–3706
28. Wu SJ, Li FT, Wang HT, Fu L, Zhang BR, Li GT (2010) Effects of poly (vinyl alcohol) (PVA) content on preparation of novel thiol-functionalized mesoporous PVA/SiO_2 composite nanofiber membranes and their application for adsorption of heavy metal ions from aqueous solution. Polymer 51:6203–6211

29. Brettmann BK, Tsang S, Forward KM, Rutledge GC, Myerson AS, Trout BL (2012) Free surface electrospinning of fibers containing microparticles. Langmuir 28:9714–9721
30. Deitzel JM, Kleinmeyer J, Harris D, Tan NCB (2001) The effect of processing variables on the morphology of electrospun nanofibers and textiles. Polymer 42:261–272
31. Reneker DH, Yarin AL, Fong H, Koombhongse S (2000) Bending instability of electrically charged liquid jets of polymer solutions in electrospinning. J Appl Phys 87:4531–4547
32. Kim GM, Wutzler A, Radusch HJ, Michler GH, Simon P, Sperling RP, Parak WJ (2005) One-dimensional arrangement of gold nanoparticles by electrospinning. Chem Mater 17: 4949–4957
33. Zhang ZP, Zhang LD, Wang SX, Chen W, Lei Y (2001) A convenient route to polyacrylonitrile/silver nanoparticle composite by simultaneous polymerization-reduction approach. Polymer 42:8315–8318
34. Tijing LD, Ruelo MTG, Amarjargal A, Pant HR, Park CH, Kim CS (2012) One-step fabrication of antibacterial (silver nanoparticles/poly(ethylene oxide))-Polyurethane bicomponent hybrid nanofibrous mat by dual-spinneret electrospinning. Mater Chem Phys 134:557–561
35. Jin M, Zhang XT, Nishimoto S, Liu ZY, Tryk DA, Murakami T, Fujishima A (2007) Large-scale fabrication of Ag nanoparticles in PVP nanofibres and net-like silver nanofibre films by electrospinning. Nanotechnology 18:075605
36. Jin WJ, Jeon HJ, Kim JH, Youk JH (2007) A study on the preparation of poly(vinyl alcohol) nanofibers containing silver nanoparticles. Synth Met 157:454–459
37. Shi Q, Vitchuli N, Nowak J, Noar J, Caldwell JM, Breidt F, Bourham M, McCord M, Zhang XW (2011) One-step synthesis of silver nanoparticle-filled nylon 6 nanofibers and their antibacterial properties. J Mater Chem 21:10330
38. Xu XY, Yang QB, Wang YZ, Yu HZ, Chen XS, Jing XB (2006) Biodegradable electrospun poly(L-lactide) fibers containing antibacterial silver nanoparticles. Eur Polym J 42: 2081–2087
39. Chen JF, Yang PP, Wang CJ, Zhan SM, Zhang LJ, Huang ZH, Li WW, Wang C, Jiang ZJ, Shao C (2011) Ag nanoparticles/PPV composite nanofibers with high and sensitive opto-electronic response. Nanoscale Res Lett 6:121
40. Penchev H, Paneva D, Manolova N, Rashkov I (2009) Electrospun hybrid nanofibers based on chitosan or N-carboxyethylchitosan and silver nanoparticles. Macromol Biosci 9:884–894
41. Penchev H, Paneva D, Manolova N, Rashkov I (2010) Hybrid nanofibrous yarns based on N-carboxyethylchitosan and silver nanoparticles with antibacterial activity prepared by self-bundling electrospinning. Carbohydr Res 345:2374–2380
42. Yang QB, Li DM, Hong YL, Li ZY, Wang C, Qiu SL, Wei Y (2003) Preparation and characterization of a PAN nanofibre containing Ag nanoparticles via electrospinning. Synth Met 137:973–974
43. Saquing CD, Manasco JL, Khan SA (2009) Electrospun nanoparticle–nanofiber composites via a one-step synthesis. Small 5:944–951
44. Jin WJ, Lee HK, Jeong EH, Park WH, Youk JH (2005) Preparation of polymer nanofibers containing silver nanoparticles by using poly(N-vinylpyrrolidone). Macromol Rapid Commun 26:1903–1907
45. Zhuang XP, Cheng BW, Kang WM, Xu XL (2010) Electrospun chitosan/gelatin nanofibers containing silver nanoparticles. Carbohydr Polym 82:524–527
46. Mahapatra A, Garg N, Nayak BP, Mishra BG, Hota G (2012) Studies on the synthesis of electrospun PAN-Ag composite nanofibers for antibacterial application. J Appl Polym Sci 124:1178–1185
47. He D, Hu B, Yao QF, Wang K, Yu SH (2009) Large-scale synthesis of flexible free-standing SERS substrates with high sensitivity: electrospun PVA nanofibers embedded with controlled alignment of silver nanoparticles. ACS Nano 3:3393–4002
48. Maity S, Downen LN, Bochinski JR, Clarke LI (2011) Embedded metal nanoparticles as localized heat sources: an alternative processing approach for complex polymeric materials. Polymer 52:1674–1685

49. Bai J, Li YX, Yang ST, Du JS, Wang SG, Zheng JF, Wang YZ, Yang QB, Chen XS, Jing XB (2007) A simple and effective route for the preparation of poly(vinylalcohol) (PVA) nanofibers containing gold nanoparticles by electrospinning method. Solid State Commun 141:292–295
50. Kim JK, Ahn H (2008) Fabrication and characterization of polystyrene/gold nanoparticle composite nanofibers. Macromol Res 16:163–168
51. Wang YZ, Li YX, Sun GE, Zhang GL, Liu H, Du JS, Yang ST, Bai J, Yang QB (2007) Fabrication of Au/PVP nanofiber composites by electrospinning. J Appl Polym Sci 105: 3618–3622
52. Fischer KDM, Freeman JW (2010) Characterization of electrospun poly(L-lactide) and gold nanoparticle composite scaffolds for skeletal muscle tissue engineering. J Tissue Eng Regen M 5:560
53. Gries K, Vieker H, Gölzhäuser A, Agarwal S, Greiner A (2012) Preparation of continuous gold nanowires by electrospinning of high-concentration aqueous dispersions of gold nanoparticles. Small 8:1436–1441
54. Li JY, Chen C, Wang XM, Gu ZZ, Chen BA (2011) Novel strategy to fabricate PLA/Au nanocomposites as an efficient drug carrier for human leukemia cells in vitro. Nanoscale Res Lett 6:29
55. Sharma N, McKeown SJ, Ma X, Pochan DJ, Cloutier SG (2010) Structure property correlations in hybrid polymer nanoparticle electrospun fibers and plasmonic control over their dichroic behavior. ACS Nano 4:5551–5558
56. Fang X, Ma H, Xiao SL, Shen MW, Guo R, Cao XY, Shi XY (2011) Facile immobilization of gold nanoparticles into electrospun polyethyleneimine/polyvinyl alcohol nanofibers for catalytic applications. J Mater Chem 21:4493
57. Li XF, Cao MH, Zhang H, Zhou L, Cheng S, Yao JL, Fan LJ (2012) Surface-enhanced Raman scattering-active substrates of electrospun polyvinyl alcohol/gold–silver nanofibers. J Colloid Interface Sci 382:28–35
58. Ji LW, Jung KH, Medford AJ, Zhang XW (2009) Electrospun polyacrylonitrile fibers with dispersed Si nanoparticles and their electrochemical behaviors after carbonization. J Mater Chem 19:4992–4997
59. Wu H, Pan W (2006) Preparation of zinc oxide nanofibers by electrospinning. J Am Ceram Soc 89:699–701
60. Guan HY, Shao CL, Wen SB, Chen B, Gong J, Yang XH (2003) A novel method for preparing Co_3O_4 nanofibers by using electrospun PVA/cobalt acetate composite fibers as precursor. Mater Chem Phys 82:1002–1006
61. Barakata NAM, Abadirc MF, Sheikhd FA, Kanjwale MA, Park SJ, Kim HY (2010) Polymeric nanofibers containing solid nanoparticles prepared by electrospinning and their applications. Chem Eng J 156:487–495
62. Li ZY, Huang HM, Wang C (2006) Electrostatic forces induce poly (vinyl alcohol)-protected copper nanoparticles to form copper/poly (vinyl alcohol) nanocables via electrospinning. Macromol Rapid Commun 27:152–155
63. Ji LW, Zhang XW (2009) Electrospun carbon nanofibers containing silicon particles as an energy-storage medium. Carbon 47:3219–3226
64. Kanehatal M, Ding B, Shiratori S (2007) Nanoporous ultra-high specific surface inorganic fibres. Nanotechnology 18:315602
65. Jiang L, Zhao Y, Zhai J (2004) A lotus-leaf-like superhydrophobic surface: s porous microsphere/nanofiber composite film prepared by electrohydrodynamics. Angew Chem Int Ed 43:4338–4341
66. Drew C, Wang XY, Samuelson LA, Kumar J (2003) The effect of viscosity and filler on electrospun fiber morphology. J Macromol Sci A 40:1415
67. Chuangchote S, Jitputti J, Sagawa T, Yoshikawa S (2009) Photocatalytic activity for hydrogen evolution of electrospun TiO_2 nanofibers. ACS Appl Mater Interfaces 1:1140–1143

68. Nakane K, Yasuda K, Ogihara T, Ogata N, Yamaguchi S (2007) Formation of poly(vinyl alcohol)-titanium lactate hybrid nanofibers and properties of TiO_2 nanofibers obtained by calcination of the hybrids. J Appl Polym Sci 104:1232–1235
69. Choi SK, Kim S, Lim SK, Park H (2010) Photocatalytic comparison of TiO_2 nanoparticles and electrospun TiO_2 nanofibers: effects of mesoporosity and interparticle charge transfer. J Phys Chem C 114:16475–16480
70. Roso M, Sundarrajan S, Pliszka D, Ramakrishna S, Modesti M (2008) Multifunctional membranes based on spinning technologies: the synergy of nanofibers and nanoparticles. Nanotechnology 19:285707
71. Li D, Xia YN (2003) Fabrication of Titania nanofibers by electrospinning. Nano Lett 3: 555–560
72. Linh NTB, Lee KH, Lee BT (2011) Fabrication of photocatalytic PVA-TiO_2 nano-fibrous hybrid membrane using the electro-spinning method. J Mater Sci 46:5615–5620
73. Wang C, Yan E, Huang ZH, Zhao Q, Xin Y (2007) Fabrication of highly photoluminescent TiO_2/PPV hybrid nanoparticle-polymer fibers by electrospinning. Macromol Rapid Commun 28:205–209
74. Song MY, Kim DK, Ihn KJ, Jo MS, Kim DY (2004) Electrospun TiO_2 electrodes for dye-sensitized solar cells. Nanotechnology 15:1861–1865
75. Asmatulu R, Ceylan M, Nuraje N (2010) Study of superhydrophobic electrospun nanocomposite fibers for energy systems. Langmuir 27:504–507
76. Lee S (2009) Multifunctionality of layered fabric systems based on electrospun polyurethane/zinc oxide nanocomposite fibers. J Appl Polym Sci 114:3652–3658
77. Munir MM, Iskandar F, Yun KM, Okuyama K, Abdullah M (2008) Optical and electrical properties of indium tin oxide nanofibers prepared by electrospinning. Nanotechnology 19:145603–145608
78. Yang XH, Shao CL, Liu YC (2007) Fabrication of Cr_2O_3/Al_2O_3 composite nanofibers by electrospinning. J Mater Sci 42:8470–8472
79. Andrew JS, Clarke DR (2008) Enhanced ferroelectric phase content of polyvinylidene difluoride fibers with the addition of magnetic nanoparticles. Langmuir 24:8435–8438
80. Stanishevsky A, Chowdhury S, Chinoda P, Thomas V (2007) Hydroxyapatite nanoparticle loaded collagen fiber composites: microarchitecture and nanoindentation study. J Biomed Mater Res A 86:873–882
81. Nie H, Wang CH (2007) Fabrication and characterization of PLGA/HAp composite scaffolds for delivery of BMP-2 plasmid DNA. J Control Release 120:111–121
82. Mei F, Zhong JS, Yang XP, Ouyang XY, Zhang S, Hu XY, Ma Q, Lu JG, Ryu S, Deng XL (2007) Improved biological characteristics of poly(L-Lactic acid) electrospun membrane by incorporation of multiwalled carbon nanotubes/hydroxyapatite nanoparticles. Biomacromolecules 8:3729–3735
83. Zhang YZ, Venugopal JR, Turki AE, Ramakrishna S, Su B, Lim CT (2008) Electrospun biomimetic nanocomposite nanofibers of hydroxyapatite/chitosan for bone tissue engineering. Biomaterials 29:4314–4322
84. Yeom BY, Shim E, Pourdeyhimi B (2010) Boehmite nanoparticles incorporated electrospun nylon-6 nanofiber web for new electret filter media. Macromol Res 18:884–890
85. Fujihara K, Kotaki M, Ramakrishna S (2005) Guided bone regeneration membrane made of polycaprolactone/calcium carbonate composite nanofibers. Biomaterials 26:4139–4147
86. Pedicini A, Farris RJ (2004) Thermally induced color change in electrospun fiber mats. J Polym Sci B Polym Phys 4:752–757
87. Hwang J, Muth J, Ghosh T (2006) Electrical and mechanical properties of carbon-black-filled, electrospun nanocomposite fiber webs. J Appl Polym Sci 104:2410–2417
88. Chuangchote S, Sirivat A, Supaphol P (2007) Mechanical and electro-rheological properties of electrospun poly(vinylalcohol) nanofiber mats filled with carbon black nanoparticles. Nanotechnology 18:145705–145712

89. Dror Y, Salalha W, Khalfin RL, Cohen Y, Yarin AL, Zussman E (2003) Carbon nanotubes embedded in oriented polymer nanofibers by electrospinning. Langmuir 19:7012–7020
90. Ge JJ, Hou HQ, Li Q, Graham MJ, Greiner A, Reneker DH, Harris FW, Cheng SZD (2004) Assembly of well-aligned multiwalled carbon nanotubes in confined polyacrylonitrile environments: electrospun composite nanofiber sheets. J Am Chem Soc 126:15754–15761
91. Klimov E, Raman V, Venkatesh R, Heckmann W, Stark R (2010) Designing nanofibers via electrospinning from aqueous colloidal dispersions: effect of cross-linking and template polymer. Macromolecules 43:6152–6155
92. Buruaga L, Sardon H, Irusta L, Gonzalez A, Berridi MJF, Iruin JJ (2010) Electrospinning of waterborne polyurethanes. J Appl Polym Sci 115:1176–1179
93. Broichsitter MB, Thieme M, Nguyen J, Schmehl T, Gessler T, Seeger W, Agarwal S, Greiner A, Kissel T (2010) Novel 'nano in nano' composites for sustained drug delivery: biodegradable nanoparticles encapsulated into nanofiber non-wovens. Macromol Biosci 10:1527–1535
94. Sun JY, Bubel K, Chen F, Kissel T, Agarwal S, Greiner A (2010) Nanofibers by green electrospinning of aqueous suspensions of biodegradable block copolyesters for applications in medicine, pharmacy and agriculture. Macromol. Rapid Commun 31:2077–2083
95. Ramakrishna S, Jose R, Archana PS, Nair AS, Balamurugan R, Venugopal J, Teo WE (2010) Science and engineering of electrospun nanofibers for advances in clean energy, water filtration, and regenerative medicine. J Mater Sci 45:6283–6312
96. Thavasi V, Singh G, Ramakrishna S (2008) Electrospun nanofibers in energy and environmental applications. Energy Environ Sci 1:205–221
97. Jose MV, Steinert BW, Thomas V, Dean DR, Abdalla MA, Price G, Janowski GM (2007) Morphology and mechanical properties of nylon 6/MWNT nanofibers. Polymer 48: 1096–1104
98. Wen S, Liu L, Zhang L, Chen Q, Zhang L, Fong H (2010) Hierarchical electrospun SiO_2 nanofibers containing SiO_2 nanoparticles with controllable surface-roughness and/or porosity. Mater Lett 64:1517–1520
99. Guo A, Liu J, Dong X, Liu M (2013) Preparation of porous silica ceramics from silica spinning solution and introduced silica particles by electrospinning. Mate Lett 95:74–77
100. Selling GW, Woods KK (2008) Improved isolation of zein from corn gluten meal using acetic acid and isolate characterization as solvent. Cereal Chem 85:202–206
101. Gopal R, Kaur S, Ma Z, Chan C, Ramakrishna S, Matsuura T (2006) Electrospun nanofibrous filtration membrane. J Membr Sci 281:581–586
102. Horzum N, Muñoz-Espí R, Glasser G, Demir MM, Landfester K, Crespy D (2012) Hierarchically structured metal oxide/silica nanofibers by colloid electrospinning. ACS Appl Mater Interfaces 4:6338–6345
103. Wessel C, Ostermann R, Dersch R, Smarsly BM (2010) Formation of inorganic nanofibers from preformed TiO_2 nanoparticles via electrospinning. J Phys Chem C 115:362–372
104. Kim YS, Nam SH, Shim HS, Ahn HJ, Anand M, Kim WB (2008) Electrospun bimetallic nanowires of PtRh and PtRu with compositional variation for methanol electrooxidation. Electrochem Commun 10:1016–1019
105. Kim S, Choi SK, Lim SK, Chang D, Park H (2010) Pt immobilization on TiO_2-embedded carbon nanofibers using photodeposition. Cryst Res Technol 45:1079–1082
106. Shui JL, Li JCM (2009) Platinum nanowires produced by electrospinning. Nano Lett 9: 1307–1314
107. Cavaliere S, Subianto S, Chevallier L, Jones DJ, Roziere J (2011) Single step elaboration of size-tuned Pt loaded titania nanofibres. Chem Commun 47:6834–6836
108. Ruotsalainen T, Turku J, Hiekkataipale P, Vainio U, Serimaa R, Brinke G, Harlin A, Ruokolainen J, Ikkala O (2007) Tailoring of the hierarchical structure within electrospun fibers due to supramolecular comb-coil block copolymers: polystyrene-block-poly(4-vinyl pyridine) plasticized by hydrogen bonded pentadecylphenol. Soft Matter 3:978–985
109. Ding B, Lin J, Wang X, Yu J, Yang J, Cai Y (2011) Investigation of silica nanoparticle distribution in nanoporous polystyrene fibers. Soft Matter 7:8376–8383

110. Wu J, Wang N, Wang L, Dong H, Zhao Y, Jiang L (2012) Electrospun porous structure fibrous film with high oil adsorption capacity. ACS Appl Mater Interfaces 4:3207–3212
111. Peer D, Karp JM, Hong S, Farokzhad OC, Margalit R, Langer R (2007) Nanocarriers as an emerging platform for cancer therapy. Nat Nanotechnol 2:751–760
112. Zhang CL, Lv KP, Huang HT, Cong HP, Yu SH (2012) Co-assembly of Au nanorods with Ag nanowires within polymer nanofiber matrix for enhanced SERS property by electrospinning. Nanoscale 4:5348–5355
113. Wang W, Feng Z, Jiang W, Zhan J (2012) Electrospun porous CuO-Ag nanofibers for quantitative sensitive SERS detection. CrystEngComm 15:1339–1344
114. Pinto N, Ramos I, Rojas R, Wang P Jr, Johnson AT (2008) Electric response of isolated electrospun polyaniline nanofibers to vapors of aliphatic alcohols. Sens Actuat B Chem 129:621–627
115. Ding B, Wang M, Yu J, Sun G (2009) Gas sensors based on electrospun nanofibers. Sensors 9:1609–1624
116. Kumar A, Jose R, Fujihara K, Wang J, Ramakrishna S (2007) Structural and optical properties of electrospun TiO_2 nanofibers. Chem Mater 19:6536–6542
117. Yi X, Li JL (2010) Synthesis and optical property of $NaTaO_3$ nanofibers prepared by electrospinning. J Sol-Gel Sci Technol 53:480–484
118. Viswanathamurthi P, Bhattarai N, Kim HY, Lee DR (2004) The photoluminescence properties of zinc oxide nanofibres prepared by electrospinning. Nanotechnology 15:320–323
119. Barakat NAM, Woo KD, Kanjwal MA, Choi KE, Khil MS, Kim HY (2008) Surface plasmon resonances, optical properties, and electrical conductivity thermal hysteresis of silver nanofibers produced by the electrospinning technique. Langmuir 24:11982–11987
120. Wang XY, Drew C, Lee SH, Senecal KJ, Kumar J, Samuelson LA (2002) Electrospun nanofibrous membranes for highly sensitive optical sensors. Nano Lett 2:1273–1275
121. Chae SK, Park H, Yoon J, Lee CH, Ahn DJ, Kim JM (2007) Polydiacetylene supramolecules in electrospun microfibers: fabrication, micropatterning, and sensor applications. Adv Mater 19:521–524
122. Kuo CC, Lin CH, Chen WC (2007) Morphology and photophysical properties of light-emitting electrospun nanofibers prepared from poly(fluorene) derivative/PMMA blends. Macromolecules 40:6959–6966
123. Liu HQ, Edel JB, Bellan LM, Craighead HG (2006) Electrospun polymer nanofibers as subwavelength optical waveguides incorporating quantum dots. Small 2:495–499
124. Hou ZY, Chai RT, Zhang ML, Zhang CM, Chong P, Xu ZH, Li GG, Lin J (2009) Fabrication and luminescence properties of one-dimensional $CaMoO_4$: Ln^{3+} (Ln = Eu, Tb, Dy) nanofibers via electrospinning process. Langmuir 25:12340–12348

Chapter 17
Applications of Electrospun Nanofibers in Oil Spill Cleanup

Aikifa Raza, Jianlong Ge, Yang Si, Jianyong Yu, Gang Sun, and Bin Ding

Abstract Oil is one of the important sources of energy in the modern industrial world which has to be transported from the source of production to many places across the globe through oceans and inland transport. During transportation the chance of oil spillage over the water body occurs due to accidents or by deliberate action during wartime that causes severe environmental pollution. Nanofibers, mainly fabricated by electrospinning, have exhibited great potential for many emerging environmental applications including oil spill cleanup. They can be considered as one of the safest nanomaterials due to their extremely long length (can be up to hundreds of kilometers) and their ability to be embedded within other media. Their high surface-to-volume ratio, large porosity (up to over 80 %), and adjustable functionality are also much more effective than conventional nonwoven and polymeric membranes in particulate oil sorption and oil–water separation. For sustainable environment, disposal of used sorbents is a major issue. In this context, the naturally available biodegradable materials have great potential than the synthetic ones. This chapter reviews about oil spill cleanup with special emphasis on the wetting phenomenon used for oil absorption and cross-flow filtration methods

A. Raza • J. Ge
State Key Laboratory for Modification of Chemical Fibers and Polymer Materials, College of Materials Science and Engineering, Donghua University, Shanghai 201620, China

Y. Si • J. Yu • B. Ding (✉)
State Key Laboratory for Modification of Chemical Fibers and Polymer Materials, College of Materials Science and Engineering, Donghua University, Shanghai 201620, China

Nanomaterials Research Center, Modern Textile Institute, Donghua University, Shanghai 200051, China
e-mail: binding@dhu.edu.cn

G. Sun
Nanomaterials Research Center, Modern Textile Institute, Donghua University, Shanghai 200051, China

B. Ding and J. Yu (eds.), *Electrospun Nanofibers for Energy and Environmental Applications*, Nanostructure Science and Technology,
DOI 10.1007/978-3-642-54160-5_17,

of oil spill cleanup as well as focus on the characteristics of nanofibrous oil sorbent materials, fluid flow through nanofibrous materials, and types of nanofibrous materials envisaged for making sorbents cross flow filtration membranes.

17.1 Introduction

An oil spill is the release of a liquid petroleum hydrocarbon into the environment, especially in marine areas due to the human activity into the ocean or coastal waters, but spills could also occur on land. Oil spills may also be due to releases of crude oil from tankers, offshore platforms, drilling rigs, and wells, as well as spills of refined petroleum products (such as gasoline, diesel, etc.) and their by-products, heavier fuels used by large ships such as bunker fuel, or the spill of any oily refuse or waste oil [1–4]. In recent years, crude oil and refined fuel spills from tanker ship accidents have damaged natural ecosystems in Alaska, the Gulf of Mexico, the Galapagos Islands, France, and many other places in the world (Fig. 17.1). The quantity of

Fig. 17.1 (**a**) An oil slick from the Montara oil spill in the Timor Sea [5]. (**b**) Volunteers cleaning up the aftermath of the Prestige oil spill [6]. (**c**) Oil spill cleanup in the Gulf [7]. (**d**) Kelp after an oil spill [8]. (**e**) An oiled bird from oil spill in San Francisco Bay [9] ((**a**) Reprinted with permission from Ref. [5]. Public domain file. (**b**) Reprinted with permission from Ref. [6]. Copyright © 2005 GFDL. (**c**) Reprinted with permission from Ref. [7]. Copyright © 2012 The Consumer Warning Network. (**d**) Reprinted with permission from Ref. [8]. Copyright © National Oceanic and Atmospheric Administration. (**e**) Reprinted with permission from Ref. [9]. Copyright © 2005 GNU)

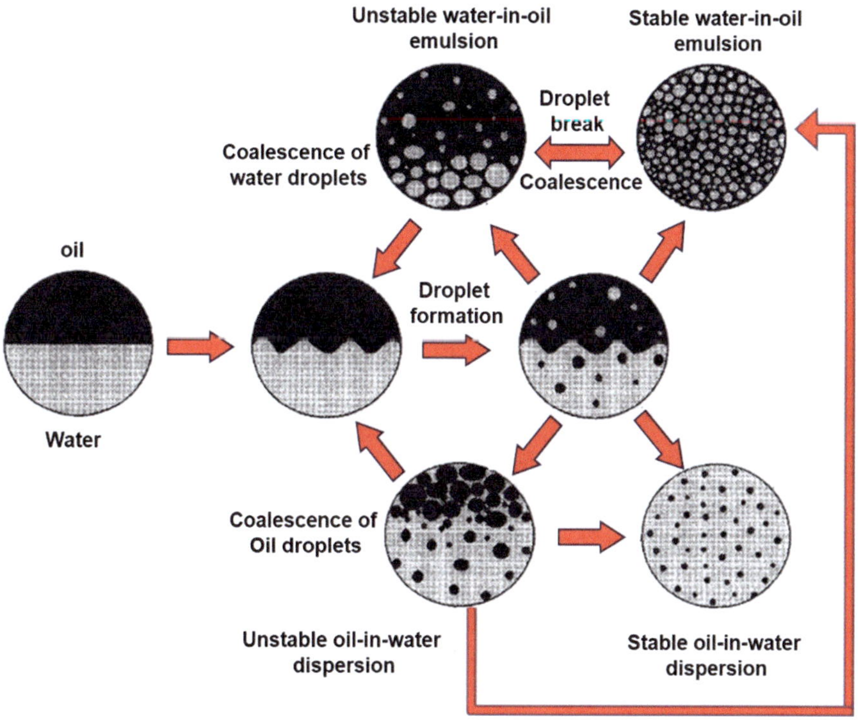

Fig. 17.2 A schematic diagram for the formation of stable and unstable emulsions and dispersions at sea (Reprinted with permission from Ref. [11]. Copyright © 1996, Elsevier)

oil spilled during accidents has ranged from a few hundred tons to several hundred thousand tons. Smaller spills also have already proven to have a great impact on ecosystems, such as the Exxon Valdez oil spill because of the remoteness of the site or the difficulty of an emergency environmental response.

Cleanup and recovery from an oil spill is difficult and depends upon many factors, including the type of oil spilled, the temperature of the water (affecting evaporation and biodegradation), and the types of shorelines and beaches involved [10]. A complicating factor in the separation of recovered spilled oil is that crude oil and refined products change their physical and chemical characteristics significantly over time. This is due to evaporation and formation of water–oil emulsions and oil-in-water dispersions [11]. Figure 17.2 illustrated the emulsification processes and formation of stable and unstable dispersions and emulsions at sea. Methods for cleaning up include bioremediation in which microorganisms or biological agents are used to break down or remove oil [12–16]. Controlled burning is also used to reduce the amount of oil in water, if done properly, but it can only be done in low wind and can cause air pollution [17, 18]. Dispersants can be used to dissipate oil slicks by breaking them up and form water-soluble micelles that are rapidly diluted. However, laboratory experiments showed that dispersants increased toxic hydrocarbon levels in fish by a factor of up to 100 and may kill fish eggs [19, 20]. Skimming is another method but it requires calm waters at all times during the process [21].

Oil spill cleanup using sorbent materials and cross-flow filtration via membranes is attractive for some applications because of the possibility of collection and complete removal of the oil from the oil spill site [22, 23]. Some properties of good absorbent materials include hydrophobicity and oleophilicity, high uptake capacity, high rate of uptake, retention over time, oil recovery from absorbents, and the reusability and biodegradability of the absorbents [24]. Similarly, membranes used also have some special wetting (hydrophobic or hydrophilic) characteristics for effective oil–water separation [25]. Over the past two decades, the sorbents used for oil sorption mainly included natural materials, inorganic mineral products, and organic synthetic fibers [23, 26, 27]. Among these sorbent materials, nonwoven polypropylene (PP) fibers, as synthetic fibers, have been widely used as sorbents in oil cleanup due to their hydrophobic–oleophilic properties, good oil–water selectivity, high buoyancy, and scalable fabrication. However, they suffered from a low oil sorption capacity (15–30 g g^{-1}) because of their low porosity and solid fibers with large diameters [24].

With continuous growth in the field of nanotechnology, various nanostructures in the form of nanoparticles, nanofibers, nanowires, nanobelts, and nanotubes have been fabricated in the past few years [28]. The appropriate choice of nanofibrous membrane with selective wettability and suitable posttreatment through the use of a nonpolar material with low surface energy on these nanomaterials enabled them to exhibit hydrophobicity–oleophilicity and hydrophilicity–oleophobicity properties, which make these materials particularly suited for applications in oil cleanup from water [29, 30]. Electrospinning, an efficient and simple method for generating polymeric fibers with diameters both on the micro- and nanoscale, has gained increasing attention as it can not only control the fiber diameter easily but also manipulate the individual fiber structures via variable tuning [28, 31–33]. More recently, electrospun nanofibrous membranes were used as oil sorbents and oil or water filters, which showed many advantages compared to the conventional fibers and porous membranes [34, 35]. The aim of this review is then to give a detailed overview of recent advances in the application of sorbents and membranes issued from electrospinning in oil spill cleaning.

17.2 Electrospun Nanofibrous Membranes for Oil Spill Cleanup

The versatility of electrospinning provides extremely suitable conditions to prepare both membranes and absorbents with the benefits of smaller, nanostructured geometries, selective wettability, and reduced manufacturing costs [33, 36–39].

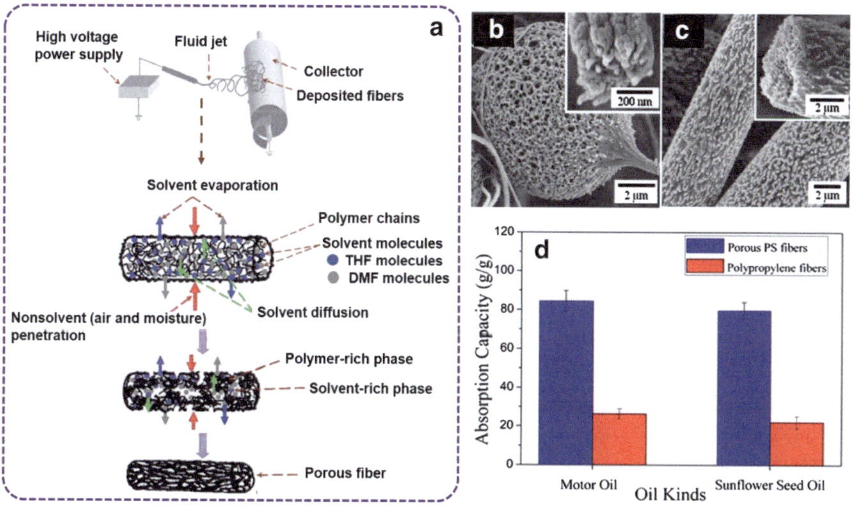

Fig. 17.3 (**a**) Schematic diagram illustrating the formation process of porous fibers during electrospinning. Scanning electron microscopic (SEM) images of the PS (Mw = 208,000 g mol^{-1}) fibers formed from (**b**) 10 wt% and (**c**) 30 wt% PS in THF–DMF with a weight ratio of 20:80. (**d**) Maximum absorption capacities of porous PS fibrous mats and commercial PP nonwoven fibers for motor oil and sunflower seed oil (Reprinted with permission from Ref. [58]. Copyright © 2011, Royal Society of Chemistry)

Electrospinning has the advantages of easiness, efficiency, low cost, high yield, and high degree of reproducibility of the obtained materials. Due to its inherent adaptability and applicability, the electrospinning technique can be applied to the preparation of different components for oil spill cleaning, from nonpolar polymeric absorbents to nonpolar and polar polymeric selective filtration membranes [40–44].

17.2.1 Oil Spill Cleanup via Absorption

Absorbent materials are attractive for some applications because of the possibility of collection and complete removal of the oil from the oil spill site. The important characteristics of good absorbent materials include hydrophobicity and oleophilicity, high uptake capacity, high rate of uptake, retention over time, oil recovery from absorbents, and the reusability of the absorbents [45–52]. Electrospun nanofibrous mats with and without appropriate posttreatment by using nonpolar materials with low surface energy enabled them to exhibit hydrophobicity–oleophilicity properties, which make these materials particularly suited for applications in oil cleanup from water [53–55] (Fig. 17.3).

17.2.1.1 Absorption Using Hydrophobic–Oleophilic Nanofibrous Membranes

Organic synthetic fibers, such as PP, which was conventionally used in oil spill cleanup because of their oleophilic–hydrophobic properties, high oil–water selectivity, and low density, were replaced by electrospun polystyrene (PS) nanofibers. Recently, PS nanofibrous mats with superhydrophobic and oleophilic features attributed to the nonpolarity and low surface energy are extensively investigated for selective oil absorption applications [56]. These PS and polyvinyl chloride nanofibrous sorbents have shown very low density (2–3 mg/cm^3) and very high porosity ($\sim$99.7 %) as compared to commercial PP microfibers. The results have proven the effectiveness of electrospinning method in the fabrication of oil sorbents with high capacity [57]. Lin et al. performed detailed experiments utilizing a variety of nanofibrous mats for oil absorption applications. PS fibers with different morphologies, such as smooth and rough surfaces, were fabricated and examined for oil sorption capacities to determine their potential use in oil spill cleanup. The porous PS fibers with smooth surfaces and small fiber diameters have presented the highest oil sorption capacities, followed by the porous PS fibers with rough surfaces and large fiber diameters. The reported sorption capacities were approximately 3–4 times higher than those of commercialized PP fibers and were attributed to the adsorption, capillary action, or a combination of both [34].

Hydrophobicity–oleophilicity of porous PS nanofibrous mats and their use as oil sorbents were investigated by varying the molecular weights of the polymers with different sources, solvent compositions, solution concentration, as well as the influence of micro- and nanostructures of electrospun PS fibers (Fig. 17.4a–c). The porous PS fibrous mats have shown the absorption capacities of 84.41 and 79.62 g g^{-1} for motor oil and sunflower seed oil, respectively, which are nearly three times larger than that of commercial PP nonwoven fabric [58]. Intra-fiber porosity and interfiber voids were controlled by electrospinning the PS solution, and to improve the mechanical properties, in particular the elasticity and recoverability of the as-prepared fibrous mats, electrospun polyurethane (PU) fibers were added via multi-nozzles electrospinning setup. The fibrous mats with low PU content have shown better oil sorption capacities of 30.81 and 24.36 g g^{-1} for motor oil and sunflower seed oil, respectively, with good reusability [35].

The poor nanofiber strength and resiliency after oil sorption, which results in poor reusability, is a challenging problem. The core–shell configuration of fibers via coaxial electrospinning provides the potential for achieving unique properties from a product that is difficult to obtain from the constituent materials if spun separately. Recently, Lin et al. reported composite PS–PU fibers with a high specific surface area prepared via coaxial electrospinning directly for use as an oil sorbent (Fig. 17.4d–f). The composite fibrous mats showed much higher oil sorption capacities for motor oil (64.40 g g^{-1}) and sunflower seed oil (47.48 g g^{-1}), approximately 2–3 times that of conventional PP fibers for these two same oils. Even after five sorption cycles, the fibrous mats still maintained a higher oil sorption capacity [59].

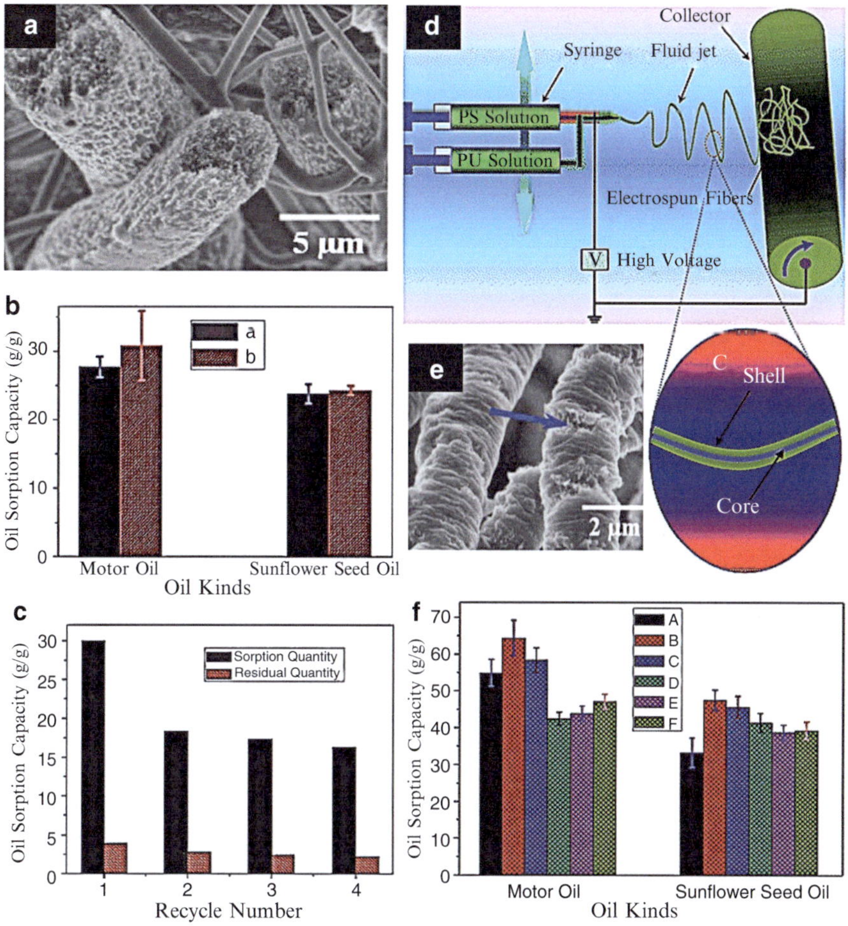

Fig. 17.4 (**a**) FE-SEM images of fibrous mats formed at an RH of 45 % with a 4/1 PS–PU nozzle ratio from 30 wt% PS solution. (**b**) Oil sorption capacities of the 4/1 PS–PU from 20 and 30 wt% PS fibers. (**c**) Reusability of the fibrous mat shown in (**a**). (**d**) Schematic representation showing the core–shell electrospinning setup. (**e**) SEM image of stretched PS–PU fibers fabricated by core–shell electrospinning. (**f**) Sorption capacities of the composite PS–PU fibrous mats for motor oil and sunflower seed oil ((**a–c**) Reprinted with permission from Ref. [35]. Copyright © 2012, Royal Society of Chemistry. (**d–f**) Reprinted with permission from Ref. [59]. Copyright © 2013, Royal Society of Chemistry)

17.2.2 Oil Spill Cleanup via Cross-Flow Filtration

Cross-flow filtration using membrane technology is highly efficient spatially in the fields of engineering and environmental protection. Recently, much attention is given to the fabrication of extremely efficient nanofibrous filtration membranes with advantages of high filtration efficiency due to very high surface area to volume ratios and resulting high surface cohesion, as well as the possibility of recycling electrospun membranes. For oil–water separation, major focus has been shown toward the fabrication of superhydrophobic and superhydrophilic membranes depending upon the type of medium to be separated. In this section, we will focus on different nanofibrous membranes with varying wetting properties used for oil–water cross-flow filtration for oil spill cleanup.

17.2.2.1 Filtration Using Hydrophobic Nanofibrous Membranes

The use of superhydrophobic nanofibrous membranes, characterized by a high water contact angle (WCA $>150°$) and low contact angle hysteresis, in many potential applications like self-cleaning and protective clothing, micro fluid dynamics, and so on is not new, but their utility as oil–water separating membranes is challenging and a point of interest [60–63]. Yang et al. reported the polydopamine-coated PS nanofibrous mats by simply immersing the PS mats into dopamine alkaline tris buffer solution for 24 h. As demonstrated, oil–water separation membranes were fabricated by using the Michael reaction of polydopamine coating with undecanethiol or 11-mercaptoundecanoic acid, respectively. The long alkyl chains of UT anchored on the membrane surface increased the contact angle of water distinctly, and water was blockaded completely, while oil smoothly passed through the membrane [64] (Fig. 17.5a–c). Recently, Shang et al. fabricated unique superhydrophobic and superoleophilic nanofibrous membranes exhibiting robust oil–water separation by a facile combination of electrospun cellulose acetate (CA) nanofibers and a novel in situ polymerized fluorinated polybenzoxazine (F-PBZ) functional layer that incorporated silica nanoparticles (SiO_2 NPs) (Fig. 17.5d). By employing the F-PBZ/SiO_2 NPs modification, the pristine hydrophilic CA nanofibrous membranes were endowed with a superhydrophobicity with the WCA of 161° and a superoleophilicity with the oil contact angle of 3°. The as-prepared membranes exhibited fast and efficient separation for oil–water mixtures and excellent stability over a wide range of pH conditions as shown in Fig. 17.5e, which would make them a good candidate in industrial oil-polluted water treatments and oil spill cleanup [65].

17.2.2.2 Filtration Using Hydrophilic Nanofibrous Membranes

By considering forthcoming new environmental regulations, nanofiltration and ultrafiltration using hydrophilic membranes have become even more important for applications in water treatment such as oil–water emulsions. Conventional membranes are effective in the removal of oily microemulsions from water, but they often suffer from low flux due to limited permeability and surface fouling [66, 67]. Membrane surface hydrophilicity is widely accepted as a dominant factor that

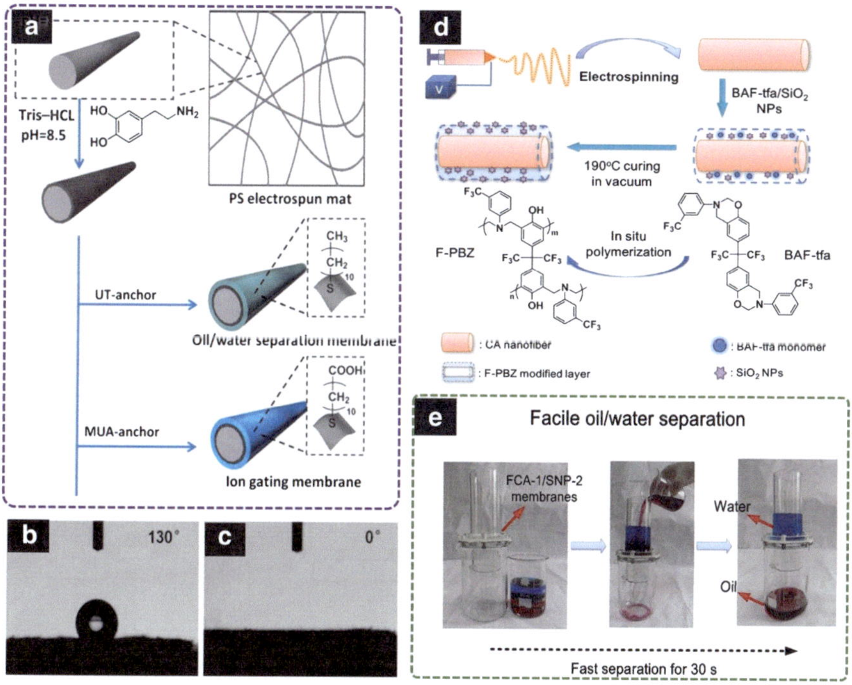

Fig. 17.5 (**a**) Schematic illustration of the preparation of porous functional membranes using a polydopamine-coated PS nanofibrous membrane as a platform. Optical images of the WCA with (**c**) as-synthesized and (**d**) polydopamine-coated PS nanofibrous membranes. (**d**) Schematic for the strategy using the in situ polymerization approach to the synthesis of F-PBZ/SiO_2 NPs modified CA nanofibrous membranes. (**e**) The facile oil–water separation using FCA-1/SNP-2 membranes; the water and oil were dyed by methyl *blue* and oil *red*, respectively ((**a–c**) Reprinted with permission from Ref. [64]. Copyright © 2012, Royal Society of Chemistry. (**d–e**) Reprinted with permission from Ref. [65]. Copyright © 2012, Royal Society of Chemistry)

governs fouling development on the membrane. A hydrophilic membrane surface generally has higher fouling resistance compared with hydrophobic membranes and to improve the fouling problems, extensive efforts have been attempted [68].

For the first time, Wang et al. fabricated high flux filtration medium, consisting of a three-tier composite structure, i.e., a nonporous hydrophilic polyether-*b*-polyamide hydrophilic nanocomposite coating top layer, an electrospun polyvinyl alcohol (PVA) nanofibrous substrate mid-layer, and a conventional nonwoven microfibrous support for oil–water emulsion separations [69–71]. When a thin hydrophilic, highly water-permeable coating layer is deposited on the nanofibrous membrane, the permeate flux of the composite membrane is often reduced due to the hydraulic resistance of the coating layer (Fig. 17.6a–b). Yoon et al. described a membrane system containing polyacrylonitrile (PAN) electrospun nanofibers as the mid-layer scaffold and cross-linked PVA as the top-layer coating. The critical

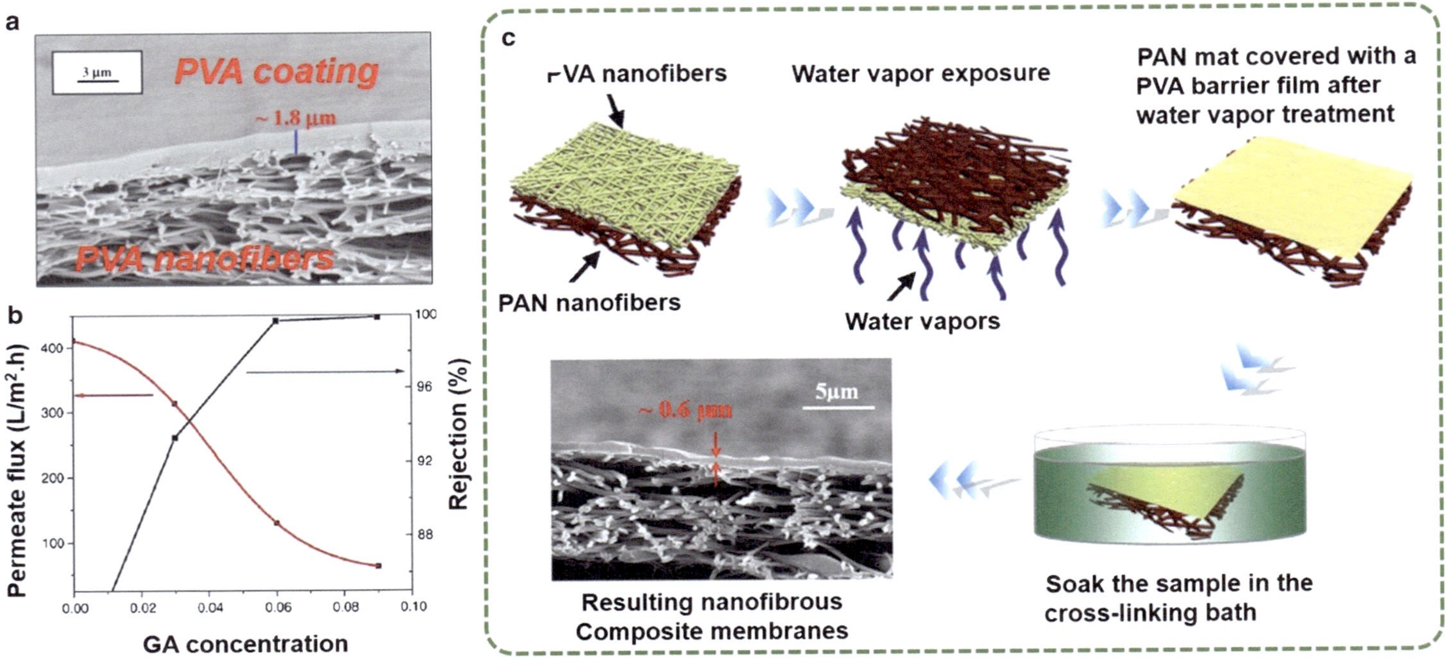

Fig. 17.6 (**a**) Typical SEM cross-sectional image of PVA nanofibrous composite membrane. (**b**) Relations of permeate flux and solute rejection of the nanofibrous composite membranes with the degree of cross-linking in the PVA hydrogel coating for separation of oil–water emulsion. (**c**) Schematic illustration of the fabrication process for thin-film nanofibrous composite membranes based on PAN electrospun nanofibrous substrate and cross-linked PVA barrier layer ((**a–b**) Reprinted with permission from Ref. [70]. Copyright © 2006, Elsevier. (**c**) Reprinted with permission from Ref. [74]. Copyright © 2010, Elsevier)

parameters, such as the thicknesses of the coating layer and mid-layer, were optimized to achieve a higher flux than that of the conventional membranes with rejection ratio (>99.5 %) for separation of oil–water mixture (1,500 ppm in water) over a long time period (tested up to 190 h) in a practical pressure range (up to 130 psig) [72]. The same group also investigated the oil–water separation characteristics of nanofibrous membranes consisting of UV-cured PVA hydrogel barrier layer over electrospun nanofibrous PVA mid-layer scaffold and polyethylene terephthalate (PET) nonwoven substrate (Fig. 17.6c) [73, 74]. Cellulose dissolved in ionic liquids deposited on the PAN nanofibrous scaffold has also shown high permeation flux in the filtration of an emulsified oil–water mixture with high rejection ratio (above 99.5 %) after prolonged operation [75]. Similarly, ultrafine polysaccharide nanofibers (i.e., cellulose and chitin) with 5–10 nm diameters were also employed as barrier layers to form thin-film nanofibrous composite membranes for oil–water separation with 10-fold higher permeation flux with above 99.5 % rejection ratio [76].

17.3 Concluding Remarks and Future Perspectives

Electrospun nanofibrous membranes have exhibited great advantages over conventional media in environmental applications, such as in oil spill cleaning; however, there remain a large number of challenges. Apart from other avenues, one-dimensional nanostructured materials have significantly contributed to meeting some of these challenges as the structural control of nanomaterials can be as important as the composition of the materials themselves. New approaches in structure manipulation, such as smaller fiber formation, porous nanofibers, interfiber adhesion, and modification of nanofiber surfaces, have been rapidly demonstrated. In particular, the processes of fabricating multifunctional nanofibers with various properties, core–shell nanofibers and bi-/multicomponent nanofibers, may be of particular importance for oil spill cleaning applications. These new processes and new chemistries may all be incorporated in the fabrication of better (e.g., higher flux, more efficient, and stronger) filtration membranes. Without question, electrospun functional nanofibrous membranes will be further explored for many different absorption/filtration applications because of the intrinsic versatility of electrospinning. In conclusion, the convenient adaptability of the nonwoven nanofibrous mats to the conventional absorption/filtration media format allows functional nanofibers to become one of the most critical elements for improvement of the existing conventional methods for environmental applications.

Acknowledgments This work is supported by the National Basic Research Program of China (973 Program, 2011CB606103 and 2012CB525005), the National Natural Science Foundation of China (No. 51173022, U1232116, and 51273038), the Shanghai Nano Special Projects (11 nm0502900), the Shanghai Committee of Science and Technology (No. 12JC1400101), the Huo Yingdong Foundation (131070), and the Program for New Century Excellent Talents of the University in China.

References

1. Dominguez J, Vidal M (2009) Factors affecting plumage oiling levels in a Kentish Plover *Charadrius alexandrinus* population after a major oil spill. Acta Ornithol 44(2):119–126. http://dx.doi.org/10.3161/000164509X482696
2. Hasr AH, Ahmed MH (2007) Detecting oil spills in the offshore Nile delta coast using image processing of ERS SAR data. WSEAS Trans Signal Process 3(1):102–107
3. Lee RF, Page DS (1997) Petroleum hydrocarbons and their effects in subtidal regions after major oil spills. Mar Pollut Bull 34(11):928–940. doi:10.1016/s0025-326x(97)00078-7
4. Liu K, Gan W, Huang M, Deng J, Yang X (2012) Risk assessment of ship oil spill based on Bayesian network. Navigation China 35(1):85–89, 104
5. NASA Earth Observatory (2009) An oil slick from the Montara oil spill in the Timor Sea. http://en.wikipedia.org/wiki/File:Oil_Slick_in_the_Timor_Sea_September-2009.jpg
6. Viajero (2005) Volunteers cleaning up the aftermath of the Prestige oil spill. http://en.wikipedia.org/wiki/File:PrestigeVolunteersInGaliciaCoast.jpg
7. Smiljanich T (2010) Who pays for the massive oil spill clean-up in the Gulf? http://www.consumerwarningnetwork.com/2010/05/07/who-pays-for-the-massive-oil-spill-clean-up-in-the-gulf/
8. Svdmolen (2004) Kelp after an oil spill. http://en.wikipedia.org/wiki/Oil_spill
9. Brocken I (2007) An oiled bird from oil spill in San Francisco Bay. http://en.wikipedia.org/wiki/File:Oiled_bird_3.jpg
10. Vandermeulen JH, Ross CW (1995) Oil-spill response in fresh-water-assessment of the impact of cleanup as a management tool. J Environ Manage 44(4):297–308. doi:10.1016/s0301-4797(95)90338-0
11. Nordvik AB, Simmons JL, Bitting KR, Lewis A, StromKristiansen T (1996) Oil and water separation in marine oil spill clean-up operations. Spill Sci Technol Bull 3(3):107–122. doi:10.1016/s1353-2561(96)00021-7
12. Atlas RM (1995) Bioremediation of petroleum pollutants. Int Biodeter Biodegr 35(1–3):317–327. doi:10.1016/0964-8305(95)00030-9
13. Atlas RM (1995) Petroleum biodegradation and oil spill bioremediation. Mar Pollut Bull 31(4–12):178–182. doi:10.1016/0025-326x(95)00113-2
14. Margesin R, Schinner F (1999) Biological decontamination of oil spills in cold environments. J Chem Technol Biotechnol 74(5):381–389, 10.1002/(sici)1097-4660(199905)74:5<381::aid-jctb59>3.0.co;2-0
15. Mills MA, Bonner JS, McDonald TJ, Page CA, Autenrieth RL (2003) Intrinsic bioremediation of a petroleum-impacted wetland. Mar Pollut Bull 7:887–899. doi:10.1016/s0025-326x(02)00367-3
16. Prince RC (1993) Petroleum spill bioremediation in marine environments. Crit Rev Microbiol 19(4):217–242. doi:10.3109/10408419309113530
17. Mullin JV, Champ MA (2003) Introduction/overview to in situ burning of oil spills. Spill Sci Technol Bull 8(4):323–330. doi:10.1016/s1353-2561(03)00076-8
18. PradoJatar MA, Brown MT (1997) Interface ecosystems with an oil spill in a Venezuelan tropical savannah. Ecol Eng 8(1):49–78
19. Brandvik PJ, Daling PS (1998) Optimisation of oil spill dispersant composition by mixture design and response surface methods. Chemometr Intell Lab 42(1–2):63–72. doi:10.1016/s0169-7439(98)00009-4
20. Zolfaghari-Baghbaderani A, Emtyazjoo M, Poursafa P, Mehrabian S, Bijani S, Farkhani D, Mirmoghtadaee P (2012) Effects of three types of oil dispersants on biodegradation of dispersed crude oil in water surrounding two Persian gulf provinces. J Environ Public Health 2012:981365–981365. doi:10.1155/2012/981365
21. Gubaydullin MG, Kalashnikov AV, Konyukhov DA, Masyagin AI, Savinov RA (2011) Method of skimming and removal of oil from water reservoir surface, device for its implementation. Russian Patent RU2475590-C2, 20 Feb 2013

22. Lee GY, Park CC (2000) Water-dispersible oil gelling agent for solidifying oil spilt at sea. KR2001073581-A; KR336175-B, 01 Aug 2001
23. Bayat A, Aghamiri SF, Moheb A, Vakili-Nezhaad GR (2005) Oil spill cleanup from sea water by sorbent materials. Chem Eng Technol 28(12):1525–1528. doi:10.1002/ceat.200407083
24. Ceylan D, Dogu S, Karacik B, Yakan SD, Okay OS, Okay O (2009) Evaluation of butyl rubber as sorbent material for the removal of oil and polycyclic aromatic hydrocarbons from seawater. Environ Sci Technol 43(10):3846–3852. doi:10.1021/es900166v
25. Thavasi V, Singh G, Ramakrishna S (2008) Electrospun nanofibers in energy and environmental applications. Energy Environ Sci 1(2):205–221. doi:10.1039/b809074m
26. Choi H, Moreau JP, Srinivasan M (1994) Cleanup of agrochemical spills using cotton sorbents. J Environ Sci Heal A 29(10):2151–2168
27. Choi HM, Kwon HJ, Moreau JP (1993) Cotton nonwovens as oil-spill cleanup sorbents. Text Res J 63(4):211–218. doi:10.1177/004051759306300404
28. Wang X, Ding B, Yu J, Wang M (2011) Engineering biomimetic superhydrophobic surfaces of electrospun nanomaterials. Nano Today 6(5):510–530. doi:10.1016/j.nantod.2011.08.004
29. Korhonen JT, Kettunen M, Ras RHA, Ikkala O (2011) Hydrophobic nanocellulose aerogels as floating, sustainable, reusable, and recyclable oil absorbents. ACS Appl Mater Interfaces 3(6):1813–1816. doi:10.1021/am200475b
30. Ding Y, Bikson B (2007) Superhydrophobic nanoporous membrane for bioethanol production. In: Abstracts of papers of the American Chemical Society 234, Boston
31. Wang X, Ding B, Sun G, Wang M, Yu J (2013) Electro-spinning/netting: a strategy for the fabrication of three-dimensional polymer nano-fiber/nets. Prog Mater Sci 58(8):1173–1243. doi:http://dx.doi.org/10.1016/j.pmatsci.2013.05.001
32. Lin J, Wang X, Ding B, Yu J, Sun G, Wang M (2012) Biomimicry via electrospinning. Crit Rev Solid State 37(2):94–114. doi:10.1080/10408436.2011.627096
33. Ding B, Wang M, Wang X, Yu J, Sun G (2010) Electrospun nanomaterials for ultrasensitive sensors. Mater Today 13(11):16–27
34. Lin J, Shang Y, Ding B, Yang J, Yu J, Al-Deyab SS (2012) Nanoporous polystyrene fibers for oil spill cleanup. Mar Pollut Bull 64(2):347–352. doi:10.1016/j.marpolbul.2011.11.002
35. Lin J, Tian F, Shang Y, Wang F, Ding B, Yu J (2012) Facile control of intra-fiber porosity and inter-fiber voids in electrospun fibers for selective adsorption. Nanoscale 4(17):5316–5320. doi:10.1039/c2nr31515g
36. Ding B, Kimura E, Sato T, Fujita S, Shiratori S (2004) Fabrication of blend biodegradable nanofibrous nonwoven mats via multi-jet electrospinning. Polymer 45(6):1895–1902. doi:10.1016/j.polymer.2004.01.026
37. Ding B, Li C, Hotta Y, Kim J, Kuwaki O, Shiratori S (2006) Conversion of an electrospun nanofibrous cellulose acetate mat from a super-hydrophilic to super-hydrophobic surface. Nanotechnology 17(17):4332–4339. doi:10.1088/0957-4484/17/17/009
38. Ding B, Lin J, Wang X, Yu J, Yang J, Cai Y (2011) Investigation of silica nanoparticle distribution in nanoporous polystyrene fibers. Soft Matter 7(18):8376–8383. doi:10.1039/c1sm05791j
39. Ding B, Ogawa T, Kim J, Fujimoto K, Shiratori S (2008) Fabrication of a super-hydrophobic nanofibrous zinc oxide film surface by electrospinning. Thin Solid Films 516(9):2495–2501. doi:10.1016/j.tsf.2007.04.086
40. Lin J, Cai Y, Wang X, Ding B, Yu J, Wang M (2011) Fabrication of biomimetic superhydrophobic surfaces inspired by lotus leaf and silver ragwort leaf. Nanoscale 3(3):1258–1262. doi:10.1039/c0nr00812e
41. Lin J, Ding B, Yang J, Yu J, Al-Deyab SS (2012) Mechanical robust and thermal tolerant nanofibrous membrane for nanoparticles removal from aqueous solution. Mater Lett 69:82–85. doi:10.1016/j.matlet.2011.11.046
42. Wang J, Raza A, Si Y, Cui L, Ge J, Ding B, Yu J (2012) Synthesis of superamphiphobic breathable membranes utilizing SiO_2 nanoparticles decorated fluorinated polyurethane nanofibers. Nanoscale 4(23):7549–7556

43. Wang N, Wang X, Ding B, Yu J, Sun G (2012) Tunable fabrication of three-dimensional polyamide-66 nano-fiber/nets for high efficiency fine particulate filtration. J Mater Chem 22(4):1445–1452. doi:10.1039/c1jm14299b
44. Zhao F, Wang X, Ding B, Lin J, Hu J, Si Y, Yu J, Sun G (2011) Nanoparticle decorated fibrous silica membranes exhibiting biomimetic superhydrophobicity and highly flexible properties. Rsc Adv 1(8):1482–1488. doi:10.1039/c1ra00605c
45. Gao B, Yang LY, Wang XR, Zhao JC, Sheng GY (2000) Influence of modified soils on the removal of diesel fuel oil from water and the growth of oil degradation micro-organism. Chemosphere 41(3):419–426. doi:10.1016/s0045-6535(99)00287-8
46. Liang Y, Liu D, Chen H (2013) Poly(methyl methacrylate-co-pentaerythritol tetraacrylate-co-butyl methacrylate)-g-polystyrene: synthesis and high oil-absorption property. J Polym Sci Pol Chem 51(9):1963–1968. doi:10.1002/pola.26576
47. Thanikaivelan P, Narayanan NT, Pradhan BK, Ajayan PM (2012) Collagen based magnetic nanocomposites for oil removal applications. Sci Rep 2. doi:10.1038/srep00230
48. Gui X, Zeng Z, Lin Z, Gan Q, Xiang R, Zhu Y, Cao A, Tang Z (2013) Magnetic and highly recyclable macroporous carbon nanotubes for spilled oil sorption and separation. ACS Appl Mater Interfaces 5(12):5845–5850. doi:10.1021/am4015007
49. Peng D, Lan Z, Guo C, Yang C, Dang Z (2013) Application of cellulase for the modification of corn stalk: leading to oil sorption. Bioresour Technol 137:414–418. doi:10.1016/j.biortech.2013.03.178
50. Sakthivel T, Reid DL, Goldstein I, Hench L, Seal S (2013) Hydrophobic high surface area zeolites derived from fly ash for oil spill remediation. Environ Sci Technol 47(11):5843–5850. doi:10.1021/es3048174
51. Teli MD, Valia SP (2013) Acetylation of jute fiber to improve oil absorbency. Fiber Polym 14(6):915–919. doi:10.1007/s12221-013-0915-8
52. Wu G, He L, Chen D (2013) Sorption and distribution of asphaltene, resin, aromatic and saturate fractions of heavy crude oil on quartz surface: molecular dynamic simulation. Chemosphere 92(11):1465–1471. doi:10.1016/j.chemosphere.2013.03.057
53. Yuan J, Liu X, Akbulut O, Hu J, Suib SL, Kong J, Stellacci F (2008) Superwetting nanowire membranes for selective absorption. Nat Nanotechnol 3(6):332–336. doi:10.1038/nnano.2008.136
54. Liang HW, Guan QF, Chen LF, Zhu Z, Zhang WJ, Yu SH (2012) Macroscopic-scale template synthesis of robust carbonaceous nanofiber hydrogels and aerogels and their applications. Angew Chem Int Edit 51(21):5101–5105. doi:10.1002/anie.201200710
55. Zou R, Zhang Z, Yu L, Tian Q, Wu J, Sun Y, Chen Z, Hu J (2010) Oriented free-standing ammonium vanadium oxide nanobelt membranes: highly selective absorbent materials. Chem-Eur J 16(48):14307–14312. doi:10.1002/chem.201002228
56. Wu J, Wang N, Wang L, Dong H, Zhao Y, Jiang L (2012) Electrospun porous structure fibrous film with high oil adsorption capacity. ACS Appl Mater Interfaces 4(6):3207–3212. doi:10.1021/am300544d
57. Zhu H, Qiu S, Jiang W, Wu D, Zhang C (2011) Evaluation of electrospun polyvinyl chloride/polystyrene fibers as sorbent materials for oil spill cleanup. Environ Sci Technol 45(10):4527–4531. doi:10.1021/es2002343
58. Lin J, Ding B, Yang J, Yu J, Sun G (2012) Subtle regulation of the micro- and nanostructures of electrospun polystyrene fibers and their application in oil absorption. Nanoscale 4(1):176–182. doi:10.1039/c1nr10895f
59. Lin J, Tian F, Shang Y, Wang F, Ding B, Yu J, Guo Z (2013) Co-axial electrospun polystyrene/polyurethane fibres for oil collection from water surface. Nanoscale 5(7): 2745–2755. doi:10.1039/c3nr34008b
60. Junping Z, Seeger S (2011) Polyester materials with superwetting silicone nanofilaments for oil/water separation and selective oil absorption. Adv Funct Mater 21(24):4699–4704. doi:10.1002/adfm.201101090

61. Raza A, Si Y, Ding B, Yu J, Sun G (2013) Fabrication of superhydrophobic films with robust adhesion and dual pinning state via in situ polymerization. J Colloid Interface Sci. doi:http://dx.doi.org/10.1016/j.jcis.2012.11.030
62. Raza A, Si Y, Wang X, Ren T, Ding B, Yu J, Al-Deyab SS (2012) Novel fluorinated polybenzoxazine-silica films: chemical synthesis and superhydrophobicity. Rsc Adv 2(33):12804–12811
63. Guangming G, Juntao W, Jingang L, Na S, Yong Z, Lei J (2012) Bio-inspired adhesive superhydrophobic polyimide mat with high thermal stability. J Mater Chem 22(17):8257–8262. doi:10.1039/c2jm16503a
64. Yang H, Lan Y, Zhu W, Li W, Xu D, Cui J, Shen D, Li G (2012) Polydopamine-coated nanofibrous mats as a versatile platform for producing porous functional membranes. J Mater Chem 22(33):16994–17001. doi:10.1039/c2jm33251e
65. Shang Y, Si Y, Raza A, Yang L, Mao X, Ding B, Yu J (2012) An in situ polymerization approach for the synthesis of superhydrophobic and superoleophilic nanofibrous membranes for oil–water separation. Nanoscale 4:7847–7854. doi:10.1039/C2NR33063F
66. Blanpain-Avet P, Migdal JF, Benezech T (2004) The effect of multiple fouling and cleaning cycles on a tubular ceramic microfiltration membrane fouled with a whey protein concentrate-membrane performance and cleaning efficiency. Food Bioprod Process 82(C3):231–243. doi:10.1205/fbio.82.3.231.44182
67. Marshall AD, Munro PA, Tragardh G (1993) The effect of protein fouling in microfiltration and ultrafiltration on permeate flux, protein retention and selectivity-a literature-review. Desalination 91(1):65–108. doi:10.1016/0011-9164(93)80047-q
68. Zhao Z, Zheng J, Wang M, Zhang H, Han CC (2012) High performance ultrafiltration membrane based on modified chitosan coating and electrospun nanofibrous PVDF scaffolds. J Membr Sci 394–395:209–217. doi:10.1016/j.memsci.2011.12.043
69. Wang XF, Chen XM, Yoon K, Fang DF, Hsiao BS, Chu B (2005) High flux filtration medium based on nanofibrous substrate with hydrophilic nanocomposite coating. Environ Sci Technol 39(19):7684–7691. doi:10.1021/es050512j
70. Hsiao BS, Chu B, Xuefen W, Dufei F, Kyunghwan Y (2006) High performance ultrafiltration composite membranes based on poly(vinyl alcohol) hydrogel coating on crosslinked nanofibrous poly(vinyl alcohol) scaffold. J Membr Sci 278(1–2):261–268. doi:10.1016/j.memsci.2005.11.009
71. You H, Yang Y, Li X, Zhang K, Wang X, Zhu M, Hsiao BS (2012) Low pressure high flux thin film nanofibrous composite membranes prepared by electrospraying technique combined with solution treatment. J Membr Sci 394–395:241–247. doi:10.1016/j.memsci.2011.12.047
72. Yoon Y, Hsiao BS, Chu B (2009) High flux ultrafiltration nanofibrous membranes based on polyacrylonitrile electrospun scaffolds and crosslinked polyvinyl alcohol coating. J Membr Sci 338(1–2):145–152. doi:10.1016/j.memsci.2009.04.020
73. Tang Z, Wei J, Yung L, Ji B, Ma H, Qiu C, Yoon K, Wan F, Fang D, Hsiao BS, Chu B (2009) UV-cured poly(vinyl alcohol) ultrafiltration nanofibrous membrane based on electrospun nanofiber scaffolds. J Membr Sci 328(1–2):1–5. doi:10.1016/j.memsci.2008.11.054
74. Wang X, Zhang K, Yang Y, Wang L, Zhou Z, Zhu M, Hsiao BS, Chu B (2010) Development of hydrophilic barrier layer on nanofibrous substrate as composite membrane via a facile route. J Membr Sci 356(1–2):110–116. doi:10.1016/j.memsci.2010.03.039
75. Ma H, Yoon K, Rong L, Mao Y, Mo Z, Fang D, Hollander Z, Gaiteri J, Hsiao BS, Chu B (2010) High-flux thin-film nanofibrous composite ultrafiltration membranes containing cellulose barrier layer. J Mater Chem 20(22):4692–4704. doi:10.1039/b922536f
76. Ma H, Burger C, Hsiao BS, Chu B (2011) Ultrafine polysaccharide nanofibrous membranes for water purification. Biomacromolecules 12(4):970–976. doi:10.1021/bm1013316

Chapter 18
Electrospun Superhydrophobic Self-Cleaning Materials

Yong Zhao and Nü Wang

Abstract In this chapter, we introduce the wettability of electrospinning products. Especially, we concentrate on the fabrication, characteristics, and applications of the electrospun self-cleaning materials. Self-cleaning materials are typical nature-inspired artificial materials learning from such as lotus leaf and many other plants or animals. Self-cleaning materials usually rely on a superhydrophobic surface, which should be of low surface free energy as well as large surface roughness. Electrospinning method is such a method that could facilely shape various hydrophobic polymers into ultrathin fibers with tunable surface microstructures. It means the electrospun products are of very large specific area, which satisfy the two basic conditions in preparing superhydrophobic surfaces. Therefore, in the last decade, scientists put forward a good few of elegant approaches to fabricate superhydrophobic materials by electrospinning. These methods can be generally classified into two routes. One is a direct route that creates superhydrophobic electrospun films from hydrophobic materials. Another is an indirect route that decorates electrospun nanofibers (no matter hydrophobic or hydrophilic) with hydrophobic chemicals. We first introduce some representative works on the fabrication of superhydrophobic self-cleaning materials by electrospinning method. Then we show some applications of these superhydrophobic materials. Finally, we give a brief personal perspective on this area.

18.1 Introduction

Self-cleaning material is a kind of typical bio-inspired artificial materials which get inspiration from lotus leaf. Natural lotus leaf has an admirable property that could clean its surface by flushing water without the need of any other aid. It is regarded

---

Y. Zhao (✉) • N. Wang
School of Chemistry and Environment, Beihang University, Beijing, China
e-mail: zhaoyong@buaa.edu.cn

B. Ding and J. Yu (eds.), *Electrospun Nanofibers for Energy and Environmental Applications*, Nanostructure Science and Technology,
DOI 10.1007/978-3-642-54160-5_18,

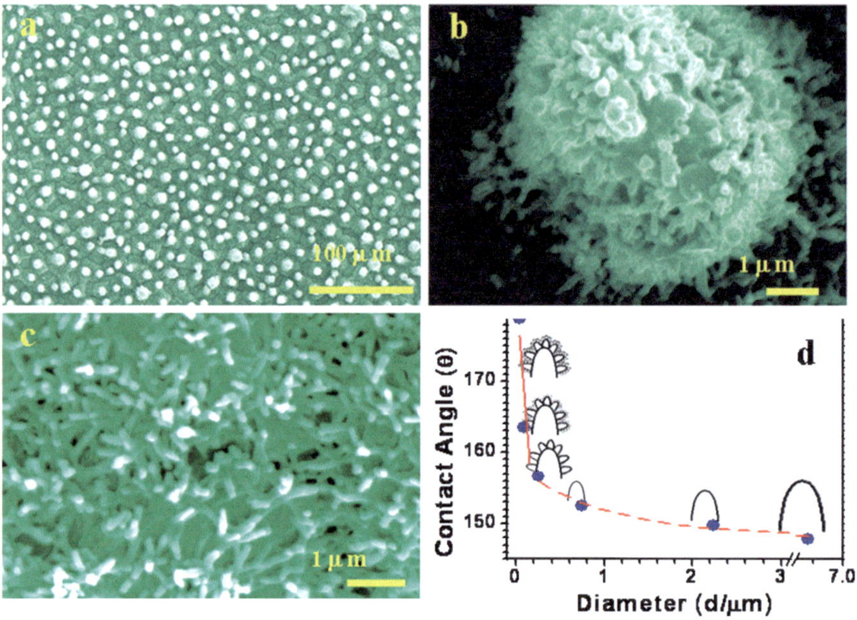

Fig. 18.1 Two-tier micro-/nanostructures of the lotus leaves. (**a**) Large-area SEM image of a lotus leaf, which has a lot of micropapillae; (**b**) enlarged view of a single papilla from (**a**); it shows there are nanobranches on the micropapillae; (**c**) SEM image of the groove area between micropapillae; there are also a large number of nanobranches; (**d**) the relationship between superhydrophobicity and micro-/nanostructures (Reproduced from Ref. [3] by permission of John Wiley & Sons Ltd)

as a symbol of sacred purity in Asian religions over 2000 years for the reputation of "unsoiled by the mud from which it springs." No doubt, lotus leaf is the most representative self-cleaning material. This is because the lotus leaf has a superhydrophobic surface that water cannot wet the leaf. Thanks to the unique non-wettable surface, dust on the lotus leaf could be stuck on the rolling water droplets on the leaf and then flow away. This phenomenon is known as famous "lotus effect." As a matter of fact, there are still many other plant leaves with such superhydrophobic self-cleaning property such as taro, lady's mantle, rice leaf, etc. This is a very attractive function that means we might not need to wash our windows or clothes anymore if we could fabricate such a superhydrophobic self-cleaning surface. People have known the self-cleaning ability of lotus leaf for thousands of years; however, the underlying mechanism was revealed just a few decades ago. Barthlott and Neinhuis firstly reported that the lotus effect is a combination of microscale surface roughness caused by papillae and the hydrophobic wax [1, 2]. In 2002, Jiang et al. further revealed that micro-/nanoscale two-tier hierarchical structures play an important role in superhydrophobicity of lotus leaf [3]. They found that there are microsized papillae distributed on the lotus leaf surface; moreover, there are numerous nanoscale branches on papillae as well as grooves between papillae (Fig. 18.1). Such composite structures make for the excellent

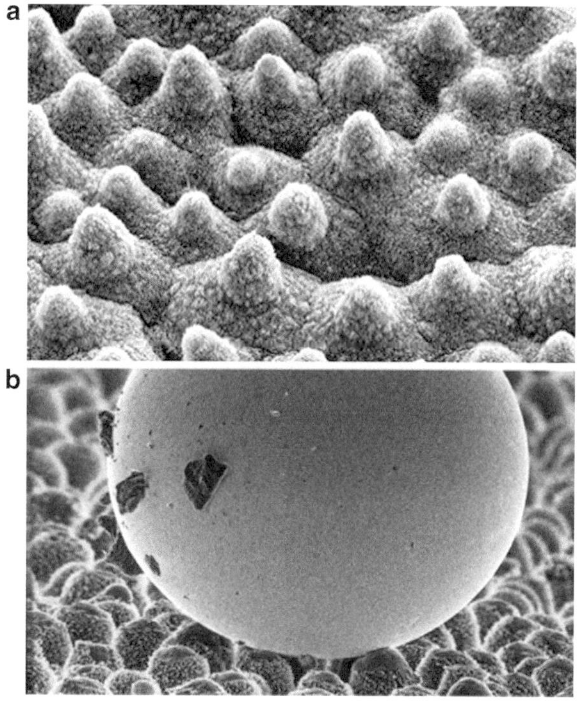

Fig. 18.2 Lotus effect: (**a**) the self-cleaning surfaces of lotus leaf; (**b**) contaminating particles are removed from the leaf when the droplet rolls (Reprinted with the permission from Ref. [5] Copyright 2004 American Chemical Society)

water-repellent self-cleaning performance of lotus leaf (Fig. 18.2) [4, 5]. These findings spurred great interests on the researches of superhydrophobic self-cleaning materials in the last decade [6–10].

18.2 Biomimetic Self-Cleaning Materials Fabricated via Electrospinning

For the better understanding of how to create a superhydrophobic self-cleaning surface, it is necessary to give a brief introduction to some surface-wetting phenomena. Wetting is a common interface phenomenon that occurs when a kind of liquid falls or condenses onto a solid substrate. In most cases, such liquid is water because water is the most important liquid to both industry and daily life. Therefore, hereinafter, wetting we discuss refers to water wetting unless otherwise specified. According to the water-repellent capability, surfaces could be classified into two categories, i.e., hydrophobic and hydrophilic, which are measured by water contact angle. The contact angle (CA) refers to the angle at which a liquid/vapor interface contacts the solid surface as Fig. 18.3 shows. Inspired by self-cleaning lotus leaf, scientists put forward two new terms that are superhydrophobic and superhydrophilic. Superhydrophobic means a surface is extremely repellent to water

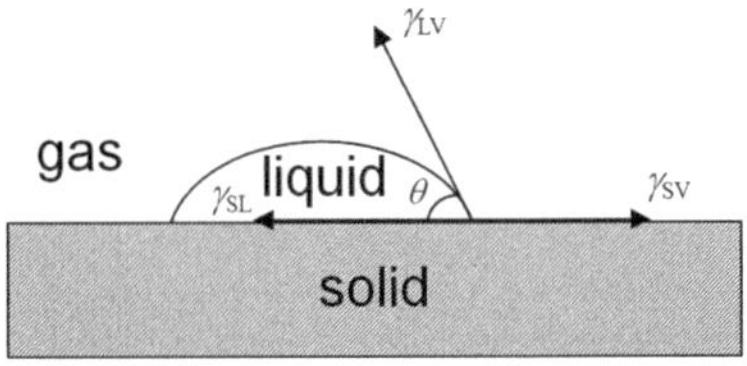

Fig. 18.3 Definition of the contact angle. It refers to the angle when a liquid droplet contacts the solid surface

with a CA larger than 150°, while superhydrophilic refers to the opposite case with CA smaller than 10°. By investigating lotus leaf, scientists have demonstrated that there are two essential conditions for fabricating a superhydrophilic surface: one is low surface free energy and the other is the micro-/nano-two-tier structures. So far as we know, the chemical material with the lowest surface energy is perfluorinated silane, which has a CA of only 119° [11]. It means that it is virtually impossible to get a superhydrophobic surface by merely changing the chemical composition. Accordingly, researchers paid considerable attention to designing and fabricating various surface micro-/nanostructures by diverse chemical or physical strategies in order to acquire the superhydrophobic surfaces such as heterogeneous nucleation method, plasma treatment method, etching method, solgel method, chemical vapor deposition method, electrochemical method, layer-by-layer method, template method, and solvent-nonsolvent methods [12–15]. Among these methods, electrospinning has been proved to be a simple and versatile approach for generating superhydrophobic surfaces from a broad variety of materials [8, 16–20].

Electrospinning is a simple technique that can stretch viscous liquids from polymer solutions into ultrafine fibers using electrostatic force. As far as we know, this is the only universal method that can be applied to produce nanofibers from a very broad range of materials, from polymers to hybrids to inorganics. Moreover, the prominent advantage of the electrospinning approach lies in that it has huge potential to tune the configuration of the spinning setup, as well as abundant posttreatment methods for the electrospun products. It thereby leaves vast scope for researchers to tailor the structures and properties of electrospun fibers.

Inspired by the fascinating hierarchical structures of lotus leaf [21], by making use of simple electrospinning technique and employing the low-cost polystyrene (PS) as raw material, Jiang et al. fabricated a superhydrophobic film with a novel composite structure consisting of porous microspheres and nanofibers (Fig. 18.4c, d) [22]. The PS morphologies can be controlled by the concentration of solvents. Different from routine electrospun products with uniform fiber structures, they found that simple nanofiber film could not become superhydrophobic because the surface roughness is not large enough. Then they fabricated a porous particle film by electrospray of a very dilute PS solution. The results showed that such film was rough enough to be superhydrophobic. However, such film is unstable because there is no linkage between individual particles. Therefore, to increase the surface roughness, they fabricated a composite structured film that combined porous microspheres and nanofibers, which showed excellent superhydrophobic property.

Fig. 18.4 (**a**) Lotus leaf with self-cleaning effect. (**b**) SEM image of lotus leaf. The morphologies are composited of microscale papillae and cilium-like nanostructures. (**c**) Electrospun PS three-dimensional network porous film (inset: water droplet on the film (CA = 160.4°)). The fibrous mats were composited by the bead-on-fiber structured fibers. (**d**) High-magnified SEM image of the surface nanostructure of a single porous microsphere (Reproduced from Ref. [22] by permission of John Wiley & Sons Ltd)

Jiang et al. further fabricated a composite polymer film by electrospinning method, which exhibited conductive and excellent superhydrophobic properties in a whole pH range (Fig. 18.2) [23]. They used a mixture of conductive PANI and hydrophobic PS as electrospinning materials to fabricate composite film which had a self-cleaning property. The outer surface of the film was composed of carbon element that possessed low surface energy, while the nitrogen atoms in the PANI which played an important role in doping and undoping were distributed in the inner membrane. It can keep the superhydrophobicity and conductivity in the acidic and basic environment. Ordinary surface could easily cause static-charge accumulation, which possibly leads to sparking under dry conditions. So the conductive and superhydrophobic coatings would be used effectively in many fields.

Rutledge et al. showed that fiber mats composed solely of uniform fibers could be obtained by electrospinning a hydrophobic material (i.e., poly(styrene-block-dimethylsiloxane) block copolymer) blended with homopolymer polystyrene (PS) [24]. The roughness of the nonwoven mat, resulting from the small diameters of the fibers (150–400 nm), combined with the enrichment of the dimethylsiloxane component at the fiber surfaces, was sufficient to yield materials with a contact angle of 163° and a hysteresis value of 15°. Besides, the lotus leaflike structure could be also fabricated by electrospun polyaniline/polystyrene as well as poly(ε-caprolactone) [23, 25]. The PCL nanofibers exhibited bead-on-string structures. Such beaded fibers had been considered to be the undesirable structures in electrospinning process; however, many extensive studies proved that such structures played a great role in increasing the hydrophobic property by increasing the surface roughness [26].

Compared with such one-step method to fabricate lotus leaf fibrous electrospun film, endowing the electrospun fibers with a secondary morphology in order to obtain the hierarchical structure is also an effective way. Kang et al. reported the fabrication of PS fibers with nanoscaled protuberances. The water contact angle was $154.2 \pm 0.7°$, which was increased dramatically compared with smooth PS film [27]. Furthermore, Rutledge et al. used two approaches involving either directly introducing nanometer-scale pore structures onto electrospun fiber surfaces or decorating the electrospun fiber with fine nanoparticles using layer-by-layer deposition techniques. Both approaches produced highly stable superhydrophobicity fabrics [28].

In addition to lotus leaf, scientists find that there are many other plant leaves such as silver ragwort leaf, Chinese watermelon, etc., that also show water-repellent and self-cleaning property [29]. Take silver ragwort leaf as an example [26, 30]; it can be seen that in Fig. 18.5a, the silver ragwort presents superhydrophobicity with water contact angle about 150° [26, 30]. Further to examine the silver ragwort leaf by SEM, it is obvious that the leaf is composed with curved fibers and multicellular grooves from epidermal tissues. The grooves in nanoscale are the secondary structures which are found along the fiber axis (Fig. 18.5b). The as-shown hierarchical micro- and nanostructure is the primary cause for the high hydrophobicity. Inspired by the silver ragwort leaf structure, many works concentrated on the fabrication of such structured surface via electrospinning. Ding and coworker demonstrated the fabrication of silver ragwort leaf by electrospinning polystyrene solution [19]. These fibers showed rough surface with nano-sized grooves which were caused by the various phase separation rates of the solvents from PS fibers during electrospinning. Such hierarchical surface and the low surface free energy of PS yielded a stable superhydrophobicity with water contact angle as high as 159.5° for 12 mg water droplet, which exceeded that of silver ragwort leaf (147°). Further, the same group blended PS with polyamide 6 by four-jet electrospinning technique in order to enhance the mechanical property of PS fibrous film [31]. Although the hydrophobicity of the resulting PS/PA6 fibrous mats decreased with decreasing the number ratios of jets of PS/PA6, the fibrous mats formed with the number ratios of jets of 2/2 (PS/PA6) showed a water contact angle of 150° with a three times increased tensile strength compared with that of the pure fibrous PS mats, which provided wide applications of these self-cleaning materials (Fig. 18.5c, d).

Thanks to the development of electrospinning technology, some polymer systems, which are normally not electrospinnable by single spinneret setup, can be electrospun to fibers via a coaxial electrospinning technique. What's more,

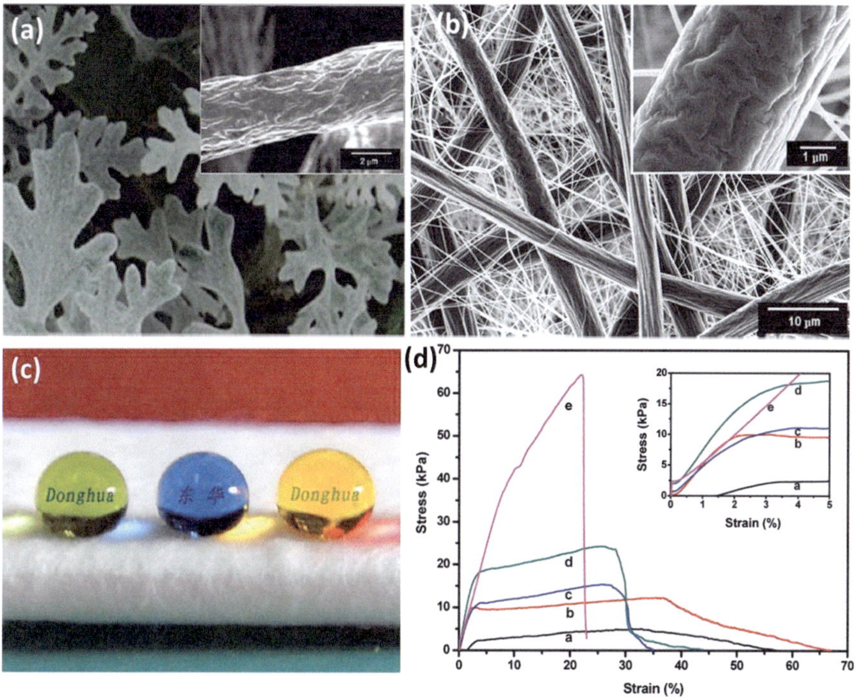

Fig. 18.5 (**a**) Photography of superhydrophobic silver ragwort leaves. Inset is the SEM image of silver ragwort leaf. It is consisted of fiber with nanogrooves. (**c**) Biomimetic PS/PA6 composite electrospun nanofibrous. (**c**) The PS/PA6 nanofibrous mats show superhydrophobic property. (**d**) Stress-strain behavior of PS/PA6 composite fibrous mats with different ratios: (**a**) 4/0, (**b**) 3/1, (**c**) 2/2, (**d**) 1/3, and (**e**) 0/4 [30]

such excellent characteristics can also be maintained. Steckl et al. put coaxial electrospinning into practice to fabricate the non-electrospinnable Teflon AF. The Teflon AF acted as the sheath material, and poly(ε-caprolactone) (PCL) was the core material (Fig. 18.6) [32]. By using this coaxial electrospinning method, superhydrophobic and oleophobic membranes have been successfully produced combined with good mechanical tensile endowed by PCL. Using such coaxial electrospinning, the two-component fibrous mats show superhydrophobicity with a water contact angle of 158° as well as ~5° rolling angle and oleophobic with an oil contact angle >150° (oil surface tension higher than ~50 mN/m).

Water strider is a kind of insect that can walk freely on water surfaces, whose legs are the most sensitive organs in its body. They can feel the struggle of water strider when it is falling into water and can help the water strider to jump 30–40 cm far and high. Water strider by sliding between a pair of legs can quickly slide on water at a speed of 1.5 m/s, owning the reputation of "skaters in the pond." Jiang et al. reported that the water strider's legs are superhydrophobic for the combination of the unique hierarchical surface structure of needle-shaped micro-setae with nanogrooves as well as the covered hydrophobic wax layer (Fig. 18.7b) [33]. By investigating the

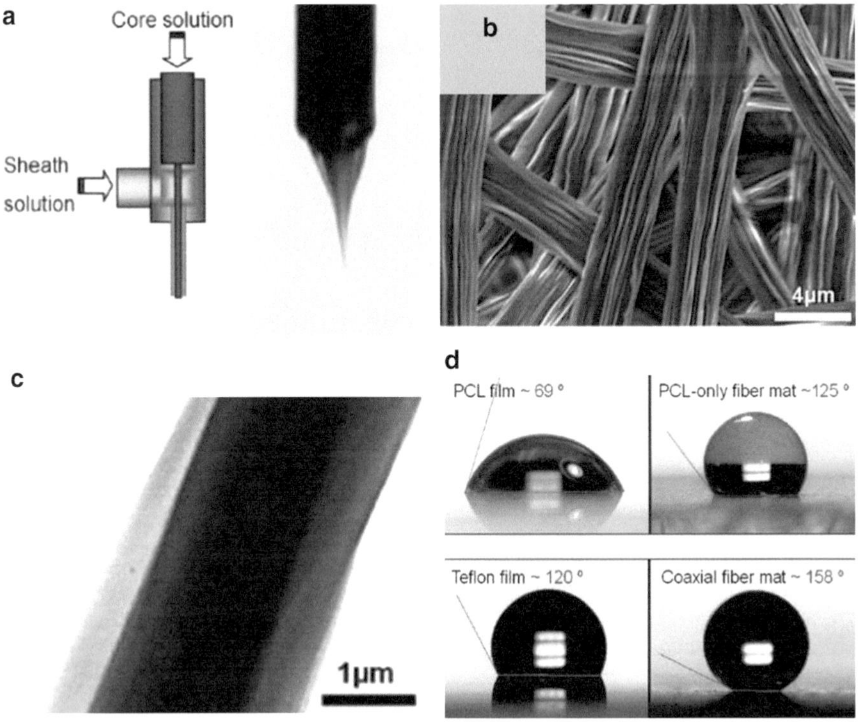

Fig. 18.6 (**a**) Coaxial electrospinning operation. (**b**) SEM image of PCL/Teflon fibers obtained with PCL 10 wt% with 1.5 mL/hand Teflon 1 wt% with 1.0 mL/h. (**c**) PCL/Teflon fiber TEM cross section. (**d**) Water droplet (2 μL) on a PCL film, Teflon film, PCL-only fiber membrane, and coaxial PCL/Teflon fiber membrane; dodecane droplet (2 μL) on fiber membranes of PCL-only and coaxial PCL/Teflon (Reprinted with the permission from Ref. [32]. Copyright 2009 American Chemical Society)

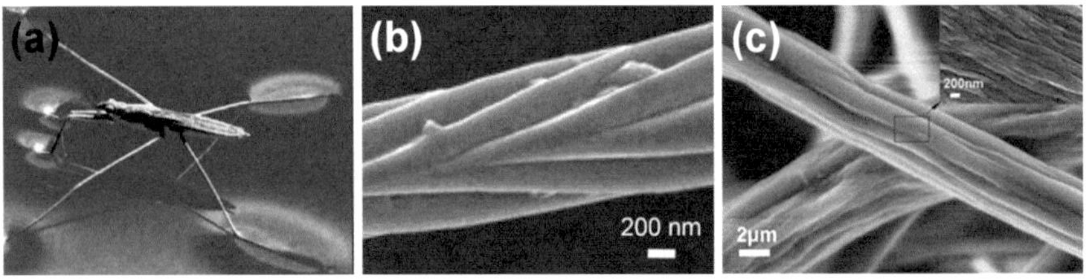

Fig. 18.7 The superhydrophobic water strider legs. (**a**) Photograph of a water strider standing on the water surface. (**b**) SEM image of the leg with oriented spindly setae and the nanogrooves on a single seta (Reprinted by permission from Macmillan Publishers Ltd: Ref. [33]. Copyright 2004) (**c**) SEM images of biomimetic electrospun water strider's leg (Reprinted with the permission from Ref. [35]. Copyright 2009 Royal Society of Chemistry)

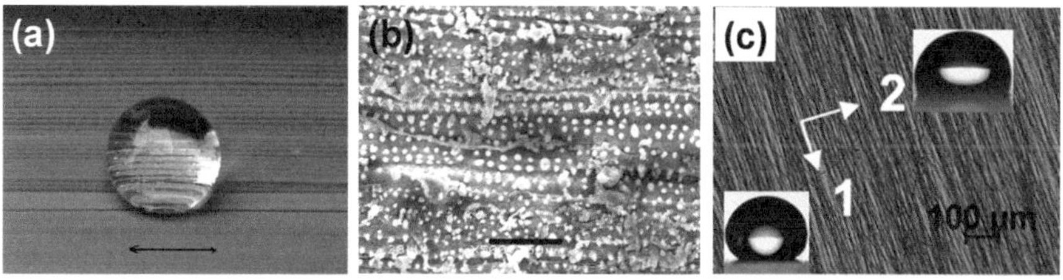

Fig. 18.8 (**a**) Optical photo of a water droplet floating on a rice leaf. (**b**) SEM image of rice leaf. (**c**) Optical micrograph of a glass slide covered by a low density of aligned PAN electrospun fibers. Insets show the photos of the droplet on the surface coated with polymerized perfluoroalkyl ethyl methacrylate taken from directions 1 and 2 [17]

topography of water strider's legs, it was found that the maximal supporting force of a single leg against water is about 15 times the total of its body weight [34].

Inspired by water strider's legs, Lin et al. successfully prepared "water strider's leg" by electrospinning a polyhedral oligomeric silsesquioxane-poly(methyl methacrylate) copolymer (Fig. 18.7c). Uniform electrospun fibers from the copolymer showed a water contact angle as high as 165° with a sliding angle as low as 6°. Interestingly, the bead-free electrospun fibers showed a nano-fibrillated structure under electron microscopes and an ordered surface molecular arrangement under an atomic force microscope. Such features have not been reported for any other electrospun fibers [35]. Mimicking water striders provides a facile method to develop microrobots to stand and maneuver on water surfaces. Pan et al. reported the biomimetic fabrication of the water strider's leg structure by using specially designed nanofiber collectors, which contained a spiculate metal needle perpendicular to a rectilinear metal strip as the fiber collector instead of two parallel electrodes during electrospinning [36]. The artificial water strider's legs with parallel grooves and maximal supporting force of more than 200 dyn/cm were facilely fabricated.

Learning from nature is a powerful method of creating new techniques and advanced materials. In nature, there are many surfaces exhibiting great anisotropic wetting. Recently, a lot of attentions have been focused on such anisotropic surfaces for their wide applications. The most common example of anisotropic wetting in organisms is rice leaves, which exhibit hierarchical structures that are similar to the lotus leaves with a water contact angle about 157° [29]. Meanwhile, different to lotus leaves, the fine structure of rice leaves reveals that the papillae are arranged in one-dimensional order parallel to the edge of the leaf edge. It endows the rice leaf with anisotropic wettability that water droplet can roll off along the parallel direction freely but hard to move along the perpendicular direction (Fig. 18.8a, b). In order to mimic rice leaves, Rutledge et al. fabricated polyacrylonitrile (PAN) fibers with parallel directions by controlling the collector with parallel electrodes. After obtaining the paralleled fibers, they were coated with polymerized perfluoroalkyl ethyl methacrylate using CVD [17]. A water contact angle of 153°and sliding angle

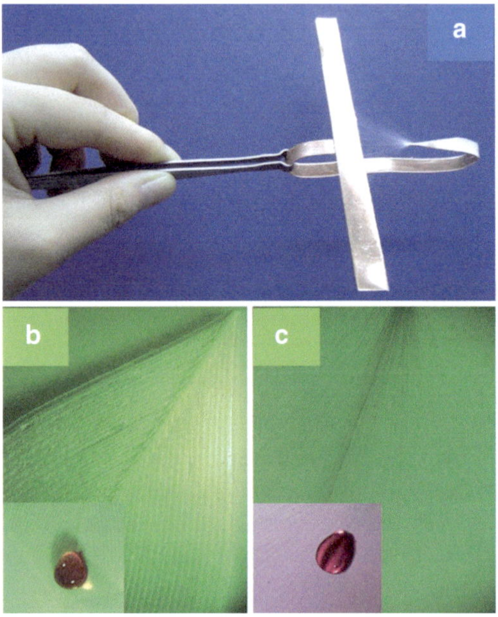

Fig. 18.9 (**a**) Digital photograph of fan-shaped radiating nanofiber pattern collected by a spiculate copper needle perpendicular to a rectilinear copper strip. (**b**) Water droplet placed on both goose feather and (**c**) fan-shaped radiating nanofiber pattern. Obviously, water shows the similar streamlined shape on the above two surfaces (Reproduced from Ref. [36] by permission of The Royal Society of Chemistry)

of 8° in the direction parallel to fiber axis exhibited superhydrophobicity. However, in the perpendicular direction, the water contact angle was 119° (Fig. 18.8c).

In nature, different from lotus leaf with hydrotropic surface, some plants, like waterfowl feathers, show special anisotropic wetting property. This anisotropic wetting property is attributed to the alignment of feather veins. If electrospun fibers can be aligned as the feather structure, it may have such anisotropic wetting phenomenon. Ding et al. electrospun aligned poly(vinyl butyral) (PVB) fibrous membranes by collecting the fibers between the metal needle and the straight strip (Fig. 18.9a) [36]. In this situation, by taking advantage of this easy-controlled collector, after the fiber deposition, goose feather-like biomimetic fan-shaped nanofiber pattern fibrous mat can be obtained. When water droplets are set on the surfaces of the goose feather and biomimetic fan-shaped radiating nanofiber pattern, respectively, both the droplets have a streamlined shape (Fig. 18.9b, c). Meanwhile, replacing the collector with two parallel copper strips, the bamboo-like leaf surface can also be mimicked.

The research route of biomimetic materials is a process that derives from nature but beyond nature ultimately. Scientists first discover the special property of bio-system and give a scientific explanation of the mechanism behind the property. Based on the understanding, researchers mimic the chemical composition and microstructures of the bio-system to recur the desired bio-function. Then abundant artificial fabrication approaches can be used to create new materials that possess better performance or more functionalities than natural materials. McKinley and Cohen et al. synthesized a series of fluorinated polyhedral oligomeric silsesquioxanes (fluoroPOSS) that were of very low free energy (Fig. 18.10a). Because pure fluoroPOSS was not spinnable, they blended the fluoroPOSS with

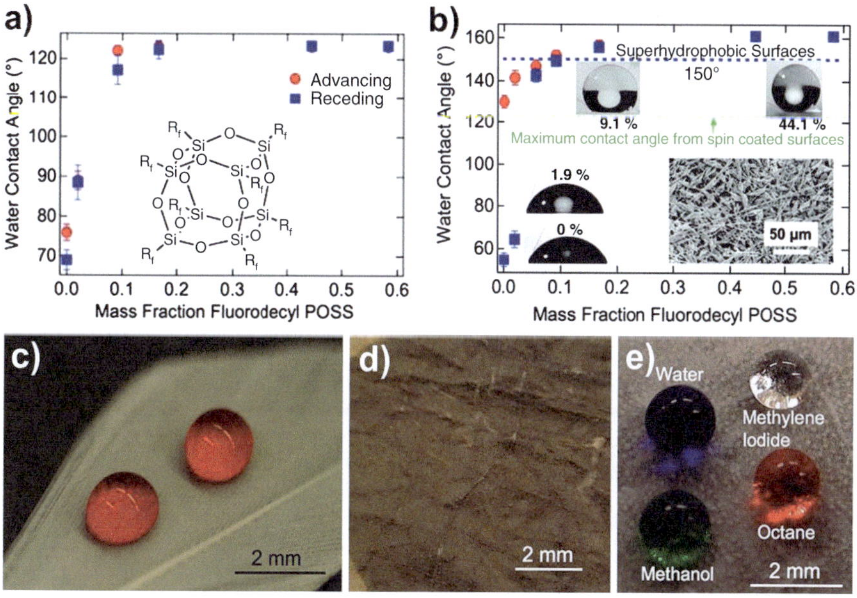

Fig. 18.10 (**a**) CA as a function of the mass fraction of fluoroPOSS in fluoroPOSS/PMMA blends. The inset is the general molecular structure of fluoroPOSS. R_f have the general molecular formula $-CH_2CH_2(CF_2)_nCF_3$, where $n = 0, 3, 5$, or 7. (**b**) CA on the electrospun surfaces. The inset shows a bead-on-a-string structure of electrospun surface containing 9.1 wt% POSS. It reaches superhydrophobicity (From [14]. Reprinted with permission from AAAS). (**c**) Rapeseed oil droplet on a duck feather dip coated with fluoroPOSS. (**d**) Droplets of octane droplet on a lotus leaf dip coated with fluoroPOSS. (**e**) Droplets of water, methylene iodide, methanol, and octane on a lotus leaf covered with electrospun fibers of PMMA and fluoroPOSS (44 wt%) [37]

PMMA, a well spinnable polymer, and fabricated a "bead-on-a-string" nanofibrous film by electrospinning (Fig. 18.10b). With the increase of fluoroPOSS ratio, the nanofibrous film showed excellent superamphiphobic property; in other words, the film was both superhydrophobic and superoleophobic (Fig. 18.10c–e) [14, 37]. These works demonstrated the successful creation of new artificial materials beyond the performance of natural biomaterials.

With the advance technology, intelligent materials with stimuli-responsive properties have inspired great interests for its intelligent multifunctions. The wettability of smart surface could be tuned with varying of temperature [38, 39]. With the ever-increasing practical needs of smart surfaces, simplified production process and low cost of such smart materials are urgently needed. Zhao et al. reported the pioneering work of the fabrication of stimuli-responsive wettable surface by electrospinning. A novel composite film of poly(N-isopropylacrylamide) (PNIPAAm) blended with polystyrene(PS) was obtained. By altering the PS concentration in the electrospinning system, obvious wettability changes can be realized near the lower critical solution temperature (LCST). Using a raw electrospinning solution composed of PNIPAAm, PS, and tetrahydrofuran (THF) in the ratio 2:10:90 (w/w/w), the wettability of the as-prepared fibrous mat could be switched between superhydrophilicity and

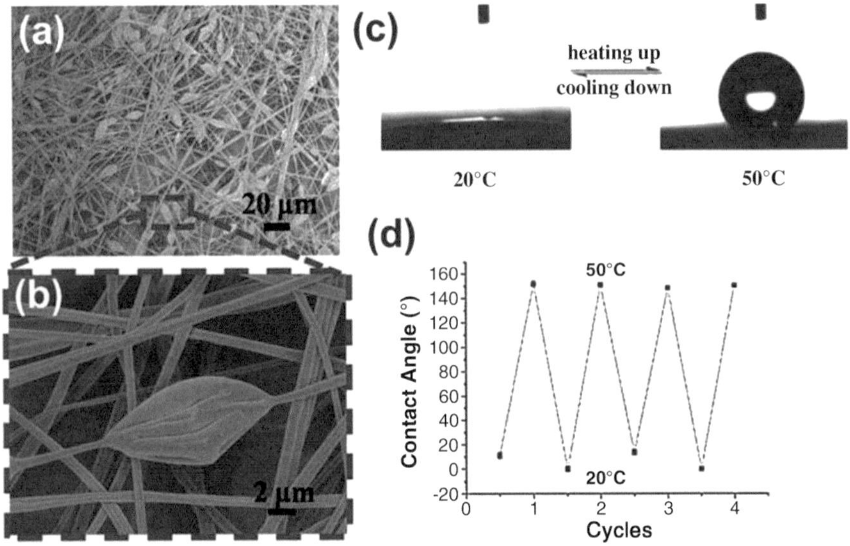

Fig. 18.11 (**a–b**) SEM image of PNIPAAm-PS composite film. (**c**) Blue dotted square magnification of the PNIPAAm-PS composite film. It shows thermoresponsive properties, with a switch between superhydrophobicity at low temperature (20 °C) and superhydrophilicity at high temperature (50 °C). (**d**) Reversibility of water contact angle transition of the PNIPAAm-PS composite film at two different temperatures. Half cycles, 20 °C, and integral cycles, 50 °C (Reproduced from Ref. [40] by permission of John Wiley & Sons Ltd)

superhydrophobicity with a change in temperature from 20 to 50 °C (Fig. 18.11). Accordingly, many similar studies have been concerned with such temperature-dependent wettability properties. As examples, Gu et al., Sigmund et al., and Linhardt et al. produced PNIPAAm-poly-L-lactide and PNIPAAm-PLLA composited fibrous mats, PS-PNIPAAm blended cross-linked PNIPAAm-PMA electrospun mats, and PNIPAAm-cellulose nanofibers, respectively. All abovementioned cases are demonstrated to have reversibly switchable surface wettability [40–44].

Apart from stimuli-responsive wettability surface, anisotropic wettability surface with different wettability also exhibits intelligent function. Zhao et al. designed and fabricated an electrospun composite fibrous film with heterogeneous wettability in the normal direction [45]. The hydrophobic polyurethane (PU) film and hydrophilic cross-linked poly(vinyl alcohol) (c-PVA) film were coupled seamlessly (Fig. 18.12a–f). By taking advantage of the distinct hydrophobic/hydrophilic difference, an interesting unidirectional water-penetration phenomenon has been successfully realized (Fig. 18.12g, h). Water can spontaneously penetrate from the hydrophobic side to the hydrophilic side; however, it is not the case for the reverse direction. Meanwhile, the unidirectional water-penetration phenomenon also cannot be realized for the single PU or c-PVA film. This is attributed to the hydrophobicity and hydrophilicity synergistic effect. Furthermore, by controlling the thickness of both hydrophobic fibrous film and the hydrophilic film, water penetration can be regulated.

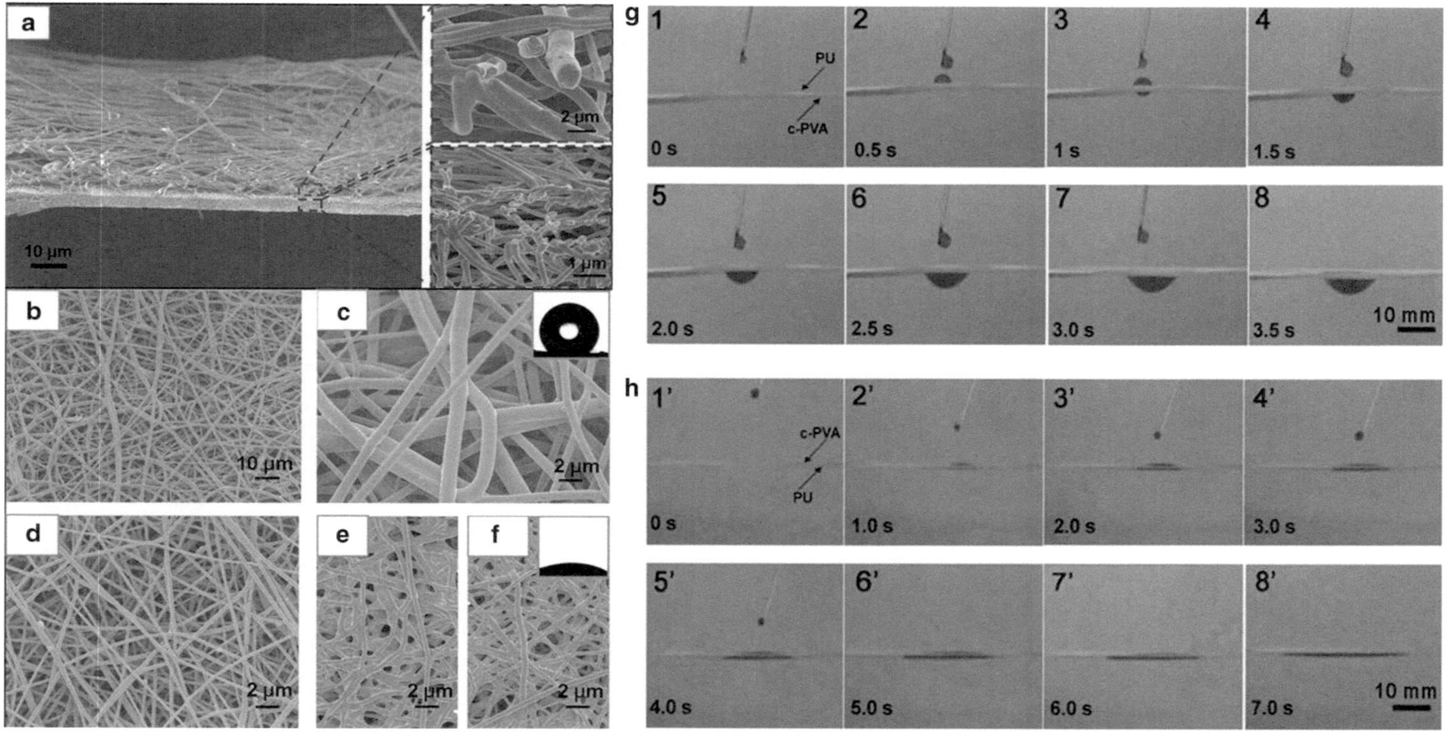

Fig. 18.12 (**a–f**) SEM images and wettability of PU/c-PVA composite fibrous film. (**a**) Cross section of PU/c-PVA film, which is of two-layer structure. (**b**, **c**) PU microfiber layer with high water repellency. (**d–f**) c-PVA nanofibers' layer with hydrophilic property. (**g**, **h**) Unidirectional water-penetration phenomenon of the hydrophobic/hydrophilic composite film. (**g**) Water drops onto the upward hydrophobic PU side and penetrates spcntaneously. (**h**) Water drops onto the upward hydrophilic c-PVA film side and spreads instead of penetration (Reproduced from Ref. [45] by permission of The Royal Society of Chemistry)

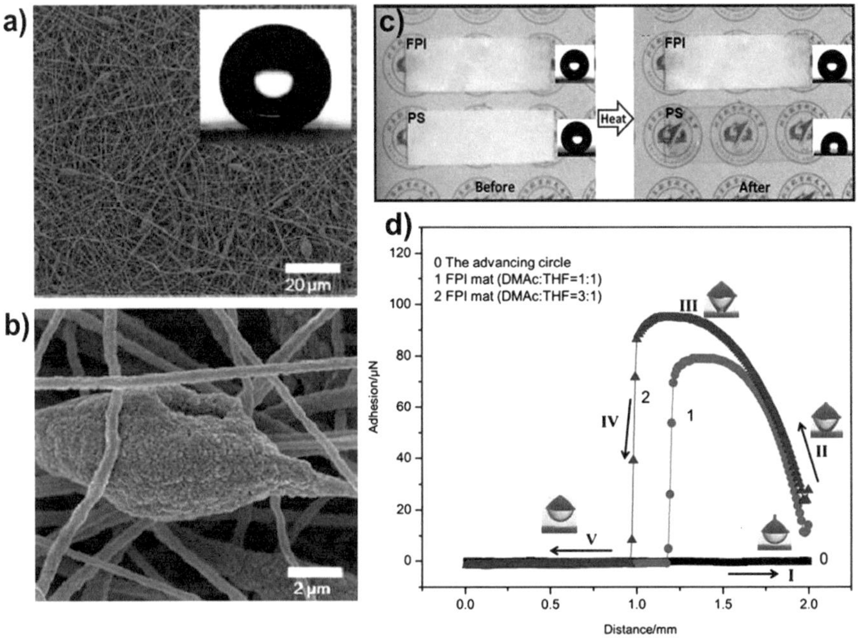

Fig. 18.13 The high-adhesive superhydrophobic FPI mat with high thermal stability. (**a**, **b**) Bead-on-string-structured FPI electrospun nanofibers' film that shows superhydrophobic property. (**c**) An annealing comparison of the thermal stability of two polymers, FPI and PS. The samples were heated at 150 °C. After heating, the PS sample turned into transparent, which means the microstructures of the PS film have been destroyed, while the FPI sample has no change. (**d**) Adhesion curve of FPI superhydrophobic electrospun film; it means that the film is high adhesive to water (Reproduced from Ref. [49] by permission of The Royal Society of Chemistry)

Surfaces possess a strong adhesive force to water means water presents a water contact angle larger than 150° and will not roll off at any tilted angles, that is, superhydrophobic surface with high adhesion. Such unique property can be found in nature, e.g., rose petals, which is named the "petal effect." Researchers found that surface with petal effect had potential for applications in many fields, such as microfluid non-loss transportation, biochemical separation, and localized chemical reactions [46–48]. Wu et al. proposed a facile electrospinning approach to produce a fluorinated polyimide (FPI) fibrous surface with excellent thermal stability as well as petal effect [49]. By adjusting the concentration of FPI electrospinning solution and tuning the ratio of the blended solvents, beaded nanofibers with highly porous surface structures can be obtained. Such FPI mat exhibited large water contact angle of 157.8° and strong adhesion to water 98.3 μN. More importantly, such film also possessed high thermal stability that could endure high temperature up to 300 °C. Such mat with petal effect as well as thermal stability properties may open broader applications in harsh environments (Fig. 18.13).

18.3 Self-Cleaning Materials Fabricated via Modifying Electrospun Fibers

As mentioned above, the wettability of fibrous materials is affected by both fiber structure and the chemical components. In this section, we will briefly introduce other techniques reported on modifications of the surface chemistry, which mainly include the applications of low-surface-energy materials and hydrothermal treatments on preformed textured fiber surfaces.

Modification of low-surface-energy materials consists of two aspects: taking advantage of the intrinsic roughness of electrospun fibrous mats and providing the low-surface-energy hydrophobic coating. The most important step for the realization of superhydrophobicity is the nature of the coating and its nanometer-scale thickness which ensures that the surface retains the desired fiber-like structure. A number of hydrophobic coatings have been modified on electrospun fibers to make a superhydrophobic surface. Rutledge et al. fabricated poly(caprolactone) (PCL) electrospun fibers and coated them with hydrophobic polymerized perfluoroalkyl ethyl methacrylate (PPFEMA) by initiating chemical vapor deposition (iCVD) [50]. The hierarchical surface roughness in the PCL fibers and the extremely low-surface-free-energy coating PPFEMA endow the mat stable superhydrophobicity with a water contact angle of 175° and a threshold sliding angle less than 2.5° for 20 mg water droplet. Hoven et al. reported an alternative multistep process to construct a superhydrophobic fibrous surface, which was achieved by the reaction of poly(vinyl alcohol) (PVA) fiber mats with multiple cycles of $SiCl_4/H_2O$ treatment followed by silanization with (1H,1H,2H,2H-perfluorooctyl)trichlorosilane (PFOTS) [51]. The $SiCl_4/H_2O$ treatment maintained the physical integrity of the PVA fibers and thus yielded PVA-silanol fiber mats. It is postulated that the strong pinning effect and hence the water adhesion originated from the collapse of the underlying fibrous structures and the removal of air pockets. The addition of silica to the PVA matrix improved the rigidity and thus prevented the fibers from collapsing, allowing air to remain trapped within the fibrous structure and giving the surface greater water repellency. The superhydrophobic fibrous membrane had a water contact angle of 168° and hysteresis of 0°. Besides PPFEMA and PFOTS, (heptadecafluoro-1,1,2,2-tetrahydrodecyl)trichlorosilane (FDTS) [52] and FAS [36, 53] are also usually used as low-free-energy coating materials in modification of electrospun fibers.

Recently, with the developed technique, the hydrothermal synthesis has been established as an efficient way to fabricate multifunctional materials with different morphologies. Thanks to the advantages of hydrothermal synthesis, various kinds of micro-/nanofibers with versatile secondary hierarchical structures can be obtained easily. Chang designed a facile and effective strategy that combines electrospinning method with hydrothermal process to prepare firecracker-shaped ZnO/polyimide (PI) hybrid nanofibers (Fig. 18.14) [54]. Firstly, PI nanofibers were prepared. In this case, the polyamic acid (PAA) solution produced from pyromellitic dianhydride

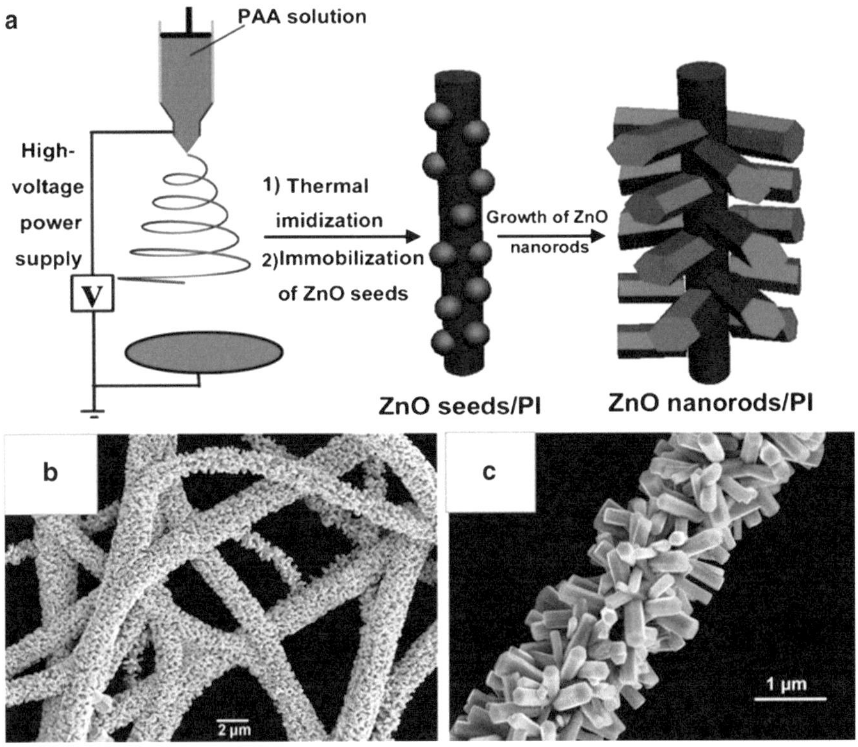

Fig. 18.14 (**a**) Strategy for growing ZnO nanorods onto PI nanofiber surface using hydrothermal method. In general, three processes are needed in sequence for the preparation of ZnO/PI hybrid nanofibers: PI nanofiber preparation, immobilization of ZnO seeds onto PI nanofibers, and growth of ZnO nanorods from the ZnO seeds. (**b**) SEM image of firecracker-shaped ZnO/polyimide(*PI*) hybrid nanofibers. (**c**) High image of firecracker-shaped ZnO/polyimide (*PI*) nanofibers (Reproduced from Ref. [54] by permission of The Royal Society of Chemistry)

and 4,4′-oxydianilineindimethylacetamide (DMAc) was synthesized by polycondensation. PI nanofibers were prepared via electrospinning a PAA solution followed by thermal imidization in a high-temperature furnace. Then the ZnO nanorods were grown on PI nanofibers via immobilization of ZnO seeds onto PI nanofibers, followed by a hydrothermal process. Finally, ZnO nanorods could be grown from ZnO seeds. Kim et al. successfully prepared cedar leaflike Ag nanostructures [55].

He et al. reported that large-area superhydrophobic titanic membranes with high adhesive forces were fabricated by electrospinning, followed by calcinations and surface modification with 1H, 1H, 2H, 2H-perfluorooctyltriethoxysilane (POTS) [56]. Various surface structures could be realized by combination of electrospinning and hydrothermal treatment. Taking advantage of hydrothermal treatment, rodlike nanoparticles deposited on fibers' surface and the surface showed a further enhanced self-cleaning property with water contact angle of 155°. Hydrothermal treatment provided the possible access to generate hierarchical nanomaterials with reasonable shape and size. However, such hierarchical structures could have better superhydrophobic property but led to worse mechanical integrity. After solving this challenging problem, this technology may find greater use in fabricating robust superhydrophobic materials.

18.4 Functions and Applications of Electrospun Self-Cleaning Materials

Steckl et al. have fabricated photocatalytic self-cleaning fibrous mats by coaxial electrospinning. It is achieved by using cellulose acetate as the core material and nanocrystalline TiO_2 as the sheath materials. As controls, they also fabricated pure cellulose, TiO_2 surface-loaded cellulose nanofibers, cellulose core and TiO_2/cellulose sheath nanofibers, and cellulose core and TiO_2 sheath nanofibers. After electrospinning, they adopted a simple deacetylation step to generate self-cleaning textile fibers. The results showed that the cellulose core and TiO_2 sheath nanofibers had better photocatalytic property than other three kinds of products (Fig. 18.15). This work provided a possibility of creating self-cleaning and environmental-friendly materials by the facile coaxial electrospinning method [57].

Besides self-cleaning property of superhydrophobic surfaces, non-wettable materials are also beneficial for protecting the underneath materials since water is hard to penetrate the superhydrophobic surface. This could be used in areas such as anticorrosion. For example, Jerome et al. fabricated a novel low-free-energy fluorinated diblock copolymer solution onto the aluminum surfaces by electrospinning, which showed excellent corrosion-resistant ability. The diblock copolymer was made of poly(heptadecafluorodecylacrylate-co-acrylic acid) (PFDA-co-AA) and polyacrylonitrile (PAN). This design endowed the copolymer with two functions that the fluorinated block increased the hydrophobicity and the carboxylic acid anchored the polymer film onto the aluminum surface after annealing, while the PAN block made the material possess a good structural stability during annealing. After an electrospinning process, they obtained a highly rough copolymer superhydrophobic coating that was firmly adhered on to an aluminum plate surface. Furthermore, an acetic acid salt spray test showed that the copolymer coating gave a good anticorrosion ability to the metallic plate [58].

In most cases, the electrospun superhydrophobic nanofibers are oleophilic because they are fabricated from hydrophobic but oleophilic raw materials. By using this property, some researchers demonstrated that the superhydrophobic nanofibers can be used to remove oil pollution from water or separate oil/water mixture. Zhao et al. reported such oil absorption materials by electrospinning. They fabricated a series of PS fibers with different diameters as well as porous structures on fibers' surfaces [32]. By investigating the wettability of PS fibrous mats with different morphologies, they found that the fiber surface morphology played important roles

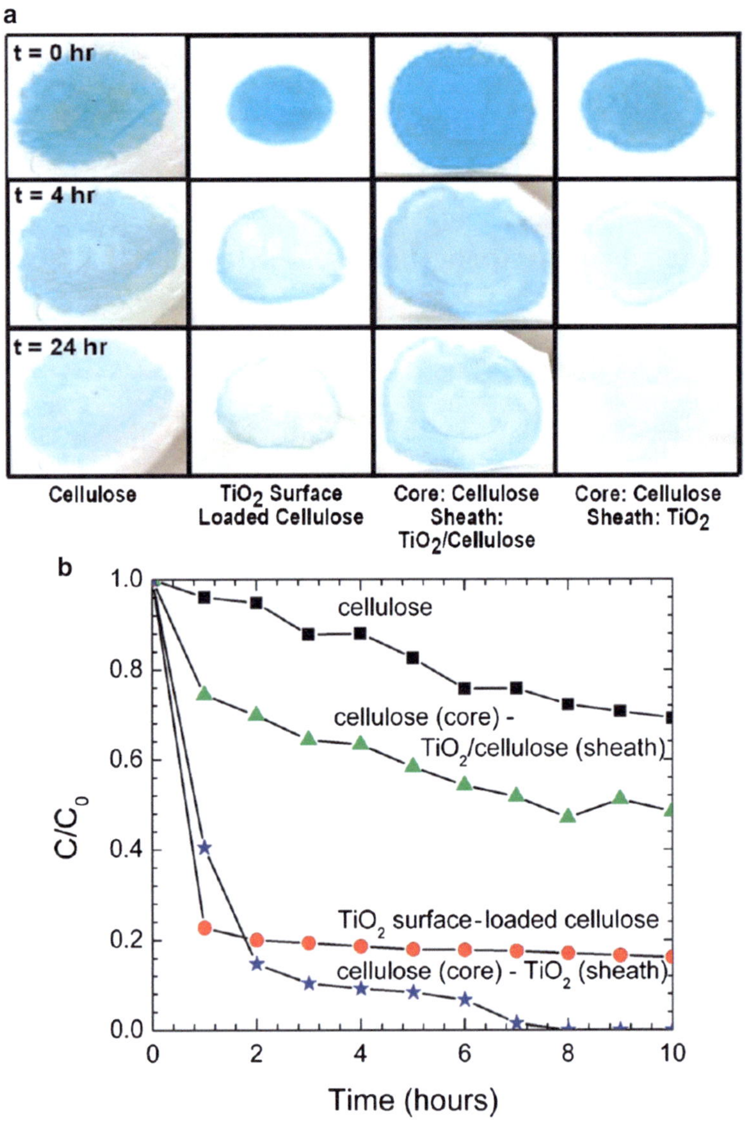

Fig. 18.15 Discoloration of Keyacid Blue (0.1 wt%) stain in electrospun nonwoven mats. (**a**) Exposed to halogen light (~13 mW/cm^2) over a 24 h. (**b**) Optical absorption spectrum exposed to halogen light (~27 mW/cm^2) as a function of time. The results show that the cellulose core and TiO_2 sheath nanofibers' mat exhibited the best photodegradation performance (Reprinted with the permission from Ref. [57]. Copyright 2010 American Chemical Society)

in the wettability of both water and oil. The oil adsorption capacity of PS mat with small fiber diameter and porous surface structure was higher than normal fibrous sorbent without any porous structures. The thinner porous PS oil sorbent also had excellent oil/water selectivity in the cleanup of oil from water. The results showed

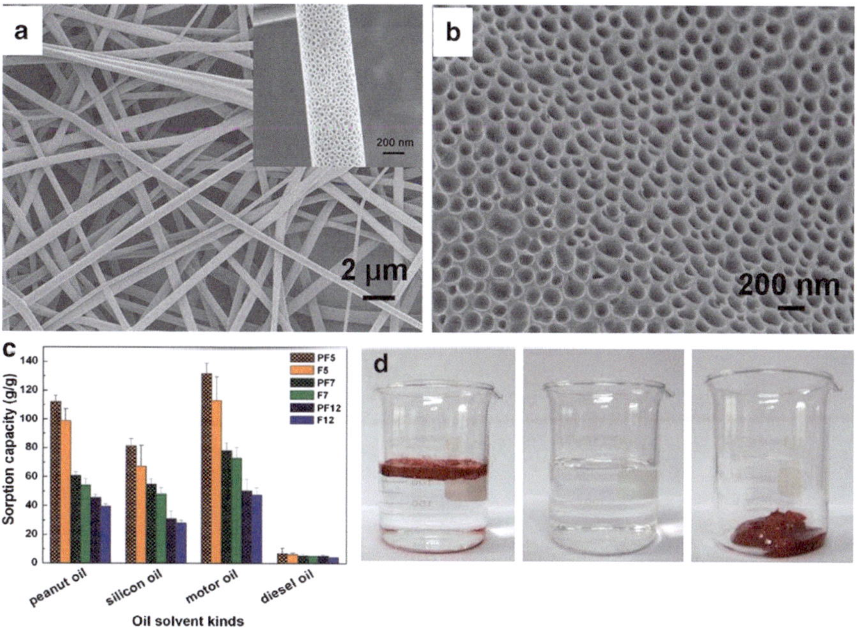

Fig. 18.16 (**a–b**) Electrospun PS nanofibers with porous surface structures (Reprinted with the permission from Ref. [32]. Copyright 2009 American Chemical Society). (**c**) The oil adsorption capacity comparison between smooth nanofibers and porous structured nanofibers with similar diameters; the nanofibers with porous surface structures show better absorption capacity. The adsorption capacity of peanut oil, silicon oil, motor oil, and diesel oil is 112.3, 81.4, 131.6, and 7.13 g/g, respectively. (**d**) Process of removing oil from water by porous nanofibers (Reprinted from Ref. [59], Copyright 2013, with permission from Elsevier)

that oil adsorption capacity of PS oil sorbent film with small diameter and porous surface structure for diesel oil, silicon oil, peanut oil, and motor oil was approximate to 7.13, 81.40, 112.30, and 131.63 g/g, respectively. In other words, the electrospun PS nanofibers with nanoporous surface structures could absorb over 100 times oil than its own weight (Fig. 18.16). This work should be of great significance in removing oil spill pollution that occurred during oil exploration, transportation, and storage [59, 60].

Another usage of superhydrophobic nanofibers is drug release. Recently, Grinstaff et al. fabricated superhydrophobic films from biocompatible materials and used them as controllable drug release systems [61]. They fabricated a novel poly(glycerol monostearate-co-ε-caprolactone) (PGC-C18)-doped poly(ε-caprolactone) (PCL) nanofibrous film by electrospinning. The doping of the PGC-C18 greatly increased the hydrophobicity of the film that enabled the film to be superhydrophobic. The increase of hydrophobicity delayed the rate at which water penetrated into the porous network and displaced entrapped air (Fig. 18.17). It thereby could be used for long-term drug delivery [62].

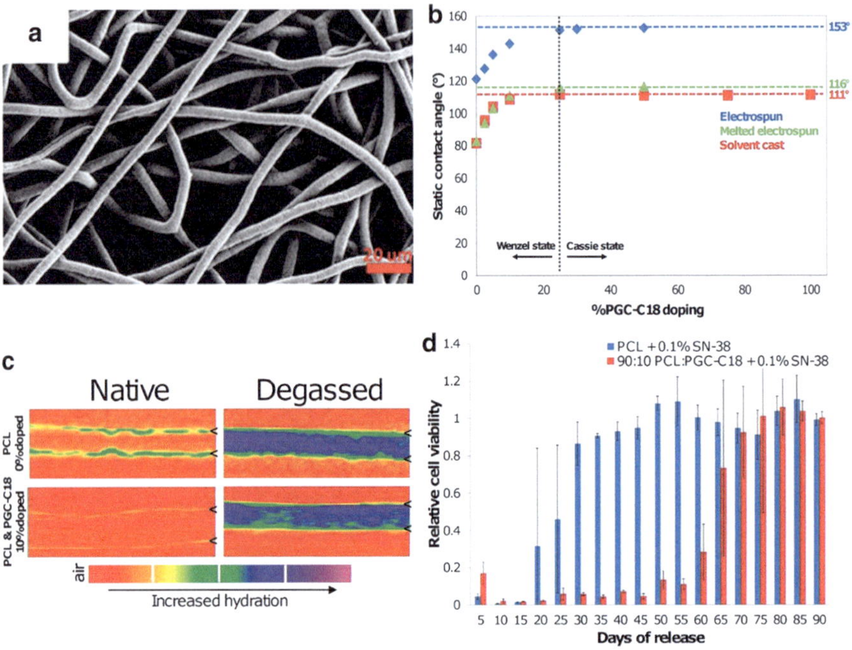

Fig. 18.17 (**a**) PGC-C18-doped PCL microfibers' film by electrospinning. (**b**) The electrospun film shows superhydrophobic property. The microfibers' structure plays an important role in increasing hydrophobicity because either solvent cast film or melted electrospun film cannot reach superhydrophobic. (**c**) CT image of electrospun films after incubation with the contrast agent Hexabrix for 2 h. Degassed meshes show complete water penetration, while native and melted meshes exhibit low surface concentration of water. (**d**) Cytotoxicity test upon incubation of LLC cells with pure PCL meshes and 10 % PGC-C18-doped PCL meshes. The PCL meshes are cytotoxic to LLC cells for 25 days, while the 10 % PGC-C18-doped PCL meshes show cytotoxicity for 65 days (Reprinted with the permission from Ref. [61]. Copyright 2012 American Chemical Society)

18.5 Summary and Outlook

Superhydrophobic self-cleaning materials and electrospun nanofibrous materials are two classes of new materials with "old" history. Natural superhydrophobic surfaces like lotus leaf have been known for thousands of years, while electrospinning is also an old technique invented about 100 years ago. However, both of these subjects were silent for quite a long time until very recent years. The booming of these two areas should thank the flourishment of nanotechnology because nanotechnology provides the powerful approaches for the manufacturing and characterization of nanomaterials. Superhydrophobic self-cleaning materials require a combination of low surface free energy and micro-/nanoscale surface protuberant structures. Coincidently, electrospinning is such a technique that could shape polymer materials (many are hydrophobic materials) into nanofibrous films with large micro-/nanoroughness. Therefore, electrospinning is facilely applied to generate superhydrophobic self-cleaning materials without obstacles. The electrospun superhydrophobic materials have found wide applications such as anticorrosion, oil/water separation, drug release, and others. Although researchers have made great progress in the fabrication of superhydrophobic self-cleaning materials by electrospinning, there still exist some shortages. Firstly, the electrospun products are normally nanofibers, which will form a continuous three-phase contact line with water drop. This is the reason why nanofiber film often has a relatively large water hysteresis that is not favorable for the self-cleaning performance. To conquer this problem, researchers adopted some secondary decoration methods to further increase the roughness of the surface. Nevertheless, such posttreatment inevitably enlarges the inconvenience of the procedure and weakens the one-step advantage of the electrospinning. Secondly, the nanofibers of an electrospun product are naturally stacked in most cases, which means that it is a weak physical contact rather than a strong chemical bond link between nanofibers. So the mechanical property or wear durability of the electrospun film is not good enough. This drawback largely limits the real applications of electrospun products. Thirdly, the productivity of electrospinning technique still needs to be further improved. Therefore, there are still many problems from fundamental research to real production and application to be resolved, which may need multidisciplinary cooperation such as chemistry, materials, physics, engineering, and mechanics. We believe that a number of exciting achievements in electrospun superhydrophobic self-cleaning materials will emerge in the very near future.

References

1. Barthlott W, Neinhuis C (1997) Purity of the sacred lotus, or escape from contamination in biological surfaces. Planta 202(1):1–8. doi:10.1007/s004250050096
2. Neinhuis C, Barthlott W (1997) Characterization and distribution of water-repellent, self-cleaning plant surfaces. Ann Bot 79(6):667–677. doi:10.1006/anbo.1997.0400
3. Feng L, Li S, Li Y, Li H, Zhang L, Zhai J, Song Y, Liu B, Jiang L, Zhu D (2002) Super-hydrophobic surfaces: from natural to artificial. Adv Mater 14(24):1857–1860. doi:10.1002/adma.200290020
4. Marmur A (2004) The lotus effect: superhydrophobicity and metastability. Langmuir 20(9):3517–3519. doi:10.1021/la036369u
5. Patankar NA (2004) Mimicking the lotus effect: influence of double roughness structures and slender pillars. Langmuir 20(19):8209–8213. doi:10.1021/la048629t
6. Lafuma A, Quere D (2003) Superhydrophobic states. Nat Mater 2(7):457–460. doi:10.1038/nmat924
7. Li XM, Reinhoudt D, Crego-Calama M (2007) What do we need for a superhydrophobic surface? A review on the recent progress in the preparation of superhydrophobic surfaces. Chem Soc Rev 36(8):1350–1368. doi:10.1039/b602486f
8. Wang X, Ding B, Yu J, Wang M (2011) Engineering biomimetic superhydrophobic surfaces of electrospun nanomaterials. Nano Today 6(5):510–530. doi:10.1016/j.nantod.2011.08.004
9. Sun TL, Feng L, Gao XF, Jiang L (2005) Bioinspired surfaces with special wettability. Acc Chem Res 38(8):644–652. doi:10.1021/ar040224c

10. Xia F, Jiang L (2008) Bio-inspired, smart, multiscale interfacial materials. Adv Mater 20(15):2842–2858. doi:10.1002/adma.200800836
11. Nishino T, Meguro M, Nakamae K, Matsushita M, Ueda Y (1999) The lowest surface free energy based on $-CF_3$ alignment. Langmuir 15(13):4321–4323. doi:10.1021/la981727s
12. Quéré D (2008) Wetting and roughness. Annu Rev Mater Res 38(1):71–99. doi:10.1146/annurev.matsci.38.060407.132434
13. Erbil HY, Demirel AL, Avci Y, Mert O (2003) Transformation of a simple plastic into a superhydrophobic surface. Science 299(5611):1377–1380. doi:10.1126/science.1078365
14. Tuteja A, Choi W, Ma M, Mabry JM, Mazzella SA, Rutledge GC, McKinley GH, Cohen RE (2007) Designing superoleophobic surfaces. Science 318(5856):1618–1622. doi:10.1126/science.1148326
15. Zhang W, Shi Z, Zhang F, Liu X, Jin J, Jiang L (2013) Superhydrophobic and superoleophilic PVDF membranes for effective separation of water-in-oil emulsions with high flux. Adv Mater 25(14):2071–2076. doi:10.1002/adma.201204520
16. Lu X, Wang C, Wei Y (2009) One-dimensional composite nanomaterials: synthesis by electrospinning and their applications. Small 5(21):2349–2370. doi:10.1002/smll.200900445
17. Ma M, Hill RM, Rutledge GC (2008) A review of recent results on superhydrophobic materials based on micro- and nanofibers. J Adhes Sci Technol 22(15):1799–1817. doi:10.1163/156856108x319980
18. Nuraje N, Khan WS, Lei Y, Ceylan M, Asmatulu R (2013) Superhydrophobic electrospun nanofibers. J Mater Chem A 1(6):1929–1946. doi:10.1039/c2ta00189
19. Lin J, Wang X, Ding B, Yu J, Sun G, Wang M (2012) Biomimicry via electrospinning. Crit Rev Solid State Mater Sci 37(2):94–114. doi:10.1080/10408436.2011.627096
20. Zhao Y, Jiang L (2009) Hollow micro/nanomaterials with multilevel interior structures. Adv Mater 21(36):3621–3638. doi:10.1002/adma.200803645
21. Zhang J, Wang J, Zhao Y, Xu L, Gao X, Zheng Y, Jiang L (2008) How does the leaf margin make the lotus surface dry as the lotus leaf floats on water? Soft Matter 4(11):2232–2237. doi:10.1039/b807857b
22. Jiang L, Zhao Y, Zhai J (2004) A lotus-leaf-like superhydrophobic surface: a porous microsphere/nanofiber composite film prepared by electrohydrodynamics. Angew Chem 116(33):4438–4441. doi:10.1002/ange.200460333
23. Zhu Y, Zhang J, Zheng Y, Huang Z, Feng L, Jiang L (2006) Stable, superhydrophobic, and conductive polyaniline/polystyrene films for corrosive environments. Adv Funct Mater 16(4):568–574. doi:10.1002/adfm.200500624
24. Ma ML, Hill RM, Lowery JL, Fridrikh SV, Rutledge GC (2005) Electrospun poly(styrene-block-dimethylsiloxane) block copolymer fibers exhibiting superhydrophobicity. Langmuir 21(12):5549–5554. doi:10.1021/la047064y
25. Yoon H, Park JH, Kim GH (2010) A superhydrophobic surface fabricated by an electrostatic process. Macromol Rapid Commun 31(16):1435–1439. doi:10.1002/marc.201000131
26. Lin J, Cai Y, Wang X, Ding B, Yu J, Wang M (2011) Fabrication of biomimetic superhydrophobic surfaces inspired by lotus leaf and silver ragwort leaf. Nanoscale 3(3):1258–1262. doi:10.1039/C0NR00812E, 10.1039/C0NR00812E#_blank#Link to landing page via DOI
27. Kang M, Jung R, Kim H-S, Jin H-J (2008) Preparation of superhydrophobic polystyrene membranes by electrospinning. Colloids Surf A Physicochem Eng Asp 313–314:411–414, doi: http://dx.doi.org/10.1016/j.colsurfa.2007.04.122
28. Zhao F, Wang X, Ding B, Lin J, Hu J, Si Y, Yu J, Sun G (2011) Nanoparticle decorated fibrous silica membranes exhibiting biomimetic superhydrophobicity and highly flexible properties. RSC Adv 1(8):1482–1488. doi:10.1039/c1ra00605c
29. Guo Z, Liu W (2007) Biomimic from the superhydrophobic plant leaves in nature: binary structure and unitary structure. Plant Sci 172(6):1103–1112. doi:10.1016/j.plantsci.2007.03.005
30. Yasuhiro M, Bin D, Seimei S (2006) Fabrication of a silver-ragwort-leaf-like superhydrophobic micro/nanoporous fibrous mat surface by electrospinning. Nanotechnology 17(20):5151. doi:10.1088/0957-4484/17/20/019

31. Li X, Ding B, Lin J, Yu J, Sun G (2009) Enhanced mechanical properties of superhydrophobic microfibrous polystyrene mats via polyamide 6 nanofibers. J Phys Chem C 113(47): 20452–20457. doi:10.1021/jp9076933
32. Han D, Steckl AJ (2009) Superhydrophobic and oleophobic fibers by coaxial electrospinning. Langmuir 25(16):9454–9462. doi:10.1021/la900660v
33. Gao X, Jiang L (2004) Biophysics: water-repellent legs of water striders. Nature 432(7013):36. doi:10.1038/432036a
34. Feng XQ, Gao X, Wu Z, Jiang L, Zheng QS (2007) Superior water repellency of water strider legs with hierarchical structures: experiments and analysis. Langmuir 23(9):4892–4896. doi:10.1021/la063039b
35. Xue Y, Wang H, Yu D, Feng L, Dai L, Wang X, Lin T (2009) Superhydrophobic electrospun POSS-PMMA copolymer fibres with highly ordered nanofibrillar and surface structures. ChemCommun (42):6418–6420. doi:10.1039/B911509A, 10.1039/B911509A#_blank#Link to landing page via DOI
36. Wu H, Zhang R, Sun Y, Lin D, Sun Z, Pan W, Downs P (2008) Biomimetic nanofiber patterns with controlled wettability. Soft Matter 4(12):2429–2433. doi:10.1039/B805570J, 10.1039/B805570J#_blank#Link to landing page via DOI
37. Tuteja A, Choi W, Mabry JM, McKinley GH, Cohen RE (2008) Robust omniphobic surfaces. Proc Natl Acad Sci 105(47):18200–18205. doi:10.1073/pnas.0804872105
38. Jeong S, Lee Y, Lee J, Shin Y, Shin H, Lim Y, Nho Y (2008) Preparation and characterization of temperature-sensitive poly(N-isopropylacrylamide)-g-poly(L-lactide-co-ε-caprolactone) nanofibers. Macromol Res 16(2):139–148. doi:10.1007/bf03218843
39. Tian Y, Chen C-Y, Yip H-L, Wu W-C, Chen W-C, Jen AKY (2009) Synthesis, nanostructure, functionality, and application of polyfluorene-block-poly(N-isopropylacrylamide)s. Macromolecules 43(1):282–291. doi:10.1021/ma9019619
40. Wang N, Zhao Y, Jiang L (2008) Low-cost, thermoresponsive wettability of surfaces: poly(*N*-isopropylacrylamide)/polystyrene composite films prepared by electrospinning. Macromol Rapid Commun 29(6):485–489. doi:10.1002/marc.200700785
41. Muthiah P, Hoppe SM, Boyle TJ, Sigmund W (2011) Thermally tunable surface wettability of electrospun fiber mats: polystyrene/poly(N-isopropylacrylamide) blended versus crosslinked poly[(N-isopropylacrylamide)-co-(methacrylic acid)]. Macromol Rapid Commun 32(21):1716–1721. doi:10.1002/marc.201100373
42. Webster M, Miao J, Lynch B, Green DS, Jones-Sawyer R, Linhardt RJ, Mendenhall J (2012) Tunable thermo-responsive poly(N-vinylcaprolactam) cellulose nanofibers: synthesis, characterization, and fabrication. Macromol Mater Eng 298(4):447–453. doi:10.1002/mame.201200081
43. Chen M, Besenbacher F (2011) Light-driven wettability changes on a photoresponsive electrospun mat. ACS Nano 5(2):1549–1555. doi:10.1021/nn103577g
44. Chen M, Dong M, Havelund R, Regina VR, Meyer RL, Besenbacher F, Kingshott P (2010) Thermo-responsive core − sheath electrospun nanofibers from poly (N-isopropylacrylamide)/polycaprolactone blends. Chem Mater 22(14):4214–4221. doi:10.1021/cm100753r
45. Wu J, Wang N, Wang L, Dong H, Zhao Y, Jiang L (2012) Electrospun porous structure fibrous film with high oil adsorption capacity. ACS Appl Mater Interfaces 4(6):3207–3212. doi:10.1021/am300544d
46. Cho WK, Choi IS (2008) Fabrication of hairy polymeric films inspired by geckos: wetting and high adhesion properties. Adv Funct Mater 18(7):1089–1096. doi:10.1002/adfm.200701454
47. Winkleman A, Gotesman G, Yoffe A, Naaman R (2008) Immobilizing a drop of water: fabricating highly hydrophobic surfaces that pin water droplets. Nano Lett 8(4):1241–1245. doi:10.1021/nl080317d
48. Zhu S, Li Y, Zhang J, Lü C, Dai X, Jia F, Gao H, Yang B (2010) Biomimetic polyimide nanotube arrays with slippery or sticky superhydrophobicity. J Colloid Interface Sci 344(2):541–546, doi: http://dx.doi.org/10.1016/j.jcis.2009.12.047

49. Gong G, Wu J, Liu J, Sun N, Zhao Y, Jiang L (2012) Bio-inspired adhesive superhydrophobic polyimide mat with high thermal stability. J Mater Chem 22:8257–8262. doi:10.1039/c2jm16503a
50. Ma ML, Mao Y, Gupta M, Gleason KK, Rutledge GC (2005) Superhydrophobic fabrics produced by electrospinning and chemical vapor deposition. Macromolecules 38(23): 9742–9748. doi:10.1021/ma0511189
51. Pisuchpen T, Chaim-ngoen N, Intasanta N, Supaphol P, Hoven VP (2011) Tuning hydrophobicity and water adhesion by electrospinning and silanization. Langmuir 27(7):3654–3661. doi:10.1021/la104978e
52. Lim J-M, Yi G-R, Moon JH, Heo C-J, Yang S-M (2007) Superhydrophobic films of electrospun fibers with multiple-scale surface morphology. Langmuir 23(15):7981–7989. doi:10.1021/la700392w
53. Pant HR, Pandeya DR, Nam KT, W-I B, Hong ST, Kim HY (2011) Photocatalytic and antibacterial properties of a TiO2/nylon-6 electrospun nanocomposite mat containing silver nanoparticles. J Hazard Mater 189(1–2):465–471. doi:10.1016/j.jhazmat.2011.02.062
54. Chang Z (2011) "Firecracker-shaped" ZnO/polyimide hybrid nanofibers via electrospinning and hydrothermal process. Chem Commun 47(15):4427–4429. doi:10.1039/c0cc05634k
55. Sheikh FA, Barakat NAM, Kanjwal MA, Chaudhari AA, Jung IH, Lee JH, Kim HY (2009) Electrospun antimicrobial polyurethane nanofibers containing silver nanoparticles for biotechnological applications. Macromol Res 17(9):688–696. doi:10.1007/BF03218929
56. Tang H, Wang H, He J (2009) Superhydrophobic titania membranes of different adhesive forces fabricated by electrospinning. J Phys Chem C 113(32):14220–14224. doi:10.1021/jp904221f
57. Bedford NM, Steckl AJ (2010) Photocatalytic self cleaning textile fibers by coaxial electrospinning. ACS Appl Mater Interfaces 2(8):2448–2455. doi:10.1021/am1005089
58. Grignard B, Vaillant A, de Coninck J, Piens M, Jonas AM, Detrembleur C, Jerome C (2011) Electrospinning of a functional perfluorinated block copolymer as a powerful route for imparting superhydrophobicity and corrosion resistance to aluminum substrates. Langmuir 27(1):335–342. doi:10.1021/la102808w
59. Viswanadam G, Chase GG (2013) Water-diesel secondary dispersion separation using superhydrophobic tubes of nanofibers. Sep Purif Technol 104:81–88. doi:10.1016/j.seppur.2012.11.020
60. Wang L, Yang S, Wang J, Wang C, Chen L (2011) Fabrication of superhydrophobic TPU film for oil–water separation based on electrospinning route. Mater Lett 65(5):869–872, doi: http://dx.doi.org/10.1016/j.matlet.2010.12.024
61. Yohe ST, Colson YL, Grinstaff MW (2012) Superhydrophobic materials for tunable drug release: using displacement of air to control delivery rates. J Am Chem Soc 134(4):2016–2019. doi:10.1021/ja211148a
62. Yohe ST, Herrera VLM, Colson YL, Grinstaff MW (2012) 3D superhydrophobic electrospun meshes as reinforcement materials for sustained local drug delivery against colorectal cancer cells. J Control Release 162(1):92–101. doi:10.1016/j.jconrel.2012.05.047

Chapter 19
Adsorbents Based on Electrospun Nanofibers

Xueqin Wang, Jianlong Ge, Yang Si, and Bin Ding

Abstract In the past few decades, removing or controlling the concentration levels of the pollutants including toxic gases, heavy metal ions, and organic contaminants in environmental systems has attracted tremendous attention. Among the numerous approaches, adsorption is considered to be one of the most versatile and promising approach in removing pollutants due to its convenience, ease of operation, simplicity of design, and universal in use. Electrospun nanofibers have unique properties such as large surface area, tailored pore structure, high porosity, and flexibility of surface functionalization, therefore could be used as advanced adsorbents for contaminant removal. Importantly, nanofiber-based adsorbents are expected to possess strong adsorption capacity, fast kinetics, and good reproducibility due to the unique structure of nanofibers. In this chapter, we summarize recent progress in the development of electrospun nanofibrous membrane-based adsorbent for the removal of toxic gases and pollutants in aqueous solution (heavy metal ions and organic contaminants), describe the design of the nanofibrous materials, and discuss their adsorption performance in detail. This chapter might trigger further development and evolution of adsorption based on electrospun nanofibers as one potential to ease the environmental pollution problem.

X. Wang
State Key Laboratory for Modification of Chemical Fibers and Polymer Materials,
College of Materials Science and Engineering, Donghua University, Shanghai 201620, China

J. Ge
Nanomaterials Research Center, Modern Textile Institute, Donghua University,
Shanghai 200051, China

Y. Si • B. Ding (✉)
State Key Laboratory for Modification of Chemical Fibers and Polymer Materials,
College of Materials Science and Engineering, Donghua University,
Shanghai 201620, China

Nanomaterials Research Center, Modern Textile Institute, Donghua University,
Shanghai 200051, China
e-mail: binding@dhu.edu.cn

B. Ding and J. Yu (eds.), *Electrospun Nanofibers for Energy and Environmental Applications*, Nanostructure Science and Technology,
DOI 10.1007/978-3-642-54160-5_19,

19.1 Introduction

Till recent years, the surge of industrial, agricultural, and domestic activities has intensified more environmental problems as seen, for example, in the deterioration of several ecosystems due to the accumulation of dangerous pollutants such as heavy metal ions (Hg(II), Pb(II), Cu(II), Cr(IV), etc.), organic contaminant (dyes, pesticides, etc.), and toxic gases (formaldehyde, benzene, NO_x, SO_x, etc.) [1–3]. Such pollutants pose a great threat to the environment and human health; it can cause a broad range of adverse effects, including respiratory system damage, immune system disorder, cardiovascular system disease, or even cancer [4, 5]. Hence, it is imperative to remove or control the concentration levels of these contaminants in environmental systems.

Up to now, various conventional methods including precipitation, filtration, ion exchange, flotation, and adsorption have been used to remove the environmental contaminant [6, 7]. Among the various methods, adsorption is considered to be one of the most versatile and promising approaches for the reasons of convenience, ease of operation, simplicity of design, and universal in use [8–10]. Several materials such as activated carbon, zeolite, silica gel, activated alumina, and metal organic framework have been used for pollution control [11–13]. It is worth noting that high efficiency, fast adsorption kinetics, and reusability are of great importance to adsorbents. However, zeolites, silica, and metal organic framework often are limited in their widespread utilization because of structural instability and inefficiency [14–16]. In addition, the adsorption capacity of these materials is remarkably diminished in humid conditions or after regeneration [17, 18]. Therefore, researchers have carried out investigation for new promising adsorbents with high adsorption capacity, fast adsorption kinetics, and good reusability and which are suitable to operate in different conditions (temperature, humidity, pressure, etc.).

As the development of the nanoscience and nanotechnology, huge amounts of nanomaterials have been prepared and used as adsorbents because of their large surface area, including nanoparticles, nanotubes, and nanofibers [19, 20]. Porous nanofibers have evoked extensive research as adsorbents to remove toxic contaminants due to its large surface area, high porosity, good interconnectivity, and flexibility of surface modification with functional groups (carboxyl, hydroxyl, mercapto groups) in the latest years [21]. A number of processing techniques such as template synthesis, phase separation, self-assembly, and electrospinning have been used to prepare polymer nanofibers [22, 23]. Among the various approaches, electrospinning appears to be the most straightforward and versatile technique for generating nanofibers with diameters ranging from several nanometers to a few micrometers facilely, effectively, and economically [24]. Numerous electrospun nanofibers have been fabricated to adsorb environmental contaminants such as activated carbon nanofibers (ACNFs), SiO_2, polyamide-6 (PA-6), poly(vinyl alcohol) (PVA), polyacrylonitrile (PAN), poly(vinylpyrrolidone) (PVP), poly(ethylene oxide) (PEO), and poly(acrylic acid) (PAA) (listed in Table 19.1), and we focus on adsorbents based on the electrospun nanofibers in this chapter.

Table 19.1 The application of electrospun nanofibers for environmental contaminant adsorption

Reference	q_m[a] (mg g^{-1})	Extraction time (h)	C_0[b] (mg L^{-1})	Pore size (nm)	S_{BET}[c] (m2 g^{-1})	Diameter (nm)	Absorbent	Adsorbate
[25]	–	–[d]	0.0135	1	802	800	ACNF	Formaldehyde
[26]	108	10.5	–	2.3	1,193	250	ACNF	Benzene
[27]	1	10	0.31	0.8	2,201	380	ACNF	NO
[28]	10.3	–	–	0.6–0.7	2,420	80–250	ACNF	H_2
[29]	66.7	0.5	20	–	–	600–800	PVA/Fe^{3+}	As(III)
[30]	263.4	96	400	–	–	100–300	PAN-oxime	Pb(II)
[31]	15.8	18	20	3.1	248	226	TiO_2/ZrO_2	Pb(II)
[32]	18.6	2	64	3.4	120	1,000	Thiol-SiO_2	Cu(II)
[33]	12.8	5	20	5.7	148	90–105	m-TiO_2	Cu(II)
[34]	504.9	0.7	640	7.13	414	300–500	Mercapto-PVA/SiO_2	Cu(II)
[35]	215.2	72	1	–	–	–	Amidoxime-PAN	Cu(II)
[36]	97	0.3	100	2.86	873	100–700	Amino-PVP/SiO_2	Cr(III)
[37]	6.8	1	10	–	–	200–600	Al_2O_3	Cr(VI)
[38]	854.6	3	20	2.79	218	200–300	Thioether-PVP/SiO_2	Hg(II)
[39]	57.5	0.5	–	–	–	–	SiO_2	Hg(II)
[40]	63.7	1	20	5.2	9.6	200–500	Fe_2O_3/Al_2O_3	Hg(II)
[41]	84	–	21.6	–	5.29	300–500	Thiol/amine-PVA	Au
[42]	357.1	2	100	4.12	312	98	PEO/Chitosan	Ni(II)
[43]	493	0.25	700	0.78, 4.26	1,642	525	ACNF	Acid red 1
[44]	495	0.75	100	2.9	356	200–300	Cyclodextrin-PVA/SiO_2	Indigo carmine
[45]	523.1	1.5	141	3.9	514.9	400–800	PAA/SiO_2	Indigo carmine
[46]	161.6	3	2,000	–	1,080	200–400	Cibacron blue/PAN	Bromelain

[a] q_m: the maximum adsorption capacity
[b] C_0: the initial concentration of the adsorbate
[c] S_{BET}: Brunauer-Emmett-Teller (BET) surface area
[d] –: not provided

19.2 Adsorbents Based on Electrospun Nanofibrous Membranes

Adsorbents based on electrospun nanofibrous membranes offer unique properties such as large surface area, tailored pore structure, flexibility of surface functionalization, and self-standing, which make them ideal candidates to pollutant adsorption. The large surface area of the fibrous membranes provides abundant active sites for materials to adsorb and consequently improve the absorption capacity, while the tailored pore structures lead the nanofibrous membranes to adsorb different materials selectively. The functionalized nanofibers could adsorb materials selectively depending on the specific force between adsorbent and adsorbate; in addition, the nanofibrous membrane-based adsorbents are easy to recycle and would not cause secondary pollution which is much more suitable than nanoparticles. In the following part, we will mainly introduce two catalogs of adsorbents: toxic gases and contaminants in liquid (heavy metal ions, organic contaminants), which are classified according to the adsorbates in different environments.

19.2.1 Adsorption of Toxic Gases

Air pollution has recently become a major concern, especially related with human health. Air pollutants such as SO_x, NO_x, and volatile organic compounds (VOCs) cause adverse health impacts on occupants, leading to sick building syndrome, immune system damage, neurological disorder, or even lung cancer. Due to the acute toxicity and volatility of these pollutants, it is strongly required to remove and sense this pollution. Several methods have been developed, including adsorption, filtration, and thermal oxidation, while adsorption is the preferred technique for controlling gaseous contaminants at low concentration levels because it is efficiency, economical, and easy-to-use.

ACNFs have been widely used as gas adsorbents due to the abundant micropores, large surface area, and excellent adsorption capacity [47–49]. The forces between gas molecules and adsorbents are the attractive dispersion force and the short-range repulsive forces; in addition, specific molecular interactions (polarization, field dipole, field gradient quadrupole) usually occur as a result of particular geometric and electronic properties of the adsorbent and adsorptive [50, 51]; through these forces, the gases could be adsorbed onto the adsorbents. The nitrogen isotherms of ACNF obtained from the experimental data are commonly identified as Type I on the basis of IUPAC classification that represents the long plateau, indicating the microporous adsorbents [52]. The adsorption properties of such materials could be explained by the following reasons: firstly, the primary filling in very narrow micropores; secondly, the formation of a monolayer in wider micropores and small mesopores; and finally, the filling of the larger micropores and mesopores by

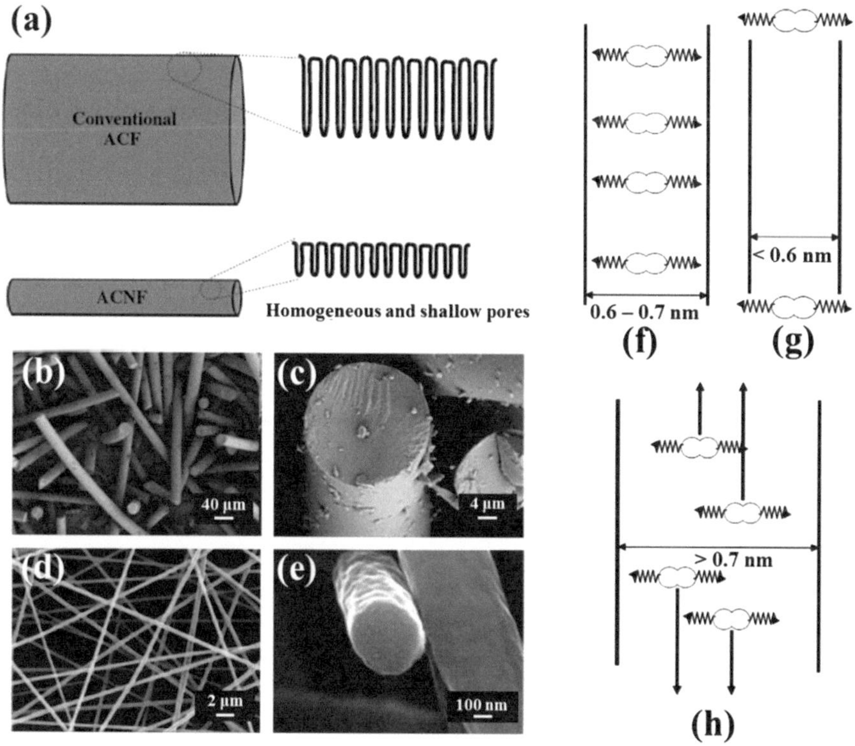

Fig. 19.1 (**a**) Pore structure of conventional thick ACF and thin ACNF. Low-magnification and cross-sectional FE-SEM image of (**b** and **c**) ACFs and (**d** and **e**) ACNF. The mechanism of hydrogen adsorption: (**f**) 0.6 nm < pore width < 0.7 nm; (**g**) pore width < 0.6 nm; and (**h**) pore width > 0.7 nm ((**a**) Reprinted with permission from Ref. [25]. © 2010 Elsevier. (**b**–**e**) Reprinted with permission from Ref. [26]. © 2006 Wiley Periodicals, Inc. (**f**–**h**) Reprinted with permission from Ref. [28]. © 2008 Elsevier)

capillary condensation process [53]. Extensive researches have focused on studying the influence factors on adsorption and promoting the adsorption capacity of the ACNF. By carbonizing and steam activating the electrospun PAN nanofibers (uniform diameter of 800 nm), ACNF with tailored microporosity and abundant nitrogen-containing functional groups was produced. The as-prepared ACNF demonstrated a much higher adsorption amount of formaldehyde in low concentration (11 ppm) which is almost twice as much for conventional activated carbon fibers (ACFs) in the dry conditions, even though the BET surface areas of two samples are similar. The higher adsorption capacity of ACNF might be explained as follows: firstly, the much more nitrogen content of ACNF could offer more available adsorption sites for formaldehyde compared with ACFs even in humid conditions; secondly, the shallower micropore of ACNF (Fig. 19.1a) would not only enable faster and easier diffusions of formaldehyde molecules to adsorption sites but also the water molecules from adsorption sites to the outside [25]. On the other hand, the smaller diameters of ACNF (about 100 times smaller than ACFs) directly related to the gas adsorption kinetics, and the morphologies of the ACNF and ACFs were characterized by the field emission scan electronic microscopy (FE-SEM), shown in Fig. 19.1b–e. Shim et al. [26] selected well-known homogeneous surface diffusion model to describe the adsorption of benzene vapor since the highly favorable isotherms are obtained from adsorption equilibrium data. The results showed that the surface diffusivity of ACNF is one order of magnitude greater than that of ACFs. Moreover, it should be noted that the pore structure of the ACNF is one of the main factors that affect the adsorption capacity; researchers have made effort to study the effect of pore structure on ACNF for gas adsorption [54, 55]. Through the Monte Carlo simulation [56], Im et al. [28] calculated that the appropriate pore width of hydrogen adsorption was suggested as 0.6 nm, and the hydrogen adsorption mechanism was schematically illustrated as shown in Fig. 19.1f–h. The results revealed that the dynamic hydrogen molecule could be trapped properly in case of pore width 0.6–0.7 nm considering the diameter of hydrogen molecule (0.4059 nm). The experiment also confirmed that through regulating the pore structure, the highest breakthrough of hydrogen adsorption capacity was in the range of pore width from 0.6 to 0.7 nm as it is predicted by Monte Carlo simulation, and the pore volume was the most effective factor which can attract the hydrogen adsorption capacity positively. Song et al. [27] synthesized the ultrafine porous carbon fibers by chemical activation of electrospun PAN nanofibers and used it to adsorb NO. However, through regulating the process parameter, the pore size could only be centered at 0.8 nm which was much broader than the NO molecule (0.124 nm) and consequently made the adsorption amount relatively low (but adsorptive properties are still better than the commercial).

Besides regulating the ACNF structure, researchers have attempted to prepare the hierarchal fabrics to improve the gas adsorption capacity. Katepalli et al. [57] synthesized a set of hierarchal fabrics: preoxidized electrospun PAN nanofibers on ACF (ACF-PANS), PAN ACNF on ACF (ACF-PANC), and carbon nanofibers on metal-impregnated ACF (ACF-CNF). The adsorption experiments implied that the stabilized PAN nanofiber-based hierarchical fabric (ACF-PANS) showed a superior performance (higher adsorption amount and faster adsorption speed) to SO_2, NO_x, and toluene attributed to the larger surface area, higher porosity, and the presence of nitrogen-based functional groups.

In brief, the tailored pore structure of electrospun nanofibers endows the adsorbents the ability to sorb different adsorbates (formaldehyde, SO_x, benzene, hydrogen, etc.) in relatively high speed, and the large surface area ensures the high gas adsorption capacity. In further researches, the adsorption capacity, especially in humid conditions, should be promoted and the adsorb selectivity should be improved. Moving forward, electrospun nanofibers could be potentially used to control and remove toxic gases for industry and potential inhabitant.

19.2.2 Adsorption of Contaminants in Water

19.2.2.1 Heavy Metal Ions

Nowadays, the water pollution caused by the indiscriminate disposal of heavy metal ions has become one of the most serious and widespread consequences. Due to the toxicity, accumulation in the food chain, and persistence in nature, heavy metal ions have been posing a significant threat to the environment and public health. If the presence of heavy metal ions is beyond a certain limit, it could bring serious hazards to living organisms such as Wilson disease, chronic asthma, kidney damage, and Minamata disease [58–60]. The removal of these toxic heavy metal ions from aqueous solutions has been traditionally carried out by chemical precipitation, ion exchange, reverse osmosis, membrane separation, adsorption processes, etc. [22, 61]. Among the various methods, adsorption technology is one of the most popular methods to remove these heavy metal ions. Yet, the adsorption efficiency, selectivity, equilibrium time, regeneration, and stability usually depend on the material characterization of the adsorbents. Therefore, it is necessary to develop new efficient materials to meet the requirement for removal of heavy metal ions from wastewater.

Compared with the traditional adsorbent such as activated carbon, zeolites, and silica gel, electrospun nanofibers are good candidates for heavy metal ion adsorption due to its large surface area, tailored pore structure, good interconnectivity of pores, and potential to incorporate active chemistry or functionality on nanoscale [62, 63]. Moreover, recycle is of great importance in the field of water treatment; taking this aspect into consideration, the nanofiber-based adsorbents are more suitable compared with powdered nanoporous materials.

Over the recent years, researchers have investigated the adsorption capacity of various nanofibers and explored approaches to improve the adsorption amount, equilibrium speed, selectivity, etc. To increase the adsorption of heavy metals on nanofiber materials, the following methods have attracted the most attention: (1) utilizing polymers which contain a molecular functional group as resources to electrospinning, such as cellulose acetate (CA), silk fibroin (SF), wool keratin (WK), and PA-6 [15, 64–67]; (2) surface modification with functional group (the removal efficiency of heavy metal ions increases significantly after the adsorbents have been modified by functional group of $–NH_2$, –SH, $–SO_3H$, $–COO^-$, etc.) [35, 62, 68]; and (3) increasing the surface areas of nanofibers [31].

The preparation of the natural polymeric nanofibers has been intensively studied due to their abundance in nature, cheap availability, and potential industrial applications such as separation, scaffolding materials, and adsorption. Ki et al. [69] investigated the adsorption performance of electrospun WK/SF blend nanofibrous membranes toward heavy metal ions (Cu(II) as a model). The results showed that the WK/SF blend nanofibrous membranes exhibited a higher adsorption capacity (2.88 mg g^{-1}) than SF membranes (1.65 mg g^{-1}) with similar specific surface area and porosity. The rise of the adsorption can be explained by a different chemical composition: WK contain more cysteine (can be oxidized to $–SO_3H$) and hydrophilic amino acid (has polar side residues) in the side or terminal groups. Furthermore, the adsorption capacity was steadily maintained above 90 % after six recycles, and the nanofibrous membranes showed much higher Cu(II) adsorption due to the extremely large specific surface area of the nanofibrous membranes (about 50 times higher than wool silver and filter paper) and the ionizable groups which are the binding sites for the positive metal ions. Chitosan (CS) is one of the most abundant and promising classes of the functional natural polymeric materials having special characteristics, such as hydrophilicity and nontoxic for use in wastewater treatment. The electrospun CS neutralized with potassium carbonate were found effective in adsorption of Cu(II) and Pb(II); the equilibrium adsorption capacities for Cu(II) and Pb(II) were 485.44 mg g^{-1} and 263.15 mg g^{-1}, respectively [70]. The adsorption equilibrium data is analyzed with the Langmuir isotherm (Eq. 19.1) [71], in which the saturation capacity is supposed to coincide with saturation of a fixed number of identical surface sites:

$$q_e = \frac{K_L q_L C_e}{1 + K_L C_e} \tag{19.1}$$

where q_e is the equilibrium quantity of the metal ions adsorbed onto the chitosan nanofiber mat (mg g^{-1}), C_e is the equilibrium concentration (mg L^{-1}), and q_L and K_L (L mg^{-1}) are the Langmuir constants related to the saturation adsorption capacity and binding energy (affinity), respectively.

In the past few decades, heavy metal ion adsorbents based on natural polymeric nanofibers developed fast, but further study should be intensively carried out to improve the adsorption capacity and simplified the pretreatment process.

Besides natural polymeric nanofibers, porous nanofibers modified by the functional groups have attracted much attention because of their properties such as (1) high surface-to-volume ratio, (2) uniform pore size distribution, (3) convenient recycling, and (4) high equilibrium adsorption capacities of heavy metal ions. So porous nanofibrous membranes are a kind of high-efficiency adsorbent in the adsorption of heavy metal ions; the mechanism of the adsorption and desorption [72] is shown in Fig. 19.2a. Stephen and coworkers [73] utilized oxolane-2,5-dione-functionalized electrospun cellulose nanofibers (deacetylation of CA) as adsorbent for the adsorption of Cd(II) and Pb(II) from wastewater samples. The experimental data displayed that adsorption capacities for functionalized cellulose nanofibers are 207.2 and 327.1 mg g^{-1} in 120 and 60 min for Pb(II) and Cd(II), respectively, compared to 0.41 and 0.22 mg g^{-1} for raw cellulose. This might be caused by the chelation and entrapment of metal ions on the ligand as well as the pores on the surface of the functionalized nanofibrous cellulose membranes. Moreover, the metal ion adsorption uptake data is in great agreement with the Freundlich adsorption isotherm with a correlation coefficient (R^2) of 0.99 [74]. Unlike the Langmuir isotherm as mentioned above, the Freundlich isotherm model is an empirical equation (Eq. 19.2) that has been widely used for heterogeneous surface energies:

$$q_e = K_F C_e^{\frac{1}{n}} \tag{19.2}$$

where q_e is the amount adsorbed, K_F and n are the Freundlich constants related to the adsorption capacity, and C_e is the equilibrium concentration. Wu et al. [72] synthesized thiol-functionalized mesoporous PVA/SiO_2 composite nanofibrous membranes by electrospinning, the as-spun membranes owned the adsorption capacity of 481.9 mg g^{-1} for Cu(II), and the experimental data was fitted well with the Redlich-Peterson models which are another widely used isotherm, expressed as Eq. 19.3:

$$q_e = \frac{K_R C_e}{1 + \alpha_R C_e^{\beta}} \tag{19.3}$$

where K_R and α_R are the Redlich-Peterson isotherm constant and β is the exponent that lies between 0 and 1. Wu et al. [75] prepared the thiol-functionalized polystyrene (PS)/SiO_2 nanofibrous membranes based on the combination of the electrospinning and evaporation-induced self-assembly mesopore-forming technique for Cu(II) adsorption. The adsorption capacity of membranes was 16.3 mg g^{-1}. However, the adsorption capacity could be estimated as up to 65.12 mg g^{-1} by adding the mesoporous coating layers. Moreover, the Langmuir, Freundlich, and Redlich-Peterson equations are used to analyze the experimental data (Fig. 19.2b). It demonstrated that the adsorption data fitted the Freundlich isotherm best with an R^2 of 0.9995, followed by Redlich-Peterson isotherm, which indicated that the Freundlich isotherm is suitable for heterogeneous adsorption, while Langmuir isotherm is suitable for monolayer adsorption, and Redlich-Peterson isotherm incorporated the characteristics of Langmuir and Freundlich isotherm. Importantly, in different testing systems (different heavy metal ions and different conditions), we should use different isotherms to analyze the adsorption capacities of the adsorbents so that it can suitably fit the isotherm [76, 77].

Abbasizadeh and coworkers [78] synthesized a PVA/TiO_2 nanofiber adsorbent modified with mercapto groups by electrospinning for the removal of U(VI) and Th(IV) from water. Due to the large surface area, uniform surface, and regular pore structure of the nanofibrous membranes, the maximum sorption capacities of U(VI) and Th(IV) by Langmuir isotherm reached 196.1 and 238.1 mg g^{-1} under optimized conditions. Compared with the Freundlich and Dubinin-Radushkevich isotherms, Langmuir isotherm fits the experimental data better; this might be caused by the homogeneous distribution of active sites on mercapto-functionalized PVA/TiO_2 nanofibrous membranes. Meanwhile, the experimental data showed the sorption kinetics are fitting well with the double exponential kinetics, indicated that the adsorption is physisorption and the sorption process occurs in two steps:

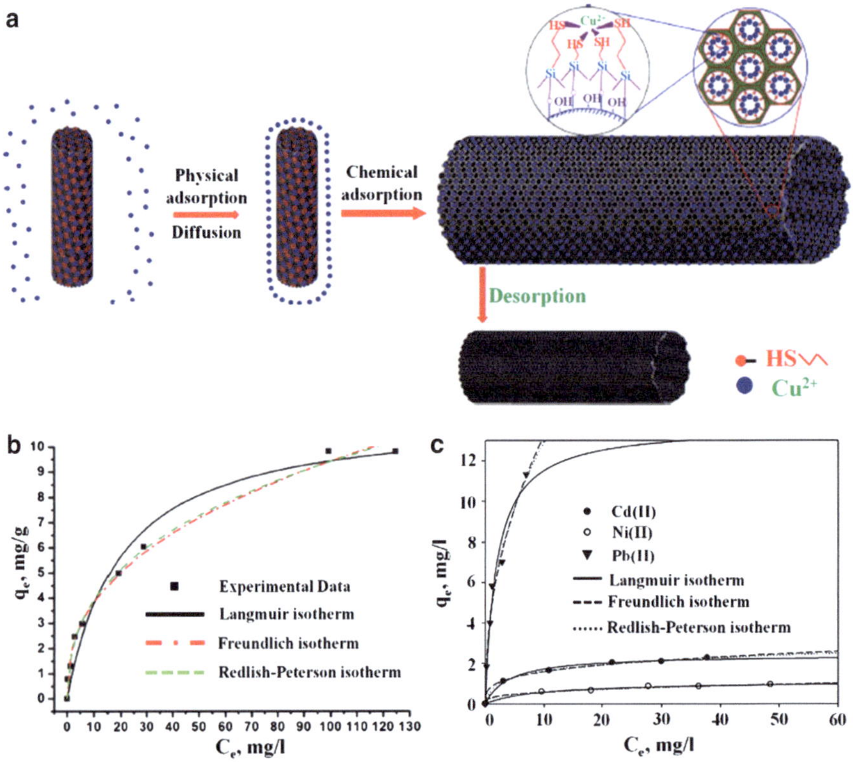

Fig. 19.2 (**a**) The mechanism of the adsorption of Cu(II) on PVA/SiO_2 composite nanofibers. (**b**) Adsorption isotherm of Cu(II) on thiol-functionalized PS/SiO_2 membranes and its fitting curves by Langmuir, Freundlich, and Redlich-Peterson isotherms. (**c**) Langmuir, Freundlich, and Redlich-Peterson isotherms for the adsorption of different heavy metal ions ((**a**) Reprinted with permission from Ref. [74]. © 2010 Elsevier. (**b**) Reprinted with permission from Ref. [75]. © 2012 The Royal Society of Chemistry. (**c**) Reprinted with permission from Ref. [77]. © 2013 Elsevier)

first, involving the external diffusion (transport of the metal ion to the external surface of the adsorbent) which is a rapid phase, while the second step involves the pore diffusion (sorption of the metal ion on the interior surface of the adsorbent) which is a slow phase. In order to improve the selectivity of the adsorbent, Wang et al. [79] prepared naphthalimide-functionalized nanofibrous membranes by copolymerization and electrospinning and then used it to adsorb Cu(II). The as-prepared fibrous membranes obtained an adsorption capacity of 10.39 mg g^{-1} (Langmuir isotherm) toward Cu(II) with negligible effect on other metal ions in aqueous solution.

Furthermore, scattered researches have been focused on immobilized metallic compound on nanofibers' surface; the metallic compounds are hydrolyzed during soaking into the wastewater and hence adsorb the negative heavy metal ion group such as $Cr_2O_7^{2-}$, $HCrO_4^-$, and $HgCl_2^-$ (Fig. 19.3a). For example, Lin et al. [80]

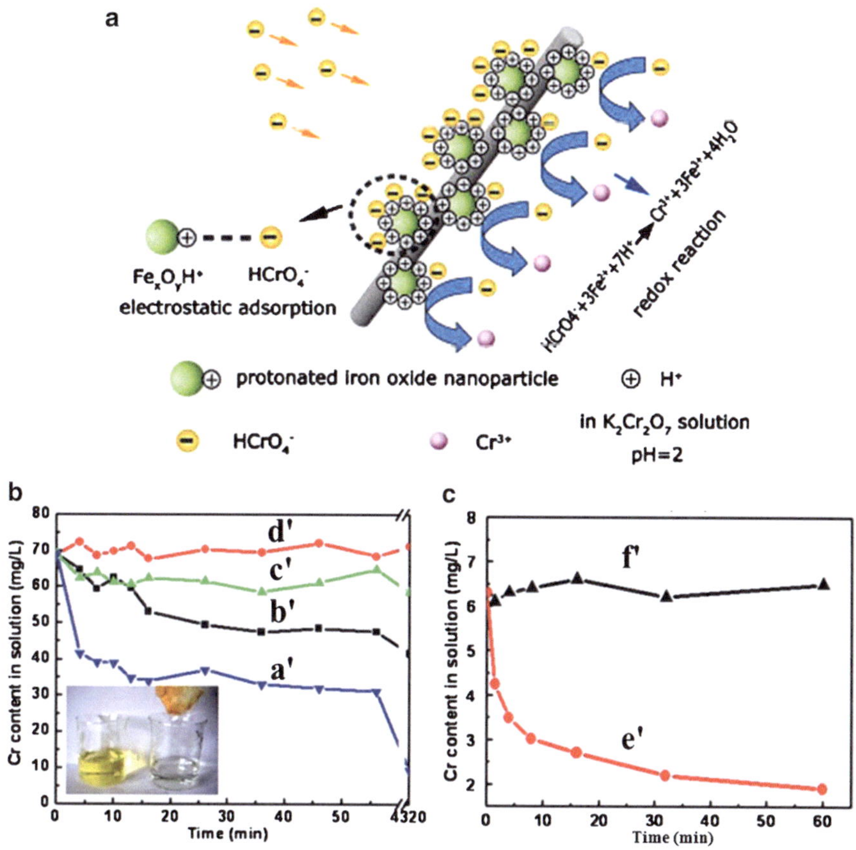

Fig. 19.3 (**a**) Schematic illustration for Cr(VI) removal mechanism. (**b**) The Cr(VI) removal measurement with Cr contents of 69.1 mg L^{-1} after the addition of different materials: (*a′*) PAN/$FeCl_2$ nanofibers, (*b′*) cast films of PAN/$FeCl_2$, (*c′*) the mixture of $FeCl_2$ powders with PAN nanofibers, and (*d′*) pure PAN nanofibers. (**c**) The Cr(VI) removal measurement with Cr contents of 6.3 mg L^{-1} after addition of (*e′*) PAN/$FeCl_2$ and (*f′*) pure PAN nanofibers. Inset: the photograph before and after the addition of the PAN/$FeCl_2$ nanofibers into the $K_2Cr_2O_7$ solution ((**a**) Reprinted with permission from Ref. [81]. © 2013 Elsevier. (**b**) and (**c**) Reprinted with permission from Ref. [80]. © 2010 The Royal Society of Chemistry)

prepared PAN/ferrous chloride ($FeCl_2$) composite porous nanofibrous membranes by electrospinning technology for removal of $Cr_2O_7{}^{2-}$ in one step. The nitrile group in PAN could form a weak complex with cations such as the hydrolysates of Fe^{2+} ($Fe(OH)^+$, $Fe(OH)_2{}^+$ $Fe(OH)^{2+}$, etc.), and then the $Cr_2O_7{}^{2-}$ could be adsorbed onto the complex by electrostatic forces; on the other hand, the Fe^{2+} could reduce the $Cr_2O_7{}^{2-}$ to Cr^{3+} which would alleviate the Cr-induced toxicity. Benefiting from the large surface area and the coordination bond between PAN and Fe^{2+}, the adsorption capacity of the as-prepared nanofibrous membranes is 11.7 mg g^{-1}, which is higher than PAN nanofibers, cast film of PAN/Fe^{2+}, and the mixture of $FeCl_2$ powder with PAN nanofibers (Fig. 19.3b–c). Li and coworkers [81] prepared PA-6/Fe_xO_y composite nanofibrous membranes by combination of electrospinning technology and hydrothermal strategy. The adsorption mechanism is similar to Lin's work [80], and the adsorption capacity can reach 150 mg g^{-1} (Langmuir isotherm). However, the adsorption capacity would decrease when the redox reaction take place, because the concentration of the negative heavy metal ions ($HCrO_4^-$) in solution was decreased.

Before beginning the experiment to estimate the adsorption capacity, the influence of environmental conditions and properties of the solutions should be investigated. Predicting the rate of adsorption for a given system is among the most important factors in adsorption design, as the system's kinetics determines adsorbate residence time and the reactor dimensions [82]. Several kinetic models have been established to understand the adsorption kinetics and rate-limiting steps; among them, the most popular adsorption kinetic models are pseudo-first-order and pseudo-second-order kinetics [6, 71]. The Lagergren first-order rate expression based on solid capacity is generally expressed as follows (Eq. 19.4) [83]:

$$\text{Pseudo-first-order}: \ q = q_e\left(1 - e^{-k_1 t}\right) \tag{19.4}$$

where q is the amount adsorbed at time t, q_e is amount of solute adsorbed at equilibrium condition, and k_1 is the rate constant of first-order adsorption. However, some researches have showed that the predicted q_e values seem to be lower than the experimental values, which were probably caused by the presence of boundary layers or external resistance controlling at the beginning of the sorption process [84]. Pseudo-second-order model is derived on the basis of the sorption capacity of the solid phase, expressed as Eq. 19.5:

$$\text{Pseudo-second-order}: q = \frac{k_2 {q_e}^2 t}{1 + k_2 q_e t} \tag{19.5}$$

where q is the amount adsorbed at time t, q_e is amount of solute adsorbed at equilibrium condition, and k_2 is the rate constant for pseudo-second-order model. As such, in comparison to pseudo-first-order kinetic, this model is considered more appropriate to represent the kinetic data in sorption systems. This tendency comes as an indication that the rate-limiting step in the sorption of heavy metal ions is chemisorption involving valence force through the sharing or enhancing of electrons between sorbent and sorbate, complexation, coordination, and chelation. In other words, in chemisorption process, the pseudo-second order is superior to the pseudo-first-order model (Fig. 19.4b) [37, 85]. It is worth noting that almost all the adsorption experiments are conducted in the acidic conditions to prevent the heavy metal ion precipitation. However, the heavy metal ions are in different states depending on solution pH, and the active component in nanofibers is unstable or dissolves under strong acidity, which would significantly influence the removal capacity, so the initial pH value of the solution should be optimized (Fig. 19.4c) [31, 40, 68]. Importantly, the initial concentration and the extraction time also

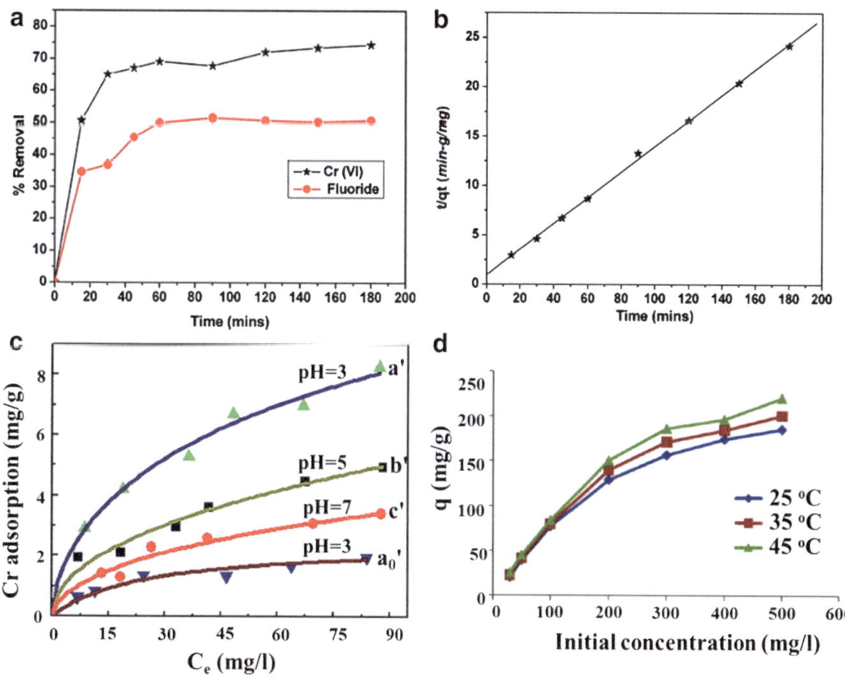

Fig. 19.4 (**a**) Percentage removal of Cr(VI) ions as a function of time (min) by Al_2O_3 nanofibers and (**b**) pseudo-second-order plot for Cr(VI) ion removal by Al_2O_3 nanofibers. (**c**) Adsorption isotherms of the α-AlOOH nanopowders (a_0) and PAN/α-AlOOH composite fibers (*a*', *b*', *c*') at different pH values. (**d**) Effect of initial concentration and temperature on Th(IV) sorption onto the mercapto-modified PVA/TiO_2 nanofiber ((**a**) and (**b**) Reprinted with permission from Ref. [37]. © 2013 American Chemical Society. (**c**) Reprinted with permission from Ref. [86]. © 2012 The Royal Society of Chemistry. (**d**) Reprinted with permission from Ref. [78]. © 2013 Elsevier)

have great impact on the adsorption capacity. In general, the adsorption capacity increases when the concentration of the heavy metal ions increases, the adsorption amount increases remarkably at the beginning, and then the adsorption process becomes gradually slow, finally reaching the equilibrium (Fig. 19.4a). It is essential to mention that the adsorption capacity increases with increasing temperature generally due to the acceleration of some slow sorption steps or creation of new vacant active sites on the nanofibrous membranes' surface (Fig. 19.4d) [78, 86].

19.2.2.2 Organic Contaminants

Dyes, persistent organic pollutant (POP), and other organic contaminants in water are harmful to the environment and human health. Various approaches for the removal of the organic contaminants in water have been investigated, including adsorption, chemical coagulation, and photodegradation. Among the technologies, adsorption based on porous nanofibrous membranes has attracted a great deal of interest due to the large surface area, high porosity, tailored pore structure, and flexible surface modification.

The extremely large surface area and high pore volume of ACNF lead the nanofibers to a strong adsorption amount and rapid adsorption rate, which make it an ideal candidate to adsorb dyes from water and received growing attention in recent years. Si and coworkers [87] synthesized hierarchical porous magnetic Fe_3O_4/ACNF based on polybenzoxazine (PBZ) precursors by a combination of electrospinning and in situ polymerization for the use of removing dyes from aqueous solution (Fig. 19.5a). The hierarchical porous structures of the Fe_3O_4/ACNF were analyzed based on 2-dimensional nonlocal density functional theory, Frenkel-Halsey-Hill theory, and synchrotron radiation small-angle X-ray scattering data. The results revealed that the pore size distribution mainly centered at 2.83 nm and 4.92 nm and the as-fabricated nanofibers owned a typically surface fractal feature. Benefiting from the extremely large surface area (1,885 $m^2\ g^{-1}$), high porosity (2.3 $cm^3\ g^{-1}$), and enhanced porous structure, the as-prepared Fe_3O_4/ACNF exhibits efficient adsorption for methylene blue and Rhodamine B in water, and all samples could achieve complete adsorption within 10 and 15 min (Fig. 19.5b). More importantly, as shown in Fig. 19.5c, the aqueous suspension after adsorption could be separated facilely (<60 s) by an external magnet without a tedious separation process, which is of great importance for real applications. In further research, Si et al. [88, 89] utilized the electrospun PAN/PBZ as composite carbon precursor to investigate the dye adsorption capacities. Teng et al. [43] prepared the mesoporous ACNF by sol–gel/electrospinning process by using phenolic resin precursor as a carbon source and triblock polymer Pluronic F127 as a template and then carbonization and removal of SiO_2. The mesoporous ACNFs were used as highly efficient adsorbents for large dye molecules (methyl thionyl chloride, methyl orange, acid red 1). The results suggested that the adsorption of different dyes depended on an appropriate pore size distribution in addition to the surface area. The CNFs C-3 which exhibited the highest surface area and the largest mesopore volume than others showed great performance in adsorption of large dye molecules, for example, the maximum adsorption capacity for methyl thionyl chloride can reach 567 $mg\ g^{-1}$ in less than 20 min (equilibrium time).

Other electrospun polymeric nanofibers have also been used to adsorb dyes from water. Li et al. [90] synthesized an absorbent by electrospinning the thiol-functionalized PVA/SiO_2 composite nanofibrous membranes and investigated the adsorption performance of indigo carmine and acid red. Due to the large surface area (140.1 $m^2\ g^{-1}$), electrostatic interaction, and hydrogen bond interaction between dyes and membranes (Fig. 19.6a–b), the maximum adsorptions (Langmuir isotherm) for indigo carmine and acid red were 266.77 and 211.74 $mg\ g^{-1}$. Moreover, the adsorptions of dyes were fitted better with Redlich-Peterson model compared with the Langmuir and Freundlich isotherm, while the adsorption kinetics were fitted with pseudo-second-order and pseudo-first-order model, relatively. By combination of electrospinning and hydrothermal reaction, the hierarchical SiO_2/γ-AlOOH (boehmite) core/sheath nanofibers are fabricated to adsorb Congo red from

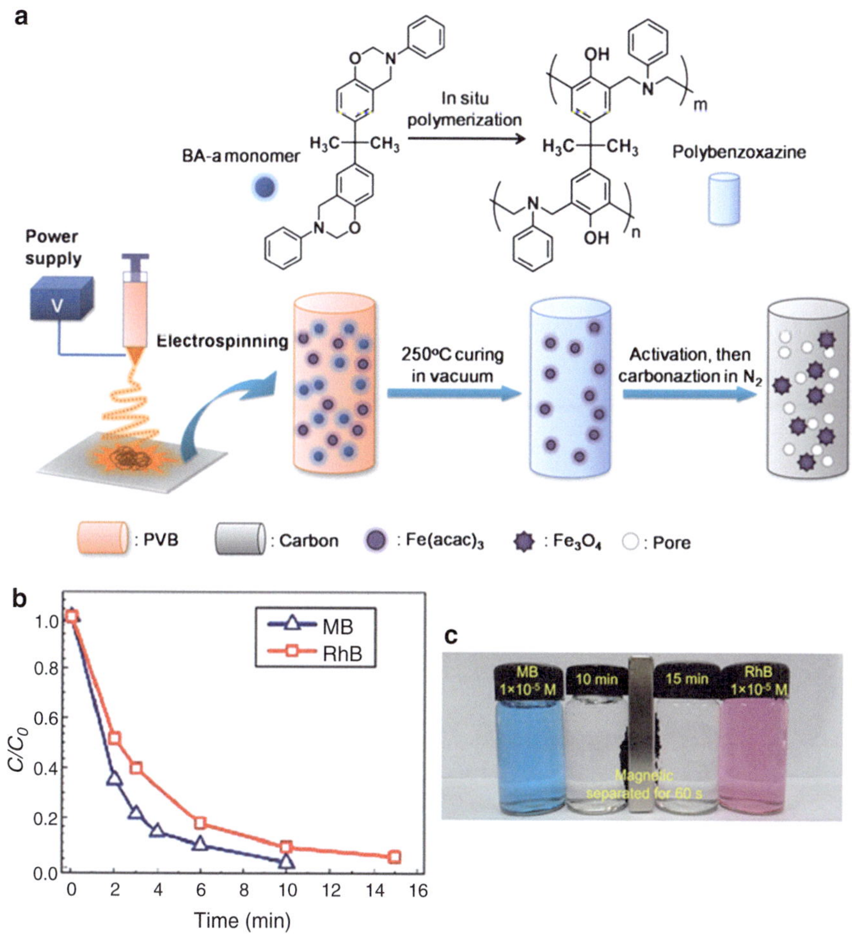

Fig. 19.5 (**a**) Illustration showing the synthesis of Fe_3O_4/ACNF by a combination of electrospinning and in situ polymerization. (**b**) The dye removal measurement after addition of Fe_3O_4/ACNF and (**c**) optical photographs showing the magnetic responsive performance (60 s) of Fe_3O_4/ACNF after adsorption of methylene blue (10 min) and Rhodamine B (15 min) ((**a**) Reprinted with permission from Ref. [87]. © 2012 Elsevier. (**b**) and (**c**) Reprinted with permission from Ref. [88]. © 2012 The Royal Society of Chemistry)

the water [91]. Compared to the boehmite powder, the self-standing membranes thus prepared are highly flexible, easy to handle, and retrievable (Fig. 19.6c), while the adsorption capacity of nanofibers (21.3 mg g^{-1}) was more than two times higher than that of boehmite powder (10.4 mg g^{-1}) which might be caused by the large surface area and improved distribution of boehmite nanoplatelets (Fig. 19.6d). Significantly, the mechanism of Congo red adsorption onto the nanofibers was similar as Li's work. Xu et al. [76] prepared the vinyl-modified mesoporous PAA/SiO_2 composite nanofibers through a sol–gel electrospinning process for the

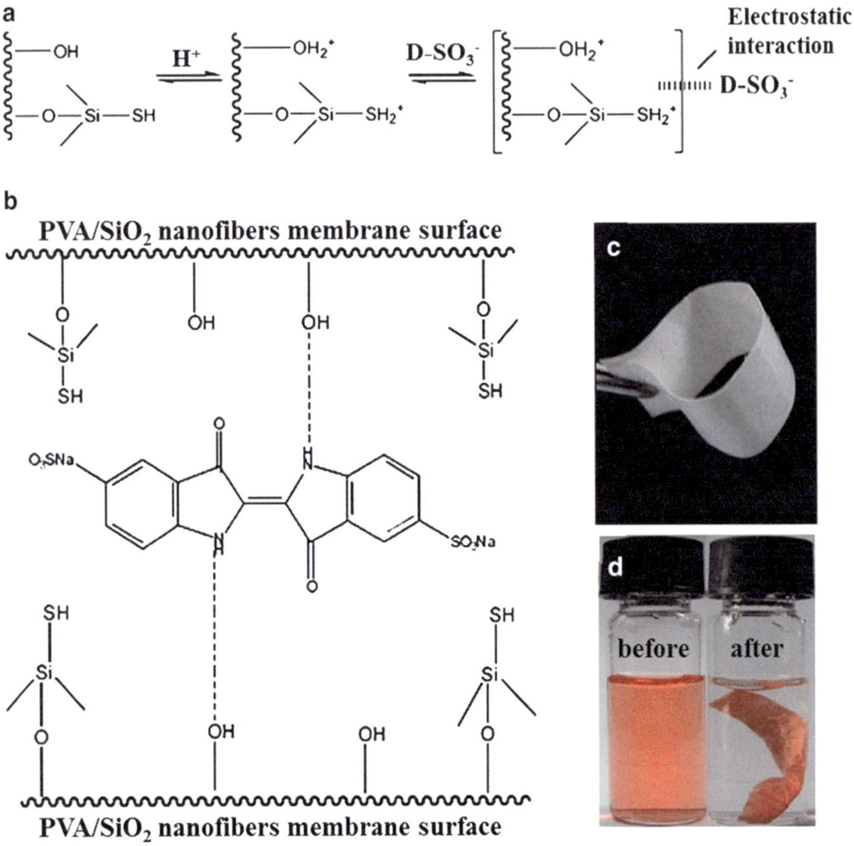

Fig. 19.6 (**a**) Electrostatic and (**b**) hydrogen-bonding mechanism of dye adsorption onto nanofibrous membranes. (**c**) Optical photograph of the self-standing fiber membranes and the performance of dye adsorption ((**a**) and (**b**) Reprinted with permission from Ref. [90]. © 2011 The Royal Society of Chemistry. (**c**) and (**d**) Reprinted with permission from Ref. [91]. © 2012 American Chemical Society)

use of adsorption of malachite green in the water. The strong adsorption capacity (220.49 mg g^{-1}) of the as-spun nanofibers may be caused by three main aspects as follows: firstly, the electrostatic interaction between negatively charged nanofibrous membrane surface (–Si-O^- of silica, –COO^- of PAA) and the positively charged malachite green molecule (–N^+); secondly, physical adsorption might play a great contribution to the dye adsorption due to the large surface area (523.84 m^2 g^{-1}) and high surface energy of nanofibers; and, finally, the conjugation effect between the delocalized π bonds of the dye molecule and vinyl (or carbonyl) group. Furthermore, the mesoporous PAA/SiO_2 nanofibrous membranes showed good generation properties in the first three cycles.

Besides adsorbing dyes from wastewater, electrospun nanofibers have also been utilized for the efficient removal of other organic contaminants such as

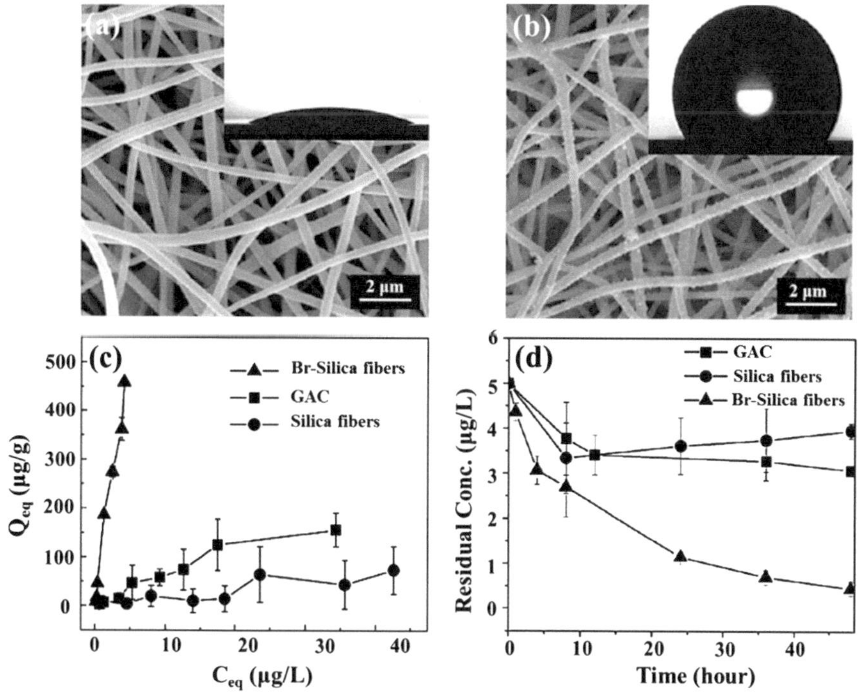

Fig. 19.7 FE-SEM images of electrospun SiO_2 nanofibers (**a**) before and (**b**) after modified with bromopropyl group. Insets are contact angles of the corresponding fibrous membranes. (**c**) Adsorption isotherm and residual concentrations as a function of time for dieldrin on SiO_2 nanofibers, Br-SiO_2 nanofibers, and GAC ((**a–d**) Reprinted with permission from Ref. [77]. © 2012 Elsevier)

dieldrin, estrogen, and phenol. Yue and coworkers [92] described the fabrication of the bromopropyl-functionalized silica nanofibrous membranes by combination of electrospinning and surface modification. Because of the strong hydrophobic interaction between dieldrin and bromopropyl groups based on the principle of "like prefers like," the large specific surface area, and the large amount of pores, the bromopropyl-modified silica nanofibers were found to be highly effective for the removal of hydrophobic POPs in low concentration (Fig. 19.7). The experimental adsorption isotherms were fitted well with the Freundlich model, and the maximum removal rate of the as-prepared adsorbent was evaluated to be 91 %, which was much higher than that of commercial granular activated carbon (GAC, 38.6 %) and silica nanofibers (20.9 %). Xu et al. [93] used electrospun PA-6 nanofibrous membranes for the extraction and enrichment of three estrogens (estradiol, ethinyl estradiol, and estrone) in environmental water samples as solid membrane extraction adsorbents. As a result, the estrogens can be quantitatively adsorbed on 1.5 mg PA-6 due to the high specific surface area (18.41 $m^2\ g^{-1}$) and the enhanced interaction between the nanofibers and estrogens. The further research confirmed that the analytical performance of PA-6 was better in terms of recovery, reproducibility, and adsorption amount compared to commercial PA-6 microporous membranes.

It is essential to mention that the pH value, initial concentration of the organic contaminants, equilibrium time, temperature, etc. have great impact on the adsorption capacities of the nanofiber-based adsorbents; the effects were similar to adsorption of heavy metal ions.

19.3 Summary and Perspectives

In summary, removal of environmental pollutants including toxic gases, heavy metal ions, dyes, and other organic contaminants in water has been extensively investigated in recent years. Among the large amount of approaches, adsorption with nanomaterial-based adsorbent is emerging to be one of the most promising pollutant strategies. In this chapter, we reviewed the recent progress in developing electrospun nanofibrous membrane-based adsorbents with potential application for environmental contaminant removal. In general, nanofibrous membranes possess advantages such as large surface area, high porosity, tailored pore structure, ease of design and synthesis, and flexibility of surface modification with functional groups, which make them highly attractive for the development of adsorbents with high adsorption capacity.

Despite the numerous advantages toward the development of nanofibrous adsorbents for removal of environmental pollutants, several challenges still remained. First, some nanofibrous membranes still do not have the adsorption ability after surface modification with functional groups, or the adsorption capacity of the as-prepared adsorbent reduced after surface modification since the undesired blocking of fibers leads to pore structure change and unavoidably reduces the adsorption capacity and kinetics of the adsorbents. Further research could be focused on exploring effective surface modification strategies, which could not change the pore structure of the nanofibrous membranes. Second, the adsorption capacity of the adsorbents reduces dramatically after regeneration, implying that a portion of the binding sites are occupied by the adsorbate; accordingly, desorption approaches should be optimized to decrease the loss of the adsorption capacity. Third, the toxic gases, heavy metal ions, and organic contaminants always involve water in practical pollutant adsorption; however, the water could reduce the adsorption capacity as mentioned above; therefore, nanofiber-based adsorbents need a high tolerance to water or superhydrophobicity. Although there are numerous challenges, further research on improving the performance of the adsorbent under the realistic conditions would allow for the realization of nanofibrous membrane-based adsorbent in practical pollutant removal process in the future.

Acknowledgment This work is supported by the National Basic Research Program of China (973 Program, 2011CB606103 and 2012CB525005), the National Natural Science Foundation of China (No. 51173022, U1232116 and 51273038), the Shanghai Nano Special Projects (11 nm0502900), the Shanghai Committee of Science and Technology (No. 12JC1400101), the Huo Yingdong Foundation (131070), and the Program for New Century Talents of the University in China.

References

1. Grimm NB, Faeth SH, Golubiewski NE, Redman CL, Wu J, Bai X, Briggs JM (2008) Global change and the ecology of cities. Science 319(5864):756–760. doi:10.1126/science.1150195
2. Jones AP (1999) Indoor air quality and health. Atmos Environ 33(28):4535–4564. doi:10.1016/s1352-2310(99)00272-1
3. Pearce CI, Lloyd JR, Guthrie JT (2003) The removal of colour from textile wastewater using whole bacterial cells: a review. Dye Pigment 58(3):179–196. doi:10.1016/s0143-7208(03)00064-0
4. Harada M (1995) Minamata disease-methylmercury poisoning in Japan caused by environmental-pollution. Crit Rev Toxicol 25(1):1–24. doi:10.3109/10408449509089885
5. Landrigan PJ, Schechter CB, Lipton JM, Fahs MC, Schwartz J (2002) Environmental pollutants and disease in American children: estimates of morbidity, mortality, and costs for lead poisoning, asthma, cancer, and developmental disabilities. Environ Health Perspect 110(7):721–728. doi:10.1289/ehp.02110721
6. Azizian S (2004) Kinetic models of sorption: a theoretical analysis. J Colloid Interf Sci 276(1):47–52. doi:10.1016/j.jcis.2004.03.048
7. Chen GH (2004) Electrochemical technologies in wastewater treatment. Sep Purif Technol 38(1):11–41. doi:10.1016/j.seppur.2003.10.006
8. Gupta VK, Carrott PJM, Ribeiro Carrott MML, Suhas (2009) Low-cost adsorbents: growing approach to wastewater treatment a review. Crit Rev Environ Sci Technol 39(10):783–842. doi:10.1080/10643380801977610
9. Chen W, Duan L, Zhu DQ (2007) Adsorption of polar and nonpolar organic chemicals to carbon nanotubes. Environ Sci Technol 41(24):8295–8300. doi:10.1021/es071230h
10. Gupta VK, Saleh TA (2013) Sorption of pollutants by porous carbon, carbon nanotubes and fullerene- an overview. Environ Sci Pollut Res Int 20(5):2828–2843. doi:10.1007/s11356-013-1524-1
11. Dettmer K, Engewald W (2002) Adsorbent materials commonly used in air analysis for adsorptive enrichment and thermal desorption of volatile organic compounds. Anal Bioanal Chem 373(6):490–500. doi:10.1007/s00216-002-1352-5
12. Forgacs E, Cserhati T, Oros G (2004) Removal of synthetic dyes from wastewaters: a review. Environ Int 30(7):953–971. doi:10.1016/j.envint.2004.02.001
13. Li JR, Kuppler RJ, Zhou HC (2009) Selective gas adsorption and separation in metal-organic frameworks. Chem Soc Rev 38(5):1477–1504. doi:10.1039/b802426j
14. Meng LY, Park SJ (2012) MgO-templated porous carbons-based CO_2 adsorbents produced by KOH activation. Mater Chem Phys 137(1):91–96. doi:10.1016/j.matchemphys.2012.08.043
15. Aluigi A, Tonetti C, Vineis C, Tonin C, Mazzuchetti G (2011) Adsorption of copper(II) ions by keratin/PA6 blend nanofibres. Eur Polym J 47(9):1756–1764. doi:10.1016/j.eurpolymj.2011.06.009
16. Rao GP, Lu C, Su F (2007) Sorption of divalent metal ions from aqueous solution by carbon nanotubes: a review. Sep Purif Technol 58(1):224–231. doi:10.1016/j.seppur.2006.12.006
17. Kizzie AC, Wong-Foy AG, Matzger AJ (2011) Effect of humidity on the performance of microporous coordination polymers as adsorbents for CO_2 capture. Langmuir 27(10):6368–6373. doi:10.1021/la200547k
18. Werner MD (1985) The effects of relative-humidity on the vapor-phase adsorption of trichlorethylene by activated carbon. Am Ind Hyg Assoc J 46(10):585–590. doi:10.1202/0002-8894(1985)046<0585:teorho>2.3.co;2
19. Zhang LD, Fang M (2010) Nanomaterials in pollution trace detection and environmental improvement. Nano Today 5(2):128–142. doi:10.1016/j.nantod.2010.03.002
20. Khin MM, Nair AS, Babu VJ, Murugan R, Ramakrishna S (2012) A review on nanomaterials for environmental remediation. Energ Environ Sci 5(8):8075–8109. doi:10.1039/c2ee21818f
21. Ma HY, Hsiao BS, Chu B (2013) Electrospun nanofibrous membrane for heavy metal ion adsorption. Curr Org Chem 17(13):1361–1370

22. Huang ZM, Zhang YZ, Kotaki M, Ramakrishna S (2003) A review on polymer nanofibers by electrospinning and their applications in nanocomposites. Compos Sci Technol 63(15):2223–2253. doi:10.1016/s0266-3538(03)00178-7
23. Ramakrishna S, Fujihara K, Teo WE, Yong T, Ma ZW, Ramaseshan R (2006) Electrospun nanofibers: solving global issues. Mater Today 9(3):40–50. doi:10.1016/s1369-7021(06)71389-x
24. Li D, McCann JT, Xia YN (2006) Electrospinning: a simple and versatile technique for producing ceramic nanofibers and nanotubes. J Am Ceram Soc 89(6):1861–1869. doi:10.1111/j.1551-2916.2006.00989.x
25. Kyung L, Shiratori N, Gang L, Miyawaki J, Mochida I, Seong Y, Jyongsik J (2010) Activated carbon nanofiber produced from electrospun polyacrylonitrile nanofiber as a highly efficient formaldehyde adsorbent. Carbon 48(15):4248–4255. doi:10.1016/j.carbon.2010.07.034
26. Shim WG, Kim C, Lee JW, Yun JJ, Jeong YI, Moon H, Yang KS (2006) Adsorption characteristics of benzene on electrospun-derived porous carbon nanofibers. J Appl Polym Sci 102(3):2454–2462. doi:10.1002/app.24554
27. Song XF, Wang C, Zhang DJ (2009) Surface structure and adsorption properties of ultrafine porous carbon fibers. Appl Surf Sci 255(7):4159–4163. doi:10.1016/j.apsusc.2008.10.113
28. Im JS, Park SJ, Kim TJ, Kim YH, Lee YS (2008) The study of controlling pore size on electrospun carbon nanofibers for hydrogen adsorption. J Colloid Interf Sci 318(1):42–49. doi:10.1016/j.jcis.10.024
29. Mahanta N, Valiyaveettil S (2013) Functionalized poly(vinyl alcohol) based nanofibers for the removal of arsenic from water. RSC Adv 3(8):2776–2783. doi:10.1039/c2ra22768a
30. Saeed K, Haider S, Oh TJ, Park SY (2008) Preparation of amidoxime-modified polyacrylonitrile (PAN-oxime) nanofibers and their applications to metal ions adsorption. J Membr Sci 322(2):400–405, doi: http://dx.doi.org/10.1016/j.memsci.2008.05.062
31. Choi J, Ide A, Truong YB, Kyratzis IL, Caruso RA (2013) High surface area mesoporous titanium-zirconium oxide nanofibrous web: a heavy metal ion adsorbent. J Mater Chem A 1(19):5847–5853. doi:10.1039/c3ta00030c
32. Wu Y, Li F, Wu Y, Jia W, Hannam P, Qiao J, Li G (2011) Formation of silica nanofibers with hierarchical structure via electrospinning. Colloid Polym Sci 289(11):1253–1260. doi:10.1007/s00396-011-2455-3
33. Vu D, Li Z, Zhang H, Wang W, Wang Z, Xu X, Dong B, Wang C (2012) Adsorption of Cu(II) from aqueous solution by anatase mesoporous TiO_2 nanofibers prepared via electrospinning. J Colloid Interf Sci 367:429–435. doi:10.1016/j.jcis.2011.09.088
34. Wu S, Li F, Wu Y, Xu R, Li G (2010) Preparation of novel poly(vinyl alcohol)/SiO_2 composite nanofiber membranes with mesostructure and their application for removal of Cu^{2+} from waste water. Chem Commun 46(10):1694–1696. doi:10.1039/b925296g
35. Huang F, Xu Y, Liao S, Yang D, Hsieh YL, Wei Q (2013) Preparation of amidoxime polyacrylonitrile chelating nanofibers and their application for adsorption of metal ions. Materials 6(3):969–980. doi:10.3390/ma6030969
36. Taha AA, Qiao J, Li F, Zhang B (2012) Preparation and application of amino functionalized mesoporous nanofiber membrane via electrospinning for adsorption of Cr^{3+} from aqueous solution. J Environ Sci 24(4):610–616, doi: http://dx.doi.org/10.1016/S1001-0742(11)60806-1
37. Mahapatra A, Mishra BG, Hota G (2013) Studies on electrospun alumina nanofibers for the removal of chromium(VI) and fluoride toxic ions from an aqueous system. Ind Eng Chem Res 52(4):1554–1561. doi:10.1021/ie301586j
38. Teng M, Wang H, Li F, Zhang B (2011) Thioether-functionalized mesoporous fiber membranes: sol-gel combined electrospun fabrication and their applications for Hg^{2+} removal. J Colloid Interf Sci 355(1):23–28. doi:10.1016/j.jcis.2010.11.008
39. Li S, Yue X, Jing Y, Bai S, Dai Z (2011) Fabrication of zonal thiol-functionalized silica nanofibers for removal of heavy metal ions from wastewater. Colloid Surf A 380(1–3):229–233, doi: http://dx.doi.org/10.1016/j.colsurfa.2011.02.027

40. Mahapatra A, Mishra BG, Hota G (2013) Electrospun Fe_2O_3-Al_2O_3 nanocomposite fibers as efficient adsorbent for removal of heavy metal ions from aqueous solution. J Hazard Mater 258–259:116–123. doi:10.1016/j.jhazmat.2013.04.045
41. Mahanta N, Valiyaveettil S (2011) Surface modified electrospun poly(vinyl alcohol) membranes for extracting nanoparticles from water. Nanoscale 3(11):4625–4631. doi:10.1039/c1nr10739a
42. Aliabadi M, Irani M, Ismaeili J, Piri H, Parnian MJ (2013) Electrospun nanofiber membrane of PEO/Chitosan for the adsorption of nickel, cadmium, lead and copper ions from aqueous solution. Chem Eng J 220:237–243, doi: http://dx.doi.org/10.1016/j.cej.2013.01.021
43. Qiao JL, Li FT, Bera PK (2012) Electrospun mesoporous carbon nanofibers produced from phenolic resin and their use in the adsorption of large dye molecules. Carbon 50(8):2877–2886. doi:10.1016/j.carbon.2012.02.056
44. Teng M, Li FT, Zhang BR, Taha AA (2011) Electrospun cyclodextrin-functionalized mesoporous polyvinyl alcohol/SiO_2 nanofiber membranes as a highly efficient adsorbent for indigo carmine dye. Colloid Surf A 385(1–3):229–234. doi:10.1016/j.colsurfa.2011.06.020
45. Xu R, Jia M, Li F, Wang H, Zhang B, Qiao J (2012) Preparation of mesoporous poly(acrylic acid)/SiO_2 composite nanofiber membranes having adsorption capacity for indigo carmine dye. Appl Phys A-Mater 106(3):747–755. doi:10.1007/s00339-011-6697-1
46. Zhang HT, Yu DG, Wu CY, Zhang YL, Christopher J, Branford W (2010) Surface modification of electrospun polyacrylonitrile nanofiber towards developing an affinity membrane for bromelain adsorption. Desalination 256:141–147. doi:10.1016/j.desal.2010.01.026
47. Fierro V, Zhao W, Izquierdo MT, Aylon E, Celzard A (2010) Adsorption and compression contributions to hydrogen storage in activated anthracites. Int J Hydrog Energ 35(17):9038–9045. doi:10.1016/j.ijhydene.2010.06.004
48. Cuervo MR, Asedegbega E, Diaz E, Vega A, Ordonez S, Castillejos E, Rodriguez-Ramos I (2008) Effect of carbon nanofiber functionalization on the adsorption properties of volatile organic compounds. J Chromatogr A 1188(2):264–273. doi:10.1016/j.chroma.2008.02.061
49. Takagi H, Hatori H, Yamada Y (2005) Reversible adsorption/desorption property of hydrogen on carbon surface. Carbon 43(14):3037–3039. doi:10.1016/j.carbon.2005.06.032
50. Lunde PJ, Kester FL (1975) Chemical and physical gas adsorption in finite multimolecular layers. Chem Eng Sci 30(12):1497–1505. doi:10.1016/0009-2509(75)85027-5
51. Brunauer S, Emmett PH, Teller E (1938) Adsorption of gases in multimolecular layers. J Am Chem Soc 60(2):309–319. doi:10.1021/ja01269a023
52. Sing KSW, Everett DH, Haul RAW, Moscou L, Pierotti RA, Rouquerol J, Siemieniewska T (1985) Reporting physisorption data for gas solid systems with special reference to the determination of surface-area and porosity (Recommendations 1984). Pure Appl Chem 57(4):603–619. doi:10.1351/pac198557040603
53. Li ZJ, Kruk M, Jaroniec M, Ryu SK (1998) Characterization of structural and surface properties of activated carbon fibers. J Colloid Interf Sci 204(1):151–156. doi:10.1006/jcis.1998.5515
54. Im JS, Kang SC, Lee SH, Lee YS (2010) Improved gas sensing of electrospun carbon fibers based on pore structure, conductivity and surface modification. Carbon 48(9):2573–2581. doi:10.1016/j.carbon.2010.03.045
55. Hong SE, Kim DK, Jo SM, Kim DY, Chin BD, Lee DW (2007) Graphite nanofibers prepared from catalytic graphitization of electrospun poly(vinylidene fluoride) nanofibers and their hydrogen storage capacity. Catal Today 120(3–4):413–419. doi:10.1016/j.cattod.2006.09.013
56. Bogdanovic B, Schwickardi M (1997) Ti-doped alkali metal aluminium hydrides as potential novel reversible hydrogen storage materials. J Alloy Compd 253:1–9. doi:10.1016/s0925-8388(96)03049-6
57. Katepalli H, Bikshapathi M, Sharma CS, Verma N, Sharma A (2011) Synthesis of hierarchical fabrics by electrospinning of PAN nanofibers on activated carbon microfibers for environmental remediation applications. Chem Eng J 171(3):1194–1200. doi:10.1016/j.cej.2011.05.025
58. Shanker AK, Cervantes C, Loza-Tavera H, Avudainayagam S (2005) Chromium toxicity in plants. Environ Int 31(5):739–753. doi:10.1016/j.envint.2005.02.003

59. Babel S, Kurniawan TA (2003) Low-cost adsorbents for heavy metals uptake from contaminated water: a review. J Hazard Mater 97(1–3):219–243. doi:10.1016/s0304-3894(02)00263-7
60. Beyersmann D, Hechtenberg S (1997) Cadmium, gene regulation, and cellular signalling in mammalian cells. Toxicol Appl Pharm 144(2):247–261. doi:10.1006/taap.1997.8125
61. Bailey SE, Olin TJ, Bricka RM, Adrian DD (1999) A review of potentially low-cost sorbents for heavy metals. Water Res 33(11):2469–2479. doi:10.1016/s0043-1354(98)00475-8
62. Li S, Yue X, Jing Y, Bai S, Dai Z (2011) Fabrication of zonal thiol-functionalized silica nanofibers for removal of heavy metal ions from wastewater. Colloid Surf A 380(1–3):229–233. doi:10.1016/j.colsurfa.2011.02.027
63. Khajeh M, Laurent S, Dastafkan K (2013) Nanoadsorbents: classification, preparation, and applications (with emphasis on aqueous media). Chem Rev. doi:10.1021/cr400086v
64. Doo Hyun B, Chang Seok K, In Chul U, Young Hwan P (2007) Metal ion adsorbability of electrospun wool keratose/silk fibroin blend nanofiber mats. Fiber Polym 8(3):271–277. doi:10.1007/bf02877269
65. Zhou W, He J, Cui S, Gao W (2011) Preparation of electrospun silk fibroin/cellulose acetate blend nanofibers and their applications to heavy metal ions adsorption. Fiber Polym 12(4):431–437. doi:10.1007/s12221-011-0431-7
66. Zhou W, He J, Cui S, Gao W (2011) Nanofibrous membrane of silk fibroin/cellulose acetate blend for heavy metal ion adsorption. In: Liu XH, Jiang ZY, Han JT (eds) Manufacturing processes and systems, Pts 1-2, vol 148–149. Advanced Materials Research, pp 1431–1435. doi:10.4028/www.scientific.net/AMR.148-149.1431
67. Darko G, Sobola A, Adewuyi S, Okonkwo JO, Torto N (2012) Pre-concentration of toxic metals using electrospun amino-functionalized nylon-6 nanofibre sorbent. S Afr J Chem-S Afr T 65:14–22
68. Tian Y, Wu M, Liu R, Li Y, Wang D, Tan J, Wu R, Huang Y (2011) Electrospun membrane of cellulose acetate for heavy metal ion adsorption in water treatment. Carbohyd Polym 83(2):743–748. doi:10.1016/j.carbpol.2010.08.054
69. Ki CS, Gang EH, Um NC, Park YH (2007) Nanofibrous membrane of wool keratose/silk fibroin blend for heavy metal ion adsorption. J Membr Sci 302(1–2):20–26. doi:10.1016/j.memsci.2007.06.003
70. Haider S, Park SY (2009) Preparation of the electrospun chitosan nanofibers and their applications to the adsorption of Cu(II) and Pb(II) ions from an aqueous solution. J Membr Sci 328(1–2):90–96. doi:10.1016/j.memsci.2008.11.046
71. Febrianto J, Kosasih AN, Sunarso J, Ju YH, Indraswati N, Ismadji S (2009) Equilibrium and kinetic studies in adsorption of heavy metals using biosorbent: a summary of recent studies. J Hazard Mater 162(2–3):616–645. doi:10.1016/j.jhazmat.2008.06.042
72. Wu S, Li F, Wang H, Fu L, Zhang B, Li G (2010) Effects of poly (vinyl alcohol) (PVA) content on preparation of novel thiol-functionalized mesoporous PVA/SiO_2 composite nanofiber membranes and their application for adsorption of heavy metal ions from aqueous solution. Polymer 51(26):6203–6211. doi:10.1016/j.polymer.2010.10.015
73. Stephen M, Catherine N, Brenda M, Andrew K, Leslie P, Corrine G (2011) Oxolane-2,5-dione modified electrospun cellulose nanofibers for heavy metals adsorption. J Hazard Mater 192(2):922–927. doi:10.1016/j.jhazmat.2011.06.001
74. Appel J (1973) Freundlich's adsorption isotherm. Surf Sci 39(1):237–244, doi: http://dx.doi.org/10.1016/0039-6028(73)90105-2
75. Wu Y, Zhang B, Li F, Zhu W, Xu D, Hannam P, Li G (2012) Electrospun fibrous mats as a skeleton for fabricating hierarchically structured materials as sorbents for Cu^{2+}. J Mater Chem 22(11):5089–5097. doi:10.1039/c2jm13874c
76. Xu R, Jia M, Zhang Y, Li F (2012) Sorption of malachite green on vinyl-modified mesoporous poly(acrylic acid)/SiO_2 composite nanofiber membranes. Micropor Mesopor Mat 149(1):111–118. doi:10.1016/j.micromeso.2011.08.024
77. Heidari A, Younesi H, Mehraban Z, Heikkinen H (2013) Selective adsorption of Pb(II), Cd(II), and Ni(II) ions from aqueous solution using chitosan–MAA nanoparticles. Int J Biol Macromol 61:251–263, doi: http://dx.doi.org/10.1016/j.ijbiomac.2013.06.032

78. Abbasizadeh S, Keshtkar AR, Mousavian MA (2013) Preparation of a novel electrospun polyvinyl alcohol/titanium oxide nanofiber adsorbent modified with mercapto groups for uranium(VI) and thorium(IV) removal from aqueous solution. Chem Eng J 220:161–171. doi:10.1016/j.cej.2013.01.029
79. Wang W, Yang Q, Sun L, Wang H, Zhang C, Fei X, Sun M, Li Y (2011) Preparation of fluorescent nanofibrous film as a sensing material and adsorbent for Cu^{2+} in aqueous solution via copolymerization and electrospinning. J Hazard Mater 194:185–192. doi:10.1016/j.jhazmat.2011.07.083
80. Lin YX, Cai WP, Tian XY, Liu XL, Wang GZ, Liang CH (2011) Polyacrylonitrile/ferrous chloride composite porous nanofibers and their strong Cr-removal performance. J Mater Chem 21(4):991–997. doi:10.1039/c0jm02334e
81. Li CJ, Li YJ, Wang JN, Cheng J (2013) PA6@Fe_xO_y nanofibrous membrane preparation and its strong Cr (VI)-removal performance. Chem Eng J 220:294–301. doi:10.1016/j.cej.2013.01.060
82. Ho YS (2006) Review of second-order models for adsorption systems. J Hazard Mater 136(3):681–689. doi:10.1016/j.jhazmat.2005.12.043
83. Ho YS (2004) Citation review of Lagergren kinetic rate equation on adsorption reactions. Scientometrics 59(1):171–177. doi:10.1023/B:SCIE.0000013305.99473.cf
84. Vijayaraghavan K, Palanivelu K, Velan M (2006) Biosorption of copper(II) and cobalt(II) from aqueous solutions by crab shell particles. Bioresour Technol 97(12):1411–1419, doi: http://dx.doi.org/10.1016/j.biortech.2005.07.001
85. Wang F, Ge M (2013) Fibrous mat of chitosan/polyvinyl alcohol/containing cerium(III) for the removal of chromium(VI) from aqueous solution. Text Res J 83(6):628–637. doi:10.1177/0040517512454188
86. Lin YX, Cai WP, He H, Wang XB, Wang GZ (2012) Three-dimensional hierarchically structured PAN@gamma-AlOOH fiber films based on a fiber templated hydrothermal route and their recyclable strong Cr(VI)-removal performance. RSC Adv 2(5):1769–1773. doi:10.1039/c2ra00945e
87. Si Y, Ren T, Li Y, Ding B, Yu JY (2012) Fabrication of magnetic polybenzoxazine-based carbon nanofibers with Fe_3O_4 inclusions with a hierarchical porous structure for water treatment. Carbon 50(14):5176–5185. doi:10.1016/j.carbon.2012.06.059
88. Si Y, Ren T, Ding B, Yu JY, Sun G (2012) Synthesis of mesoporous magnetic Fe_3O_4@carbon nanofibers utilizing in situ polymerized polybenzoxazine for water purification. J Mater Chem 22(11):4619–4622. doi:10.1039/c2jm00036a
89. Ren T, Si Y, Yang J, Ding B, Yang X, Hong F, Yu J (2012) Polyacrylonitrile/polybenzoxazine-based Fe_3O_4@carbon nanofibers: hierarchical porous structure and magnetic adsorption property. J Mater Chem 22(31):15919–15927. doi:10.1039/c2jm33214k
90. Li M, Wang H, Wu S, Li F, Zhi P (2012) Adsorption of hazardous dyes indigo carmine and acid red on nanofiber membranes. RSC Adv 2(3):900–907. doi:10.1039/c1ra00546d
91. Miao YE, Wang R, Chen D, Liu Z, Liu T (2012) Electrospun self-standing membrane of hierarchical SiO_2@gamma-AlOOH (Boehmite) core/sheath fibers for water remediation. ACS Appl Mater Interfaces 4(10):5353–5359. doi:10.1021/am3012998
92. Yue X, Feng S, Li S, Jing Y, Shao C (2012) Bromopropyl functionalized silica nanofibers for effective removal of trace level dieldrin from water. Colloid Surf A 406:44–51. doi:10.1016/j.colsurfa.2012.04.047
93. Xu Q, Wu SY, Wang M, Yin XY, Wen ZY, Ge WN, Gu ZZ (2010) Electrospun nylon6 nanofibrous membrane as spe adsorbent for the enrichment and determination of three estrogens in environmental water samples. Chromatographia 71(5–6):487–492. doi:10.1365/s10337-009-1453-9

Chapter 20
Application of Electrospun Nanofibers in Electromagnetic Interference Shielding

Qing-Qing Ni, Hong Xia, Xudong Jin, and Fan Liu

Abstract The challenges of developing multifunctional nano-composite fibers are reviewed in this chapter. Nano-composite fibers made from polyacrylonitrile (PAN) or polyvinyl alcohol (PVA) containing the components, such as multiwalled CNTs (MWCNT), magnetic nanoparticles Fe_3O_4, and $CoFe_2O_4$, were fabricated by electrospinning technology. $CoFe_2O_4$/PAN nanofibers with various $CoFe_2O_4$ loadings were prepared, and the fiber diameter was controlled between 200 and 400 nm. For developed PAN nano-composite fibers containing CNTs and $CoFe_2O_4$ nanoparticles, the presence of $CoFe_2O_4$ nanoparticles and CNTs in the PAN matrix was confirmed by XRD patterns, Raman spectra, and SEM and TEM observations. The PVA-based nano-composite fibers with Fe_3O_4 or MWCNT were also fabricated. Polyacrylic acid (PAA) was used as a polymer surfactant to stabilize Fe_3O_4 NPs aqueous suspension and facilitate their dispersal in PVA solutions. PAA was also used as a cross-linking agent to enhance the water resistance of the magnetic composite nanofibers through heat treatment, which provides a simple and convenient process to prepare homogenous NPs/PVA composite system. Incorporating CNTs, Fe_3O_4, and $CoFe_2O_4$ nanoparticles, the results showed that the developed electrospun nanofibers were suitable for the applications in electromagnetic interference (EMI) shielding effectiveness (SE) field and their performance can be tailored by adjusting fillers loading.

20.1 Introduction

The widespread use of electrical instruments in commercial and military settings as well as in our daily lives means that electromagnetic interference (EMI) has become a serious problem in modern society [1] and drawn the wide attention all over the

Q.-Q. Ni (✉) • H. Xia • X. Jin • F. Liu
Faculty of Textile Science and Technology, Shinshu University, Ueda 386-8567, Japan
e-mail: niqq@shinshu-u.ac.jp

B. Ding and J. Yu (eds.), *Electrospun Nanofibers for Energy and Environmental Applications*, Nanostructure Science and Technology,
DOI 10.1007/978-3-642-54160-5_20,

world. To avoid EMI, there is increased interest on the development of shielding materials. Novel shielding materials with superior properties to those currently available are needed, and a significant body of research has been performed with this goal. This includes applications of intrinsic conducting polymers [2–6], studies on materials with unique structures such as multilayers [7, 8] and foam [9–11], and preparation of polymer-based conductive or magnetic composites [12–17].

EMI shielding refers to the surface reflection, multi-reflection, and/or adsorption of electromagnetic radiation by a material, which thereby acts as a shield against the penetration of the radiation through it [18]. Conductivity is considered to be a critical factor for electromagnetic shielding; thus metals, such as steel, copper, aluminum, etc., are the most common materials used for EMI shielding. Since metal shielding has shortcomings of heavy weight, corrosion, and physical rigidity, polymer-based composites with discontinuous conducting fillers, such as metal particles, carbon particles, and carbon fiber, are extensively employed in EMI shielding [19–24]. Although these composites are not strong enough for most structural applications, they are attractive because of their superior molding and more dependable lightweight. For these composites, the EMI shielding effectiveness increases with the increasing fraction and/or aspect ratio of the fillers [21, 25].

Electrospinning can be used to produce novel fibers with the diameters in the range from 100 nm to 10 μm. As the solvent evaporates, the polymer molecules can come together by either a phase separation through a spinodal reaction or through classic nucleation and growth of the crystalline phase. As a result, the structure in the polymer deposited on the collector may form a totally amorphous, a part of oriented, spherulitic, or textured fibrillar structure.

Among the approaches reported [13] and some mentioned in Chap. 1, electrospinning technology has proved to be a simple, efficient, and versatile way to fabricate nano-composites fibers, which are suitable for EMI SE applications. A number of novel electrospun nano-composite fibers have consequently been researched and developed [26, 27]. Electrospun nano-composite fibers combining poly(vinyl pyrrolidone) and rare-earth complexes and having strong photoluminescence properties were prepared by Zhang et al. [15]. Several types of magnetic polymer fiber were fabricated by incorporating magnetic nanoparticles such as Fe_3O_4 and FePt into a polymer matrix and then electrospinning [16–18]. The incorporation of a component such as Fe_3O_4 or CNTs could clearly introduce new functionality to a nano-composite fiber. However, usually, nanofibers are limited to a single component, and there are relatively few publications on multicomponent nano-composite fibers for EMI SE applications.

CNT is a prospective novel material for multicomponent nano-composite fibers, as they have been widely applied as composite components because of their excellent optical, mechanical, and electrical properties [21, 28–31]. Magnetic nanoparticles (MNPs) of Fe_3O_4 and cobalt ferrite ($CoFe_2O_4$) are important magnetic materials used in high-density data storage, ferrofluids, and electronic devices [32, 33]. Thus, in this chapter, nano-composite fibers made from polyacrylonitrile (PAN) or polyvinyl alcohol (PVA) containing these components, such as multi walled CNTs, Fe_3O_4, and $CoFe_2O_4$ nanoparticles, were fabricated by electrospinning.

Incorporating CNTs, Fe_3O_4, and $CoFe_2O_4$ nanoparticles was expected to give rise to fibers with multifunctionality and EMI shielding properties, which are helpful for EMI SE applications.

20.2 PAN Based Nano-composite Fibers for EMI SE [34, 35]

Polyacrylonitrile (PAN) with an average molecular weight of 150,000 was used as matrix and multiwalled CNTs with diameter of 20–30 nm were used as fillers. All other reagents in this work were of analytical grade and used as received without further purification.

20.2.1 Fabrication of $CoFe_2O_4$/PAN Nanofibers by Electrospinning

20.2.1.1 Treatment of CNTs

Pristine CNTs were first dispersed into nitric acid by ultrasonication. This CNT suspension was then transferred into a round-bottom flask equipped with magnetic stirring bar and a reflux condenser and refluxed with vigorous stirring. The treated CNTs were washed repeatedly with deionized water and ethanol to remove any residual nitric acid. The CNTs were collected by filtration and dried in a vacuum drying oven.

20.2.1.2 Fabrication of $CoFe_2O_4$/PAN Nanofibers

$CoFe_2O_4$ nanoparticles were synthesized by hydrothermal method. The synthesized $CoFe_2O_4$ particles were dispersed in DMF by ultrasonication, and the PAN was added into this solution. The PAN concentration was kept at a specified content. After complete stirring, the solution was used to spin the nanofibers by electrospinning. As a comparison, nanofibers with several $CoFe_2O_4$ loadings were prepared.

20.2.2 Fabrication of PAN/CNT-$CoFe_2O_4$ Nanofibers by Electrospinning

The treated CNTs and the synthesized $CoFe_2O_4$ nanoparticles as mentioned above were added to DMF and dispersed by ultrasonication. PAN was then dissolved into this suspension with magnetic stirring. After that, a more vigorous agitation was carried out to make sure the electrospinning solution was completely mixed.

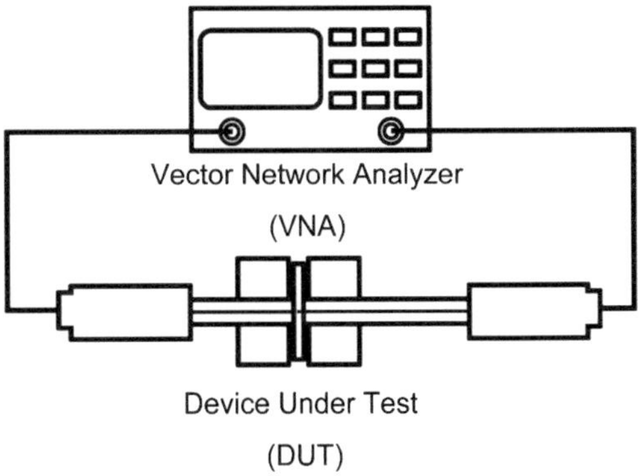

Fig. 20.1 Schematic diagram showing the system used to measure EMI SE (Reproduced from Ref. [67] by permission of SAGE Publications)

The mass fractions of CNTs and $CoFe_2O_4$ nanoparticles to dissolved PAN were set at CNT:PAN = 5 %, 10 %; and $CoFe_2O_4$:PAN = 5 %, 10 %. The concentration of PAN to DMF (PAN: DMF) was adjusted between 7 and 8 %.

The electrospinning process was carried out. A voltage of 8–12 kV was applied at ambient temperature and a polymer jet was formed under the influence of the external electric field, and the fabricated nano-composite fibers were collected using a rotating disk collector. The fiber samples will be referred to by their mass fractions (CNT:PAN and $CoFe_2O_4$:PAN), and two nano-composite fibers, PAN/5CNT-10$CoFe_2O_4$ (CNT:PAN = 5 %, $CoFe_2O_4$:PAN = 10 %) and PAN/10CNT-5$CoFe_2O_4$ (CNT:PAN = 10 %, $CoFe_2O_4$:PAN = 5 %), were prepared. As a control, single-component PAN fibers (PAN/5CNT, PAN/10$CoFe_2O_4$) were also prepared.

20.2.3 Evaluation Methods

Raman spectra were recorded using a HoloLab series 5,000 Raman spectroscope (Kaiser Optical Systems, Inc., USA) at a 514 nm laser excitation. Morphologies of as-prepared $CoFe_2O_4$ nanoparticles and the electrospun fibers were observed by FESEM (S-5000, Hitachi) and scanning electron microscope (SEM, S-3400 N, Hitachi, Japan) with an operating voltage of 20.0 kV. Analyses of X-ray diffraction (XRD) patterns were carried out on a Rigaku Geigerflex 2028 diffractometer (Rigaku, Japan) with Cu Kα radiation (1.5406 Å). Transmission electron microscopy (TEM) observations were performed on JEM-2010 electron microscope at 200 kV accelerating voltage, and the electrospun fibers were directly collected on a copper grid with carbon film during the electrospinning process.

EMI SE was estimated by using an EMI SE test system as shown in Fig. 20.1. Samples with a diameter of 12 mm were used to match the holder of the equipment. The measurable frequency region is from 0.5 to 18 GHz. Magnetic property was measured on a vibrating sample magnetometer (VSM) with a maximum magnetic field of $\pm$10 kOe at room temperature.

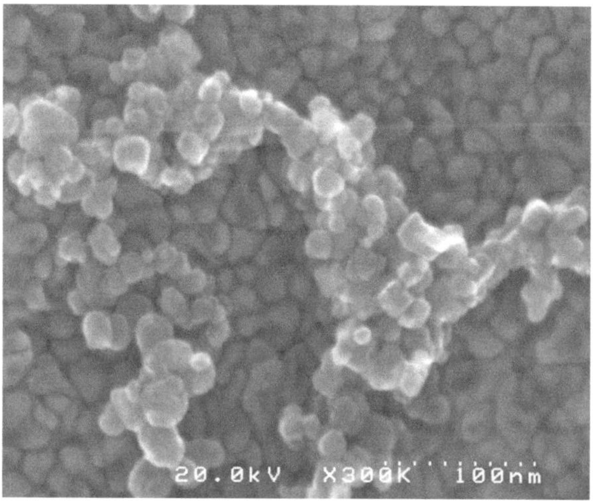

Fig. 20.2 FESEM image of $CoFe_2O_4$ particles synthesized by hydrothermal method [35]

20.2.4 Characterization of $CoFe_2O_4$/PAN Nanofibers

Prior to fabrication of PAN fibers containing carbon nanotubes and cobalt ferrite, the $CoFe_2O_4$/PAN nanofibers were first fabricated here.

The FESEM image of synthesized $CoFe_2O_4$ particles is shown in Fig. 20.2. The particles with diameter about 10–25 nm are observed, and the irregular shape of the particle could be attributed to the aggregation and surfactant-free condition.

Figure 20.3 shows the microscope images of the $CoFe_2O_4$/PAN nanofibers with various $CoFe_2O_4$ loadings (1, 4, 7, 10 wt%). It could be observed that the $CoFe_2O_4$/PAN nanofibers were prepared, and the diameter distribution was controlled at 200–400 nm for all of the $CoFe_2O_4$ loadings.

As shown in Fig. 20.4, obvious peaks could be observed in the XRD pattern of $CoFe_2O_4$, which is in a good agreement with the standard pattern of the $CoFe_2O_4$. It confirmed that the $CoFe_2O_4$ particles were synthesized by hydrothermal method successfully in this work. In addition, the prepared $CoFe_2O_4$/PAN nanofibers with 10 wt% $CoFe_2O_4$ loading were also characterized by XRD. Due to the effect of polymer PAN, the intensity of the peaks decreased significantly. Although some weak peaks were covered, the main characteristic peaks could be easily identified, and all of them still matched with the pattern of $CoFe_2O_4$. It suggested that the $CoFe_2O_4$ were embedded in the PAN.

20.2.5 PAN/CNT-$CoFe_2O_4$ Nanofibers

A uniform dispersion of CNTs in the electrospinning solution is critical for fabricating nano-composite fibers. Poorly dispersed CNTs always yield beaded morphologies, even at low concentrations [36]. However, van der Waals' forces

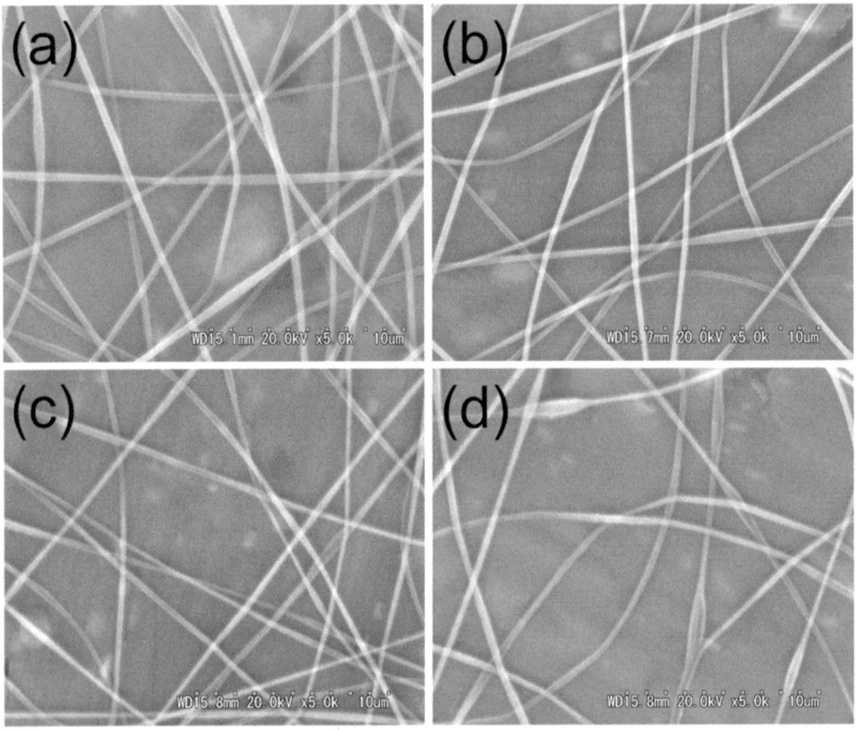

Fig. 20.3 SEM images of $CoFe_2O_4$/PAN nanofibers with various $CoFe_2O_4$ loadings: (**a**) 1 wt%, (**b**) 4 wt%, (**c**) 7 wt%, (**d**) 10 wt% [35]

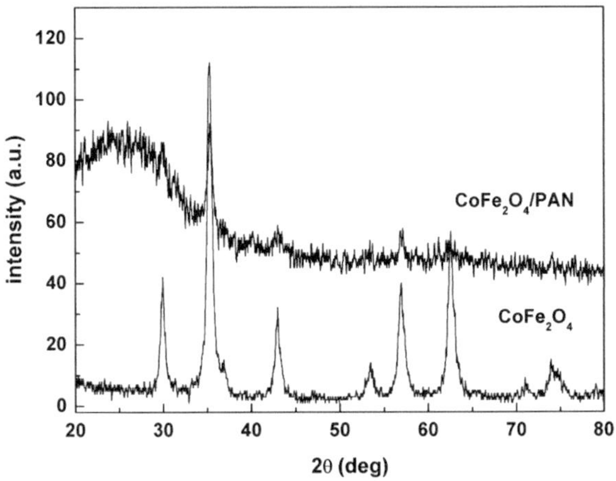

Fig. 20.4 XRD patterns of $CoFe_2O_4$ and $CoFe_2O_4$/PAN nanofibers [35]

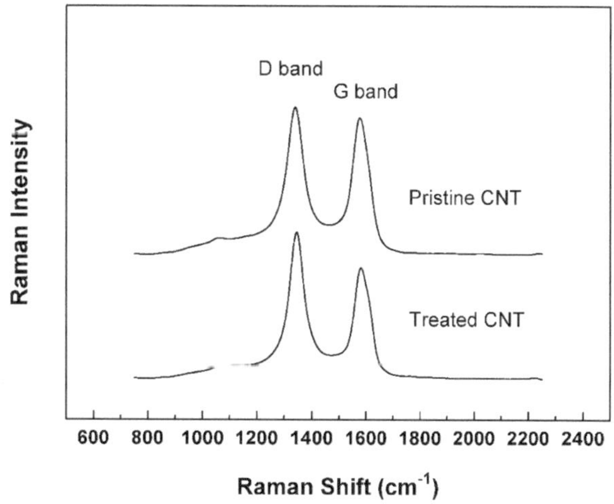

Fig. 20.5 Raman spectra of pristine and treated CNTs (Reproduced from Ref. [34] by permission of John Wiley & Sons Ltd)

make these CNTs readily aggregate, which is the reason for the difficulty in dispersing CNTs in liquid [37, 38]. To overcome this problem, the CNTs were treated with concentrated nitric acid, which has proved to be an effective method for improving the dispersion of CNTs in liquid [38, 39].

Raman spectroscopy was used to investigate the structures of the CNTs before and after treatment, and the spectra are shown in Fig. 20.5. For both pristine and treated CNTs, the patterns display two obvious bands centered around 1,343 cm^{-1} and 1,580 cm^{-1}, which may correspond to the D and G bands. The D band originates from disordered structures or defects in graphite carbon, and the G band could be associated with tangential stretching vibrations of the C-C bonds present in graphite [40]. Treatment by concentrated nitric acid may introduce functional groups (such as carboxyl, hydroxyl) onto the CNTs, the presence of which is considered to lead to an increase in disordered structures. Therefore, the intensity ratio of D and G band ($R = I_D/I_G$) is an index for disorder density [41] and was used to evaluate the effectiveness of the treatment on the CNTs. From Fig. 20.5, the R value is 1.03 for pristine CNTs, while the treated CNTs show a value of 1.32. The increase of R value indicates the presence of an increased disordered structure in the CNTs after treatment, presumably owing to the introduction of functional groups. We assume that the treatment was limited to the CNT surface because of the similar Raman spectra patterns for pristine and treated CNTs shown in Fig. 20.5.

Photographs of several kinds of electrospun fiber sheets (PAN, PAN/5CNT, PAN/10$CoFe_2O_4$, PAN/5CNT-10$CoFe_2O_4$) are shown in Fig. 20.6. The photographs provide visual evidence that regular sheets of nano-composite fibers were fabricated by electrospinning. Compared with the white PAN fiber sheet (Fig. 20.6a), the PAN/5CNT and PAN/10$CoFe_2O_4$ fiber sheets (Fig. 20.6b and 6c) are deep black and light brown in color, while the PAN/5CNT-10$CoFe_2O_4$ fiber sheet (Fig. 20.6d) shows a mixture of colors from Fig. 20.6b, c. These variations in the color of the electrospun fiber sheets could be attributed to the incorporation of CNTs and

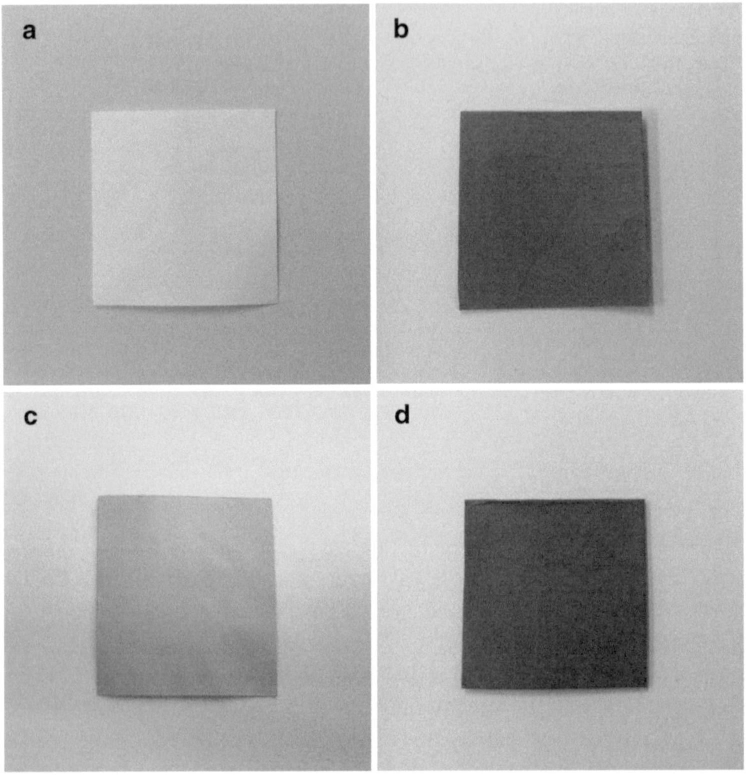

Fig. 20.6 Photographs of electrospun fiber sheets: (**a**) PAN; (**b**) PAN/5CNT; (**c**) PAN/10$CoFe_2O_4$; (**d**) PAN/5CNT-10$CoFe_2O_4$ (Reproduced from Ref. [34] by permission of John Wiley & Sons Ltd)

$CoFe_2O_4$ nanoparticles into PAN. Color also indicates the presence of CNTs and $CoFe_2O_4$ nanoparticles in the PAN fibers.

The SEM images (Fig. 20.7) give further detailed information about the electrospun fibers. The PAN fibers with 200–400 nm diameters display a smooth surface, and no beads are observed (Fig. 20.7a). Compared with PAN fibers, there is no obvious difference for PAN/10$CoFe_2O_4$ fibers (Fig. 20.7b), except for a slight increase in diameter. This suggests that adding of $CoFe_2O_4$ nanoparticles to a PAN electrospinning solution hardly affected the appearance and size of the electrospun fibers. Figure 20.7c shows PAN/5CNT fibers with a further increased diameter of about 400–700 nm. This could be attributed to the increase of solution viscosity because of the introduction of CNTs [42]. The regular morphology of the fibers also suggests a uniform CNT dispersion through the PAN solution even at high CNT concentrations. Figure 20.8d illustrates PAN/5CNT-10$CoFe_2O_4$ fibers fabricated by electrospinning; the larger diameter of about 700–1,000 nm may be the result of a further increase in viscosity when CNTs and $CoFe_2O_4$ nanoparticles were added into the electrospinning solution. In addition, we found that the surface of the fibers

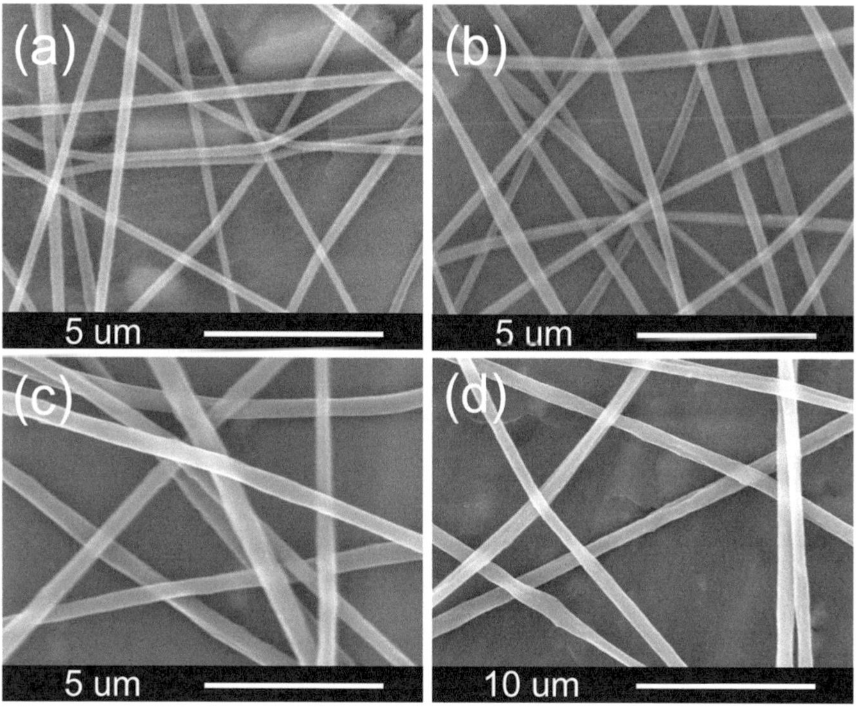

Fig. 20.7 SEM images of electrospun fibers: (**a**) PAN; (**b**) PAN/10$CoFe_2O_4$; (**c**) PAN/5CNT; (**d**) PAN/5CNT-10$CoFe_2O_4$ (Reproduced from Ref. [34] by permission of John Wiley & Sons Ltd)

become a little rough in some places; this could be attributed to a slight aggregation of CNTs and $CoFe_2O_4$ nanoparticles inside the fibers.

To investigate the crystal structures present in the nano-composite fibers, XRD patterns from PAN/5CNT and PAN/5CNT-10$CoFe_2O_4$ fibers were taken and are shown in Fig. 20.8. For PAN/5CNT fibers, two strong diffraction peaks are observed (Fig. 20.8a). The peak around $2\theta = 17^{\circ}$ could correspond to orthorhombic PAN (110) [26]. The other peak at $2\theta = 26^{\circ}$ is assigned to the (002) plane, which corresponds to the graphene layers of CNT [42]. Therefore, the appearance of these two characteristic diffraction peaks proves the existence of PAN/CNT composites. As shown in Fig. 20.9b, besides the two diffraction peaks shown in Fig. 20.8a, several other strong diffraction peaks are also observed. These peaks may be assigned to the (220), (311), (222), (400), (422), (511), and (440) crystal planes, respectively, which are a good match with the spinel structure of $CoFe_2O_4$ (JCPDS 22-1086). The presence of the spinel structure confirms that the $CoFe_2O_4$ was synthesized by a hydrothermal process. The average particle size of $CoFe_2O_4$ was further estimated using the Scherrer formula, and the calculation result based on the peak indexed (311) was about 17 nm. From the XRD pattern shown in Fig. 20.9b, we conclude that both CNTs and nanoparticles were incorporated into the PAN matrix.

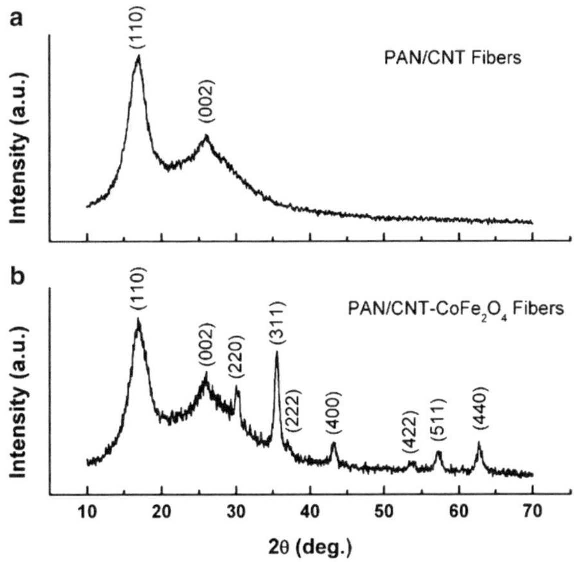

Fig. 20.8 XRD patterns of (**a**) PAN/5CNT fibers and (**b**) PAN/5CNT-10$CoFe_2O_4$ fibers (Reproduced from Ref. [34] by permission of John Wiley & Sons Ltd)

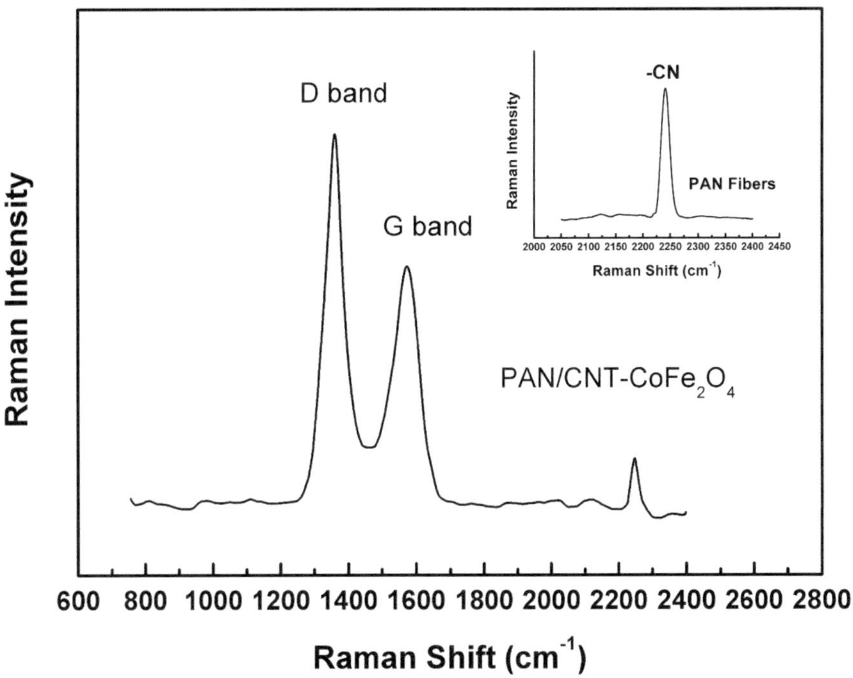

Fig. 20.9 Raman spectrum of PAN/5CNT-10$CoFe_2O_4$ fibers; Inset shows the Raman spectrum of PAN fibers [35]

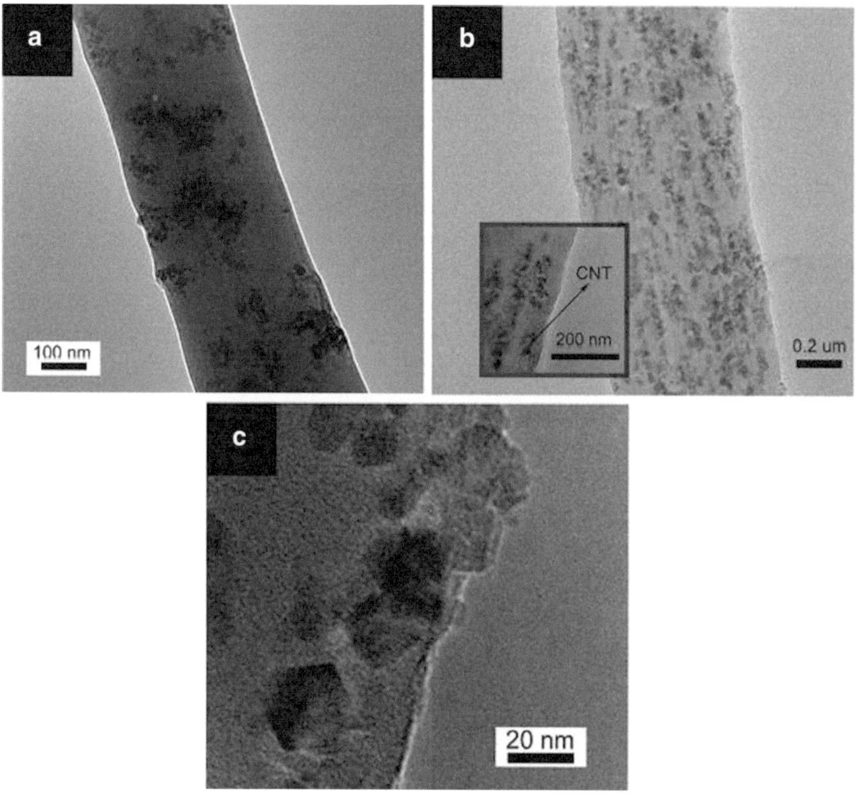

Fig. 20.10 TEM images of (**a**) PAN/10$CoFe_2O_4$ fiber, (**b**) PAN/5CNT-10$CoFe_2O_4$ fiber, and (**c**) $CoFe_2O_4$ nanoparticles on the surface of composite fiber (Reproduced from Ref. [34] by permission of John Wiley & Sons Ltd)

The presence of CNTs inside the nano-composite fibers was further confirmed using Raman spectroscopy, and the obtained spectrum of PAN/5CNT-10$CoFe_2O_4$ fibers is shown in Fig. 20.9, where three obvious bands are observed. By comparing with the spectrum of PAN fibers (shown in the inset) we can confirm that the band centered at 2,240 cm^{-1} comes from PAN and corresponds to the nitrile group (–CN) [43]. The other two bands that could be assigned to D and G bands come from CNTs – referring to the spectra in Fig. 20.5. Therefore, the Raman spectra (as shown in Fig. 20.9) prove the incorporation of CNTs into the PAN matrix. Compared with Fig. 20.5, we also note that there is a slight shift of Raman bands and an increase of $R = I_D/I_G$ for the CNTs incorporated in PAN that may be attributed to the interactions between CNTs and PAN [44].

TEM observations were performed to study the distribution of CNTs and $CoFe_2O_4$ nanoparticles inside the nano-composite fibers. For PAN/10$CoFe_2O_4$ fibers (Fig. 20.10a), it could be clearly observed that $CoFe_2O_4$ nanoparticles are embedded in PAN, although the distribution is not homogeneous due to the aggregation of $CoFe_2O_4$ nanoparticles. The nano-scale size, high concentration, and lack of subsequent surface modification of these as-prepared $CoFe_2O_4$ particles may be the reasons for aggregation during the electrospinning process.

Figure 20.10b shows a PAN/5CNT-10$CoFe_2O_4$ fiber. The $CoFe_2O_4$ nanoparticles inside the PAN fiber exhibit a general ribbon arrangement along the fiber axis, which is very different from the distribution of $CoFe_2O_4$ nanoparticles observed in Fig. 20.10a. In contrast to the obvious presence of $CoFe_2O_4$ nanoparticles, the CNTs are hardly seen. This may be partly attributed to the relative larger size of the fiber. Here, we further assume that the $CoFe_2O_4$ nanoparticles absorb onto the CNTs, and many of the CNTs were encapsulated by these $CoFe_2O_4$ nanoparticles. The functional groups (such as carboxyl) on the surface of CNTs, which was introduced by nitric acid treatment, were considered to serve as attachment points to anchor the other nanoparticles [45]. CNTs have been observed to be well arranged along the axis in polymer fibers by electrospinning [46, 47]; this also explains the ribbon arrangement of $CoFe_2O_4$ nanoparticles inside the PAN fibers. As shown in the inset in Fig. 20.10b, a part of a CNT without encapsulation by $CoFe_2O_4$ nanoparticles is observed, which further supports the assumption. Due to the occasional defects appearing on the surface of the nano-composite fibers, some $CoFe_2O_4$ nanoparticles incompletely embedded inside the PAN fibers are observed – as shown in Fig. 20.10c. The high magnification image clearly shows the synthesized $CoFe_2O_4$ nanoparticles, with a particle size about 10–25 nm, which is in agreement with the estimation result from XRD pattern.

The incorporation of CNTs into a polymer matrix can increase conductivity [48, 49], which means these composites could be possibly used for electromagnetic shielding. Therefore, the PAN/CNT-$CoFe_2O_4$ fibers fabricated in this work were also expected to exhibit shielding properties. To test this, EMI SE of PAN/10CNT-5$CoFe_2O_4$ fiber sheets was evaluated. SE is defined as the attenuation of electromagnetic waves and is expressed as

$$SE = 10 \log \frac{P_O}{P_T} \tag{20.1}$$

where P_O and P_T are power of the incident and transmitted electromagnetic waves through a shielding material [49, 50].

Figure 20.11a shows the shielding curve of a PAN/10CNT-5$CoFe_2O_4$ fiber sheet over 8–12 GHz, and a curve from a PAN/5CNT-10 $CoFe_2O_4$ fiber sheet is presented for comparison. The areal density was controlled at about 9.7 mg/cm^2 for both of the samples. Compared with the PAN/5CNT-10$CoFe_2O_4$ fiber sheet, the PAN/10CNT-5$CoFe_2O_4$ fiber sheet exhibits higher SE. This indicates that the SE of the nano-composite fibers depends on the concentration of CNTs and increases depending on the amount of CNTs in the fiber. Conductivity is considered to be an important factor for shielding, and a higher conductivity should bring a higher SE. Therefore, the increased SE observed in the nano-composite fibers could be related

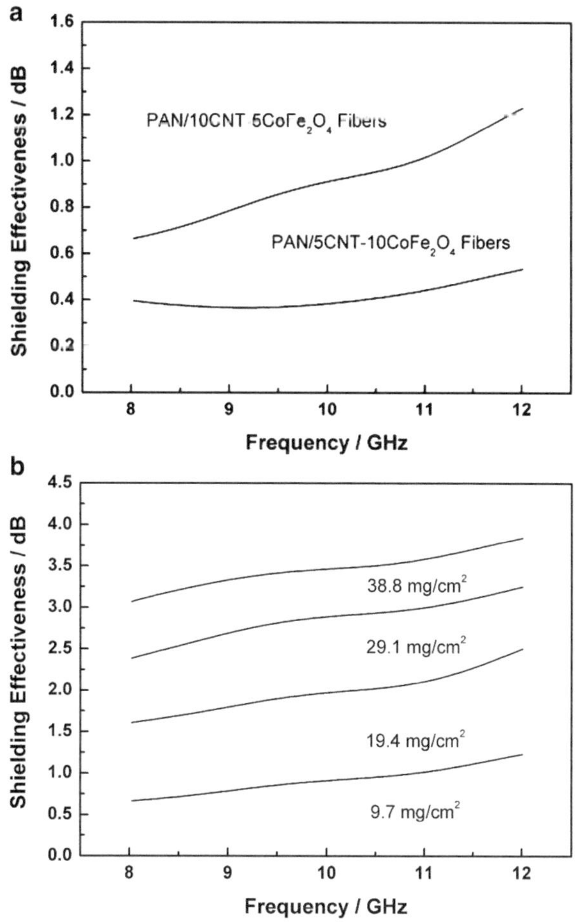

Fig. 20.11 SE curves of (**a**) PAN/5CNT-10$CoFe_2O_4$ and PAN/10CNT-5$CoFe_2O_4$ fiber sheets, (**b**) PAN/10CNT-5$CoFe_2O_4$ fiber sheets with different areal density (Reproduced from Ref. [34] by permission of John Wiley & Sons Ltd)

to the improvement of conductivity owing to the formation of a CNT conductive network. No obvious SE effect from the $CoFe_2O_4$ nanoparticles is observed over this frequency range. In Fig. 20.11b, the SE curves (PAN/10CNT-5$CoFe_2O_4$ fiber sheets) corresponding to different areal densities (9.7, 19.4, 29.1, 38.8 mg/cm^2) are arranged to confirm the effect of areal density.

It is observed that the SE significantly increases accompanied with the increase in areal density. For the sample with an areal density of 38.8 mg/cm^2, an SE about 3.9 dB is achieved at 12 GHz. From the results of our SE evaluation, we can conclude that the SE of PAN/CNT-$CoFe_2O_4$ fibers depend on CNT concentration and areal density: higher CNT concentration and areal density yield higher SE. Thus, there is a possible application for nano-composite fibers in electromagnetic shielding.

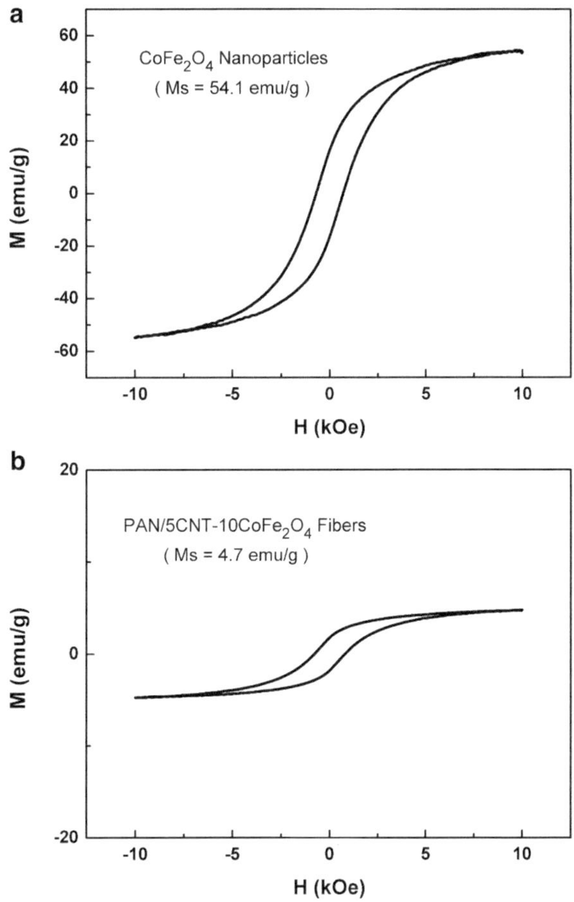

Fig. 20.12 Magnetization hysteresis loops of (**a**) $CoFe_2O_4$ nanoparticles and (**b**) PAN/5CNT-10$CoFe_2O_4$ fibers [35]

To illustrate the magnetic properties of the fibers, the room temperature magnetization curves of PAN/5CNT-10$CoFe_2O_4$ fibers and as-prepared $CoFe_2O_4$ nanoparticles are shown in Fig. 20.12, and typical hysteresis loops are observed for both. For the PAN/5CNT-10$CoFe_2O_4$ fibers, the saturation magnetization (Ms) and coercivity (H_C) are 4.7 emu/g and 0.8 kOe, respectively. The magnetic properties of the PAN/5CNT-10$CoFe_2O_4$ fibers clearly come from the $CoFe_2O_4$ nanoparticles incorporated into the fibers. Based on the $CoFe_2O_4$ loading in PAN/5CNT-10$CoFe_2O_4$ fibers, the calculated value of pure $CoFe_2O_4$ nanoparticles is about Ms = 54 emu/g, which is very close to the experimental value Ms = 54.1 emu/g according to the curve taken from $CoFe_2O_4$ nanoparticles. This indicates that the Ms of the nano-composite fibers is well matched with the proportion of $CoFe_2O_4$ nanoparticles in the nano-composite fibers. That means the magnetic properties of these nano-composite fibers could be accurately tailored by adjusting the $CoFe_2O_4$ nanoparticles loading.

20.3 PVA-Based Nano-composite Fibers for EMI SE [51, 52]

Polyvinyl alcohol (PVA) is one of the most popular polymers used for electrospun composite fiber production due to its thermostability, excellent biocompatibility, and high mechanical performance [53–56]. However, its applications are limited by its high hydrophilicity, through which it dissolves immediately on contact with water. Therefore, PVA nanofibers have usually been modified by either chemical or physical cross-linking to improve their water resistance and mechanical properties [57].

This part describes the fabrication of magnetite nanoparticles (MNPs)/PAA/PVA composite nanofibers with homogeneous dispersion of Fe_3O_4 nanoparticles (NPs) and high water resistance. In our experiments, polyacrylic acid (PAA) was used as a polymer surfactant [58, 59] to stabilize Fe_3O_4 NPs aqueous suspension and facilitate their dispersal in PVA solutions, since to PAA could provide electrostatic and steric repulsion against particle aggregation. Furthermore, PAA was also used as a cross-linking agent to enhance the water resistance of the magnetic composite nanofibers through heat treatment [60]. This method provides a simple and convenient process to prepare homogenous NPs/PVA composite system with high stability and can greatly improve the practical application of NPs/PVA nano-composites.

20.3.1 *Fabrication of Fe_3O_4 NPs/PAA/PVA Nanofibers*

PVA HC with a degree of hydrolysis (DH = 99.9 mol%) and degree of polymerization (DP = 1,700) was kindly provided by Kuraray Co. Ltd, Tokyo, Japan. PAA ($M_w = 5{,}000$), iron (II) sulfate heptahydrate ($FeSO_4 \cdot 7H_2O$), iron (III) chloride anhydrous ($FeCl_3$), ethanol, HCl solution, and ammonia solution (28 % NH_3 in water) were purchased from Wako Pure Chemical Industries, Ltd., Japan. All chemicals were of chemical grade and were used without further purification. Distilled water was used as the solvent.

20.3.1.1 Preparation of Magnetic Fe_3O_4 NPs

Magnetic Fe_3O_4 NPs were prepared by coprecipitating Fe^{2+} and Fe^{3+} ions in aqueous ammonia solution [61]. Next, a mixture of $FeCl_3$ and $FeSO_4$ at pH 2 was prepared under N_2. Then, NH_4OH aqueous solution was dropped into it with violent stirring until the pH of the solution rose to 10. The resulting precipitate was stirred. After triple washes with deionized water and ethanol, the precipitates were resuspended in deionized water. The pure magnetic NPs were isolated from the solution by a magnet bar and dried in vacuum.

20.3.1.2 Preparation of Fe_3O_4 NPs/PVA and Fe_3O_4 NPs/PAA/PVA Suspension

Fe_3O_4 NPs were dispersed in distilled water and PAA aqueous solution, respectively, then both suspensions was treated by ultrasonic mixing to get two kinds of Fe_3O_4 NPs suspension for reserve. PVA chips were dissolved in distilled water in a static rotary mixer and then rapidly heated to a specified temperature, and this temperature was maintained with high-speed stirring (1,000 rpm) to obtain a PVA solution. Varying amounts of Fe_3O_4 NP suspension, PAA solution, and water were then added to the as-prepared PVA solution to adjust the Fe_3O_4 NPs concentration in the PVA to 5, 10, and 15 wt% (weight of solid PVA content) while maintaining the PVA concentration at 8 wt% (PVA/PAA = 2:1) in each composite suspension with high-speed stirring. Finally, two kinds of suspension were obtained: PVA/Fe_3O_4 NPs and PVA/PAA/Fe_3O_4 NPs.

20.3.1.3 Electrospinning

The electrospinning setup (Kato Tech, Kyoto, Japan) used in this study consists of a syringe with a flat-end metal needle (1.20 mm × 38 mm), a syringe pump for controlling the feeding rate, a grounded cylindrical stainless steel mandrel, and a high-voltage direct current (DC) power supply. For the electrospinning procedure, the suspensions were loaded into a syringe coupled with nozzles and electrospun under a high DC voltage of 10 kV, with a distance from the needle tip to the collector of 15 cm and a feeding speed of 0.5 mL/h.

20.3.2 Fabrication of Magnetic Fe_3O_4 NPs/PAA/PVA Nanofibers

The preparation of magnetic nanoparticles/polymer composite systems has been intensively pursued because of these systems' broad applications, including magnetic storage media, ferrofluids, magnetic resonance imaging, and magnetically guided drug delivery. However, one of the main challenges facing almost all of these applications is the ability of the composite systems to deliver well-dispersed nanoparticles with the desired composition, structure, and uniformity as well as the prevention of aggregation. The principal cause of aggregation is the short-range forces –van der Waals attraction – between the two particles [62]. To counteract these attractive interactions and promote stability, equally short-range repulsive forces are required. These are enforced either by electrostatic repulsion between the particles or by coating the particles with organic long-chain molecules.

For fully hydrolyzed PVA aqueous solution, the viscosity is very high even in low concentration (<10 wt%); therefore, it is extremely difficult to prepare

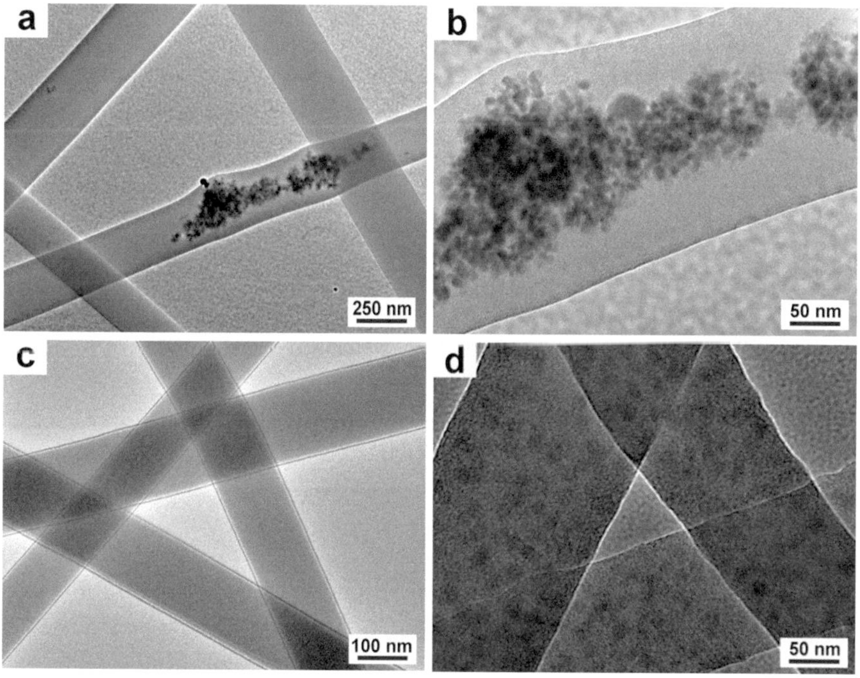

Fig. 20.13 TEM images of Fe_3O_4 NPs/polymer composite nanofibers. (**a**) and (**b**): Fe_3O_4 NPs/PVA composite nanofibers; **c** and **d**: Fe_3O_4 NPs/PAA/PVA composite nanofibers) (Reproduced from Ref. [51] by permission of SAGE Publications)

homogeneously dispersed NPs/PVA composite materials. In our experiments, PAA was used as a polymer surfactant to stabilize NPs aqueous suspension, which could provide electrostatic and steric repulsion against particle aggregation. As seen in Fig. 20.13a, b, Fe_3O_4 NPs randomly formed irregular aggregation in PVA electrospun nanofibers (diameter is about 250 nm), which will cause heterogeneity of magnetism distribution in the magnetic nanofiber mat. But Fe_3O_4 NPs, which were formed prior to dispersal in the PAA solution, can homogenously disperse in PVA composite nanofibers (Fig. 20.13c, d).

The XRD pattern of Fe_3O_4 NPs/PAA/PVA composite nanofibers revealed the characteristic peaks of both the Fe_3O_4 and the PVA. A wide peak was observed at 19.4°, which is in agreement with the XRD peak of the crystalline pure PVA nanofibers, while the other peaks are identical to Fe_3O_4 NPs. The intensity of the peak at 19.4° became stronger along with the content of Fe_3O_4 NPs, which indicated that the existence of Fe_3O_4 NPs facilitated the crystallization of polymer.

As a magnetic material, the magnetic property of the Fe_3O_4/polymer nanofiber nonwoven is important to its applications. The magnetization curves at 25 °C of the Fe_3O_4/polymer nanofiber nonwoven are shown in Fig. 20.14a. The results in

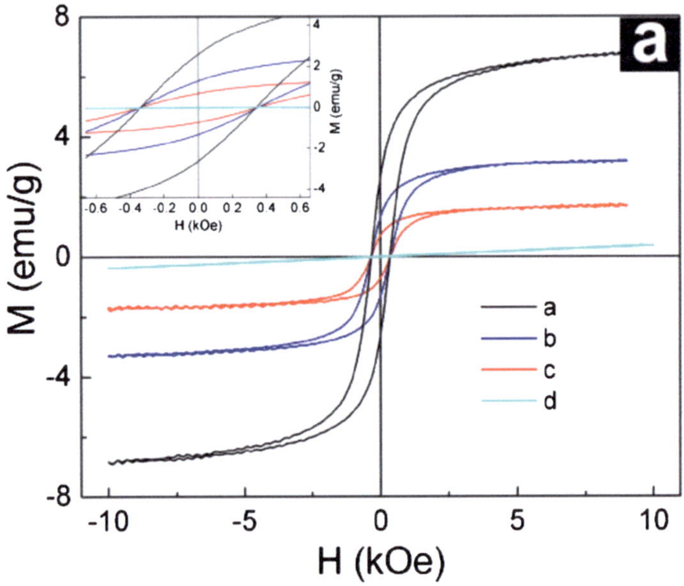

Fig. 20.14 Hysteresis loops of the polymer/Fe_3O_4 hybrids measured at 25 °C (the weight rate of Fe_3O_4-NPs/polymer in Fe_3O_4-NPs/PAA/PVA nanofiber nonwoven: *a* 14.23 wt%, *b* 8.47 wt%, *c* 3.94 wt%, *d* PVA/PAA only) (Reproduced from Ref. [51] by permission of SAGE Publications)

Table 20.1 Content of Fe_3O_4 in magnetic nanofibers and saturation magnetization

Fe_3O_4/polymer (wt%)		
Preparation content	Content (calculated based on TG data)	Saturation magnetization (emu/g)
5	3.94	1.72
10	8.47	3.17
15	14.23	6.77

Table 20.1 show that the magnetic nanofiber nonwoven has saturation magnetization of 1.72 to 6.77 emu/g, which is lower than that of the corresponding bulk (92 emu/g) resulting from the existence of polymer [63].

The coercivity is also about 300 Oe at 25 °C (see the inset of Fig. 20.14). Therefore, the products show a soft ferromagnetic behavior, which is the same as Fe_3O_4. The magnetism of the products was demonstrated by holding a commercial magnet close to the sample and then the magnetic nanofiber nonwoven was attracted to the magnetite, indicating that the nanofiber nonwoven had strong magnetism.

20.3.3 Fabrication of MWCNTs/PVA Nanofibers

20.3.3.1 Modification of CNTs

Another practical method to improve the strength of electrospun PVA nanofibers and enhance the EMI SE performance is to produce composite nanofibers with CNTs. Here, we use the one-step process using ammonium persulfate (APS) as oxidant to prepare water-soluble multiwalled carbon nanotubes (MWCNTs), which is a low-cost, eco-friendly, facile method without the need for organic solvents and acids. Then the APS-treated MWCNTs were mixed with PVA solution to prepare homogenous MWCNTs/PVA aqueous solutions. The solutions had undergone electrospinning using electrospinning equipment with an ultra-high-speed rotating cylinder as a collector to fabricate aligned electrospun nanofibers.

20.3.3.2 Aligned MWCNTs/PVA Nanofibers

One of the biggest problems that limit the application of electrospun nanofibers is the weakness in strength. A practical method to deal with this issue is to fabricate aligned nanofibers. Here, we adjusted the rotating speed of collector during electrospinning process, which is simple and convenient to prepare aligned electrospun nanofibers. SEM images of our samples were shown in Fig. 20.15. From Fig. 20.15a–d, rolling speed of the collector increased from 0 to 2,000 rpm, and we can see clearly that the fibers aligned gradually along with the rolling direction during this process. Especially, when the speed became 2,000 rpm, the fibers were almost unidirectionally aligned and the average angle between each fiber is only about 10°. Although the EMI SE has not been estimated for these currently developed nanofibers, their SE performance is expected to be improved due to the loading of CNTs and aligned nanofibers.

20.4 Design of EMI SE Structures

Here, the composite film of shape memory polymer (SMP) and CNTs are designed for improving EMI SE performance. Although the electrospun fibers are not used in the example, the nanofiber sheets may be expected to be important for an effective design in replace of the SMP/CNTs film for EMI SE applications.

In the fabrication of SMP/CTNs film, CNTs were dispersed in DMF by ultrasonication and a solution of SMP in DMF was added to the suspension of CNTs. The resulting mixture was stirred mechanically and then ultrasonicated. The completely dispersed mixture was then cast in a mold and dried. Samples with various CNT loading (3, 7, and 9 wt%) were prepared.

Based on the as-prepared CNT/SMP film with the thickness of 0.18 mm, the laminated sheet with different thicknesses could be fabricated by simply stacking layer by layer, and then pressed using a hot pressboard.

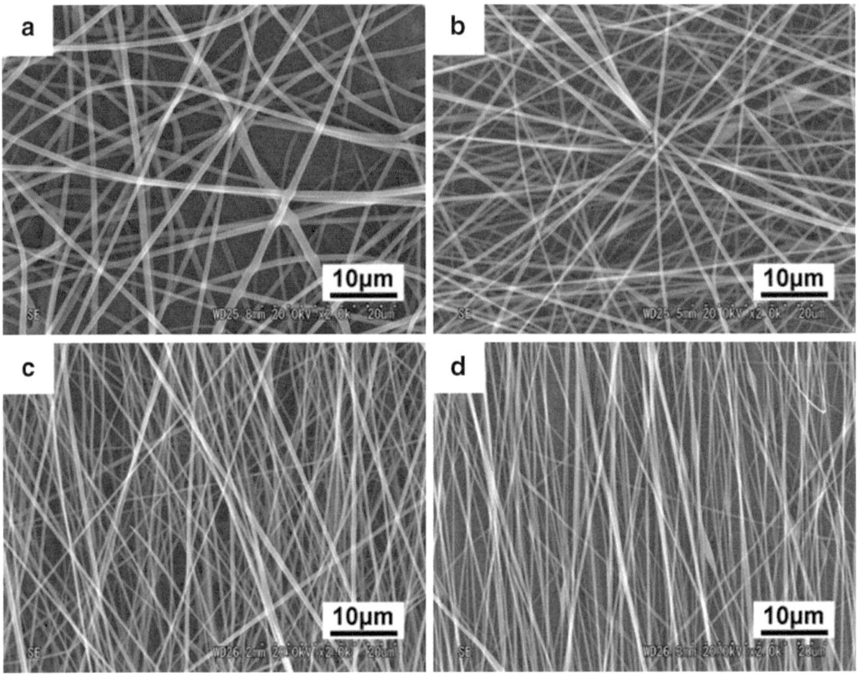

Fig. 20.15 SEM images of APS-treated MWCNTs/PVA electrospun fibers (MWCNTs/PVA = 0.1 wt%). The rolling speeds of collector are (**a**) 0 rpm; (**b**) 500 rpm; (**c**) 1,000 rpm; (**d**) 2,000 rpm [52]

20.4.1 Theory of EMI SE for Sheet Materials

According to Maxwell's equations, an electromagnetic wave is composed of electric and magnetic fields. An alternating electric field generates an alternating magnetic field, which in turn generates an alternating electric field. Therefore, an electromagnetic wave can be considered as a self-propagating transverse alternating wave of electric and magnetic fields. There are two cases that should be considered for EMI shielding, the near-field and far-field. The far-field is defined as distances further than $\lambda/2\pi$ between the radiation source and shielding material, where λ is the wavelength of the electromagnetic wave [64]. Here, the electromagnetic wave used to test shielding materials was limited to a plane wave in the far-field. The SE is expressed as

$$\mathrm{SE} = 10\log\frac{P_O}{P_T} = 20\log\frac{|H_O|}{|H_T|} = 20\log\frac{|E_O|}{|E_T|} \tag{20.2}$$

where P_O, H_O, and E_O are the power, magnetic, and electric field intensity when there is no shielding material; P_T, H_T, E_T are the power, magnetic, and electric field intensity in the presence of a shielding material.

When an electromagnetic wave propagates to a shielding material, a typical process is as follows: part of the wave is reflected on the surface of the material, which strongly depends on the impedance matching; part of the wave is absorbed and dissipated as heat; part of the wave is multi-reflected inside the shielding material; and the others penetrate through the material. Based on the description above, electromagnetic wave shielding is mainly dominated by three functions: reflection (SE_R), absorption (SE_A), and multi-reflection (SE_M), giving Eq. 20.3:

$$SE = SE_{R+} SE_{A+} SE_M \tag{20.3}$$

These three functions further relate to the intrinsic permittivity, permeability, conductivity, and thickness of the material and the frequency of the electromagnetic wave.

The skin depth (δ) is defined as the distance at which the electric field decreases to $1/e$ of its incident strength, which is expressed as follows:

$$\delta = 1/\sqrt{\pi f \mu \sigma} \tag{20.4}$$

where f is the wave frequency, μ is the permeability, and σ is the electrical conductivity in S/m. The prepared CNT/SMP composites can be regarded as nonmagnetic, thus $\mu = \mu_0 = 4\pi \times 10^{-7}$ Hm^{-1}, where μ_0 is the permeability in free space.

When the shielding material is thicker than the skin depth, the multi-reflection is effectively reduced, especially in the case when $SE_A \geq 15$ dB [65]. Therefore, for a very approximate analysis, the multi-reflection can be ignored, and Eq. 20.3 simplifies to

$$SE = SE_{R+} SE_A \tag{20.5}$$

According to Al-Saleh et al. [66], for conductive materials, SE_R and SE_A can be described as follows:

$$SE_R = 39.5 + 10 \log \frac{\sigma}{2\pi f \mu} \tag{20.6}$$

$$SE_A = 8.7 \frac{d}{\delta} = 8.7d\sqrt{\pi f \mu \sigma} \tag{20.7}$$

where d is the thickness of the sample. According to Eqs. 20.6 and 20.7, it is apparent that when f and μ are defined, the reflection depends on the conductivity, while the absorption is determined by both the thickness and conductivity of the material. Therefore, the total SE is a function of the frequency, thickness, and conductivity and is expressed as

$$SE = 39.5 + 10 \log \frac{\sigma}{2\pi f \mu} + 8.7d\sqrt{\pi f \mu \sigma} \tag{20.8}$$

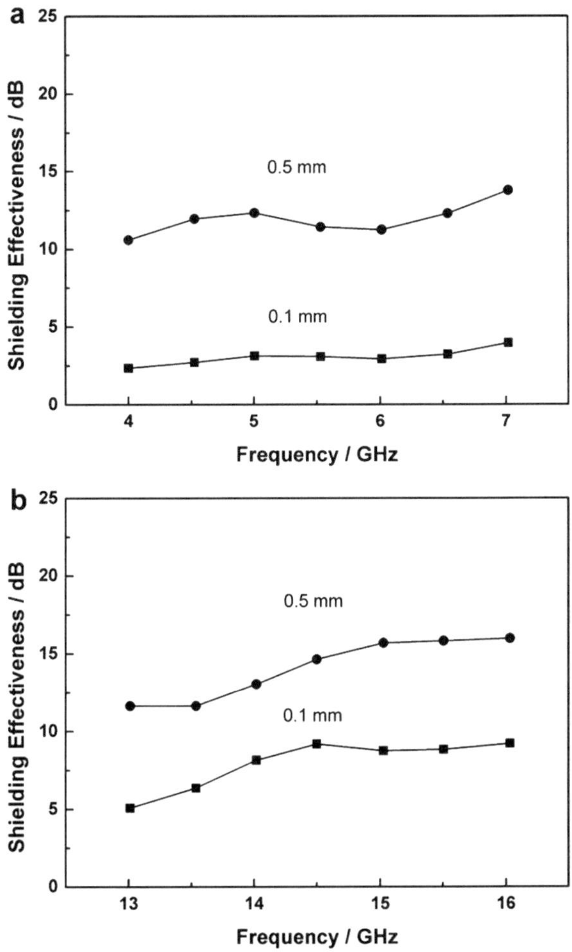

Fig. 20.16 SE of CNT/SMP composites containing 9 wt% CNTs with thicknesses of 0.1 and 0.5 mm over frequency ranges of (**a**) 4–7 GHz and (**b**) 13–16 GHz (Reproduced from Ref. [67] by permission of SAGE Publications)

20.4.2 Shielding Effectiveness (SE) of CNT/SMP Composites Sheets [35, 67]

The EMI SE performance of CNT/SMP composite sheets with various CNT loadings and thicknesses was measured over the frequency ranges of 4–7 and 13–16 GHz. The SE was plotted as a function of frequency to investigate the effect of material thickness (Fig. 20.16). The SE curves were recorded for composites containing 9 wt% CNTs with thicknesses of 0.1 and 0.5 mm.

As shown in Fig. 20.16a, the SE of the sample with a thickness of 0.5 mm is greater over the entire frequency range of 4–7 GHz than that of the sample that

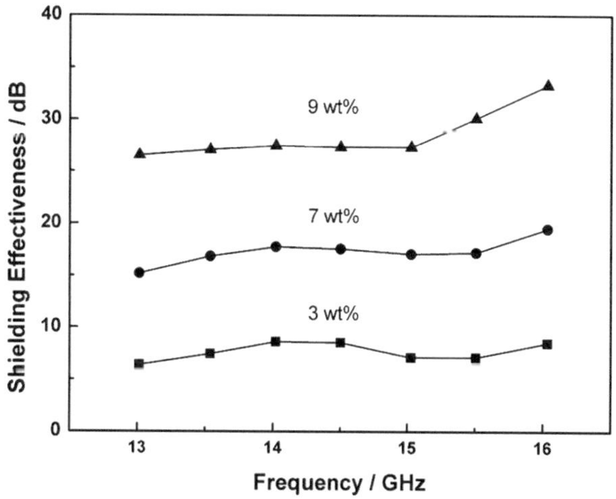

Fig. 20.17 SE of CNT/SMP composites with a thickness of 2 mm with various CNT loadings (3, 7, and 9 wt%) over 13–16 GHz (Reproduced from Ref. [67] by permission of SAGE Publications)

is 0.1 mm thick. Almost the same pattern is observed for the SE curves of the same samples between 13 and 16 GHz (Fig. 20.16b). These results demonstrate that the thickness of the sample had a great effect on the SE, with the thicker sample exhibiting an increased SE. According to Eq. 20.4, the thickness of the samples (0.1 and 0.5 mm) is less than the skin depth, so the influence of multi-reflection cannot be ignored. Therefore, the dependence of the SE on the thickness could be attributed to both the increased absorption and the influence of multi-reflection as the thickness increases.

The SE of samples with a thickness of 2 mm and CNT loadings of 3, 7, and 9 wt% was measured over the frequency range of 13–16 GHz (Fig. 20.17). The SE increases significantly as the CNT loading increases. For the sample containing 3 wt% CNTs, the average SE over 13–16 GHz was 7.68 GHz, while the samples containing 7 and 9 wt% CNTs showed average SE of 17.31 and 28.46 dB, respectively. These results show that the composites with higher conductivity achieved higher SE. A maximum SE of almost 35 dB was realized for the sample containing 9 wt% CNTs at a frequency of 16 GHz.

The samples containing 7 and 9 wt% CNTs are assumed to behave as conductive materials. According to Eq. 20.4, a thickness of 2 mm is greater than the skin depth. Therefore, the SE of these two samples was calculated using Eq. 20.8 to compare with the experiment results. As shown in Fig. 20.18, the calculated SE exhibits a fairly similar trend to the experiment results for the samples containing 7 and 9 wt% CNTs. However, the difference between the calculated and observed results cannot be ignored. Figure 20.18 shows that Eq. 20.8 provides approximate predictions for the SE of 2 mm thick samples containing 7 and 9 wt% CNTs over 13–16 GHz. Without regard to precise measurements, the difference present between the experimental and calculated SE may be attributed to the neglect of multi-reflection and the assumption that the composites are conductive materials.

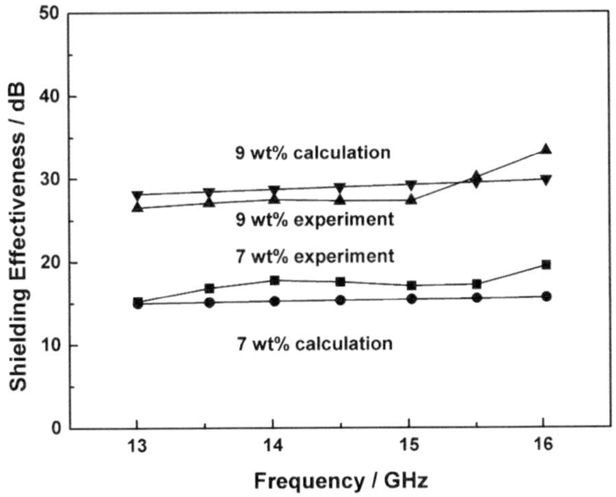

Fig. 20.18 Comparison between calculated and experiment results for the SE of CNT/SMP composites with a thickness of 2 mm and CNT loadings of 7 and 9 wt% (Reproduced from Ref. [67] by permission of SAGE Publications)

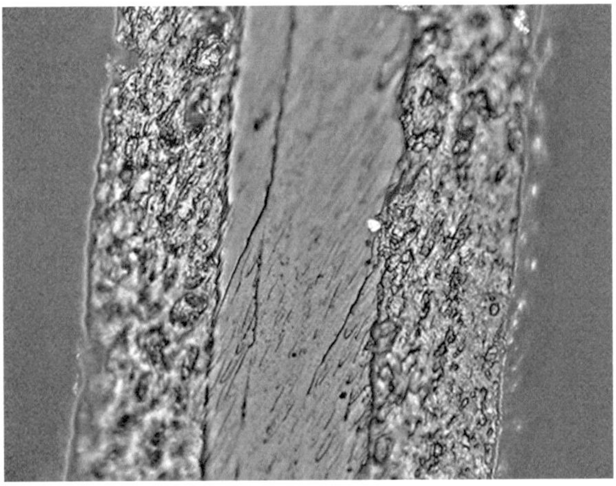

Fig. 20.19 Optical micrograph of layered structure [35]

20.4.3 CNT/SMP Composites Sheet with Layered Structure [35, 67]

As shown in Fig. 20.19, the image presents a clear section of CNT/SMP composites with layered structure, and it could be observed that an SMP layer was inserted between two layers of CNT/SMP composites.

The SE curve is shown in Fig. 20.20, and two obvious peaks are observed. This enhancement of SE over specific frequency range could be attributed to the insertion of SMP layer. It is supposed that the inserted SMP layer (an insulated medium with certain thickness) between CNT/SMP composites layers (as shown in Fig. 20.19) maybe lead to the neutralization of the electromagnetic wave with specific wave length by overlapping with each other, and this resulted in a functional EMI shielding at a specified frequency region.

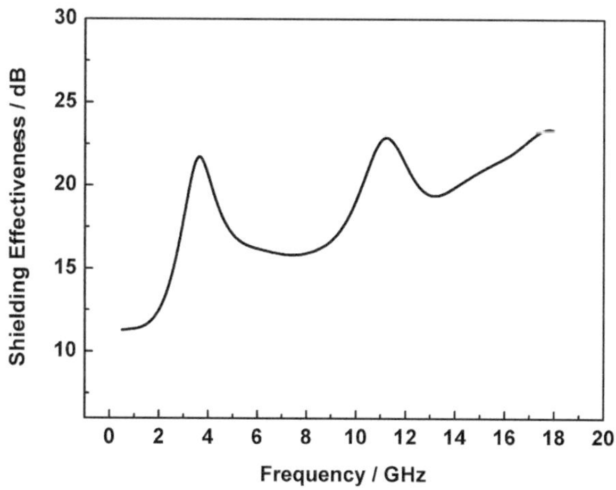

Fig. 20.20 SE of SI as function of frequency in 0.5–18 GHz [35]

20.5 Summary

In this chapter, nano-composite fibers made from polyacrylonitrile (PAN) or polyvinyl alcohol (PVA) containing the components such as multiwalled CNTs, magnetic nanoparticles Fe_3O_4, and $CoFe_2O_4$ were fabricated by electrospinning. The results showed the developed electrospun nanofibers incorporating CNTs, Fe_3O_4, and $CoFe_2O_4$ nanoparticles were suitable for the applications in EMI SE field. The theoretical analysis has shown reasonable results with the experiments. A layered structure may lead to a flexible and optimum design in EMI shielding applications.

References

1. Dexmer J, Leroy CM, Binet L, Heresanu V, Launois P, Steunou N, Coulon C, Maquet J, Brun N, Livage J, Backov R (2008) Vanadium oxide-PANI nanocomposite-based macroscopic fibers: 1D alcohol sensors bearing enhanced toughness. Chem Mater 20(17):5541–5549. doi:10.1021/cm800886v
2. Guo H, Tao S (2007) Silver nanoparticles doped silica nanocomposites coated on an optical fiber for ammonia sensing. Sens Actuat B Chem 123(1):578–582. doi:10.1016/j.snb.2006.09.055
3. Luoh R, Hahn HT (2006) Electrospun nanocomposite fiber mats as gas sensors. Compos Sci Technol 66(14):2436–2441. doi:10.1016/j.compscitech.2006.03.012
4. Yeo SY, Jeong SH (2003) Preparation and characterization of polypropylene/silver nanocomposite fibers. Polym Int 52(7):1053–1057. doi:10.1002/pi.1215
5. Yeo SY, Lee HJ, Jeong SH (2003) Antibacterial effect of nanosized silver colloidal solution on textile fabrics. J Mater Sci 38(10):2199–2204. doi:10.1023/A:1023736416361

6. Li Y, Chung T (2008) Exploratory development of dual-layer carbon-zeolite nanocomposite hollow fibre membranes with high performance for oxygen enrichment and natural gas separation. Microporous Mesoporous Mater 113(1–3):315–324. doi:10.1016/j.micromeso.2007.11.038
7. Alshebani A, Pera-Titus M, Landrivon E, Schiestel T, Miachon S, Dalmon JA (2008) Nanocomposite MFI—ceramic hollow fibres: prospects for CO_2 separation. Microporous Mesoporous Mater 115(1–2):197–205. doi:10.1016/j.micromeso.2007.11.050
8. Jiang LY, Chung TS, Kulprathipanja S (2006) Fabrication of mixed matrix hollow fibers with intimate polymer–zeolite interface for gas separation. AIChE J 52(8):2898–2908. doi:10.1002/aic.10909
9. Wolf H, Pajkic Z, Gerdes T, Willert-Porada M (2009) Carbon-fiber-silicon-nanocomposites for lithium-ion battery anodes by microwave plasma chemical vapor deposition. J Power Sources 190(1):157–161. doi:10.1016/j.jpowsour.2008.07.035
10. Lu HW, Li D, Sun K, Li YS, Fu ZW (2009) Carbon nanotube reinforced NiO fibers for rechargeable lithium batteries. Solid State Sci 11(5):982–987. doi:10.1016/j.solidstatesciences.2009.02.014
11. Uddin MJ, Cesano F, Bonino F, Bordiga S, Spoto G, Scarano D, Zecchina A (2007) Photoactive TiO_2 films on cellulose fibres: synthesis and characterization. J Photochem Photobiol A 189(2–3):286–294. doi:10.1016/j.jphotochem.2007.02.015
12. Jin M, Zhang X, Emeline AV, Liu Z, Tryk DA, Murakami T, Fujishima A (2006) Fibrous TiO_2–SiO_2 nanocomposite photocatalyst. Chem Commun 43:4483–4485. doi:10.1039/B611580B
13. Zhang D, Karki AB, Rutman D, Young DP, Wang A, Cocke D, Ho TH, Guo Z (2009) Electrospun polyacrylonitrile nanocomposite fibers reinforced with Fe3O4 nanoparticles: fabrication and property analysis. Polymer 50(17):4189–4198. doi:10.1016/j.polymer.2009.06.062
14. Chen D, Liu T, Zhou X, Tjiu WC, Hou H (2009) Electrospinning fabrication of high strength and toughness polyimide nanofiber membranes containing multiwalled carbon nanotubes. J Phys Chem B 113(29):9741–9748. doi:10.1021/jp9025128
15. Zhang H, Song H, Yu H, Bai X, Li S, Pan G, Dai Q, Wang T, Li W, Lu S, Ren X, Zhao H (2007) Electrospinning preparation and photoluminescence properties of rare-earth complex/polymer composite fibers. J Phys Chem C 111(17):6524–6527. doi:10.1021/jp0684123
16. Wang S, Wang C, Zhang B, Sun Z, Li Z, Jiang X, Bai X (2010) Preparation of Fe_3O_4/PVA nanofibers via combining in-situ composite with electrospinning. Mater Lett 64(1):9–11. doi:10.1016/j.matlet.2009.09.043
17. Wang B, Sun Y, Wang H (2010) Preparation and properties of electrospun PAN/Fe_3O_4 magnetic nanofibers. J Appl Polym Sci 115(3):1781–1786. doi:10.1002/app.31288
18. Song T, Zhang YZ, Zhou TJ, Lim CT, Ramakrishna S, Liu B (2005) Encapsulation of self-assembled FePt magnetic nanoparticles in PCL nanofibers by coaxial electrospinning. Chem Phys Lett 415(4–6):317–322. doi:10.1016/j.cplett.2005.09.035
19. Chung DLL (2001) Electromagnetic interference shielding effectiveness of carbon materials. Carbon 39(2):279–285. doi:10.1016/S0008-6223(00)00184-6
20. Shinagawa S, Kumagai Y, Urabe K (1999) Conductive papers containing metallized polyester fibers for electromagnetic interference shielding. J Porous Mater 6(3):185–190. doi:10.1023/A:1009619711017
21. Heiser JA, King JA, Konell JP, Sutter LL (2004) Shielding effectiveness of carbon-filled nylon 6,6. Polym Compos 25(4):407–416. doi:10.1002/pc.20034
22. Lee BO, Woo WJ, Park HS, Hahm HS, Wu JP, Kim MS (2002) Influence of aspect ratio and effect on EMI shielding of coating materials fabricated with carbon nanofiber/PVDF. J Mater Sci 37(9):1839–1843. doi:10.1023/A:1014970528482
23. Das NC, Chaki TK, Khastgir D (2001) Electromagnetic interference shielding effectiveness of conductive carbon black and carbon fiber-filled composites based on rubber and rubber blends. Adv Polym Technol 20(3):226–236. doi:10.1002/adv.1018
24. Das NC, Khasgir D, Chaki TK, Chakraborty A (2000) Electromagnetic interference shielding effectiveness of carbon black and carbon fiber filled EVA and NR based composites. Compos Part A 31(10):1069–1081. doi:10.1016/S1359-835X(00)00064-6

25. Luo X, Chung DDL (1999) Electromagnetic interference shielding using continuous carbon-fiber carbon-matrix and polymer-matrix composites. Compos Part B 30(3):227–231. doi:10.1016/S1359-8368(98)00065-1
26. Heikkilä P, Harlin A (2009) Electrospinning of polyacrylonitrile (PAN) solution: effect of conductive additive and filler on the process. Expr Polym Lett 3(7):437–445. doi:10.3144/expresspolymlett.2009.53
27. Chen J, Li Z, Chao D, Zhang W, Wang C (2008) Synthesis of size-tunable metal nanoparticles based on polyacrylonitrile nanofibers enabled by electrospinning and microwave irradiation. Mater Lett 62(4–5):692–694. doi:10.1016/j.matlet.2007.06.047
28. Ma CCM, Huang YL, Kuan HC, Chiu YS (2005) Preparation and electromagnetic interference shielding characteristics of novel carbon-nanotube/siloxane/poly-(urea urethane) nanocomposites. J Polym Sci Polym Phys 43(4):345–358. doi:10.1002/polb.20330
29. Chae HG, Sreekumar TV, Uchida T, Kumar A (2005) A comparison of reinforcement efficiency of various types of carbon nanotubes in polyacrylonitrile fiber. Polymer 46(24):10925–10935. doi:10.1016/j.polymer.2005.08.092
30. Vivien L, Lançon P, Riehl D, Hache F, Anglaret E (2002) Carbon nanotubes for optical limiting. Carbon 40(10):1789–1797. doi:10.1016/S0008-6223(02)00046-5
31. Meincke O, Kaempfer D, Weickmann H, Friedrich C, Vathauer M, Warth H (2004) Mechanical properties and electrical conductivity of carbon-nanotube filled polyamide-6 and its blends with acrylonitrile/butadiene/styrene. Polymer 45(3):739–748. doi:10.1016/j.polymer.2003.12.013
32. Khedr MH, Omar AA, Abdel-Moaty SA (2006) Magnetic nanocomposites: preparation and characterization of Co-ferrite nanoparticles. Colloid Surf A 281(1–3):8–14. doi:10.1016/j.colsurfa.2006.02.005
33. Ju YW, Park JH, Jung HR, Cho SJ, Lee WJ (2008) Fabrication and characterization of cobalt ferrite ($CoFe_2O_4$) nanofibers by electrospinning. Mater Sci Eng B-Solid 147(1):7–12. doi:10.1016/j.mseb.2007.10.018
34. Jin XD, Ni QQ, Fu YQ, Zhang L, Natsuki T (2012) Electrospun nanocomposite polyacrylonitrile fibers containing carbon nanotubes and cobalt ferrite. Polym Compos 33(3):317–323. doi:10.1002/pc.21251
35. Jin XD (2012) Doctoral thesis, development and characterization of carbon nanotube based functional nanocomposites. Shinshu University, Ueda
36. Ra EJ, An KH, Kim KK, Jeong SY, Lee YH (2005) Anisotropic electrical conductivity of MWCNT/PAN nanofiber paper. Chem Phys Lett 413(1–3):188–193. doi:10.1016/j.cplett.2005.07.061
37. Chen IH, Wang CC, Chen CY (2010) Fabrication and structural characterization of polyacrylonitrile and carbon nanofibers containing plasma-modified carbon nanotubes by electrospinning. J Phys Chem C 114(32):13532–13539. doi:10.1021/jp103993b
38. Sahoo NG, Rana S, Cho JW, Li L, Chan SH (2010) Polymer nanocomposites based on functionalized carbon nanotubes. Prog Polym Sci 35(7):837–867. doi:10.1016/j.progpolymsci.2010.03.002
39. Xie XL, Mai YW, Zhou XP (2005) Dispersion and alignment of carbon nanotubes in polymer matrix: a review. Mater Sci Eng R 49(4):89–112. doi:10.1016/j.mser.2005.04.002
40. McNally T, Pötschke P, Halley P, Murphy M, Martin D, Bell SEJ, Brennan GP, Bein D, Lemoine P, Quinn JP (2005) Polyethylene multiwalled carbon nanotube composites. Polymer 46(19):8222–8232. doi:10.1016/j.polymer.2005.06.094
41. Osswald S, Havel M, Gogotsi Y (2007) Monitoring oxidation of multiwalled carbon nanotubes by Raman spectroscopy. J Raman Spectrosc 38(6):728–736. doi:10.1002/jrs.1686
42. Wang GX, Shen XP, Yao J, Wexler D, Ahn JH (2009) Hydrothermal synthesis of carbon nanotube/cobalt oxide core-shell one-dimensional nanocomposite and application as an anode material for lithium-ion batteries. Electrochem Commun 11(3):546–549. doi:10.1016/j.elecom.2008.12.048
43. Hou H, Ge JJ, Zeng J, Li Q, Reneker DH, Greiner A, Cheng SZD (2005) Electrospun polyacrylonitrile nanofibers containing a high concentration of well-aligned multiwall carbon nanotubes. Chem Mater 17(5):967–973. doi:10.1021/cm0484955

44. Zhang H, Xu L, Yang F, Geng L (2010) The synthesis of polyacrylonitrile/carbon nanotube microspheres by aqueous deposition polymerization under ultrasonication. Carbon 48(3):688–695. doi:10.1016/j.carbon.2009.10.013
45. Wepasnick KA, Smith BA, Schrote KE, Wilson HK, Diegelmann SR, Fairbrother DH (2011) Surface and structural characterization of multi-walled carbon nanotubes following different oxidative treatments. Carbon 49(1):24–36. doi:10.1016/j.carbon.2010.08.034
46. Ge JJ, Hou H, Li Q, Graham MJ, Greiner A, Reneker DH, Harris FW, Cheng SZD (2004) Assembly of well-aligned multiwalled carbon nanotubes in confined polyacrylonitrile environments: electrospun composite nanofiber sheets. J Am Chem Soc 126(48):15754–15761. doi:10.1021/ja048648p
47. Saeed K, Park S (2010) Preparation and characterization of multiwalled carbon nanotubes/polyacrylonitrile nanofibers. J Polym Res 17(4):535–540. doi:10.1007/s10965-009-9341-4
48. Yang T, Wu D, Lu L, Zhou W, Zhang M (2011) Electrospinning of polylactide and its composites with carbon nanotubes. Polym Compos 32(8):1280–1288. doi:10.1002/pc.21149
49. Li Y, Chen CX, Zhang S, Ni Y, Huang J (2008) Electrical conductivity and electromagnetic interference shielding characteristics of multiwalled carbon nanotube filled polyacrylate composite films. Appl Surf Sci 254(18):5766–5771. doi:10.1016/j.apsusc.2008.03.077
50. Im JS, Kim JG, Lee SH, Lee YS (2010) Enhanced adhesion and dispersion of carbon nanotube in PANI/PEO electrospun fibers for shielding effectiveness of electromagnetic interference. Colloids Surf A 364(1–3):151–157. doi:10.1016/j.colsurfa.2010.05.015
51. Liu F, Ni QQ, Murakami Y (2013) Preparation of magnetic polyvinyl alcohol composite nanofibers with homogenously dispersed nanoparticles and high water resistance. Text Res J 83(5):510–518. doi:10.1177/0040517512444334
52. Liu F (2012) Doctoral thesis, development of functional polyvinyl alcohol nanofibers by electrospinning technology. Shinshu University, Ueda
53. Chen C, Zhu Y, Bao H, Yang X, Li C (2010) Physically controlled cross-linking in gelated crystalline colloidal array photonic crystals. ACS Appl Mater Interf 2(5):1499–1504. doi:10.1021/am100130n
54. Jensen BEB, Smith AAA, Fejerskov B, Postma A, Senn P, Reimhult E, Pla-Roca M, Isa L, Sutherland DS, Städler B, Zelikin AN (2011) Poly(vinyl alcohol) physical hydrogels: noncryogenic stabilization allows nano- and microscale materials design. Langmuir 27(16):10216–10223. doi:10.1021/la201595e, http://pubs.acs.org/doi/abs/10.1021/la201595e
55. Ossipov DA, Piskounova S, Hilborn J (2008) Poly(vinyl alcohol) cross-linkers for in vivo injectable hydrogels. Macromolecules 41(11):3971–3982. doi:10.1021/ma800332c
56. Cavalieri F, Chiessi E, Villa R, Vigano L, Zaffaroni N, Telling MF, Paradossi G (2008) Novel PVA-based hydrogel microparticles for doxorubicin delivery. Biomacromolecules 9(7):1967–1973. doi:10.1021/bm800225v
57. Kim CK, Kim BS, Sheikh FA, Lee US, Khil MS, Kim HY (2007) Amphiphilic poly(vinyl alcohol) hybrids and electrospun nanofibers incorporating polyhedral oligosilsesquioxane. Macromolecules 40(14):4823–4828. doi:10.1021/ma070056e
58. Lin CL, Lee CF, Chiu WY (2005) Preparation and properties of poly(acrylic acid) oligomer stabilized superparamagnetic ferrofluid. J Colloid Interface Sci 291(2):411–420. doi:10.1016/j.jcis.2005.05.023
59. Ma YH, Wu SY, Wu T, Chang YJ, Hua MY, Chen JP (2009) Magnetically targeted thrombolysis with recombinant tissue plasminogen activator bound to polyacrylic acid-coated nanoparticles. Biomaterials 30(19):3343–3351. doi:10.1016/j.biomaterials.2009.02.034
60. Park JC, Ito T, Kim KO, Kim KW, Kim BS, Khil MS, Kim HK, Kim IS (2010) Electrospun poly(vinyl alcohol) nanofibers: effects of degree of hydrolysis and enhanced water stability. Polym J 42(3):273–276. doi:10.1038/pj.2009.340
61. Molday RS (1984) Magnetic iron-dextran microspheres. US Patent 4452773
62. Hamaker HC (1937) The London-van der Waals attraction between spherical particles. Physica 4(10):1058–1072. doi:10.1016/S0031-8914(37)80203-7
63. Han DH, Wang JP, Luo HL (1994) Crystallite size effect on saturation magnetization of fine ferrimagnetic particles. J Magn Magn Mater 136(1–2):176–182. doi:10.1016/0304-8853(94)90462-6

64. Liu Z, Bai G, Huang Y, Ma Y, Du F, Li F, Guo T, Chen Y (2007) Reflection and absorption contributions to the electromagnetic interference shielding of single-walled carbon nanotube/polyurethane composites. Carbon 45(4):821–827. doi:10.1016/j.carbon.2006.11.020
65. Zhang CS, Ni QQ, Fu SY, Kurashiki K (2007) Electromagnetic interference shielding effect of nanocomposites with carbon nanotube and shape memory polymer. Compos Sci Technol 67(14):2973–2980. doi:10.1016/j.compscitech.2007.05.011
66. Al-Saleh MH, Sundararaj U (2009) Electromagnetic interference shielding mechanisms of CNT/polymer composites. Carbon 47(7):1738–1746. doi:10.1016/j.carbon.2009.02.030
67. Jin X, Ni QQ, Natsuki T (2011) Composites of multi-walled carbon nanotubes and shape memory polyurethane for electromagnetic interference shielding. J Compos Mater 45(11):2547–2554. doi:10.1177/0021998311401106